S0-BHS-049

A Physicist's Desk Reference

The Second Edition of
Physics Vade Mecum

Herbert L. Anderson
Editor in Chief

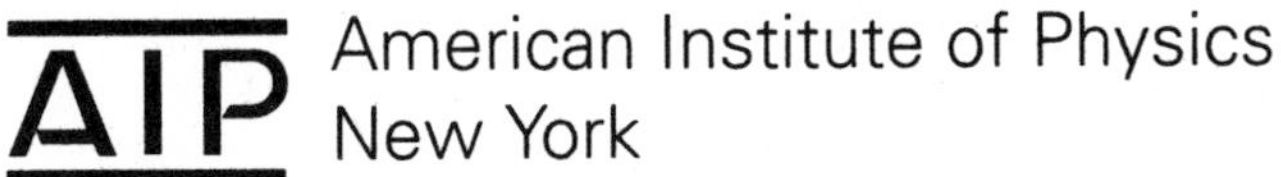

American Institute of Physics
New York

335 East 45th Street, New York, NY 10017

Printed in the United States of America

Library of Congress Cataloging-in-Publication Data

A physicist's desk reference: The second edition of physics vade mecum/Herbert L. Anderson, editor-in-chief.—2nd. ed.

p. cm.
Rev. ed. of: AIP 50th anniversary physics vade mecum. c1981.
Includes index.
1. Physics—Handbooks, manuals, etc.2. Astronomy—Handbooks, manuals, etc. I. Anderson,Herbert Lawrence. II. American Institute of Physics. III. AIP 50th anniversary physics vade mecum.
QC61.P49 1989 89-15201
530—dc20
ISBN 0-88318-629-2
ISBN 0-88318-610-1 (pbk.)

The Second Edition of the Vade Mecum is dedicated to the memory of

Herbert L. Anderson

Foreword

To commemorate the fiftieth anniversary of the American Institute of Physics in 1981, H. William Koch, former Executive Director of the Institute, proposed to publish a handbook that "presents the essence of concepts and numerical data contained primarily in archival journals in which physicists and astronomers publish." Herbert L. Anderson served as Editor in Chief of this Handbook, which appeared in 1981 as *Physics Vade Mecum*. Anderson's preface to the original edition is reproduced to the right.

Four years ago, Herbert Anderson was asked if he might be willing to assume responsibility for an update. He favored an update, but his poor health prevented him from working on it. Both he and the AIP staff members responsible for books were pleased when Fay Ajzenberg-Selove and Henry H. Barschall agreed to serve as Co-Editors of the revision.

Unfortunately, three of the original authors, Herbert L. Anderson, R. Bruce Lindsay, and Yeram S. Touloukian, are no longer living. Their sections have been updated by E. Richard Cohen, Robert T. Beyer, and David R. Lide, respectively. Other new authors involved in the project include David Bodansky, Robert J. Mowris, and Thomas G. Trippe.

The Co-Editors have followed the spirit and form established by Anderson in preparing this new edition. Authors have checked and updated each chapter as necessary, and portions have been changed substantially. The sections in the first chapter on physical constants and recommended symbols, for example, have been revised and expanded. New material appears on high-T_c superconductors, and the chapters on energy, elementary particles, and surface physics have undergone major revisions. In addition, the index has been recompiled and more than doubled in size, making the volume even more functional.

Finally, for the benefit of those who navigate more comfortably in English than in Latin, we have entitled this new edition *A Physicist's Desk Reference* and have retained *Physics Vade Mecum* as part of the subtitle.

I thank Fay Ajzenberg-Selove and Heinz Barschall for their dedicated effort in bringing out this new edition, and Rita Lerner of AIP for shepherding the project to completion. All of us hope that, even in this online world, this reference material on paper will prove useful to students and working scientists.

Kenneth W. Ford
Executive Director
American Institute of Physics

Preface to the First Edition

Those of us who were fortunate enough to watch Enrico Fermi at work marveled at the speed and ease with which he could produce a solution to almost any problem in physics brought to his attention. When he had heard enough to know what the problem was, he proceeded to the blackboard and let the solution flow out of his chalk. He kept in trim by doing a lot of problems, either for the courses that he taught, the talks he gave, or the papers that he wrote. Most frequently, he worked out his own solutions to problems he heard about, in seminars, or in discussions with those who came to talk physics with him. Fermi's solutions were almost always simpler and easier to understand than the ones obtained by the person who raised the question in the first place.

As he grew older, Fermi became concerned about his speed in solving problems. He claimed that his memory was becoming less reliable and less certain and that he had to spend more and more time rederiving even the simple formulas he could previously write down without hesitation. To save time, he decided to organize an "artificial memory" kept close at hand and available for ready reference as needed. Unfortunately, he died before his artificial memory was well developed. However, the idea remained that a handy collection of useful physics formulas and data would help us all deal with the problems in physics that crossed our paths.

The occasion of the 50th anniversary of the American Institute of Physics seemed to be an opportunity to make a start in that direction. A committee was appointed under the chairmanship of David R. Lide, Jr. The members were Herbert L. Anderson, Hans Frauenfelder,

Laurence W. Fredrick, John N. Howard, Robert H. Marks, and Murray Strasberg. The task was to assemble a carry-with-you compendium that would be useful to the wide spectrum of physicists associated with the AIP through its member societies. For our Vade Mecum we chose 22 subjects broadly representative of the fields physicists are supposed to know something about. Each subeditor was charged to compile within 10 pages the most useful information, formulas, numerical data, definitions, and references most physicists would like to have at hand. In the General Section there are collected the fundamental constants, the SI units and prefixes, conversion factors, magnitudes, basic mathematical and physics formulas, formulas useful in practical physics applications, and a list of physics data centers. Although the formulary is brief, it is intended to serve as a starting point.

There were a number of good examples of what could be done in special fields. The shirt-pocket-size booklet *Particle Properties* distributed by the Particle Data Group at LBL and CERN is used actively by many high-energy physicists. Plasma physicists like to keep with them the NRL *Plasma Formulary*, compiled by David L. Book of the Naval Research Laboratory. Another useful compendium is *Astrophysical Quantities* by C. W. Allen, whose style and content influenced the present work. There is, of course, the AIP *Physics Handbook*, but this goes far beyond what we had in mind in size. We refer to it as a backup for more detail in the tables and other material we present. We refer also to the *Handbook of Optics* recently issued by the Optical Society of America. The section on medical physics was drawn from the recently published *Handbook for Medical Physics*. An excellent little compendium is W.H.J. Childs, *Physical Constants*. There are also "pocket" handbooks that come close to ours in general coverage but are larger in size. Among them are H. Ebert, *Physics Handbook*, and B. Yavorsky and A. Detlef, *Handbook of Physics*. We have excerpted some material from all of these.

The handbook is arranged for quick and easy location of the material being sought. The idea is to make each formula and each table usable with a minimum of reading. The explanatory material is placed nearby. Unlike a text book, it is not necessary to read the first page to understand what is written later on. As far as possible, each page is made complete in itself with the name of the principal contributor on its heading. The book should serve as an aid to quick recall of the information physicists most need to have readily available. In future editions it should be possible to remove pages and add new ones without disturbing the utility of the rest. In this way revisions can easily be made.

We wish to thank those who were kind enough to read the manuscripts in their specialty and make useful comments. In particular, David R. Lide, Jr., Emilio Segrè, D. Allan Bromley, Robert T. Beyer, Murray Strasberg, Stefan L. Wipf, Albert Petschek, Nicolas C. Metropolis, John R. Cunningham, and George W. Smith gave good advice, which we were happy to have.

We hope to make the Vade Mecum grow in value but not in size with each passing year.

Herbert L. Anderson,
Editor in Chief, 1981

Contents

1.00. General section

HERBERT L. ANDERSON[†] *Los Alamos National Laboratory* and E. RICHARD COHEN *Rockwell International*

CONTENTS

[†] Deceased.

1.01. Table: PRECISE PHYSICAL CONSTANTS

Physical quantities are listed in this table in both SI and cgs electrostatic units. In cgs units "permittivity of vacuum" μ_0 and "permeability of vacuum" ϵ_0 are dimensionless unit quantities; in SI they have the values $\mu_0 = 4\pi \times 10^{-7}$ m·kg·s^{-2}·A$^{-2} = 4\pi \times 10^{-7}$ N·A$^{-2} = 4\pi \times 10^{-7}$ T·A^{-1} and $\epsilon_0 = 1/\mu_0 c^2$. When the formula for a quantity is given as the product of two factors, the factor in square brackets is to be omitted to obtain the expression in cgs units. The numerical values are based on the 1986 CODATA recommendations. [*The 1986 adjustment of the fundamental physical constants*, E. Richard Cohen and Barry N. Taylor, *Reviews of Modern Physics*, **59**, 1121 (1987); CODATA Bulletin #**63**, November, 1986 (Pergamon, Oxford/New York).] Since the uncertainties in this table are correlated, the full variance matrix of the least-squares solution must be used to evaluate the uncertainty associated with further quantities calculated from these entries.

Quantity	Symbol (expression)	Value, SI unit (cgs unit)	Error (ppm)
1. Speed of light in vacuum	c	$2.997\,924\,58 \times 10^{8}$ m·s^{-1} (10^{10} cm·sec^{-1})	(exact)
2. Gravitational constant	G	$6.672\,59 \times 10^{-11}$ m^3·kg^{-1}·s^{-2} (10^{-8} cm^3·gm^{-1}·sec^{-2})	128
3. Elementary charge	e	$1.602\,177\,33 \times 10^{-19}$ C (10^{-20} emu)	0.30
		($4.803\,2068 \times 10^{-10}$ esu)	0.30
4. Planck constant	h	$6.626\,0755 \times 10^{-34}$ J·s (10^{-27} erg·sec)	0.60
	$\hbar = h/2\pi$	$1.054\,572\,66 \times 10^{-34}$ J·s (10^{-27} erg·sec)	0.60
5. Avogadro constant	N_A	$6.022\,1367 \times 10^{23}$ mol^{-1}	0.59
6. Faraday constant	$F = N_A e$	$9.648\,5309 \times 10^{4}$ C·mol^{-1} (10^3 emu·mol^{-1})	0.30
7. Electron mass	m_e	$9.109\,3897 \times 10^{-31}$ kg (10^{-28} gm)	0.59
8. Rydberg constant	$R_\infty = m_e c \alpha^2/2h$	$1.097\,373\,1534 \times 10^{7}$ m^{-1} (10^5 cm^{-1})	0.0012
9. Fine structure constant	$\alpha = [4\pi\epsilon_0]^{-1} e^2/\hbar c$	$7.297\,353\,08 \times 10^{-3}$	0.045
	α^{-1}	137.035 9895	0.045
10. Classical electron radius	$r_e = (\hbar/m_e c)\alpha$	2.817 940 92 fm (10^{-13} cm)	0.13
11. Electron Compton wavelength	$\lambda_C = h/m_e c$	2.426 310 58 pm (10^{-10} cm)	0.089
12. Bohr radius	$a_0 = r_e \alpha^{-2}$	$5.291\,772\,49 \times 10^{-11}$ m (10^{-9} cm)	0.045
13. Atomic mass unit	$m_u = \frac{1}{12} m(^{12}\mathrm{C})$	$1.660\,5402 \times 10^{-27}$ kg (10^{-24} gm)	0.59
14. Proton mass	m_p	$1.672\,6231 \times 10^{-27}$ kg (10^{-24} gm)	0.59
15. Neutron mass	m_n	$1.674\,9286 \times 10^{-27}$ kg (10^{-24} gm)	0.59
16. Magnetic flux quantum	$\Phi_0 = [c]^{-1} hc/2e$	$2.067\,834\,61 \times 10^{-15}$ Wb (10^{-7} Gs·cm^2)	0.30
17. Quantum of circulation	$h/2m_e$	$3.636\,948\,07 \times 10^{-4}$ m^2·s^{-1} (cm^2·sec^{-1})	0.089
18. Specific electron charge	$-e/m_e$	$-1.758\,819\,62 \times 10^{11}$ C·kg^{-1} (10^7 emu·gm^{-1})	0.30
19. Bohr magneton	$\mu_B = [c]e\hbar/2m_e c$	$9.274\,0154 \times 10^{-24}$ J·T^{-1} (10^{-21} erg·Gs^{-1})	0.34
20. Electron magnetic moment	μ_e	$9.284\,7701 \times 10^{-24}$ J·T^{-1} (10^{-21} erg·Cs^{-1})	0.34
21. Electron magnetic moment ratio	$\mu_e/\mu_B = \frac{1}{2} g_e$	1.001 159 652 193	10^{-5}
22. Nuclear magneton	$\mu_N = [c]e\hbar/2m_p c$	$5.050\,7866 \times 10^{-27}$ J·T^{-1} (10^{-24} erg·Gs^{-1})	0.34
23. Proton magnetic moment	μ_p	$1.410\,607\,61 \times 10^{-26}$ J·T (10^{-23} erg·Gs^{-1})	0.34
24. Proton gyromagnetic ratio	γ_p	$2.675\,221\,28 \times 10^{8}$ rad·s^{-1}·T^{-1} (10^4 rad·sec^{-1}·Gs^{-1})	0.30
25. Quantum Hall resistance	R_H	25 812.8056 Ω	0.045
26. Molar gas constant	R	8.314 510 J·mol^{-1}·K^{-1} (10^7 erg·mol^{-1}·K^{-1})	8.4
27. Boltzmann constant	$k = R/N_A$	$1.380\,658 \times 10^{-23}$ J·K^{-1} (10^{-16} erg·K^{-1})	8.5
28. Molar volume,			
$T_0 = 273.15$ K, $p_0 = 10^5$ Pa	$V_M = RT_0/p_0$	$22.711\,08 \times 10^{-3}$ m^3·mol^{-1} (10^3 cm^3·mol^{-1})	8.4
$T_0 = 273.15$ K, $p_0 = 1$ atm (1 atm ≡ 101 325 Pa)		$22.414\,10 \times 10^{-3}$ m^3·mol^{-1} (10^3 cm^3·mol^{-1})	8.4
29. Stefan–Boltzmann constant	$\sigma = (\pi^2/60)k^4/\hbar^3 c^2$	$5.670\,51 \times 10^{-8}$ W·m^{-2}·K^{-4} (10^{-5} erg·sec·cm^{-2}·K^{-4})	34
30. First radiation constant	c_1	$3.741\,7749 \times 10^{-16}$ W·m^2 (10^{-5} erg·sec^{-1}·cm^2)	0.60
31. Second radiation constant	c_2	$1.438\,769 \times 10^{-2}$ m·K (cm·K)	8.4
32. Wien displacement constant	$b = \lambda_{max} T = c_2/4.965\,114$	$2.897\,756 \times 10^{-3}$ m·K (10^{-1} cm·K)	8.4

Energy Equivalents

Quantity	Symbol	Value	Error (ppm)
Atomic mass unit	u	931.494 32 MeV	0.30
Electron mass	m_e	0.510 999 06 MeV	0.30
Muon mass	m_μ	105.658 389 MeV	0.32
Proton mass	m_p	938.272 31 MeV	0.30
Neutron mass	m_n	939.565 63 MeV	0.30
Electron volt	1 eV	$1.602\,177\,33 \times 10^{-19}$ J	0.30
	1 eV/hc	8065.5410 cm^{-1}	0.30
	1 eV/h	241.798 836 THz	0.30
Electron volt per particle	1 eV/k	11 604.45 K	8.4
Planck constant	$\hbar$	$6.582\,1220 \times 10^{-22}$ MeV·s	0.30
	$\hbar c$	197.327 053 MeV·fm	0.30
	$(\hbar c)^2$	0.389 379 66 GeV^2·mb	0.59
Rydberg constant	$R_\infty hc$	13.605 6981 eV	0.30
Voltage–wavelength product	$V \cdot \lambda$	12 398.4244 eV·Å	0.30
Gas constant	R	1.987 216 cal·mol^{-1}	8.4

1.02. THE INTERNATIONAL SYSTEM OF UNITS (SI)

A. Table: SI base units

From *The International System of Units,* Natl. Bur. of Stand. (US) Spec. Pub. **330** (1986 Edition).

Name	Symbol	Definition
meter	m	"The meter is the length of path travelled by light in vacuum during a time interval of 1/299 792 458 of a second."
mass	kg	"The kilogram is the unit of mass, it is equal to the mass of the international prototype of the kilogram." (The international prototype is a platinum-iridium cylinder kept at the BIPM in Sèvres (Paris), France.)
second	s	"The second is the duration of 9 192 631 770 periods of the radiation corresponding to the transition between the two hyperfine levels of the ground state of the cesium-133 atom."
ampere	A	"The ampere is that constant current which, if maintained in two straight parallel conductors of infinite length, of negligible circular cross section, and placed 1 meter apart in vacuum, would produce between these conductors a force equal to 2×10^{-7} newton per meter of length."
kelvin	K	"The kelvin, unit of thermodynamic temperature, is the fraction 1/273.16 of the thermodynamic temperature of the triple point of water." (The Celsium temperature scale is defined by the equation $t=T-T_0$, where T is the thermodynamic temperature in kelvins and $T_0=273.15$ K.)
mole	mol	"The mol is the amount of substance of a system which contains as many elementary entities as there are atoms in 0.012 kg of carbon-12."
candela	cd	"The candela is the luminous intensity, in a given direction, of a source that emits monochromatic radiation of frequency 540×10^{12} hertz and that has a radiant intensity in that direction of (1/683) watt per steradian."

B. Table: SI derived units

Name	Symbol (dimensions)	Definition	Value in cgs units ($\xi=2.997\ 924\ 58$)
Absorbed dose	Gy ($m^2{\cdot}s^{-2}$)	The *gray* is the absorbed dose when the energy per unit mass imparted to matter by ionizing radiation is one joule per kilogram. (The gray is also used for the ionizing radiation quantities: specific energy imparted, kerma, and absorbed dose index, which have the SI unit joule per kilogram.) 1 rad $=10^{-2}$ Gy.	10^4 erg/gm
Activity	Bq (s^{-1})	The *becquerel* is the activity of a radionuclide decaying at the rate of one spontaneous nuclear transition per second. 1 Ci(curie) $=3.7\times10^{10}$ Bq.	1 sec^{-1}
Dose equivalent	Sv ($m^2{\cdot}s^{-2}$)	The *sievert* is the dose equivalent when the absorbed dose of ionizing radiation multiplied by the dimensionless factors Q (quality factor) and N (product of any other multiplying factors) stipulated by the International Commission on Radiological Protection is one joule per kilogram.	10^4 erg/gm
Electric capacitance	F ($m^{-2}{\cdot}kg^{-1}{\cdot}s^4{\cdot}A^2$)	The *farad* is the capacitance of a capacitor between the plates of which there appears a difference of potential of one volt when it is charged by a quantity of electricity equal to one coulomb.	$\xi^2\times10^{11}$ esu
Electric conductance	S ($m^{-2}{\cdot}kg^{-1}{\cdot}s^3{\cdot}A^2$)	The *siemens* is the electric conductance of a conductor in which a current of one ampere is produced by an electric potential difference of one volt.	$\xi^2\times10^{11}$ esu (cm/sec)

B. Table—*Continued*

Name	Symbol (dimensions)	Definition	Value in cgs units ($\xi = 2.997\,924\,58$)
Electric inductance	H ($m^2 \cdot kg \cdot s^{-2} \cdot A^{-2}$)	The *henry* is the inductance of a closed circuit in which an electromotive force of one volt is produced when the electric current in the circuit varies uniformly at a rate of one ampere per second.	10^9 emu (cm)
Electric potential difference, electromotive	V ($m^2 \cdot kg \cdot s^{-3} \cdot A^{-1}$)	The *volt* (unit of electric potential difference and electromotive force) is the difference of electric potential between two points of a conductor carrying a constant current of one ampere, when the power dissipated between these points is equal to one watt.	$(0.01/\xi)$ esu ($cm^{1/2} \cdot gm^{1/2} \cdot sec^{-1}$)
Electric resistance	Ω ($m^2 \cdot kg \cdot s^{-3} \cdot A^{-2}$)	The *ohm* is the electric resistance between two points of a conductor when a constant difference of potential of one volt, applied between these two points, produces in this conductor a current of one ampere, this conductor not being the source of any electromotive force.	$\xi^{-2} \times 10^{11}$ esu ($cm^{-1} \cdot sec$)
Energy	J ($m^2 \cdot kg \cdot s^{-2}$)	The *joule* is the work done when the point of application of a force of one newton is displaced a distance of one meter in the direction of the force.	10^7 erg ($cm^2 \cdot gm \cdot sec^{-2}$)
Force	N ($m \cdot kg \cdot s^{-2}$)	The *newton* is that force which, when applied to a body having a mass of one kilogram, gives it an acceleration of one meter per second squared.	10^5 dyne ($cm^2 \cdot gm \cdot sec^{-2}$)
Frequency	Hz (s^{-1})	The *hertz* is the frequency of a periodic phenomenon of which the period is one second.	cycle/sec (sec^{-1})
Illuminance	lx ($cd \cdot sr \cdot m^{-2}$)	The *lux* is the illuminance produced by a luminous flux of one lumen uniformly distributed over a surface of one square meter.	
Luminous flux	lm ($cd \cdot sr$)	The *lumen* is the luminous flux emitted in a solid angle of one steradian by a point source having a uniform intensity of one candela.	
Magnetic flux	Wb ($m^2 \cdot kg \cdot s^{-2} \cdot A^{-1}$)	The *weber* is the magnetic flux which, linking a circuit of one turn, produces in it an electromotive force of one volt as it is reduced to zero at a uniform rate in one second.	10^8 Mx ($cm^{3/2} \cdot gm^{1/2} \cdot sec^{-1}$)
Magnetic flux density	T ($kg \cdot s^{-2} \cdot A^{-1}$)	The *tesla* is the magnetic flux density given by a magnetic flux of one weber per square meter.	10^4 Gs ($cm^{-1/2} \cdot gm^{1/2} \cdot sec^{-1}$)
Power	W ($m^2 \cdot kg \cdot s^{-3}$)	The *watt* is the power which gives rise to the production of energy at the rate of one joule per second.	10^7 erg/sec ($cm^2 \cdot gm \cdot sec^{-2}$)
Pressure or stress	Pa ($m^{-1} \cdot kg \cdot s^{-2}$)	The *pascal* is the pressure or stress of one newton per square meter.	10 dyn/cm^2 ($cm^{-1} \cdot gm \cdot sec^{-2}$)
Quantity of electricity	C	The *coulomb* is the quantity of electricity transported in one second by a current of one ampere.	$\xi \times 10^9$ esu ($cm^{3/2} \cdot gm^{1/2} \cdot sec^{-1}$)

C. Table: SI prefixes

Factor	Prefix	Symbol	Factor	Prefix	Symbol
10^{18}	exa	E	10^{-1}	deci	d
10^{15}	peta	P	10^{-2}	centi	c
10^{12}	tera	T	10^{-3}	milli	m
10^9	giga	G	10^{-6}	micro	μ
10^6	mega	M	10^{-9}	nano	n
10^3	kilo	k	10^{-12}	pico	p
10^2	hecto	h	10^{-15}	femto	f
10^1	deka	da	10^{-18}	atto	a

D. Table: Conversion to SI units

Five-figure accuracy except for exact values. The symbol ≡ indicates exact values. Taken in part from American Society for Testing and Materials Standard for Metric Practice E380-79 (1980), C. W. Allen, Astrophysical Quantities (The Athlone Press, University of London, 1976), and L. V. Judson, Natl. Bur. Stand. (U. S.) Lett. Circ. **LC 1035** (1960) (Revised, 1988).

1. Acceleration

The gal is a special unit employed in geodesy and geophysics to express the acceleration due to gravity.

1 ft/s^2	≡0.3048 m/s^2	1 gal	≡0.01 m/s^2
Standard gravity (g)	≡9.80665 m/s^2		

2. Angle

1 degree (°)	$=1.7453\times10^{-2}$ rad	1 second(″)	$=4.8481\times10^{-6}$ rad
1 minute (′)	$=2.9089\times10^{-4}$ rad		

3. Area

1 acre	=4046.9 m^2	1 in.^{2}	$\equiv6.4516\times10^{-4}$ m^2
1 are	≡100 m^2	1 square mile	$=2.5900\times10^{6}$ m^2
1 barn (b)	$\equiv1\times10^{-28}$ m^2	1 square rod (rd^2), square pole, or square perch	=25.293 m^2
1 circular mil	$=5.0671\times10^{-10}$ m^2	1 square yard (yd^2)	=0.83613 m^2
1 ft^2	=0.092 903 m^2		
1 hectare	≡10 000 m^2		

4. Density

1 grain/gal (U.S. liquid)	=0.017118 kg/m^3	1 ton (short)/yd^3	=1186.6 kg/m^3
1 oz (avoirdupois)/in.^{3}	=1730.0 kg/m^3	Density of water (4 °C)	=999.97 kg/m^3
1 lb/ft^3	=16.018 kg/m^3	Density of mercury (0 °C)	=13595 kg/m^3
1 lb/in.^{3}	=27680 kg/m^3	Solar mass/cubic parsec	$=6.770\times10^{-20}$ kg/m^3
1 lb/gal (U.S. liquid)	=119.83 kg/m^3	STP gas density for molecular weight M_0	$=0.044615M_0$ kg/m^3
1 ton (long)/yd^3	=1328.9 kg/m^3		

5. Electricity and magnetism

A = ampere, C = coulomb, F = farad, H = henry, Ω = ohm, S = siemens, V = volt, T = tesla, Wb = weber.

1 abampere	≡10 A	1 ohm centimeter	$\equiv1\times10^{-2}$ Ω m
1 abcoulomb	≡10 C	1 ohm circular-mil per foot	$=1.6624\times10^{-9}$ Ω m
1 abfarad	$\equiv1\times10^{9}$ F	1 statampere	$=3.3356\times10^{-10}$ A
1 abhenry	$\equiv1\times10^{-9}$ H	1 statcoulomb	$=3.3356\times10^{-10}$ C
1 abmho	$\equiv1\times10^{9}$ S	1 statfarad	$=1.1127\times10^{-12}$ F
1 abohm	$\equiv1\times10^{-9}$ Ω	1 stathenry	$=8.9876\times10^{11}$ H
1 abvolt	$\equiv1\times10^{-8}$ V	1 statmho	$=1.1127\times10^{-12}$ S
1 ampere hour	≡3600 C	1 statohm	$=8.9876\times10^{11}$ Ω
1 emu of capacitance	$\equiv1\times10^{8}$ F	1 statvolt	=299.79 V
1 emu of current	≡10 A	1 unit pole	$=1.2566\times10^{-7}$ Wb
1 emu of electric potential	$\equiv1\times10^{-8}$ V	Potential of electron at 1st Bohr orbit	=27.211 V
1 emu of inductance	$\equiv1\times10^{-9}$ H	Ionization potential from 1st Bohr orbit	=13.606 V
1 emu of resistance	$\equiv1\times10^{-9}$ Ω	Nuclear electric field at 1st Bohr orbit	$=5.1422\times10^{11}$ V/m
1 esu of capacitance	$=1.1127\times10^{-12}$ F	Current in 1st Bohr orbit	$=3.5051\times10^{-13}$ A
1 esu of current	$=3.3356\times10^{-10}$ A	Dipole moment of nucleus and electron in 1st Bohr orbit	$=8.4784\times10^{-30}$ C·m
1 esu of electric potential	$=2.9979\times10^{2}$ V	Magnetic field, atomic unit	=1715.3 T
1 esu of inductance	$=8.9876\times10^{11}$ H	Field at nucleus due to electron in 1st Bohr orbit	$=9.9606\times10^{6}$ A/m
1 esu of resistance	$=8.9876\times10^{11}$ Ω	Magnetic moment, atomic unit	$=2.542\times10^{-21}$ J/T
1 faraday (based on ^{12}C)	$=9.6486\times10^{4}$ C		
1 gamma	$\equiv1\times10^{-9}$ T		
1 gauss	$\equiv1\times10^{-4}$ T		
1 gilbert	$=7.9577\times10^{-1}$ A(amp. turns)		
1 maxwell	$\equiv1\times10^{-8}$ Wb		
1 mho	≡1 S		
1 oersted	=79.577 A/m		

6. Energy

Btu = British thermal unit (thermochemical), 1 Btu (International Table) = 1.000 67 Btu (thermochemical); cal = calorie (thermochemical), 1 cal (International Table) = 1.000 67 cal (thermochemical); J = joule; W = watt.

1 Btu	=1054.4 J	1 Btu (60°F)	=1054.7 J
1 Btu (mean)	=1055.9 J	1 calorie	≡4.184 J
1 Btu (39°F)	=1059.7 J	1 calorie (mean)	=4.1900 J

D. Table—*Continued*

1 calorie (15 °C)	$\equiv 4.1858$ J	1 ton (equivalent of TNT)	$\equiv 4.184 \times 10^{9}$ J
1 calorie (20 °C)	$= 4.1819$ J	1 watt hour (W·h)	$\equiv 3600$ J
1 kilocalorie	$\equiv 4184$ J	1 watt second (W·s)	$\equiv 1$ J
1 electron volt (eV)	$= 1.6022 \times 10^{-19}$ J	Energy of unit wave number (*hc*)	$= 1.9864 \times 10^{-25}$ J
1 erg	$\equiv 1 \times 10^{-7}$ J	Mass energy of unit atomic weight	$= 1.4924 \times 10^{-10}$ J
1 foot-pound (ft·lbf)	$= 1.3558$ J		
1 foot-poundal	$= 0.042140$ J		
1 kilowatt hour (kW·h)	$\equiv 3.6 \times 10^{6}$ J		
1 therm	$= 1.0551 \times 10^{8}$ J		

Note: 1 quad $\equiv 10^{15}$ Btu
1 quad per year = 0.472 million barrels of oil per day (1 barrel = 42 gallons)
= 1 trillion cubic feet of gas per year
= 44.4 million tons of coal per year (for medium heating value coal at 22.5 Btu/ton)
= 33.4 million kilowatts of electricity
= 293 billion kilowatt-hours of electricity per year at 100% efficiency
= 95.2 billion kilowatt-hours of electricity per year at 32.5% efficiency
See Chaps. 10 (Energy Demand) and 11 (Energy Supply).

7. Force

1 dyne	$\equiv 1 \times 10^{-5}$ N	1 lbf/lb (thrust/weight [mass] ratio)	$\equiv 9.80665$ m s^{-2}
1 kilogram-force	$\equiv 9.80665$ N	1 poundal	$= 0.13825$ N
1 kip (1000 lbf)	$= 4448.2$ N	1 ton-force (2000 lbf)	$= 8896.4$ N
1 ounce-force	$= 0.27801$ N	Proton-electron attraction at distance a_0	$= 8.2387 \times 10^{-8}$ N
1 pound-force (lbf)	$= 4.4482$ N		

8. Frequency

1 hertz (Hz)	= 1 cycle/s	Frequency of free electron in magnetic field B	$= (2.7992 \times 10^{10}$ Hz·T$^{-1})$ B
Rydberg frequency (cR_∞)	$= 3.2898 \times 10^{15}$ Hz	Plasma frequency associated with electron density N_e	$= (80.617$ m$^3 \cdot N_e)^{1/2}$ Hz
Frequency of 1st Bohr orbit ($2cR_\infty$)	$= 6.5797 \times 10^{15}$ Hz		

9. Heat

Btu = British thermal unit (thermochemical), 1 Btu (International Table) = 1.000 67 Btu (thermochemical); cal = calorie (thermochemical), 1 cal (International Table) = 1.000 67 cal (thermochemical); J = joule, K = kelvin; W = watt; h = hour.

Thermal conductivity k:			
1 Btu·ft/h·ft^2·°F	= 1.7926 W/m-K	1 Btu-in./s-ft^2·°F	= 518.87 W/m-K
Heat capacity:			
1 Btu-lb.°F	$\equiv$ 4184 J/kg·K	1 cal/s	$\equiv$ 4.184 W
1 cal/g.°C	$\equiv$ 4184 J/kg·K		
Thermal resistance R:		Thermal resistivity:	
1 °F·h·ft^2/Btu	= 0.176 23 K·m^2/W	1°F·h·ft/Btu·in.	= 6.9381 K·m/W
Thermal diffusivity:			
1 ft^2/h	$= 2.5806 \times 10^{-5}$ m^2/s		

10. Length

1 angstrom (Å)	$\equiv 1 \times 10^{-10}$ m	1 light year (ly)	$= 9.4605 \times 10^{15}$ m
1 atomic unit (a_0)	$= 0.529\ 18 \times 10^{-10}$ m	1 microinch	$\equiv 2.54 \times 10^{-8}$ m
1 astronomical unit (AU)	$= 1.4960 \times 10^{11}$ m	1 mil	$\equiv 2.54 \times 10^{-5}$ m
1 chain	= 20.117 m	1 nautical mile	$\equiv$ 1852 m
1 electron radius (r_e)	$= 2.8179 \times 10^{-15}$ m	1 mile	= 1609.3 m
1 fathom	$\equiv$ 1.8288 m	1 parsec (pc)	$= 3.0857 \times 10^{16}$ m
1 fermi (femtometer) (fm)	$\equiv 1 \times 10^{-15}$ m	1 pica (printer's)	$= 4.2175 \times 10^{-3}$ m
1 foot (ft)	$\equiv$ 0.3048 m	1 point (printer's)	$= 3.5146 \times 10^{-4}$ m
1 furlong	= 201.17 m	1 rod	$\equiv$ 5.0292 m
1 hand	= 0.1016 m	Wavelength of 1-eV photon (*hc*/eV)	$= 1.2398 \times 10^{-6}$ m
1 inch (in.)	$\equiv$ 0.0254 m	1 yard	$\equiv$ 0.9144 m
1 league (land)	= 4828 m		

D. Table—*Continued*

11. Light

cd = candela, lm = lumen, [1 lumen = luminous flux corresponding to (1/683) W at 540 THz], lx = lux

1 apostilb	≡1 lm/m^2 for perfectly diffusing surface	1 foot-lambert	=3.4263 cd/m^2
1 cd/in.^{2}	=1550.0 cd/m^2	1 lambert	=3183.1 cd/m^2
1 foot-candle	=10.764 lx	1 stilb (sb)	≡10 000 cd/m^2
		1 phot	≡10 000 lx

12. Mass

1 atomic mass unit (^{12}C scale) (u)	=1.6605×10^{-27} kg	1 ounce (avoirdupois)	=2.8350×10^{-2} kg
1 carat (metric)	≡2×10^{-4} kg	1 ounce (troy or apothecary)	=3.1103×10^{-2} kg
1 dram, apothecary	=3.8879×10^{-3} kg	1 pennyweight (troy)	=1.5552×10^{-3} kg
1 dram, avoirdupois	=1.7718×10^{-3} kg	1 pound (lb avoirdupois)	=0.45359 kg
1 gamma	≡1×10^{-9} kg = 1 μg	1 pound (troy or apothecary)	=0.37324 kg
1 grain	=6.4799×10^{-5} kg	1 quintal (q)	≡100 kg
1 hundredweight (gross or long)	=50.802 kg	1 scruple	=1.2960×10^{-3} kg
1 hundred weight (net or short)	=45.359 kg	1 slug	=14.594 kg
1 kgf·s^2/m	≡9.80665 kg	1 ton (long, 2240 lb)	=1016.0 kg
		1 ton (short, 2000 lb)	=907.18 kg
		1 ton (metric ton)	≡1000 kg

13. Mass per unit length

1 denier	=1.1111×10^{-7} kg/m	1 tex	≡1×10^{-6} kg/m

14. Mass per unit time

1 perm (0 °C)	=5.7214×10^{-11} kg/Pa·s·m^2	1 lb/hp·h	=1.6897×10^{-7} kg/J
1 perm·in. (0 °C)	=1.4532×10^{-12} kg/Pa·s·m	1 ton (short)/h	=0.25200 kg/s
1 lb/h	=1.2600×10^{-4} kg/s		

15. Power

1 Btu (int.)/h	=0.29307 W	1 ft·lbf/h	=3.7662×10^{-4} W
1 Btu (int.)/s	=1055.1 W	1 horsepower (550 ft·lbf/s)	=745.70 W
1 Btu (thermochem.)/h	=0.292 88 W	1 horsepower (boiler)	=9809.5 W
1 cal (thermochem.)/s	≡4.184 W	1 horsepower (electric)	≡746 W
1 force de cheval	=735.5 W	1 horsepower (metric)	=735.50 W
1 erg/s	≡1×10^{-7} W	1 horsepower (water)	=746.04 W

16. Pressure or stress (force per unit area)

1 atmosphere (standard)	≡101 325 Pa	1 inch of water (39.2 °F)	=249.08 Pa
1 atmosphere (technical) = 1 kgf/cm^2	≡98 066.5 Pa	1 kgf/cm^2	≡98 066.5 Pa
1 bar	≡100 000 Pa	1 kip/in.^{2} (ksi)	=6.8948×10^6 Pa
1 cm Hg(0 °C)	=1333.2 Pa	1 millibar	≡100 Pa
1 centimeter of water (4 °C)	=98.064 Pa	1 newton/cm^2	≡10 000 Pa
1 dyne/cm^2	≡0.1 Pa	1 poundal/ft^2	=1.4882 Pa
1 foot of water (39.2 °F)	=2989.0 Pa	1 lbf/ft^2	=47.880 Pa
1 gf/cm^2	≡98.0665 Pa	1 lbf/in.^{2} (psi)	=6894.8 Pa
1 inch of mercury (32 °F)	=3386.4 Pa	1 torr (mm Hg, 0 °C)	=133.32 Pa

17. Temperature

Degree Celsius	$T_K = t_c + 273.15$	Triple point of natural water	≡273.16 K
Degree Fahrenheit	$t_C = (t_F - 32)/1.8$	Elementary temperature ($m_e c^2/\alpha k$)	=8.1261×10^{11} K
Degree Fahrenheit	$T_K = (t_F + 459.67)/1.8$	Temperature of 1 eV	=11 604 K
Degree Rankine	$T_K = T_R/1.8$		
Kelvin	$t_C = T_K - 273.15$		

18. Time

1 day	≡86 400 s	1 minute	≡60 s
1 day (sidereal)	=86 164 s	1 second (sidereal)	=0.99727 s
1 hour	≡3600 s	1 year (365 days)	≡31 536 000 s
1 hour (sidereal)	=3590.2 s		

D. Table—*Continued*

1 year (sidereal)	$=3.1558\times10^{7}$ s	1 atomic unit ($\tau_0=\hbar/m_ec^2\alpha^2=1/4\pi R_\infty c$)	$=2.4189\times10^{-17}$ s
1 year (tropical)	$=3.1557\times10^{7}$ s	Jordan's elementary time ($\alpha^3\tau_0=r_e/c$)	$=9.3996\times10^{-24}$ s
1 second	≡9 192 631 770 ^{133}Cs cycles		
19. Torque			
1 dyne-cm	$\equiv1\times10^{-7}$ N·m	1 lbf·in.	=0.11298 N·m
1 kgf·m	≡9.80665 N·m	1 lbf·ft	=1.3558 N·m
1 ozf·in.	=0.0070616 N·m		
20. Velocity			
1 ft/s	≡0.3048 m/s	1 AU per year	=4.7406 km/s
1 in./s	≡0.0254 m/s	1 parsec per year	$=9.7781\times10^{8}$ m/s
1 km/h	=0.27778 m/s	Electron in Bohr orbit	$=2.1877\times10^{6}$ m/s
1 knot	=0.51444 m/s	1–eV electron	$=5.9309\times10^{5}$ m/s
1 mi/h	≡0.44704 m/s	Angular velocity of Earth on its axis	$=7.2921\times10^{-5}$ rad/s
1 mi/s	=1609.3 m/s	Mean angular velocity of Earth in its orbit	$=1.9910\times10^{-7}$ rad/s
1 mi/h	=1.6093 km/h		
Velocity of light (c)	$=2.9979\times10^{8}$ m/s		
21. Viscosity			
1 centipoise	$\equiv1\times10^{-3}$ Pa·s	1 lb/ft·s	=1.4882 Pa·s
1 centistokes	$\equiv1\times10^{-6}$ m^2/s	1 lbf·s/ft^2	≡47.880 Pa·s
1 ft^2/s	=0.092030 m^2/s	1 lbf·s/in.^{2}	=6894.8 Pa·s
1 poise	≡0.1 Pa·s	1 rhe	≡10 (Pa·s)$^{-1}$
1 poundal·s/ft^2	=1.4882 Pa·s	1 slug/ft·s	=47.880 Pa·s
1 lb/ft·h	$=4.1338\times10^{-4}$ Pa·s	1 stokes	$\equiv1\times10^{-4}$ m^2/s
22. Volume			
1 acre-foot	=1233.5 m^3	1 gill (U.S.)	$=1.1829\times10^{-4}$ m^3
1 barrel (petroleum, 42 gal)	=0.15899 m^3	1 in.^{3}	$=1.6387\times10^{-5}$ m^3
1 barrel (bbl) (other liquids 31.5 gal.)	=0.11924 m^3	1 liter	$\equiv1\times10^{-3}$ m^3
1 board foot	$=2.3597\times10^{-3}$ m^3	1 ounce (U.K. fluid)	$=2.8413\times10^{-5}$ m^3
1 bushel (bu), struck measure (U.S.) 2150.42 in.^{3}	$=3.5239\times10^{-2}$ m^3	1 ounce (U.S. fluid)	$=2.9574\times10^{-5}$ m^3
1 cord (cd) (firewood)	≡128 ft^3	1 cubic parsec	$=2.9380\times10^{49}$ m^3
1 cup (measuring)	$=2.3659\times10^{-4}$ m^3	1 peck (U.S.)	$=8.8098\times10^{-3}$ m^3
1 dram (U.S. fluid)	$=3.6967\times10^{-6}$ m^3	1 pint (U.S. dry)	$=5.5061\times10^{-4}$ m^3
1 fluid ounce (U.S.)	$=2.9574\times10^{-5}$ m^3	1 pint (U.S. liquid)	$=4.7318\times10^{-4}$ m^3
1 ft^3	=0.028 317 m^3	1 quart (U.S. dry)	$=1.1012\times10^{-3}$ m^3
1 gallon (U.K. liquid, 277.418 in.^{3})	$=4.5461\times10^{-3}$ m^3	1 quart (U.S. liquid)	$=9.4635\times10^{-4}$ m^3
1 gallon (U.S. dry)	$=4.4049\times10^{-3}$ m^3	1 stere	≡1 m^3
1 gallon (U.S. liquid, 231 in.^{3})	$=3.7854\times10^{-3}$ m^3	1 tablespoon	$=1.4787\times10^{-5}$ m^3
1 gill (U.K.)	$=1.4207\times10^{-4}$ m^3	1 teaspoon	$=4.9289\times10^{-6}$ m^3
		1 registry ton (100 ft^3)	=2.8317 m^3
		1 yd^3	=0.764 55 m^3

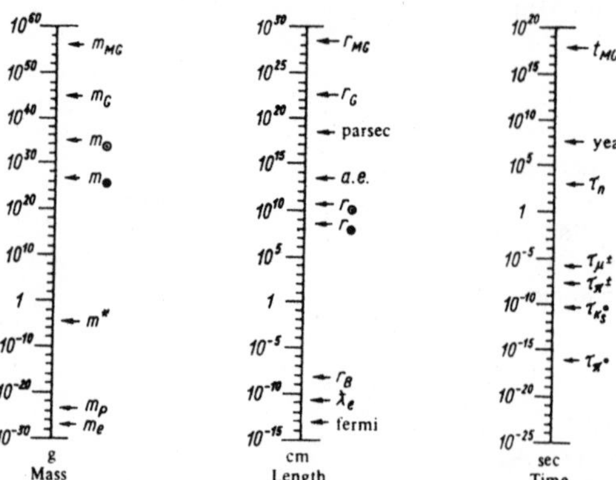

E. Figure: Characteristic mass, length, and lifetime for various physical and astrophysical objects

m_{MG}, mass of Metagalaxy; m_G, Galaxy; $m_\odot$, Sun; $m_\oplus$, Earth; m^*, Planck particle; m_p, proton; m_e, electron. r_{MG}, radius of Metagalaxy; r_G, of Galaxy; $r_\odot$, Sun; $r_\oplus$, Earth; r_B, Bohr orbit; $\bar{\lambda}_e$, electron Compton wavelength. t_{MG}, lifetime of Metagalaxy; τ_n, of neutron; τ_μ, muon; $\tau_{\pi^\pm}$, charged pions; $\tau_{K_S^0}$, short-lived neutral kaon; τ_{π^0}, neutral pion. From R. M. Muradyan, Sov. J. Part. Nucl. **8**, 73 (1977).

1.03 MATHEMATICAL FORMULAS[1]

A. Quadratic equations

If $a \neq 0$, the roots of $ax^2 + bx + c = 0$ are

$$x = \frac{-b \pm (b^2 - 4ac)^{1/2}}{2a} = \frac{-2c}{b \pm (b^2 - 4ac)^{1/2}}.$$

B. Binomial theorem

$$(1+x)^n$$
$$= 1 + nx + \frac{n(n-1)}{2!}x^2 + \frac{n(n-1)(n-2)}{3!}x^3$$
$$+ \cdots + \frac{n!}{(n-r)!r!}x^r + \cdots.$$

The coefficient of x^r is denoted by $\binom{n}{r}$ or ${}_nC_r$;

$$(x+y)^n = \sum_{r=0}^{n} {}_nC_r x^{n-r} y^r = \sum_{r=0}^{n} \binom{n}{r} x^{n-r} y^r.$$

Note that, here and elsewhere, we take $0! = 1$. If n is a positive integer, the expression consists of a finite number of terms. If n is not a positive integer, the series is convergent for $x^2 < 1$; and if $n > 0$, the series is convergent also for $x^2 = 1$. In particular,

$$(1 \pm x)^{1/2} = 1 \pm \tfrac{1}{2}x - \frac{1 \cdot 1}{2 \cdot 4}x^2 \pm \frac{1 \cdot 1 \cdot 3}{2 \cdot 4 \cdot 6}x^3$$
$$- \frac{1 \cdot 1 \cdot 3 \cdot 5}{2 \cdot 4 \cdot 6 \cdot 8}x^4 \pm \cdots,$$
$$(1 \pm x)^{-1/2} = 1 \mp \tfrac{1}{2}x + \frac{1 \cdot 3}{2 \cdot 4}x^2 \mp \frac{1 \cdot 3 \cdot 5}{2 \cdot 4 \cdot 6}x^3$$
$$+ \frac{1 \cdot 3 \cdot 5 \cdot 7}{2 \cdot 4 \cdot 6 \cdot 8}x^4 \mp \cdots.$$

C. Factorials

$0! = 1, \quad 2! = 2, \quad 5! = 120, \quad 10! = 3\,628\,800.$

D. Stirling's formula

$$\lim_{n \to \infty} \frac{n!}{n^n e^{-n} n^{1/2}} = (2\pi)^{1/2}.$$

This gives approximate values of $n!$ for large n. For $n = 12$ the approximation is too small by 0.7%; for $n = 20$ it is too small by 0.4%.

E. Series

1. Arithmetic progression

Of the first order (first differences constant), to n terms:

$$a + (a+d) + (a+2d) + (a+3d) + \cdots$$
$$+ [a + (n-1)d] = na + \tfrac{1}{2}n(n-1)d$$
$$\equiv \tfrac{1}{2}n(\text{first term} + \text{last term}).$$

2. Geometric progression

To n terms:

$$a + ar + ar^2 + ar^3 + \cdots + ar^{n-1}$$
$$\equiv a(1 - r^{n-1})/(1 - r).$$

If $r^2 < 1$, the limit of the sum of an infinite number of terms is $a/(1-r)$.

3. Harmonic progression

The reciprocals of the terms of a series in arithmetic progression of the first order. Thus

$$\frac{1}{a}, \frac{1}{a+d}, \frac{1}{a+2d}, \ldots, \frac{1}{a+(n-1)d}$$

are in harmonic progression.

4. Arithmetic mean

of n quantities:

$(1/n)(a_1 + a_2 + a_3 + \cdots + a_n).$

5. Geometric mean

of n quantities:

$(a_1 a_2 a_3 \cdots a_n)^{1/n}.$

6. Harmonic mean

of n quantities:

$$n\left(\frac{1}{a_1} + \frac{1}{a_2} + \frac{1}{a_3} + \cdots + \frac{1}{a_n}\right)^{-1}.$$

7. Summation formulas

$$1 + 2 + 3 + \cdots + n = \tfrac{1}{2}n(n+1),$$
$$1^2 + 2^2 + 3^2 + \cdots + n^2 = \tfrac{1}{6}n(n+1)(2n+1),$$
$$1^3 + 2^3 + 3^3 + \cdots + n^3 = \tfrac{1}{4}n^2(n+1)^2.$$

8. Maclaurin's series

$$f(h) = f(0) + hf'(0) + \frac{h^2}{2!}f''(0) + \frac{h^3}{3!}f'''(0) + \cdots.$$

9. Taylor's series

$$f(x+h) = f(x) + hf'(x) + \frac{h^2}{2!} f''(x) + \frac{h^3}{3!} f'''(x)$$
$$+ \cdots + \frac{h^{n-1}}{(n-1)!} f^{n-1}(x) + R_n,$$

where, for a suitable value of θ between 0 and 1,

$$R_n = \frac{h^n}{n!} f^n(x+\theta h),$$

$$f(x+h, y+k)$$
$$= f(x,y) + \left(h \frac{\partial f(x,y)}{\partial x} + k \frac{\partial f(x,y)}{\partial y}\right)$$
$$+ \frac{1}{2!}\left(h^2 \frac{\partial^2 f(x,y)}{\partial x^2} + 2hk \frac{\partial^2 f(x,y)}{\partial x \partial y} + \frac{k^2 \partial^2 f(x,y)}{\partial y^2}\right)$$
$$+ \frac{1}{3!}\left(h^3 \frac{\partial^3 f(x,y)}{\partial x^3} + 3h^2 k \frac{\partial^3 f(x,y)}{\partial x^2 \partial y} + 3hk^2 \frac{\partial^3 f(x,y)}{\partial x \partial y^2} + k^3 \frac{\partial^3 f(x,y)}{\partial y^3}\right) + \cdots .$$

F. Algebraic functions

1. Derivatives

$$\frac{d}{dx}(au) = a \frac{du}{dx},$$
$$\frac{d}{dx}(u+v) = \frac{du}{dx} + \frac{dv}{dx}, \quad \frac{d}{dx}(uv) = u\frac{dv}{dx} + v\frac{du}{dx},$$
$$\frac{d}{dx}(uvw) = uv\frac{dw}{dx} + vw\frac{du}{dx} + wu\frac{dv}{dx},$$
$$\frac{d}{dx}(x^n) = nx^{n-1}, \quad \frac{d}{dx}(x^{1/2}) = \tfrac{1}{2}x^{-1/2},$$
$$\frac{d}{dx}(1/x) = -\frac{1}{x^2}.$$

2. Some indefinite integrals

$$\int dx = x, \quad \int x\,dx = \tfrac{1}{2}x^2, \quad \int \frac{dx}{x} = \ln|x|,$$
$$\int x^n\,dx = \frac{x^{n+1}}{n+1} \quad (n \neq -1),$$
$$\int \frac{dx}{x^n} = -\frac{1}{(n-1)x^{n-1}} \quad (n \neq 1).$$

3. Integration by parts

$$\int u\,dv = uv - \int v\,du$$

or

$$\int u\,dv = uv - \int v \frac{du}{dv}\,dv.$$

4. Rational algebraic functions

$$\int \frac{dx}{1+x^2} = \tan^{-1}x$$

(the principal value of $\tan^{-1}x$ is to be taken; that is, $-\pi/2 < \tan^{-1}x < \pi/2$),

$$\int \frac{dx}{a^2+b^2x} = \frac{1}{ab}\tan^{-1}\frac{bx}{a},$$
$$\int \frac{dx}{1-x^2} = \tfrac{1}{2}\ln\left|\frac{1+x}{1-x}\right|,$$
$$\int \frac{dx}{a^2-x^2} = \frac{1}{2a}\ln\left|\frac{a+x}{a-x}\right|.$$

5. Irrational algebraic functions

$$\int \frac{dx}{(x^2+a^2)^{1/2}} = \ln(x + \surd(x^2+a^2)),$$
$$\int \frac{dx}{(a^2-x^2)^{1/2}} = \sin^{-1}\frac{x}{a}, \quad (x^2 < a^2)$$

[take $-\pi/2 \leqslant \sin^{-1}(x/a) \leqslant \pi/2$, $a \geqslant 0$],

$$\int \frac{dx}{x(x^2-a^2)^{1/2}} = (1/a)\cos|a/x|$$

[take $0 \leqslant \cos^{-1}|a/x| \leqslant \pi/2$].

G. Trigonometric functions

$$\sin^2 A + \cos^2 A = 1,$$
$$\sin(A \pm B) = \sin A \cos B \pm \cos A \sin B,$$
$$\cos(A \pm B) = \cos A \cos B \mp \sin A \sin B,$$
$$2 \sin A \cos B = \sin(A+B) + \sin(A-B),$$
$$2 \sin A \sin B = \cos(A-B) - \cos(A+B),$$
$$2 \cos A \cos B = \cos(A+B) + \cos(A-B),$$
$$\sin 2A = 2 \sin A \cos A,$$
$$\cos 2A = \cos^2 A - \sin^2 A,$$
$$\sin x = (1/2i)(e^{ix} - e^{-ix}), \quad i = +(-1)^{1/2},$$
$$\cos x = \tfrac{1}{2}(e^{ix} + e^{-ix}),$$
$$e^{ix} = \cos x + i \sin x.$$

1. Formulas for plane triangles

Let a, b, c be the sides opposite angles A, B, C.

$$a^2 = b^2 + c^2 - 2bc \cos A,$$
$$a/\sin A = b/\sin B = c/\sin C,$$
$$a = b \cos C + c \cos B,$$
$$A + B + C = \pi \text{ rad} = 180°.$$

2. Area of triangle

$$\tfrac{1}{2}ab\sin C = \frac{a^2}{2}\frac{\sin B \sin C}{\sin A},$$

$$= [s(s-a)(s-b)(s-c)]^{1/2}$$

where $s = \frac{1}{2}(a+b+c)$.

3. Trigonometric series

$$\sin x = x - \frac{x^3}{3!} + \frac{x^5}{5!} - \frac{x^7}{7!} + \cdots, \quad (x^2 < \infty),$$

$$\cos x = 1 - \frac{x^2}{2!} + \frac{x^4}{4!} - \frac{x^6}{6!} + \cdots, \quad (x^2 < \infty).$$

4. Derivatives

$$\frac{d}{dx}\sin x = \cos x, \quad \frac{d}{dx}\cot x = -\csc^2 x,$$

$$\frac{d}{dx}\cos x = -\sin x, \quad \frac{d}{dx}\sec x = \sec x \tan x,$$

$$\frac{d}{dx}\tan x = \sec^2 x, \quad \frac{d}{dx}\csc x = -\csc x \cot x.$$

5. Integrals

$$\int \sin x\, dx = -\cos x,$$

$$\int x \sin x\, dx = \sin x - x\cos x,$$

$$\int \sin^2 x\, dx = \frac{x}{2} - \frac{\sin 2x}{4}$$

$$= \frac{x}{2} - \frac{\sin x \cos x}{2},$$

$$\int \frac{\sin x\, dx}{x} = \mathrm{Si}(x) = x - \frac{x^3}{3 \cdot 3!} + \frac{x^5}{5 \cdot 5!} - \frac{x^7}{7 \cdot 7!} + \cdots,$$

$$\int \cos x\, dx = \sin x,$$

$$\int x\cos x\, dx = \cos x + x\sin x,$$

$$\int \cos^2 x\, dx = \frac{x}{2} + \frac{\sin 2x}{4}$$

$$= \frac{x}{2} + \frac{\sin x \cos x}{2},$$

$$\int \frac{\cos x}{x}\, dx = \ln|x| - \frac{x^2}{2 \cdot 2!} + \frac{x^4}{4 \cdot 4!} - \frac{x^6}{6 \cdot 6!} + \cdots,$$

$$\int \tan x\, dx = -\ln|\cos x|,$$

$$\int \tan^2 x\, dx = \tan x - x.$$

6. Derivatives of inverse trigonometric functions

$$\frac{d}{dx}\sin^{-1}\frac{x}{a}$$

$$= \frac{1}{(a^2 - x^2)^{1/2}} \quad [-\pi/2 < \sin^{-1}(x/a) < \pi/2],$$

$$(a > 0)$$

$$\frac{d}{dx}\cos^{-1}\frac{x}{a}$$

$$= \frac{-1}{(a^2 - x^2)^{1/2}} \quad [0 < \cos^{-1}(x/a) < \pi],$$

$$(a > 0)$$

$$\frac{d}{dx}\tan^{-1}\frac{x}{a} = \frac{a}{a^2 + x^2}$$

$$\frac{d}{dx}\cot^{-1}\frac{x}{a} = \frac{-a}{a^2 + x^2}.$$

7. Integrals of inverse trigonometric functions

$$\int \sin^{-1}\frac{x}{a}\, dx = x\sin^{-1}\frac{x}{a} + (a^2 - x^2)^{1/2}$$

$$\int \cos^{-1}\frac{x}{a}\, dx = x\cos^{-1}\frac{x}{a} - (a^2 - x^2)^{1/2}$$

$$\int \tan^{-1}\frac{x}{a}\, dx = x\tan^{-1}\frac{x}{a} - \frac{a}{2}\ln(a^2 + x^2)$$

H. Exponential functions

1. Series

$$e^{\pm x} = 1 \pm \frac{x}{1!} + \frac{x^2}{2!} \pm \frac{x^3}{3!} + \cdots + (\pm 1)^n \frac{x^n}{n!} + \cdots,$$

$$(x^2 < \infty)$$

$$e^{\sin u} = 1 + u + \frac{u^2}{2!} - \frac{3u^4}{4!} - \frac{8u^5}{5!}$$

$$- \frac{3u^6}{6!} + \frac{56u^7}{7!} + \cdots, \quad (u^2 < \infty)$$

2. Derivatives

$$\frac{d}{dx}e^x = e^x, \quad \frac{d}{dx}e^{ax} = ae^{ax}$$

$$\frac{d}{dx}a^x = a^x \ln a, \quad (a \text{ is a constant})$$

$$\frac{d}{dx}x^y = yx^{y-1} + x^y(\ln x)\frac{dy}{dx}$$

$$\frac{d}{dx}x^x = x^x(1 + \ln x)$$

3. Integrals

$$\int e^x dx = e^x, \quad \int e^{ax} dx = \frac{1}{a}e^{ax}$$

$$\int xe^{ax} dx = e^{ax}\left(\frac{x}{a} - \frac{1}{a^2}\right)$$

I. Logarithmic functions

1. Series

$$\ln(1+x) = x - \frac{x^2}{2} + \frac{x^3}{3} - \frac{x^4}{4} + \frac{x^5}{5} - \cdots, \quad (x^2<1,\ x=1)$$

$$\ln x = (x-1) - \frac{(x-1)^2}{2} + \frac{(x-1)^3}{3} - \frac{(x-1)^4}{4} + \cdots, \quad (0<x\leqslant 2)$$

$$\ln x = 2\left(\frac{x-1}{x+1} + \frac{(x-1)^3}{3(x+1)^3} + \frac{(x-1)^5}{5(x+1)^5} + \cdots\right), \quad (x>0)$$

2. Integrals

$$\int \ln x\, dx = x \ln x - x$$

$$\int x \ln x\, dx = \frac{x^2}{2}\ln x - \frac{x^2}{4}$$

$$\int x^n \ln x\, dx = \frac{x^{n+1}}{n+1}\left(\ln x - \frac{1}{n+1}\right)$$

J. Definite integrals:

$$\int_0^\infty x^{n-1}e^{-x}dx = \Gamma(n),$$

where $\Gamma(n)$ is the gamma function. The integral is finite when $n>0$.

$$\Gamma(n+1) = n\Gamma(n), \quad \Gamma(n)\Gamma(1-n) = \pi/\sin n\pi.$$

$\Gamma(n) = (n-1)!$, when n is an integer >0.

$$\Gamma(1) = \Gamma(2) = 1, \quad \Gamma(\tfrac{1}{2}) = \pi.$$

$$\int_1^\infty \frac{dx}{x^m} = \frac{1}{m-1}, \quad (m>1),$$

$$\int_0^\infty \frac{a\,dx}{a^2+x^2} = \pi/2, \quad (a>0),$$

$$= 0, \quad (a=0),$$

$$= -\pi/2, \quad (a<0),$$

$$\int_0^\pi \sin^2 x\, dx = \int_0^\pi \cos^2 x\, dx = \frac{\pi}{2},$$

$$\int_0^\infty e^{-a^2x^2}dx = \frac{\pi^{1/2}}{2a}, \quad (a>0),$$

$$\int_0^\infty x e^{-x^2}dx = \tfrac{1}{2},$$

$$\int_0^\infty x^2 e^{-x^2}dx = \frac{\pi^{1/2}}{4},$$

$$\int_0^\infty x^{2a}e^{-px^2}dx = \frac{1\cdot 3\cdot 5\cdots(2a-1)}{2^{a+1}p^a}\sqrt{\frac{\pi}{p}}.$$

K. Vector algebra[2]

1. Notation

A, B, etc., are scalars; **A**, **B**, etc., are vectors.

2. Scalar inner product

$$\mathbf{A}\cdot\mathbf{B} = (\mathbf{A},\mathbf{B}) = |\mathbf{A}||\mathbf{B}|\cos\phi,$$

ϕ being the angle between **A** and **B**.

3. Vector product

$\mathbf{A}\times\mathbf{B} = [\mathbf{A},\mathbf{B}]$ is a vector which is perpendicular to the plane determined by **A** and **B**, has absolute value $|\mathbf{A}||\mathbf{B}|\sin\phi$, and a direction such that **A**, **B**, [**A**,**B**] form a right-handed screw.

In Cartesian coordinates with $\mathbf{A} = A_x\mathbf{i} + A_y\mathbf{j} + A_z\mathbf{k}$,

$$\mathbf{A}\times\mathbf{B} = (A_yB_z - A_zB_y)\mathbf{i} + (A_zB_x - A_xB_z)\mathbf{j} + (A_xB_y - A_yB_x)\mathbf{k}$$

$$= \begin{vmatrix} \mathbf{i} & \mathbf{j} & \mathbf{k} \\ A_x & A_y & A_z \\ B_x & B_y & B_z \end{vmatrix}.$$

4. Vector equations

$$\mathbf{X}\cdot\mathbf{A} = p, \quad \mathbf{X}\times\mathbf{A} = \mathbf{B}.$$

Solution:

$$\mathbf{X} = \frac{\mathbf{A}p}{A^2} + \frac{\mathbf{A}\times\mathbf{B}}{A^2}.$$

$$\mathbf{X}\cdot\mathbf{A} = p \quad \mathbf{X}\cdot\mathbf{B} = q, \quad \mathbf{X}\cdot\mathbf{C} = r, \quad (\mathbf{A}\cdot\mathbf{B}\times\mathbf{C} \neq 0).$$

Solution:

$$\mathbf{X} = \frac{p\mathbf{B}\times\mathbf{C} + q\mathbf{C}\times\mathbf{A} + r\mathbf{A}\times\mathbf{B}}{\mathbf{A}\cdot\mathbf{B}\times\mathbf{C}}.$$

5. Differential operators

The *gradient* of a scalar $\phi(x, y, z)$ gives the absolute value and direction of the most rapid change of ϕ. It is perpendicular to the surface $\phi = $ const. In Cartesian coordinates,

$$\mathrm{grad}\phi = \nabla\phi = \frac{\partial\phi}{\partial x}\mathbf{i} + \frac{\partial\phi}{\partial y}\mathbf{j} + \frac{\partial\phi}{\partial z}\mathbf{k}.$$

The *divergence* of a vector **A** is the excess of the outflowing over the inflowing vector flux through the surface of a volume element divided by the volume. In Cartesian coordinates,

$$\mathrm{div}\mathbf{A} = \nabla\cdot\mathbf{A} = \frac{\partial A_x}{\partial x} + \frac{\partial A_y}{\partial y} + \frac{\partial A_z}{\partial z}.$$

The *curl* of a vector **A** gives the absolute value and direction of the maximum circulation about a fixed point around a surface element containing the fixed point, divided by the area enclosed by the element. The direction of the curl forms a right-handed screw with the sense of rotation of the circulation. In Cartesian coordinates,

$$\text{curl}\mathbf{A} = \nabla\times\mathbf{A} = \begin{vmatrix} \mathbf{i} & \mathbf{j} & \mathbf{k} \\ \partial/\partial x & \partial/\partial y & \partial/\partial z \\ A_x & A_y & A_z \end{vmatrix}.$$

Formal definition of grad, div, and curl:

$$\nabla\phi = \lim_{V\to 0}\frac{1}{V}\int \phi\, d\mathbf{S},$$

$$\nabla\cdot\mathbf{A} = \lim_{V\to 0}\frac{1}{V}\int \mathbf{A}\cdot d\mathbf{S},$$

$$\nabla\times\mathbf{A} = \lim_{V\to 0}\frac{1}{V}\int d\mathbf{S}\times\mathbf{A}.$$

6. Vector identities[3]

f, g, etc., are scalars; $\mathbf{A}$, $\mathbf{B}$, etc., are vectors; $\mathbf{T}$ is a tensor.

$$\mathbf{A}\cdot(\mathbf{B}\times\mathbf{C}) = (\mathbf{A}\times\mathbf{B})\cdot\mathbf{C} = \mathbf{B}\cdot(\mathbf{C}\times\mathbf{A}) = (\mathbf{B}\times\mathbf{C})\cdot\mathbf{A} = \mathbf{C}\cdot(\mathbf{A}\times\mathbf{B}) = (\mathbf{C}\times\mathbf{A})\cdot\mathbf{B}, \quad (1)$$

$$\mathbf{A}\times(\mathbf{B}\times\mathbf{C}) = (\mathbf{C}\times\mathbf{B})\times\mathbf{A} = (\mathbf{A}\cdot\mathbf{C})\mathbf{B} - (\mathbf{A}\cdot\mathbf{B})\mathbf{C}, \quad (2)$$

$$\mathbf{A}\times(\mathbf{B}\times\mathbf{C}) + \mathbf{B}\times(\mathbf{C}\times\mathbf{A}) + \mathbf{C}\times(\mathbf{A}\times\mathbf{B}) = 0, \quad (3)$$

$$(\mathbf{A}\times\mathbf{B})\cdot(\mathbf{C}\times\mathbf{D}) = (\mathbf{A}\cdot\mathbf{C})(\mathbf{B}\cdot\mathbf{D}) - (\mathbf{A}\cdot\mathbf{D})(\mathbf{B}\cdot\mathbf{C}), \quad (4)$$

$$(\mathbf{A}\times\mathbf{B})\times(\mathbf{C}\times\mathbf{D}) = (\mathbf{A}\times\mathbf{B}\cdot\mathbf{D})\mathbf{C} - (\mathbf{A}\times\mathbf{B}\cdot\mathbf{C})\mathbf{D}, \quad (5)$$

$$\nabla(fg) = \nabla(gf) = f\nabla g + g\nabla f, \quad (6)$$

$$\nabla\cdot(f\mathbf{A}) = f\nabla\cdot\mathbf{A} + \mathbf{A}\cdot\nabla f, \quad (7)$$

$$\nabla\times(f\mathbf{A}) = f\nabla\times\mathbf{A} + \nabla f\times\mathbf{A}, \quad (8)$$

$$\nabla\cdot(\mathbf{A}\times\mathbf{B}) = \mathbf{B}\cdot(\nabla\times\mathbf{A}) - \mathbf{A}\cdot(\nabla\times\mathbf{B}), \quad (9)$$

$$\nabla\times(\mathbf{A}\times\mathbf{B}) = \mathbf{A}(\nabla\cdot\mathbf{B}) - \mathbf{B}(\nabla\cdot\mathbf{A}) + (\mathbf{B}\cdot\nabla)\mathbf{A} - (\mathbf{A}\cdot\nabla)\mathbf{B}, \quad (10)$$

$$\mathbf{A}\times(\nabla\times\mathbf{B}) = (\nabla\mathbf{B})\cdot\mathbf{A} - \mathbf{A}\cdot\nabla\mathbf{B}, \quad (11)$$

$$\nabla(\mathbf{A}\cdot\mathbf{B}) = \mathbf{A}\times(\nabla\times\mathbf{B}) + \mathbf{B}\times(\nabla\times\mathbf{A}) + (\mathbf{A}\cdot\nabla)\mathbf{B} + (\mathbf{B}\cdot\nabla)\mathbf{A}, \quad (12)$$

$$\nabla^2 f = \nabla\cdot\nabla f, \quad (13)$$

$$\nabla^2\mathbf{A} = \nabla(\nabla\cdot\mathbf{A}) - \nabla\times\nabla\times\mathbf{A}, \quad (14)$$

$$\nabla\times\nabla f = 0, \quad (15)$$

$$\nabla\cdot\nabla\times\mathbf{A} = 0. \quad (16)$$

If $\mathbf{e}_1$, $\mathbf{e}_2$, $\mathbf{e}_3$ are orthonormal unit vectors, a second-order tensor $\mathbf{T}$ can be written in the dyadic form

$$\mathbf{T} = \sum_{i,j} T_{ij}\mathbf{e}_i\mathbf{e}_j. \quad (17)$$

In Cartesian coordinates the divergence of a tensor is a vector with components

$$(\nabla\cdot\mathbf{T})_i = \sum_j(\partial T_{ji}/\partial x_j). \quad (18)$$

[This definition is required for consistency with Eq. (28).] In general,

$$\nabla\cdot(\mathbf{AB}) = (\nabla\cdot\mathbf{A})\mathbf{B} + (\mathbf{A}\cdot\nabla)\mathbf{B}, \quad (19)$$

$$\nabla\cdot(f\mathbf{T}) = \nabla f\cdot\mathbf{T} + f\nabla\cdot\mathbf{T}. \quad (20)$$

Let $\mathbf{r} = \mathbf{i}x + \mathbf{j}y + \mathbf{k}z$ be the radius vector of magnitude r, from the origin to the point x,y,z. Then

$$\nabla\cdot\mathbf{r} = 3, \quad (21)$$

$$\nabla\times\mathbf{r} = 0, \quad (22)$$

$$\nabla r = \mathbf{r}/r, \quad (23)$$

$$\nabla(1/r) = -\mathbf{r}/r^3, \quad (24)$$

$$\nabla(\mathbf{r}/r^3) = 4\pi\delta(\mathbf{r}). \quad (25)$$

If V is a volume enclosed by a surface S and $d\mathbf{S} = \mathbf{n}dS$, where $\mathbf{n}$ is the unit normal outward from V,

$$\int_V dV\,\nabla f = \int_S d\mathbf{S}f, \quad (26)$$

$$\int_V dV\,\nabla\cdot\mathbf{A} = \int_S d\mathbf{S}\cdot\mathbf{A}, \quad (27)$$

$$\int_V dV\,\nabla\cdot\mathbf{T} = \int_S d\mathbf{S}\cdot\mathbf{T}, \quad (28)$$

$$\int_V dV\,\nabla\times\mathbf{A} = \int_S d\mathbf{S}\times\mathbf{A}, \quad (29)$$

$$\int_V dV(f\nabla^2 g - g\nabla^2 f) = \int_S d\mathbf{S}\cdot(f\nabla g - g\nabla f), \quad (30)$$

$$\int_V dV(\mathbf{A}\cdot\nabla\times\nabla\times\mathbf{B} - \mathbf{B}\cdot\nabla\times\nabla\times\mathbf{A}) = \int_S d\mathbf{S}\cdot(\mathbf{B}\times\nabla\times\mathbf{A} - \mathbf{A}\times\nabla\times\mathbf{B}). \quad (31)$$

If S is an open surface bounded by the contour C of which the line element is $d\mathbf{l}$,

$$\int_S d\mathbf{S}\times\nabla f = \oint_C d\mathbf{l}f, \quad (32)$$

$$\int_S d\mathbf{S}\cdot\nabla\times\mathbf{A} = \oint_C d\mathbf{l}\cdot\mathbf{A}, \quad (33)$$

$$\int_S (d\mathbf{S}\times\nabla)\times\mathbf{A} = \oint_C d\mathbf{l}\times\mathbf{A}, \quad (34)$$

$$\int_S d\mathbf{S}\cdot(\nabla f\times\nabla g) = \oint_C f\,dg = -\oint_C g\,df. \quad (35)$$

7. Vector differential equations

a. Cylindrical coordinates. Divergence:

$$\nabla\cdot\mathbf{A} = \frac{1}{r}\frac{\partial}{\partial r}(rA_r) + \frac{1}{r}\frac{\partial A_\phi}{\partial\phi} + \frac{\partial A_z}{\partial z}.$$

Gradient:

$$(\nabla f)_r = \frac{\partial f}{\partial r},\quad (\nabla f)_\phi = \frac{1}{r}\frac{\partial f}{\partial\phi},\quad (\nabla f)_z = \frac{\partial f}{\partial z}.$$

Curl:

$$(\nabla\times\mathbf{A})_r = \frac{1}{r}\frac{\partial A_z}{\partial\phi} - \frac{\partial A_\phi}{\partial z},$$

$$(\nabla\times\mathbf{A})_\phi = \frac{\partial A_r}{\partial z} - \frac{\partial A_z}{\partial r},$$

$$(\nabla\times\mathbf{A})_z = \frac{1}{r}\frac{\partial}{\partial r}(rA_\phi) - \frac{1}{r}\frac{\partial A_r}{\partial\phi}.$$

Laplacian:

$$\nabla^2 f = \frac{1}{r}\frac{\partial}{\partial r}\left(r\frac{\partial f}{\partial r}\right) + \frac{1}{r^2}\frac{\partial^2 f}{\partial \phi^2} + \frac{\partial^2 f}{\partial z^2}.$$

Laplacian of vector:

$$(\nabla^2 \mathbf{A})_r = \nabla^2 A_r - \frac{2}{r^2}\frac{\partial A_\phi}{\partial \phi} - \frac{A_r}{r^2},$$

$$(\nabla^2 \mathbf{A})_\phi = \nabla^2 A_\phi + \frac{2}{r^2}\frac{\partial A_r}{\partial \phi} - \frac{A_\phi}{r^2},$$

$$(\nabla^2 \mathbf{A})_z = \nabla^2 A_z.$$

Components of $(\mathbf{A}\cdot\nabla)\mathbf{B}$:

$$(\mathbf{A}\cdot\nabla\mathbf{B})_r = A_r\frac{\partial B_r}{\partial r} + \frac{A_\phi}{r}\frac{\partial B_r}{\partial \phi} + A_z\frac{\partial B_r}{\partial z} - \frac{A_\phi B_\phi}{r},$$

$$(\mathbf{A}\cdot\nabla\mathbf{B})_\phi = A_r\frac{\partial B_\phi}{\partial r} + \frac{A_\phi}{r}\frac{\partial B_\phi}{\partial \phi} + A_z\frac{\partial B_\phi}{\partial z} + \frac{A_\phi B_r}{r},$$

$$(\mathbf{A}\cdot\nabla\mathbf{B})_z = A_r\frac{\partial B_z}{\partial r} + \frac{A_\phi}{r}\frac{\partial B_z}{\partial \phi} + A_z\frac{\partial B_z}{\partial z}.$$

Divergence of tensor:

$$(\nabla\cdot\mathbf{T})_r = \frac{1}{r}\frac{\partial}{\partial r}(rT_{rr}) + \frac{1}{r}\frac{\partial}{\partial \phi}(T_{\phi r}) + \frac{\partial T_{zr}}{\partial z} - \frac{1}{r}T_{\phi\phi},$$

$$(\nabla\cdot\mathbf{T})_\phi = \frac{1}{r}\frac{\partial}{\partial r}(rT_{r\phi}) + \frac{1}{r}\frac{\partial T_{\phi\phi}}{\partial \phi} + \frac{\partial T_{z\phi}}{\partial z} + \frac{1}{r}T_{\phi r},$$

$$(\nabla\cdot\mathbf{T})_z = \frac{1}{r}\frac{\partial}{\partial r}(rT_{rz}) + \frac{1}{r}\frac{T_{\phi z}}{\partial \phi} + \frac{\partial T_{zz}}{\partial z}.$$

b. Spherical coordinates. Divergence:

$$\nabla\cdot\mathbf{A} = \frac{1}{r^2}\frac{\partial}{\partial r}(r^2 A_r) + \frac{1}{r\sin\theta}\frac{\partial}{\partial \theta}(A_\theta \sin\theta) + \frac{1}{r\sin\theta}\frac{\partial A_\phi}{\partial \phi}.$$

Gradient:

$$(\nabla f)_r = \frac{\partial f}{\partial r},\quad (\nabla f)_\theta = \frac{1}{r}\frac{\partial f}{\partial \theta},\quad (\nabla f)_\phi = \frac{1}{r\sin\theta}\frac{\partial f}{\partial \phi}.$$

Curl:

$$(\nabla\times\mathbf{A})_r = \frac{1}{r\sin\theta}\frac{\partial}{\partial \theta}(A_\phi \sin\theta) - \frac{1}{r\sin\theta}\frac{\partial A_\theta}{\partial \phi},$$

$$(\nabla\times\mathbf{A})_\theta = \frac{1}{r\sin\theta}\frac{\partial A_r}{\partial \phi} - \frac{1}{r}\frac{\partial}{\partial r}(rA_\phi),$$

$$(\nabla\times\mathbf{A})_\phi = \frac{1}{r}\frac{\partial}{\partial r}(rA_\theta) - \frac{1}{r}\frac{\partial A_r}{\partial \theta}.$$

Laplacian:

$$\nabla^2 f = \frac{1}{r^2}\frac{\partial}{\partial r}\left(r^2\frac{\partial f}{\partial r}\right) + \frac{1}{r^2\sin\theta}\frac{\partial}{\partial \theta}\left(\sin\theta\frac{\partial f}{\partial \theta}\right) + \frac{1}{r^2\sin^2\theta}\frac{\partial^2 f}{\partial \phi^2}.$$

Laplacian of vector:

$$(\nabla^2\mathbf{A})_r = \nabla^2 A_r - \frac{2A_r}{r^2} - \frac{2}{r^2}\frac{\partial A_\theta}{\partial \theta} - \frac{2}{r^2\sin\theta}\frac{\partial A_\phi}{\partial \phi},$$

$$(\nabla^2\mathbf{A})_\theta = \nabla^2 A_\theta + \frac{2}{r^2}\frac{\partial A_r}{\partial \theta} - \frac{A_\theta}{r^2\sin^2\theta} - \frac{2\cos\theta}{r^2\sin^2\theta}\frac{\partial A_\phi}{\partial \phi},$$

$$(\nabla^2\mathbf{A})_\phi = \nabla^2 A_\phi - \frac{A_\phi}{r^2\sin^2\theta} + \frac{2}{r^2\sin\theta}\frac{\partial A_r}{\partial \phi} + \frac{2\cos\theta}{r^2\sin^2\theta}\frac{\partial A_\theta}{\partial \phi}.$$

Components of $(\mathbf{A}\cdot\nabla)\mathbf{B}$:

$$(\mathbf{A}\cdot\nabla\mathbf{B})_r = A_r\frac{\partial B_r}{\partial r} + \frac{A_\theta}{r}\frac{\partial B_r}{\partial \theta} + \frac{A_\phi}{r\sin\theta}\frac{\partial B_r}{\partial \phi} - \frac{A_\theta B_\theta + A_\phi B_\phi}{r},$$

$$(\mathbf{A}\cdot\nabla\mathbf{B})_\theta = A_r\frac{\partial B_\theta}{\partial r} + \frac{A_\theta}{r}\frac{\partial B_\theta}{\partial \theta} + \frac{A_\phi}{r\sin\theta}\frac{\partial B_\theta}{\partial \phi} + \frac{A_\theta B_r}{r} - \frac{A_\phi B_\phi \cot\theta}{r},$$

$$(\mathbf{A}\cdot\nabla\mathbf{B})_\phi = A_r\frac{\partial B_\phi}{\partial r} + \frac{A_\theta}{r}\frac{\partial B_\phi}{\partial \theta} + \frac{A_\phi}{r\sin\theta}\frac{\partial B_\phi}{\partial \phi} + \frac{A_\phi B_r}{r} + \frac{A_\phi B_\theta \cot\theta}{r}.$$

Divergence of tensor:

$$(\nabla\cdot\mathbf{T})_r = \frac{1}{r^2}\frac{\partial}{\partial r}(r^2 T_{rr}) + \frac{1}{r\sin\theta}\frac{\partial}{\partial \theta}(T_{\theta r}\sin\theta) + \frac{1}{r\sin\theta}\frac{\partial T_{\phi r}}{\partial \phi} - \frac{T_{\theta\theta} + T_{\phi\phi}}{r},$$

$$(\nabla\cdot\mathbf{T})_\theta = \frac{1}{r^2}\frac{\partial}{\partial r}(r^2 T_{r\theta}) + \frac{1}{r\sin\theta}\frac{\partial}{\partial \theta}(T_{\theta\theta}\sin\theta) + \frac{1}{r\sin\theta}\frac{\partial T_{\phi\theta}}{\partial \phi} + \frac{T_{\theta r}}{r} - \frac{\cot\theta}{r}T_{\phi\phi},$$

$$(\nabla\cdot\mathbf{T})_\phi = \frac{1}{r^2}\frac{\partial}{\partial r}(r^2 T_{r\phi}) + \frac{1}{r\sin\theta}\frac{\partial}{\partial \theta}(T_{\theta\phi}\sin\theta) + \frac{1}{r\sin\theta}\frac{\partial T_{\phi\phi}}{\partial \phi} + \frac{T_{\phi r}}{r} + \frac{\cot\theta}{r}T_{\phi\theta}.$$

L. Fourier series[4]

A function $f(x)$ in the interval $-a/2 \leqslant x \leqslant a/2$ may be expanded in the series

$$f(x) = \tfrac{1}{2}A_0 + \sum_{m=1}^{\infty}\left[A_m \cos\left(\frac{2\pi mx}{a}\right) + B_m \sin\left(\frac{2\pi mx}{a}\right)\right],$$

where

$$A_m = \frac{2}{a}\int_{-a/2}^{a/2} f(x)\cos\left(\frac{2\pi mx}{a}\right)dx,$$

$$B_m = \frac{2}{a}\int_{-a/2}^{a/2} f(x)\sin\left(\frac{2\pi mx}{a}\right)dx.$$

The functions

$$\sqrt{\frac{2}{a}}\sin\left(\frac{2\pi mx}{a}\right) \quad \text{and} \quad \sqrt{\frac{2}{a}}\cos\left(\frac{2\pi mx}{a}\right)$$

form an orthonormal set.

If the interval becomes infinite the expansion becomes the *Fourier integral*

$$f(x) = \frac{1}{(2\pi)^{1/2}}\int_{-\infty}^{\infty} A(k)e^{ikx}dk,$$

where

$$A(k) = \frac{1}{(2\pi)^{1/2}}\int_{-\infty}^{\infty} e^{-ikx}f(x)dx.$$

The orthogonality condition is

$$\frac{1}{2\pi}\int_{-\infty}^{\infty} e^{i(k-k')x}dx = \delta(k-k'),$$

while the completeness relation is

$$\frac{1}{2\pi}\int_{-\infty}^{\infty} e^{ik(x-x')}dk = \delta(x-x'),$$

where $\delta(k-k')$ and $\delta(x-x')$ are δ functions.

M. Dirac δ function[4]

Properties:

$$\delta(x-a) = 0 \quad \text{for } x \neq a,$$

and

$$\int \delta(x-a)dx = 1$$

if the region of integration includes $x=0$ and is zero otherwise.

For an arbitrary function $f(x)$,

$$\int f(x)\delta(x-a)dx = f(a),$$

$$\int f(x)\delta'(x-a)dx = -f'(a),$$

where the prime denotes differentiation with respect to the argument.

If the δ function has as an argument a function $f(x)$ of the independent variable x, it can be transformed according to the rule

$$\delta(f(x)) = \sum_i \frac{1}{|f'(x_i)|}\delta(x-x_i).$$

where $f(x)$ is assumed to have only simple zeros, located at $x=x_i$.

N. Spherical harmonics[4,5]

Legendre polynomials

$$P_l(x) = \frac{1}{2^l l!}\frac{d^l}{dx^l}(x^2-1)^l,$$

$$(1-x^2)P_l'' - 2xP_l' + l(l+1)P_l = 0,$$

$$\int_{-1}^{1} P_l^2(x)dx = \frac{2}{2l+1},$$

$$\int_{-1}^{1} P_l(x)P_{l'}(x)dx = 0 \quad \text{for } l \neq l',$$

$$P_l = \frac{2l-1}{l}xP_{l-1} - \frac{l-1}{l}P_{l-2},$$

$$P_0 = 1, \quad P_1 = x, \quad P_2 = \tfrac{3}{2}x^2 - \tfrac{1}{2},$$

$$P_3 = \tfrac{5}{2}x^3 - \tfrac{3}{2}x, \quad P_4 = \tfrac{35}{8}x^4 - \tfrac{15}{4}x^2 + \tfrac{3}{8},$$

$$P_5 = \tfrac{63}{8}x^5 - \tfrac{35}{4}x^3 + \tfrac{15}{8}x; \quad P_l(1) = 1.$$

Alternate definition:

$$\frac{1}{(1-2rx+r^2)^{1/2}} = \sum_{l=0}^{\infty} P_l(x)r^l,$$

$$\int_{-1}^{1} xP_l(x)P_{l'}(x)dx$$

$$= \frac{(l+l'+1)}{(l+l')(l+l'+2)} \quad \text{for } |l-l'| = 1$$

$$= 0 \quad \text{otherwise.}$$

Potential at **r** due to a unit point charge at **r**′:

$$\frac{1}{|\mathbf{r}-\mathbf{r}'|} = \sum_{l=0}^{\infty}\frac{r_<^l}{r_>^{l+1}}P_l(\cos\gamma),$$

where $r_<$ ($r_>$) is the smaller (larger) of $|\mathbf{r}|$ and $|\mathbf{r}'|$, and γ is the angle between **r** and **r**′.

$$Y_{lm}(\theta,\phi) = \frac{1}{N_{lm}}e^{im\phi}\sin^{|m|}\theta\frac{d^{|m|}P^l(\cos\theta)}{d(\cos\theta)^{|m|}},$$

$$\frac{1}{N_{lm}} = \pm\frac{1}{(2\pi)^{1/2}}\sqrt{\frac{2l+1}{2}\frac{(l-|m|)!}{(l+|m|)!}}$$

for $m \leqslant 0$ (+ sign) and $m > 0$ (− sign).

Normalization

$$\int_{4\pi} Y_{lm}^* Y_{l'm'}d\Omega = \delta_{ll'}\delta_{mm'},$$

where the asterisk denotes the complex conjugate.

Differential equation

$$\Lambda Y_{lm} + l(l+1)Y_{lm} = 0,$$

$$\Lambda = \frac{1}{\sin\theta}\frac{\partial}{\partial\theta}\left(\sin\theta\frac{\partial}{\partial\theta}\right) + \frac{1}{\sin^2\theta}\frac{\partial^2}{\partial\phi^2},$$

$$\nabla^2(r^l Y_l) = 0, \quad \nabla^2(r^{-l-1}Y_l) = 0 \quad \text{(except origin)},$$

$$\nabla^2 = \frac{\partial^2}{\partial r^2} + \frac{2}{r}\frac{\partial}{\partial r} + \frac{1}{r^2}\Lambda,$$

$$Y_{00} = \frac{1}{(4\pi)^{1/2}}, \quad Y_{10} = \sqrt{\frac{3}{4\pi}}\cos\theta,$$

$$Y_{1,\pm1} = \mp\sqrt{\frac{3}{8\pi}}\sin\theta\, e^{\pm i\phi},$$

$$Y_{20} = \sqrt{\frac{5}{4\pi}}\left(\tfrac{3}{2}\cos^2\theta - \tfrac{1}{2}\right),$$

$$Y_{2,\pm1} = \mp\sqrt{\frac{15}{8\pi}}\sin\theta\cos\theta\, e^{\pm i\phi},$$

$$Y_{2,\pm2} = \frac{1}{4}\sqrt{\frac{15}{2\pi}}\sin^2\theta\, e^{\pm 2i\phi},$$

$$Y_{30} = \sqrt{\frac{7}{4\pi}}\left(\tfrac{5}{2}\cos^3\theta - \tfrac{3}{2}\cos\theta\right),$$

$$Y_{3,\pm1} = \mp\frac{1}{4}\sqrt{\frac{21}{4\pi}}\sin\theta\,(5\cos^2\theta - 1)e^{\pm i\phi},$$

$$Y_{3,\pm2} = \frac{1}{4}\sqrt{\frac{105}{2\pi}}\sin^2\theta\cos\theta\, e^{\pm 2i\phi},$$

$$Y_{3,\pm3} = \mp\frac{1}{4}\sqrt{\frac{35}{4\pi}}\sin^3\theta\, e^{\pm 3i\phi}.$$

Completeness relation

$$\sum_{l=0}^{\infty}\sum_{m=-l}^{l} Y^*_{lm}(\theta',\phi')Y_{lm}(\theta,\phi) = \delta(\phi-\phi')\delta(\cos\theta - \cos\theta').$$

Potential problems

The general solution for a boundary-value problem in spherical coordinates can be written

$$\Phi(r,\theta,\phi) = \sum_{l=0}^{\infty}\sum_{m=-l}^{l}(A_{lm}r^l + B_{lm}r^{-l-1})Y_{lm}(\theta,\phi).$$

If the potential on a spherical surface is given by $f(\theta,\phi)$ and there are no charges at the origin so that $B_{lm} = 0$, then

$$A_{lm} = \int_{4\pi} d\Omega\, Y^*_{lm}(\theta,\phi)f(\theta,\phi).$$

Addition theorem

Given two vectors **r** and **r**′ with spherical coordinates (r,θ,ϕ) and (r',θ',ϕ'), respectively, with angle γ between them, then

$$P_l(\cos\gamma) = \frac{4\pi}{2l+1}\sum_{m=-l}^{l} Y^*_{lm}(\theta',\phi')Y_{lm}(\theta,\phi),$$

where $\cos\gamma = \cos\theta\cos\theta' + \sin\theta\sin\theta'\cos(\phi - \phi')$.

O. Clebsch-Gordan coefficients

The Clebsch-Gordan coefficients are the values of $\langle T,T_3|I',I'_3;I'',I''_3\rangle$ which govern the transformation

$$|T,T_3\rangle = \sum_{I',I''}\langle T,T_3|I',I'_3;I'',I''_3\rangle|I',I'_3\rangle|I'',I''_3\rangle,$$

$$T = I' + I'', \quad I' + I'' - 1, \quad I' + I'' - 2, \ldots, |I' - I''|,$$

$$T_3 = I'_3 + I''_3.$$

The Clebsch-Gordan coefficients are designed so that if the initial set of wave functions is complete and orthornormal, the new set will also be complete and orthonormal. The following table applies to the case $I' = 1$, $I'' = 1/2$.

$I',I'_3;I'',I''_3$	T,T_3: $\frac{3}{2},\frac{3}{2}$	$\frac{3}{2},\frac{1}{2}$	$\frac{3}{2},-\frac{1}{2}$	$\frac{3}{2},-\frac{3}{2}$	$\frac{1}{2},\frac{1}{2}$	$\frac{1}{2},-\frac{1}{2}$
$1,+1;\frac{1}{2},\frac{1}{2}$	1					
$1,0;\frac{1}{2},\frac{1}{2}$		$\sqrt{\frac{2}{3}}$			$-\sqrt{(\frac{1}{3})}$	
$1,-1;\frac{1}{2},\frac{1}{2}$			$\sqrt{\frac{1}{3}}$			$-\sqrt{(\frac{2}{3})}$
$1,+1;\frac{1}{2},-\frac{1}{2}$		$\sqrt{\frac{1}{3}}$			$\sqrt{\frac{2}{3}}$	
$1,0;\frac{1}{2},-\frac{1}{2}$			$\sqrt{\frac{2}{3}}$			$\sqrt{\frac{1}{3}}$
$1,-1;\frac{1}{2},-\frac{1}{2}$				1		

The same coefficients apply to the inverse transformation. Thus, from the table, reading down expresses the total spin wave functions in terms of the individual wave functions:

$$|\tfrac{3}{2},-\tfrac{1}{2}\rangle = (\sqrt{\tfrac{1}{3}})|1,-1\rangle|\tfrac{1}{2},\tfrac{1}{2}\rangle + (\sqrt{\tfrac{2}{3}})|1,0\rangle|\tfrac{1}{2},-\tfrac{1}{2}\rangle,$$

and reading across expresses the individual spin wave functions in terms of the total spin wave functions:

$$|1,0\rangle|\tfrac{1}{2},-\tfrac{1}{2}\rangle = (\sqrt{\tfrac{2}{3}})|\tfrac{3}{2},-\tfrac{1}{2}\rangle + (\sqrt{\tfrac{1}{3}})|\tfrac{1}{2},-\tfrac{1}{2}\rangle.$$

1. Table: Clebsch-Gordan coefficients[6]

Sign convention is that of E. P. Wigner, *Group Theory* (Academic, New York, 1959), also used by E. U. Condon and G. H. Shortley, *The Theory of Atomic Spectra* (Cambridge University, New York, 1953), M. E. Rose, *Elementary Theory of Angular Momentum* (Wiley, New York, 1957), and E. R. Cohen, *Tables of the Clebsch-Gordan Coefficients* (North American Rockwell Science Center, Thousand Oaks, CA, 1974). The signs and numbers in the current tables have been calculated by computer programs written independently by Cohen and at LBL.

Note: a $\sqrt{}$ is to be understood over every coefficient, e.g., for $-8/15$ read $-\sqrt{8/15}$.

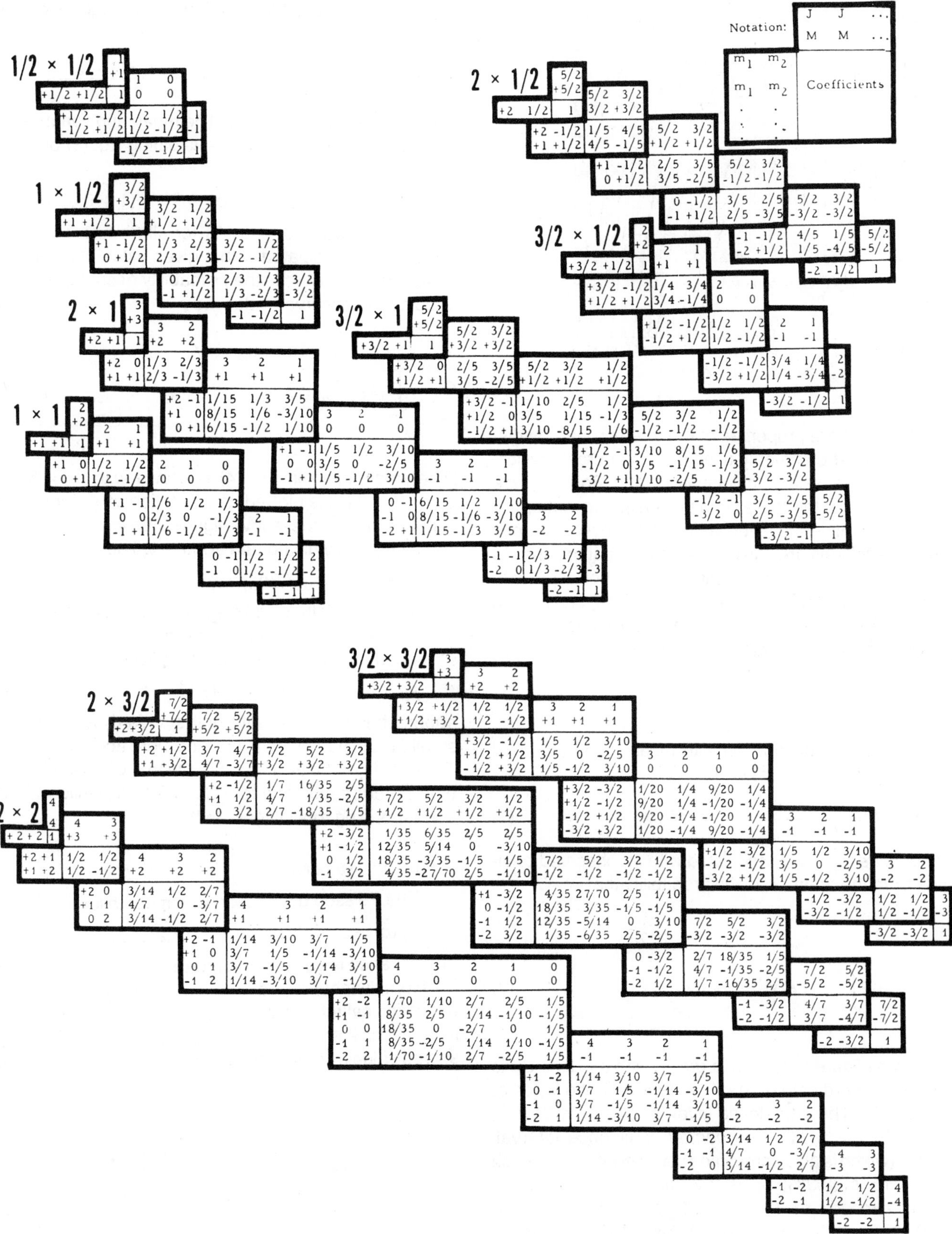

1.04. PROBABILITY AND STATISTICS[6]

A. Figure: Probability distributions and confidence levels

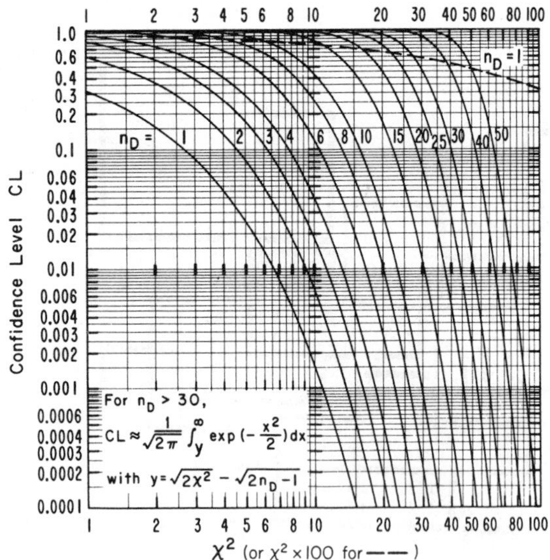

We give here properties of the three probability distributions most commonly used in high-energy physics: normal (or Gaussian), chi-squared, and Poisson. We *warn* the reader the there is no universal convention for the term "confidence level" as used by physicists; thus explicit definitions are given for each distribution, and we have attempted to choose definitions that correspond to common usage. It is explained below how confidence levels for all three distributions can be extracted from the above figure, which shows the χ^2 confidence level versus χ^2 for n_D degrees of freedom.

1. Normal distribution

The normal distribution with mean $\bar{x}$ and standard deviation σ (variance σ^2) is

$$P(x)dx = \frac{1}{(2\pi)^{1/2}\sigma} e^{-(x-\bar{x})^2/2\sigma^2} dx. \tag{1}$$

The *confidence level* associated with an observed deviation from the mean, δ, is the probability that $|x - \bar{x}| > \delta$, i.e.,

$$\mathrm{CL} = 2\int_{\bar{x}+\delta}^{\infty} dx\, P(x). \tag{2}$$

[The small figure in Eq. (2) is drawn with $\delta = 2\sigma$.] CL is given by the ordinate of the $n_D = 1$ curve in the figure at $\chi^2 = (\delta/\sigma)^2$. The confidence level for $\delta = 1\sigma$ is 31.7%; 2σ, 4.6%; 3σ, 0.3%. The *central* confidence interval, 1 − CL (which is also sometimes called confidence level) for $\delta = 1\sigma$ is 68.3%; 2σ, 95.4%; 3σ, 99.7%. The *odds* against exceeding δ, (1 − CL)/CL, for $\delta = 1\sigma$ are 2.15:1; 2σ, 21:1; 3σ, 370:1; 4σ, 16 000:1; 5σ, 1 700 000:1. Relations between σ and other measures of the *width*: probable error (CL = 0.5 deviation) = 0.67σ; mean absolute deviation = 0.80σ; rms deviation = σ; half-width at half maximum = 1.18σ.

2. Chi-squared distribution

The chi-squared distribution for n_D degrees of freedom is

$$P_{n_D}(\chi^2)d\chi^2 = \frac{1}{2^h\Gamma(h)} (\chi^2)^{h-1} e^{-\chi^2/2} d\chi^2, \quad (\chi^2 \geqslant 0) \tag{3}$$

where h (for "half") = $n_D/2$. The mean and variance are n_D and $2n_D$, respectively. In evaluating Eq. (3) one may use *Stirling's approximation:*

$$\Gamma(h) = (h-1)! \approx 2.507 e^{-h} h^{h-1/2}(1 + 0.0833/h),$$

which is accurate to $\pm 0.1\%$ for all $h \geqslant 1/2$. The confidence level associated with a given value of n_D and an observed value of χ^2 is the probability of chi-squared exceeding the observed value, i.e.,

$$\mathrm{CL} = \int_{\chi^2}^{\infty} d\chi^2 P_{n_D}(\chi^2). \tag{4}$$

[The small figure in Eq. (4) is drawn with $n_D = 5$ and CL = 10%.] CL is plotted as a function of χ^2 for several values of n_D in the above figure. For *large* n_D, χ^2 becomes normally distributed about n_D. Thus

$$y_1 = (\chi^2 - n_D)/(2n_D)^{1/2} \tag{5}$$

becomes normally distributed with unit standard deviation. A better approximation, due to Fisher,[7] is that χ, not χ^2, becomes normally distributed; specifically

$$y_2 = (2\chi^2)^{1/2} - (2n_D - 1)^{1/2} \tag{6}$$

approaches normality with unit standard deviation. For small CL's in particular, y_2 is much more accurate than y_1. Thus, for $n_D = 50$ and $\chi^2 = 80$, the true CL = 0.45%, but y_1 is 3.0 corresponding to a CL of 0.13%, while y_2 is 2.7 corresponding to a CL of 0.35%.

3. Poisson distribution

The Poisson distribution with mean $\bar{n}$ is

$$P_{\bar{n}}(n) = e^{-\bar{n}}(\bar{n})^n/n! \quad (n = 0,1,2,\ldots). \tag{7}$$

The variance is equal to the mean. *Confidence levels* for Poisson distributions are usually defined in terms of quantities called *upper limits* as follows: The confidence level associated with a given upper limit N and an observed value n_0 of n is the probability that $n > n_0$ if $\bar{n} = N$, i.e.,

$$\mathrm{CL} = \sum_{n=n_0+1}^{\infty} P_N(n) = 1 - \sum_{n=0}^{n_0} P_N(n). \tag{8}$$

[The small figure in Eq. (8) is drawn with $n_0 = 2$ and CL = 90%.] A useful relation between Poisson and chi-squared confidence levels allows one to look up this quantity on the above figure. Specifically, the quantity $1 - \text{CL}$ is given by the ordinate of the $n_D = 2(n_0 + 1)$ curve at $\chi^2 = 2N$. Thus 90% confidence level upper limits for $n_0 = 0, 1$, and 2 are given by half the χ^2 value corresponding to an ordinate of 0.1 on the $n_D = 2$, 4, and 6 curves, respectively; the values are $N = 2.3, 3.9$, and 5.3.

Tables of confidence levels for all three of these distributions, the relation between Poisson and chi-squared confidence levels, and numerous other useful tables and relations may be found in Ref. 8.

B. Statistics

We consider here the situation in which one is presented with N independent data, $y_n \pm \sigma_n$, and it is desired to make some *inference* about the "true" value of the quantity represented by these data. For this purpose we interpret each datum y_n as a single sample point drawn randomly (and independently of the other data) from a distribution having mean $\bar{y}_n$ (which we wish to estimate) and variance σ_n^2. (Identification of the true σ_n with the σ_n datum is an *approximation* which may become seriously inaccurate when σ_n is an appreciable fraction of y_n.) Some methods of estimation commonly used in high-energy physics are given below; see Ref. 9 for numerous applications. Section 1.04.B.1 deals with the case in which all $\bar{y}_n$ are the same, e.g., several different measurements of the same quantity; Sec. 1.04.B.2 deals with the case in which $\bar{y}_n = \bar{y}(x_n)$, where x_n represents some set of independent variables, e.g., cross-section measurements at various values of energy and angle, $x_n = \{E_n, \theta_n\}$.

1. Single mean and variance estimates

(1) If the y_n represent a set of values all supposedly drawn from a single distribution with mean $\bar{y}$ and variance σ^2 (i.e., the σ_n are all the same, but their common value is unknown), then

$$\bar{y}_e = \frac{1}{N}\sum_n y_n \tag{9}$$

and

$$\sigma_e^2 = \frac{1}{N-1}\sum_n (y_n - \bar{y}_e)^2 = \frac{N}{N-1}\left[\left(\overline{y^2}\right)_e - \bar{y}_e^2\right] \tag{10}$$

are unbiased estimates of $\bar{y}$ and σ^2. The variance of $\bar{y}_e$ is σ^2/N. If the parent distribution is normal and N is large, the variance of σ_e^2 is $2\sigma^4/N$.

(2) If the $\bar{y}_n$ all have the common value $\bar{y}$ and the σ_n are known, then the weighted average

$$\bar{y}_e = \frac{1}{w}\sum_n w_n y_n, \tag{11}$$

where $w_n = 1/\sigma_n^2$ and $w = \Sigma w_n$, is an appropriate unbiased estimate of $\bar{y}$. This choice of weighting factors in Eq. (11) minimizes the variance of the estimate; the variance is $1/w$.

2. Linear least-squares fit

A least-squares fit of the function $y(x) = \Sigma_i a_i f_i(x)$ to independent data $y_n \pm \sigma_n$ (e.g., a Legendre fit in which the f_i are Legendre polynomials and the a_i are Legendre coefficients) gives the following estimates of the parameters a_i:

$$a_{e,i} = \sum_{j,n} V_{ij} f_j(x_n) y_n/\sigma_n^2. \tag{12}$$

Here V is the covariance matrix of the fitted parameters

$$V_{ij} = \overline{(a_{e,i} - \bar{a}_{e,i})(a_{e,j} - \bar{a}_{e,j})}, \tag{13}$$

which is given by

$$(V^{-1})_{ij} = \sum_n f_i(x_n) f_j(x_n)/\sigma_n^2. \tag{14}$$

The variance of an interpolated or extrapolated value of y at point x, $y_e = \Sigma a_{e,i} f_i(x)$, is

$$\overline{(y_e - \bar{y}_e)^2} = \sum_{ij} V_{ij} f_i(x) f_j(x). \tag{15}$$

For the case of a *straight-line fit*, $y(x) = a + bx$, one obtains the following estimates of a and b:

$$\begin{aligned} a_e &= (S_y S_{xx} - S_x S_{xy})/D, \\ b_e &= (S_1 S_{xy} - S_x S_y)/D, \end{aligned} \tag{16}$$

where

$$S_1, S_x, S_y, S_{xx}, S_{xy} = \sum (1, x_n, y_n, x_n^2, x_n y_n)/\sigma_n^2, \tag{17}$$

$$D = S_1 S_{xx} - S_x^2.$$

The convariance matrix of the fitted parameters is

$$\begin{pmatrix} V_{aa} & V_{ab} \\ V_{ab} & V_{bb} \end{pmatrix} = \frac{1}{D}\begin{pmatrix} S_{xx} & -S_x \\ -S_x & S_1 \end{pmatrix}. \tag{18}$$

The variance of an interpolated or extrapolated value of y at point x is

$$\overline{(y_e - \bar{y}_e)^2} = \frac{1}{S_1} + \frac{S_1}{D}\left(x - \frac{S_x}{S_1}\right)^2. \tag{19}$$

C. Error propagation

We consider here the situation in which one wishes to calculate the value and error of a function of some other quantities with the errors, e.g., in a Monte Carlo program. Let $\{y\}$ be a set of random variables with means $\{\bar{y}\}$ and convariance matrix V. Then the mean and variance of a function of these variables are approximately (to second order in $\{y - \bar{y}\}$)

$$\bar{f} \approx f(\{y\}) + \tfrac{1}{2}\sum_{mn} V_{mn}\left(\frac{\partial^2 f}{\partial y_m \partial y_n}\right)_{\{y\}=\{\bar{y}\}}, \tag{20}$$

$$\overline{(f - \bar{f})^2} = \sum_{mn} V_{mn}\left(\frac{\partial f}{\partial y_m}\right)_{\{y\}=\{\bar{y}\}}\left(\frac{\partial f}{\partial y_n}\right)_{\{y\}=\{\bar{y}\}}. \tag{21}$$

E.g., the mean and variance of a function of a *single variable* with mean $\bar{y}$ and variance σ^2 are

$$\bar{f} \approx f(\bar{y}) + \tfrac{1}{2}\sigma^2 f''(\bar{y}), \tag{22}$$

$$\overline{(f-\bar{f})^2} = \sigma^2 f'(\bar{y})^2. \quad (23)$$

Note that these equations will usually be applied by substituting some measured quantities, $\{\tilde{y}\}$ say, for the true means, $\{\bar{y}\}$. If, as is often the case, $\tilde{y}_n - \bar{y}_n$ is order $(V_{nn})^{1/2}$, then there is no point in keeping the second-order terms in Eq. (20) or (22) since the substitution itself introduces first-order errors.

1.05. PHYSICS FORMULARY

A. Mechanics

1. Newtonian dynamics of particles

Newton's laws. (1) If $\mathbf{F}=0$, $\mathbf{v}=$ const in an inertial frame (defines the inertial frame). (2) The acceleration $\mathbf{a}=\mathbf{F}/m$, defines the force

$$\mathbf{F}=\frac{d}{dt}(m\mathbf{v}).$$

(3) Action of particle 1 on particle 2 is equal and opposite to that of particle 2 on 1, $\mathbf{F}_{12}=-\mathbf{F}_{21}$.

Statics. Forces in balance, no acceleration:

$$\sum_i \mathbf{F}_a = 0.$$

Coriolis force. When a particle moves in a moving frame Σ' rotating with angular velocity ω with respect to an inertial frame Σ, and second law becomes

$$\mathbf{F}=m\frac{d^2\mathbf{r}}{dt^2}=m\frac{d'^2\mathbf{r}}{dt^2}+2m\left(\boldsymbol{\omega}\times\frac{d'\mathbf{r}}{dt}\right)+m\boldsymbol{\omega}\times(\boldsymbol{\omega}\times\mathbf{r}).$$

The term $2\boldsymbol{\omega}\times d'\mathbf{r}/dt$ is the *Coriolis acceleration.* Here $\mathbf{r}$ is the position vector of the particle with respect to the moving axis and d'/dt is in the moving frame.

The impulse of a force

$$\mathbf{I}=\int_{t_0}^{t}\mathbf{F}\,dt$$

is related to the change in momentum: $\mathbf{I}=m\mathbf{v}-m\mathbf{v}_0$.

The work done is given by

$$W=\int_{s_0}^{s_1}\mathbf{F}\cdot d\mathbf{s},$$

where s_0, s_1 are two positions along a path s. The change in kinetic energy is

$$W=\tfrac{1}{2}mv_1^2-\tfrac{1}{2}mv_0^2.$$

Potential energy. If the work done does not depend on the path but only on initial and final positions, the force is said to be conservative. In this case $\nabla\times\mathbf{F}=0$ and $\mathbf{F}=\nabla\phi$ defines the potential energy ϕ. The total mechanical energy is

$$U=\tfrac{1}{2}mv^2+\phi(x,y,z).$$

Rigid body. The distance between any two points within a rigid body remains fixed. Six numbers suffice to specify its positions, the position of the origin and the orientation of a coordinate systems fixed in the body. Two types of motion: translation and rotation. Total mass:

$$M=\int_V \rho\,dV,$$

where ρ is density and V volume. Center or mass:

$$\mathbf{R}=\frac{1}{M}\int_V \rho\mathbf{r}\,dV.$$

Translation motion:

$$\mathbf{F}=M\ddot{\mathbf{R}},$$

where $\mathbf{F}$ is the resultant of the applied external forces,

$$\mathbf{F}=\sum_i \mathbf{F}_i.$$

For an instantaneous angular velocity $\boldsymbol{\omega}$, the velocity of any point in the rigid body with position $\mathbf{r}_p$ relative to an origin on the axis is given by

$$\mathbf{v}_p=\boldsymbol{\omega}\times\mathbf{r}_p.$$

The *angular acceleration*

$$\boldsymbol{\alpha}=\dot{\boldsymbol{\omega}}.$$

The *total moment of momentum* or *angular momentum* with respect to some fixed origin of coordinates whether inside or outside the body is given in Cartesian coordinates by

$$\begin{aligned}\mathbf{L}=&\,\mathbf{i}(\omega_x I_{xx}-\omega_y I_{xy}-\omega_z I_{xz})\\&+\mathbf{j}(-\omega_x I_{yx}+\omega_y I_{yy}-\omega_z I_{yz})\\&+\mathbf{k}(-\omega_x I_{zx}-\omega_y I_{zy}+\omega_z I_{zz}).\end{aligned}$$

Moments of inertia:

$$I_{xx}=\int_V (y^2+z^2)\rho\,dV, \quad \text{etc.}$$

Products of inertia:

$$I_{xy}=\int_V xy\rho\,dV, \quad \text{etc.}$$

By proper choice of axes

$$I_{xy}=I_{xz}=I_{yx}=\cdots=0 \quad \text{and} \quad I_{xx}=MR^2,$$

where

$$R^2=\frac{1}{M}\int_V (y^2+z^2)\rho\,dV$$

is the square of the *radius of gyration* about the x axis.

The *torque* is

$$\mathbf{T}=\sum_i \mathbf{r}_i\times\mathbf{F}_i,$$

where $\mathbf{r}_i$ is the position of application of the force $\mathbf{F}_i$ with respect to the fixed origin.

Newton's second law for the motion of a rigid body about a fixed origin is

$$\mathbf{T}=\dot{\mathbf{L}}.$$

If the fixed origin is taken as the center of mass, the translational motion is simply that of the center of mass.

For motion about a point fixed in space, I_{xx}, I_{xy}, etc., and ω change in time. However, with axes fixed in the body I_{xx}, I_{xy}, etc., remain constant. Then, for motion about a fixed point the axes rotate, and we have

$$\dot{\mathbf{L}}=\mathbf{i}\dot{L}_x+\mathbf{j}\dot{L}_y+\mathbf{k}\dot{L}_z+\boldsymbol{\omega}\times\mathbf{L},$$

where $\dot{L}_x, \dot{L}_y, \dot{L}_z$ are taken about the moving axes and ω is taken about the instantaneous axis of rotation. If we choose principal axes, $I_{xy} = I_{xz} = \cdots = 0$, we have *Euler's equations*

$$\mathbf{T} = \mathbf{i}[I_{xx}\dot{\omega}_x + (I_{zz} - I_{yy})\omega_y\omega_z] + \mathbf{j}[I_{yy}\dot{\omega}_y + (I_{xx} - I_{zz})\omega_z\omega_x] + \mathbf{k}[I_{zz}\dot{\omega}_z + (I_{yy} - I_{xx})\omega_x\omega_y].$$

2. Elastic constants

In general, if the displacement of any particle in a solid body from its equilibrium position is denoted by the vector

$$\mathbf{s} = \mathbf{i}\xi + \mathbf{j}\eta + \mathbf{k}\zeta,$$

where the displacement components ξ, η, ζ are functions of both space and time, the effective *strain* is denoted by the covariant tensor of the second order written in matrix form as

$$D = \begin{pmatrix} \partial\xi/\partial x & \frac{1}{2}(\partial\xi/\partial y + \partial\eta/\partial x) & \frac{1}{2}(\partial\xi/\partial z + \partial\zeta/\partial x) \\ \frac{1}{2}(\partial\eta/\partial x + \partial\xi/\partial y) & \partial\eta/dy & \frac{1}{2}(\partial\eta/\partial z + \partial\zeta/\partial y) \\ \frac{1}{2}(\partial\zeta/\partial x + \partial\xi/\partial z) & \frac{1}{2}(\partial\zeta/\partial y + \partial\eta/\partial z) & \partial\zeta/\partial z \end{pmatrix}.$$

In symbolic form

$$D = \begin{pmatrix} \epsilon_{xx} & \frac{1}{2}\epsilon_{xy} & \frac{1}{2}\epsilon_{xz} \\ \frac{1}{2}\epsilon_{yx} & \epsilon_{yy} & \frac{1}{2}\epsilon_{yz} \\ \frac{1}{2}\epsilon_{zx} & \frac{1}{2}\epsilon_{zy} & \epsilon_{zz} \end{pmatrix}.$$

In single crystals depending on the symmetry, Hooke's law becomes

$$\epsilon_{xx} = S_{11}\sigma_x + S_{12}\sigma_y + S_{13}\sigma_z + S_{14}\tau_{yz} + S_{15}\tau_{zx} + S_{16}\tau_{xy}$$

$$\epsilon_{yy} = S_{21}\sigma_x + S_{22}\sigma_y + S_{23}\sigma_z + S_{24}\tau_{yz} + S_{25}\tau_{zx} + S_{26}\tau_{xy}$$

$$\epsilon_{zz} = S_{31}\sigma_x + \cdots, \quad \epsilon_{zx} = S_{41}\sigma_x + \cdots,$$

$$\epsilon_{zy} = S_{51}\sigma_x + \cdots, \quad \epsilon_{xy} = S_{61}\sigma_x + \cdots,$$

etc. The 36 coefficients $S_{11}, S_{12}, \ldots, S_{ij}, \ldots, S_{ii}$ are called the *elastic constants.* The corresponding coefficients C_{ij} in equations for the stresses in terms of the strains are called the *elastic coefficients.* For any i, j, $S_{ij} = S_{ji}$ and $C_{ij} = C_{ji}$.

Examples. C_{ij} in units of 10^{10} Pa (10^{11} dyn/cm^2), S_{ij} in units of 10^{-12} Pa^{-1} (10^{-13} cm^2/dyn).

NaCl, a cubic crystal:

$$C_{11} = 4.9, \quad C_{12} = 1.2, \quad C_{44} = 1.3;$$

$$S_{11} = 22.9, \quad S_{12} = -4.7, \quad S_{44} = 79.4.$$

$BaTiO_3$, hexagonal:

$$C_{11} = 16.8, \quad C_{33} = 18.9, \quad C_{44} = 5.5,$$

$$C_{12} = 7.8, \quad C_{13} = 7.1;$$

$$S_{11} = 8.2, \quad S_{33} = 6.8, \quad S_{44} = 18.3,$$

$$S_{12} = -3.0, \quad S_{13} = -2.0.$$

For a more complete table see Ref. 10.

3. Lorentz transformations

Transformation from an inertial frame to reference K to a frame K' which travels with respect to K at a velocity βc along $+x$ direction, $\gamma = (1 - \beta^2)^{-1/2}$:

$$x' = \gamma(x - \beta ct), \quad y' = y, \quad z' = z,$$

$$t' = \gamma[t - (\beta/c)x].$$

Transformation from K' to K: change the sign of β.

Velocities. A body moves with velocity $\mathbf{v}$ in K and $\mathbf{v}'$ in K':

$$v_x = \frac{v_x' + \beta c}{1 + (\beta/c)v_x'},$$

$$v_y = \frac{v_y'}{\gamma[1 + (\beta/c)v_x']},$$

$$v_z = \frac{v_z'}{\gamma[1 + (\beta/c)v_x']}.$$

B. Thermodynamics[11]

1. Equation of state

Relates the pressure, volume, and temperature of a given amount of substance in a system:

$$f(p, V, T) = 0.$$

The *work done* in a transformation from an initial state A to a final state B:

$$L = \int_A^B p\,dV.$$

In an *isochore* transformation no work is done. This is the case for $V =$ const. In *isobaric* transformations the pressure is constant; in *isothermal* transformations the temperature is constant.

2. Ideal gas

The equation of state for a mass m of a gas whose molar mass is M is

$$pV = (m/M)RT.$$

R is a universal constant: $R = 8.314$ J/K.

For 1 mol of gas

$$L = RT\ln(V_2/V_1) = RT\ln(p_1/p_2).$$

3. First law

Principle of conservation of energy. In a transformation of a system between state A and state B for which the energies are U_A and U_B, respectively,

$$\Delta U + L = Q,$$

where $\Delta U = U_B - U_A$, and Q is the amount of energy received by the system in forms other than work. For an infinitesimal transformation,

$$dU + dL = dQ.$$

For an ideal gas, $dL = p\,dV$.

4. Molecular heat

The amount of heat required to raise the temperature of 1 mol of a substance 1 deg. [*Specific heat* refers to either 1 g (cgs) or 1 kg (SI) of substance.] At constant volume

$$C_V = \left(\frac{\partial Q}{\partial T}\right)_V = \left(\frac{\partial U}{\partial T}\right)_V.$$

At constant pressure

$$C_p = \left(\frac{\partial Q}{\partial T}\right)_p = \left(\frac{\partial U}{\partial T}\right)_p + p\left(\frac{\partial V}{\partial T}\right)_p.$$

For an ideal gas, and assuming $C_V = \text{const}$,

$$C_p = C_V + R.$$

From kinetic theory, in good agreement with experiment,

$$C_V = \tfrac{3}{2}R \quad \text{for monatomic gas}$$
$$= \tfrac{5}{2}R \quad \text{for diatomic gas,}$$
$$\gamma = C_p/C_V = \tfrac{5}{3} \quad \text{for monatomic gas}$$
$$= \tfrac{7}{5} \quad \text{for diatomic gas.}$$

In an *adiabatic* transformation the thermodynamic system is thermally isolated; the transformation is reversible, no heat is exchanged between it and its environment. For an ideal gas

$$pV^{\gamma} = \text{const}, \quad TV^{\gamma-1} = \text{const}.$$

5. Second law

Kelvin postulate. A transformation whose only final result is to transform into work heat extracted from a source which is at the same temperature throughout is impossible.

Clausius postulate. A transformation whose only result is to transfer heat from a body at a given temperature to a body at a higher temperature is impossible.

6. Carnot cycle

A cycle in which a fluid undergoes a series of transformations, isothermal at temperature T_2 from A to B, adiabatic from B to C, isothermal at temperature T_1 from C and D, and adiabatic back to A. If the cycle is carried out reversibly the efficiency is

$$\eta = (T_2 - T_1)/T_2,$$

and is the highest possible efficiency that an engine working between the temperatures T_1 and T_2 can have.

A Carnot cycle operated in the reverse sense can be used to extract an amount of heat Q_1 from a source at the low temperature T_1 by absorbing an amount of work L. Thus, if T_2 is the temperature of the receiver,

$$Q_1 = LT_1/(T_2 - T_1).$$

An actual refrigerator will always have a lower efficiency because irreversible processes are always involved in refrigerating devices.

7. Entropy

In a reversible transformation, e.g. one whose successive states differ by infinitesimals from equilibrium states,

$$S(B) - S(A) = \int_A^B \frac{dQ}{T},$$

where $S(B)$ and $S(A)$ are the entropies of states B and A, respectively; the integral is independent of the path over which the transformation goes from A to B. In differential form,

$$dS = dQ/T.$$

For an irreversible transformation

$$S(B) - S(A) \geqslant \int_A^B \frac{dQ}{T}.$$

Unlike dQ, dS is a perfect differential. For an ideal gas

$$S = C_V \ln T + R \ln V + a,$$

where a is a constant of integration.

8. Clapeyron equation

Deals with a system composed of a liquid and its vapor in equilibrium. Liquid and vapor can coexist at temperatures below the critical temperature T_c. In this region liquid transforms to vapor in an isothermal increase in volume, and the pressure remains constant. Above the critical temperature only vapor exists. At the critical temperature the pressure and volume at which the two phases, liquid and vapor, no longer coexist are the critical pressure p_c and critical volume V_c, respectively. Clapeyron's equation gives the increase in pressure with temperature for a system in which liquid and vapor coexist:

$$\frac{dp}{dT} = \frac{\lambda}{T(v_2 - v_1)},$$

where v_1 and v_2 are the specific volumes (inverse of densities) for vapor and liquid phases, respectively; λ is the *latent heat of vaporization,* the amount of heat required to vaporize a unit weight of the substance, gram, mole, or kilogram as the case may be.

9. Thermodynamic potentials

Free energy:

$$F = U - TS.$$

For a system whose initial and final states are at the same temperature as the environment with which it exchanges only heat, the work that can be done in transformation from A to B is

$$L \leqslant F(A) - F(B) = -\Delta F.$$

The free energy of 1 mole of an ideal gas is given by

$$F = C_V T + W - T(C_V \ln T + R \ln V + a),$$

where W represents the energy left in the gas at absolute zero temperature, and a is a constant of integration.

Enthalpy:

$$H = U + pV.$$

Thermodynamic potential at constant pressure. In thermodynamical transformations in which both pressure and temperature do not change, but instead remain equal to the pressure and temperature of the environment,

$$G = U - TS + pV,$$

also called the *Gibbs free energy.*

If temperature and pressure of a system are kept constant, the state of the system for which G is a minimum is a state of stable equilibrium.

10. Third law

Nernst postulate. In any isothermal process carried out at absolute zero temperature, the entropy change of the system is zero:

$$\Delta S_{T=0} = 0.$$

Principle of unattainability of absolute zero. It is impossible to accomplish a process as a result of which the temperature of a body is reduced to $T = 0$ K.

11. Tables: Thermodynamic symbols, definitions, and formulas

From *American Institute of Physics Handbook,* 2nd ed., edited by Dwight E. Gray (McGraw-Hill, New York, 1957).

a. Thermodynamic systems and coordinates

System	Intensive coordinate		Extensive coordinate	
Chemical system	Pressure	p	Volume	V
Stretched wire	Tension	$\mathscr{F}$	Length	l
Surface film	Surface tension	$\mathscr{S}$	Area	A
Electric cell	emf	$\mathscr{E}$	Charge	Z
Capacitor	Electric intensity	E	Polarization	p
Magnetic substance	Magnetic intensity	$\mathscr{H}$	Magnetization	I

b. Work done by thermodynamic systems

System	Intensive quantity (generalized force)	Extensive quantity (generalized displacement)	Work
Chemical system	p in newtons/m^2	V in m^3	$p\,dV$ in joules
Stretched wire	$\mathscr{F}$ in newtons	l in m	$-\mathscr{F}\,dl$ in joules
Surface film	$\mathscr{S}$ in newtons/m	A in m^2	$-\mathscr{S}\,dA$ in joules
Electric cell	$\mathscr{E}$ in volts	Z in coulombs	$-\mathscr{E}\,dZ$ in joules
Capacitor	E in newtons/coulomb	p in coulombs m	$-E\,dp$ in joules
Magnetic substance	$\mathscr{H}$ in amperes/m	I in weber m	$-\mathscr{H}\,dI$ in joules

c. Definitions and symbols for thermal quantities

Thermal quantity	Symbol	Definition
Heat	Q	
Internal energy	U	
Entropy	S	
Enthalpy (also called heat content, heat function, total heat)	H	$U + pV$
Helmholtz function (also called free energy or work function)	F, A	$U - TS$
Gibbs function (also called free energy, free enthalpy, or thermodynamic potential)	G	$H - TS$
Volume expansivity (coefficient of volume expansion)	β	$\frac{1}{V}\left(\frac{\partial V}{\partial T}\right)_p$
Isothermal bulk modulus	B	$-V\left(\frac{\partial p}{\partial V}\right)_T$
Adiabatic bulk modulus	B_S	$-V\left(\frac{\partial p}{\partial V}\right)_S$
Isothermal compressibility	k	$-\frac{1}{V}\left(\frac{\partial V}{\partial p}\right)_T$
Adiabatic compressibility	k_s	$-\frac{1}{V}\left(\frac{\partial V}{\partial p}\right)_S$
Heat capacity at constant volume	C_V	$\left(\frac{dQ}{dT}\right)_V = T\left(\frac{\partial S}{\partial T}\right)_V$
Heat capacity at constant pressure	C_p	$\left(\frac{dQ}{dT}\right)_p = T\left(\frac{\partial S}{\partial T}\right)_p$
Ratio of heat capacities	γ	$\frac{C_p}{C_V}$
Joule coefficient	η	$\left(\frac{\partial T}{\partial V}\right)_U$
Joule-Thomson (Kelvin) coefficient	μ	$\left(\frac{\partial T}{\partial p}\right)_H$
Linear expansivity	α	$\frac{1}{l}\left(\frac{\partial l}{\partial T}\right)_{\mathscr{F}}$
Isothermal Young's modulus	Y	$\frac{l}{A}\left(\frac{\partial \mathscr{F}}{\partial l}\right)_T$
Adiabatic Young's modulus	Y_S	$\frac{l}{A}\left(\frac{\partial \mathscr{F}}{\partial l}\right)_S$

11. Tables—Continued

d. Thermodynamic formulas for a chemical system of constant mass

First law of thermodynamics:

$$Q = U_2 - U_1 + L$$
$$dQ = dU + dL$$

Second law of thermodynamics:

$$dQ = T\,dS$$

Third law of thermodynamics:

$$\lim_{T\to 0} \Delta S_T = 0$$

$$dU = T\,dS - p\,dV$$
$$dH = T\,dS + V\,dp$$
$$dF = -S\,dT - p\,dV$$
$$dG = -S\,dT + V\,dp$$

Maxwell's equations:

$$\left(\frac{\partial T}{\partial V}\right)_S = -\left(\frac{\partial p}{\partial S}\right)_V$$
$$\left(\frac{\partial T}{\partial p}\right)_S = \left(\frac{\partial V}{\partial S}\right)_p$$
$$\left(\frac{\partial S}{\partial V}\right)_T = \left(\frac{\partial p}{\partial T}\right)_V$$
$$\left(\frac{\partial S}{\partial p}\right)_T = -\left(\frac{\partial V}{\partial T}\right)_p$$

Basic thermodynamic equations:

$$C_V = \left(\frac{\partial U}{\partial T}\right)_V = T\left(\frac{\partial S}{\partial T}\right)_V = -T\left(\frac{\partial p}{\partial T}\right)_V\left(\frac{\partial V}{\partial T}\right)_S$$
$$C_P = \left(\frac{\partial H}{\partial T}\right)_p = T\left(\frac{\partial S}{\partial T}\right)_p = T\left(\frac{\partial V}{\partial T}\right)_p\left(\frac{\partial p}{\partial T}\right)_S$$
$$T\,dS = C_p\,dT - T\left(\frac{\partial V}{\partial T}\right)_p dp = C_p\,dT - V\beta T\,dp$$
$$T\,dS = C_V\left(\frac{\partial T}{\partial p}\right)_V dp + C_p\left(\frac{\partial T}{\partial V}\right)_p dV = \frac{C_V k}{\beta}\,dp + \frac{C_p}{\beta V}\,dV$$
$$T\,dS = C_V\,dT + T\left(\frac{dp}{\partial T}\right)_V dV = C_V\,dT + \frac{\beta T}{k}\,dV$$
$$\left(\frac{\partial C_V}{\partial V}\right) = T\left(\frac{\partial^2 p}{\partial T^2}\right)_V \qquad \left(\frac{\partial C_p}{\partial p}\right)_T = -T\left(\frac{\partial^2 V}{\partial T^2}\right)_p$$
$$C_p - C_V = T\left(\frac{\partial p}{\partial T}\right)_V\left(\frac{\partial V}{\partial T}\right)_p = -T\left(\frac{\partial V}{\partial T}\right)_p^2\left(\frac{\partial p}{\partial V}\right)_T = \frac{TV\beta^2}{k}$$
$$\gamma = \frac{C_p}{C_V} = \frac{(\partial p/\partial V)_S}{(\partial p/\partial V)_T} = \frac{k}{k_s}$$
$$\mu = \left(\frac{\partial T}{\partial p}\right)_H = \frac{1}{C_p}\left[T\left(\frac{\partial V}{\partial T}\right)_p - V\right] = \frac{V}{C_p}(\beta T - 1)$$
$$\eta = \left(\frac{\partial T}{\partial V}\right)_U = -\frac{1}{C_V}\left[T\left(\frac{\partial p}{\partial T}\right)_V - p\right] = -\frac{1}{C_V}\left(\frac{\beta T}{k} - p\right)$$

e. First derivatives of T, p, V, and S

$\left(\frac{\partial T}{\partial p}\right)_V = \frac{k}{\beta}$	$\left(\frac{\partial p}{\partial T}\right)_V = \frac{\beta}{k}$	$\left(\frac{\partial V}{\partial T}\right)_p = V\beta$	$\left(\frac{\partial S}{\partial T}\right)_V = \frac{C_V}{T}$
$\left(\frac{\partial T}{\partial p}\right)_S = \frac{V\beta T}{C_p} = \frac{(\gamma-1)k}{\gamma\beta}$	$\left(\frac{\partial p}{\partial T}\right)_S = \frac{C_p}{V\beta T} = \frac{\gamma\beta}{(\gamma-1)k}$	$\left(\frac{\partial V}{\partial T}\right)_S = -\frac{C_V k}{\beta T} = -\frac{V\beta}{\gamma-1}$	$\left(\frac{\partial S}{\partial T}\right)_p = \frac{C_p}{T}$
$\left(\frac{\partial T}{\partial V}\right)_p = \frac{1}{V\beta}$	$\left(\frac{\partial p}{\partial V}\right)_T = -\frac{1}{Vk}$	$\left(\frac{\partial V}{\partial p}\right)_T = -Vk$	$\left(\frac{\partial S}{\partial p}\right)_T = -V\beta$
$\left(\frac{\partial T}{\partial V}\right)_S = -\frac{\beta T}{C_V k} = -\frac{\gamma-1}{V\beta}$	$\left(\frac{\partial p}{\partial V}\right)_S = -\frac{\gamma}{Vk}$	$\left(\frac{\partial V}{\partial p}\right)_S = -\frac{Vk}{\gamma}$	$\left(\frac{\partial S}{\partial p}\right)_V = \frac{C_V k}{\beta T} = \frac{V\beta}{\gamma-1}$
$\left(\frac{\partial T}{\partial S}\right)_p = \frac{T}{C_p}$	$\left(\frac{\partial p}{\partial S}\right)_T = -\frac{1}{V\beta}$	$\left(\frac{\partial V}{\partial S}\right)_T = \frac{k}{\beta}$	$\left(\frac{\partial S}{\partial V}\right)_p = \frac{C_p}{V\beta T} = \frac{\gamma\beta}{(\gamma-1)k}$
$\left(\frac{\partial T}{\partial S}\right)_V = \frac{T}{C_V}$	$\left(\frac{\partial p}{\partial S}\right)_V = \frac{\beta T}{C_V k} = \frac{\gamma-1}{V\beta}$	$\left(\frac{\partial V}{\partial S}\right)_p = \frac{V\beta T}{C_p} = \frac{(\gamma-1)k}{\gamma\beta}$	$\left(\frac{\partial S}{\partial V}\right)_T = \frac{\beta}{k}$

f. First derivatives of U, H, F, and G with respect to T, p, V, and S

Internal energy U	Enthalpy H	Free energy F	Gibbs function G
$\left(\frac{\partial U}{\partial T}\right)_p = C_p - pV\beta$	$\left(\frac{\partial H}{\partial T}\right)_p = C_p$	$\left(\frac{\partial F}{\partial T}\right)_p = -S - pV\beta$	$\left(\frac{\partial G}{\partial T}\right)_p = -S$
$\left(\frac{\partial U}{\partial T}\right)_V = C_V$	$\left(\frac{\partial H}{\partial T}\right)_V = C_V + \frac{V\beta}{k}$	$\left(\frac{\partial F}{\partial T}\right)_V = -S$	$\left(\frac{\partial G}{\partial T}\right)_V = -S + \frac{V\beta}{k}$
$\left(\frac{\partial U}{\partial T}\right)_S = \frac{pC_V k}{\beta T} = \frac{pV\beta}{\gamma-1}$	$\left(\frac{\partial H}{\partial T}\right)_S = \frac{C_p}{\beta T}$	$\left(\frac{\partial F}{\partial T}\right)_S = -S + \frac{pV\beta}{\gamma-1} = -S + \frac{PC_V k}{\beta T}$	$\left(\frac{\partial G}{\partial T}\right)_S = -S + \frac{C_p}{\beta T} = -S + \frac{V\gamma\beta}{(\gamma-1)k}$
$\left(\frac{\partial U}{\partial p}\right)_T = -V\beta T + Vkp$	$\left(\frac{\partial H}{\partial p}\right)_T = V(1-\beta T) = -C_p\mu$	$\left(\frac{\partial F}{\partial p}\right)_T = pVk$	$\left(\frac{\partial G}{\partial p}\right)_T = V$
$\left(\frac{\partial U}{\partial p}\right)_V = \frac{C_V k}{\beta} = \frac{V\beta T}{\gamma-1}$	$\left(\frac{\partial H}{\partial p}\right)_V = V + \frac{C_V k}{\beta}$	$\left(\frac{\partial F}{\partial p}\right)_V = -\frac{Sk}{\beta}$	$\left(\frac{\partial G}{\partial p}\right)_V = -\frac{Sk}{\beta} + V$
$\left(\frac{\partial U}{\partial p}\right)_S = \frac{pVk}{\gamma}$	$\left(\frac{\partial H}{\partial p}\right)_S = V$	$\left(\frac{\partial F}{\partial p}\right)_S = -\frac{SV\beta T}{C_p} + \frac{pVk}{\gamma} = -\frac{Sk(\gamma-1)}{\gamma\beta} + \frac{pVk}{\gamma}$	$\left(\frac{\partial G}{\partial p}\right)_S = -\frac{SV\beta T}{C_p} + V = -\frac{S(\gamma-1)k}{\gamma\beta} + V$

11. Tables—Continued

$\left(\frac{\partial U}{\partial V}\right)_T=\frac{\beta T}{k}-p$	$\left(\frac{\partial H}{\partial V}\right)_T=\frac{\beta T}{k}-\frac{1}{k}$	$\left(\frac{\partial F}{\partial V}\right)_T=-p$	$\left(\frac{\partial G}{\partial V}\right)_T=-\frac{1}{k}$
$\left(\frac{\partial U}{\partial V}\right)_p=\frac{C_p}{V\beta}=\frac{\gamma\beta T}{(\gamma-1)k}$	$\left(\frac{\partial H}{\partial V}\right)_p=\frac{C_p}{V\beta}$	$\left(\frac{\partial F}{\partial V}\right)_p=-\frac{S}{V\beta}-p$	$\left(\frac{\partial G}{\partial V}\right)_p=-\frac{S}{V\beta}$
$\left(\frac{\partial U}{\partial V}\right)_S=-p$	$\left(\frac{\partial H}{\partial V}\right)_S=-\frac{\gamma}{k}$	$\left(\frac{\partial F}{\partial V}\right)_S=\frac{S\beta T}{C_V k}-p$ $=\frac{S(\gamma-1)}{V\beta}-p$	$\left(\frac{\partial G}{\partial V}\right)_S=\frac{S\beta T}{C_V k}-\frac{\gamma}{k}$ $=\frac{S(\gamma-1)}{V\beta}-\frac{\gamma}{k}$
$\left(\frac{\partial U}{\partial S}\right)_T=T-\frac{pk}{\beta}$	$\left(\frac{\partial H}{\partial S}\right)_T=T-\frac{1}{\beta}$	$\left(\frac{\partial F}{\partial S}\right)_T=-\frac{pk}{\beta}$	$\left(\frac{\partial G}{\partial S}\right)_T=\frac{1}{\beta}$
$\left(\frac{\partial U}{\partial S}\right)_p=T-\frac{pV\beta T}{C_p}$ $=T-\frac{(\gamma-1)pk}{\gamma\beta}$	$\left(\frac{\partial H}{\partial S}\right)_p=T$	$\left(\frac{\partial F}{\partial S}\right)_p=-\frac{ST}{C_p}-\frac{pV\beta T}{C_p}$ $=-\frac{ST}{C_p}-\frac{pk(\gamma-1)}{\gamma\beta}$	$\left(\frac{\partial G}{\partial S}\right)_p=-\frac{ST}{C_p}$
$\left(\frac{\partial U}{\partial S}\right)_V=T$	$\left(\frac{\partial H}{\partial S}\right)_V=T+\frac{\gamma-1}{\beta}$	$\left(\frac{\partial F}{\partial S}\right)_V=-\frac{ST}{C_V}$	$\left(\frac{\partial G}{\partial S}\right)_V=-\frac{ST}{C_V}+\frac{V\beta T}{C_V k}$ $=-\frac{ST}{C_V}+\frac{V(\gamma-1)}{V\beta}$

C. Electricity and magnetism[4]

1. Maxwell's equations

SI units	Gaussian units
$\nabla\cdot\mathbf{D}=\rho$	$\nabla\cdot\mathbf{D}=4\pi\rho$
$\nabla\times\mathbf{H}=\mathbf{j}+\frac{\partial\mathbf{D}}{\partial t}$	$\nabla\times\mathbf{H}=\frac{4\pi}{c}\mathbf{j}+\frac{1}{c}\frac{\partial\mathbf{D}}{\partial t}$
$\nabla\times\mathbf{E}+\frac{\partial\mathbf{B}}{\partial t}=0$	$\nabla\times\mathbf{E}+\frac{1}{c}\frac{\partial\mathbf{B}}{\partial t}=0$
$\nabla\cdot\mathbf{B}=0$	$\nabla\cdot\mathbf{B}=0$
$\mathbf{D}=\epsilon_0\mathbf{E}+\mathbf{P}$	$\mathbf{D}=\mathbf{E}+4\pi\mathbf{P}$
$\mathbf{H}=(1/\mu_0)\mathbf{B}-\mathbf{M}$	$\mathbf{H}=\mathbf{B}-4\pi\mathbf{M}$
$\epsilon_0=10^7/4\pi(c^2)\mathrm{F\cdot m^{-1}}$	
$\mu_0=4\pi\times10^{-7}\ \mathrm{H\cdot m^{-1}}$	

2. Continuity

$$\frac{\partial\rho}{\partial t}+\nabla\cdot\mathbf{j}=0$$

3. Lorentz force

$\mathbf{F}=q(\mathbf{E}+\mathbf{v}\times\mathbf{B})$ $\qquad$ $\mathbf{F}=q[\mathbf{E}+(1/c)\mathbf{v}\times\mathbf{B}]$

Boundary conditions between two media:

SI units	Gaussian units
$(\mathbf{D}_2-\mathbf{D}_1)\cdot\mathbf{n}=\sigma$	$(\mathbf{D}_2-\mathbf{D}_1)\cdot\mathbf{n}=4\pi\sigma$
$(\mathbf{B}_2-\mathbf{B}_1)\cdot\mathbf{n}=0$	$(\mathbf{B}_2-\mathbf{B}_1)\cdot\mathbf{n}=0$
$\mathbf{n}\times(\mathbf{E}_2-\mathbf{E}_1)=0$	$\mathbf{n}\times(\mathbf{E}_2-\mathbf{E}_1)=0$
$\mathbf{n}\times(\mathbf{H}_2-\mathbf{H}_1)=\mathbf{j}_{\mathrm{surf}}$	$\mathbf{n}\times(\mathbf{H}_2-\mathbf{H}_1)=(4\pi/c)\ \mathbf{j}_{\mathrm{surf}}$

n is unit normal to the surface drawn from medium 1 to 2.

4. Conservation of energy and momentum

For a single charge the rate of doing work is

$$W=q\mathbf{v}\cdot\mathbf{E},$$

where **v** is the velocity of the charge.

For a continuous distribution of charge and current in a volume τ,

$$W=\int_\tau \mathbf{j}\cdot\mathbf{E}\,d\tau.$$

The total energy density is given by

$$u=\tfrac{1}{2}(\mathbf{E}\cdot\mathbf{D}+\mathbf{B}\cdot\mathbf{H})\quad\text{(SI units)}$$
$$=(1/8\pi)(\mathbf{E}\cdot\mathbf{D}+\mathbf{B}\cdot\mathbf{H})\quad\text{(Gaussian).}$$

The vector **S**, representing energy flow, is called the *Poynting vector.* It is given by

$$\mathbf{S}=(\mathbf{E}\times\mathbf{H})\quad\text{(SI units)}$$
$$=(c/4\pi)(\mathbf{E}\times\mathbf{H})\quad\text{(Gaussian),}$$

and has dimensions energy/(area $\times$ time).

The electromagnetic momentum density is given by

$$\mathbf{g}=(1/c^2)(\mathbf{E}\times\mathbf{H})\quad\text{(SI units)}$$
$$=(1/4\pi c)(\mathbf{E}\times\mathbf{H})\quad\text{(Gaussian).}$$

a. Lorentz transformations for electromagnetic field. The Lorentz transformation formulas for the components of the electric and magnetic fields **E** and **B** in transforming to a stationary coordinate system K from a system K', moving at uniform velocity $V=\beta c$ along the x axis [using $\gamma=(1-\beta^2)^{-1/2}$]:

Gaussian units		SI units	
$E_x = E_x'$	$B_x = B_x'$	$E_x = E_x'$	$B_x = B_x'$
$E_y = \gamma(E_y' + \beta B_z')$	$B_y = \gamma(B_y' - \beta E_z')$	$E_y = \gamma(E_y' + \beta c B_z')$	$B_y = \gamma[B_y' - (\beta/c)E_z']$
$E_z = \gamma(E_z' - \beta B_y')$	$B_z = \gamma(B_z' + \beta E_y')$	$E_z = \gamma(E_z' - \beta c B_y')$	$B_z = \gamma[B_z' + (\beta/c)E_y']$

For the inverse transformation interchange prime and unprimed quantities and change the sign of β.

5. Table: Conversions for symbols and formulas

The symbols for mass, length, time, force, and other not specifically electromagnetic quantities are unchanged. To convert any equation in Gaussian variables to the corresponding equation in SI quantities, on both sides of the equation replace the relevant symbols listed below under "Gaussian" by the corresponding SI symbols listed on the right. The reverse transformation is also allowed. Since the length and time symbols are unchanged, quantities which differ dimensionally from one another only by powers of length and/or time are grouped together where possible. From Ref. 4.

Quantity	Gaussian	SI	Quantity	Gaussian	SI
Velocity of light	c	$(\mu_0\epsilon_0)^{-1/2}$	Magnetic field	$\mathbf{H}$	$(4\pi\mu_0)^{1/2}\mathbf{H}$
Electric field (potential,voltage)	$\mathbf{E}(\Phi,V)$	$(4\pi\epsilon_0)^{1/2}\mathbf{E}(\Phi,V)$	Magnetization	$\mathbf{M}$	$(\mu_0/4\pi)^{1/2}\mathbf{M}$
Displacement	$\mathbf{D}$	$(4\pi/\epsilon_0)^{1/2}\mathbf{D}$	Conductivity	σ	$\sigma/4\pi\epsilon_0$
Charge density (charge, current density, current, polarization)	$\rho(q,\mathbf{J},I,\mathbf{P})$	$(4\pi\epsilon_0)^{-1/2}\rho(q,\mathbf{J},I,\mathbf{P})$	Dielectric constant	ϵ	ϵ/ϵ_0
Magnetic induction	$\mathbf{B}$	$(4\pi/\mu_0)^{1/2}\mathbf{B}$	Permeability	μ	μ/μ_0
			Resistance (impedance)	$R(Z)$	$4\pi\epsilon_0 R(Z)$
			Inductance	L	$4\pi\epsilon_0 L$[a]
			Capacitance	C	$(1/4\pi\epsilon_0)C$

[a]Here L is in esu. For L in emu the conversion is $(4\pi/\mu_0)\,L$.

6. Table: Transformation properties of various physical quantities under rotations, spatial inversion, and time reversal

For quantities that are functions of $\mathbf{x}$ and t, it is necessary to be very clear what is meant by evenness or oddness under space inversion or time reversal. For example, the magnetic induction is such that under space inversion, $\mathbf{B}(\mathbf{x},t) \to \mathbf{B}_I(\mathbf{x},t) = +\mathbf{B}(-\mathbf{x},t)$, while under time reversal, $\mathbf{B}(\mathbf{x},t) \to \mathbf{B}_T(\mathbf{x},t) = -\mathbf{B}(\mathbf{x},-t)$. From Ref. 4.

Physical quantity		Rotation (rank of tensor)	Space inversion (name)	Time reversal
Mechanical				
Coordinate	$\mathbf{x}$	1	Odd (vector)	Even
Velocity	$\mathbf{v}$	1	Odd (vector)	Odd
Momentum	$\mathbf{p}$	1	Odd (vector)	Odd
Angular momentum	$\mathbf{L} = \mathbf{x}\times\mathbf{p}$	1	Even (pseudovector)	Odd
Force	$\mathbf{F}$	1	Odd (vector)	Even
Torque	$\mathbf{N} = \mathbf{x}\times\mathbf{F}$	1	Even (pseudovector)	Even
Kinetic energy	$p^2/2m$	0	Even (scalar)	Even
Potential energy	$U(\mathbf{x})$	0	Even (scalar)	Even
Electromagnetic				
Charge density	ρ	0	Even (scalar)	Even
Current density	$\mathbf{j}$	1	Odd (vector)	Odd
Electric field	$\mathbf{E}$	1	Odd (vector)	Even
Polarization	$\mathbf{P}$	1	Odd (vector)	Even
Displacement	$\mathbf{D}$	1	Odd (vector)	Even
Magnetic induction	$\mathbf{B}$	1	Even (pseudovector)	Odd
Magnetization	$\mathbf{M}$	1	Even (pseudovector)	Odd
Magnetic field	$\mathbf{H}$	1	Even (pseudovector)	Odd
Poynting vector	$\mathbf{S} = (c/4\pi)(\mathbf{E}\times\mathbf{H})$	1	Odd (vector)	Odd
Maxwell stress tensor	$T_{\alpha\beta}$	2	Even (tensor)	Even

7. Electrostatics[12]

Expressions are in SI units. For cgs electrostatic units omit the factor $4\pi\epsilon_0$ wherever it appears. $\epsilon_0 = 8.8542\times10^{-12}\,\mathrm{C^2\cdot N^{-1}\cdot m^{-2}} = 8.8542\times10^{-12}\,\mathrm{F/m}$ is the permittivity of free space.

Coulomb's law for two point charges:

$$\mathbf{F}_{12} = \frac{1}{4\pi\epsilon_0}\frac{q_1q_2}{r_{12}^3}\mathbf{r}_{12}.$$

Electric field due to a point charge:

$$\mathbf{E} = \frac{1}{4\pi\epsilon_0}\frac{q}{r^3}\mathbf{r},$$

where $\mathbf{r}$ is the radius vector drawn from the point charge to the field point. In an isotropic homogeneous medium with relative dielectric constant ϵ,

$$\mathbf{E} = \frac{1}{4\pi\epsilon_0}\frac{1}{\epsilon}\frac{q}{r^3}\mathbf{r}.$$

For an *array of charges*

$$\mathbf{E} = \frac{1}{4\pi\epsilon_0}\sum_i \frac{q_i}{r_i^2}\frac{\mathbf{r}_i}{r_i}.$$

Electric potential:

$$V = \frac{1}{4\pi\epsilon_0}\sum_i \frac{q_i}{r_i}.$$

For a *charge distribution* with surface density σ and volume density ρ,

$$V = \frac{1}{4\pi\epsilon_0}\left(\int \frac{\rho}{r}\,d\tau + \int \frac{\sigma}{r}\,dS\right),$$

where r is the distance from the elements of volume $d\tau$ and of surface dS to the field point. In a medium that is homogeneous and isotropic with relative dielectric constant ϵ, the potential compared to that in a vacuum is $V_\epsilon = (1/\epsilon)V$.

$$\mathbf{E} = -\nabla V.$$

Poisson's equation:

$$\nabla^2 V = -(1/4\pi\epsilon_0)4\pi\rho.$$

For a *dipole*, $-q$ and $+q$ separated by $\mathbf{l}$, the electric dipole moment is $\mathbf{p} = q\mathbf{l}$. The vector $\mathbf{l}$ is directed from the negative to the positive charge.

$$\mathbf{E} = \frac{1}{4\pi\epsilon_0}\left(\frac{3(\mathbf{p}\cdot\mathbf{r})\mathbf{r}}{r^5} - \frac{\mathbf{p}}{r^3}\right),$$

$$E_r = \frac{1}{4\pi\epsilon_0}\frac{2p\cos\theta}{r^3}, \quad E_\theta = \frac{1}{4\pi\epsilon_0}\frac{p\sin\theta}{r^3},$$

$$E = \frac{1}{4\pi\epsilon_0}\frac{p}{r^3}(3\cos^2\theta + 1)^{1/2},$$

$$V = \frac{1}{4\pi\epsilon_0}\frac{\mathbf{p}\cdot\mathbf{r}}{r^3}.$$

$\mathbf{r}$ is the radius vector drawn from the middle of the dipole to the field point.

Cylindrical cylinder of infinite length charged uniformly at a density ζ per unit length:

$$\mathbf{E} = \frac{1}{4\pi\epsilon_0}\frac{2\zeta}{r}\frac{\mathbf{r}}{r},$$

$$V - V_0 = \frac{1}{4\pi\epsilon_0}2\zeta\ln\frac{r_0}{r},$$

where r_0 is the radius of the cylinder, and $\mathbf{r}$ the radius vector to the field point. Inside the cylinder, $r \leqslant r_0$, $\mathbf{E} = 0$, $V = V(r_0)$.

Infinite plane charged uniformly at a density σ per unit area, $\mathbf{n}$ a unit vector perpendicular to the plane:

$$\mathbf{E} = (1/4\pi\epsilon_0)2\pi\sigma\mathbf{n},$$

$$V_1 - V_2 = (1/4\pi\epsilon_0)2\pi\sigma(x_2 - x_1),$$

where x_1 and x_2 are the distances from the plane to the points 1 and 2.

Between *two infinite parallel planes* uniformly and oppositely charged,

$$\mathbf{E} = (1/4\pi\epsilon_0)4\pi\sigma\mathbf{n}.$$

The unit vector $\mathbf{n}$ is perpendicular to the planes and points from the positive to the negative charges. The potential difference between the planes is

$$V_1 - V_2 = (1/4\pi\epsilon_0)4\pi\sigma d,$$

where d is the distance between the planes.

Sphere of radius R, whose charge q is uniformly distributed over its surface:

$$\mathbf{E}(\mathbf{r}) = \frac{1}{4\pi\epsilon_0}\frac{q}{r^2}\frac{\mathbf{r}}{r}, \quad V(\mathbf{r}) = \frac{1}{4\pi\epsilon_0}\frac{q}{r},$$

the same as for a point charge q placed at the center of the sphere. Inside the sphere $E = 0$, $V = V(R)$.

Sphere of radius R charged uniformly throughout its volume with volume density ρ: for $r \geqslant R$,

$$\mathbf{E}(\mathbf{r}) = \frac{1}{4\pi\epsilon_0}\frac{4\pi}{3}\rho\left(\frac{R}{r}\right)^3\mathbf{r},$$

$$V(R) - V(\mathbf{r}) = \frac{1}{4\pi\epsilon_0}q\left(\frac{1}{R} - \frac{1}{r}\right);$$

for $r \leqslant R$,

$$\mathbf{E}(\mathbf{r}) = \frac{1}{4\pi\epsilon_0}\frac{4\pi}{3}\rho\mathbf{r},$$

$$V(R) - V(\mathbf{r}) = \frac{1}{4\pi\epsilon_0}\frac{2\pi}{3}\rho(r^2 - R^2),$$

where $\mathbf{r}$ is the radius vector from the center of the sphere. $q = (4\pi/3)\rho R^3$ is the total charge of the sphere.

In a *dielectric medium,*

$$\mathbf{D} = \epsilon_0\mathbf{E} + \mathbf{P} \quad \text{(SI units)}$$

$$= \mathbf{E} + 4\pi\mathbf{P} \quad \text{(cgs esu)},$$

where

$$\mathbf{P} = \lim_{\Delta\tau\to 0}\left(\frac{1}{\Delta\tau}\sum_{\Delta\tau}\mathbf{p}_i\right)$$

is the polarization, and $\Delta\tau$ an element of volume, and $\mathbf{P}_i$ is the dipole moment of the ith molecule (or atom).

Gauss's theorem. The displacement flux

$$\Phi = \oint_s \mathbf{D}\cdot\mathbf{n}\, dS = Q \quad \text{(SI units)}$$

$$= \oint_s \mathbf{D}\cdot\mathbf{n}\, dS = 4\pi Q \quad \text{(cgs esu)},$$

where Q is the total charge enclosed by the surface S, and **n** is a unit normal to the surface element dS.

Energy. The energy of electrostatic interaction of an array of point charges

$$W = \frac{1}{2}\sum_i q_i V_i .$$

V_i is the potential at the position of q_i due to all other charges except q_i.

For a capacitor,

$$W = \tfrac{1}{2} C V^2,$$

where C is the capacitance and V the potential difference between the plates.

In terms of surface and volume density of charge,

$$W = \frac{1}{2}\int \rho V\, d\tau + \frac{1}{2}\int_s \sigma V\, dS,$$

$$= \frac{1}{2}\int_\tau \mathbf{D}\cdot\mathbf{E}\, d\tau \quad \text{(SI units)}$$

$$= \frac{1}{8\pi}\int_\tau \mathbf{D}\cdot\mathbf{E}\, d\tau \quad \text{(cgs esu)}.$$

8. Kirchhoff's laws

First law. The algebraic sum of the currents I_k which meet at any junction must be zero:

$$\sum_k I_k = 0.$$

Second law. In any mesh with a circuit in which each segment j has a current I_j, a resistance R_j, and an electromotive force $\mathscr{E}_j$,

$$\sum_j I_j R_j = \sum_j \mathscr{E}_j .$$

A definite direction around the mesh, clockwise or counterclockwise, must be chosen. The currents I_j in that direction are taken positive and the emf's $\mathscr{E}_k$ of the current soures are taken positive if they produce a current in the chosen direction around the loop.

The following procedure is employed in analyzing a complex dc circuit:

(a) The directions of the currents are arbitrarily assigned in all elements of the circuit.

(b) For m junctions, $m-1$ independent equations are written on the basis of the first law.

(c) In applying the mesh rule, select the meshes so that each new mesh contains at least one element of the circuit not included in the previously considered meshes. For m junctions with p branches, the number of independent equations based on the second law equals $p-m+1$.

9. Magnetostatics

Except when indicated otherwise, expressions are given in SI units. For Gaussian units replace $\mu_0/4\pi$ wherever it appears by $1/c$. Here $\mu_0 = 4\pi\times10^{-7}$ H·m^{-1} is the permeability of empty space; $c \simeq 3\times10^8$ m/s is the conversion constant from electromagnetic to electrostatic units, equal to the velocity of light.

a. Ampere's law:

$$d\mathbf{F} = I\, d\mathbf{l}\times\mathbf{B} \quad \text{(SI units)}$$

$$= (1/c) I\, d\mathbf{l}\times\mathbf{B} \quad \text{(Gaussian)},$$

defines the magnetic induction **B**. Here $d\mathbf{l}$ is a vector element of the conductor pointing in the direction in which the current flows. Note: $d\mathbf{F}$ is in the direction a right-handed screw would move if turned from $d\mathbf{l}$ toward **B**.

b. Biot-Savart law:

$$d\mathbf{B} = \frac{\mu_0}{4\pi}\frac{I}{r^3}\, d\mathbf{l}\times\mathbf{r},$$

$$\mathbf{B} = \frac{\mu_0}{4\pi}\int \frac{\mathbf{j}(\mathbf{r}')\times\mathbf{r}'}{r'^3}\, d\tau,$$

where $\mathbf{j}(\mathbf{r}')$ is the current density and $d\tau$ is the volume element in which the current flows.

c. Vector potential:

$$\mathbf{B}(\mathbf{r}) = \nabla\times\mathbf{A}(\mathbf{r}).$$

In the Coulomb gauge, $\Psi =$ constant,

$$\mathbf{A}\rightarrow\mathbf{A} + \nabla\Psi \quad \text{and} \quad \nabla\cdot\mathbf{A} = 0.$$

The vector potential satisfies the Poisson equation,

$$\nabla^2\mathbf{A} = -(\mu_0/4\pi)4\pi\mathbf{j},$$

with solution

$$\mathbf{A}(\mathbf{r}) = \frac{\mu_0}{4\pi}\int \frac{\mathbf{j}(\mathbf{r}')}{|\mathbf{r}-\mathbf{r}'|}\, d\tau,$$

This is the basis of useful expressions to calculate magnetic fields from given current distributions. (See Ref. 4.)

d. Magnetic moment. For current distributions localized in space small compared to scale of length involved in observations.

Magnetization:

$$\mathbf{M}(\mathbf{r}) = (\mu_0/4\pi)\tfrac{1}{2}[\mathbf{r}\times\mathbf{j}(\mathbf{r})].$$

Magnetic moment:

$$\mathbf{m} = \frac{\mu_0}{4\pi}\frac{1}{2}\int \mathbf{r}'\times\mathbf{j}(\mathbf{r}')\, d\tau.$$

If the current is confined to a plane, but otherwise arbitrary, $\mathbf{m} = (\mu_0/4\pi)\oint \mathbf{r}\times d\mathbf{l}$, and $|\mathbf{m}| = (\mu_0/4\pi)IS$, where S is the area of the loop.

Torque:

$$\mathbf{N} = \mathbf{m}\times\mathbf{B} \quad \text{(SI or Gaussian units)}.$$

Potential energy of dipole:

$$U = -\mathbf{m}\cdot\mathbf{B} \quad \text{(SI or Gaussian units)}.$$

Note: this is *not* the total energy. It does not include the work that must be done to keep the current **j** which produces **m** constant.

e. Magnetization in matter:

$$\mathbf{M}(\mathbf{r}) = \sum_i N_i \langle m_i \rangle,$$

where $\langle m_i \rangle$ is the average molecular magnetic moment in a small volume at the point **r**; N_i is the average number of molecules or atoms per unit volume.

Effective current density due to magnetization:

$$\mathbf{j}_M = (4\pi/\mu_0)\nabla\times\mathbf{M}.$$

The magnetic field **H** is defined by

$$\mathbf{H} = (1/\mu_0)\mathbf{B} - \mathbf{M} \quad \text{(SI units)}$$

$$= \mathbf{B} - 4\pi\mathbf{M} \quad \text{(Gaussian)}.$$

f. Permeability:

$$\mathbf{B} = \mu\mu_0\mathbf{H} \quad \text{(SI units)}$$

$$= \mu\mathbf{H} \quad \text{(Gaussian)}.$$

For paramagnetic substances $\mu - 1 \simeq 10^{-4}$–10^{-5}. for diamagnetic substances $1 - \mu \simeq 10^{-5}$. For ferromagnetic substances μ is large but depends on **H** and the history of magnetization. The **B** vs **H** relation exhibits hysteresis.

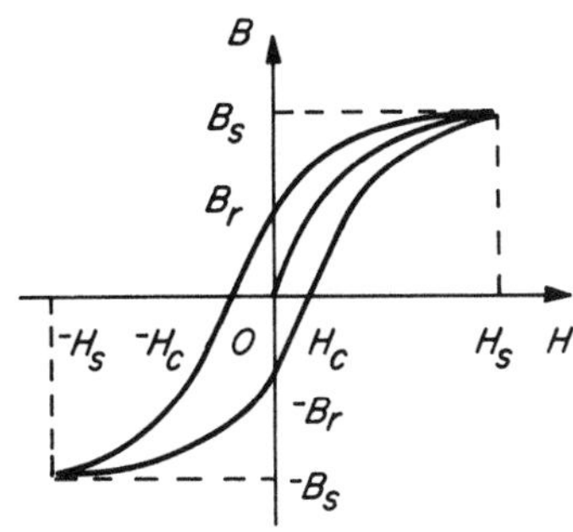

B_{sat} and H_{sat} are *saturation* values. B_r is the remanant magnetic induction. H_c is the coercive force.

g. Table: Properties of high-permeability materials [a]

Name	Composition (%)	Permeability Initial	Permeability Maximum	Coercivity H_e (Oe)	Retentivity B_r (G)	B(max) (G)	Resitivity ($\mu\Omega$ cm)
Iron							
Pure (lab. conditions)	Annealed	25 000	350 000	0.05	12 000	14 000	9.7
Swedish		250	5 500	1.0	13 000	21 000	10
Cast		100	600	4.5	5 300	20 000	30
Silicon	4 Si, bal. Fe (hot rolled)	500	7 000	0.3	7 000	20 000	50
Rhometal	36 Ni, bal. Fe	1 000	5 000	0.5	3 600	10 000	90
Permalloy 45	45 Ni, bal. Fe	2 500	25 000	0.3		16 000	45
Mumetal	71–78 Ni, 4.3–6 Cu, 0–2 Cr, bal. Fe	20 000	100 000	0.05	6 000	7 200	25–50
Supermalloy	79 Ni, 5 Mo, bal. Fe	100 000	1 000 000	0.002		8 000	60
HyMu80	80 Ni, bal. Fe	20 000	100 000	0.05		8 700	57
Alfenol	16 Al, bal. Fe	3 450	116 000	0.025	3 800	7 825	150
Permendur	50 Co, 1–2 V, bal. Fe	800	4 500	2.0	14 000	24 000	26
Sendust	10 Si, 5 Al, bal. Fe (cast)	30 000	120 000	0.05	5 000	10 000	60–80
Ferroxcube 3	Mn-Zn-ferrite	1 000	1 500	0.1	1 000	3 000	$>10^6$
Ferroxcube 101	Ni-Zn-ferrite	1 100		0.18	1 100	2 300	$>10^5$

[a]Expressed in cgs electromagnetic system (typical values). Values are approximate only and vary with heat treatment and mechanical working of the material. From *Reference Data for Radio Engineers* (Sams, Indianapolis, 1977).

10. Magnetic fields due to currents [12]

a. Straight wire:

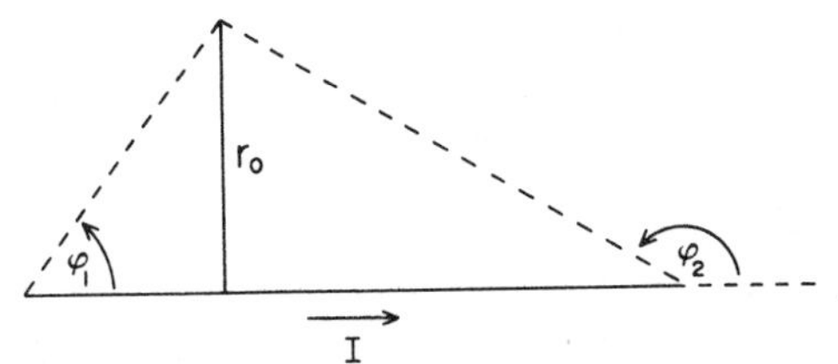

$$\mathbf{B} = (\mu_0/4\pi)(I/r_0)(\cos\phi_1 - \cos\phi_2)\mathbf{n},$$

where **n** is a unit normal to the plane of r_0 and I in the direction $\mathbf{I}\times\mathbf{r}_0$.

b. Straight wire, infinitely long:

$$\mathbf{B} = (\mu_0/4\pi)(2I/r_0)\mathbf{n}.$$

c. Circular turn. Field on the axis a distance h above the plane of the current loop.

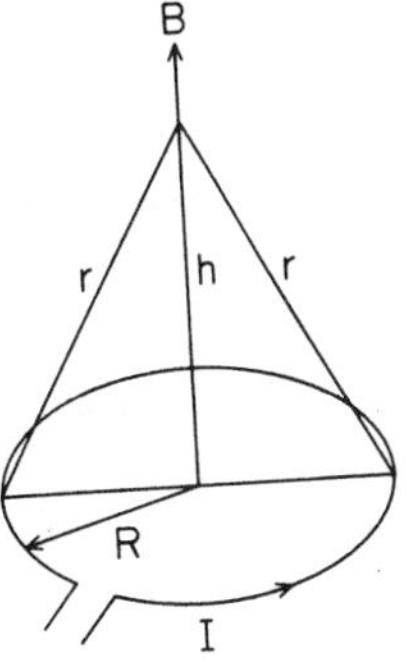

$$\mathbf{B} = \frac{\mu_0}{4\pi}\frac{2IS}{(R^2+h^2)^{3/2}}\mathbf{n}.$$

Here $S=\pi R^2$, **n** is unit normal to the plane of the loop. Right-hand rule: if the fingers of the right hand follow the current, the **B** follows the tumb.

d. Toroid:

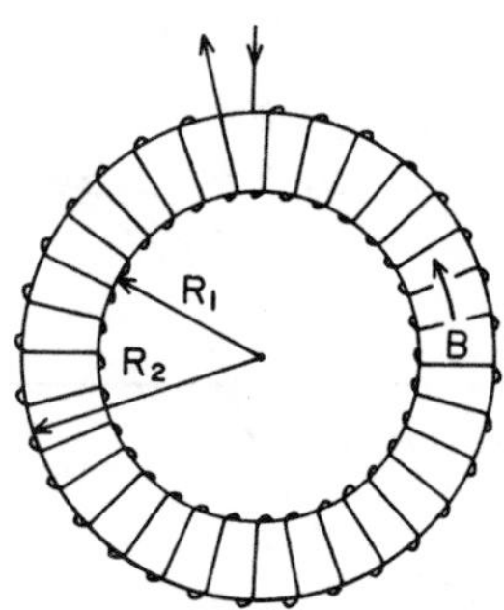

$$B=(\mu_0/4\pi)\mu 2NI/r,$$
$$H=NI/2\pi r \quad \text{(SI units)}$$
$$=(1/c)NI/r \quad \text{(Gaussian)},$$
$$R_1 \leqslant r \leqslant R_2.$$

Here N is the total number of turns, μ the relative permeability of the toroid core.

e. Solenoid:

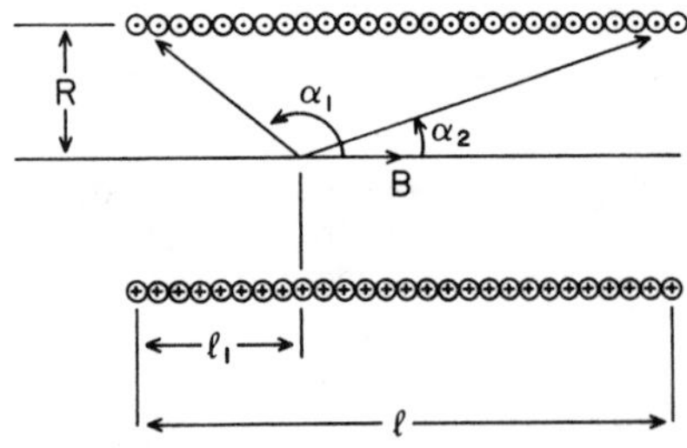

$$B=(\mu_0/4\pi)2\pi nI(\cos\alpha_2-\cos\alpha_1).$$

For an infinitely long solenoid,

$$B=(\mu_0/4\pi)4\pi nI,$$

where n is number of turns per unit length.

Magnetic dipole:

$$\mathbf{B}=\frac{\mu_0}{4\pi}\left(\frac{3(\mathbf{m}\cdot\mathbf{r})\mathbf{r}}{r^5}-\frac{\mathbf{m}}{r^3}\right), \quad \text{(SI units)}.$$

$$\mathbf{B}=\frac{3(\mathbf{m}\cdot\mathbf{r})\mathbf{r}}{r^5}-\frac{\mathbf{m}}{r^3} \quad \text{(Gaussian)}$$

11. Capacitance[17]

Dimensions in cm. For the dielectric between conductors, ϵ is the dielectric constant; $\epsilon\simeq 1$ for air. Constants given are correct to 0.1%. Edge effects are neglected. A is the area, d the separation, R, R_1, R_2 the radii of circles, spheres, cylinders; for wires, a is the radius, l the length, h the height above ground. Capacitance in picofarads (1 pF $=10^{-12}$ F).

a. Parallel plates:

$$C=0.0885\epsilon A/d \text{ pF}$$

b. Isolated disk:

$$C=0.708R.$$

c. Isolated sphere:

$$C=1.112R.$$

d. Concentric spheres:

$$C=1.112\epsilon R_1R_2/(R_2-R_1).$$

e. Coaxial cylinders:

$$C=0.2416\epsilon l\,[\log_{10}(R_2/R_1)]^{-1}.$$

f. Single long wire parallel to the ground:

$$C=0.2416l\left(\log_{10}\frac{2h}{a}+\log_{10}\frac{l+(l^2+a^2)^{1/2}}{l+(l^2+4h^2)^{1/2}}\right)^{-1}.$$

g. Two horizontal parallel wires above the ground:

$$C=0.1208l\,[\log_{10}(d/a)-d^2/8h^2]^{-1}.$$

$(a/l \ll 1, \quad d/l \ll 1).$

12. Inductance[17]

Inductance Formulas are given for low frequencies. See Ref. 17 for frequency effect. Formulas assume no iron in vicinity of the conductors, no iron wires. Dimensions in cm, accuracy about 0.1%. Edge effects are neglected. For wires, l is the length, a the radius, d the separation, h the distance above ground return. For coils, radii are R, R_1, R_2. L is the self-inductance M the mutual inductance. Inductance in microhenries (μH).

a. Circular ring of circular section (R is the mean radius of ring):

$$L=0.01257R[\ln(8R/a)-1.75].$$

b. Circular coil of circular cross section (coil of n fine wires; R is the mean radius of the turns, a the radius of winding cross section):

$$L=0.01257Rn^2[\ln(8R/a)-1.75].$$

c. Torus with single-layer winding (R is the distance from axis to center of cross section of winding, a the radius of the turns of the winding, n the number of turns of winding):

$$L=0.01257n^2[R-(R^2-a^2)^{1/2}].$$

d. Single-layer solenoid (n is the number of turns, R the radius of coil measured from axis to center of any wire, l the length of coil):

$$L=(0.03948R^2n^2/l)K.$$

For $l \gg R$, $K=1$. For $2R/l=0.00$, 0.2, 0.5, 1.0, 2.0, 5.0, and 10.0, $K=1.0000$, 0.9201, 0.8181, 0.6884, 0.5255, 0.3198, and 0.2033, respectively.

e. Straight round wire:

$$L=0.002l\,[\ln(2l/a)-\tfrac{3}{4}+a/l]\ \mu\text{H}.$$

At infinite frequency L decreases by 6% for $l/a=50$, 2% for $l/a=200\,000$. If the return conductor is not far away, the mutual inductances have to be taken into account.

f. Parallel wires, opposite currents:

$$L = 0.004l\,[\ln(d/a) - d/l + \tfrac{1}{4}].$$

g. Return circuit of rectangular wires (cross section of wires b by c):

$$L = 0.004l\left(\ln\frac{d}{b+c} + \frac{3}{2} - \frac{d}{l} + 0.2235\,\frac{b+c}{l}\right).$$

h. Square of round wires (side of square has length s):

$$L = 0.008s[\ln(s/a) + a/s - 0.774 + \delta].$$

i. Grounded horizontal wire (earth is return circuit):

$$L = 0.004605l\left(\log_{10}\frac{2h}{a} + \log_{10}\frac{l+(l^2+a^2)^{1/2}}{l+(l^2+4h^2)^{1/2}}\right) + 0.002[(l^2+4h^2)^{1/2} - (l^2+a^2)^{1/2} + \delta l - 2h + a].$$

In the above formulas $\delta = 0.25$ at low frequencies, $\delta = 0.0$ at infinite frequency.

j. Mutual inductance of two parallel wires stretched horizontally with ends grounded (earth forms the return circuit):

$$M = 0.004605l\left(\log_{10}\frac{(4h^2+d^2)^{1/2}}{d} + \log_{10}\frac{l+(l^2+d^2)^{1/2}}{l+(l^2+d^2+4h^2)^{1/2}}\right) + 0.002l\,[(l^2+d^2+4h^2)^{1/2} - (l^2+d^2)^{1/2} + d - (d^2+4h^2)^{1/2}].$$

13. ac impedance

Ohm's law:

$$V = ZI, \quad V = V_0 e^{i\omega t}.$$

Impedance of self-inductance L:

$$Z = i\omega L.$$

Impedance of a capacitance C:

$$Z = 1/i\omega C.$$

Impedance of a flat conductor of width w at high frequency:

$$Z = (1+i)\rho/\omega\delta,$$

where ρ is the resistivity in $10^{-8}\ \Omega\cdot$m:

~1.7 for Cu,	~5.5 for W,
~2.4 for Au,	~73 for SS 304,
~2.8 for Al,	~100 for Nichrome

(Al alloys may have up to double the stated value); δ is the effective skin depth:

$$\delta = \sqrt{\frac{\rho}{\pi\nu\mu}} \approx \frac{6.6\ \text{cm}}{\nu^{1/2}} \quad \text{for Cu}$$

(ν = the frequency in hertz, μ = magnetic susceptibility).

Impedance of free space:

$$Z = (\mu_0/\epsilon_0)^{1/2} = 376.7\ \Omega.$$

14. Transmission lines (no loss) (SI):

For flat plates of width w, separated by $d \ll w$, $C = \epsilon\epsilon_0 w/d$, $L = \mu\mu_0 d/w$. For coax cable of inside and outside radii r_1 and r_2, $C = 2\pi\epsilon\epsilon_0/\ln(r_2/r_1)$, $L = (\mu\mu_0/2\pi)\ln(r_1/r_2)$.

$$\text{velocity} = v = (LC)^{-1/2} = (\mu\epsilon)^{-1/2}c,$$

$$\text{impedance} = (L/C)^{1/2}.$$

L and C are inductance and capacitance per unit length. $\epsilon = 2$ to 6 for plastics, 4 to 8 for porcelain, glasses.

15. Synchrotron radiation (cgs):

$$\text{energy loss/revolution} = \frac{4\pi}{3}\frac{e^2}{\rho}\beta^3\gamma^4,$$

where ρ is the orbit radius.

For electrons ($\beta \approx 1$),

$$\frac{\Delta E\ (\text{MeV})}{\text{rev}} = \frac{0.0885[E\ (\text{GeV})]^4}{\rho\ (\text{m})}.$$

Critical frequency:

$$\omega_c = 3\gamma^3 c/\rho.$$

Frequency spectrum (for $\gamma \gg 1$):

$$I(\omega) \cong 4.8\,\frac{e^2\gamma}{c}\left(\frac{\omega}{\omega_c}\right)^{1/3}, \quad \omega \ll \omega_c;$$

$$I(\omega) \cong (3\pi)^{1/2}\frac{e^2\gamma}{c}\left(\frac{\omega}{\omega_c}\right)^{1/2} e^{-2\omega/\omega_c}, \quad \omega \gtrsim 2\omega_c;$$

$$I(\omega) \cong (1.0,\ 1.6,\ 1.6,\ 0.5,\ 0.08)e^2\gamma/c$$

at $\omega/\omega_c = 0.01, 0.1, 0.2, 1.0, 2.0$, respectively.

The radiation is confined to angles $\lesssim 1/\gamma$ relative to the instantaneous direction of motion.

For more formulas and details, see Ref. 4.

16. ac circuits[18]

For a resistance R, inductance L, and capacitance C in series with a voltage source $V = V_0\exp(j\omega t)$, where $j = \surd(-1)$, the current $I = dq/dt$, where q satisfies

$$L\frac{d^2q}{dt^2} + R\frac{dq}{dt} + \frac{q}{C} = V.$$

Solutions are $q(t) = q_s + q_t$, $I(t) = I_s + I_t$, where the steady state is $I_s = j\omega q_s = V/Z$ in terms of the impedance $Z = R + j(\omega L - 1/\omega C)$. For initial conditions $q(0) = q_0 = \bar{q}_0 + q_s(0)$, $I(0) = I_0$, the transients can be of three types, depending on $\Delta = R^2 - 4L/C$:

(a) Overdamped, $\Delta > 0$:

$$q_t = [(I_0 + \gamma_+\bar{q}_0)e^{-\gamma_- t} - (I_0 + \gamma_-\bar{q}_0)e^{-\gamma_+ t}]/(\gamma_+ - \gamma_-),$$

where $\gamma_\pm = (R \pm \Delta^{1/2})/2L$.

(b) Critically damped, $\Delta = 0$:

$$q_t = [\bar{q}_0 + (I_0 + \gamma_R\bar{q}_0)t\,]e^{-\gamma_R t},$$

where $\gamma_R = R/2L$.

(c) Underdamped, $\Delta < 0$:

$$q_t = [\omega_1^{-1}(\gamma_R\bar{q}_0 + I_0)\sin\omega_1 t + \bar{q}_0\cos\omega_1 t\,]e^{-\gamma_R t}.$$

Here $\omega_1 = \omega_0(1 - R^2C/4L)^{1/2}$, where $\omega_0 = (LC)^{-1/2}$ is the resonant frequency. The quality of the circuit is $Q = \omega_0 L/R$. At $\omega = \omega_0$, $Z = R$. Instability results when L, R, C are not all of the same sign.

17. Table: Electromagnetic frequency/wavelength bands[3]

Nomenclature: E, Extreme; F, Frequency; H, High; L, Low; M, Medium; S, Super; U, Ultra; V, Very.

	Frequency range		Wavelength range	
Designation	Lower	Upper	Lower	Upper
ULF[a]		10 Hz	3 Mm	
ELF[a]	10 Hz	3 kHz	100 km	3 Mm
VLF	3 kHz	30 kHz	10 km	100 km
LF	30 kHz	300 kHz	1 km	10 km
MF	300 kHz	3 MHz	100 m	1 km
HF	3 MHz	30 MHz	10 m	100 m
VHF	30 MHz	300 MHz	1 m	10 m
UHF	300 MHz	3 GHz	10 cm	1 m
SHF[b]	3 GHz	30 GHz	1 cm	10 cm
S	2.6	3.95	7.6	11.5
G	3.95	5.85	5.1	7.6
J	5.3	8.2	3.7	5.7
H	7.05	10.0	3.0	4.25
X	8.2	12.4	2.4	3.7
M	10.0	15.0	2.0	3.0
P	12.4	18.0	1.67	2.4
K	18.0	26.5	1.1	1.67
R	26.5	40.0	0.75	1.1
EHF	30 GHz	300 GHz	1 mm	1 cm
Submillimeter	300 GHZ	3 THz	100 μ	1 mm
Infrared	3 THz			100 μ

[a]The boundary between ULF and ELF is variously defined.
[b]The SHF (microwave) band is further subdivided approximately as shown.

D. Hydrogenlike atoms[5,13]

Schrödinger equation with central forces:

$$\nabla^2 u + (2m/\hbar^2)[E - U(r)]u = 0,$$

where m is the reduced mass,

$$u = \sum_{l,m} R_l(r) Y_{l,m}(\theta,\phi), \quad R_l(r) = \frac{1}{r} V_l(r).$$

For hydrogenlike atom, Coulomb potential:

$$V_l''(r) + \frac{2m}{\hbar^2}\left(E + \frac{Ze^2}{r} - \frac{\hbar^2}{2m}\frac{l(l+1)}{r^2}\right)V_l(r) = 0.$$

Put

$$x = \frac{2r}{r_0}, \quad r_0 = \sqrt{\frac{\hbar^2}{2m|E|}}, \quad A = \frac{Ze^2}{2r_0|E|}.$$

For $E<0$, only acceptable solutions have A a positive integer, and

$$E_n = \frac{-mZ^2e^4}{2\hbar^2 n^2}, \quad n = l+1, l+2, \ldots$$

$$R_\infty hc = me^4/2\hbar^2 = 21.799\times 10^{-12}\ \text{erg}$$

$$= 13.606\ \text{eV} = 109\,737\ \text{cm}^{-1}.$$

Solutions expressible in Laguerre polynomials:

$$L_k(x) = e^x \frac{d^k}{dx^k}(x^k e^{-x}), \quad L_k^{(j)} = \frac{d^j}{dx^j} L_k(x),$$

$$L_0 = 1, \quad L_1 = 1 - x, \quad L_2 = 2 - 4x + x^2,$$

$$L_3 = 6 - 18x + 9x^2 - x^3.$$

Normalized eigenfunctions for hydrogenlike atoms:

$$u_{nlm} = R_{nl}(r) Y_{lm}(\theta,\phi),$$

$$R_{ne} = \sqrt{\frac{4(n-l-1)!}{a^3 n^4 [(n+l)!]^3}}\, e^{-r/na}\left(\frac{2r}{na}\right)^l L_{n+l}^{(2l+1)} \frac{2r}{na},$$

$$a = \frac{\hbar^2}{me^2}\frac{1}{Z},$$

where $\hbar^2/me^2$ = Bohr radius = $0.529\,177\times 10^{-8}$ cm.

$$u(1s) = \frac{1}{(\pi a^3)^{1/2}} e^{-r/a},$$

$$u(2s) = \frac{(2 - r/a)e^{-r/2a}}{4(2\pi a^3)^{1/2}},$$

$$u(2p) = \frac{(r/a)e^{-r/2a}}{8(\pi a^3)^{1/2}} \begin{cases} -\sin\theta\, e^{i\phi} \\ (\sqrt{2})\cos\theta \\ \sin\theta\, e^{-i\phi}. \end{cases}$$

E. Phase-shift analysis of scattering

Plane wave incident, spinless particles, neglect Coulomb.

Wave equation:

$$\nabla^2\psi + k^2\psi = 0 \quad \text{at large } r.$$

Solution:

$$\psi = \psi_{\text{inc}} + \psi_{\text{scat}}.$$

Differential cross section:

$$\frac{d\sigma}{d\Omega} = \frac{v_{\text{scat}}|\psi_{\text{scat}}(\theta)|^2 r^2}{v_{\text{inc}}|\psi_{\text{inc}}|^2},$$

$$\psi_{\text{inc}} = e^{ikz} = e^{ikr\cos\theta} = \sum_l A_l(kr) Y_{l0}(\theta,\phi),$$

$$A_l(kr) = \int e^{ikr\cos\theta} Y_{l0}^*(\theta,\phi)\, d\Omega$$

$$= (4\pi)^{1/2}(2l+1)^{1/2} i^l j_l(kr),$$

$$j_l(kr) = (\pi/kr)^{1/2} J_{l+1/2}(kr)$$

$$\underset{r\to\infty}{\longrightarrow} \sin(kr - l\pi/2),$$

$$e^{ikz} \underset{r\to\infty}{\longrightarrow} \frac{2\pi^{1/2}}{kr}\sum_l (2l+1)^{1/2} i^l Y_{l0}(\cos\theta)$$

$$\times \sin(kr - l\pi/2).$$

The phase shift is introduced in

$$\psi \underset{r\to\infty}{\longrightarrow} \frac{2\pi^{1/2}}{kr}\sum_l (2l+1)^{1/2} i^l Y_{l0}(\cos\theta) A_l$$

$$\times \sin(kr - l\pi/2 + \delta_l),$$

$$\psi_{\text{scat}} \underset{r\to\infty}{\longrightarrow} \frac{\pi^{1/2}}{ikr}\sum_l (2l+1)^{1/2}(e^{2i\delta_l} - 1) Y_{l0}(\cos\theta).$$

For elastic scattering and given l,

$$\left(\frac{d\sigma}{d\Omega}\right)_l = \frac{4\pi}{k^2}(2l+1)[Y_{l0}(\cos\theta)]^2 \sin^2\delta_l,$$

$$\int \left(\frac{d\sigma}{d\Omega}\right)_l d\Omega = 4\pi\lambda\!\!\!^{-2}(2l+1)\sin^2\delta_l.$$

1.06. PRACTICAL PHYSICS

A. Friction

The coefficient of *static friction* is the ratio of the force F required to start sliding between two surfaces and the normal force N which presses them together:

$$f = F/N.$$

The coefficient of *sliding friction* is the same ratio to maintain sliding once started. The *angle of repose* is the angle θ of a surface with respect to the horizontal upon which sliding will begin: $\theta = \tan^{-1} f$.

Belt friction opposes the slipping of a belt on a pulley. When power is transmitted the tension T_1 on the driving side is greater than T_2 on the driven side. Neglecting centrifugal force,

$$T_1/T_2 = e^{f\alpha},$$

where α is the angle in radians of the contact between belt and pulley. The power transmitted is

$$P = (T_1 - T_2)v,$$

where v is the velocity of the pulley surface.

Greasy lubrication. The lubricant is thin but adheres strongly enough to the rubbing surfaces that the layers of lubricant slip over one another.

Viscous lubrication. The lubricant is thick enough to carry the entire hydrostatic or hydrodynamic pressure.

1. Table: Coefficients of static and sliding friction

From Theodore Baumeister, Mark's *Mechanical Engineer's Handbook,* 7th ed. (McGraw-Hill, New York, 1967).

Materials, lubricant	Static	Sliding
Hard steel on hard steel		
Dry	0.78	0.42
Oleic acid	0.11	0.12
Light mineral oil	0.23	
Castor oil	0.15	0.081
Stearic acid	0.0052	0.029
Hard steel on graphite		
Dry	0.21	
Oleic acid	0.09	
Mild steel on lead		
Dry	0.95	0.95
Medium mineral oil	0.5	0.3
Aluminum on mild steel, dry	0.61	0.47
Aluminum on aluminum, dry	1.05	1.4
Glass on glass		
Dry	0.94	0.40
Ricinoleic acid	0.005	...
Oleic acid	...	0.09
Teflon on Teflon, dry	0.04	
Teflon on steel, dry	0.04	

B. Bending of beams

Solid bodies resist the deformation caused by external forces by producing internal forces. The force per unit area on a surface element within the body is called the stress. These are normal (σ) or shear (τ) stresses depending on whether they act perpendicularly or tangentially to the surface element. In a crystalline substance the response may be anisotropic and both the stress and the strain, which describes the deformation, will be tensors. Practical metals are generally microcrystalline and exhibit simple strain and simple stress. A rectangular bar of length l_1 will stretch under tension an amount e_1. At the same time its width l_2 will contract an amount e_2. Thus the normal strains are

$$\epsilon_1 = e_1/l_1, \quad \epsilon_2 = e_2/l_2.$$

The forces of shear act tangentially to the surface and cause a relative slippage δ of parallel surfaces separated by a distance l. The shear strain is the displacement angle

$$\gamma = \delta/l.$$

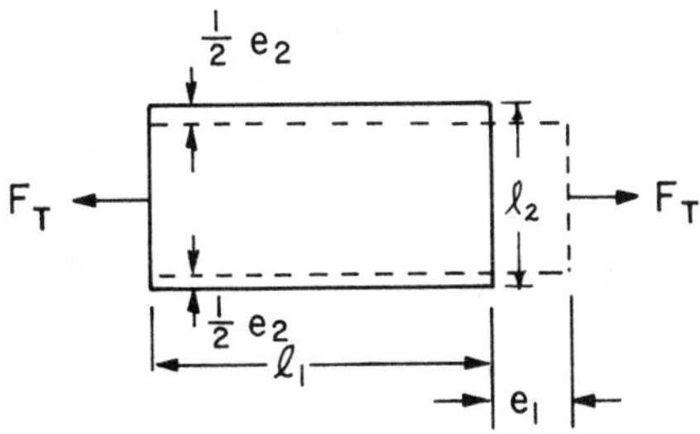

normal strains $\epsilon_1 = e_1/l_1, \quad \epsilon_2 = e_2/l_2$

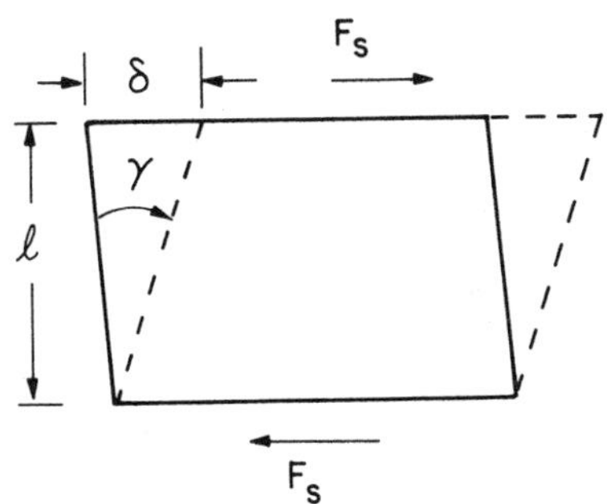

shearing strain $\gamma = \delta/l$

Modulus of elasticity (Young's modulus):

$$E = \frac{\sigma}{\epsilon_1} = \frac{F_T/A}{e_1/l_1} \quad (F_T = \text{force},\ A = \text{area}).$$

Shear modulus:

$$G = \frac{\tau}{\gamma} = \frac{F_s/A}{\delta/l}.$$

Poisson's ratio:

$$\mu = \epsilon_1/\epsilon_2.$$

Bulk modulus:

$$K = -\frac{F_T/A}{\Delta V/V} \quad (V = \text{volume}).$$

Hooke's law:

$$\epsilon = \sigma/E, \quad \gamma = \tau/G.$$

Following relations hold:

$$E = 3K(1 - 2\mu) = 2G(1 + \mu).$$

C. Deflections in uniform beams

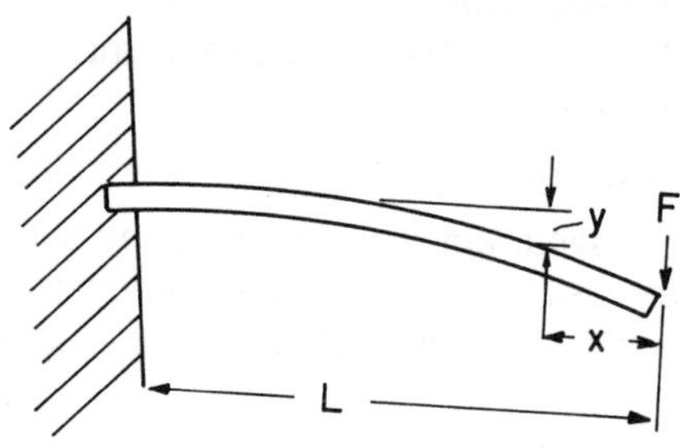

In a cantilevered uniform beam of length L with a force F at one end, the section at a distance x from F has a bending moment

$$M = Fx.$$

For several forces, taken to be in the y direction,

$$M = \sum_i F_i x_i.$$

More generally,

$$\mathbf{M} = \sum_i (\mathbf{r}_i \times \mathbf{F}_i).$$

Moments tending to turn the section clockwise are positive; counterclockwise, negative.

Across the section the fiber stress increases linearly from the neutral axis. At a distance ξ from the neutral axis the stress is given by

$$\sigma = M\xi/I,$$

where I is the moment of inertia of the section. At any point x along the beam the deflection y is governed by the differential equation

$$EI = \frac{d^2y}{dx^2} = M,$$

where E is Young's modulus.

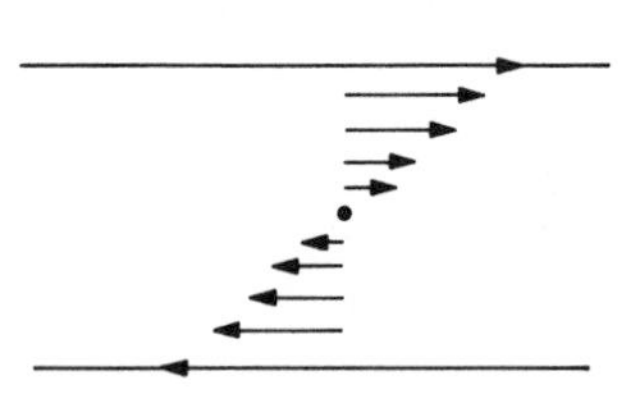

Distribution of stress within uniform beam

The maximum fiber stress at any section occurs at a point $\xi = c$, most remote from the neutral axis. Hence

$$\sigma_b = M_b/(I/c).$$

I/c is called the *section modulus* of the beam. The maximum fiber stress in the beam occurs at the section of greatest bending moment. Care should be taken to see that the maximum fiber stress is within the elastic range.

The shearing force is related to the bending moment by

$$F_s = \frac{dM}{dx}.$$

The maximum shear stress τ_b occurs at the section of greatest vertical shear. The maximum shear stress at any section occurs at the neutral axis provided the net width b is as small there as anywhere else. The maximum shear stress

$$\tau_b = \alpha F_s/A,$$

where F_s/A is the *average* shear on the section and α is a factor that depends on the shape of the section. For a rectangular section, $\alpha = 3/2$; for a solid circular section, $\alpha = 4/3$.

1. Figure: Deflection of beams [14]

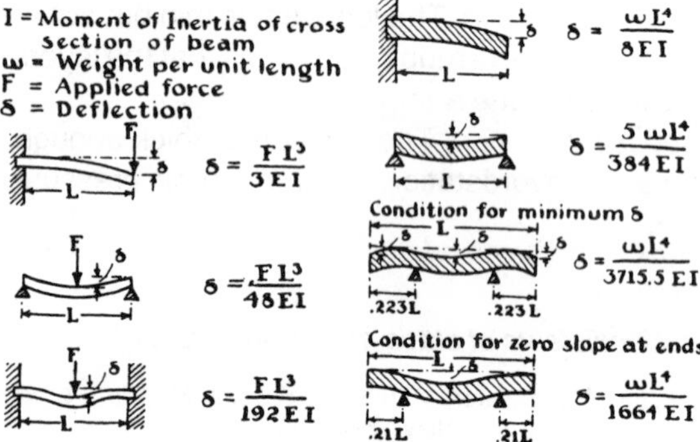

2. Figure: Vacuum vessels [14]

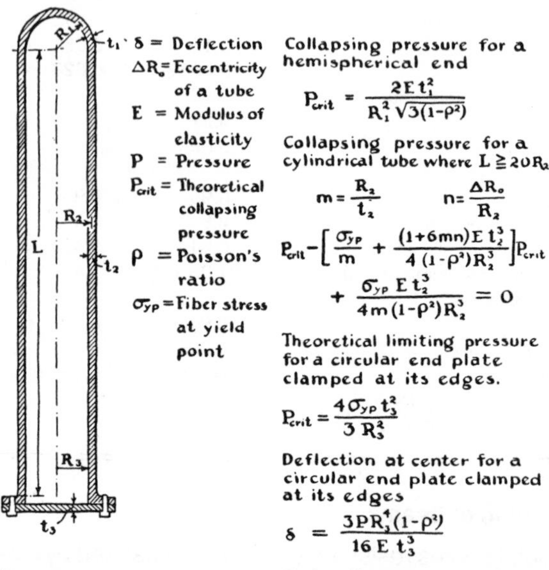

3. Figure: Moments of inertia[14]

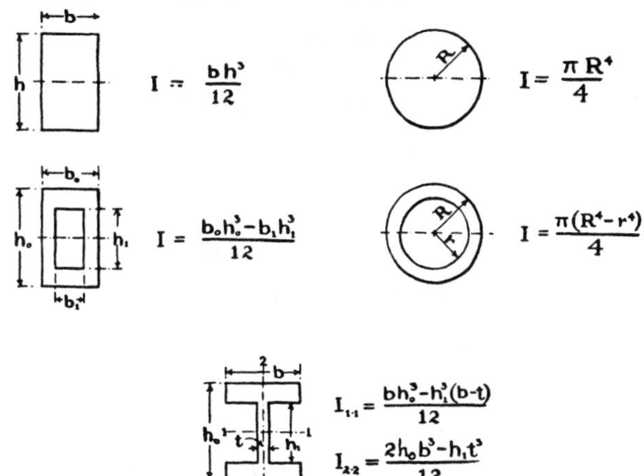

D. Table: Elastic constants of solids

The elastic constants of a sample depend to a great extent on its past history, crystalline structure, etc. The values below can only be taken as approximate. (10^{10} Pa $= 1.45\times10^{6}$ lbs/in.^{2}). Mainly from W. H. J. Childs, *Physical Constants* (Methuen, London, 1951).

Material	Bulk modulus (10^{10} Pa)	Young's modulus (10^{10} Pa)	Shear modulus (10^{10} Pa)	Poisson's ratio
Aluminum	7.5	7.0	2.5	0.34
Al alloys				
AlSiMg		6.8–7.2	2.6–2.8	0.34
AlCuMg		7.2–7.4	2.8	0.33–0.34
Bismuth	3.0	3.2	1.2	0.33
Cadmium	4.2	5.0	2.0	0.30
Copper	13.5	11.0	4.4	0.34
Gold	16.5	8.0	2.8	0.42
Iron				
Wrought	16.0	21.0	7.7	0.28
Cast	9.5	11.0	5.0	0.27
Lead	4.1	1.6	0.6	0.44
Magnesium	3.3	4.1	1.7	$\simeq$0.3
Nickel	17.0	21.0	7.8	0.30
Platinum	24.5	17.0	6.3	0.39
Silver	10.5	7.7	2.8	0.37
Steel				
Cast	17.0	20.0	7.5	0.28
Mild	16.0	22.0	8.0	0.28
Ni steel				
>5% Ni		20.3–20.5	7.8–7.9	0.31
>25% Ni		18.2	7.0	0.3
>36% Ni		15.6	6.0	0.3
Tantalum		19.0		
Tin	5.3	5.3	1.9	0.33
Tungsten	30.0	39.0	15.0	
Zinc	3.5	8.0	3.6	0.23
Brass	6.0	9.0	3.5	0.35
Bronze	9.0	10.5	3.7	0.36
Phosphor bronze		12.0	4.4	0.38
German silver	15.0	11.0	4.5	0.37
Glass, crown	5.0	6.0	2.5	0.25
India rubber		0.05	0.000 15	0.48
Quartz fiber	1.5	5.4	3.0	
Wood				
Oak		1.3		
Teak		1.7		

E. Strength of materials

Yield stress. In a tensile test a specimen rod of the material is gradually stretched and the force and elongation are both measured. The yield stress is the applied load at which the stress-strain relationship departs appreciably from linearity. *Ultimate strength* is the maximum stress reached before rupture. *Creep strength,* usually appreciable only at elevated temperatures, is the stress under constant load at which the deformation increases with time.

1. Table: Yield stress of steels—Representative types

10^7 Pa $= 1.45\times10^3$ pounds/in.^{2} From *Ryerson Steel and Aluminum Data Book* (Ryerson, Chicago, 1975).

Material	Yield strength (10^7 Pa)
AISI M1020 low carbon steel	24
Hot-rolled bars	
AISI 1141 low carbon steel bars, oil quenched	70
AISI 4340 medium carbon steel bars, oil quenched	112
AISI 1095 high carbon steel bars, water quenched	95
Cor-Ten high-strength plates and sheets	34
Jalloy alloy plates and sheets	97
AISI 1010 carbon steel tubing	
Hot rolled	21
Cold drawn	34
Stainless steel 304 (18% Cu, 8% Ni)	24

2. Table: Yield stress of aluminum and aluminum alloys—Representative types

10^7 Pa $= 1.45\times10^3$ pounds/in.^{2} From *Ryerson Steel and Aluminum Data Book* (Ryerson, Chicago, 1975). See also *Alcoa Structural Handbook* and *Alcoa Aluminum Handbook* (Aluminum Company of America, Pittsburgh, 1960).

Material	Yield stress (10^7 Pa)
EC aluminum 99.45% Al, electrical grade	
Annealed	2.8
Tempered	16.5
2011 5–6% Cu, tempered	31
2024 4–5% Cu, 0.3–0.9% Mn, 1.5% Mg	
Annealed	7.6
Tempered	29–46
3004 1.0–1.5% Mn, 0.8–1.3% Mg	
Annealed	7
Tempered	17–25
Brass, rolled (70% Cu, 30% Zn)	17
Bronze, case	14
Copper, hard drawn	26

F. Heat transfer[15]

1. Heat conduction

Heat flow across an area A, in a material with thermal conductivity K where temperature gradient is dT/dx:

$$\Phi = -KA\frac{dT}{dx}.$$

For a rectangular parallelipiped with opposite ends at T_1, T_2 spaced x apart,

$$\Phi = -(KA/x)(T_2 - T_1).$$

For two concentric isothermal cylindrical surfaces of radii r_1, r_2 maintained at T_1, T_2 and with length $l \gg r$,

$$\Phi = -\frac{2.73l}{\log_{10}(r_2/r_1)}K(T_2 - T_1).$$

For two concentric spheres,

$$\Phi = -\frac{4\pi}{1/(r_1 - r_2)}K(T_2 - T_1).$$

The *time dependent* differential equation for heat conduction is

$$\frac{dT}{dt} = \frac{K}{\rho c}\nabla^2 T,$$

where ρ is the density, c the specific heat, and $h^2 = K/\rho c$ the *thermal diffusivity*. The relaxation time is defined by $\tau = 4x_0^2/\pi^2h^2$.

Here we consider only the one-dimensional case of a plane parallel slab with dimensions large compared to its thickness. The slab is initially at temperature T_0 *and is* immersed into an environment at fixed temperature $T_1 < T_0$. The faces of the slab coincide with the planes $x = +x_0$ and $-x_0$. Newton's law of cooling applies with the heat loss per unit area of surface given by

$$W = N(T_{x_0} - T_1) \text{ cal/s·cm}^2.$$

For slow cooling, $Nx_0/K \ll 1$;

$$T = T_1 + (T_0 - T_1)\exp(-Nh^2t/Kx_0).$$

The relaxation time is defined by

$$\tau = 4x_0^2/\pi^2h^2.$$

For air,

$$N \simeq 11.7 \text{ W/m}^2\text{·K}.$$

For a cylinder of radius r_0 and $Nr_0/K \ll 1$, the temperature wil be practically uniform throughout the cylinder:

$$T = T_1 + (T_0 - T_1)\exp(-2Nh^2t/Kr_0).$$

See Refs. 15 and 16 for a more complete treatment.

2. Table: Thermal conductivity κ, diffusivity h^2, and relaxation time τ for an infinite slab of 2 cm thickness

Materials are at room temperature (298.2 K) unless otherwise specified (1 W cm^{-1} K^{-1} = 0.239 cal s^{-1} cm^{-1} °C^{-1}, 1 cm^2 s^{-1} = 10^{-4} m^2 s^{-1}).

Material	κ (W cm^{-1} K^{-1})	h^2 (cm^2 s^{-1})	τ (s)
Aluminum[a]	2.37	0.969	0.418
Bismuth (polycrystalline)[a]	0.079	0.065	6.25
Brass (yellow)[b]	1.05	0.350	1.16
Chromium[a]	0.939	0.291	1.39
Constantan (60 Cu, 40 Ni)[c]	0.243	0.069	5.87
Copper[a]	4.01	1.17	0.346
Gold[a]	3.18	1.28	0.317
Invar[b]	0.139	0.037	11.0
Iron[d]	0.791	0.228	1.78
Armco iron[a]	0.728	0.207	1.96
Carbon steel[b]	0.605	0.145	2.80
Lead[a]	0.353	0.243	1.67
Mercury (liquid)[a]	0.083	0.044	9.21
Monel metal[b]	0.227	0.065	6.24
Nichrome[b]	0.129	0.040	10.1
Nickel[a]	0.916	0.230	1.76
Platinum[a]	0.716	0.252	1.61
Silver[a]	4.29	1.74	0.233
Titanium (polycrystalline)[a]	0.219	0.093	4.36
Tungsten[a]			
298.2 K	1.74	0.664	0.610
1500 K	1.07	0.354	0.533
3000 K	0.914	0.263	1.54
Bonded silicon carbide[e]	0.400	0.020	20
TiC[e]	0.315	0.120	3.38
Diamond (Type I)[a]	9.07	5.09	0.080
Graphite (ATJ, along molding pressure)[a]	0.982	0.80	0.51
Quartz glass[f]			
298.2 K	0.0138	0.0083	48.8
500 K	0.0162	0.0077	52.6
1000 K	0.0287	0.0114	35.6
Limestone (Bedford, Indiana)[g]	0.0220	0.010	41
Soapstone[e]	0.0335	0.015	27
MgO[f]	0.487	0.137	2.96
Al_2O_3 (sapphire)[e]	0.464	0.149	2.72
Al_2O_3 (polycrystalline)[e]	0.363	0.118	3.43
Mica[e]	0.0079	0.0031	131
Glass (Pyrex)[f]	0.0110	0.0064	63

2. Table: Continued

Material	κ (W cm^{-1} K^{-1})	h^2 (cm^2 s^{-1})	τ (s)
Air[h]	0.00026	0.202	2.01
Asbestos (loose fiber)[e]	0.0016	0.0033	121
Firebrick[e]	0.0075	0.0033	121
Concrete (cement, sand, & gravel aggregate)[e]	0.0182	0.0090	45
Cork[e]	0.00042	0.0014	289
Paraffin[e]	0.00454	0.00197	206
Water (saturated liquid)[i]	0.00609	0.00141	287
Water (saturated vapor)[i]	0.000186	4.08	0.10
Mahogany[e]			
Across grain	0.00161	0.0023	176
With grain	0.00310	0.0044	92

[a] C. Y. Ho, R. W. Powell, and P. E. Liley, *Thermal Conductivity of the Elements: A Comprehensive Review*, J. Phys. Chem. Ref. Data **3**, Suppl. 1 (1974). Most thermal diffusivity values are from Y. S. Touloukian, R. W. Powell, C. Y. Ho, and M. C. Nicolaou, *Thermal Diffusivity*, Vol. 10 of Thermophysical Properties of Matter: The TPRC Data Series (IFI/Plenum, New York, 1973).
[b] Y. S. Touloukian, R. W. Powell, C. Y. Ho, and P. G. Klemens, *Thermal Conductivity: Metallic Elements and Alloys*, Vol. 1 of Thermophysical Properties of Matter: The TPRC Data Series (IFI/Plenum, New York, 1970).
[c] C. Y. Ho, M. W. Ackerman, K. Y. Wu, S. G. Oh, and T. N. Havill, "Thermal Conductivity of Ten Selected Binary Alloy Systems," J. Phys. Chem. Ref. Data **7**(3), 959 (1978).
[d] Y. S. Touloukian, C. Y. Ho, R. H. Bogaard, P. D. Desai, H. H. Li *et al.*, *Properties of Selected Ferrous Alloying Elements*, Vol. III-1 of CINDAS Data Series on Material Properties (McGraw–Hill, New York, 1981).
[e] Y. S. Touloukian, R. W. Powell, C. Y. Ho, and P. G. Klemens, *Thermal Conductivity: Nonmetallic Solids*, Vol. 2 of Thermophysical Properties of Matter: The TPRC Data Series (IFI/Plenum, New York, 1970).
[f] R. W. Powell, C. Y. Ho, and P. E. Liley, "Thermal Conductivity of Selected Materials," National Standard Reference Data Series, National Bureau of Standards, NSRDS-NBS 8 (1966).
[g] P. D. Desai, R. A. Navarro, S. E. Hasan, C. Y. Ho, D. P. DeWitt, and T. R. West, "Thermophysical Properties of Selected Rocks," Purdue University, TPRC Report 23 (1974).
[h] Y. S. Touloukian, P. E. Liley, and S. C. Saxena, *Thermal Conductivity: Nonmetallic Liquids and Gases*, Vol. 3 of Thermophysical Properties of Matter: The TPRC Data Series (IFI/Plenum, New York, 1970).
[i] P. E. Liley, T. Makita, and Y. Tanaka, *Properties of Inorganic and Organic Fluids*, Vol. V-1 of CINDAS Data Series on Material Properties (Hemisphere, New York, 1988).

3. Heat transfer by radiation

The energy emitted per unit area A radiating a heat spectrum between wavelengths λ and $\lambda + \Delta\lambda$ is

$$\Phi(\lambda) = \pi\epsilon_\lambda J_\lambda \Delta\lambda.$$

This represents an integration over the hemisphere (2π steradians); ϵ_λ is the emissivity of the surface the ratio of the emission of the surface to that of a "black body" at the same temperature. Planck's expression

$$J_\lambda = (c_1/\lambda^5)\left(\frac{1}{e^{c_2/\lambda T} - 1}\right)$$

gives the distribution of energy of the heat spectrum:

$$c_1 = 3.74\times10^{-16}\ \text{W·m}^2,\quad c_2 = 1.439\times10^{-2}\ \text{m·K}.$$

The total heat lost per unit area of a "blackbody" with emissivity ϵ_T at temperature T is given by Stefan's formula:

$$\Phi = \epsilon_T\sigma T^4,$$

where $\sigma = 5.67\times10^{-8}\ \text{W·m}^{-2}\text{·K}^{-4}$.

The heat emitted by a flat surface of area A into a cone defined by a solid angle $d\Omega$ is

$$d\Phi = A\cos\theta\,(d\Omega/\pi)\epsilon_T\sigma T^4.$$

ϵ_T is an emissivity averaged over all wavelengths. It is ordinarily a slowly varying function of temperature. For porous nonmetallic substances it is very nearly unit, regardless of the color of the material. ϵ_T for aluminum paints around room temperature varies between 0.3 and 0.5. For nonmetallic pigment paints $\epsilon_T \simeq 1$. For clean metals ϵ_T varies with temperature in such a way that the total emission is conveniently given by an expression of the form

$$\Phi = M(T/1000\ \text{K})^m$$

a. Table: Radiation constants of metals (610–980 K)[14]

Metal	M (W/m^2)	m
Silver	6 000	4.1
Platinum	23 000	5.0
Nickel	8900	4.65
Iron	14 300	5.55
Nichrome	36 000	4.1

The heat transfer per unit area by radiation between two parallel black isothermal surfaces at absolute temperatures T_1 and T_2 separated by a small distance is

$$W = \sigma(T_2^4 - T_1^4)\ \text{W/s·m}^2.$$

Baffles held at intermediate temperatures may be used to reduce the amount of heat transfer.

4. Heat transfer by free convection

a. Langmuir formula. The heat transfer from a vertical surface at temperature T_1 to the ambient air at temperature T_2 is

$$W = -(\phi_2 - \phi_1)/0.45 \text{ cal/s·cm}^2$$

$$= -93\,000(\phi_2 - \phi_1) \text{ W/m}^2.$$

Langmuir found that heat losses by free convection from a horizontal surface facing upward are 10% greater than they are from a vertical surface and 50% less from a surface facing downward.

b. Table: Values of ϕ for air[14]

T (K)	ϕ (cal/s·cm)	T (K)	ϕ (cal/s·cm)
0	0	1100	0.1017
100	0.000 98	1300	0.1376
200	0.004 01	1500	0.1776
300	0.009 24	1700	0.222
400	0.016 0	1900	0.271
500	0.024 3	2100	0.325
700	0.045 1	2300	0.384
900	0.070 9	2500	0.447

1.07. PROPERTIES OF MATERIALS

A. Table: Periodic table of the elements[18]

PERIODIC TABLE OF THE ELEMENTS

Group: New notation / Previous IUPAC form / CAS version

KEY TO CHART: Atomic Number → 50; Symbol → Sn; 1983 Atomic Weight → 118.71; Oxidation States ← +2 +4; Electron Configuration ← 18 18 4

1	2	3	4	5	6	7	8	9	10	11	12	13	14	15	16	17	18	Orbit
IA	IIA	IIIA	IVA	VA	VIA	VIIA	VIIIA	VIIIA	VIIIA	IB	IIB	IIIB	IVB	VB	VIB	VIIB	VIIIA	
		IIIB	IVB	VB	VIB	VIIB	VIII	VIII	VIII			IIIA	IVA	VA	VIA	VIIA		
1 +1 −1 H 1.00794 1																	2 0 He 4.00260 2	K
3 +1 Li 6.941 2-1	4 +2 Be 9.01218 2-2											5 +3 B 10.81 2-3	6 +2 +4 −4 C 12.011 2-4	7 +1 +2 +3 +4 +5 −1 −2 −3 N 14.0067 2-5	8 −2 O 15.9994 2-6	9 −1 F 18.9984 2-7	10 0 Ne 20.179 2-8	K-L
11 +1 Na 22.9898 2-8-1	12 +2 Mg 24.305 2-8-2											13 +3 Al 26.9815 2-8-3	14 +2 +4 −4 Si 28.0855 2-8-4	15 +3 +5 −3 P 30.9738 2-8-5	16 +4 +6 −2 S 32.06 2-8-6	17 +1 +5 +7 −1 Cl 35.453 2-8-7	18 0 Ar 39.948 2-8-8	K-L-M
19 +1 K 39.0983 -8-8-1	20 +2 Ca 40.08 -8-8-2	21 +3 Sc 44.9559 -8-9-2	22 +2 +3 +4 Ti 47.88 -8-10-2	23 +2 +3 +4 +5 V 50.9415 -8-11-2	24 +2 +3 +6 Cr 51.996 -8-13-1	25 +2 +3 +4 +7 Mn 54.9380 -8-13-2	26 +2 +3 Fe 55.847 -8-14-2	27 +2 +3 Co 58.9332 -8-15-2	28 +2 +3 Ni 58.69 -8-16-2	29 +1 +2 Cu 63.546 -8-18-1	30 +2 Zn 65.39 -8-18-2	31 +3 Ga 69.72 -8-18-3	32 +2 +4 Ge 72.59 -8-18-4	33 +3 +5 −3 As 74.9216 -8-18-5	34 +4 +6 −2 Se 78.96 -8-18-6	35 +1 +5 −1 Br 79.904 -8-18-7	36 0 Kr 83.80 -8-18-8	-L-M-N
37 +1 Rb 85.4678 -18-8-1	38 +2 Sr 87.62 -18-8-2	39 +3 Y 88.9059 -18-9-2	40 +4 Zr 91.224 -18-10-2	41 +3 +5 Nb 92.9064 -18-12-1	42 +6 Mo 95.94 -18-13-1	43 +4 +6 +7 Tc (98) -18-13-2	44 +3 Ru 101.07 -18-15-1	45 +3 Rh 102.906 -18-16-1	46 +2 +4 Pd 106.42 -18-18-0	47 +1 Ag 107.868 -18-18-1	48 +2 Cd 112.41 -18-18-2	49 +3 In 114.82 -18-18-3	50 +2 +4 Sn 118.71 -18-18-4	51 +3 +5 −3 Sb 121.75 -18-18-5	52 +4 +6 −2 Te 127.60 -18-18-6	53 +1 +5 +7 −1 I 126.905 -18-18-7	54 0 Xe 131.29 -18-18-8	-M-N-O
55 +1 Cs 132.905 -18-8-1	56 +2 Ba 137.33 -18-8-2	57* +3 La 138.906 -18-9-2	72 +4 Hf 178.49 -32-10-2	73 +5 Ta 180.948 -32-11-2	74 +6 W 183.85 -32-12-2	75 +4 +6 +7 Re 186.207 -32-13-2	76 +3 +4 Os 190.2 -32-14-2	77 +3 +4 Ir 192.22 -32-15-2	78 +2 +4 Pt 195.08 -32-16-2	79 +1 +3 Au 196.967 -32-18-1	80 +1 +2 Hg 200.59 -32-18-2	81 +1 +3 Tl 204.383 -32-18-3	82 +2 +4 Pb 207.2 -32-18-4	83 +3 +5 Bi 208.980 -32-18-5	84 +2 +4 Po (209) -32-18-6	85 At (210) -32-18-7	86 0 Rn (222) -32-18-8	-N-O-P
87 +1 Fr (223) -18-8-1	88 +2 Ra 226.025 -18-8-2	89** +3 Ac 227.028 -18-9-2	104 +4 Unq (261) -32-10-2	105 Unp (262) -32-11-2	106 Unh (263) -32-12-2	107 Uns (262) -32-13-2												OPQ

															Orbit
*Lanthanides	58 +3 +4 Ce 140.12 -20-8-2	59 +3 Pr 140.908 -21-8-2	60 +3 Nd 144.24 -22-8-2	61 +3 Pm (145) -23-8-2	62 +2 +3 Sm 150.36 -24-8-2	63 +2 +3 Eu 151.96 -25-8-2	64 +3 Gd 157.25 -25-9-2	65 +3 Tb 158.925 -27-8-2	66 +3 Dy 162.50 -28-8-2	67 +3 Ho 164.930 -29-8-2	68 +3 Er 167.26 -30-8-2	69 +3 Tm 168.934 -31-8-2	70 +2 +3 Yb 173.04 -32-8-2	71 +3 Lu 174.967 -32-9-2	NOP
**Actinides	90 +4 Th 232.038 -18-10-2	91 +5 +4 Pa 231.036 -20-9-2	92 +3 +4 +5 +6 U 238.029 -21-9-2	93 +3 +4 +5 +6 Np 237.048 -22-9-2	94 +3 +4 +5 +6 Pu (244) -24-8-2	95 +3 +4 +5 +6 Am (243) -25-8-2	96 +3 Cm (247) -25-9-2	97 +3 +4 Bk (247) -27-8-2	98 +3 Cf (251) -28-8-2	99 +3 Es (252) -29-8-2	100 +3 Fm (257) -30-8-2	101 +2 +3 Md (258) -31-8-2	102 +2 +3 No (259) -32-8-2	103 +3 Lr (260) -32-9-2	OPQ

Numbers in parentheses are mass numbers of most stable isotope of that element

From *Chemical and Engineering News*, 63(5), 27, 1985. This format numbers the groups 1 to 18.

B. Table: Densities[19]

Substance	Density (g cm^{-3})	Substance	Density (g cm^{-3})	Substance	Density (g cm^{-3})
Aluminum	2.70	Samarium	7.75	Teak	0.58
Antimony	6.62	Scandium	3.02		
Arsenic (metallic)	5.73	Selenium (amorph.)	4.82	**Liquids (15°C)**	
Barium	3.5	Silicon (amorph.)	2.42	Acetone	0.792
Beryllium	1.85	Silver	10.49	Alcohol	
Bismuth	9.78	Sodium	0.97	Ethyl	0.791
Boron	2.53	Strontium	2.56	Methyl	0.810
Cadmium	8.65	Sulfur (amorph.)	1.92	Aniline	1.02
Calcium	1.55	Tantalum	16.6	Benzene	0.899
Carbon		Tellurium (cryst.)	6.25	Ether	0.736
Graphite	2.25*	Thallium	11.86	Glycerine	1.26
Diamond	3.514	Thorium	11.3	Oil	
Cerium	6.90	Tin	7.3	Lubricating	0.90–0.92
Cesium	1.87	Titanium	4.5	Olive	0.92
Chromium	7.14	Tungsten (wire)	19.3	Paraffin	0.8
Cobalt	8.71	Uranium	18.7	Seawater	1.025
Copper	8.96	Vanadium	5.98	Turpentine	0.87
Gallium	5.93	Yttrium	3.8	Normal $\frac{1}{2}H_2SO_4$	1.0304
Germanium	5.46	Zinc	7.1	Normal HCl	1.0162
Gold	19.3	Zirconium	6.4	Normal HNO_3	1.0322
Hafnium	13.3			Normal NaOH	1.0414
Indium	7.28	**Alloys**		Normal NaCl	1.0388
Iodine	4.94	Bell metal	8.7	Normal KOH	1.048
Iridium	22.42	Brass	8.4–8.7	Normal KCl	1.0446
Iron		Bronze	8.8–8.9		
Pure	7.88	Phosphor	8.8		(g/liter at NTP)
Wrought	7.85	Constantan	8.88	**Gases**	
Cast	7.6	Invar	8.00	Air	1.2928
Steel	7.83	Magnalium	2.0–2.5	Ammonia	0.7708
Lanthanum	6.15	Manganin	8.50	Argon	1.783
Lead	11.34	Steel	7.8	Bromine	7.139
Lithium	0.534	Wood's metal	9.5–10.5	Carbon monoxide	1.2502
Magnesium	1.74			Carbon dioxide	1.9768
Manganese	7.41	**Miscellaneous**		Chlorine	3.220
Mercury		Asbestos	2.0–2.8	Fluorine	1.69
(solid, −39°C)	14.19	Celluloid	1.4	Helium	0.1785
Molybdenum	10.22	Cork	0.22–0.26	Hydrochloric acid	1.639
Neodymium	6.96	Ebonite	1.15	Hydrogen	0.0899
Nickel	8.88	Glass	2.4–2.8	Hydrogen sulfide	1.539
Osmium	22.5	Ice, 0°C	0.917	Krypton	3.68
Palladium	12.16	Mica	2.6–3.2	Methane	0.7167
Phosphorus		Fused silica	2.1–2.2	Neon	0.900
Red	2.20	Paraffin wax	0.9	Nitrogen	1.251
Yellow	1.83	Woods (oven dry)		Oxygen	1.429
Platinum	21.45	Ash	0.52–0.64	Xenon	5.85
Potassium	0.86	Balsa	0.12–0.20		
Praseodymium	6.48	Beech	0.65–0.67		
Rhodium	12.44	Elm	0.55–0.67		
Rubidium	1.53	Mahogany	0.54–0.67		
Ruthenium	12.1	Oak	0.67–0.98		

T (°C)	Water (g cm^{-3})	Mercury (g cm^{-3})	T (°C)	Water (g cm^{-3})	Mercury (g cm^{-3})
0	0.999 841	13.5951	50	0.988 04	13.4725
1 or 7	0.999 902	...	60	0.983 21	13.4482
2 or 6	0.999 941	...	70	0.977 79	13.4240
3 or 5	0.999 965	...	80	0.971 80	13.4000
4	0.999 973	...	90	0.965 31	13.3758
10	0.999 700	13.5704	100	0.958 35	13.3518
20	0.998 203	13.5458	150	0.917 3	13.2326
30	0.995 646	13.5213	200	0.862 8	13.1144
40	0.992 21	13.4970			

*Crystalline graphite, bulk density ~1.60–1.65 g cm^{-3}.

C. Table: Viscosities[19]

The unit throughout is 10^{-5} poise $= 10^{-6}$ Pa·s. 1 poise = 1 g cm^{-1} s^{-1}. For water, glycerol, and air, see Tables 12.04.A, A.1, B.

Water

T (°C)	Viscosity	*T* (°C)	Viscosity
0	1793	60	469
10	1309	70	406
20	1006	80	357
30	800	90	315
40	657	100	284
50	550		

Mercury

T (°C)	Viscosity
0	1690
50	1410
100	1220

Miscellaneous liquids

Substance	*T* (°C)	Viscosity
Acetone	20	330
Alcohol		
Ethyl	20	1192
Methyl	20	591
Benzene	20	649
Carbon disulfide	20	367
Ether	20	234
Glycerine	20	83×10^4
Nitric acid	10	1770
Oils		
Castor	20	986×10^3
Linseed	30	331×10^3
Olive	20	84×10^3
Sulfuric acid	20	22×10^3
Turpentine	20	1490
Xylol		
Ortho	20	807
Meta	20	615
Para	20	643

Gases

Substance	*T* (°C)	Viscosity
Air	20	18.1
Ammonia	20	10.8
Argon	0	21.0
Carbon monoxide	20	18.4
Carbon dioxide	20	16.0
Chlorine	20	14.7
Helium	0	18.9
Hydrogen	20	9.5
Hydrogen sulfide	20	13.0
Hydrogen chloride	0	14.0
Krypton	0	23.3
Methane	20	12.0
Neon	0	29.8
Nitric oxide	20	18.6
Nitrogen	20	18.4
Oxygen	20	20.9
Sulfur dioxide	20	13.8
Water vapor	15	9.8
Xenon	0	21.1

D. Table: Hydrometer units[19]

Baumé	Density (15°C) in g·cm^{-3} = 144.3/(144.3 − *B*) (*B* = Baumé degrees)
Twaddell	Density (60°F) = 1 + *T*/200 (*T* = Twaddell degrees)

E. Table: Surface tension[19]

Measured in dynes per cm at 20°C. In all cases the liquid is assumed to be in contact with air. 1 dyne/cm = 10^{-3} N·m^{-1}.

Substance	Surface tension	Substance	Surface tension
Acetone	23.7	Olive oil	33
Alcohol		Paraffin oil	26
Ethyl	22.3	Turpentine	27
Methyl	22.6	Water (°C)	
Amyl acetate	24.7	5	74.92
Aniline	43	10	74.22
Benzene	28.9	15	73.49
Chloroform	27.2	20	72.75
Glycerine	64	25	71.97
Mercury	475[a]	30	71.18
Solutions			
Conc. H_2SO_4	55		
HNO_3	41		
$CaCl_2$	0.29[c]	KOH	0.32[c]
$CuSO_4$	0.11[c]	NaCl	0.28[c]
KCl	0.19[c]	NaOH	0.50[c]
		NH_4Cl	0.26[c]

[a]In vacuum.
[b]In air; decrease with age, 500–400.
[c]For every 1 g of anhydrous salt per 100 cm^3 of water, add this (approximate) value to the value for pure water.

F. Table: Solubility of common gases in water[19]

The value given is the volume of gas, at NTP, absorbed by 1 cm^3 of water at 20°C under a pressure of 1 atm.

Gas	Volume absorbed	Gas	Volume absorbed
Air	0.019	Hydrochloric acid	440
Ammonia	700	Neon	0.017
Argon	0.038	Nitrogen	0.0154
Carbon dioxide	0.88	Oxygen	0.031
Chlorine	2.3	Sulfur dioxide	39
Helium	0.0138		
Hydrogen	0.0181		
Hydorgen sulfide	2.6		

G. Table: Solubility of common chemical compounds in water[19]

Solubilities are expressed in g of anhydrous substance per 100 g of water.

Substance	Formula	Solubility at 15°C	100°C
Ammonium			
chloride	NH_4Cl	35.3	77.3
nitrate	NH_4NO_3	173	871
Sulfate	$(NH_4)_2SO_4$	74.2	103.3
Boric acid	H_3BO_3	4.3	40
Copper sulfate	$CuSO_45H_2O$	19.0	75.4
Ferrous sulfate	$FeSO_47H_2O$	23.5	37.3 (90°)
Lead nitrate	$Pb(NO_3)_2$	52.4	138.8
Potash alum	$Al_2(SO_4)_3K_3SO_424H_2O$	5.0	109 (90°)
Potassium			
chloride	KCl	32.5	56.7
bichromate	$K_2Cr_2O_7$	9	95
hydroxide	KOH	107	178
iodide	KI	140	208
nitrate	KNO_3	26.3	246
permanganate	$KMnO_4$	5.4	22.2 (60°)
Silver nitrate	$AgNO_2$	196	952
Sodium			
carbonate	$Na_2CO_310H_2O$	17	45
bicarbonate	$NaHCO_3$	8.9	16.4 (60°)
chloride	$NaCl$	35.9	39.8
hydroxide	$NaOH$	105	340
sulfate	$Na_2SO_410H_2O$	13.1	42
thiosulfate	$Na_2S_2O_35H_2O$	65.5	266
Zinc sulfate	$ZnSO_47H_2O$	50.7	...
Cane sugar	$C_{12}H_{22}O_{11}$	197	487

H. Table: Compressibility of liquids[19]

The quantity given is $(1/V_0)\partial V/\partial P$, where V_0 is the volume at 0 °C and 1 atm. The unit of pressure is 1 kg cm^{-2}. 1 kg $cm^{-2} = 0.9807\times10^5$ $N\cdot m^{-2}$ = 0.9807 bar = 0.967 atm. The value of V/V_0 is also given, in parentheses. All values (except mercury) for 20 °C.

Substance	Pressure 1	500	1000
Acetone	120×10^{-6} (1.0279)	61×10^{-6} (0.9829)	51×10^{-6} (0.9553)
Alcohol			
Amyl	89×10^{-6} (1.0181)	60×10^{-6} (0.9800)	45×10^{-6} (0.9526)
Ethyl	104×10^{-6} (1.0212)	62×10^{-6} (0.9794)	53×10^{-6} (0.9506)
Methyl	113×10^{-6} (1.0283)	64×10^{-6} (0.9823)	53×10^{-6} (0.9530)
Propyl	91×10^{-6} (1.0173)	65×10^{-6} (0.9780)	46×10^{-6} (0.9498)
Carbon disulfide	91×10^{-6} (1.0235)	57×10^{-6} (0.9865)	47×10^{-6} (0.9586)
Ether	184×10^{-6} (1.0315)	83×10^{-6} (0.9681)	60×10^{-6} (0.9363)
Mercury (22°C)	3.95×10^{-6} (1.003 98)	3.89×10^{-6} (1.002 02)	3.83×10^{-6} (1.000 07)
Water	45.3×10^{-6} (1.0016)	38.1×10^{-6} (0.9808)	33.6×10^{-6} (0.9630)

I. Table: Vapor pressures (mm Hg)[19]

See also Table 22.03 for vapor pressure of the elements.

T (°C)	Water	Mercury
Liquid air	...	2×10^{-27}
Solid CO_2	6×10^{-4}	3×10^{-9}
−20	0.79 (ice)	...
−10	1.97 (ice)	...
0	4.58	0.0004
10	9.2	...
20	17.5	0.0013
30	31.8	...
40	55.3	0.006
50	92.5	...
60	149.4	0.03
70	234	...
80	355	0.09
90	526	...
100	760	0.28
150	3 580	2.9
200	11 700	18
300	...	247

Substance	Vapor pressure (T = 20°C)
Acetic acid	11.6
Acetone	185
Alcohol	
Isoamyl	2.3
Ethyl	44.5
Methyl	88.7
Benzene	74.6
Bromine	172
Carbon disulfide	298
Carbon dioxide	42.9×10^3
Carbon tetrachloride	91
Chloroform	161
Ethyl acetate	72.8
Ethyl ether	440
Hydrogen sulfide	14×10^3
Sulfur dioxide	2.46×10^3

J. Table: Wet- and dry-bulb hygrometer tables (ventilated-type hygrometer)[19]

The tabulated values are relative humidities.

Depression of wet bulb (°C)	Dry-bulb temperature (°C) 0	5	10	15	20	25	30	35	40
1	81	87	88	89	90	92	93	93	94
2	64	72	76	80	82	85	86	87	88
3	46	59	66	71	74	77	79	81	82
4	29	45	55	62	66	70	73	75	76
5	13	33	44	53	59	63	67	70	72
6	...	21	34	44	52	57	61	64	66
7	...	9	25	36	45	50	55	59	61
8	...	...	15	28	38	44	50	54	56
9	...	...	6	20	30	38	44	50	52
10	...	...	...	13	24	33	39	44	48

K. Table: Thermal constants of common substances[19]

See Chap. 22.00 for thermal properties of the elements.

Melting points (°C)

Substance	mp	Substance	mp
Beeswax	62	Gun metal	1010
Naphthalene	80	Invar	1500
Paraffin wax	50–60	Magnalium	610
Quartz (fused)	1700	Solder	
Potassium nitrate	335	Hard	≈900
Sodium chloride	801	Soft	≈180
Brass	900	Steel	1400
Constantan	1290	Wood's metal	65

Boiling points (atm. pressure) (°C)

Substance	bp	Substance	bp
Acetone	56.7	Carbon disulfide	46.2
Aniline	184.2	Carbon tetrachlor.	76.7
Alcohol		Chloroform	61.2
Ethyl	78.3	Ether	34.6
Methyl	64.7	Glycerine	290
Benzene	80.2	Turpentine	161

Melting and boiling points of gaseous compounds (°C)

	mp	bp		mp	bp
Ammonia	− 78	− 33.5	Nitric oxide	− 167	− 150
Carbon dioxide	− 57	− 78.5	Nitrogen peroxide	− 10	21
Carbon monoxide	− 199	− 190	Sulfur dioxide	− 72	− 10
Hydrogen chloride	− 111	− 85	Sulfuretted hydrogen	− 83	− 61.5
Methane	− 184	− 161.5			

Expansion coefficients (per °C)

Solids (linear coeff., 10^{-6})		Liquids (volume coeff., 10^{-3})	
Bakelite	22	Alcohol	
Brass	19	Ethyl	1.10
Brick	9.5	Methyl	1.22
Celluloid	110	Benzene	1.24
Constantan	16	Carbon disulfide	1.21
Duralumin	22.6	Carbon tetrachlor.	1.23
Ebonite	84	Ether	1.63
German silver	18.4	Glycerine	0.53
Glass	8.5	Mercury	0.182
Gun metal	18.1	Oil	
Invar	1	Olive	0.72
Magnalium	24	Paraffin	0.90
Nichrome	12	Hydrochloric acid	0.57
Quartz (fused)	0.42	Sulfuric acid	0.57
Solder	25	Water (°C)	
		10–20	0.15
		20–30	0.25
		30–40	0.35
		40–60	0.46
		60–80	0.59
		80–100	0.70

Specific heats (cal·g^{-1}·°C^{-1})

Solids		Liquids	
Brass	0.092	Alcohol	
Building stone	0.18–0.23	Ethyl	0.58
Constantan	0.10	Methyl	0.60
Ebonite	0.40	Aniline	0.514
Glass	0.16	Benzene	0.41
Gun metal	0.086	Brine (25 wt. % NaCl)	0.785
Ice (0°C)	0.50	Carbon disulfide	0.24
Invar	0.12	Carbon tetrachlor.	0.20
Magnalium	0.22	Chloroform	0.234
Marble	0.21	Ether	0.56
Paraffin wax	0.69	Glycerine	0.58
Quartz	0.174	Paraffin oil	≈0.52
Rubber	0.40	Turpentine	0.42
Solder	0.043		
Steel	0.107		
Wood's metal	0.035		
Wood	0.4		

Gases and vapors	C_p	C_p/C_v
Acetylene	0.402	1.24
Air	0.241	1.40
Ammonia	0.52	1.31
Argon	0.127	1.65
Carbon monoxide	0.250	1.40
Carbon dioxide	0.202	1.30
Chlorine	0.124	1.36
Helium	1.25	1.66
Hydrogen	3.41	1.41
Methane	0.53	1.31
Nitrogen	0.249	1.40
Oxygen	0.218	1.40
Sulfur dioxide	0.154	1.29
Sulfuretted hydrogen	0.26	1.34

Latent heats (cal·g^{-1})

Fusion		Vaporization	
Acetic acid	44	Alcohol	
Beeswax	42	Ethyl	205
Benzene	31	Methyl	267
Bismuth	14.1	Benzene	93
Glycerine	42	Carbon disulfide	83.8
Ice	80	Carbon tetrachlor.	46
Lead	5.9	Ether	88.4
Sulfur	9.4	Mercury	65
Tin	14	Nitrogen	47.7
		Oxygen	51
		Water	539

L. Table: Thermal conductivities (10^{-4} cal·cm^{-1}·s^{-1}·°C^{-1})[19]

See also Tables 22.07 and 22.08 for thermal conductivity of the elements.

Substance	Bulk density (g cm^{-3})	Thermal conductivity	Substance	Bulk density (g cm^{-3})	Thermal conductivity
Solids					
Asbestos paper, wool	0.5	3.5	Wood		
Brass	8.6	2600	Balsa	0.14	1.3
Brick	1.6	12	Common, perpendicular		
Celluloid	1.4	5.0	to grain	...	2.6–3.8
Concrete	2.2	≈25.0	Common, parallel		
Constantan	8.9	540	to grain	...	5.3–8.4
Cork	0.20	1.2	Sawdust	0.20	1.4
Cotton wool	0.08	1.0			
Ebonite	1.2	3.8	**Liquids**		
Felt (hair)	0.27	0.9	Alcohol		
Glass wool	0.22	0.9	Ethyl		4.2
Ice	0.92	53	Methyl		5.0
Kapok	0.015	0.8	Aniline		4.0
Paper	...	1.2	Benzene		3.3
Quartz	2.2	24	Paraffin oil		3.6
Rubber (pure)	...	3.1	Turpentine		3.3
Sheepswool	0.08	1.0	Water		14.0
Silk fabric	...	1.0			
Slag wool	0.20	1.1	**Gases (0°C)**		
Slate	...	33	Air		0.57
Snow	0.25	3.6	Carbon dioxide		0.33
Soda glass	2.6	17	Hydrogen		3.30
Soil (dry)	...	4.3	Methane		0.74
Steel	7.8	1100	Nitrogen		0.55
Steel wool	0.10	1.9	Oxygen		0.56

M. Table: Selected properties of coolants[19]

Property	Helium	Water	Lithium	Sodium	Molten salt[a]
Melting point (K)	...	273	459	370	415
Boiling point (K)	4.2	373	1590	1151	...
Temperature (K)	400	338	473	473	422
Pressure (MPa)	7.09	0.344	0.101	0.101	0.101
Density (g·cm^{-3})	0.00833	0.980	0.515	0.903	1.98
Viscosity (g·cm^{-1}·s^{-1})	0.000245	0.00427	0.00566	0.00457	0.0017 at 700 K
Thermal conductivity (W·cm^{-1}·K^{-1})	0.00191	0.0066	0.4602	0.816	0.00605
Prandtl No.	0.664	2.71	0.0514	0.0074	4.39
Specific heat (J·g^{-1}·K^{-1})	5.192	4.184	4.142	1.326	1.561

[a] $NaNO_2$, $NaNO_3$, KNO_3, eutectic mixture.

N. Table: Properties of shielding materials[19]

γ-ray linear attenuation coefficient μ for various materials.

Material	Density ρ (g/cm^3)	Linear attenuation coefficient (cm^{-1}) 1 MeV	3 MeV	6 MeV
Air	0.001 294	0.000 076 6	0.000 043 0	0.000 030 4
Aluminum	2.7	0.166	0.095 3	0.071 8
Ammonia (liquid)	0.771	0.061 2	0.032 2	0.022 1
Beryllium	1.85	0.104	0.057 9	0.039 2
Beryllium carbide	1.9	0.112	0.062 7	0.042 9
Beryllium oxide (hot-pressed blocks)	2.3	0.140	0.078 9	0.055 2
Bismuth	9.80	0.700	0.409	0.440
Boral	2.53	0.153	0.086 5	0.067 8
Boron (amorphous)	2.45	0.144	0.079 1	0.067 9
Boron carbide (hot pressed)	2.5	0.150	0.082 5	0.067 5
Bricks				
Fire clay	2.05	0.129	0.073 8	0.054 3
Kaolin	2.1	0.132	0.075 0	0.055 2
Silica	1.78	0.113	0.064 6	0.047 3
Carbon	2.25	0.143	0.080 1	0.055 4
Clay	2.2	0.130	0.080 1	0.059 0
Cements				
Colemanite borated	1.95	0.128	0.072 5	0.052 8
Plain (1 Portland cement: 3 sand mixture)	2.07	0.133	0.076 0	0.055 9
Concretes				
Barytes	3.5	0.213	0.127	0.110
Barytes-boron frits	3.25	0.199	0.119	0.101
Barytes-limonite	3.25	0.200	0.119	0.099 1
Barytes-lumonite-colemanite	3.1	0.189	0.112	0.093 9
Iron-Portland[a]	6.0	0.364	0.215	0.181
MO (ORNL mixture)	5.8	0.374	0.222	0.184
Portland (1 cement: 2 sand: 4 gravel mixture)	2.2	0.141	0.080 5	0.059 2
	2.4	0.154	0.087 8	0.064 6
Flesh[b]	1	0.0699	0.0393	0.0274
Fuel oil (medium weight)	0.89	0.0716	0.0350	0.0239
Gasoline	0.739	0.0537	0.0299	0.0203
Glass				
Borosilicate	2.23	0.141	0.0805	0.0591
Lead (Hi-D)	6.4	0.439	0.257	0.257
Plate (avg)	2.4	0.152	0.0862	0.0629
Iron	7.86	0.470	0.282	0.240
Lead	11.34	0.797	0.468	0.505
Lithium hydride (pressed powder)	0.70	0.0444	0.0239	0.0172
Lucite (polymethyl methacrylate)	1.19	0.0816	0.0457	0.0317
Paraffin	0.89	0.0646	0.0360	0.0246
Rocks				
Granite	2.45	0.155	0.0887	0.0654
Limestone	2.91	0.187	0.109	0.0824
Sandstone	2.40	0.152	0.0871	0.0641
Rubber				
Copolymer	0.915	0.0662	0.0370	0.0254
Natural	0.92	0.0652	0.0364	0.0248
Neoprene	1.23	0.0813	0.0462	0.0333
Sand	2.2	0.140	0.0825	0.0587
Type 347 stainless steel	7.8	0.462	0.279	0.236
Steel (1% carbon)	7.83	0.460	0.276	0.234
Uranium	18.7	1.46	0.813	0.881
Uranium hydride	11.5	0.903	0.504	0.542
Water	1.0	0.0706	0.0396	0.0277
Wood				
Ash	0.51	0.0345	0.0193	0.0134
Oak	0.77	0.0521	0.0293	0.0203
White pine	0.67	0.0452	0.0253	0.0175

[a]Elemental composition (wt. %): hydrogen, 1.0; oxygen, 52.9; silicon, 33.7; aluminum, 3.4; iron, 1.4; calcium, 4.4; magnesium, 0.2; carbon, 0.1; sodium, 1.6; potassium, 1.3.
[b]Composition (wt. %): oxygen, 65.99; carbon, 18.27; hydrogen, 10.15; nitrogen, 3.05; calcium, 1.52; phosphorus, 1.02.

O. Table: Radiation damage doses for typical electrical insulating materials[19]

Material	Threshold damage dose (Mrads)	25% damage dose (Mrads)
Teflon	0.017	0.037
Formvar	16	82
Mylar	30	120
Typical epoxies, unfilled	200	3200
Mica	2000	~15 000
Epoxy, mineral filled with glass fiber reinforcing	8000	50 000

P. Table: Properties of common materials used in nuclear physics experiments[20]

See also Table 9.05.F for atomic and nuclear properties of materials.

Material	Chemical name or type	Composition	Density (g/cm^3)	Volume resistivity (Ω cm)	Manufacturer
Aquadag	Colloidal graphite dispersion in isopropyl alcohol	C, H, O (varies)			Acheson Colloids
Celluloid	Cellulose acetate	$(C_9H_{13}O_7)_n$	1.23–1.34	10^{10}–10^{14}	
Cymel	Melamine and formaldehyde condensation product	C, H, O, N (varies)			American Cyanamid
Formvar	Polyvinyl {alcohol, acetate, formal}	C, H, O (varies)	1.21–1.23		Monsanto
Grafoil	graphite film	C	1.6–1.8		Union Carbide
Havar		Fe (17.9%), Co (42.5%), Cr (20.0%), Ni (13.0%), W (2.8%), Mo, Mn, C, Be (traces)	8.3		Hamilton Watch
Hevimet		W (90%), Ni (7.5%), Cu (2.5%)	16.9–17.2		General Electric
Kapton[a]	Polyamide film	$(C_{22}H_{10}N_2O_4)_n$	1.08–1.14		Du Pont
Kel-F	Chlorotrifluoroethylene polymer	$(CF_2CHCl)_n$	2.10–2.13	1.2×10^{18}	3M
Lucite (Plexiglas)	Acrylic (methyl methacrylate) resins	$(C_5H_8O_2)_n$	1.18–1.19	$>10^{14}$	Du Pont, Rohm and Haas
Mica[b]	Muscovite (white mica)	$K_2O{\cdot}3Al_2O_3{\cdot}6SiO_2{\cdot}2H_2O$	2.76–3.0		
Mylar[c]	Polyester film	$(C_{10}H_8O_4)_n$	1.38–1.40	10^{14}	Du Pont
Nylon	Polyamides	$(C_{12}H_{22}N_2O_2)_n$	1.08–1.14	10^{12}–10^{15}	Du Pont
Polyethylene (polythene)		$(CH_2{:}CH_2)_n$	0.910–0.965	$>10^{16}$	
Polystyrene (styron)		$(C_6H_5CH{:}CH_2)_n$	0.98–1.10	$>10^{16}$	Dow Chemical
Teflon	Tetrafluoroethylene resin	$(CF_2)_n$	2.1–2.2	$>10^{18}$	Du Pont
VYNS	Polyvinylchloride-acetate copolymer (solution)	CH_2CHCl (90%), $CH_2CHO_2CCH_3$ (10%)	1.36		Union Carbide
Zapon	Nitrocellulose in a blend of hydrocarbon solvents	C, H, O, N (varies)			Glidden

[a]Formerly known as H-Film. This material is ≈ 50 times more resistant to radiation damage than Mylar.[c]
[b]Many types of mica are known; the one listed is merely typical.
[c]Mylar suffers radiation damage at doses greater than about 5×10^8 rads: A. M. Koehler *et al.*, Nucl. Instrum. Methods **33**, 341 (1965).

1.08. Table: RECOMMENDED SYMBOLS FOR PHYSICAL QUANTITIES

This table presents a listing of the most commonly used symbols for physical quantities. It is not intended to be complete, and the absence of a symbol should not necessarily prohibit its use. Many of the symbols listed are general; they may be made more specific by adding superscripts or subscripts or by using both lower and upper case forms if there is no ambiguity or conflict with other symbols. The expression given with the name of a symbol should be considered as a description rather than as a definition.

Symbol	Quantity
a	annihilation operator
	length
	relative chemical activity
$a^\dagger$	creation operator
b	annihilation operator
	breadth
	impact parameter
	mobility ratio: μ_n/μ_p
	photon annihilation operator
$b^\dagger$	phonon creation operator
c	concentration: $c = n/V$
	velocity of sound
c_g	group velocity
c_{ijkl}	elasticity tensor: $\tau_{ij} = c_{ijkl}\epsilon_{lk}$
c_l	velocity of longitudinal waves
c_t	velocity of transverse waves
$\bar{c}$	average speed
$\hat{c}$	most probable speed

1.08. Table: *Continued*

Symbol	Quantity
$\langle c\rangle$	average speed
d	diameter: $2r$
	lattice plane spacing
	relative density
	thickness
d_{hkl}	lattice plane spacing
e	linear strain
f	frequency
$f(c)$	velocity distribution function
g	acceleration of free fall
	g-factor: μ/μ_N
h	height
h,k,l	Miller indices
i	electric current
j	electric current density
j_i	total angular momentum quantum number
k	angular wave number
k_T	thermal diffusion ratio
l	length
	mean free path
l_i	orbital angular momentum quantum number
l_e	mean free path of electrons
l_{ph}	mean free path of phonons
$\mathrm{d}l$	element of path
m	mass
	molality of solution
m^*	effective mass
m_a	atomic mass
m_i	magnetic quantum number
m_r	reduced mass: $m_1m_2/(m_1+m_2)$
m_u	atomic mass unit: $\frac{1}{12}m_a(^{12}\mathrm{C})$
m_N	nuclear mass
n	amount of substance
	number density of particles: (N/V)
	order of reflection
	principal quantum number
n_i	principal quantum number
n_a	acceptor number density
n_d	donor number density
n_i	intrinsic number density: $(np)^{1/2}$
n_n	electron number density
n_p	hole number density
n_+	hole number density
n_-	electron number density
p	acoustic pressure
	pressure
q	electric charge
q_D	Debye angular wave number
r	molar ratio of solution
	radius
s	long range order parameter
	symmetry number
s_i	spin quantum number
s_{klji}	compliance tensor: $\epsilon_{kl}=s_{klji}\tau_{ij}$
$\mathrm{d}s$	element of path
t	time
u	average speed
	electromagnetic energy density
	speed: $\mathrm{d}s/\mathrm{d}t$
v	specific volume
	speed: $\mathrm{d}s/\mathrm{d}t$
	vibrational quantum number
	volume
v_{dr}	drift velocity
$\bar{v}$	average speed
$\hat{v}$	most probable speed
$\langle v\rangle$	average speed
w	electromagnetic energy density
	mass fraction
x	molar fraction
z	ionic charge
	reduced activity: $(2\pi mkT/h^2)^{3/2}\lambda$
A	nucleon number, mass number
	area
	chemical affinity
	activity (radioactivity)
	Richardson constant: $j=AT^2\exp(-\Phi/kT)$
$A^\dagger$	Hermitian conjugate of A: $(A^\dagger)_{ij}=(A_{ji})^*$
A_{ij}	matrix element: $\int\phi_i^*(A\phi_j)\mathrm{d}\tau$
A_E	Ettinghausen coefficient
A_H	Hall coefficient
A_N	Ettinghausen–Nernst coefficient
A_r	relative atomic mass: m_a/m_u
A_{RL}	Righi–Leduc coefficient
B	susceptance
C	capacitance
D	Debye–Waller factor
	diffusion coefficient
D_{td}	thermal diffusion coefficient
E	electromotive force
	energy
	Young's modulus
E_a	acceptor ionization energy
E_{ab}	thermo-electromotive force
E_d	donor ionization energy
E_g	energy gap
E_k	kinetic energy
E_p	potential energy
E_F	Fermi energy
F	hyperfine quantum number
F_m	magnetomotive force
G	conductance
	shear modulus
H	Boltzmann function
	Hamiltonian: $\Sigma_i p_i\dot{q}_i - L$
H_c	superconductor critical field strength
I	electric current
	moment of inertia
	nuclear spin quantum number (atomic physics)
J	action integral: $\oint p\,\mathrm{d}q$
	electric current density
	exchange integral
	nuclear spin quantum number (nuclear physics)
	rotational quantum number
	total angular momentum quantum number
K	bulk modulus
	equilibrium constant
	kinetic energy
	relative permittivity
	rotational quantum number
L	Lagrangian: $T(q_i,\dot{q}_i)-V(q_i,\dot{q}_i)$
	length
	orbital angular momentum quantum number
	self-inductance
L_p	sound pressure level
L_N	loudness level
L_W	sound power level
L_{12}	mutual inductance
M	magnetic quantum number
	molar mass
	mutual inductance
M_r	relative atomic mass: m_a/m_u
	relative molar mass
N	neutron number: $A-Z$
	number of particles
N_E	density of states: $\mathrm{d}N(E)/\mathrm{d}E$
N_ω	(spectral) density of vibrational modes
P	power: $\mathrm{d}E/\mathrm{d}t$
	pressure
	probability density

1.08. Table: *Continued*

Symbol	Meaning
P_E	Ettinghausen coefficient
Q	quadrupole moment
	quantity of electricity
	reaction energy, disintegration energy
R	resistance
R_l	linear range
R_H	Hall coefficient
S	action integral: $\oint p\,dq$
	area
	Hamilton's characteristic function: $2\int T\,dt$
	spin quantum number
S_a	atomic stopping power
S_{ab}	Seebeck coefficient
S_l	linear stopping power
S_p	Hamilton's principal function: $\int L\,dt$
S_{RL}	Righi–Leduc coefficient
T	kinetic energy
	period, periodic time
T_c	superconductor critical transition temperature
T_C	Curie temperature
T_N	Néel temperature
$T_{1/2}$	half-life
U	potential difference
	potential energy
U_m	magnetic potential difference
V	electric potential
	potential difference
	potential energy
	volume
W	energy
	Hamilton's principal function: $\int L\,dt$
	sound energy flux, acoustic power
	weight
	work: $\int \mathbf{F}\cdot d\mathbf{s}$
X	reactance
Y	admittance: $Y = 1/Z = G + jB$
	Young's modulus
Z	canonical partition function
	impedance: $R + jX$
	proton number, atomic number
(h,k,l)	single plane or set of parallel planes in a lattice
$\{h,k,l\}$	full set of equivalent lattice planes
$[u,v,w]$	direction in a lattice
$\langle u,v,w\rangle$	full set of equivalent lattice directions
$\langle A\rangle$	expectation value of A: $\mathrm{Tr}(A)$
$[A,B]$	commutator of A and B: $AB - BA$
$[A,B]_+$	anticommutator of A and B: $AB + BA$
$[A,B]_-$	commutator of A and B: $AB - BA$
$\mathbf{a}$	acceleration
$\mathbf{b}$	Burgers' vector
$\mathbf{c}$	velocity
$\mathbf{c}_O$	average velocity
$\langle\mathbf{c}\rangle$	average velocity
$\mathbf{e}$	polarization vector
$\mathbf{k}$	angular wave vector
	propagation vector
	(particle) propagation vector
$\mathbf{k}_F$	Fermi angular wave vector
$\mathbf{m}$	magnetic dipole moment
$\mathbf{p}$	electric dipole moment
	momentum: $m\mathbf{v}$
$\mathbf{p}, p_i$	generalized momentum: $\partial L/\partial q_i$
$\mathbf{q}$	(phonon) propagation vector
$\mathbf{q}, q_i$	generalized coordinate
$\mathbf{r}$	position vector
$\mathbf{s}$	position vector
$\mathbf{u}$	velocity
$\mathbf{v}$	velocity
$\mathbf{v}_O$	average velocity
$\langle\mathbf{v}\rangle$	average velocity
$\mathbf{A}$	magnetic vector potential
$\mathbf{B}$	magnetic induction
$\mathbf{D}$	electric displacement
$\mathbf{E}$	electric field
$\mathbf{F}$	force
$\mathbf{G}$	(circular) reciprocal lattice vector
	$\mathbf{G}\cdot\mathbf{R} = 2\pi m$, m = integer
$\mathbf{H}$	angular impulse: $\int \mathbf{M}\,dt$
	magnetic field (strength)
$\mathbf{I}$	impulse: $\int \mathbf{F}\,dt$
$\mathbf{J}$	angular momentum: $\mathbf{r}\times\mathbf{p}$
$\mathbf{L}$	angular momentum: $\mathbf{r}\times\mathbf{p}$
$\mathbf{M}$	magnetization
	moment of force
$\mathbf{P}$	dielectric polarization
$\mathbf{R}$	lattice vector
$\mathbf{S}$	Poynting vector
$\mathbf{T}$	torque, moment of a couple
$\mathscr{E}$	electromotive force
$\mathscr{H}$	Hamiltonian: $\Sigma_i p_i \dot{q}_i - L$
$\mathscr{L}$	Lagrangian: $T(q_i,\dot{q}_i) - V(q_i,\dot{q}_i)$
α	absorption coefficient
	angular acceleration
	annihilation operator
	internal conversion coefficient
	Madelung constant
	plane angle
	polarizability
	recombination coefficient
$\alpha^\dagger$	creation operator
α_a	absorption coefficient
α_T	thermal diffusion factor
β	annihilation operator
	plane angle
$\beta^\dagger$	creation operator
γ	conductivity: $1/\rho$
	growth rate
	gyromagnetic ratio: ω/B
	plane angle
	polarizability
	shear strain
	surface tension
δ	damping coefficient
	loss angle: arctan X/R
	thickness
ϵ	linear strain
	permittivity
ϵ_{ij}	strain tensor
ϵ_r	relative permittivity
ϵ_F	Fermi energy
η	viscosity
θ	plane angle
	scattering angle
ϑ	Bragg angle
	scattering angle
κ	bulk modulus
	Landau–Ginzburg parameter
λ	absolute activity: $\exp(\mu/kT)$
	damping coefficient
	decay constant, disintegration constant
	mean free path
	wavelength
λ_{ik}	thermal conductivity tensor
λ_C	Compton wavelength: h/mc
λ_L	London penetration depth
μ	chemical potential
	permeability
	Poisson ratio
	reduced mass: $m_1 m_2/(m_1 + m_2)$
	shear modulus

1.08. Table: *Continued*

μ_l	linear attenuation coefficient	$\tau_{1/2}$	half life
μ_a	atomic attenuation coefficient	χ	magnetic susceptibility
μ_m	mass attenuation coefficient	χ_e	electric susceptibility
μ_r	relative permeability: μ/μ_0	χ_m	magnetic susceptibility
ν	amount of substance	ω	angular frequency: $2\pi f$
	frequency		solid angle
	kinematic velocity: η/ρ	ω_D	Debye angular frequency
	Poisson ratio	ω_L	Larmor circular frequency
ν_B	stoichiometric number of substance B	Δ	superconductor energy gap
ξ	coherence length	Γ	level width
ϕ	electric potential	Θ	characteristic temperature
	osmotic coefficient	Θ_{ab}	thermo-electromotive force
	plane angle	Θ_{rot}	characteristic rotational temperature: $h^2/8\pi^2kI$
	volume fraction	Θ_{vib}	characteristic vibrational temperature: $h\nu/k$
$\psi(\mathbf{r})$	one-electron wave function	Θ_D	Debye temperature: $h\nu_D/k$
ρ	charge density	Θ_E	Einstein temperature: $h\nu_E/k$
	reflection coefficient	Θ_W	characteristic (Weiss) temperature
	resistivity	Λ	logarithmic decrement
	(mass) density		mean free path of phonons
ρ_{ik}	resistivity tensor	Π	osmotic pressure
ρ_R	residual resistivity	Π_{ab}	Peltier coefficient
σ	conductivity: $1/\rho$	Σ	macroscopic cross section: $n\sigma$
	cross section	Φ	magnetic flux
	normal stress		potential energy
	short range order parameter	Ξ	grand canonical partition function
	surface charge density	Ψ	electric flux
	surface tension		wave function
	wave number	Ω	microcanonical partition function
σ_{ik}	electrical conductivity tensor		solid angle
τ	mean life		volume in γ phase space
	relaxation time	$\boldsymbol{\mu}$	magnetic dipole moment
	shear stress	$\boldsymbol{\sigma}$	wave vector
	transmission coefficient	$\boldsymbol{\omega}$	angular velocity: $d\phi/dt$
τ_{ij}	stress tensor	$\langle\cdots\vert$	Dirac bra vector
τ_m	mean life	$\vert\cdots\rangle$	Dirac ket vector

Pauli matrices: $\boldsymbol{\sigma}, I$

$$\sigma_x=\begin{pmatrix}0&1\\1&0\end{pmatrix},\ \sigma_y=\begin{pmatrix}0&-i\\i&0\end{pmatrix},\ \sigma_z=\begin{pmatrix}1&0\\0&-1\end{pmatrix}$$

$$I=\begin{pmatrix}1&0\\0&1\end{pmatrix}\ \text{(unit matrix)}$$

Dirac matrices: $\boldsymbol{\alpha},\beta$

$$\alpha_x=\begin{pmatrix}0&\sigma_x\\\sigma_x&0\end{pmatrix},\ \sigma_y=\begin{pmatrix}0&\sigma_y\\\sigma_y&0\end{pmatrix},\ \alpha_z=\begin{pmatrix}0&\sigma_z\\\sigma_z&0\end{pmatrix}$$

$$\beta=\begin{pmatrix}I&0\\0&-I\end{pmatrix}$$

Fundamental translation vectors for the crystal lattice:

$\mathbf{R}=n_1\mathbf{a}_1+n_2\mathbf{a}_2+n_3\mathbf{a}_3$ (n_1,n_2,n_3, integers).

Fundamental translation vectors for the reciprocal lattice:

$\mathbf{b}_1,\mathbf{b}_2,\mathbf{b}_3$ $\mathbf{a}^*,\mathbf{b}^*,\mathbf{c}^*$

(In solid-state physics, $\mathbf{a}_i\cdot\mathbf{b}_k=2\pi\delta_{ik}$; in crystallography, however, $\mathbf{a}_i\cdot\mathbf{b}_k=\delta_{ik}$, where δ_{ik} is the Kronecker symbol).

1.09. SOURCES OF PHYSICAL DATA
A. Journals, publications, and databases

The following journals, publication series, and computerized databases provide evaluated physical property data of general interest.

Journal of Physical and Chemical Reference Data (published by the American Institute of Physics and American Chemical Society for the National Institute of Standards and Technology)

Atomic Data and Nuclear Data Tables (Academic Press)

Nuclear Data Sheets (Academic Press)

Physikdaten/Physics Data (Fachinformationszentrum Energie, Physik, Mathematik GmbH)

Landolt-Bornstein, Numerical Data and Functional Relationships in Science and Technology (Springer-Verlag)

National Standard Reference Database Series (National Institute of Standards and Technology)—A series of computerized databases of physical and chemical properties.

Information on sources for specific data may be obtained from:

Office of Standard Reference Data
National Institute of Standards and Technology
Gaithersburg, MD 20899
Telephone: (301) 975-2208

B. Data centers

A number of specialized data centers in the United States compile and evaluate data of interest to physicists on a regular basis. Most of these centers can respond to inquiries from the public. In the list below, those data centers for which no address is given are located at the National Institute of Standards and Technology, Gaithersburg, MD 20899.

Fundamental physical constants

Fundamental Constants Data Center. (301) 975-4220. Current information on precision measurements and fundamental physical constants.

Nuclear and particle physics data

National Nuclear Data Center, Brookhaven National Laboratory, Upton, NY 11973. (516) 282-2902. Neutron cross sections and nuclear structure data.

Isotopes Project, Lawrence Berkeley Laboratory, University of California, Berkeley, CA 94720. (415) 486-6152. Nuclear structure and radioactivity data.

Nuclear Data Project, Oak Ridge National Laboratory, P.O. Box X, Oak Ridge, TN 37870. (675) 574-4691. Nuclear structure data.

Photon and Charged Particle Data Center. (301) 975-5551. X- and γ-ray cross sections and absorption coefficients, electron stopping power.

Idaho Engineering Laboratory, P.O. Box 1625, Idaho Falls, ID 83401. (208) 526-1185. γ-ray spectrum catalog.

Fundamental Particle Data Center, Lawrence Berkeley Laboratory, University of California, Berkeley, CA 94720. (415) 486-5885. Fundamental particle properties, cross sections for high-energy processes.

Atomic and molecular properties, including spectra

Atomic Collision Cross Section Data Center, Joint Institute for Laboratory Astrophysics, University of Colorado, Boulder, CO 80390. (303) 492-8089. Collision cross sections for electron and photons, energy transfer data for atoms and small molecules.

Atomic Energy Levels Data Center. (301) 975-3212. Energy levels, ionization potentials, and spectra.

Atomic Transition Probabilities Data Center. (301) 975-3200. Transition probabilities, line shapes, and shifts.

Ion Energetics Data Center. (301) 975-2562. Ionization and appearance potentials of molecules.

Molecular Spectra Data Center. (301) 975-2385. Microwave and high-resolution infrared spectra.

Controlled Fusion Atomic Data Center, Oak Ridge National Laboratory, P.O. Box X, Bldg. 6003, Oak Ridge, TN 37830. (615) 574-4704. Atomic collision cross sections and other data relevant to fusion research.

Structural, optical, and electrical properties of solids

Alloy Data Center. (301) 975-6040. Phase diagrams and other alloy properties.

Crystal Data Center. (301) 975-6254. Lattice constants and other properties of single crystals.

Diffusion in Metals Data Center. (301) 975-6157. Diffusion constants for various materials in metals and alloys.

Phase Diagrams for Ceramists Data Center. (301) 975-6119. Phase diagrams for nonmetallic inorganic systems.

Center for Information and Numerical Data Analysis and Synthesis (CINDAS), Purdue University, 2595 Yeager Road, West Lafayette, IN 47906. (317) 494-9393; Thermal, electrical, and optical properties of solids.

Thermodynamics and chemical kinetics

Chemical Thermodynamics Data Center. (301) 975-2526. Enthalpy, entropy, Gibbs energy, and heat capacity of inorganic compounds; properties of aqueous electrolytes.

Fluid Mixtures Data Center, National Institute of Standards and Technology, Boulder, CO 80303. (303) 497-3257. Thermodynamic and transport properties of pure fluids and mixtures.

Molten Salts Data Center, Rensselaer Polytechnic Institute, Troy, NY 12181. (518) 276-6337. Density, viscosity, electrical conductance, and surface tension of molten salts.

Thermodynamics Research Center, Texas A&M University, Department of Chemistry, College Station, TX 77843. (409) 845-4940. Physical and thermodynamic properties of organic compounds.

Chemical Kinetics Information Center. (301) 975-2569. Rate constants and photochemistry of gas phase reactions.

Radiation Chemistry Data Center, Radiation Laboratory, University of Notre Dame, Notre Dame, IN 46556. (219) 239-6527. Effects of ionizing and photo-optical radiation on chemical systems in solution.

1.10. REFERENCES

[1]Except as noted (Refs. 2–5), the material in Sec. 1.05 is from H. B. Dwight, *Tables of Integrals and Other Mathematical Data* (Macmillan, New York, 1949).

[2]In part from H. Ebert, *Physics Pocketbook* (Oliver and Boyd, Edinburgh, 1967).

[3]David L. Book, *NRL Plasma Formulary* (Naval Research Laboratory, Washington, DC, 1980). See also P. M. Morse and H. Feshbach, *Methods of Theoretical Physics* (McGraw-Hill, New York, 1953).

[4]J. D. Jackson, *Classical Electrodynamics* (Wiley, New York, 1975).

[5]Enrico Fermi, *Notes on Quantum Mechanics* (University of Chicago, Chicago, 1961).

[6]Particle Data Group, *Review of Particle Properties*, Rev. Mod. Phys. **52**, No. 2, Pt. II (1980).

[7]R. A. Fisher, *Statistical Methods for Research Workers* (Oliver and Boyd, Edinburgh, 1958).

[8]*Handbook of Mathematical Functions*, edited by M. Abramovitz and I. Stegun (National Bureau of Standards, Washington, DC, 1964), Appl. Math. Ser. 55.

[9]W. T. Eadie, D.Drijard, F. E. James, M. Roos, and B. Sadoulet, *Statistical Methods in Experimental Physics* (North-Holland, Amsterdam, 1971).

[10]*American Institute of Physics Handbook*, 3rd ed., edited by Dwight E. Gray (McGraw-Hill, New York, 1972).

[11]Enrico Fermi, *Thermodynamics* (Prentice-Hall, Englewood Cliffs, NJ, 1937).

[12]B. Yavorsky and A. Detlaf, *Handbook of Physics* (Mir. Moscow, 1975).

[13]H. A. Bethe and E. E. Salpeter, *Quantum Mechanics of One- and Two-Electron Atoms* (Academic, New York, 1957).

[14]John Strong, *Procedures in Experimental Physics* (Prentice-Hall, Englewood Cliffs, NJ, 1938).

[15]H. S. Carslaw, *Introduction to the Mathematical Theory of the Conduction of Heat in Solids*, 2nd ed. (MacMillan, London, 1921).

[16]W. E. Byerly, *Fourier Series and Spherical Harmonics* (Ginn, Boston, 1893).

[17]Natl. Bur. Stand. (U.S.) Circ. **74** (1924).

[18]Chem. Eng. News **63**(5), 27 (1985).

[19]W. H. Childs, *Physical Constants* (Methuen, London, 1951).

[20]J. B. Marion and F. C. Young, *Nuclear Reaction Analysis* (Wiley, New York, 1968). See also R. C. Reid, J. M. Prausnitz, and T. K. Sherwood, *The Properties of Gases and Liquids* (McGraw-Hill, New York, 1977).

2.00. Acoustics

R. BRUCE LINDSAY* AND ROBERT T. BEYER

Brown University

CONTENTS

*Deceased.

2.01. IMPORTANT ACOUSTICAL UNITS

Acoustic ohm. Unit of acoustic impedance. An acoustic impedance (including acoustic resistance and reactance) has a magnitude of 1 acoustic ohm (cgs) when a sound pressure of 1 μbar produces a volume velocity of 1 cm^3/s. The specific acoustic impedance (impedance for a unit area) of a plane wave is the product of the density of the medium and the velocity of sound. Typical values are: 41.5 $g{\cdot}cm^{-2}{\cdot}s^{-1}$; water, 145 000 $g{\cdot}cm^{-2}{\cdot}s^{-1}$; iron, 4×10^6 $g{\cdot}cm^{-2}{\cdot}s^{-1}$.

Cent. The interval between two musical sounds having as a basic frequency ratio the 1200th root of 2. The number of cents between frequencies f_1 and f_0 is

$$1200 \log_2(f_1/f_0) = 3986 \log_{10}(f_1/f_0).$$

Decibel. Unit expressing the magnitude of the ratio of two sound power or intensity levels. The number of decibels between powers P_1 and P_2 is $10 \log_{10}(P_1/P_2)$. For pressure the number of decibels is $20 \log_{10}(P_1/P_2)$. Abbreviation: dB. Ordinary sound intensities in air: 10^{-16}–10^{-4} $W{\cdot}cm^{-2}$. Audible threshold is 10^{-16} $W{\cdot}cm^{-2}$, 2×10^{-4} μbars.

Mechanical ohm. Unit of mechanical impedance. A mechanical impedance has a magnitude of 1 mechanical ohm (cgs) when an applied pressure of 1 dyn/cm^2 produces a velocity of 1 cm/s.

Mel. Unit of subjectively estimated pitch. The pitch of a 1000-Hz tone at 40 dB above threshold is taken to be 1000 mels. The pitch of any sound judged to be double that pitch is taken as 2000 mels, etc.

Microbar. Unit of pressure commonly used in acoustics. 1 μbar = 1 dyn/cm^2. The SI unit is the micropascal, 1 μPa = 10^{-5} μbars.

Neper. Unit expressing the ratio between two amplitudes as a natural (Nasperian) logarithm. 1 Np = 8.687 dB.

Noy. Unit of noisiness related to the preceived noise level in PNdB through the relation

$$10 \log_2(\text{noy}) = \text{PNdB} - 40,$$

where PNdB, the perceived noise level in dB, is defined as the sound-pressure level of a reference sound of the order of one octave wide centered at 1000 Hz, which is subjectively judged to be equally noisy as the sound being measured.

Phon. Unit of loudness level. The loudness level of a given sound is the sound-pressure level in dB of a pure tone of frequency 1000 Hz (relative to 2×10^{-4} μbars) which is assessed by normal observers as being equally loud as the sound in question. Thus phons are effectively expressed in dB. The value at the audible threshold is 0 phons, the threshold of feeling is about 140 phons. A loudness level of 74 phons corresponds to a sound pressure of 1 μbar of a 7000-Hz tone.

Rayl. Unit of specific acoustic impedance. A specific acoustic impedance has a magnitude of 1 rayl (cgs) when a sound pressure of 1 μbar produces a linear velocity of 1 cm/s. See *Acoustic ohm.*

Sabin. Unit of absorption of surface covering in room acoustics equal to that of 1 ft^2 of perfectly absorbing material. The unit is effectively the square foot in the English system. The metric sabin is the square meter.

Sone. Unit of subjective loudness. If S is the loudness in sones and P the loudness level in phons,

$$\log_{10} S = 0.0301P - 1.204.$$

2.02. OSCILLATIONS OF A LINEAR SYSTEM

The equation of motion of a dissipative system with one degree of freedom subject to a linear restoring force, with equivalent localized mass m, resistance (damping factor) R, and stiffness k, subject to no external force is

$$m\ddot{\xi} + R\dot{\xi} + k\xi = 0, \tag{1}$$

where ξ is the displacement from equilibrium. The general solution is

$$\xi = e^{-Rt/2m}\left(Ae^{+(R^2/4m^2 - k/m)^{1/2}t} + Be^{-(R^2/4m^2 - k/m)^{1/2}t}\right), \tag{2}$$

where A and B are arbitrary constants to be fixed by the initial conditions. If

$$R^2/4m^2 < k/m, \tag{3}$$

the motion is a damped oscillation with frequency

$$f = (1/2\pi)(k/m - R^2/4m^2)^{1/2} \tag{4}$$

and amplitude which varies with the time through the term $e^{-Rt/2m}$. The logarithmic decrement (the logarithm to the base e of the ratio of successive amplitudes) is

$$\delta = R/2mf. \tag{5}$$

If

$$R^2/4m^2 \geqslant k/m, \tag{6}$$

no oscillations take place. The case of the equals sign is called "critical damping."

If the system is subject to a harmonic force $F_0e^{i\omega t}$, with the angular frequency $\omega = 2\pi f$, the displacement in the steady state

$$\xi = \frac{F_0e^{i\omega t}}{i\omega R + k - \omega^2 m}\,. \tag{7}$$

The velocity ($\dot{\xi}$) amplitude has its maximum for the angular frequency

$$\omega = (k/m)^{1/2}, \tag{8}$$

which is called the *resonance* frequency of the system.

Acoustic example of a lumped oscillating system: Helmholtz resonator, a hollow air-filled sphere of volume V, having an inlet opening for sound and an outlet placed in the external ear, is shown in Figure 2.02.A.

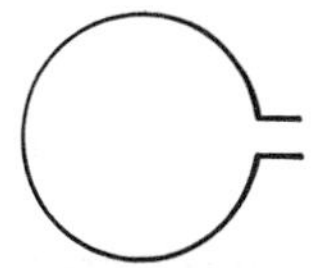

FIGURE 2.02.A. Helmholtz resonator

It is used as a resonant cavity in sound analysis. The acoustic elements of the oscillator are as follows: m is the equivalent mass of moving air in the opening,

$$m = \rho_0 S^2/c_0, \quad (9)$$

R is the resistance due to radiation of sound from the opening,

$$R = 2\pi\rho_0 f^2 S^2, \quad (10)$$

and k is the equivalent stiffness,

$$k = \rho_0 c^2/V, \quad (11)$$

with ρ_0 the equilibrium density of the fluid in the resonator, V the volume of the resonator cavity, c the velocity of sound in the fluid medium, S the area of the resonator opening, and c_0 the acoustic conductivity. The approximate value of the resonance frequency is

$$f_r = (c/2\pi)(c_0/V)^{1/2}. \quad (12)$$

For an inlet consisting of a tube of length l and cross-sectional radius R, the acoustic conductivity becomes

$$c_0 = \frac{\pi R^2}{l + \pi R/2}.$$

For very small l compared with R, c_0 reduces to $2R$.

For resonance frequencies and normal modes of other oscillating systems see Secs. 2.10 and 2.14.

Analogies. In the normal classical electromechanical analogy between an oscillating electrical circuit of inductance L, resistance R, and capacitance C subject to an alternating voltage E_0, L is analogous to m, R to the mechanical resistance R, and C to the reciprocal of the stiffness k.

In the so-called mobility analogy, electric current corresponds to mechanical force, electromotive force to mechanical velocity, mechanical resistance to the reciprocal of electric resistance, and mechanical compliance (reciprocal of stiffness) to electric inductance.

Mechanical impedance is the ratio of the mechanical force to the velocity.

2.03. GENERAL LINEAR ACOUSTICS; WAVE PROPAGATION IN FLUIDS

A. Plane waves

A plane wave of acoustic excess pressure p in the x direction with velocity c is subject to the equation

$$\frac{\partial^2 p}{\partial x^2} = \frac{1}{c^2}\frac{\partial^2 p}{\partial t^2}, \quad (13)$$

with the general solution

$$p = \phi_1(x - ct) + \phi_2(x + ct), \quad (14)$$

where ϕ_1 and ϕ_2 are arbitrary continuous and differentiable functions. For a harmonic wave of frequency f in the positive x direction,

$$\phi_1 = p_0 \cos[(2\pi f/c)(x - ct) + \theta_0], \quad (15)$$

where p_0 is the pressure amplitude and θ_0 the initial phase at $x = 0$.

The acoustic impedance is defined as the complex ratio of the sound pressure on a given surface lying in a wave front to the volume velocity through unit area in that surface. The real part of the impedance is the acoustic *resistance* and the imaginary part is the acoustic *reactance*. The acoustic *stiffness* for a fluid medium traversed by a harmonic wave of frequency f is $2\pi f$ times the acoustic resistance. The acoustic *compliance* is the reciprocal of the acoustic stiffness. For a plane harmonic wave the acoustic impedance is real and has the value

$$Z = \rho_0 c/S, \quad (16)$$

where S is the area of the wave front and ρ_0 the equilibrium density of the fluid. The specific impedance, or the impedance for unit area of wave front, is

$$Z_S = \rho_0 c. \quad (17)$$

The acoustic intensity defined as the average rate at which wave energy is transmitted in a specified direction at a given point through unit area normal to the direction at this point is

$$I = p_0^2/2Z_S. \quad (18)$$

The absolute unit is the W/m^2. But see *Decibel* in Sec. 2.01.

The velocity of sound in a fluid medium of density ρ_0 and adiabatic bulk modulus B is

$$c = (B/\rho_0)^{1/2}. \quad (19)$$

For an ideal gas this becomes

$$c = (\gamma p/\rho)^{1/2} = (\gamma R_m T)^{1/2} = c_0(T/273)^{1/2}, \quad (20)$$

where γ is the ratio of specific heat at constant pressure to that at constant volume, T the absolute temperature, R_m the gas constant per gram. The velocity of sound at 0°C is c_0. (See Sec. 2.05.)

Classical acoustic wave attenuation is shown in the equation

$$p = Ae^{-ax}\cos[(\omega/c)(x - ct) + \theta_0], \quad (21)$$

with the attenuation coefficient α given by

$$\alpha = \frac{2\omega^2\eta}{3\rho_0 c^3} + \left(\frac{\gamma - 1}{\gamma}\right) \times \frac{\omega^2 M_k}{2\rho_0 c^3 C_v}, \quad (22)$$

where η is the coefficient of viscosity, κ the thermal conductivity, M the molecular weight, C_v the molar specific heat at constant volume, and γ the ratio of specific heat at constant pressure to that at constant volume.

For relaxation attenuation see Sec. 2.08.

B. Spherical waves

A spherical wave from a monopole source at the origin is governed by

$$\frac{\partial^2}{\partial r^2}(rp) = \frac{1}{c^2}\frac{\partial^2(rp)}{\partial t^2}. \tag{23}$$

The specific acoustic impedance is

$$Z_s = Z_{s_1} + iZ_{s_2}, \tag{24}$$

with Z_{s_1} the specific acoustic resistance,

$$Z_{s_1} = \frac{\omega^2 \rho_0 r^2}{c(1+\omega^2 r^2/c^2)} \tag{25}$$

and Z_{s_2} the specific acoustic reactance,

$$Z_{s_2} = Z_{s_1}\left(\frac{c}{\omega r}\right). \tag{26}$$

C. Reflection and transmission

The reflection and transmission of a plane wave at a plane interface are given by

$$\frac{I_r}{I_i} = \frac{(Z_2\cos\theta_i - Z_1\cos\theta_t)^2}{(Z_2\cos\theta_i + Z_1\cos\theta_t)^2}, \tag{27}$$

where I_i is the intensity of the incident wave, I_r the reflected intensity, θ_i the angle of incidence, θ_t the angle of refraction, Z_1 the specific acoustic resistance of the incident wave, and Z_2 the specific acoustic resistance of the transmitted wave.

Similarly,

$$\frac{I_t}{I_i} = \frac{4Z_2/Z_1}{(Z_2/Z_1 + \cos\theta_t/\cos\theta_i)^2}, \tag{28}$$

where I_t is the intensity of the transmitted wave. The law of refraction is

$$\frac{\sin\theta_i}{\sin\theta_t} = \frac{c_1}{c_2}, \tag{29}$$

where c_1 and c_2 are the sound velocities in the two media, respectively.

D. Velocity and attenuation of sound

The following tables present relevant acoustical data on the velocity and attenuation of sound.

TABLE 2.03.D.1. Velocity of sound in selected gases at sonic frequencies. From Ref. 1.

Gas	Formula	Velocity at 0°C (m/s)
Air (dry)	...	331.45
Ammonia	NH_3	415
Argon	Ar	319
Carbon monoxide	CO	338
Carbon dioxide	CO_2	259.0
Carbon disulfide	CS_2	189
Chlorine	Cl_2	206
Ethylene	C_2H_4	317
Helium	He	965
Hydrogen	H_2	1284
Illuminating gas (coal)	...	453
Methane	CH_4	430
Neon	Ne	435
Nitric oxide (10°C)	NO	325
Nitrogen	N_2	334
Nitrous oxide	N_2O	263
Oxygen	O_2	316
Steam (134°C)	H_2O	494

TABLE 2.03.D.2. Velocity of sound in pure water and in sea water (salinity 3.5%) at 1 atm. From Ref. 1.

Temperature (°C)	Velocity (m/s) Pure water	Sea water
0	1402.3	1449.4
10	1447.2	1490.4
20	1482.3	1522.2
30	1509.0	1546.2
50	1542.5	
70	1554.7	
100	1543.0	

TABLE 2.03.D.3. Velocity of sound in selected liquids. From Ref. 1.

Liquid	Velocity (m/s)
Mercury (20°C, 1 atm)	1451
Pentane (20°C, 1 atm)	1008
CS_2 (25°C, 1 atm, 2 mHz)	1140
CCl_4 (25°C, 1 atm, 4.85 mHz)	930
Ether (25°C, 1 atm)	976
Acetone (20°C, 1 atm)	1203
Ethyl alcohol (20°C, 1 atm)	1161.8
Methyl alcohol (20°C, 1 atm)	1121.2
Liquid helium I (4 K, 15 mHz, 1 atm)	211

TABLE 2.03.D.4. Elastic constants, sound velocities, and specific impedances of selected solids. Y_0 is Young's modulus, μ the shearing modulus, and λ the Lamé coefficient $B - \frac{2}{3}\mu$, where B is the bulk modulus; V_l is the longitudinal compressional wave velocity, V_s the shear wave velocity, and V_{ext} the velocity of a wave in a thin rod. From Ref. 1.

Materials	Y_0 (10^{10} N/m²)	μ (10^{10} N/m²)	λ (10^{10} N/m²)	Poisson's ratio σ	$V_l = [(\lambda + 2\mu)/\rho]^{1/2}$ (m/s)	$V_s = (\mu/\rho)^{1/2}$ (m/s)	$V_{ext} = (Y_0/\rho)^{1/2}$ (m/s)	$Z_l = [\rho(\lambda + 2\mu)]^{1/2}$ (10^6 kg/s m²)	$Z_s = (\rho\mu)^{1/2}$ (10^6 kg/s m²)
Aluminum, rolled	6.8–7.1	2.4–2.6	6.1	0.355	6 420	3 040	5 000	17.3	8.2
Beryllium	30.8	14.7	1.6	0.05	12 890	8 880	12 870	24.1	16.6
Brass, yellow, 70 Cu, 30 Zn	10.4	3.8	11.3	0.374	4 700	2 110	3 480	40.6	18.3
Constantan	16.1	6.09	11.4	0.327	5 177	2 625	4 270	45.7	23.2
Copper, rolled	12.1–12.8	4.6	13.1	0.37	5 010	2 270	3 750	44.6	20.2
Duralumin 17S	7.15	2.67	5.44	0.335	6 320	3 130	5 150	17.1	8.5
Gold, hard-drawn	8.12	2.85	15.0	0.42	3 240	1 200	2 030	62.5	23.2
Iron, cast	15.2	5.99	6.92	0.27	4 994	2 809	4 480	37.8	21.35
Iron electrolytic	20.6	8.2	11.3	0.29	5 950	3 240	5 120	46.4	25.3
Armco	21.2	8.24	11.35	0.29	5 960	3 240	5 200	46.5	25.3
Lead, rolled	1.5–1.7	0.54	3.3	0.43	1 960	690	1 210	22.4	7.85
Magnesium, drawn, annealed	4.24	1.62	2.56	0.306	5 770	3 050	4 940	10.0	5.3
Monel metal	16.5–18	6.18–6.86	12.4	0.327	5 350	2 720	4 400	47.5	24.2
Nickel	21.4	8.0	16.4	0.336	6 040	3 000	4 900	53.5	26.6
Nickel silver	10.7	3.92	11.2	0.37	4 760	2 160	3 575	40.0	18.1
Platinum	16.7	6.4	9.9	0.303	3 260	1 730	2 800	69.7	37.0
Silver	7.5	2.7	8.55	0.38	3 650	1 610	2 680	38.0	16.7
Steel, K9	21.6	8.29	10.02	0.276	5 941	3 251	5 250	46.5	25.4
347 stainless steel	19.6	7.57	11.3	0.30	5 790	3 100	5 000	45.7	24.5
Tin, rolled	5.5	2.08	4.04	0.34	3 320	1 670	2 730	24.6	11.8
Titanium	11.6	4.40	7.79	0.32	6 070	3 125	5 090	27.3	14.1
Tungsten, drawn	36.2	13.4	31.3	0.35	5 410	2 640	4 320	103	50.5
Tungsten carbide	53.4	21.95	17.1	0.22	6 655	3 984	6 240	91.8	55.0
Zinc, rolled	10.5	4.2	4.2	0.25	4 210	2 440	3 850	30	17.3
Fused silica	7.29	3.12	1.61	0.17	5 968	3 764	5 760	13.1	8.29
Pyrex glass	6.2	2.5	2.3	0.24	5 640	3 280	5 170	13.1	7.6
Heavy silicate flint	5.35	2.18	1.77	0.224	3 980	2 380	3 720	15.4	9.22
Light borate crown	4.61	1.81	2.2	0.274	5 100	2 840	4 540	11.4	6.35
Lucite	0.40	0.143	0.562	0.4	2 680	1 100	1 840	3.16	1.3
Nylon 6-6	0.355	0.122	0.511	0.4	2 620	1 070	1 800	2.86	1.18
Polyethylene	0.076	0.026	0.288	0.458	1 950	540	920	1.75	0.48
Polystyrene	0.360	0.133	0.319	0.353	2 350	1 120	1 840	2.49	1.19

TABLE 2.03.D.5. Attenuation of sound in air as a function of temperature, humidity, and frequency, in dB/100 m for an atmospheric pressure of 1.013×10^5 Pa (1 atm). From Ref. 2.

T	Relative humidity (%)	Frequency (Hz) 125	250	500	1000	2000	4000
	10	0.09	0.19	0.35	0.82	2.6	8.8
	20	0.06	0.18	0.37	0.64	1.4	4.4
30°C	30	0.04	0.15	0.38	0.68	1.2	3.2
(86°F)	50	0.03	0.10	0.33	0.75	1.3	2.5
	70	0.02	0.08	0.27	0.74	1.4	2.5
	90	0.02	0.06	0.24	0.70	1.5	2.6
	10	0.08	0.15	0.38	1.21	4.0	10.9
	20	0.07	0.15	0.27	0.62	1.9	6.7
20°C	30	0.05	0.14	0.27	0.51	1.3	4.4
(68°F)	50	0.04	0.12	0.28	0.50	1.0	2.8
	70	0.03	0.10	0.27	0.54	0.96	2.3
	90	0.02	0.08	0.26	0.56	0.99	2.1
	10	0.07	0.19	0.61	1.9	4.5	7.0
	20	0.06	0.11	0.29	0.94	3.2	9.0
10°C	30	0.05	0.11	0.22	0.61	2.1	7.0
(50°F)	50	0.04	0.11	0.20	0.41	1.2	4.2
	70	0.04	0.10	0.20	0.38	0.92	3.0
	90	0.03	0.10	0.21	0.38	0.81	2.5
	10	0.10	0.30	0.89	1.8	2.3	2.6
	20	0.05	0.15	0.50	1.6	3.7	5.7
0°C	30	0.04	0.10	0.31	1.08	3.3	7.4
(32°F)	50	0.04	0.08	0.19	0.60	2.1	6.7
	70	0.04	0.08	0.16	0.42	1.4	5.1
	90	0.03	0.08	0.15	0.36	1.1	4.1

TABLE 2.03.D.6. Classical attenuation α/f^2 for selected gases at 20°C and 1 atm, where α is the attenuation coefficient in cm^{-1} and f the frequency. From Ref. 3.

Gas	$10^{13}\alpha/f^2$ (s²/cm) Due to heat conduction	Due to shear viscosity	Total
He	0.216	0.309	0.525
Ar	0.77	1.08	1.85
H_2	0.052	0.117	0.169
N_2	0.39	0.94	1.33
O_2	0.49	1.16	1.65
Air	0.38	0.99	1.37
CO_2	0.31	1.09	1.40

TABLE 2.03.D.7. Attenuation in selected gases due to vibrational relaxation in diatomic molecules $[(\alpha c/f)_{max}]$. f_r is the relaxation frequency; α is the attenuation coefficient in m^{-1}, c the velocity of sound in the gas, and f the frequency. From Ref. 3.

Gas	T (K)	f_r (kHz)	$(\alpha c/f)_{max}$
N_2	300	10^{-2}	0.03
NO	300	400	0.14
O_2	288	5×10^{-2}	0.56
F_2	301	84	4.3
Cl_2	293	39	8.3
Br_2	301	240	11.8

TABLE 2.03.D.8. Attenuation of sound in H_2O at 1 atm. α is the attenuation coefficient in m^{-1} and f the frequency. From Ref. 1.

Temperature (°C)	$10^{15}\alpha/f^2$ (s^2/m)
0	56.9
10	36.1
20	25.3
50	12.0
80	7.9

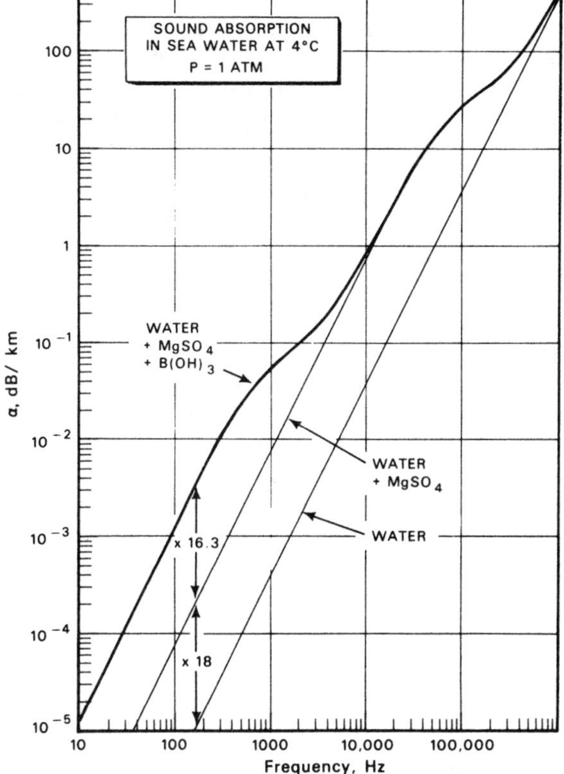

PLOT 2.03.D.9. Attenuation of sound in sea water. These curves were calculated from laboratory acoustic measurements by V. P. Simmons from 6 to 350 kHz in a 200-liter spherical resonator using Lyman and Fleming sea water of salinity = 3.5% and pH = 8.0. This refers to artificial sea water as defined by Lyman and Fleming,[4] which is sufficiently close to most actual sea water to be illustrative. From Ref. 5.

Table 2.03.D.10. Attenuation of sound in selected liquids at 1 atm. α is the attenuation coefficient in m^{-1} and f the frequency.

Liquid	T(°C)	$10^{15}\alpha/f^2$(s^2/m)
Mercury[a]	25	5.7
Sodium[a]	100	12
Lead[a]	340	9.4
Tin[a]	240	5.6
Methyl alcohol[a]	30	30.2
Ethyl alcohol[a]	30	48.6
Benzene[b]	25	870
CCl_4[b]	25	540
CS_2[b]	25	5700

[a] Reference 1.
[b] Reference 3.

2.04. MACROSONICS; HIGH-INTENSITY SOUNDS

A. Lagrangian form of wave equation

Given below is the Lagrangian form of the wave equation for transmission in one direction in a fluid in which the small-amplitude sound velocity is $c_v = (\gamma p_0/\rho_0)^{1/2}$, where γ is the usual specific heat ratio and p_0 and ρ_0 are the equilibrium values of the fluid pressure and density, respectively. This scheme follows the fate of a particle of fluid which was at the particular point $x = a$ at time $t = 0$ and has reached any other point x at time t. Then x is a function of a and t, which are the independent variables. Putting $x = a + \xi$, we have the wave equation

$$\frac{\partial^2\xi}{\partial t^2} = \frac{c_0^2}{(1 + \partial\xi/\partial a)^{2 + B/A}} \frac{\partial^2\xi}{\partial a^2}, \tag{30}$$

where

$$A = \rho_0\left(\frac{\partial p}{\partial \rho}\right)_{S,\,\rho=\rho_0} = \rho_0 c_0^2, \tag{31}$$

$$B = \rho_0^2\left(\frac{\partial^2 p}{\partial \rho^2}\right)_{S,\,\rho=\rho_0}, \tag{32}$$

where the derivatives are taken at constant entropy. For a gas,

$$2 + B/A = \gamma + 1. \tag{33}$$

The ratio B/A is known as the parameter of nonlinearity.

B. Solutions of Eulerian wave equation

Earnshaw's implicit solution of the Eulerian wave equation associated with Eq. (30) for harmonic waves with frequency f is

$$u(x,f) = u_0 \sin\left[\omega t - \frac{\omega x}{c_0}\left(1 + \frac{B}{2A}\frac{u}{c_0}\right)^{-2A/B-1}\right]. \tag{34}$$

From Eq. (34) the Fubini explicit solution can be obtained in the form

$$u = 2u_0 \sum_{n=1}^{\infty} \frac{J_n(nx/l)}{nx/l} \sin n\left(\omega t - \frac{\omega x}{c_0}\right). \tag{35}$$

Table 2.04.B.1 presents data relevant to nonlinear acoustics (B/A values).

TABLE 2.04.B.1. Nonlinear parameters for selected liquids. Values of B/A at 1 atm (see text). From Ref. 6.

Liquid	T(°C)	B/A
Water (distilled)	0	4.2
	20	5.0
Sea water (3.5% sal.)	0	4.9[a]
	20	5.25[a]
Methyl alcohol	20	9.6
Ethyl alcohol	20	10.5
Acetone	20	9.2
Benzene	20	9.0
Mercury	30	7.8
Sodium	110	2.7
Bismuth	318	7.1

[a] Reference 6a.

As a macrosonic wave progresses through a dissipative medium the wave front gradually assumes a sawtooth character, with the front part of the wave profile very steep. It then becomes a shock wave, characterized by a relatively large change in excess pressure across a very small region of space. The well-known sonic boom is a good example of an acoustic shock wave.

C. Radiation pressure

The Rayleigh radiation pressure p_R is the difference between the time average of the pressure at any point in a fluid traversed by a compressional wave and that which would have existed in a fluid of the same mean density at rest. For an ideal gas, we have

$$p_R = \tfrac{1}{4}(\gamma + 1)\bar{E}, \tag{36}$$

with γ the usual specific heat ratio and $\bar{E}$ the average energy density in the sound wave. For a liquid,

$$p_R = \tfrac{1}{2}(1 + B/2A)\bar{E}, \tag{37}$$

where B and A have the meanings in Eqs. (31) and (32).

The Langevin radiation pressure p_L is the difference between the average pressure at a reflecting or absorbing wall and that behind the wall in the same acoustic medium at rest:

$$p_L = \bar{E}. \tag{38}$$

2.05. ATMOSPHERIC ACOUSTICS

A. Velocity of sound in air

The velocity of sound in air is given by Eq. (20). For normal air, $\gamma \simeq 1.4$ and $R_m \simeq 0.288$ J/kg·K.

B. Refraction in a fluid medium

For a nonhomogeneous stratified fluid medium like the atmosphere the general ray equation for refraction is

$$c \sec\theta - c_0 \sec\theta_0 = u_{x0} - u_x \tag{39}$$

for a ray in the xz plane, where the stratification is in the z direction. The sound velocity in any particular layer for still air is c, with initial value c_0. The ray makes the angle θ with the x axis (horizontal), θ_0 being the initial angle. The instantaneous large-scale velocity of the air in the direction is u_x.

C. Attenuation in atmosphere

For attenuation in the atmosphere see Sec. 2.03.

D. Doppler effect

This phenomenon is the change in the observed frequency of a sound wave caused by a time rate of change in the effective length of the path of travel between the sound source and place of observation.

The fundamental formula is

$$f_0 = \frac{1 + v_0/c}{1 - v_s/c} f_s, \tag{40}$$

where f_0 is the observed frequency, f_s the frequency at the source, v_0 the component of velocity (relative to the medium) of the observation point toward the source, v_s the component of velocity (relative to the medium) of the source toward the observation point, and c the velocity of sound in a stationary medium.

2.06. UNDERWATER SOUND

The *Lloyd mirror effect* is the interference between the direct sound in the water from source to receiver and that reflected on the way by the sea water surface. The fundamental equation is

$$I_r/I_d = 1 + R^2 - 2R\cos(\delta_1 - \delta_2), \tag{41}$$

where I_d is the intensity produced at the receiver by the direct radiation from the source located below the water surface; I_r is the resultant intensity at the receiver due to the combination of the direct sound and that reflected from the surface; R is the ratio of acoustic pressure due to the reflected radiation to that due to the direct radiation; $\delta_1 = 2\pi f l/c$, the phase difference between the disturbance at the receiver and that at the source, where l is the direct distance from source to receiver, c the velocity of sound in the water, and f the frequency; and δ_2 is the corresponding phase difference for the reflected sound.

Figure 2.06.A shows underwater sound rays in a thermocline (sound velocity approximately uniform from the surface to C, decreases from C to C', and at lower depths is more or less constant save for change of pressure and density with depth). S is the source of sound. The shaded region between rays 2 and 3 is an acoustic shadow zone.

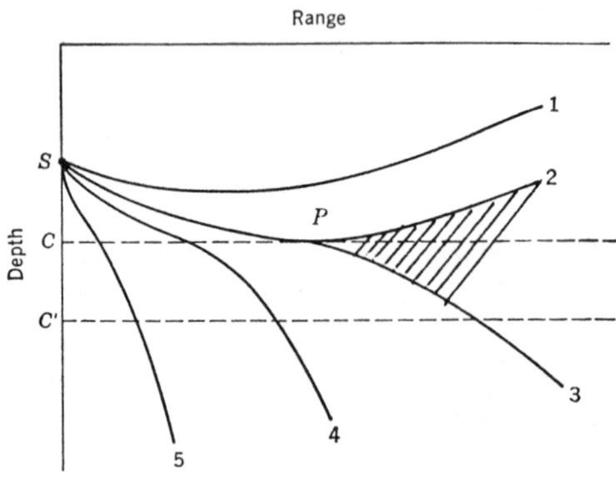

FIGURE 2.06.A. Underwater sound rays; shadow zone. From Ref. 7.

The ray equations mentioned are most readily obtained from the solution of the eikonal equation. For a stratified ocean medium in which the index of refraction $n(z)$ (ratio of the sound velocity at a given depth to some standard sound velocity) is a function of the depth only (z axis), this has the form

$$\left(\frac{\partial\psi}{\partial x}\right)^2 + \left(\frac{\partial\psi}{\partial z}\right)^2 = [n(z)]^2. \tag{42}$$

ψ is the eikonal. When this equation is solved for $\psi(x,z)$ the differential equation for the rays in the xz plane is

$$\frac{dz}{dx}=\frac{\partial\psi/\partial z}{\partial\psi/\partial x}. \tag{43}$$

In Figures 2.06.B, the second figure shows the sound-channeling effect in the sea due to the velocity profile shown in the first figure.

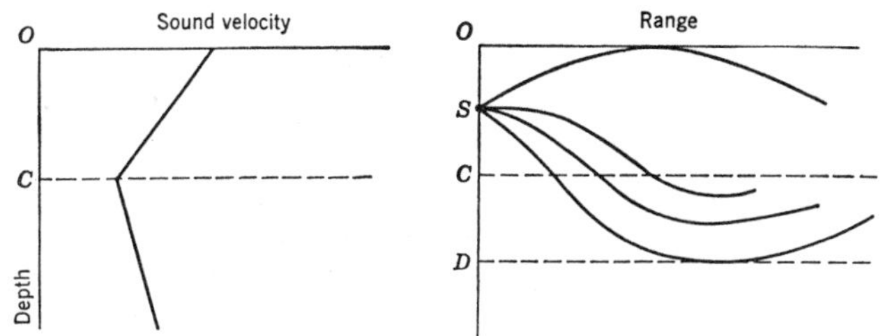

FIGURES 2.06.B. Underwater sound-channeling effect. From Ref. 7.

2.07. ACOUSTIC TRANSMISSION IN SOLIDS

A. Velocity in an extended polycrystalline solid

The velocity of a compressional (sound) wave in an extended polycrystalline solid is given by

$$c=\left(\frac{B+\frac{4}{3}\mu}{\rho}\right)^{1/2}, \tag{44}$$

where B is the bulk modulus, μ the shear modulus, and ρ the density. For the special case of a thin solid rod,

$$c=(Y/\rho)^{1/2}, \tag{45}$$

where Y is Young's modulus. For shear waves in a solid,

$$c_s=(\mu/\rho)^{1/2}. \tag{46}$$

B. Configurational dispersion in a solid rod

The configurational dispersion in a solid rod is the change in compressional (sound) wave velocity with frequency due to the interaction between the longitudinal wave and the associated lateral vibrations. The ratio of the longitudinal wave velocity in such a rod to the shear wave velocity increases with frequency. The effect is most noted in the ultrasonic range.

C. Attenuation of sound in solids

The attenuation of sound in solids depends on many factors, including temperature, heat flow, magnetization, and dislocations. In general, the attenuation may be considered exponential [see Eq. (21)], with the attenuation coefficient varying directly either with the frequency or its square. This is a very complicated subject, and the elaborate special literature should be consulted for details. For finite solids the attenuation is usually best represented by the logarithmic decrement [see Eq. (5)]. Table 2.07.C.1 lists some typical values of the decrement.

TABLE 2.07.C.1. Log decrement for selected solids (resonant frequencies in the range 10–50 kHz). From Ref. 8.

Solid	$10^3\delta$
Aluminum (annealed)	
200 K	0.03
275 K	0.1
Copper (unannealed)	
100 K	13.5
250 K	0.65
Lead	
225 K	3.15
250 K	9.5
Silver (annealed), 250 K	
(little change with temperature)	0.04
Polystyrene, room temperature	48

2.08. MOLECULAR ACOUSTICS; RELAXATION PROCESSES IN FLUIDS

A. Propagation of sound

The propagation of sound is fundamentally a molecular process, and the interaction between elastic wave propagation and molecular behavior has a significant effect on sound dispersion and attenuation, particularly at ultrasonic frequencies.

If a sound wave in a fluid disturbs any particular equilibrium molecular aggregation (involving, for example, transfer of translational energy into internal energy modes of the polyatomic molecules), it takes a certain time τ, called the relaxation time, for the original state to be restored after the passage of the crest of the wave. The process is usually called thermal relaxation.

For a polyatomic gas it is found that the quantity $\alpha/\pi f$, where α is the attenuation coefficient [see Eq. (21)] and f the frequency, when plotted as a function of frequency has a maximum at

$$f=1/2\pi\tau \quad \text{or} \quad \omega t=1 \tag{47}$$

and that the course of the attenuation as a function of frequency is given by

$$\frac{\alpha}{2\pi f^2}=2\left(\frac{\alpha}{\pi f}\right)_{\max}\frac{\tau}{1+\omega^2\tau^2}. \tag{48}$$

The value of τ is usually evaluated in practice experimentally from Eq. (47). For rotational relaxation in hydrogen at 20°C the value of τ is of the order of 10^{-8} s. For oxygen vibrational relaxation prevails, with τ of the order of 10^{-3} s. The values of α discussed here are, of course, in excess of those due to viscosity and heat conduction (Sec. 2.03).

B. Excess sound attenuation in liquids

Excess sound attenuation in liquids is attributable to a variety of relaxation processes connected with changes in the structural aggregation of molecules and with chemical reactions and dissociations. Chemical dissociation (as affected by sound propagation) connected with dissolved magnesium sulfate in sea water has been invoked to explain the abnormally high sound attenuation in this medium. The literature must be consulted for details.

C. Ultrasonic propagation at very low temperatures

There are numerous anomalies in ultrasonic propagation at very low temperatures. This is particularly evident in superconductors, as the following figure shows.

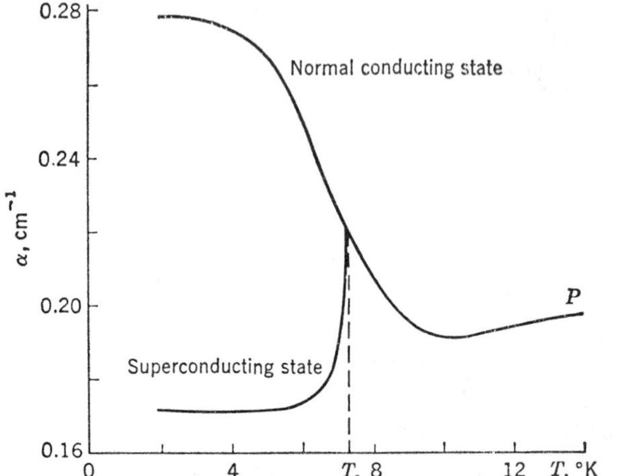

FIGURE 2.08.C.1. Variation of acoustic attenuation with temperature for a superconductor. From Ref. 7.

In superfluid liquid helium four different varieties of sound and sound velocity in addition to the normal kind have been detected.

Normal sound in liquid helium II is called first or zero sound. Second sound in this liquid is a temperature wave propagating at a speed of about 20 m/s between 1 and 2 K, decreasing to zero at the λ transition point. Third sound is a propagation of thickness variation in a thin film of liquid helium II. The velocity depends on the film thickness, ranging from about 10 m/s for 15 atomic layers to 60 m/s for 5 atomic layers. Fourth sound is a compressional wave in liquid helium II in narrow pores in which the normal component is locked in and only the superfluid component participates in the wave propagation. The velocity varies with the temperature, ranging from about 220 m/s at 1.2 K to 100 m/s at 2.1 K. Fifth sound results if in experiments leading to second sound one immobilizes the normal component and applies a pressure-release boundary condition. The velocity is a function of temperature, reaching a maximum of 10 m/s at 2 K and vanishing at the λ point. See Ref. 9.

2.09. NOISE AND ITS CONTROL

Noise is unwanted sound. Ambient noise is the composite of all unwanted sound in a given environment. Common sources of noise are machinery operating inside and outside buildings, transportation vehicles (in particular, aircraft), and human beings and animals. Intense noise can produce temporary and even permanent hearing loss, particularly if the exposure is long continued.

It can be seen from Fig. 2.11.B.1 that the human ear is far less sensitive to sound frequencies below 200 Hz than it is over higher frequencies. A weighting scheme that takes this into account is known as *A*-weighting, and the decibel readings with such a weighting are called dBa. This weighting, incorporated into the sound level meter, matches its sensitivity as far as possible with that of the human ear.

The following tables and figure relate to noise and its effects, as well as to recommended levels for safety and comfort.

TABLE 2.09.A. Typical noise levels due to various sources (approximate values at the source). Mainly from Ref. 2.

Sound power (dB) relative to 10^{-12} W	Source
200	Large rocket engine
160	Aircraft turbo jet engine
140	Light airplane, cruising
115	Crawler tractor, 150 hp
105	100-hp electric motor at 2600 rpm
100	Pneumatic drill
90	Subway with train passing
85	Vacuum cleaner
75	Busy traffic
65	Conversational speech
40	Whispered speech

TABLE 2.09.B. Recommended acceptable average noise levels in unoccupied rooms. The levels are "weighted," i.e., measured with a standard sound-level meter incorporating an A (40-dB) frequency-weighting network. From Ref. 10.

Rooms	Noise level (dB)
Radio, recording, and television studios	25–30
Music rooms	30–35
Legitimate theaters	30–35
Hospitals	35–40
Motion-picture theaters, auditoriums	35–40
Churches	35–40
Apartments, hotels, homes	35–45
Classrooms, lecture rooms	35–40
Conference rooms, small offices	40–45
Courtrooms	40–45
Private offices	40–45
Libraries	40–45
Large public offices, banks, stores, etc.	45–55
Restaurants	50–55

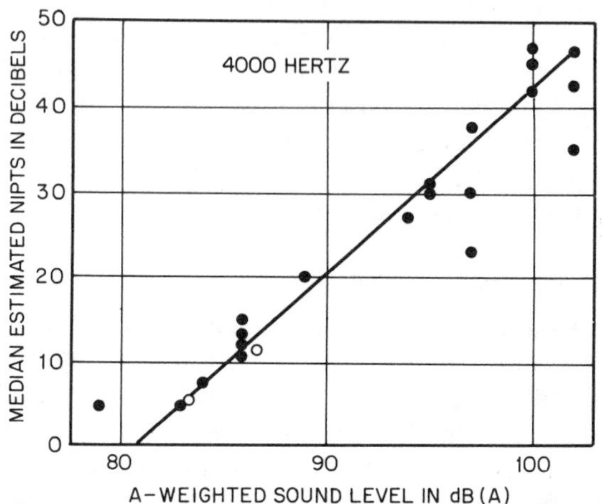

PLOT 2.09.C. NIPTS (noise-induced permanent threshold shift) at 4000 Hz for population median after approximately 10 yr of daily exposure to noises having A-weighted sound levels given along the horizontal axis. Open circles indicate more recent data. From Ref. 2.

2.10. ROOM AND ARCHITECTURAL ACOUSTICS

A. Sabine's formula

In room acoustics the reverberation time is the time after a source of given frequency has been stopped for the sound intensity level in the room to decrease by 60 dB. Sabine's formula for the reverberation time is

$$T = 0.049\ V/A, \tag{49}$$

where T is the reverberation time in seconds, V is the volume of the room in cubic feet, and A is the absorption in sabins (for V in cubic meters and A in metric sabins the coefficient 0.049 is replaced by 0.161).

Sabine's reverberation time formula has been modified by C. F. Eyring to the form

$$T = \frac{kV}{-S\ln(1-\bar{\alpha}) + 4mV}, \tag{50}$$

where V is the volume of the room, S the total surface area of the room, and $\bar{\alpha}$ the average absorption coefficient of the room surfaces. The correction term $4mV$ is due to the absorption of the air in the room. The Eyring formula has been considered particularly applicable to "dead" rooms. k is a constant with value depending on the choice of units.

B. Practical architectural acoustics

The following figure and table present material relating to practical architectural acoustics.

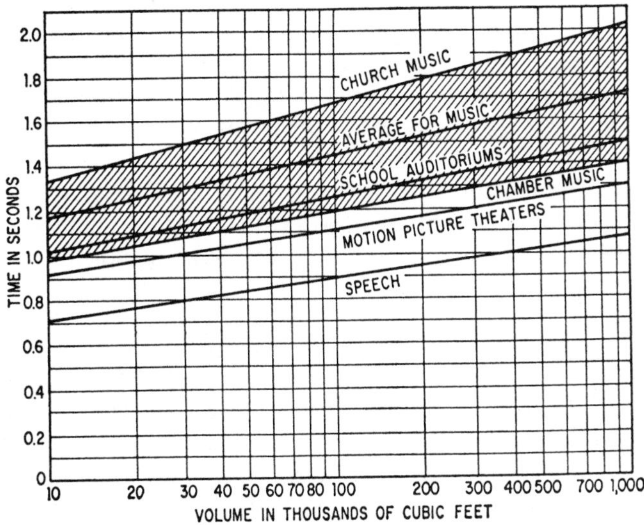

FIGURE 2.10.B.1. Optimum reverberation time at 500 Hz for different types of rooms as a function of room volume. From Ref. 1 (after Ref. 10).

TABLE 2.10.B.2.. Absorption coefficients (in sabins) for building materials. From Ref. 11.

Materials	Frequency (Hz)					
	125	250	500	1000	2000	4000
Brick, unglazed	0.03	0.03	0.03	0.04	0.05	0.07
Painted	0.01	0.01	0.02	0.02	0.02	0.03
Carpet, heavy, on concrete	0.02	0.06	0.14	0.37	0.60	0.65
On 40-oz hairfelt or foam rubber	0.08	0.24	0.57	0.69	0.71	0.73
With impermeable latex backing on 40-oz hairfelt or foam rubber	0.08	0.27	0.39	0.34	0.48	0.63
Concrete block						
Coarse	0.36	0.44	0.31	0.29	0.39	0.25
Painted	0.10	0.05	0.06	0.07	0.09	0.08
Fabrics						
Light velour, 10 oz/yd^2 hung straight, in contact with wall	0.03	0.04	0.11	0.17	0.24	0.35
Medium velour, 14 oz/yd^2, draped to half area	0.07	0.31	0.49	0.75	0.70	0.60
Heavy velour, 18 oz/yd^2, draped to half area	0.14	0.35	0.55	0.72	0.70	0.65
Floors						
Concrete or terrazzo	0.01	0.01	0.015	0.02	0.02	0.02
Linoleum, asphalt, rubber, or cork tile on concrete	0.02	0.03	0.03	0.03	0.03	0.02
Wood	0.15	0.11	0.10	0.07	0.06	0.07
Wood parquet in asphalt on concrete	0.04	0.04	0.07	0.06	0.06	0.07
Glass						
Large panes of heavy plate glass	0.18	0.06	0.04	0.03	0.02	0.02
Ordinary window glass	0.35	0.25	0.18	0.12	0.07	0.04
Gypsum board, $\frac{1}{2}$ in. nailed to 2×4's 16 in. o.c.	0.29	0.10	0.05	0.04	0.07	0.09
Marble or glazed tile	0.01	0.01	0.01	0.01	0.02	0.02
Openings						
Stage, depending on furnishing				0.25–0.75		
Deep balcony, upholstered seats				0.50–1.00		
Grills, ventilating				0.15–0.50		
Plaster, gypsum, or lime						
Smooth finish on tile or brick	0.013	0.015	0.02	0.03	0.04	0.05
Rough finish on lath	0.02	0.03	0.04	0.05	0.04	0.03
Smooth finish on lath	0.02	0.02	0.03	0.04	0.04	0.03
Plywood paneling, $\frac{3}{8}$ in. thick	0.28	0.22	0.17	0.09	0.10	0.11
Water surface, as in a swimming pool	0.008	0.008	0.013	0.015	0.020	0.025

2.11. PHYSIOLOGICAL AND PSYCHOLOGICAL ACOUSTICS

Physiological acoustics is the study of the physiological response in animals, including man, to acoustic stimuli. It includes reference to the anatomy of the outer, middle, and inner ear, with the physiology of the cochlea and the auditory central nervous system. Physics enters essentially in the mechanics of the middle ear and cochlea and in the electrical transmissions of nerve impulses from the cochlea to the brain.

Psychological acoustics is the study of the psychological response of animals, including man, to acoustic stimuli. It includes reference to the following topics.

A. Loudness and loudness level

Loudness is the subjective impression of the intensity of sound. The unit is the sone (see Sec. 2.01). Loudness level is the sound-pressure level in decibels of a pure tone of frequency 1000 Hz (relative to 2×10^{-4} dyn/cm^2) which is assessed by normal observers as being equally loud as the sound being measured. The unit is the phon (see Sec. 2.01).

The loudness S of a sound of 1000 Hz measured in sones is given by

$$\log_{10}S = 0.0301P - 1.204, \qquad (51)$$

where P is the loudness level in phons.

B. Auditory sensation area

The following figure plots the average minimum audible field threshold for young adults as a function of frequency.

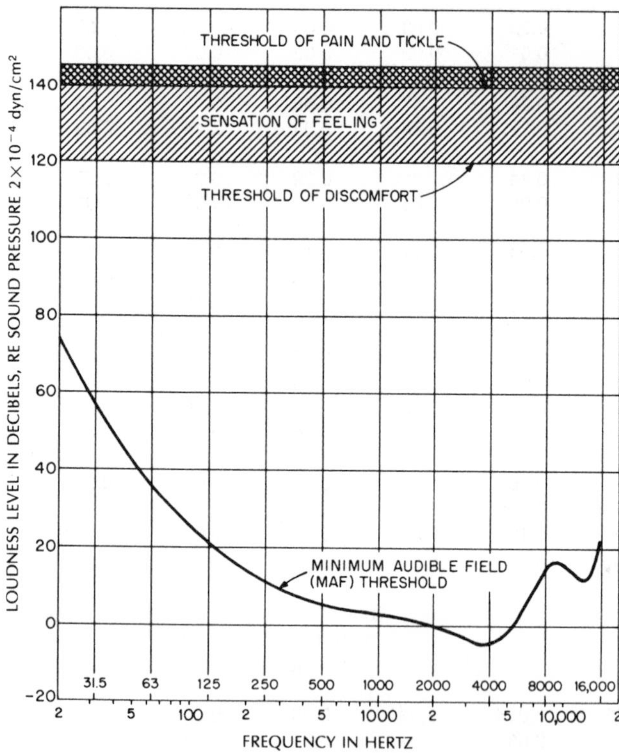

FIGURE 2.11.B.1. Average minimum audible field threshold as a function of frequency. Upper limits are set by the thresholds of discomfort, feeling, pain, and tickle. Lower limit is minimum audible field (MAF) threshold for young adults. From Ref. 2.

C. Masking

Masking is the process by which the threshold of hearing of one sound is raised due to the presence of another sound. It is primarily in general due to noise and is measured in decibels. An important quantity in this connection is the *signal-to-noise* ratio, the ratio of the signal intensity (peak or root-mean-square) to that of the noise, expressed in decibels.

D. Temporary threshold shift (TTS)

The TTS is the temporary increase in the hearing threshold induced by noise. In such a case hearing ultimately returns to normal. The following figure is illustrative. For permanent threshold shift see Sec. 2.09.

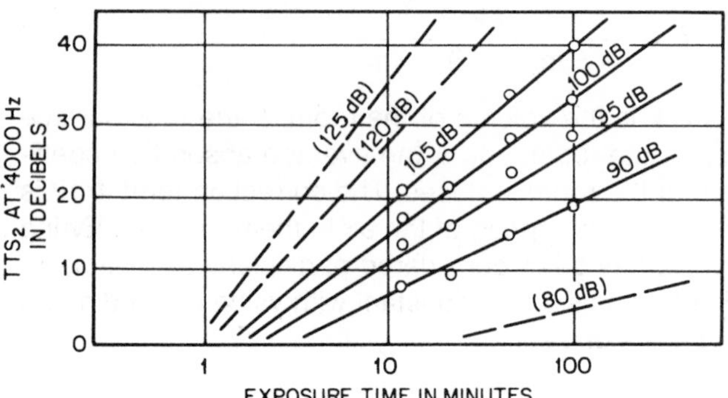

FIGURE 2.11.D.1. Temporary threshold shift at 4000 Hz measured 2 min after exposure to an octave band of noise centered at 1700 Hz at the sound-pressure levels and durations indicated. From Ref. 2.

E. Pitch

Pitch is the subjective estimate of a tone as higher or lower on a scale. It is measured in mels (see Sec. 2.01). Its relation to sound frequency is indicated in the following table.

TABLE 2.11.E.1. Pitch of a pure tone (in mels) as a function of frequency. From Ref. 1.

Frequency	Mels	Frequency	Mels	Frequency	Mels
20	0	350	460	1750	1428
30	24	400	508	2000	1545
40	46	500	602	2500	1771
60	87	600	690	3000	1962
80	126	700	775	3500	2116
100	161	800	854	4000	2250
150	237	900	929	5000	2478
200	301	1000	1000	6000	2657
250	358	1250	1154	7000	2800
300	409	1500	1296	10000	3075

F. Binaural hearing

Binaural hearing is hearing by use of the two ears, through which the direction of the source of sound may be determined. It also assists in the extraction of the signal from disturbing noise.

G. Audiogram

An audiogram is a graph showing hearing loss as a

function of frequency. It is produced by the use of an instrument called an audiometer. Hearing loss due to aging is called presbycusis. It mainly affects tones of higher frequency.

2.12. SPEECH COMMUNICATION

Speech is the result of the alteration in frequency and intensity of voiced and unvoiced sounds produced by the lungs and larynx. Modification of the cavities of the mouth and nose serve to introduce spectral content into speech. The normal range of speech frequencies is approximately 400–6000 Hz.

Total average radiated conversational speech power has been estimated at about 30 μW.

Among important acoustical aspects of speech are: spectral analysis of speech, speech synthesis (artificial speech), perception of speech by individuals and machines, speech transmission systems, and phonetics (speech and linguistics). Some important quantities in speech communication are:

Articulation. The percentage of speech units spoken by a speaker correctly identified by a listener.

Formant of a speech sound. The frequency range of the spectrum of the sound within which the partials (harmonics) have relatively large amplitudes. The central frequency within the formant is called the formant frequency.

Phoneme. The smallest phonetic unit of a speech sound.

2.13. BIOACOUSTICS

Bioacoustics is the study of the acoustical characteristics of biological media. It also includes the effects of sound on living systems, the generation and detection of sound by animals, as well as medical diagnosis and therapy with acoustical radiation. The role of ultrasonic radiation is particularly stressed. The following table is illustrative.

TABLE 2.13.A. Ultrasonic propagation properties of some tissues at 1 MHz, listed in order of increasing attenuation coefficient. Except for bone, the attenuation coefficient follows the relation $\alpha = \alpha_1 F^{1.1}$, where α_1 is the attenuation coefficient at 1 MHz (from table) and F the frequency in MHz. Condensed from Ref. 12.

Tissue	Attenuation coefficient (Np/cm)	Velocity (m/s)	Density (g/cm^3)	Impedance (10^5 rayl)	Trends
Blood	0.014	1566	1.04	1.63	increasing structural protein content ↓; decreasing water content
Fat	0.07	1478	0.92	1.36	
Nerve	0.10				
Muscle	0.14	1552	1.04	1.62	
Blood vessel	0.20	1530	1.08	1.65	
Skin	0.31	1519		1.58	
Tendon	0.56	1750			
Cartilage	0.58	1665			
Bone	1.61	3445	1.82	6.27	

2.14. MUSICAL ACOUSTICS

A. Resonance frequencies for an organ pipe

Resonance frequencies of an open organ pipe of length l (open at both ends) are given by

$$f = \frac{nc}{2(l + x_1 + x_2)}, \tag{52}$$

where c is the velocity of sound in the medium; x_1 = end correction = $0.3d$ for the unimpeded end, where d is the diameter of the pipe; $x_2 = 1.4d$ for the mouth of the pipe; and n is an integer giving the particular harmonic of the pipe.

For an organ pipe closed at one end,

$$f = \frac{nc}{4(l + x)}, \tag{53}$$

where $x = 1.4d$.

B. Resonance frequencies for a rectangular membrane

Resonance frequencies for a rectangular membrane with sides a_1 and a_2 are given by

$$f = \frac{c}{2}\left[\left(\frac{n_1}{a_1}\right)^2 + \left(\frac{n_2}{a_2}\right)^2\right]^{1/2}, \tag{54}$$

where n_1 and n_2 are any integers, and c is the velocity of transverse waves in the membrane, with

$$c = (T/\sigma)^{1/2}, \tag{55}$$

where T is the surface tension and σ the surface density.

C. Resonance frequencies for a circular membrane

Resonance frequencies for a circular membrane of radius a, clamped at the periphery, are given by

$$J_n(2\pi f a/c) = 0, \tag{56}$$

where c is given by Eq. (55). J_n is the Bessel function of order n. The lowest resonance frequencies of the clamped circular membrane are

$$0.7655c/2a, \quad 1.2197c/2a, \quad 1.6347c/2a. \tag{57}$$

2.15. ACOUSTICAL MEASUREMENTS AND INSTRUMENTS

All acoustical measurements demand the availability of sound sources capable of producing both continuous and pulsed radiation with controllable frequency and intensity. Equally suitable sound receivers (microphones) are also necessary. These are all referred to as acoustic transducers. The most common transducers are based on the use of the piezoelectric and ferroelectric effects, electrostatic and electrodynamic effects (loudspeakers), and magnetostriction. Sources of this kind are usually coupled to plane diaphragms (so-called piston sources), radiating sound that varies in intensity with the distance from the source as well as the direction. For a given piston a directivity factor can be calculated. This is the ratio of the intensity at a given distance from the source on a presented axis from the transducer to the intensity that would be produced at the same point by a spherical source centered at the transducer and radiating the same total acoustic power into a free field. By the use of appropriate baffles a piston source may be made to produce approximate plane radiation at a sufficiently large distance from the source.

Most transducers can be adapted to the production of short pulses of radiation, which are useful in many measurements.

A. Absolute measurement of sound intensity

This can be carried out by means of a Rayleigh disk, which is a solid circular disk suspended in a tube so that it can rotate about a diametral axis normal to the axis of the tube. The torque M on such a disk of radius a with angle θ between the normal to the disk and the direction of propagation of the sound wave is given by

$$M=\tfrac{4}{3}\rho a^3u^2\sin 2\theta, \quad (58)$$

where ρ is the density of the medium and u the rms particle velocity in the sound wave, which can be measured by the disk with the use of this equation, thus providing a direct measurement of the intensity of the sound wave.

B. Calibration of microphones

This is now usually carried out by a method based on the acoustical reciprocity principle. In its simplest form this says that in a region containing a simple source such a source at point A produces the same sound pressure at another point B in the medium as would have been produced at A had the source been located at B. The specific literature should be consulted for details.

C. Frequency measurement

Acoustical frequency standardization is based on a comparison with standard radio frequencies accurate to one part in 10^7.

The precision electrically driven tuning fork is also in use as a secondary frequency standard. The cathode ray oscilloscope is used for the comparison of two frequencies by means of Lissajous figures, which are the resultant patterns on an oscilloscope screen when sound is detected at two separate microphones connected separately to the x and y plates of the oscilloscope. Phase differences between the sounds at the microphones give rise to different geometric figures. This provides a means of measuring phase differences (and hence sound velocity) or of comparing the frequencies of two sounds.

D. Acoustic filter

An acoustical structure with periodic variations in acoustical properties or mode of confinement of the medium constitutes an acoustic filter in the sense that (disregarding irreversible dissipation) sound radiation of certain frequencies is transmitted through the structure (passbands), whereas for other frequencies there is complete attenuation (attenuation bands). As an example consider a fluid-filled tube of circular cross-sectional area S with side branches spaced a distance l apart and with branch impedance equal to Z_b for each branch. The transmission through such a structure (assumed infinite in extent) is governed by

$$\cos W=\cos\frac{4\pi f l}{c}+\frac{iZ}{2Z_b}\sin\frac{4\pi f l}{c}, \quad (59)$$

where f is the frequency, c is the velocity of sound in the fluid, and $Z=\rho_0 c/S$, the acoustic resistance of a plane wave progressing along the tube. For $|\cos W|\leqslant 1$ there is complete transmission; for $|\cos W|>1$ there is complete attenuation. Thus the plot of $|\cos W|$ as a function of frequency gives the pass and attenuation bands.

2.16. REFERENCES

[1] *American Institute of Physics Handbook*, 3rd ed., edited by Dwight E. Gray (McGraw-Hill, New York, 1972).

[2] *Handbook of Noise Control*, 2nd ed., edited by Cyril M. Harris (McGraw-Hill, New York, 1979).

[3] R. T. Beyer and S. V. Letcher, *Physical Ultrasonics* (Academic, New York, 1969).

[4] J. Lyman and R. H. Fleming, J. Mar. Res. **3**, 134 (1940).

[5] F. H. Fisher and V. P. Simmons, J. Acoust. Soc. Am. **62**, 558 (1977).

[6] R. T. Beyer, *Nonlinear Acoustics* (Naval Sea Systems Command, Washington, DC, 1974).

[6a] A. B. Coppens *et al.*, J. Acoust. Soc. Am. **38**, 797 (1965).

[7] R. B. Lindsay, *Mechanical Radiation* (McGraw-Hill, New York, 1960).

[8] P. G. Bordoni, J. Acoust. Soc. Am. **26**, 495 (1954).

[9] I. Rudnick, J. Acoust. Soc. Am. **68**, 36 (1980).

[10] V. O. Knudsen and C. M. Harris, *Acoustical Designing in Architecture* (Wiley, New York, 1950). Reprinted 1978 by American Institute of Physics for Acoustical Society of America.

[11] Acoust. Mater. Assoc. Bull. **29** (1969).

[12] S. A. Goss, R. L. Johnston, and F. Dunn, J. Acoust. Soc. Am. **64**, 423 (1978).

[13] P. M. Morse and K. U. Ingard, *Theoretical Acoustics* (McGraw-Hill, New York, 1968).

3.00. Astronomy and astrophysics

Laurence W. Fredrick

University of Virginia

CONTENTS

3.01. ASTRONOMICAL CONSTANTS

Solar mass:

$$M_\odot = 1.989 \times 10^{33} \text{ g.}$$

Solar radius:

$$R_\odot = 6.9599 \times 10^{10} \text{ cm.}$$

Solar luminosity:

$$L_\odot = 3.826 \times 10^{33} \text{ ergs s}^{-1}.$$

Gravity at solar surface:

$$g_\odot = 2.74 \times 10^4 \text{ cm s}^{-2}.$$

Earth mass:

$$M_E = 5.976 \times 10^{27} \text{ g.}$$

Earth equatorial radius:

$$R_E = 6378.164 \text{ km.}$$

Gravity at Earth's surface:

$$g_E = 980.665 \text{ cm s}^{-2}.$$

Astronomical unit:

$$\text{AU} = 1.495\,978\,70 \times 10^{13} \text{ cm.}$$

Solar effective temperature:

$$T_e = 5800 \text{ K.}$$

Solar magnitude:

$$M_{pv}(\text{Sun}) = +4.84, \quad m_{pv}(\text{Sun}) = -26.73.$$

Solar constant (1976):

$$= 0.1353 \text{ W cm}^{-2}.$$

Julian century:

$$= 36\,525 \text{ ephemeris days.}$$

Tropical year (1900.0):

$$= 365.242 \text{ days}$$

$$= 3.1557 \times 10^7 \text{ ephemeris seconds.}$$

Ephemeris day:

$$= 86\,400 \text{ ephemeris seconds.}$$

Solar motion:

$$\text{velocity} = 19.7 \text{ km s}^{-1},$$

apex: $\alpha = 18^h4^m$, $\delta = +30°$; $l^{II} = 57°$, $b^{II} = +22°$.

Stellar luminosity, $M_{bol} = 0$:

$$= 2.97 \times 10^{35} \text{ ergs s}^{-1}.$$

Julian date for 1 January 1981:

$$= 2\,444\,605.5 \text{ days.}$$

3.02. UNIT CONVERSIONS

1 parsec $= 3.261\,633$ light years

$= 3.085\,678 \times 10^{18}$ cm

1 light year $= 9.460\,530 \times 10^{17}$ cm

1 xu $= 1.002\,08 \times 10^{-11}$ cm

1 angstrom $\equiv 1 \times 10^{-8}$ cm

1 rayleigh $\equiv (1/4\pi) \times 10^6$ photons cm^{-2} s^{-1} sr^{-1}

1 *Uhuru* ct/s

$= 1.7 \times 10^{-11}$ ergs cm^{-2} s^{-1} (2–6 keV)

$= 2.4 \times 10^{-11}$ ergs cm^{-2} s^{-1} (2–10 keV)

1 flux unit $\equiv 10^{-26}$ W m^{-2} Hz^{-1} $\equiv$ 1 jansky

3.03. ASTRONOMICAL COORDINATE TRANSFORMATIONS

A. Horizon-equatorial (celestial) systems

$$\cos a \sin A = -\cos\delta \sin h,$$

$$\cos a \cos A = \sin\delta \cos\phi - \cos\delta \cos h \sin\phi,$$

$$\sin a = \sin\delta \sin\phi + \cos\delta \cos h \cos\phi,$$

$$\cos\delta \cos h = \sin a \cos\phi - \cos a \cos A \sin\phi,$$

$$\sin\delta = \sin a \sin\phi + \cos a \cos A \cos\phi,$$

where $h =$ local sidereal time $- \alpha$, $A =$ azimuth, $a =$ altitude, $\phi =$ observer's latitude, $h =$ hour angle, $\alpha =$ right ascension, $\delta =$ declination.

B. Ecliptic-equatorial (celestial) systems

$$\cos\delta \cos\alpha = \cos\beta \cos\lambda,$$

$$\cos\delta \sin\alpha = \cos\beta \sin\lambda \cos\epsilon - \sin\beta \sin\epsilon,$$

$$\sin\delta = \cos\beta \sin\lambda \sin\epsilon + \sin\beta \cos\epsilon,$$

$$\cos\beta \sin\lambda = \cos\delta \sin\alpha \cos\epsilon + \sin\delta \sin\epsilon,$$

$$\sin\beta = \sin\delta \cos\epsilon - \cos\delta \sin\alpha \sin\epsilon,$$

where $\lambda =$ ecliptic longitude, $\beta =$ ecliptic latitude, $\epsilon =$ obliquity of the ecliptic $= 23°27'\ 8''.26 - 46''.845T - 0''.0059T^2 + 0''.00181\,T^3$, where T is the time in centuries from 1900, $\alpha =$ right ascension, $\delta =$ declination.

C. Galactic-equatorial (celestial) systems

$$\cos b^{II} \cos(l^{II} - 33°) = \cos\delta \cos(\alpha - 282°.25),$$

$$\cos b^{II} \sin(l^{II} - 33°) = \cos\delta \sin(\alpha - 282°.25) \times \cos 62°.6 + \sin\delta \sin 62°.6,$$

$$\sin b^{II} = \sin\delta \cos 62°.6 - \cos\delta \sin(\alpha - 282°.25) \sin 62°.6,$$

$$\cos\delta \sin(\alpha - 282°.25) = \cos b^{II} \sin(l^{II} - 33°) \times \cos 62°.6 - \sin b^{II} \sin 62°.6,$$

$$\sin\delta = \cos b^{II} \sin(l^{II} - 33°) \sin 62°.6 + \sin b^{II} \cos 62°.6,$$

where $l^{II} =$ new galactic longitude, $b^{II} =$ new galactic latitude, $\alpha =$ right ascension (1950.0), $\delta =$ declination (1950.0). Galactic center:

$$l^{II} = b^{II} = 0, \quad \alpha = 17^h42^m.4, \quad \delta = -28°55'.$$

North galactic pole:

$$\alpha = 12^h48^m, \quad \delta = +27°.4.$$

D. Reduction for precession—Approximate formulas*

Approximate formulas for the reduction of coordinates and orbital elements referred to the mean equinox and equator or ecliptic of date t are as follows.

For reduction to 1950.0:

$$\alpha_0 = \alpha - M - N \sin\alpha_m \tan\delta_m,$$
$$\delta_0 = \delta - N\cos\alpha_m,$$
$$\lambda_0 = \lambda - a + b\cos(\lambda + c')\tan\beta_0,$$
$$\beta_0 = \beta - b\sin(\lambda + c'),$$
$$\Omega_0 = \Omega - a + b\sin(\Omega + c')\cot i_0,$$
$$i^0 = i - b\cos(\Omega + c'),$$
$$\omega_0 = \omega - b\sin(\Omega + c')\csc i_0.$$

For reduction from 1950.0:

$$\alpha = \alpha_0 + M + N\sin\alpha_m \tan\delta_m,$$
$$\delta = \delta_0 + N\cos\alpha_m,$$
$$\lambda = \lambda_0 + a - b\cos(\lambda_0 + c)\tan\beta,$$
$$\beta = \beta_0 + b\sin(\lambda_0 + c),$$
$$\Omega = \Omega_0 + a - b\sin(\Omega_0 + c)\cot i,$$
$$i = i_0 + b\cos(\Omega_0 + c),$$
$$\omega = \omega_0 + b\sin(\Omega_0 + c)\csc i.$$

The subscript zero refers to epoch 1950.0 and α_m, δ_m refer to the mean epoch; with sufficient accuracy, $\alpha_m = \alpha - \frac{1}{2}(M + N\sin\alpha\tan\delta)$, etc. The precessional constants M, N, etc., are given by

$$M = 1^\circ\!.280\,526\,7T + 0^\circ\!.000\,387\,5T^2 + 0^\circ\!.000\,010\,0T^3,$$
$$N = 0^\circ\!.556\,737\,6T - 0^\circ\!.000\,118\,3T^2 - 0^\circ\!.000\,011\,7T^3,$$
$$a = 1^\circ\!.396\,319T + 0^\circ\!.000\,308\,3T^2,$$
$$b = 0^\circ\!.013\,076T - 0^\circ\!.000\,009\,7T^2,$$
$$c = 5^\circ\!.592\,58 + 0^\circ\!.241\,743T + 0^\circ\!.000\,154\,2T^2,$$
$$c' = 5^\circ\!.592\,58 - 1^\circ\!.154\,576T - 0^\circ\!.000\,154\,2T^2,$$

where

$$T = (t - 1950.0)/100$$
$$= (\mathrm{JD} - 2\,433\,282.423)/36\,524.22.$$

3.04. RADIATION FORMULAS

A. Thermal radiation

Blackbody:

$$B_\nu = \frac{2h\nu^3}{c^2}\left[\exp\left(\frac{h\nu}{kT}\right) - 1\right]^{-1}.$$

Rayleigh-Jeans law ($h\nu \ll kT$):

$$I_\nu = 2\nu^2 kT/c^2.$$

Wien law ($h\nu \gg kT$):

$$I_\nu = \frac{2h\nu^3}{c^2}\exp\left(-\frac{h\nu}{kT}\right).$$

*From Astron. Almanac (1981).

B. Bremsstrahlung

Thermal bremsstrahlung emission (in cgs units):

$$\epsilon_\nu^{\rm ff} = 4\pi j_\nu^{\rm ff} = 6.8\times10^{-38} Z^2 n_e n_i T^{-1/2} \bar{g}_{\rm ff} \times \exp(-h\nu/kT)\ \mathrm{ergs\ s^{-1}\ cm^{-3}\ Hz^{-1}},$$

where $\bar{g}_{\rm ff}$ is the velocity-averaged Gaunt factor and is of order unity ($\bar{g}_{\rm ff}\sim1$ for $h\nu/kT\sim1$, increasing to $\bar{g}_{\rm ff}\sim5$ for $h\nu/kT\sim10^{-4}$); The integrated emission is

$$\epsilon^{\rm ff} = 1.4\times10^{-27} Z^2 n_e n_i T^{1/2}\bar{g}_B \times(1 + 4.4\times10^{-10}T)\ \mathrm{ergs\ s^{-1}\ cm^{-3}},$$

where $\bar{g}_B$ is a frequency average of $\bar{g}_{\rm ff}$ and is in the range 1.1–1.5. The second term in the parentheses is a relativistic correction.

Thermal bremsstrahlung absorption (in cgs units):

$$j_\nu^{\rm ff} = \alpha_\nu^{\rm ff} B_\nu = \epsilon_\nu^{\rm ff}/4\pi,$$

so that

$$\alpha_\nu^{\rm ff} = 3.7\times10^8 T^{-1/2} Z^2 n_e n_i \nu^{-3} \times[1 - \exp(-h\nu/kT)]\bar{g}_{\rm ff}\ \mathrm{cm^{-1}}.$$

C. Faraday rotation

In cgs units:

$$\phi = 8.12\times10^9\lambda^2\int_0^d n_e B_\parallel\, ds\ \mathrm{rad},$$

where $B_\parallel$ is the component of the magnetic field along the line of sight in gauss, λ is the wavelength in cm, and n_e is the electron density in $\mathrm{cm^{-3}}$.

D. Synchrotron radiation

1. Single electron

γ = electron Lorentz factor, η = pitch angle, $B_\perp = B\sin\eta$ = magnetic field (in gauss) transverse to line of sight.

Spectrum peaks at

$$\nu = 0.44\gamma^2\frac{eB_\perp}{2\pi m_e c} = 1.22\times10^6\gamma^2 B_\perp\ \mathrm{Hz}.$$

Total power emitted, with polarizations respectively perpendicular and parallel to the magnetic field:

$$W^{\perp,\parallel} = \left(\tfrac{7}{8},\tfrac{1}{8}\right)\frac{2e^4B_\perp^2}{3m^2c^3}\gamma^2 = 1.59\times10^{-15}\gamma^2B_\perp^2\ \mathrm{ergs\ s^{-1}}.$$

Emitted radiation is strongly peaked in the forward direction, with opening angle $\sim\gamma^{-3}$.

2. Power-law electron spectrum

For an isotropic power-law electron energy spectrum

$$N_e dE = \underline{K}E^{-(2\alpha+1)}dE.$$

Emissivity (power per unit volume, frequency, steradian):

$$j\nu = (2c_1)^\alpha c_5(\alpha)B_\perp^{\alpha+1}\underline{K}\nu^{-\alpha}.$$

Opacity:

$$\kappa_\nu = (2c_1)^{\alpha+5/2}c_6(\alpha)B_\perp^{\alpha+3/2}\underline{K}\nu^{-(\alpha+5/2)},$$

where

$$c_1 = \frac{3e}{4\pi m^3 c^5} = 6.26\times10^{18} \text{ cgs},$$

$$c_5 = \frac{3^{1/2}}{16\pi}\frac{e^3}{mc^2}\left(\frac{\alpha+5/3}{\alpha+1}\right) \times\Gamma\left(\frac{3\alpha+1}{6}\right)\Gamma\left(\frac{3\alpha+5}{6}\right),$$

$$c_6 = \frac{3^{1/2}\pi}{36} em^5c^{10}\left(\alpha+\frac{13}{6}\right) \times\Gamma\left(\frac{6\alpha+5}{12}\right)\Gamma\left(\frac{6\alpha+13}{12}\right).$$

In cgs units:

α	c_5	c_6
0	4.88×10^{-23}	1.18×10^{-40}
$\frac{1}{2}$	1.37×10^{-23}	8.61×10^{-41}
1	7.52×10^{-24}	7.97×10^{-41}

Polarization in optically thin, thick units:

$$\tau \ll 1:\quad \pi = \frac{I_\perp - I_\parallel}{I_\perp + I_\parallel} = \frac{\alpha+1}{\alpha+5/3};$$

$$\tau \gg 1:\quad \pi = \frac{I_\parallel - I_\perp}{I_\perp + I_\parallel} = \frac{1}{4\alpha+5}.$$

High-frequency cutoff: Assume electrons injected at time $t=0$, energy spectrum is maintained power-law and isotropic. Then as a result of energy losses, synchrotron spectrum dies exponentially above

$$\nu_c = 1.07\times10^{24}B^{-3}t^{-2} \text{ Hz}$$

(B in gauss, t in seconds).

E. Inverse Compton scattering

γ = electron Lorentz factor, ν = photon frequency, ν' = photon frequency in electron rest frame,

$$\epsilon = h\nu/m_ec^2,\quad \epsilon' = h\nu'/m_ec^2.$$

Cross section:

$$\epsilon' \ll 1:\quad \sigma = \sigma_T = \frac{8\pi}{3}\frac{e^4}{m^2c^4} \quad \text{(Thomson)};$$

$$\epsilon' \gg 1:\quad \sigma = \sigma_T\frac{3}{8\epsilon'}\left(\tfrac{1}{2} + \log_e 2\epsilon'\right) \quad \text{(Klein-Nishina)}.$$

Final frequency of photon (f = final, i = initial):

$$\nu_f \sim \gamma^2\nu_i \quad \text{(Thomson)},$$

$$\nu_f \sim \gamma m_ec^2 \quad \text{(Klein-Nishina)},$$

and in any case

$$\epsilon_f \leqslant 4\gamma^2\epsilon_i/(1+4\gamma\epsilon_i).$$

Single-electron energy loss in isotropic radiation field of energy density u_{rad}:

$$W = \tfrac{4}{3}\sigma_T\gamma^2cu_{rad} \quad \text{(Thomson)},$$

$$W = (\sigma_Tc/\epsilon^2)u_{rad} \quad \text{(Klein-Nishina)}.$$

3.05. FIGURE: ATMOSPHERIC TRANSMISSION

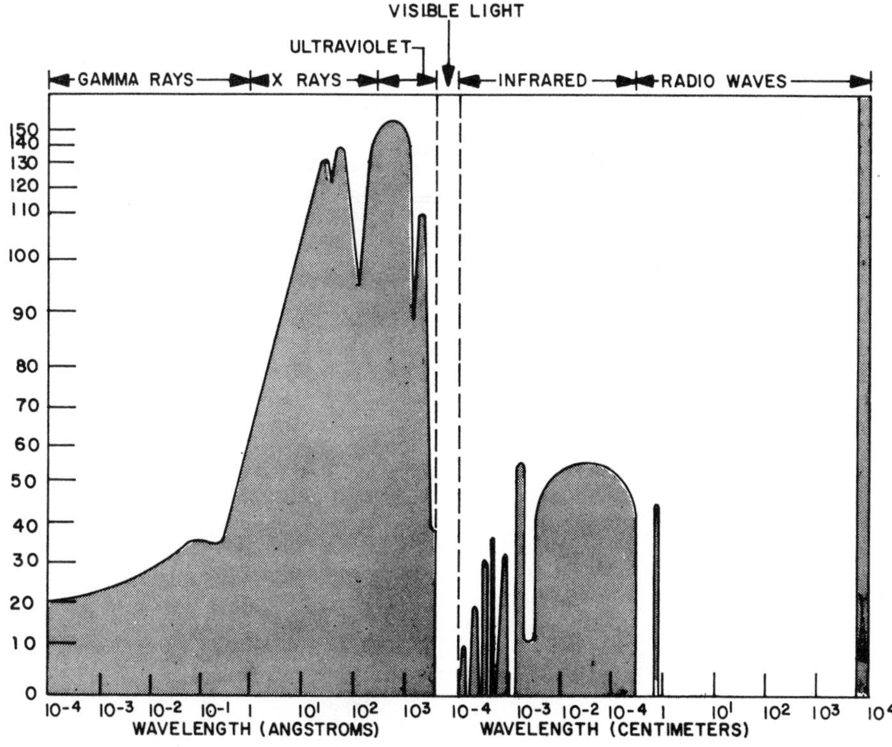

Curve specifies the altitude (in km) where the intensity of arriving radiation is reduced to half its original value. From J. M. Pasachoff, *Astronomy Now* (Saunders, Philadelphia, 1978).

3.06. TABLE: ELEMENTAL ABUNDANCES

log $A(H) = 12.0$. From J. E. Ross and L. H. Aller, Science **191**, 1223 (1976); A. G. W. Cameron, in *Explosive Nucleosynthesis*, edited by D.N. Schramm and W. D. Arnett (University of Texas, Austin, 1973), or Space Sci. Rev. **15**, 121 (1973).

		log *A*				log *A*	
Atomic number	Element	Ross-Aller solar system	Cameron solar system	Atomic number	Element	Ross-Aller solar system	Cameron solar system
2	He	10.8 ± 0.2	10.84	44	Ru	1.83 ± 0.4	1.78
3	Li	1.0 ± 0.1	3.19	45	Rh	1.40 ± 0.4	1.10
4	Be	1.15 ± 0.2	1.41	46	Pd	1.5 ± 0.4	1.61
5	B	$<2.1 \pm 0.2$	4.04	47	Ag	0.85 ± 0.10	1.15
6	C	8.62 ± 0.12	8.57	48	Cd	1.85 ± 0.15	1.67
7	N	7.94 ± 0.15	8.07	49	In	1.65 ± 0.12	0.77
8	O	8.84 ± 0.07	8.83	50	Sn	2.0 ± 0.4	2.05
9	F	4.56 ± 0.33	4.89	51	Sb	1.0 ± 0.4	1.00
10	Ne	7.57 ± 0.12	8.03	52	Te		2.31
11	Na	6.28 ± 0.05	6.28	53	I		1.53
12	Mg	7.60 ± 0.15	7.52	54	Xe		2.23
13	Al	6.52 ± 0.12	6.43	55	Cs	<1.9	1.09
14	Si	7.65 ± 0.08	7.50	56	Ba	2.09 ± 0.11	2.18
15	P	5.50 ± 0.15	5.48	57	La	1.13 ± 0.3	1.15
16	S	7.2 ± 0.15	7.20	58	Ce	1.55 ± 0.2	1.57
17	Cl	5.5 ± 0.4	5.25	59	Pr	0.66 ± 0.15	0.67
18	Ar	6.0 ± 0.20	6.57	60	Nd	1.23 ± 0.3	1.39
19	K	5.16 ± 0.10	5.12	62	Sm	0.72 ± 0.3	0.85
20	Ca	6.35 ± 0.10	6.35	63	Eu	0.7 ± 0.3	0.43
21	Sc	3.04 ± 0.07	3.04	64	Gd	1.12 ± 0.3	0.97
22	Ti	5.05 ± 0.12	4.94	65	Tb		0.24
23	V	4.02 ± 0.15	3.92	66	Dy	1.06 ± 0.3	1.05
24	Cr	5.71 ± 0.14	5.60	67	Ho		0.40
25	Mn	5.42 ± 0.16	5.47	68	Er	0.76 ± 0.4	0.85
26	Fe	7.50 ± 0.08	7.42	69	Tm	0.26 ± 0.2	0.03
27	Co	4.90 ± 0.18	4.84	70	Yb	0.9 ± 0.4	0.83
28	Ni	6.28 ± 0.09	6.18	71	Lu	0.76 ± 0.3	0.05
29	Cu	4.06 ± 0.13	4.23	72	Hf	0.8 ± 0.1	0.82
30	Zn	4.45 ± 0.15	4.59	73	Ta		-0.18
31	Ga	2.80 ± 0.15	3.18	74	W	1.7 ± 0.4	0.70
32	Ge	3.50 ± 0.08	3.56	75	Re	<-0.3	0.22
33	As		2.32	76	Os	0.7 ± 0.2	1.37
34	Se		3.32	77	Ir	0.85 ± 0.2	1.35
35	Br		2.63	78	Pt	1.75 ± 0.15	1.64
36	Kr		3.17	79	Au	0.75 ± 0.15	0.80
37	Rb	2.60 ± 0.05	2.27	80	Hg	<2.1	1.10
38	Sr	2.90 ± 0.10	2.93	81	Tl	0.90 ± 0.17	0.78
39	Y	2.10 ± 0.25	2.18	82	Pb	1.93 ± 0.12	2.10
40	Zr	2.75 ± 0.16	2.94	83	Bi	<1.9	0.65
41	Nb	1.9 ± 0.2	1.64	90	Th	0.2 ± 0.1	0.26
42	Mo	2.16 ± 0.2	2.10	92	U	<0.60	-0.08

3.07. TABLE: PROPERTIES OF THE PLANETS

Parameter	Mercury	Venus	Earth	Mars	Jupiter	Saturn	Uranus	Neptune	Pluto
Semimajor axis (mean distance from Sun)[a]									
In AU	0.387	0.723	1.000[b]	1.524	5.203	9.523	19.164	29.987	39.37
In 10^6 km	57.9	108.2	149.6	228.0	778.4	1424.6	2866.9	4486.0	5889.7
Eccentricity[a]	0.206	0.007	0.017[b]	0.093	0.049	0.053	0.046	0.012[c]	0.249
Inclination to ecliptic[a]	7°00′	3°24′	defines ecliptic	1°51′	1°18′	2°29′	0°46′	1°47′[c]	17°10′
Mean orbital velocity (km/s)	47.90	35.05	29.80	24.14	13.06	9.65	6.80	5.43	4.74
Sidereal period of revolution	88.0 days[d]	224.7 days[d]	1.000 year	1.88 years[e]	11.86 years[e]	29.46 years[e]	84.01 years[e]	164.1 years[e]	247 years[e]
Orbital angular momentum (kg km²/s)	8.99×10^{32}	1.85×10^{34}	2.66×10^{34}	3.52×10^{33}	1.9×10^{37}	7.78×10^{36}	1.69×10^{36}	2.51×10^{36}	$\sim1.8\times10^{34}$
Aphelion distance (10^6 km)	69.8	108.9	152.1	249.2	816.5	1500.1	2998.8	4539.8	7356.2
Perihelion distance (10^6 km)	46.0	107.5	147.1	206.8	740.3	1349.1	2735.0	4432.2	4423.2[f]
Mean surface magnitude at zero phase [V_{mag}/(sec arc)²]	1.45	0.85	2.1	3.9	5.6	6.9	8.2	9.6	~18.2
Mean surface brightness at zero phase (Earth = 1)	1.82	3.16	1	1.96×10^{-1}	3.98×10^{-2}	1.20×10^{-3}	3.63×10^{-2}	1.00×10^{-3}	3.63×10^{-7}
Mass (10^{24} kg)	0.330 22	4.8690	5.9742	0.641 91	1898.8	568.41	86.967	102.85	0.014[g]
Equatorial radius[h] (km)	2439	6052	6378.140	3393.4	71 398	60 000	25 400	24 300	2000[g]
Oblateness[h]	0	0	0.003 352 81	0.005 186 5	0.064 808 8	0.107 620 9	0.030	0.0259	0
Mean density (g/cm³)	5.43	5.24	5.515	3.93	1.33	0.70	1.3	1.8	~0.4[g]
Spherical harmonic J_2	$80\pm60\times10^{-6}$	$1\pm10\times10^{-6}$	1082.63×10^{-6}	1964×10^{-6}	$14\,733\times10^{-6}$	1615.6×10^{-5}	5×10^{-3}	0.005	...
Spherical harmonic J_4	...	...	-1.61×10^{-6}	-57.3×10^{-6}	-587×10^{-6}	−0.001 01	...	...	...
Equatorial surface gravity[i] (cm s⁻²)	370	890	978	371	2288	905	830	1115	~20
Rotation period									
Synodic[d]	176.0 days	116.7 days	24^h by def.	$24^h39^m35^s$	$9^h55^m33^s$[j]	$10^h39^m24^s$[j]	$\sim16^h$?	$\sim18\frac{1}{2}^h$?	$6^d09^h18^m$?
Sidereal[d]	58.6 days	243.0 days	$24^h56^m04^s.1$	$24^h37^m23^s$	$9^h55^m30^s$[j]	same	same	same	same
Inclination of equator to orbit	near 0°	~177°	23°.45	25°.2	3°.1	26°.7	97°.9	28°.8	115°[g]
Satellites (see Table 3.11)	0	0	1	2	16	16[k]	5	2	1

[a]Osculating element for 30 December 1949 except Neptune.
[b]Element of Earth-Moon barycenter (Earth-Moon system center of gravity).
[c]Osculating element for 5 January 1941.
[d]In units of mean solar time (Earth clock time).
[e]Sidereal years.
[f]Note that Pluto at perihelion is nearer the Sun than Neptune approaches.
[g]IAU circulars 3509, 3515.
[h]Recommended reference spheriods for mapping purpose (IAU, January 1980).
[i]Including centrifugal term.
[j]Rotation period of magnetic field.
[k]Liable to increase.

3.08. TABLE: EQUATORIAL FRAME COORDINATES AND VELOCITIES OF THE OUTER PLANETS

In AU and AU/day at epoch JD 2 430 000.5. From C. H. Cohen, G. C. Hubbard, and C. Oesterwriter, Astron. Papers Am. Ephemeris & Nautical Almanac **22** (1973).

Planet	x	y	z
Jupiter	$0.342\,947\,415\,189\,30\times10^{1}$	$0.335\,386\,959\,710\,90\times10^{1}$	$0.135\,494\,901\,714\,70\times10^{1}$
Saturn	$0.664\,145\,542\,549\,10\times10^{1}$	$0.597\,156\,957\,878\,30\times10^{1}$	$0.218\,231\,499\,727\,50\times10^{1}$
Uranus	$0.112\,630\,437\,207\,30\times10^{2}$	$0.146\,952\,576\,793\,50\times10^{2}$	$0.627\,960\,525\,067\,30\times10^{1}$
Neptune	$-0.301\,552\,268\,759\,30\times10^{2}$	$0.165\,699\,966\,404\,30\times10^{1}$	$0.143\,785\,752\,720\,50\times10^{1}$
Pluto	$-0.211\,251\,165\,209\,38\times10^{2}$	$0.284\,480\,700\,993\,97\times10^{2}$	$0.153\,892\,789\,502\,95\times10^{2}$
Planet	$\dot{x}$	$\dot{y}$	$\dot{z}$
Jupiter	$-0.557\,160\,570\,448\,70\times10^{-2}$	$0.505\,696\,783\,284\,30\times10^{-2}$	$0.230\,578\,543\,899\,70\times10^{-2}$
Saturn	$-0.415\,570\,776\,371\,30\times10^{-2}$	$0.365\,682\,722\,807\,70\times10^{-2}$	$0.169\,143\,213\,294\,10\times10^{-2}$
Uranus	$-0.325\,325\,669\,158\,70\times10^{-2}$	$0.189\,706\,021\,963\,50\times10^{-2}$	$0.877\,265\,322\,780\,70\times10^{-3}$
Neptune	$-0.240\,476\,254\,177\,30\times10^{-3}$	$-0.287\,659\,532\,607\,30\times10^{-2}$	$-0.117\,219\,543\,174\,10\times10^{-2}$
Pluto	$-0.176\,874\,324\,769\,98\times10^{-2}$	$-0.261\,427\,477\,414\,44\times10^{-2}$	$-0.148\,703\,713\,658\,42\times10^{-3}$

3.09. TABLE: SATELLITE DATA

Satellite	Semimajor axis (km)	Eccentricity	Inclination[a]	Sidereal period	Mean opposition magnitude V_o	Diameter (km)	Mass[i]	Density (g cm^{-3})
				Earth's satellite				
Moon	384 399	0.055	18.5°–28.5°[j]	27.321 66 days[j]	−13.0	3476[d]	0.0123	3.343
				Mars's known satellites				
Phobos	9 378.53	0.015 0	1.04°	$7^h39^m13^s.8$[b]	11.4	27.0×21.4×19.2[d]	1.5×10^{-8}	1.9 ± 0.6
Deimos	23 458.91	0.000 8	2.79°	$1^d6^h17^m54^s.9$[b]	12.5	15.0×12.0×11.0[d]	...	...
				Jupiter's known satellites				
1979 J3	128 000	...	0°	$7^h04^m30^s \pm 3^s$	...	~40	...	...
1979 J1	129 000	...	0°	$7^h09^m \pm 1^m$	...	~25	...	...
J V (Amalthea)	181 500	0.002 8[c]	0°27.3′[c]	$11^h57^m22^s.70$	14.1	270×170×155[d]	...	...
1979 J2	221 700	...	1°15′	$16^h11^m21^s.3 \pm ^s.5$	...	75 ± 5	...	...
J I(Io)	422 000	0.000 0	0°1.6′	$1^d18^h27^m33^s.51$[b]	5.0	3632[d]	4.684×10^{-5}	3.53
J II (Europa)	671 400	0.000 3	0°28.1′	$3^d13^h13^m42^s.05$[b]	5.3	3126[d]	2.523×10^{-5}	3.17
J III (Ganymede)	1 071 000	0.001 5	0°11.0′	$7^d3^h42^m33^s.35$[b]	4.6	5276[d]	7.803×10^{-5}	1.99
J IV (Callisto)	1 884 000	0.007 5	0°15.2′	$16^d16^h32^m11^s.21$[b]	5.6	4820[d]	5.661×10^{-5}	1.76
J XIII (Leda)	11 094 000	0.148	26.7°	238.7 days	~21.5	...	...	...
J VI (Himalia)	11 487 000	0.158	27.6°	250.57	14.8	170 ± 20	...	...
J VII (Elara)	11 747 000	0.207	24.8°	259.65	16.4	80 ± 20	...	...
J X (Lysithea)	11 861 000	0.130	29.0°	263.55	18.6	...	...	...
J XII (Ananke)	21 250 000	0.169	147°	631	18.8	...	...	...
J XI (Carme)	22 540 000	0.207	164°	692	18.1	...	...	...
J VIII (Pasiphae)	23 510 000	0.378	145°	739	18.8	...	...	...
J IX (Sinope)	23 670 000	0.275	153°	758	18.3	...	...	...
				Saturn's known satellites[k]				
1980 S28	137 000			0.597 days		~100		
1980 S27	139 000			0.614				
1980 S26	142 000			0.630				
1980 S3	151 000			0.692 8		~150		
1980 S1	151 000			0.693 1		~150		
Mimas	185 500	0.020 1	1°31.0′	0.942 422	12.9	400[d]	$6.59 \pm 0.15\times10^{-8}$	~1
Enceladus	238 000	0.004 44	0°01.4°	1.370 218[b]	11.7	550[d]	$1.48 \pm 0.61\times10^{-7}$	~1
Tethys	294 700	0.002 23	1°05.6′	1.887 802[b]	10.3	1040[d]	$1.095 \pm 0.022\times10^{-6}$	~1
1980 S13				1.981				
Dione	377 400	0.002 21	0°01.4′	2.736 916[b]	10.5	1000[d]	$2.039 \pm 0.053\times10^{-6}$	~2
Dione B (1980 S6)				2.722				
Rhea	527 000	0.000 98	0°21′	4.517 503[b]	9.8	1600[d]	$4.8 \pm 0.8\times10^{-6}$	1.3
Titan	1 222 000	0.029	0°20′	15.945 452[b]	8.4	5800[d]	$2.461 \pm 0.0029\times10^{-4}$	1.37
Hyperion	1 484 000	0.104	(16′–56′)[f]	21.276 665	14.2	224[d]	2.0×10^{-7}?	...
Iapetus	3 562 000	0.028 28	14.72°	79 330 82[b]	10.2–11.9	1450[d]	$3.94 \pm 1.93\times10^{-6}$	~1
Phoebe	12 960 000	0.163 26	150.05°	550.45	16.5	...	...	...
				Uranus's known satellites[l]				
Miranda	129 800	0.017	3.4	1.413 49 days	16.5	300[d]	$\sim1\times10^{-6}$	...
Ariel	190 900	0.002 8	~0	2.520 38	14.4	800[d]	$\sim15\times10^{-6}$	...
Umbriel	266 000	0.003 5	~0	4.144 18	15.3	550[d]	$\sim6\times10^{-6}$	...
Titania	436 000	0.002 4	~0	8.705 87	14.0	1000[d]	$\sim50\times10^{-6}$	...
Oberon	583 400	0.000 7	~0	13.463 25	14.2	900[d]	$\sim29\times10^{-6}$	...
				Neptune's known satellites				
Triton	355 550	0	159.945°	5.876 844 days	13.5	3200[d]	$1.3 \pm 0.3\times10^{-3}$	~3.0
Nereid	5 567 000	0.749 34	27.71°	359.881	18.7	...	...	...
				Pluto's known satellite				
Charon	~17 000	~0	115°[h]	6.387 5 days	16–17	2000[g]	...	~0.4[g]

[a]With respect to equatorial plane of planet.
[b]Known to be synchronously rotating with respect to primary.
[c]The eccentricities and inclinations for the regular satellites are slightly variable. Those for the irregular satellites are extremely variable.
[d]Recommended reference diameters for mapping purposes (IAU,January 1980).
[e]To plane of ring.
[f]Varies from 17′ to 56′.
[g]IAU Circulars 3509, 3515.
[h]To orbital plane of Pluto.
[i]Mass with respect to primary.
[j]Inclination to ecliptic = 5°9′. Synodic month = 29.5306 days.
[k]Includes early Voyager I results. Designation of 1980, S1, S3, S6, S13 follows S. M. Larson and J. W. Fountain, Sky Teles. **60**, 356 (Nov. 1980). The satellite Janus discovered in 1966 and originally assigned a period of 17.975^h was probably one of 1980 S1 or S3. 1980 S1 and S3 are also commonly designated S10, S11 (no date). The common nomenclature for the rings of Saturn is D (tenuous, unconfirmed), C (1.21–1.53R_S), B (1.53–1.95R_S), A (2.03–2.29R_S), F (2.35R_S, braided), and E (extends to ~8.5 R_S, broad maximum around 4.15–4.6R_S).
[l]Uranus has at least nine narrow rings, currently designated 6, 5, 4, α, β, η, γ, δ, and ϵ. Their respective mean radii in km are 41 980, 42 360, 42 663, 44 844, 45 799, 47 323, 47 746, 48 423, and about 51 400. For details see J. L. Elliot *et al.*, Astron. J. **83**, 980 (1978).

3.10. COMETS

Recent review papers on comets can be found in *Comets Asteroids Meteorites Interrelations, Evolution and Origins,* edited by A. H. Delsemme (University of Toledo, Toledo, 1977), and *The Study of Comets,* Proceedings of IAU Colloquium No. 25, edited by B. Donn, M. Mumma, W. Jackson, M. A. Hearn, and R. Harrington, NASA Report SP-393, 1976. Compiled by Ray Newburn.

3.11. TABLE: METEORITES

Classification and data mainly from J. T. Wasson, *Meteorites: Classification & Properties* (Springer, Berlin, 1974), but with fall data for iron meteorites from Vagn F. Buchwald, *Handbook of Iron Meteorites: Their History, Distribution, Composition & Structure* (University of California, Berkeley, 1976). Antarctic meteorites discovered since 1969 are not included in this table. Compiled by Ed Scott.

Group	Synonym	Example	Falls	Finds	Fall frequency (%)
Chondrites					
E	Enstatite chondrite	Indarch	11	5	1.5
H[a]	Bronzite chondrite	Richardton	229	230	32.3
L	Hypersthene chondrite	Bjurböle	278	192	39.3
LL	Amphoterite chondrite	St. Mesmin	51	16	7.2
CI[a]	C1 carbonaceous chondrite	Orgueil	5	0	0.71
CM	C2 carbonaceous chondrite	Mighei	14	0	2.0
CO	C3 carbonaceous chondrite	Ornans	6	1	0.85
CV	C3 carbonaceous chondrite	Allende	8	3	1.1
Other chondrites		Kakangari	2	2	0.28
Total					85
Differentiated meteorites					
Ureilites	Olivine-pigeonite achondrites	Novo Urei	3	3	0.42
Aubrites	Enstatite achondrites	Norton County	8	1	1.1
Diogenites	Hypersthene achondrites	Johnstown	8	0	1.1
Howardites	Pyroxene-plagioclase achondrites	Kapoeta	18	1	2.5
Eucrites	Pyroxene-plagioclase achondrites	Juvinas	20	4	2.8
Mesosiderites	Stony irons	Estherville	6	14	0.85
Pallasites	Stony irons	Krasnojarsk	2	33	0.28
IAB irons	[b]	Canyon Diablo	7	83	0.85[c]
IC irons		Bendego	0	10	0.09
IIAB irons		Coahuila	5	47	0.49
IIC irons		Ballinoo	0	7	0.06
IID irons		Needles	2	11	0.12
IIE irons		Weekeroo Station	1	11	0.11
IIIAB irons		Cape York	6	151	1.46
IIICD irons		Tazewell	2	10	0.11
IIIE irons		Rhine Villa	0	8	0.08
IIIF irons		Nelson County	0	5	0.05
IVA irons		Gibeon	2	38	0.38
IVB irons		Hoba	0	11	0.10
Other irons		Mbosi	6	62	0.62
Other differentiated stones		Shergotty	7	6	0.99
Total					15

[a]H, L, and LL chondrites comprise the ordinary chondrites; CI–CV are the carbonaceous chondrites.

[b]The structural classification of irons into hexahedrites, octahedrites, and ataxites does not identify all genetically related meteorites. However, most fine octahedrites belong to group IVA, medium octahedrites to IIIAB, coarse octahedrites to IA, and hexahedrites to IIA.

[c]Fall frequencies for iron meteorites are calculated on the basis of 32 observed falls allocated according to the frequencies of all classified irons in E. R. D. Scott and J. T. Wasson, Rev. Geophys. Space Phys. **73**, 530 (1975).

3.12. TABLE: ASTEROID DATA

Recent review papers on asteroids can be found in *Comets Asteroids Meteorites Interrelations, Evolution and Origins,* edited by A.H. Delsemme (University of Toledo, Toledo, 1977). Compiled by Clark Chapman and Ben Zellner from the TRIAD[a] data file.

Asteroid	Semi-major axis (AU)	Eccentricity	Inclination	Absolute magnitude $B(1,0)$ (mag.)	Color $U-B$ (mag.)	Color $B-V$ (mag.)	Albedo[b]	Diameter[c] (km)	Rotation period	Type[d]	Inferred mineralogy[e]
Asteroids larger than 200 km in diameter (in order of size)											
1 Ceres	2.767	0.0784	10°61	4.48	0.42	0.72	0.053	1020	9ʰ078	U	Silicate (olivine?) + opaque (magnetite?)
4 Vesta	2.362	0.0890	7.13	4.31	0.48	0.78	0.235	549	5.342 13	U	Clinopyroxene (+ plagioclase?)
2 Pallas	2.769	0.2353	34.83	5.18	0.26	0.65	0.079	538[f]	7.881 06	U	Salicate (olivine?) + opaque (magnetite?)
10 Hygiea	3.151	0.0996	3.81	6.50	0.31	0.69	0.041	450	18	C	Phyllo-silicate + opaque (carbonaceous?)
511 Davida	3.187	0.1662	15.81	7.36	0.35	0.71	0.033	341	5.12	C	Phyllo-silicate + opaque (carbonaceous?)
704 Interamnia	3.057	0.1553	17.31	7.24	0.25	0.64	0.035	339	8.723	C	Silicate (olivine?) + opaque (magnetite?)
31 Euphrosyne	3.154	0.2244	26.30	7.28	...	...	...	(333)	...	CM	...
451 Patientia	3.061	0.0772	15.23	8.05	0.31	0.67	0.026	327	7.11	C	...
65 Cybele	3.434	0.1154	3.55	7.99	0.28	0.68	0.022	308	...	C	...
52 Europa	3.092	0.1138	7.47	7.62	0.35	0.68	0.035	290	11.258 2	C	Phyllo-silicate + opaque (carbonaceous?)
16 Psyche	2.920	0.1390	3.09	6.88	0.25	0.70	0.093	252	4.303	M	Nickel-iron (+ enstatite?)
324 Bamberga	2.686	0.3360	11.16	8.07	0.29	0.69	0.031	251	8	C	Phyllo-silicate + opaque (carbonaceous?)
3 Juno	2.670	0.2557	12.99	6.51	0.42	0.82	0.151	248	7.213	S	Nickel-iron + olivine + pyroxene
15 Eunomia	2.642	0.1883	11.73	6.29	0.44	0.82	0.167	246	6.080 6	S	Nickel-iron + silicate (olivine>pyroxene)
13 Egeria	2.576	0.0889	16.50	8.15	0.45	0.75	0.033	241	7.045	C	...
45 Eugenia	2.721	0.0806	6.60	8.31	0.27	0.68	0.030	228	5.700	C	...
87 Sylvia	3.481	0.0985	10.85	8.12	0.24	0.69	...	(225)	...	CMEU	...
19 Fortuna	2.442	0.1576	1.56	8.45	0.38	0.75	0.030	221	7.46	C	Phyllo-silicate + opaque (carbonaceous?)
216 Kleopatra	2.793	0.2520	13.09	8.10	0.24	0.72	...	(219)	5.394	CMEU	...
532 Herculina	2.771	0.1789	16.35	8.05	0.43	0.86	0.120	217[g]	9.406	S	Pyroxene + olivine + nickel-iron + opaque?
624 Hektor	5.150	0.0248	18.26	8.65	0.26	0.76	0.038	216[h]	6.922 5	U	...
107 Camilla	3.489	0.0699	9.92	8.28	0.29	0.70	0.037	210	4.56	C	...
7 Iris	2.386	0.2303	5.50	6.84	0.47	0.83	0.160	210	7.135	S	Nickel-iron + olivine + minor pyroxene
24 Themis	3.138	0.1208	0.77	8.27	0.34	0.69	...	210	8.375	C	...
409 Aspasia	2.575	0.0733	11.26	8.31	0.31	0.71	..	208	...	C	...
88 Thisbe	2.768	0.1619	5.22	8.07	0.28	0.67	0.045	207	6.042 2	C	Phyllo-silicate + opaque (carbonaceous?)
747 Winchester	2.994	0.3438	18.15	8.84	0.32	0.71	0.024	205	8	C	...
702 Alauda	3.194	0.0347	20.54	8.29	0.31	0.66	...	205	...	C	...
165 Loreley	3.128	0.0802	11.24	8.81	0.31	0.74	...	203	...	C	...
Other interesting asteroids											
8 Flora	2.202	0.1561	5.89	7.73	0.46	0.88	0.125	153	13.6	S	Nickel-iron + clinopyroxene
25 Phocaea	2.401	0.2531	21.61	9.07	0.51	0.93	0.184	65	9.945	S	Nickel-iron + pyroxene + clinopyroxene
44 Nysa	2.422	0.1517	3.71	7.85	0.26	0.71	0.467	72	6.418	E	Enstatite?
80 Sappho	2.295	0.2008	8.66	9.22	0.50	0.92	0.113	86	> 20	U	Silicate (olivine?) + opaque (carbonaceous?)
158 Koronis	2.868	0.0559	1.00	10.95	0.38	0.84	...	36	...	S	...
221 Eos	3.014	0.0958	10.85	8.94	0.41	0.77	...	(97)	...	U	Silicate (olivine?) + opaque (carbonaceous?)
279 Thule	4.258	0.0327	2.34	9.79	0.22	0.77	...	(60)	...	MEU	...
349 Dembowska	2.925	0.0862	8.26	7.24	0.55	0.97	0.260	145	4.701 2	R	Olivine>pyroxene (+ nickel-iron?)
433 Eros	1.458	0.2220	10.83	12.40	0.50	0.88	0.180	16	5.270 3	S	Silicate (olivine = pyroxene) + minor nickel-iron
434 Hungaria	1.944	0.0736	22.51	12.45	0.24	0.70	0.300	11	...	E	...
785 Zwetana	2.576	0.2029	12.72	10.73	0.17	0.64	0.078	45	...	U	...
944 Hidalgo	5.820	0.6565	42.49	12.05	0.23	0.74	...	(39)	10.064 4	CMEU	...
1566 Icarus	1.078	0.8267	22.99	17.32	0.54	0.80	...	(1.7)	2.273 0	U	...
1580 Betulia	2.196	0.4905	52.04	15.66	0.27	0.66	...	6.5	6.130	C	...
1620 Geographos	1.244	0.3351	13.33	15.97	0.50	0.87	...	2.4	5.223 3	S	...
1685 Toro	1.368	0.4360	9.37	16.20	0.47	0.88	...	(7.6)	10.195 6	U	Pyroxene + olivine?

[a]Tucson Revised Index of Asteroid Data (TRIAD), the source of all data, except as noted in subsequent footnotes. Contributors to this computerized file are D. Bender (osculating orbital elements), E. Bowell (*UBV* colors), C. Chapman (spectral parameters), M. Gaffey (spectrophotometry), T. Gehrels (magnitudes), D. Morrison (radiometry), E. Tedesco (rotations), and B. Zellner (polarimetry). TRIAD is described in Icarus **33**, 630 (1978). To use TRIAD, contact B. Zellner, Lunar and Planetary Laboratory, University of Arizona, Tucson, AZ 85721.

[b]Albedos and geometric albedos from radiometry. They are not always consistent with tabulated diameters.

[c]Except as noted, diameters are from E. Bowell *et al.*, Icarus (Sept. 1978). Values are reliable for asteroids for which no albedo is listed in previous column. Especially unreliable diameters are listed in parentheses.

[d]Taxonomic type, related to surface composition, is from E. Bowell *et al.*, Icarus (Sept. 1978), wherein the types are defined.

[e]From M. T. Gaffey and T. B. McCord, in *Proceedings of the Eighth Lunar Science Conference* (1977), p. 113, here augmented in several cases by C. Chapman. Refers only to optically important phases.

[f]Stellar occultation diameter: H. Wasserman *et al.* and J. L. Elliot *et al.*, Bull. Am. Astron. Soc. (Oct. 1978).

[g]Stellar occultation diameter: E. Bowell *et al.*, Bull Am. Astron. Soc. (Oct. 1978).

[h]Hartmann and D. P. Cruikshank, Icarus (1979).

3.13. TABLE: REPRESENTATIVE DATA FOR MAIN-SEQUENCE STARS

Quantities tabulated are typical values for stars of approximately solar composition, from a variety of sources including mainly the following. Masses $m \leqslant 0.6 M_\odot$: data from G. J. Veeder, Astron. J. **79**, 1056 (1974); $0.8 \leqslant m \leqslant 10 M_\odot$: theoretical tracks of J. G. Mengel *et al.*, Astrophys. J. Suppl. **40**, 733 (1979), and C. Alcock and B. Paczynski, Astrophys. J. **223**, 244 (1978); $m \geqslant 20 M_\odot$: theoretical tracks of C. Chiosi *et al.*, Astron. Astrophys. **63**, 103 (1978), with maximum mass loss rates. For $m \leqslant 0.6 M_\odot$, the mean empirical main sequence is given; for $0.8 \leqslant m \leqslant 1.0 M_\odot$, points are for an age of 5 Gyr; and for more massive stars, points are for the middle of the main-sequence lifetime ($\tau_{ms}/2$). From B. M. Tinsley, Fundam. Cosmic Phys. **5**, 287 (1980).

m ($M_\odot$)	τ_{ms} (Gyr)	$\log L/L_\odot$	M_v	$\log T_{eff}$	$B-V$	Sp
0.15	...	−2.5	14.2	3.48	1.80	M7
0.25	...	−2.0	12.0	3.52	1.60	M5
0.4	...	−1.4	10.0	3.57	1.48	M1
0.6	...	−0.9	7.6	3.64	1.18	K5
0.8	25	−0.4	6.0	3.70	0.88	K1
0.9	15	−0.2	5.4	3.73	0.76	G8
1.0	10	0.0	4.9	3.76	0.64	G2
1.1	6.4	0.2	4.3	3.79	0.56	F8
1.2	4.5	0.4	3.7	3.82	0.47	F6
1.3	3.2	0.5	3.5	3.84	0.42	F5
1.4	2.5	0.7	3.0	3.86	0.36	F2
1.5	2.0	0.8	2.8	3.88	0.30	F0
2	0.75	1.3	1.4	3.98	0.00	A0
3	0.25	2.1	−0.2	4.10	−0.12	B7
4	0.12	2.6	−0.6	4.18	−0.17	B5
6	0.05	3.2	−1.5	4.30	−0.22	B3
8	0.03	3.6	−2.2	4.35	−0.25	B1
10	0.02	3.9	−2.7	4.40	−0.27	B0.5
15	0.01	4.4	−3.7	4.45	−0.29	
20	0.008	4.7	−4.3	4.48	−0.30	BO
30	0.006	5.1	−5.1	4.51	−0.31	O9.5
40	0.004	5.4	−5.7	4.53	−0.31	O9
60	0.003	5.7	−6.2	4.58	−0.32	O5

3.14. FIGURE: LUMINOSITY FUNCTION FOR STARS IN THE LOCAL SOLAR NEIGHBORHOOD

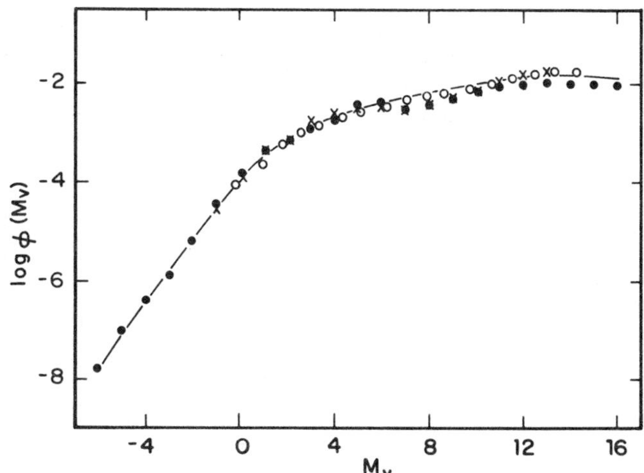

The luminosity function $\phi(M_v)$ given as the number of stars per unit absolute visual magnitude M_v and per pc^3 in the plane of the Galaxy: ●, S. W. McCuskey (1966); ○, W. J. Luyten (1968); ×, W. Wielen (1974); solid line, adopted relation. From G. E. Miller and J. M. Scalo, Astrophys. J. Suppl. **41**, 513 (1979).

3.15. STANDARD SOURCES

(1) *UBVRI* photometric standards.
(2) *uvby* and Hβ standards.
(3) Standard radial velocity stars.
(4) Radio flux calibrators.

Data on the above are given in Sec. H of Astron. Almanac (1981).

No definitive photometric standards are yet available in published form for the infrared *J*, *H*, *K*, and *L* passbands.

References to standard star photometry

D. S. Hayes and D. W. Latham, Astrophys. J. **197**, 593 (1975).

J. B. Oke, Astrophys. J. **140**, 689 (1964).

J. B. Oke and R. E. Schild, Astrophys. J. **161**, 1015 (1970).

R. E. Schild, D. M. Peterson, and J. B. Oke, Astrophys. J. **166**, 95 (1971).

R. P. S. Stone, Astrophys. J. **193**, 135 (1974); **218**, 767 (1977).

3.16. STARS NEARER THAN 5 pc

From S. L. Lippincott, Space Sci. Rev. **22**, 153 (1978).

No.	Gliese No.	Name	(1950) R.A.	Decl.	Proper motion	Position angle	Radial velocity (km s^{-1})	Parallax	Distance (lt-yr)	Visual apparent magnitude and spectrum of components A	B	C
1		Sun								−26.8 G2		
2	559	α Centauri[a]	14h36m2	−60°38′	3″68	281°	−22	0″753	4.3	−0.1 G2	1.5 Ko	11.0 M5e
3	699	Barnard's Star	17 55.4	+4 33	10.31	356	−108	0.544	6.0	9.5 M5	[b]	
4	406	Wolf 359	10 54.1	+7 19	4.71	235	+13	0.432	7.5	13.5 M8e		
5	411	BD+36°2147	11 00.6	+36 18	4.78	187	−84	0.400	8.2	7.5 M2		
6	65	Luyten 726-8	1 36.4	−18 13	3.36	80	+30	0.385	8.4	12.5 M6e	13.0 M6e	
7	244	Sirius	6 42.9	−16 39	1.33	204	−8	0.377	8.6	−1.5 A1	8.3 DA	
8	729	Ross 154	18 46.7	−23 53	0.72	103	−4	0.345	9.4	10.6 M5e		
9	905	Ross 248	23 39.4	+43 55	1.58	176	−81	0.319	10.2	12.3 M6e		
10	144	ε Eridani	3 30.6	−9 38	0.98	271	+16	0.305	10.7	3.7 K2	[c]	
11	447	Ross 128	11 45.1	+1 06	1.37	153	−13	0.302	10.8	11.1 M5		
12	866	Luyten 789-6	22 35.7	−15 36	3.26	46	−60	0.302	10.8	12.2 M6		
13	820	61 Cygni	21 04.7	+38 30	5.22	52	−64	0.292	11.2	5.2 K5e	6.0 K7	[c]
14	845	ε Indi	21 59.6	−57 00	4.69	123	−40	0.291	11.2	4.7 K5e		
15	71	τ Ceti	1 41.7	−16 12	1.92	297	−16	0.289	11.3	3.5 G8		
16	280	Procyon	7 36.7	+5 21	1.25	214	−3	0.285	11.4	0.4 F5	10.7	
17	725	Σ2398	18 42.2	+59 33	2.28	324	+5	0.284	11.5	8.9 M4	9.7 M5	
18	15	BD+43°44	0 15.5	+43 44	2.89	82	+17	0.282	11.6	8.1 M1eSB	11.0 M6e	
19	887	CD−36°15693	23 02.6	−36 09	6.90	79	+10	0.279	11.7	7.4 M2e		
20		G51-15	8 26.9	+26 57	1.26	241		0.273	11.9	14.8 m		
21	54.1	L725-32	1 10.1	−17 16	1.22	62		0.264	12.3	11.5 M5e		
22	273	BD+5°1668	7 24.7	+5 23	3.73	171	+26	0.264	12.3	9.8 M4		
23	825	CD−39°14192	21 14.3	−39 04	3.46	251	+21	0.260	12.5	6.7 M0e		
24	191	Kapteyn's Star	5 09.7	−45 00	8.89	131	+245	0.256	12.7	8.8 M0		
25	860	Krüger 60	22 26.3	+57 27	0.86	246	−26	0.254	12.8	9.7 M4	11.2 M6	
26	234	Ross 614	6 26.8	−2 46	0.99	134	+24	0.243	13.4	11.3 M5e	14.8	
27	628	BD−12°4523	16 27.5	−12 32	1.18	182	−13	0.238	13.7	10.0 M5SB:		
28	473	Wolf 424	12 30.9	+9 18	1.75	277	−5	0.234	13.9	13.2 M6e	13.4 M6e	
29	35	van Maanen's Star	0 46.5	+5 09	2.95	155	+54	0.232	14.0	12.4 DG		
30	1	CD−37°15492	0 02.5	−37 36	6.08	113	+23	0.225	14.5	8.6 M3		
31	83.1	Luyten 1159-16	1 57.4	+12 51	2.08	149		0.221	14.7	12.3 M8		
32	380	BD+50°1725	10 08.3	+49 42	1.45	249	−26	0.217	15.0	6.6 K7		
33	674	CD−46°11540	17 24.9	−46 51	1.13	147		0.216	15.1	9.4 M4		
34	832	CD−49°13515	21 30.2	−49 13	0.81	185	+8	0.214	15.2	8.7 M3		
35	682	CD−44°11909	17 33.5	−44 17	1.16	217		0.213	15.3	11.2 M5		
36	687	BD+68°946	17 36.7	+68 23	1.33	194	−22	0.213	15.3	9.1 M3.5SB	[b]	
37		G158-27	0 04.2	−7 48	2.06	204		0.212	15.4	13.7 m		
38		G208-44/45	19 53.3	+44 17	0.75	143		0.210	15.5	13.4 me	14.0 m	
39	876	BD−15°6290	22 50.6	−14 31	1.16	125	+9	0.209	15.6	10.2 M5		
40	166	40 Eridani	4 13.0	−7 44	4.08	213	−43	0.207	15.7	4.4 K0	9.5 DA	11.2 M4e
41	440	L145-141	11 43.0	−64 33	2.68	97		0.206	15.8	11.4 DG		
42	388	BD+20°2465	10 16.9	+20 07	0.49	264	+11	0.203	16.0	9.4 M4.5		
43	702	70 Ophiuchi	18 02.9	+2 31	1.13	167	−7	0.203	16.0	4.2 K1	6.0 K5	
44	873	BD+43°4305	22 44.7	+44 05	0.83	237	−2	0.200	16.3	10.2 M4.5e	[b]	
45	768	Altair	19 48.3	+8 44	0.66	54	−26	0.198	16.5	0.8 A7IV,V		
46	445	AC+79°3888	11 44.6	+78 58	0.89	57	−119	0.193	16.8	10.9 M4		
47		G9-38	8 55.4	+19 57	0.89	266		0.190	17.1	14.1 m	14.9 m	

[a]The position of α Centuri C ("Proxima") is 14h26m3, −62°28′; 2°11′ from the center of mass of α Centauri A and B. The proper motion of C is 3″84 in position angle 282°. Gliese No. 551.
[b]Unseen components.
[c]Suspected.

3.17. TABLE: PARAMETERS OF THE INTERSTELLAR GAS

Mean density ρ	3×10^{-24} g cm^{-3}
Typical particle density (cm^{-3})	
n(H I) in diffuse clouds	20
n(H I) between clouds	0.1
n_H in molecular clouds	10^3–10^6
Typical temperature T(K)	
Diffuse H I clouds	80
H I between clouds	6000
H II photo ionized regions	8000
Coronal gas between clouds	6×10^5
Root-mean-square random cloud velocity	10 km s^{-1}
Isothermal sound speed C (km s^{-1})	
H I cloud at 80 K	0.7
H II gas at 8000 K	10
Magnetic field B	2.5×10^{-6} G
Effective thickness $2H$ of H I cloud layer	250 pc

3.18. INTERSTELLAR MOLECULES OBSERVED

See in A. P. C. Mann and D. A. Williams, Nature **283**, 721 (1980).

3.19. TABLE: PARAMETERS OF THE GALAXY

Diameter	25 kpc
Width of disk	
Star system	2 kpc
Interstellar material	200 pc
Mass ($M_\odot$)	
Star system	1.4×10^{11}
Interstellar material	3×10^{9}
Age	1.2×10^{10} yr
Star density (disk) (pc^{-3})	
Solar neighborhood	0.1
Galactic nucleus	10^{2}
Number of stars	10^{11}
Absorption of optical radiation from extragalactic objects	$\Delta m=\alpha \csc b$, $\alpha=0.36$ (V), 0.47 (B)
Period of rotation	2.5×10^{8} yr

Luminosity of Galaxy (ergs s^{-1})	
Radio	3×10^{38}
Infrared	3×10^{41}
Optical	3×10^{43}
X-ray	10^{39}–10^{40}
Rotational velocity in solar neighborhood	220 ± 10 km s^{-1}
Distance of the Sun from galactic center	8.5 ± 0.5 kpc
Oort (differential galactic rotation) constants (km s^{-1} kpc^{-1})	$A=13\pm2$, $B=-13\pm2$
Height of Sun above galactic disk	8 pc
Energy density in Galaxy (10^{-12} ergs cm)	
Starlight	0.7
Turbulent gas	0.5
Cosmic rays	2
Magnetic field	2
2.7 K radiation	0.4

A. Table: Galactic classification—Revised Hubble types

From G. de Vaucouleurs, *Reference Catalogue of Bright Galaxies* (University of Texas, Austin, 1964).

Classes	Families	Varieties	Stages	Type
Ellipticals				E
			Ellipt. (0–7)	E0
			Intermediate	E0–1
			Late ellipt.	E^+
Lenticulars				S0
	Ordinary			SA0
	Barred			SB0
	Mixed			SAB0
		Inner ring		S(r)0
		S-shaped		S(s)0
		Mixed		S(rs)0
			Early	$S0^-$
			Intermediate	S0°
			Late	$S0^+$
Spirals	Ordinary			SA
	Barred			SB
	Mixed			SAB
		Inner ring		S(r)
		S-shaped		S(s)
		Mixed		S(rs)
			0/a	S0/a
			a	Sa
			ab	Sab
			b	Sb
			bc	Sbc
			c	Sc
			cd	Scd
			d	Sd
			dm	Sdm
			m	Sm
Irregulars	Ordinary			IA
	Barred			IB
	Mixed			IAB
		S-shaped		I(s)
			Magellanic	Im
			Non-Magell.	I0
Peculiars				P
Peculiarities (all types)			Peculiarity	P
			Uncertain	:
			Doubtful	?
			Spindle	sp
			Outer ring	(R)
			Pseudo outer ring	(R′)

B. Table: Mean corrected colors of galaxies

"Corrected" colors: after eight cycles of 2σ rejection of large residuals. From G. de Vaucouleurs, in *The Evolution of Galaxies and Stellar Populations,* edited by B. M. Tinsley and R. B. Larson (Yale University Observatory, New Haven, 1977).

Revised Hubble type	$\langle(B-V)\rangle$	Standard deviation	$\langle(U-B)\rangle$	Standard deviation
E	0.894	0.033	0.471	0.057
$S0^-$	0.863	0.051	0.462	0.039
S0	0.856	0.049	0.419	0.092
$S0^+$	0.808	0.071	0.374	0.078
S0/a	0.804	0.053	0.180	0.298
Sa	0.704	0.097	0.267	0.070
Sab	0.737	0.057	0.226	0.106
Sb	0.657	0.080	0.060	0.123
Sbc	0.564	0.066	−0.042	0.104
Sc	0.511	0.070	−0.072	0.103
Scd	0.441	0.080	−0.190	0.110
Sd	0.444	0.079	−0.213	0.102
Sdm	0.418	0.098	−0.218	0.085
Sm	0.353	0.087	−0.337	0.087
Im	0.370	0.079	−0.331	0.108

C. Formulas for galactic extinction of extragalactic objects

The following formulas are from G. de Vaucouleurs, A. de Vaucouleurs, and H. G. Corwin, *Second Reference Catalogue of Bright Galaxies* (University of Texas, Austin, 1976).

Blue extinction A_B as a function of new galactic coordinates l, b:

$$A_B=0.19(1+S_N\cos b)|C| \quad (b>0)$$
$$=0.21(1+S_S\cos b)|C| \quad (b<0),$$

where

$$S_N(l)=0.1948\cos l+0.0725\sin l$$
$$+0.1168\cos 2l-0.0921\sin 2l$$
$$+0.1147\cos 3l+0.0784\sin 3l$$
$$+0.0479\cos 4l+0.0847\sin 4l$$

$$S_S(l) = 0.2090 \cos l - 0.0133 \sin l$$
$$+ 0.1719 \cos 2l - 0.0214 \sin 2l$$
$$- 0.1071 \cos 3l - 0.0014 \sin 3l$$
$$+ 0.0681 \cos 4l + 0.0519 \sin 4l,$$

with for both hemispheres,

$$C = \csc[b - b_0(l)]$$
$$= \csc(b + 0\overset{\circ}{.}25 - 1\overset{\circ}{.}7 \sin l - 1\overset{\circ}{.}0 \cos 3l).$$

3.20. COSMOLOGY

A. Formulas

Redshift-distance relation:

$$z = H_0 d/c + \tfrac{1}{2}(1 + q_0)(H_0 d/c)^2 + O((H_0 d/c)^3).$$

Magnitude-redshift relation:

$$m - M = 25 - 5 \log_{10} H_0 + 5 \log_{10} c$$
$$+ 5 \log_{10}[1 - q_0 + q_0 z$$
$$+ (q_0 - 1)(2q_0 z + 1)^{1/2}]$$
$$\approx 25 - 5\log_{10} H_0 + 5 \log_{10} cz$$
$$+ 1.086(1 - q_0)z \quad \text{for } z \ll 1.$$

Here H_0 is in km s^{-1} Mpc^{-1}, c in km s^{-1}, d in Mpc.

B. Constants

Hubble constant (where h is of order unity):

$$H_0 = 100h \text{ km s}^{-1} \text{ Mpc}^{-1},$$

Hubble time:

$$1/H_0 = 9.8 \times 10^9 h^{-1} \text{ yr}.$$

Hubble distance:

$$R = c/H_0 = 3000h^{-1} \text{ Mpc}.$$

Critical density:

$$\rho_c = 3H_0^2/8\pi G = 1.9 \times 10^{-29} h^2 \text{ g cm}^{-3}.$$

Volume:

$$\tfrac{4}{3}\pi R^3 1.1 \times 10^{11} h^{-3} \text{ Mpc}^3.$$

Smoothed density of galactic material throughout universe [C. W. Allen, 3rd ed. (Athlone, Atlantic Highlands, NJ, 1973)]:

$$= 2 \times 10^{-31} \text{ g cm}^{-3}$$
$$= 1 \times 10^{-7} \text{ atom cm}^{-3}$$
$$= 3 \times 10^9 M_\odot \text{ Mpc}^{-3}.$$

Space density of galaxies:

$$= 0.02 \text{ Mpc}^{-3}.$$

Luminous emission from galaxies:

$$= 3 \times 10^8 L_\odot \text{ Mpc}^{-3}.$$

Mean sky brightness from galaxies:

$$= 1.4(m_v = 10) \text{ deg}^{-2}.$$

4.00. Atomic collision properties

CLARENCE F. BARNETT

Oak Ridge National Laboratory

CONTENTS

Most of the atomic and molecular collision formulas presented are derived from classical or semiclassical considerations and should be regarded as approximations. More precise calculations are based on quantum mechanics, and analytical expressions are usually left in open form involving integrals over matrix elements which depend on the choice of initial- and final-state wave functions and the interaction potential assumed. The reader is referred to Refs. 1–3 for a comprehensive survey of the quantum-mechanical development of atomic collisions.

The volume of experimental and theoretical data in the atomic collisions field is tremendous. Data presented in the following are representative of the many facets of atomic collisions, and the tabulated data are experimentally determined values except as noted.

4.01. ATOMIC UNITS

In making theoretical computations, atomic units are introduced to avoid handling many numerical numbers. These units are composed of combinations of the fundamental units of charge, mass, and Planck's constant $\hbar$. Experimentally, collision cross sections are usually expressed in terms of cm^2 as a function of particle energy (eV) or velocity (cm/s). Table 4.01.A presents the relationship between atomic and cgs units. (See Precise Physical Constants Table 1.01 for equivalent SI units.)

TABLE 4.01.A. Constants

	Atomic units	cgs
Length	$a=\hbar^2/me^2$	5.29×10^{-9} cm
Mass	m = mass of electron	9.108×10^{-28} gm
Charge	e = charge of electron	4.803×10^{-10} esu
Velocity	$v_0=e^2/\hbar$	2.188×10^{8} cm/s
Energy	$E=e^2/a$ = 1 hartree[a]	4.359×10^{-11} ergs (27.21 eV)
Velocity of light	137.077	2.997×10^{10} cm/s
Cross section	πa_0^2	8.797×10^{-17} cm^2
Momentum	mv_0	1.993×10^{-19} gm cm/s
Time	a/v_0	2.419×10^{-17} s
Reaction rate coefficient	$\pi a_0^2 e^2/\hbar$	6.126×10^{-9} cm^3/s
Frequency	$v_0/a=4\pi$ Ry	4.13×10^{16} s^{-1}

[a]1 hartree of energy = 2 Ry = 27.21 eV = 1 a.u.

4.02. GENERAL COLLISIONS

A. Maxwellian distribution functions

For a gas in thermodynamic equilibrium the energy distribution may be described as

$$f(E)dE=\frac{N_t 2E^{1/2}}{\pi^{1/2}(kT)^{3/2}}e^{-E/kT}dE. \qquad (1)$$

N_t is the total number of particles per cm^3 $[\int_0^\infty f(E)dE=N_t]$, k Boltzmann's constant (1.38×10^{16} ergs K^{-1}), T the gas temperature in K, and $f(E)dE$ the number of particles with energies between E and $E+dE$. For kT of 11 600 K, $E=1$ eV.

The distribution function may be expressed in terms of the particle velocity:

$$f(v)dv=N_t(2/\pi)^{1/2}(m/kT)^{3/2}v^2e^{-mv^2/2kT}dv. \qquad (2)$$

In one direction this reduces to

$$f(v_x)dv_x=N_t(m/2\pi kT)^{1/2}e^{-mv_x^2/2kT}dv_x. \qquad (3)$$

Mean velocity for Maxwellian distribution:

$$v_m=(8kT/\pi m)^{1/2}. \qquad (4)$$

Root-mean-square velocity:

$$v_r=(3kT/m)^{1/2}. \qquad (5)$$

Most probable velocity:

$$v_p=(2kT/m)^{1/2}. \qquad (6)$$

The *energy distribution of charged particles* in an external field of force is described by the Maxwell-Boltzmann distribution:

$$N(E)dE=N_0\frac{2}{(\pi)^{1/2}(kT)^{3/2}}e^{-(E+V)/kT}E^{1/2}dE. \qquad (7)$$

N_0 is the total number density at a point where $V=0$, V is the potential energy per molecule as a function of position, and the number density at each position is given by $N=N_0e^{-V/kT}$.

B. Collision cross section

The early concept of collision probability has been replaced with the more practical concept of collision cross section. The interrelationship is

$$P=L\sigma/760 \quad cm^{-1}\,(mm\ Hg)^{-1}. \qquad (8)$$

P is the collision probability, the number of collisions per cm of path length at a pressure of 1 mm Hg and $T=0$°C. L is Loschmidt's number (2.687×10^{19} cm^{-3}), and σ the total cross section in cm^2.

C. Particle beam attenuation

In passing through a gas:

$$I=I^0e^{-n\sigma x}. \qquad (9)$$

n is the gas number density, σ the cross section, and x the path length in the gas.

D. Collision frequency

$$\nu=v/\lambda=vn\sigma. \qquad (10)$$

v is the particle velocity, λ the mean free path, n the gas number density, and σ the total cross section.

E. Differential scattering cross section

For case of azimuthal symmetry:

$$d\sigma=I(\theta)\sin\theta\, d\theta\, d\phi. \qquad (11)$$

$I(\theta)$ is the intensity of scattered particles at angle θ divided by incident intensity.

F. Total cross section

$$\sigma = \int_0^{2\pi} \int_0^{\pi} I(\theta) \sin\theta \, d\theta \, d\phi. \tag{12}$$

4.03. ATOMIC COLLISION CROSS SECTIONS

Computed atomic collision cross sections are dependent on the *interaction potential* between two or more particles during the collision. Various interaction potentials have been formulated to describe the interaction. The following interaction potentials are some of the more commonly used.

A. Coulomb potential

$$V(r_{12}) = e^2 Z_1 Z_2 / r_{12}. \tag{13}$$

r_{12} is the distance between particles, and Z_1 and Z_2 the effective charges of two particles.

B. Screened Coulomb

$$V(r_{12}) = (Z_1 Z_2 e^2 / r_{12}) e^{-r_{12}/a}. \tag{14}$$

a is known as the screening length and is given by Bohr's formulation as

$$a = a_0 / (Z_1^{2/3} + Z_2^{2/3})^{1/2}, \tag{15}$$

where a_0 is the Bohr radius (5.29×10^{-9} cm).

C. Lennard-Jones

$$V(r_{12}) = a/r_{12}^m - b/r_{12}^n. \tag{16}$$

D. Lennard-Jones 6-12

$$V(r_{12}) = 4V_0[(D/r_{12})^{12} - (D/r_{12})^6]. \tag{17}$$

D is the combined radius of particle spheres 1 and 2.

E. Buckingham

$$V(r_{12}) = ae^{-br_{12}} - c/r_{12}^6 - d/r_{12}^8. \tag{18}$$

a, b, c, and d are constants.

F. Sutherland

$$V(r_{12}) = \begin{cases} \infty, & r < D \\ -a/r_{12}^n, & r > D. \end{cases} \tag{19}$$

D is the combined radius of particle spheres 1 and 2.

G. Inverse fourth-power polarization potential

$$V(r_{12}) = -\alpha Z e^2 / 2r_{12}^4. \tag{20}$$

Ze is the charge on the ion for ion-neutral collision, α the electric polarizability of the neutral molecule $[= (K-1)/4\pi n]$, K the gas dielectric constant, and n the gas number density.

H. Centrifugal

$$V_c(r_{12}) = J^2 / 2M_r r_{12}^2. \tag{21}$$

M_r is the reduced mass and J the angular momentum. $V_c(r_{12})$ represents the rotational kinetic energy of the system and can be considered as a fictitious force directed outwardly and is termed the "centrifugal potential."

I. Coulomb collisions

1. Energy transfer

Energy transferred during Coulomb collision:

$$\Delta E = \frac{4M_1 M_2}{(M_1 + M_2)^2} E \sin^2 \tfrac{1}{2}\theta. \tag{22}$$

2. Average energy loss

Average energy loss per collision for two interacting Maxwellian gases:

$$\Delta E = \frac{8}{3} \frac{M_1 M_2}{(M_1 + M_2)^2} \left(1 - \frac{T_{M_1}}{T_{M_2}}\right). \tag{23}$$

T_{M_1} and T_{M_2} are the temperatures of gases 1 and 2.

3. Differential cross section for Coulomb scattering

Using the Coulomb interaction potential $Z_1 Z_2 e^2 / r_{12}$, the differential scattering cross section can be written

$$I(\theta) d\omega = \frac{Z_1^2 Z_2^2 e^4}{4M_r^2 v_{12}^4 \sin^4 \frac{1}{2}\theta} d\omega. \tag{24}$$

M_r is the reduced mass, and v_{12} the relative velocity between particles 1 and 2. Because of the $\sin^4 \frac{1}{2}\theta$ term in this expression, the total cross section will have an infinite value. To avoid this difficulty, the screened Coulomb potential is introduced [see Eq. (16)]. Classical mechanics and quantum mechanics give the result (24).

4. Reaction rate

$$\text{reaction rate} = \langle \sigma v \rangle n_1 n_2 = \alpha n_1 n_2. \tag{25}$$

n_1 and n_2 are the densities of reactants 1 and 2, and α is the rate coefficient.

Reaction rate coefficient in terms of cross section for Maxwellian distribution:

$$\alpha = \frac{1}{(\pi M_r)^{1/2}} \left(\frac{2}{kT}\right)^{3/2} \int_0^\infty \sigma(E) E \exp^{-E/kT} dE. \tag{26}$$

$\sigma(E)$ is the collision cross section as a function of the relative energy E, kT the gas temperature, and M_r the reduced mass.

J. Massey adiabatic criterion

When $a\Delta E/\hbar v < 1$, the collision is adiabatic, and when it is $\simeq 1$, the collision cross section is a maximum. ΔE is the internal energy difference between reactants and products, v the relative velocity of approach of the particles, and a the range of the interaction or adiabatic parameter.

K. Interaction energy

In merged- or inclined-beams experiments:

$$E=\frac{M_1M_2}{M_1+M_2}\left[\frac{E_1}{M_1}+\frac{E_2}{M_2}-2\left(\frac{E_1E_2}{M_1M_2}\right)^{1/2}\cos\theta\right]. \quad (27)$$

E_1, E_2, M_1, and M_2 are the laboratory energies and masses of particles 1 and 2, and θ is the angle of intersection of the two beams.

L. Electron ionization cross section

For ions and neutrals using beams intersecting at an angle θ:

$$\sigma(E)=(RF/IJ)K. \quad (28)$$

I and J are the currents of beams 1 and 2, F is a "form factor" which corrects for nonuniform current distribution in the beams, and K given by

$$K=\frac{e_1e_2v_1v_2\sin\theta}{(v_1^2+v_2^2-2v_1v_2\cos\theta)^{1/2}},$$

where v_1, v_2, e_1, and e_2 are the velocities and charges of the particles in the beams that intersect at angle θ;

$$F=\int i(z)dz\int j(z)dz\bigg/\int i(z)j(z)dz,$$

where $i(z)$ and $j(z)$ are the currents flowing in elements of the beam of height dz.

M. Born approximation

The plane wave Born approximation provides a simple method to compute quantum mechanically both the elastic and inelastic interactions of two particles. The approximation relies on the assumption that the incident and outgoing atomic wave functions describing the relative motion between the incoming particle and target are undistorted during the collision, and the interaction may be treated as a small perturbation such that $V(r)\ll E$, where $V(r)$ is the interaction potential and E the relative energy. Applying the Born approximation to the elastic scattering amplitude results in

$$f(\theta)=-2\frac{M_r}{\hbar^2}\int_0^\infty\frac{\sin Kr}{Kr}V(r)r^2dr,$$

$$d\sigma=|f(\theta)|^2d\omega. \quad (29)$$

K is the magnitude of the momentum transferred during collision. For an exponential screened Coulomb potential, Eq. (29) reduces to

$$f(\theta)=-\frac{2Z_1Z_2e^2M_r}{\hbar^2}\frac{1}{4K^2\sin^2\frac{1}{2}\theta+1/a^2},$$

and therefore

$$\lim_{a\to\infty}f(\theta)=-\frac{Z_1Z_2e^2M_r}{2\hbar^2K^2\sin^2\frac{1}{2}\theta}. \quad (30)$$

$\hbar K$ is the linear momentum M_rv_0, and a the screening length defined in Eq. (15). For the differential scattering cross section, one obtains the same classical result as given in Eq. (24).

For electron excitation of ions and atoms, the plane wave Born approximation is valid for electron energies greater than $\sim 7E_0$, where E_0 is the threshold energy for ionization or excitation. For lower energies the Born approximation consistently overestimates the cross section. For heavy-particle collisions Born's approximation is valid at relative collision velocities large compared with that of the atomic electron involved.

4.04. ELECTRON COLLISIONS

A. Thomson cross section

Thomson classical electron ionization cross section:

$$\sigma_i=4n\left(\frac{E_H}{E_i}\right)^2\frac{E_i}{E}\left(1-\frac{E_i}{E}\right)\pi a_0^2 \quad (\text{cm}^2). \quad (31)$$

n is the number of atomic electrons in the outer shell, E_H the ionization potential of the hydrogen atom, E_i the ionization potential of the atom, and E the incident energy. Classical theory predicts the correct order of magnitude for the cross section. At low energies the Thomson formula overestimates σ by a factor of 5, while at high energies the cross sections increase as $\ln E$. (See Table 4.04.A.1 for experimental data.)

TABLE 4.04.A.1. Electron ionization cross sections of atoms, molecules, and ions. Energy in eV, cross sections in 10^{-16} cm^2. Data from Ref. 8.

Energy	H	H_2	He	He^+	O_2	N_2	N	N^{2+}	N^{4+}
2.0 E1	2.68 E − 1	2.70 E − 1			3.12 E − 1	3.19 E − 1			
4.0 E1	6.38 E − 1	8.59 E − 1	1.71 E − 1		1.50 E0	1.55 E0	9.40 E − 1		
6.0 E1	6.78 E − 1	9.80 E − 1	2.99 E − 1	1.2 E − 2	2.32 E0	2.34 E0	1.36 E0	7.80 E − 2	
8.0 E1	6.38 E − 1	9.70 E − 1	3.41 E − 1	2.3 E − 2	2.62 E0	2.61 E0	1.50 E0	1.41 E − 1	
1.0 E2	5.98 E − 1	9.23 E − 1	3.60 E − 1	3.6 E − 2	2.77 E0	2.69 E0	1.53 E0	1.67 E − 1	1.6 E − 3
2.0 E2	4.22 E − 1	7.21 E − 1	3.38 E − 1	4.5 E − 2	2.59 E0	2.40 E0	1.30 E0	1.69 E − 1	1.2 E − 2
4.0 E2	2.55 E − 1	4.71 E − 1	2.44 E − 1	3.9 E − 2	1.85 E0	1.72 E0	8.93 E − 1	1.26 E − 1	1.5 E − 2
6.0 E2	1.69 E − 1	2.74 E − 1	1.87 E − 1	3.1 E − 2	1.42 E0	1.34 E0	6.10 E − 1	9.70 E − 2	
7.5 E2	1.4 E − 1	2.90 E − 1	1.58 E − 1	2.7 E − 2	1.21 E0	1.15 E0	4.60 E − 1	8.20 E − 2	
1.0 E3	1.2 E − 1[a]	2.24 E − 1	1.22 E − 1	2.2 E − 2	9.76 E − 1	9.36 E − 1			
5.0 E3	2.8 E − 2[a]	5.00 E − 2	3.60 E − 2	6.3 E − 3	2.60 E − 1	2.44 E − 1			
1.0 E4	1.5 E − 2[a]	2.82 E − 2		3.2 E − 3	1.40 E − 1	1.23 E − 1			

[a]Theoretical.

B. Binary-encounter cross section

Refined classical binary-encounter ionization cross section:

$$\sigma_i = \frac{\pi e^4}{E}\left(\frac{1}{E_i} - \frac{1}{E}\right) + \tfrac{2}{3}E_2\left(\frac{1}{E_i^2} - \frac{1}{E_2^2}\right) - \frac{\ln(E/E_i)}{E+E_i}. \quad (32)$$

E is the incident electron energy, E_i the ionization energy of state i, and E_2 the initial energy of the bound electron. (See Table 4.04.A.1 for experimental data.)

C. Gryzinski cross section

Gryzinski's classical electron ionization cross section:

$$\sigma_i \approx 4\xi_i \pi r_0^2 \left(\frac{E_0}{E_i}\right)^2 \frac{1}{U_i} \times \left(\frac{U_i}{U_i+1}\right)^{3/2}\left(\frac{5}{3} \frac{2}{U_i}\right) \quad \text{for } E \geqslant 2E_i \ (\mathrm{cm}^2) \quad (33)$$

$$\approx 4\xi_i \pi a_0^2 \frac{4\sqrt{2}}{3}\left(\frac{E_0}{E_i}\right)^2 \frac{1}{U_i} \times \left(\frac{U_i-1}{U_i+1}\right)^{3/2} \quad \text{for } E \leqslant 2E_i \ (\mathrm{cm}^2). \quad (34)$$

E is the incident electron energy, E_i the ionization energy of state i, E_0 the ionization energy of the H atom, ξ_i the number of equivalent electrons in state i, and $U_i = E/E_i$. (See Table 4.04.A.1 for experimental data.)

D. Lotz cross section

Lotz's approximate electron ionization cross section for positive ions:

$$\sigma_i = 4.5\times 10^{-14} \sum_i \frac{\xi_i}{E_i E} \ln\frac{E}{E_i} \quad (\mathrm{cm}^2). \quad (35)$$

E is the incident electron energy in eV, ξ_i the number of electrons in level i, and E_i the ionization energy of level i in eV. Lotz's formula provides an accurate estimate of the electron ionization of positive ions except in those cases in which the electron excites an inner-shell electron which autoionizes. This process can increase the ionization by a factor of 5–10 for some elements such as are found in the transition series. (See Table 4.04.A.1 for experimental data.)

Lotz's approximate ionization rate coefficient:

$$\alpha_i = 3.0\times 10^{-6} \sum_i \frac{\xi_i}{E_i T_e^{1/2}} E_1 \frac{E_i}{T_e} \quad (\mathrm{cm}^3\,\mathrm{s}^{-1}). \quad (36)$$

T_e is the electron temperature in eV, E_i the ionization energy of level i in eV, ξ_i the number of electrons in level i, and E_1 the exponential integral of order 1.

E. Bethe-Born cross section

Bethe-Born approximation for electron ionization cross section for n, l shell and for hydrogenic wave functions:

$$\sigma_{nl} \simeq \frac{2\pi e^4}{m_e v_0^2} \frac{c_{nl}}{|E_{nl}|} \xi_i(n,l) \ln \frac{2m_e v_0^2}{C_{nl}}. \quad (37)$$

$\xi_i(n,l)$ is the number of electrons with quantum numbers n, l, E_{nl} the energy of electrons in the n, l shell, v_0 the electron impact velocity, and C_{nl} the energy approximately equal to the ionization energy of the n, l shell $|E_{nl}|$. Valid for high electron energy where $E \gg E_{nl}$. (See Table 4.04.A.1 for experimental data.)

F. Seaton's ionization rate coefficient

Seaton's[4] approximate electron ionization rate coefficient:

$$\alpha_i = 2.2\times 10^{-6} T_e^{1/2} \times \sum \xi_i(n,l) E_{nl}^{-2} E^{-E_{nl}/T_e} \quad (\mathrm{cm}^3\,\mathrm{s}^{-1}). \quad (38)$$

T_e is the electron temperature in eV, E_{nl} the binding energy or ionization potential of state nl in eV, and $\xi_i(n,l)$ the number of electrons with quantum numbers nl.

TABLE 4.04.G.1. Electron excitation and emission cross sections of atoms and positive ions:

(1) H(1s)→H(2p) emission cross section, 121.6 nm
(2) He(2 ^1S)→He(2 ^1P) emission cross section, 2058.4 nm
(3) He(1 ^1S)→He(2 ^1P) emission cross section, 58.4 nm
(4) He^+ (2 $^2P^0$)→He^+ (4 ^2D) emission cross section, 121.5 nm (4→2)
(5) C^{3+} (2s $^2S_{1/2}$)→C^{3+} (2p $^2P_{1\ 2,3\ 2}$) excitation cross section, 155 nm
(6) N^{4+} (1$s^2$2s $^2S_{1/2}$)→N^{4+} (1$s^2$2p $^2P_{1\ 2,3\ 2}$) excitation cross section, 124 nm
(7) Ba^+ (6 $^2S_{1/2}$)→Ba^+ (6 $^2P_{1/2}$) excitation cross section, 493.4 nm
(8) Ba^+ (6 $^2S_{1/2}$)→Ba^+ (6 $^2P_{3/2}$) excitation cross section, 455.4 nm
(9) Hg^+ (6s $^2S_{1/2}$)→Hg^+ (6p $^2P_{3/2}$) emission cross section, 165 nm

Energy in eV, cross sections in 10^{-16} cm^2. Data in columns (1)–(3), (7), and (8) from Ref. 9, in (5) from Ref. 10, in (6) from Ref. 11, in (9) from Ref. 12.

	Cross sections								
Energy	(1)	(2)	(3)	(4)	(5)	(6)	(7)	(8)	(9)
3.0 E0							1.7 E1	3.2 E1	
1.0 E1	4.0 E − 2						1.1 E1	1.9 E1	2.0 E1
1.5 E1	5.6 E − 1				4.0 E0	3.3 E0	9.2 E0	1.8 E1	2.2 E1
2.0 E1	6.5 E − 1				3.3 E0	2.2 E0	7.9 E0	1.5 E1	2.2 E1
2.5 E1	7.2 E − 1	1.5 E − 2			2.9 E0	1.7 E0	7.3 E0	1.4 E1	1.9 E1
3.0 E1	7.6 E − 1	3.2 E − 2			2.5 E0	1.4 E0	6.7 E0	1.3 E1	1.7 E1
5.0 E1	7.3 E − 1	9.0 E − 2	9.6 E − 2		1.4 E0	9.0 E − 1	5.1 E0	1.0 E1	1.5 E1
1.0 E2	6.0 E − 1	1.2 E − 1	8.7 E − 2	2.2 E − 1	8.2 E − 1		3.5 E0	7.1 E0	1.4 E1
1.5 E2	4.9 E − 1	1.0 E − 1	7.3 E − 2	2.9 E − 1	7.6 E − 1		2.8 E0	5.3 E0	8.7 E0
2.0 E2	4.2 E − 1	9.3 E − 2	6.5 E − 2	2.8 E − 1	6.9 E − 1		2.4 E0	4.4 E0	7.4 E0
3.0 E2		7.9 E − 2	5.6 E − 2	2.2 E − 1	5.1 E − 1		1.9 E0	3.4 E0	6.0 E0
5.0 E2			4.3 E − 2	1.6 E − 1	3.3 E − 1		1.3 E0	2.3 E0	
7.0 E2			3.5 E − 2	1.2 E − 1			9.2 E − 1	1.8 E0	
1.0 E3			2.9 E − 2	8.2 E − 2					

G. Electron excitation cross section

$$\sigma_{ij} = \frac{8\pi}{\sqrt{3}} \frac{E_0^2}{E} \frac{f_{ij}}{\Delta E_{ji}} \bar{g}(j,i) \pi a_0^2. \qquad (39)$$

E_0 is the binding energy of the electron in the ground state of the H atom, f_{ij} the oscillator strength, ΔE_{ji} the difference in binding energy of state j and i, and $\bar{g}(j,i)$ the average Gaunt factor (~ 1 for $\Delta n = 0$ transitions). (See Table 4.04.G.1 for experimental data.)

H. Electron excitation rate coefficient

For Maxwellian distribution for atoms:

$$\alpha_{ij} = \frac{1.6\times 10^{-5} f_{ij} \bar{g}(i,j)}{\Delta E_{ji} T_e^{1/2}} \times \exp\left(-\frac{\Delta E_{ji}}{T_e}\right) \quad (\text{cm}^3\ \text{s}^{-1}). \qquad (40)$$

For ions, replace 1.6×10^{-5} with 6×10^{-6}. T_e is the electron temperature in eV, ΔE_{ji} the difference in binding energy between state j and i, f_{ij} the oscillator strength, and $\bar{g}(i,j)$ the effective Gaunt factor (~ 1 for $\Delta n = 0$).

I. Cross section and collision strength

Relation between cross section and collision strength:

$$\Omega_{ij}(x) = \omega_i E \sigma_{ij}(E). \qquad (41)$$

$\sigma_{ij}(e)$ is the cross section in units of πa_0^2, x the scaled electron energy, $E/\Delta E_{ij}$, E the incident electron energy, ΔE_{ij} the threshold energy, and ω_i the statistical weight of initial state i, $\omega_i = (2L_i + 1)(2J_i + 1)$.

J. Recombination of electrons and ions in a plasma

Radiative recombination rate coefficient: for $T_e > 0.05$ keV,

$$\alpha_r = 3.5\times 10^{-15}(Z+1)^2 I_Z^{1/2}/T_e \quad (\text{cm}^3\ \text{s}^{-1}); \qquad (42)$$

for $T_e < 0.05$ keV,

$$\alpha_r = 2.8\times 10^{-13} I_Z n_0 g/T_e^{1/2} \quad (\text{cm}^3\ \text{s}^{-1}). \qquad (43)$$

Z is the charge of the ion after recombining, I_Z the ionization energy in keV of the ion of charge Z, T_e the electron temperature in keV, n_0 the ground-state principal quantum number of the ion of charge Z, and $g \simeq 4$ for Fe and Ni and $\simeq 3$ for lighter masses.

Dielectronic recombination rate coefficient: Burgess[5] has given a semiempirical formula for dielectronic recombination,

$$\alpha_d = \frac{7.6\times 10^{-14}}{T_e^{3/2}} B(Z) \times \sum_j f(i,j) A(X) e^{-\bar{E}/T_e} \quad (\text{cm}^3\ \text{s}^{-1}). \qquad (44)$$

T_e is the electron temperature in keV,

$$B(Z) = \frac{Z^{1/2}(Z+1)^{5/2}}{(Z^2 + 13.4)^{1/2}},$$

$$A(X)=\frac{X^{1/2}}{1+0.105X+0.015X^2},\quad X=(Z+1)\epsilon_{ij}$$

$$\epsilon_{ij}=1/\nu_i^2-1/\nu_j^2,$$

$$\bar{E}/T_e=\frac{0.0136(Z+1)^2E_{ij}}{1+0.015Z^3(Z+1)^{-2}T_e},$$

$f(i,j)$ is the oscillator strength of the transition $i\to j$, ν_i and ν_j are principal quantum numbers of states i and j of the recombining ion, and T_e the electron temperature in keV. For the reaction $e+C^{2+}(2s^2\,{}^1S)$ Burgess's formula yields a dielectronic rate coefficient of 5.3×10^{-11} $cm^3\,s^{-1}$ at $T_e=10^5$ K.

Three-body recombination rate coefficient $(e+e+X^{z+})$:

$$\alpha_3=5.6\times10^{-27}T_e^{-9/2}\quad (cm^6\,s^{-1}). \qquad (45)$$

T_e is the electron temperature in eV.

Three-body recombination rate coefficient $(X^{Z+}+A+e)$:

$$\alpha_3=8.75\times10^{-27}Z^3T_e^{-9/2}\quad (cm^6\,s^{-1}). \qquad (46)$$

T_e is the electron temperature in eV, and Z the charge of recombining ion.

K. Power radiated from collisions of electrons and ions in a plasma

Bremsstrahlung (free-free) radiation:

$$P_b=1.53\times10^{-32}n_en_ZZ_{\rm eff}^2T_e^{1/2}\quad (W/cm^3). \qquad (47)$$

n_e and n_Z are densities of electrons and recombining ions in cm^{-3}, Z^2 the effective charge of the recombining ion, and T_e the electron temperature in eV.

Radiative recombination (free-bound):

$$P_r=1.6\times10^{-19}n_en_Z\alpha_r(I_{Z-1}+E_Z)\quad (W/cm^3). \qquad (48)$$

n_e and n_Z are electron and ion densities in cm^{-3}, α_r is the radiative recombination rate coefficient given in Eqs. (42) and (43), I_{Z-1} the ionization energy in eV of the ion after recombining, and E_Z the energy of excitation in eV of resonant line of ion Z.

Dielectronic recombination:

$$P_d=1.6\times10^{-19}n_en_Z\frac{B(Z)}{T_e^{3/2}}\sum_j A(Z,j)\times e^{-E_{Zj}/T_e}(I_{Z-1}+E_{Zj})\quad (W/cm^3). \qquad (49)$$

$B(Z)$ and $A(Z)$ are defined in Eq. (44), summation over j is over all transitions of ion Z, n_e and n_Z are densities of electrons and ions in cm^{-3}, T_e is the electron temperature in eV, and E_{Zj} is the energy of excitation in eV for level j.

4.05. HEAVY-PARTICLE COLLISIONS

A. Electron capture cross section

Brinkman-Kramer:

$$\sigma^{BK}=(2^{18}/5)\pi a_0^2Z_1^5Z_2^5(v_0/v)^{12}\quad (cm^2). \qquad (50)$$

σ^{BK} is the Brinkman-Kramer electron capture cross section, v_0 the Bohr electron velocity in the H atom, Z_1 and Z_2 are nuclear charges of incident and target particles, and v is the velocity of the incident particle. The formula is valid only at high energies. For the reaction $H^++H(1s)\to H+H^+$ at 100 keV, the calculated value is $\sim5.5\times10^{-17}$ cm^2. (See Tables 4.05.A.1 and 4.05.A.2 for experimental data.)

TABLE 4.05.A.1. Symmetric electron capture cross sections of singly charged ions:

(1) $H^++H\to H+H^+$ (3) $O^++O\to O+O^+$ (5) $Ar^++Ar\to Ar+Ar^+$
(2) $He^++He\to He+He^+$ (4) $Ne^++Ne\to Ne+Ne^+$ (6) $H_2^++H_2\to H_2+H_2^+$

Symmetric charge changing collisions are characterized by the cross sections having maximum values as the incident particle energy approaches zero. Energy in eV, cross sections in 10^{-16} cm^2. Data from Ref. 8 and an evaluation of values found in literature.

	Cross sections					
Energy	(1)	(2)	(3)	(4)	(5)	(6)
1.0 E − 1				4.5 E1	5.1 E1	
2.0 E − 1				4.3 E1	5.0 E1	
4.0 E − 1		2.8 E1		3.9 E1	5.0 E1	
1.0 E0		2.6 E1		3.7 E1	4.7 E1	
2.0 E0	5.6 E1	2.5 E1		3.3 E1	4.5 E1	
4.0 E0	5.3 E1	2.3 E1		3.1 E1	4.2 E1	
1.0 E1	4.8 E1	2.0 E1		2.6 E1	4.1 E1	1.1 E1
4.0 E1	3.8 E1	1.7 E1		2.1 E1	3.6 E1	9.4 E0
1.0 E2	3.2 E1	1.6 E1	2.3 E1	1.7 E1	3.2 E1	8.5 E0
4.0 E2	2.2 E1	1.2 E1	1.8 E1	1.3 E1	2.9 E1	7.3 E0
1.0 E3	1.7 E1	1.0 E1	1.6 E1	1.1 E1	2.6 E1	6.6 E0
4.0 E3	1.2 E1	7.2 E0	1.4 E1	1.0 E1	2.2 E1	5.6 E0
1.0 E4	8.4 E0	5.6 E0	1.2 E1	8.5 E0	1.9 E1	4.6 E0
4.0 E4	2.0 E0	3.5 E0	1.0 E1	5.0 E0	1.5 E1	2.1 E0
1.0 E5	1.5 E − 1	2.0 E0	8.4 E0	3.1 E0	1.0 E1	4.5 E − 1
4.0 E5		2.5 E − 1	4.5 E0	1.8 E0	3.1 E0	
1.0 E6		2.4 E − 2	1.9 E0	5.9 E − 1	1.7 E0	

TABLE 4.05.A.2. Unsymmetric electron capture cross sections of ions in gases:
(1) $H^+ + H_2 \rightarrow H^0$ (2) $H^+ + H_2 \rightarrow H^-$ (3) $He^{++} + H_2 \rightarrow He^+$ (4) $He^{++} + He \rightarrow He^+$ (5) $O^+ + H \rightarrow O$ (6) $O^+ + H_2 \rightarrow O$ (7) $O^{3+} + H \rightarrow O^{2+}$ (8) $O^{3+} + H_2 \rightarrow O^{2+}$ (9) $O^{6+} + H \rightarrow O^{5+}$ (10) $O^{6+} + H_2 \rightarrow O^{5+}$
In general, the energy at which the cross section is a maximum is given approximately by the Massey adiabatic criteria. Capture by multicharged ions is predominantly into excited electronic states. Energy in eV/amu (H = 1), cross sections in 10^{-16} cm^2. Data in columns (1)–(6) from Ref. 8, in (7)–(10) from Ref. 13.

	Cross sections									
Energy	(1)	(2)	(3)	(4)	(5)	(6)	(7)	(8)	(9)	(10)
2.0 E0					1.4 E1					
5.0 E0					1.3 E1					
1.0 E1					1.2 E1	7.8 E0				
5.0 E1	2.7 E − 1				1.0 E1	4.7 E0				
1.0 E2	4.1 E − 1				9.5 E0	3.5 E0				
2.0 E2	7.5 E − 1	1.2 E − 4	4.9 E − 1	1.4 E − 1	9.0 E0	4.1 E0				
5.0 E2	2.3 E0	4.4 E − 4	1.2 E0	2.1 E − 1	8.0 E0	7.7 E0				
1.0 E3	5.2 E0	1.2 E − 3	1.6 E0	2.8 E − 1	6.1 E0	1.1 E1		8.0 E0		
2.0 E3	8.5 E0	3.6 E − 3	1.9 E0	4.1 E − 1	5.0 E0	1.1 E1		8.2 E0	3.2 E1	3.6 E1
5.0 E3	8.6 E0	1.1 E − 2	4.2 E0	7.4 E − 1	3.9 E0	7.5 E0	2.1 E1	1.2 E1	3.8 E1	3.5 E1
1.0 E4	8.6 E0	4.7 E − 2	8.5 E0	1.4 E0	2.6 E0	4.8 E0	1.9 E1	1.4 E1	3.8 E1	3.3 E1
5.0 E4	1.7 E0	6.0 E − 3	3.6 E0	2.4 E0	6.2 E − 1	1.2 E0	6.7 E0	8.4 E0	2.2 E1	1.9 E1
1.0 E5	2.8 E − 1	1.0 E − 4	9.5 E − 1	9.7 E − 1			1.1 E0	2.3 E0	8.0 E0	1.0 E1
2.0 E5	1.8 E − 1		1.1 E − 1	3.0 E − 1					1.0 E0	7.3 E − 1

First-Born-approximation-symmetric collisions ($A^+ + A \rightarrow A + A^+$):

$$\sigma^{BK} \simeq 0.661 + 5.33(m/M)^2(\hbar v/e^2)^{-6}\pi a_0^2 \quad (\text{cm}^2). \tag{51}$$

σ^{BK} is the Brinkman-Kramer cross section, m the electron mass, M the mass of the heavy particle, and v the velocity of the incident ion. The first Born approximation should be used only at high energies. For the reaction $H^+ + H(1s) \rightarrow H + H^+$ at 100 keV the calculated value is 1.08×10^{-17}. (See Table 4.05.A.1 for experimental data.)

B. Electron stripping cross section—Bohr classical

$$\sigma_s = 4\pi a_0^2 Z_1^{-2}(Z_2^2 + Z_2)(v_0/v)^2 \quad (\text{cm}^2). \tag{52}$$

v_0 is the Bohr electron velocity in the H atom, v the incident particle velocity, Z_1 the nuclear charge of the incident particle, and Z_2 the nuclear charge of the target particle (applicable to high incident particle velocities).

C. Ionization cross section—Firsov

$$\sigma_i = 39(Z_1 + Z_2)^{-2/3}(v/v_0 - 1)^2 \pi a_0^2 \quad (\text{cm}^2). \tag{53}$$

Z_1 and Z_2 are the nuclear charges of the incident particle and target atom, v is the impact velocity, and

$$1/v_0 = 0.35(Z_1 + Z_2)^{5/3}(\hbar/a_0)E_i,$$

where E_i is the binding energy of the electron in the target atom (formulation useful at small collision energies).

D. Reaction cross section

Ion-molecule or atom-molecule interchange reaction cross section (orbiting or Langevin cross section):

$$\sigma = (2\pi e/v_{12})(\alpha/M_r)^{1/2} \quad (\text{cm}^2). \tag{54}$$

v_{12} is the relative velocity, α the polarizability, and M_r the reduced mass. Orbiting occurs when impact parameter P_0 is less than or equal to $(2\alpha e^2/E)^{1/4}$, where E is the kinetic energy of relative motion.

The Langevin formula for the rate coefficient is

$$\alpha(T) = (\alpha e^2/M_r)^{1/2} \quad (\text{cm}^3\ \text{s}^{-1}).$$

(See Tables 4.05.D.1 and 4.05.D.2 for experimental values of reaction rate coefficients.)

E. Arrhenius reaction rate coefficient

For two-body reaction:

$$\alpha = P\sigma_g(8kT/\pi M_r)^{1/2} e^{-\xi_a/kT}. \tag{55}$$

P is the steric factor (~ 1), σ_g the geometric cross section, M_r the reduced mass, ξ_a the difference in internal

TABLE 4.05.D.1. Reaction rate coefficients for ion-molecule exchange reactions. Temperature or energy range in K, rate coefficient α in cm^3 s^{-1}. Data From Ref. 14.

Reaction	α	Energy range	Error (%)
$C^+ + O_2 \rightarrow CO^+ + O$ $\rightarrow O^+ + CO$	1.1×10^{-9}	300	± 30
$H^+ + D_2 \rightarrow D^+ + DH$	3.7×10^{-10}	80–278	± 50
$H_3^+ + N_2 \rightarrow HN_2^+ + H_2$	1.9×10^{-9}	464–8120	± 30
$N^+ + O_2 \rightarrow NO^+ + O$	6.1×10^{-10}	300–870	± 30
$N_2^+ + H_2 \rightarrow HN_2^+ + H$	2.1×10^{-9}	300	± 20
$N_2^+ + O \rightarrow NO^+ + N$	1.4×10^{-10}	295	$+100$ -50
$O^+ + H_2 \rightarrow HO^+ + H$	1.7×10^{-9}	300	± 20
$O^+ + N_2 \rightarrow NO^+ + N$	1.2×10^{-12}	300	± 20
$O_2^+ + N \rightarrow NO^+ + O$	1.2×10^{-10}	296	± 50
$O_2^+ + N_2 \rightarrow NO^+ + NO$	$\leqslant 1 \times 10^{-15}$	300	
$D^- + D_2O \rightarrow DO^- + D_2$	2×10^{-10}	300	± 50
$HO^- + H_2 \rightarrow H^- + H_2O$	$\leqslant 5 \times 10^{-12}$	300	
$O^- + N_2O \rightarrow NO^- + NO$	2.2×10^{-10}	278–475	± 20

TABLE 4.05.D.2 Reaction rate coefficients for atom-molecule exchange reactions. Temperature in K, reaction rate coefficient α in $cm^3 s^{-1}$. Data from Ref. 14.

Reaction	Temperature	α
$H+O_3 \rightarrow HO+O_2$	300	2.6×10^{-11}
$H+H_2O \rightarrow H_2+OH$	298	1.7×10^{-25}
$H+NO_2 \rightarrow HO+NO$	300	4.8×10^{-11}
$H+CO_2 \rightarrow CO+HO$	298	3.7×10^{-19}
$HO+H_2 \rightarrow H_2O+H$	298	7.4×10^{-15}
$HO+CO \rightarrow CO_2+H$	298	1.6×10^{-13}
$H_2O+O \rightarrow 2HO$	300	$<1\times10^{-20}$
$NO_2+O_3 \rightarrow NO_3+O_3$	298	5×10^{-17}
$O(^1S)+O_3 \rightarrow$ all products	298	5.8×10^{-10}
$O(^1D)+O_3 \rightarrow O_2+O_2$		
$\rightarrow O_2+2O(^3P)$	298	2.7×10^{-10}
$O(^3P)+O_3 \rightarrow O_2+O_2$	298	8.4×10^{-15}
$O(^1D)+H_2 \rightarrow HO+H$	298	2.9×10^{-10}
$O(^3P)+H_2 \rightarrow HO+H$	298	2.6×10^{-18}
$O_3+NO \rightarrow O_2+NO_2$	298	1.5×10^{-14}

energy of initial and post reaction molecules, and kT the gas temperature.

4.06. PHOTON COLLISIONS

A. Thomson scattering cross section

$$\sigma_{\text{Th}} = (8\pi/3)r_0^2 = 6.65\times10^{-25}\ \text{cm}^2. \tag{56}$$

r_0 is the electron radius.

B. Rayleigh scattering cross section

$$\sigma_R = \frac{8\pi^3}{3n^3}\frac{(\mu^2-1)^2}{\lambda^4}\ \text{cm}^2. \tag{57}$$

λ is the photon wavelength and μ the index of refraction at gas density n.

C. Photoionization recombination cross sections

Relation between photoionization and electron radiative recombination cross sections:

$$\sigma_i^E = m^2v^2c^2\sigma_E^i/2h^2\nu^2. \tag{58}$$

σ_i^E is the photoionization cross section of an atom in state i to produce a free electron of energy E, σ_E^i the cross section from a continuous state of energy E to a discrete state i, ν the photon frequency, v the velocity of the free electron, and m the electron mass.

D. Photoionization cross section for H-like ion

$$\sigma_i = [8\pi\alpha a_0^2 Z^4/3(\sqrt{3})n^5\omega^3]g. \tag{59}$$

g is the Kramers-Gaunt factor, which is ~ 1 at low energies and falls off toward 0 as $\omega^{-1/2}$ at high energies, α is the fine-structure constant, Z the charge of the H-like ion, n the quantum number average over l substates, and ω the photon energy.

E. Synchrotron radiation

$$\lambda_p = 23.48\times10^8\ (R/E^3). \tag{60}$$

λ_p is the wavelength in Å corresponding to the peak of the continuum produced by monoenergetic electrons, R the radius of the electron orbit in meters, and E the energy in MeV.

4.07. PARTICLE TRANSPORT COLLISIONS

The following formulas for particle transport are derivable from elementary kinetic theory and are accurate to only a factor of 2–3. For accurate results, more elaborate equations are required involving multiple integrals which usually are integrated numerically. See Refs. 6 and 7 for heavy-particle and electron transport.

A. Diffusion or momentum transfer cross section

$$\sigma_m = 2\pi\int_0^{\pi}\sigma_S(\theta,v)(1-\cos\theta)\sin\theta\ d\theta. \tag{61}$$

$\sigma_S(\theta,v)$ is the differential scattering cross section and θ the scattering angle.

Average momentum transfer cross section for electron-ion collision:

$$\langle\sigma_m\rangle = \frac{\pi}{2}\frac{eZ}{T_e}\ln\frac{3T_e}{2e^2Zn_i^{1/3}}. \tag{62}$$

T_e is the electron temperature, Z the nuclear charge, and n_i the ion density. (See Table 4.07.A.1 for experimental data.)

B. Collision frequency for momentum transfer

$$\nu_m = n\bar{v}\sigma_m. \tag{63}$$

$\bar{v}$ is the mean velocity of particles, N the gas number density, and σ_m is defined by Eq. (61).

Average ion energy of ion drifting in electric field:

$$\tfrac{1}{2}mv_i^2 = \tfrac{1}{2}mv_d^2 + \tfrac{1}{2}Mv_d^2 + \tfrac{3}{2}kT. \tag{64}$$

m and M are the ionic and molecular masses, v_i^2 the mean square of the total ion velocity, and v_d the drift velocity. The first term on the right is the field energy associated with the drift velocity, the second term is the random part of the field energy, and the third term is the thermal energy.

TABLE 4.07.A.1. Electron momentum transfer cross sections in gases. Cross sections in 10^{-16} cm^2, energy in eV. Data from Ref. 15.

	Cross sections					
Energy	H	H_2	He	O_2	CO_2	Ar
1.0 E − 2	41	7.3	5.3	0.85	168	4.2
2.0 E − 2	42	8.0	5.4	1.4	118	3.2
4.0 E − 2	42	8.9	5.6	1.9	88	1.9
1.0 E − 1	41	10.5	5.9	2.7	52	0.58
2.0 E − 1	38	12.0	6.2	3.2	30	0.095
4.0 E − 1	34	13.9	6.5	4.7	13	0.33
1.0 E0	25	17.5	7.0	7.8	5.7	1.4
2.0 E0	18	18.0	7.1	6.9	5.3	2.8
4.0 E0	13	16.2	6.6	5.4	16	5.2
1.0 E1	3.5	9.2	4.7	8.0	12	13.8

TABLE 4.07.C.1. Reduced mobilities and drift velocities of positive and negative ions in gases. E/N in Td (10^{-17} V cm^2), mobilities K_0 in cm^2/V s, drift velocities v_d in 10^4 cm/s. Data from Ref. 16.

	$H^+ + H_2$		$He^+ + He$		$N^+ + N_2$		$N_2^+ + N_2$		$K^+ + N_2$		$O^- + O_2$	
E/N	K_0	v_d	K_0	v_d	K_0	v_d	K_0	v_d	K_0	v_d	K_0	v_d
4	16.0	1.72							2.53	0.272	3.2	0.344
10	16.0	4.30	10.2	2.74	3.01	0.809	1.88	0.505	2.54	0.682	3.2	0.860
50	14.9	20.0	8.97	12.1	2.92	3.92	1.76	2.36	2.59	3.48	3.81	5.12
100	13.4	36.0	7.67	20.6	2.88	7.74	1.60	4.30	2.76	7.42	4.73	12.7
150	13.1	52.8	6.78	27.3	2.87	11.6	1.47	5.92	2.95	11.9	4.7	18.9
200	13.1	70.4	6.12	32.9	2.86	15.4	1.37	7.36	3.04	16.3	4.41	23.7
250	13.2	88.7	5.60	37.6	2.86	19.2	1.28	8.60	3.05	20.5		
300	13.3	107	5.19	41.8	3.00	23.1	1.20	9.67	3.00	24.2		
400	13.7	147	4.58	49.2	2.83	30.4	1.10	11.8	2.86	30.7		
500			4.17	56.0	2.79	37.5	1.02	13.7	2.71	36.4		
600			3.81	61.4			0.95	15.3	2.56	41.3		

C. Ion drift velocity

$$v_d = \mathscr{E} e\lambda / m\bar{v}_i. \tag{65}$$

$\mathscr{E}$ is the electric field, λ the ion mean free path, $\bar{v}_i$ the ion mean velocity, and m the ionic mass. (See Table 4.07.C.1 for experimental data.)

D. Mutual-diffusion coefficient

$$D_{12} = \tfrac{1}{3}\bar{v}_i\lambda. \tag{66}$$

$\bar{v}_i$ is the mean ion velocity and λ the ion mean free path.

E. Langevin diffusion coefficient for elastic spheres

$$D_{12} = (1/4d_{12}N_2)(2kT/\pi M_r)^{1/2}. \tag{67}$$

d_{12} is the sum of the radii of spherical molecules of m_1 and m_2, M_r the reduced mass, N_2 the density of molecules of mass m_2, and T the gas temperature.

F. Ion-diffusion coefficient in $E \times H$ field

$$D_m = \frac{D_{12}}{1 + \omega^2/\nu_c^2}. \tag{68}$$

D_{12} is the diffusion coefficient in the absence of magnetic (H) field, ω the angular frequency of the rotation of the ion (He/m), and ν_c the average collision frequency.

G. Ion diffusion and mobility

Relation between ion-diffusion coefficient and mobility at thermal equilibrium:

$$K/D_{12} = 1.16 \times 10^4/T. \tag{69}$$

T is the gas temperature in kelvin, K the mobility in cm^2/V s, and D_{12} the diffusion coefficient in cm^2/s.

H. Ion mobility

$$K = e\lambda/m\bar{v}. \tag{70}$$

m is the mass of the ion, λ the mean path of the ion, and $\bar{v}$ the mean velocity. If an ion of mass m is drifting in a molecular gas of mass M, the mobility becomes

$$K = 0.815 \frac{e}{M v_R \pi N D_{12}^2}\left(1 + \frac{M}{m}\right)^{1/2}. \tag{71}$$

The mobility and drift velocity relationship may be expressed with the use of Eq. (65):

$$v_d = K\mathscr{E}. \tag{72}$$

$\mathscr{E}$ is the electric field in V/cm and K the mobility in cm^2/V s.

Einstein relationship. At low values of E/p, the mobility is related to the ion-atom mutual diffusion coefficient D_{12} by

$$K = eD_{12}/kT. \tag{73}$$

T is the atomic or molecular gas temperature.

Reduced ion mobility. The measured mobility is usually converted to a reduced mobility K_0:

$$K_0 = K(p/760)(273/T). \tag{74}$$

The relationship between E/p and E/N is given by

$$E/N = 2.828E/p_0 = (1.035 \times 10^{-2} T)E/p, \tag{75}$$

where E/N is in 10^{-17} V cm^2, T in K, and E/p_0 in V/cm Torr. E/N is sometimes expressed in townsend, or Td, where 1 Td $= 10^{-17}$ V cm^2. (See Table 4.07.C.1 for experimental data.)

I. Ambipolar diffusion

$$D_A = (D^+K^- + D^-K^+)/(K^+ + K^-). \tag{76}$$

K^+ and K^- are the mobilities of ions and electrons, and D^+ and D^- the diffusion coefficients of ions and electrons:

$$D_A = (kT_e/e)K^+, \quad T_e \gg T_i \tag{77}$$

$$= (2kT/e)K^+, \quad T_e = T_i = T \tag{78}$$

where T_e, T_i, and T are the temperatures of electrons, ions, and gases, respectively.

J. Electron drift velocity in $\mathscr{E}$ field

$$W=\frac{e}{3mN}\overline{\left[c^{-2}\frac{d}{dc}\left(\frac{c^2}{A}\right)\right]}\frac{\mathscr{E}}{P}. \tag{79}$$

m is the electron mass, N the number of molecules in unit volume of the gas at unit pressure ($T=288$ K, $P=1$ mm Hg, $N=3.35\times10^{16}$ cm^{-3}), c the electron velocity of random motion, A the collision cross section, $\mathscr{E}$ the electric field, and P the gas pressure.

K. Electron diffusion coefficient

$$D_e=\frac{1}{3NP}\overline{\left(\frac{c}{A}\right)}. \tag{80}$$

N is the number of molecules in unit volume at unit pressure ($T=288$ K, $P=1$ mm Hg, $N=3.35\times10^{16}$ cm^{-3}), $\overline{c}$ the electron velocity of random motion, and A the collision cross section.

4.08. PARTICLE PENETRATION IN MATTER

A. Stopping cross section

$$\sigma_n=\frac{4\pi aZ_1Z_2e^2M_1}{M_1+M_2}s_n \quad (\text{eV cm}^2/10^{15}\text{ atoms}). \tag{81}$$

Z_1 and Z_2 are the nuclear charges of incident M_1 and target atom M_2, and s_n is the reduced elastic stopping cross section:

$$s_n=\frac{3.4\epsilon^{1/2}\ln(\epsilon+2.718)}{1+6.355\epsilon^{1/2}+\epsilon(6.9\epsilon^{1/2}-1.7)}, \tag{82}$$

with ϵ the reduced energy,

$$\epsilon=aEM/Z_1Z_2e^2(M_1+M_2), \tag{83}$$

where a is the Thomas-Fermi in Å and E the incident energy in eV,

$$a=0.885a_0/(Z_1^{2/3}+Z_2^{2/3})^{1/2}, \tag{84}$$

where a_0 is the Bohr radius in Å.

B. Cross section to deflect a particle through 90°

$$\sigma_{90^\circ}=Z_1^2Z_2^2e^4\pi/4E_1^2. \tag{85}$$

Z_1 and Z_2 are the nuclear charges of projectile and target atom, and E_1 is the total energy of the particle.

C. Multiple scattering cross section through 90°

$$\sigma_{90^\circ}^M\simeq8\pi\left(\frac{e^2Z_1Z_2}{M_rv_{12}^2}\right)^2\ln\frac{\rho_{\max}^2}{\rho_{\min}^2}. \tag{86}$$

$\rho_{\min}$ is the impact parameter,

$$\rho_{\min}=2e^2/M_rv_{12}^2, \tag{87}$$

and $\rho_{\max}$ the maximum or cutoff impact parameter,

$$\rho_{\max}=(kT/4\pi n_e^2)^{1/2}, \tag{88}$$

where M_r is the reduced mass, v_{12} the relative velocity between particles 1 and 2, and kT the Maxwellian temperature.

4.09. REFERENCES

[1]N. F. Mott and H. S. W. Massey, *The Theory of Atomic Collisions* (Oxford University, London, 1949).

[2]*Electronic and Ionic Impact Phenomena* (Oxford University, London): Vol. 1, *Electron Collisions with Atoms,* H. S. W. Massey and E. H. S. Burhop (1969); Vol. 2, *Electron Collisions with Molecules—Photoionization,* H. S. W. Massey (1969); Vol. 3, *Slow Collisions of Heavy Particles,* H. S. W. Massey *et al.* (1971); Vol. 4, *Recombination and Fast Collisions of Heavy Particles,* H. S. W. Massey *et al.* (1974).

[3]E. W. McDaniel, *Collision Phenomena in Ionized Gases* (Wiley, New York, 1964).

[4]M. J. Seaton, Planet Space Sci. **12**, 55 (1964).

[5]A. Burgess, Astrophys. J. **141**, 1588 (1965).

[6]E. W. McDaniel and E. A. Mason, *The Mobility and Diffusion of Ions in Gases* (Wiley, New York, 1973).

[7]R. W. Crompton and L. G. H. Huxley, *The Diffusion and Drift of Electrons in Gases* (Wiley, New York, 1974).

[8]C. F. Barnett *et al.*, "Atomic Data for Controlled Fusion Research," ORNL Report ORNL-5206 and 5207, 1977.

[9]L. J. Kieffer, At. Data **1**, 121 (1969).

[10]P. O. Taylor, D. Gregory, G. H. Dunn, R. A. Phaneuf, and D. H. Crandall, Phys. Rev. Lett. **39**, 1256 (1977).

[11]D. Gregory, G. H. Dunn, R. A. Phaneuf, and D. H. Crandall, Phys. Rev. A **20**, 410 (1979).

[12]D. H. Crandall, R. A. Phaneuf, and G. H. Dunn, Phys. Rev. A **11**, 1223 (1975).

[13]D. L. Albritton, At. Data Nucl. Data Tables **22**, 1 (1978); see also E. L. Ferguson, *ibid.* **12**, 159 (1973).

[14]L. G. Anderson, Rev. Geophys. Space Phys. **14**, 151 (1976).

[15]Y. Itikawa, At. Data Nucl. Data Tables **21**, 69 (1975); **14**, 1 (1974).

[16]H. W. Ellis, R. Y. Pai, E. W. McDaniel, E. A. Mason, and L. A. Viehland, At. Data Nucl. Data Tables **17**, 177 (1976).

5.00. Atomic spectroscopy

WOLFGANG L. WIESE AND GEORGIA A. MARTIN

National Institute of Standards and Technology

CONTENTS

5.01. DESCRIPTION OF SPECTRA, SPECTROSCOPIC NOTATION, AND ELECTRON COUPLING SCHEMES

A. One-electron spectra (hydrogen, alkalis)

Spectral lines are grouped in series, with decreasing line spacings towards series limit.

(1) *Atomic states* are described by a combination of quantum numbers:

n = principal ($n = 1,2,\ldots,\infty$),

l = orbital angular momentum ($l = 0,1,\ldots,n-1$), with customary letter designations $0 = s$, $1 = p$, $2 = d$, $3 = f$, and so on in alphabetical order (j not used),

s = spin angular momentum ($s = \frac{1}{2}$),

j = total angular momentum ($j = |\mathbf{l} + \mathbf{s}| = l \pm \frac{1}{2} \geqslant \frac{1}{2}$).

(2) *Electron transitions* are described by listing the quantum description of the lower atomic state first, followed by the upper state. Often the description is confined to giving only n and l. Thus a *typical spectral series* is $2s$-np ($n = 2,3,\ldots,\infty$)—principal series for Li-like species, where the transitions are from the fixed lower state $2s$ to upper states np.

(3) *Hydrogen* is a special case because atomic states with same n but different l coincide (l degeneracy). Principal series are:

Lyman series→1–n ($n = 2,3,\ldots$) (L_α line→1–2, L_β →1–3,...),

Balmer series→2–n ($n = 3,4,\ldots$) (H_α→2–3, H_β→2–4,...),

Paschen series→3–n ($n = 4,5,\ldots$) (P_α→3–4, P_β →3–5,...).

B. Many-electron spectra

(1) *Typical feature* is appearance of multiplets, i.e., small groups of closely spaced lines.

(2) General *quantum designations* characteristic of atomic states (in the absence of external fields) regardless of electron coupling scheme:

(a) *Parity*, determined by $\Sigma_i l_i$:

$$(-1)^{\Sigma_i l_i} = \begin{cases} +1 & \text{for even parity} \\ -1 & \text{for odd parity.} \end{cases}$$

(b) *Total angular momentum quantum number J:* resultant of vectorial coupling (see below) of single-electron spin (s) and orbital (l) angular momentum quantum numbers.

(3) *Coupling* between outer (valence) electrons occurs in several ways. For the prominent lower atomic states of most elements the *Russell-Saunders* or *LS-coupling* scheme is a good approximation. According to this most important scheme:

(a) The orbital angular momenta of the valence electrons couple among themselves vectorially: $L = |\Sigma \mathbf{l}|$.

(b) The spin momenta couple among themselves: $S = |\Sigma \mathbf{s}| \leqslant \frac{1}{2} m_0$, where m_0 is the number of electrons in unfilled subshells [Sec. 5.01.B(5)].

(c) Spin-orbit interactions (magnetic interactions) are weak (ideally: negligible).

(d) Resulting orbital angular momentum $\mathbf{L}$ and resulting spin $\mathbf{S}$ form vectorially the total angular momenta: $\mathbf{J} = \mathbf{L} + \mathbf{S}$, with $|L - S| \leqslant J \leqslant |L + S|$.

(4) *Customary designations* for L are letters (same as for l above), such that if L→0, 1, 2, 3, 4, 5, 6, 7, 8,..., the designation for L is L→$S, P, D, F, G, H, I, K, L$, and so on in alphabetical order.

(5) *Sample* spectroscopic designation of an atomic energy level in LS coupling:

$(1s^2 2s^2 2p^6 3s^2)3p^5\ {}^2P^\circ_{3/2}$ (ground state of chlorine).

Interpretation: Closed electron subshells (in sample: $1s^2 2s^2 2p^6 3s^2$) are usually not given, and are therefore listed here in parentheses ($1s^2 = 2$ electrons in filled $1s$ subshell, etc.). $3p^5$ indicates 5 equivalent electrons (i.e., having the same values of n and l) with principal quantum number $n = 3$ and orbital angular momentum quantum number $l = 1$ (p) in the unfilled $3p$ subshell; these form by vector addition a resulting state of $L = 1$ (P); superscript 2 indicates the multiplicity, $2S + 1$ (in this case $S = |\Sigma \mathbf{s}| = 1/2$); subscript 3/2 indicates the total angular momentum quantum number J (either 1/2 or 3/2 is possible for the given L and S). Finally, superscript ° (for "odd") indicates level of odd parity [Sec. 5.01.B(2)]. (No symbol when parity is even.)

Note that the Pauli exclusion principle restricts the possible values of L and S in configurations involving two or more equivalent electrons (see, e.g., Ref. 1).

(6) *Hierarchies*. Groups of atomic states and transitions, specified by the indicated sets of quantum numbers, have the designations listed in Table 5.01.B.1. These constitute a hierarchy starting with the largest, all-inclusive entity at the left.

(7) *Other coupling schemes.* For heavier atoms,

TABLE 5.01.B.1. Designations for atomic states and transitions

Atomic entity→	Configuration	Polyad	Term	Level[a]	State[b]
Required quantum number specifications[c]	n,l	n,l,S	n,l,L,S	n,l,L,S,J	n,l,L,S,J,M
Corresponding transition	Transition array	Supermultiplet	Multiplet	Spectral line	Component of line

[a]Splittings of terms into levels (and thus of multiplets into lines) constitute fine structure.
[b]Splittings of levels into states (and thus of lines into components) occur only in magnetic fields; these are identified by magnetic quantum numbers M ($|M| \leqslant J$).
[c]Designations are usually confined to n and l of the electrons in the unfilled subshell(s) (valence electrons).

TABLE 5.01.B.2. Ground state configurations, ground levels, and ionization energies of neutral atoms

Atomic number	Element	Ground state configuration[a]	Ground level	Ionization energy (eV)
1	H	$1s$	$^2S_{1/2}$	13.598
2	He	$1s^2$	$^1S_0$	24.588
3	Li	[He]$2s$	$^2S_{1/2}$	5.392
4	Be	[He]$2s^2$	$^1S_0$	9.322
5	B	[Be]$2p$	$^2P^o_{1/2}$	8.298
6	C	[Be]$2p^2$	$^3P_0$	11.260
7	N	[Be]$2p^3$	$^4S^o_{3/2}$	14.534
8	O	[Be]$2p^4$	$^3P_2$	13.618
9	F	[Be]$2p^5$	$^2P^o_{3/2}$	17.422
10	Ne	[Be]$2p^6$	$^1S_0$	21.564
11	Na	[Ne]$3s$	$^2S_{1/2}$	5.139
12	Mg	[Ne]$3s^2$	$^1S_0$	7.646
13	Al	[Mg]$3p$	$^2P^o_{1/2}$	5.986
14	Si	[Mg]$3p^2$	$^3P_0$	8.151
15	P	[Mg]$3p^3$	$^4S^o_{3/2}$	10.486
16	S	[Mg]$3p^4$	$^3P_2$	10.360
17	Cl	[Mg]$3p^5$	$^2P^o_{3/2}$	12.967
18	Ar	[Mg]$3p^6$	$^1S_0$	15.759
19	K	[Ar]$4s$	$^2S_{1/2}$	4.341
20	Ca	[Ar]$4s^2$	$^1S_0$	6.113
21	Sc	[Ca]$3d$	$^2D_{3/2}$	6.562
22	Ti	[Ca]$3d^2$	$^3F_2$	6.820
23	V	[Ca]$3d^3$	$^4F_{3/2}$	6.740
24	Cr	[Ar]$3d^5 4s$	$^7S_3$	6.766
25	Mn	[Ca]$3d^5$	$^6S_{5/2}$	7.437
26	Fe	[Ca]$3d^6$	$^5D_4$	7.870
27	Co	[Ca]$3d^7$	$^4F_{9/2}$	7.864
28	Ni	[Ca]$3d^8$	$^3F_4$	7.638
29	Cu	[Ar]$3d^{10} 4s$	$^2S_{1/2}$	7.478
30	Zn	[Ca]$3d^{10}$	$^1S_0$	9.394
31	Ga	[Zn]$4p$	$^2P^o_{1/2}$	5.999
32	Ge	[Zn]$4p^2$	$^3P_0$	7.899
33	As	[Zn]$4p^3$	$^4S^o_{3/2}$	9.81
34	Se	[Zn]$4p^4$	$^3P_2$	9.752
35	Br	[Zn]$4p^5$	$^2P^o_{3/2}$	11.814
36	Kr	[Zn]$4p^6$	$^1S_0$	13.999
37	Rb	[Kr]$5s$	$^2S_{1/2}$	4.177
38	Sr	[Kr]$5s^2$	$^1S_0$	5.695
39	Y	[Sr]$4d$	$^2D_{3/2}$	6.22
40	Zr	[Sr]$4d^2$	$^3F_2$	6.84
41	Nb	[Kr]$4d^4 5s$	$^6D_{1/2}$	6.88
42	Mo	[Kr]$4d^5 5s$	$^7S_3$	7.099
43	Tc	[Sr]$4d^5$	$^6S_{5/2}$	7.28
44	Ru	[Kr]$4d^7 5s$	$^5F_5$	7.37
45	Rh	[Kr]$4d^8 5s$	$^4F_{9/2}$	7.46
46	Pd	[Kr]$4d^{10}$	$^1S_0$	8.34
47	Ag	[Pd]$5s$	$^2S_{1/2}$	7.576
48	Cd	[Pd]$5s^2$	$^1S_0$	8.993
49	In	[Cd]$5p$	$^2P^o_{1/2}$	5.786
50	Sn	[Cd]$5p^2$	$^3P_0$	7.344
51	Sb	[Cd]$5p^3$	$^4S^o_{3/2}$	8.641
52	Te	[Cd]$5p^4$	$^3P_2$	9.009
53	I	[Cd]$5p^5$	$^2P^o_{3/2}$	10.451
54	Xe	[Cd]$5p^6$	$^1S_0$	12.130
55	Cs	[Xe]$6s$	$^2S_{1/2}$	3.894
56	Ba	[Xe]$6s^2$	$^1S_0$	5.212
57	La	[Ba]$5d$	$^2D_{3/2}$	5.577
58	Ce	[Ba]$4f 5d$	$^1G^o_4$	5.466
59	Pr	[Ba]$4f^3$	$^4I^o_{9/2}$	5.422
60	Nd	[Ba]$4f^4$	$^5I_4$	5.489
61	Pm	[Ba]$4f^5$	$^6H^o_{5/2}$	5.554
62	Sm	[Ba]$4f^6$	$^7F_0$	5.631
63	Eu	[Ba]$4f^7$	$^8S^o_{7/2}$	5.666
64	Gd	[Ba]$4f^7 5d$	$^9D^o_2$	6.141
65	Tb	[Ba]$4f^9$	$^6H^o_{15/2}$	5.852
66	Dy	[Ba]$4f^{10}$	$^5I_8$	5.927
67	Ho	[Ba]$4f^{11}$	$^4I^o_{15/2}$	6.018
68	Er	[Ba]$4f^{12}$	$^3H_6$	6.101
69	Tm	[Ba]$4f^{13}$	$^2F^o_{7/2}$	6.184
70	Yb	[Ba]$4f^{14}$	$^1S_0$	6.254
71	Lu	[Yb]$5d$	$^2D_{3/2}$	5.426
72	Hf	[Yb]$5d^2$	$^3F_2$	6.65
73	Ta	[Yb]$5d^3$	$^4F_{3/2}$	7.89
74	W	[Yb]$5d^4$	$^5D_0$	7.98
75	Re	[Yb]$5d^5$	$^6S_{5/2}$	7.88
76	Os	[Yb]$5d^6$	$^5D_4$	8.7
77	Ir	[Yb]$5d^7$	$^4F_{9/2}$	9.1
78	Pt	[Xe]$4f^{14}5d^9 6s$	$^3D_3$	9.0
79	Au	[Xe]$4f^{14}5d^{10}6s$	$^2S_{1/2}$	9.225
80	Hg	[Yb]$5d^{10}$	$^1S_0$	10.437
81	Tl	[Hg]$6p$	$^2P^o_{1/2}$	6.108
82	Pb	[Hg]$6p^2$	$^3P_0$	7.416
83	Bi	[Hg]$6p^3$	$^4S^o_{3/2}$	7.289
84	Po	[Hg]$6p^4$	$^3P_2$	8.42
85	At	[Hg]$6p^5$	$^2P^o_{3/2}$	8.8
86	Rn	[Hg]$6p^6$	$^1S_0$	10.748
87	Fr	[Rn]$7s$	$^2S_{1/2}$	3.8
88	Ra	[Rn]$7s^2$	$^1S_0$	5.279
89	Ac	[Ra]$6d$	$^2D_{3/2}$	5.17
90	Th	[Ra]$6d^2$	$^3F_2$	6.08
91	Pa	[Rn]$5f^2(^3H_4)6d7s^2$	$(4,3/2)_{11/2}$	5.89
92	U	[Rn]$5f^3(^4I^o_{9/2})6d7s^2$	$(9/2,3/2)^o_6$	6.05
93	Np	[Rn]$5f^4(^5I_4)6d7s^2$	$(4,3/2)_{11/2}$	6.19
94	Pu	[Ra]$5f^6$	$^7F_0$	6.06
95	Am	[Ra]$5f^7$	$^8S^o_{7/2}$	5.993
96	Cm	[Rn]$5f^7(^8S^o_{7/2})6d7s^2$	$(7/2,3/2)^o_2$	6.02
97	Bk	[Ra]$5f^9$	$^6H^o_{15/2}$	6.23
98	Cf	[Ra]$5d^{10}$	$^5I_8$	6.30
99	Es	[Ra]$5f^{11}$	$^4I^o_{15/2}$	6.42
100	Fm	[Ra]$5f^{12}$	$^3H_6$	6.50
101	Md	[Ra]$5f^{13}$	$^2F^o_{7/2}$	6.58
102	No	[Ra]$5f^{14}$	$^1S_0$	6.65
103	Lr	[No]$6d$	$^2D_{3/2}$	8.6

[a]Ground state configurations of elements with filled electron subshells—indicated by their respective element symbols in square brackets—are used for abbreviation. For example, [Mg] = [Ne]$3s^2$ = [Be]$2p^6 3s^2$ = [He]$2s^2 2p^6 3s^2$ = $1s^2 2s^2 2p^6 3s^2$.

highly stripped atomic ions, and higher atomic levels (larger n), spin-orbit or magnetic interactions between l and s of the same electron can become large compared to interactions between the l_i (or s_i) of different electrons. In this case, l and s of each electron are coupled:

$$\mathbf{l}_i + \mathbf{s}_i = \mathbf{j}_i \quad (j_i = l_i \pm \tfrac{1}{2} \geqslant \tfrac{1}{2}),$$

and the j_i couple to a resultant $J = |\Sigma \mathbf{j}|$ (*jj* coupling). Often the coupling within a spectrum changes from *LS* to *jj* as n increases, and there are regions of *intermediate coupling* (where electrostatic and magnetic interactions are of same magnitude).

(8) Several additional types of coupling are possible (see, e.g., Ref. 2). One of these—the $J_1 l$ (pair) coupling scheme—is a very good approximation for *noble gas spectra,* as well as others. For the noble gases, the (his-

torical) Paschen notation is often used for the more prominent lines, and the levels are characterized by an integer (related to the principal quantum number n), the orbital quantum number l, and a running number (subscript).

(9) *Ground state configurations* as well as ionization energies for all elements are compiled in Table 5.01.B.2.

5.02. PHOTON ENERGIES AND SPECTRAL LINE POSITIONS

A. Photon energy ΔE

Photon energy due to an electron transition between an upper atomic level k (of energy E_k) and a lower level i:

$$\Delta E = E_k - E_i = h\nu = hc\sigma = hc/\lambda_{\text{vac}},$$

where ν is the frequency, σ the wave number in vacuum, and λ_{vac} the wavelength in vacuum. E's are usually listed in atomic energy level tables[2–4] in wave-number units (see below). Reference point ($E = 0$) is the ground state.

B. Frequency ν

Frequencies are often used in low-frequency or high-precision spectroscopy, with MHz, GHz, and THz the customary units.

C. Wavelength λ

Conversions between customary units:

$$1\ \text{nm} = 10\ \text{Å} = 10^{-3}\ \mu\text{m} \quad (\text{Å} = \text{angstrom}).$$

Spectral ranges:

$\lambda > 7500$ Å	infrared,
$4000 < \lambda < 7500$ Å	visible,
$2000 < \lambda < 4000$ Å	near (or air) ultraviolet (uv),
$\lambda < 2000$ Å	vacuum ultraviolet (vuv).

The vuv extends to a few Å (extreme uv, or xuv), where it overlaps with the soft x-ray region (at about $\lambda < 50$ Å). In wavelength tables,[5–9] vacuum wavelengths (λ_{vac}) are usually given for $\lambda < 2000$ Å and wavelengths in air for $\lambda > 2000$ Å. Since energy levels, wave numbers, and fre-

TABLE 5.02.C.1. Wavelengths λ, upper energy levels E_k, statistical weights g_i and g_k of lower and upper levels, and transition probabilities A_{ki} for persistent spectral lines of neutral atoms

Spectrum	λ[a] (Å)	E_k (cm^{-1})	g_i	g_k	A_{ki} (10^8 s^{-1})	Accuracy[b]
Ag	3280.7g	30 473	2	4	1.4	B
	3382.9g	29 552	2	2	1.3	B
	5209.1	48 744	2	4	0.75	D
	5465.5	48 764	4	6	0.86	D
Al	3082.2g	32 435	2	4	0.63	C
	3092.7g	32 437	4	6	0.74	C
	3944.0g	25 348	2	2	0.493	C
	3961.5g	25 348	4	2	0.98	C
Ar	1048.2g	95 400	1	3	5.1	C
	4158.6	117 184	5	5	0.0145	C
	4259.4	118 871	3	1	0.0415	C
	7635.1	106 238	5	5	0.274	C
	7948.2	107 132	1	3	0.196	C
	8115.3	105 463	5	7	0.366	C
As	1890.4g	52 898	4	6	2.0	D
	1937.6g	51 610	4	4	2.0	D
	2288.1	54 605	6	4	2.8	D
	2349.8	53 136	4	2	3.1	D
Au	2428.0g	41 174	2	4	1.5	D
	2676.0g	37 359	2	2	1.1	D
B	1825.9g	54 767	2	4	2.0	C
	1826.4g	54 767	4	6	2.4	C
	2496.8g	40 040	2	2	0.85	C
	2497.7g	40 040	4	2	1.69	C
Ba	5535.5g	18 060	1	3	1.15	B
	6498.8	24 980	7	7	0.86	D
	7059.9	23 757	7	9	0.71	D
	7280.3	22 947	5	7	0.53	D
Be	2348.6g	42 565	1	3	5.56	B
	2650.6	59 696	9	9[c]	4.31	C
Bi	2228.3g	44 865	4	4	0.89	D
	2898.0	45 916	4	2	1.53	C
	2989.0	44 865	4	4	0.55	C
	3067.7g	32 588	4	2	2.07	C
Br	1488.5g	67 184	4	4	1.2	D
	1540.7g	64 907	4	4	1.4	D
	7348.5	78 512	4	6	0.12	D
C	1561.4g	64 087	5	7	1.4	D
	1657.0g	60 393	5	5	2.4	D
	1930.9	61 982	5	3	3.7	D
	2478.6	61 982	1	3	0.18	D

Spectrum	λ[a] (Å)	E_k (cm^{-1})	g_i	g_k	A_{ki} (10^8 s^{-1})	Accuracy[b]
Ca	4226.7g	23 652	1	3	2.18	B
	4302.5	38 552	5	5	1.36	C
	5588.8	38 259	7	7	0.49	D
	6162.2	31 539	5	3	0.477	C
	6439.1	35 897	7	9	0.53	D
Cd	2288.0g	43 692	1	3	5.3	C
	3466.2	59 498	3	5	1.2	D
	3610.5	59 516	5	7	1.3	D
	5085.8	51 484	5	3	0.56	C
Cl	1347.2g	74 226	4	4	4.19	C
	1351.7g	74 866	2	2	3.23	C
	4526.2	96 313	4	4	0.051	C
	7256.6	85 735	6	4	0.15	D
Co	3405.1	32 842	10	10	0.98	C
	3453.5	32 431	10	12	1.1	C
	3502.3	32 028	10	8	0.90	C
	3569.4	35 451	8	8	1.6	C
Cr	3578.7g	27 935	7	9	1.48	B
	3593.5g	27 820	7	7	1.50	B
	3605.3g	27 729	7	5	1.62	B
	4254.4g	23 499	7	9	0.315	B
	4274.8g	23 386	7	7	0.306	B
	5208.4	26 788	5	7	0.51	D
Cs	3876.1g	25 792	2	4	0.0038	C
	4555.3g	21 946	2	4	0.0188	C
	4593.2g	21 765	2	2	0.0080	C
Cu	2178.9g	45 879	2	4	0.913	B
	3247.5g	30 784	2	4	1.39	B
	3274.0g	30 535	2	2	1.37	B
	5218.2	49 942	4	6	0.75	C
F	954.83g	104 731	4	4	6.4	C
	6856.0	116 987	6	8	0.42	D
	7398.7	115 918	6	6	0.31	D
	7754.7	117 623	4	6	0.30	D
Fe	3581.2	34 844	11	13	1.02	B
	3719.9g	26 875	9	11	0.163	B
	3734.9	33 695	11	11	0.902	B
	3745.6g	27 395	5	7	0.115	B
	3859.9g	25 900	9	9	0.0970	B
	4045.8	36 686	9	9	0.75	C

TABLE 5.02.C.1—*Continued*

Spectrum	λ[a] (Å)	E_k (cm^{-1})	g_i	g_k	A_{ki} (10^8 s^{-1})	Accuracy[b]
Ga	2874.2g	34 782	2	4	1.2	C
	2943.6g	34 788	4	6	1.4	C
	4033.0g	24 789	2	2	0.49	C
	4172.0g	24 789	4	2	0.92	C
Ge	2651.6g	37 702	1	3	0.85	C
	2709.6g	37 452	3	1	2.8	C
	2754.6g	37 702	5	3	1.1	C
	3039.1	40 021	5	3	2.8	C
He	537.03g	186 209	1	3	5.66	A
	584.33g	171 135	1	3	17.99	AA
	3888.7	185 565	3	9[c]	0.09478	AA
	4026.2	193 917	9	15[c]	0.116	A
	4471.5	191 445	9	15[c]	0.246	A
	5875.7	186 102	9	15[c]	0.7053	AA
Hg	2536.5g	39 412	1	3	0.13	D
	3125.7	71 396	3	5	0.51	D
	4358.3	62 350	3	3	0.40	D
	5460.7	62 350	5	3	0.56	D
I	1782.8g	56 093	4	4	2.71	C
	1830.4g	54 633	4	6	0.16	D
In	3256.1g	32 915	4	6	1.3	D
	3039.4g	32 892	2	4	1.3	D
	4101.8g	24 373	2	2	0.56	C
	4511.3g	24 373	4	2	1.02	C
K	4044.1g	24 720	2	4	0.0124	C
	4047.2g	24 701	2	2	0.0124	C
	7664.9g	13 043	2	4	0.387	B
	7699.0g	12 985	2	2	0.382	B
Kr	5570.3	97 919	5	3	0.021	D
	5870.9	97 945	3	5	0.018	D
	7601.5	93 123	5	5	0.31	D
	8112.9	92 294	5	7	0.36	D
Li	3232.7g	30 925	2	6[c]	0.012	B
	4602.9	36 623	6	10[c]	0.236	C
	6103.6	31 283	6	10[c]	0.716	B
	6707.8g	14 904	2	6[c]	0.372	A
Mg	2025.8g	49 347	1	3	0.84	D
	2852.1g	35 051	1	3	4.95	B
	4703.0	56 308	3	5	0.255	C
	5183.6	41 197	5	3	0.575	B
Mn	2794.8g	35 770	6	8	3.7	C
	2798.3g	35 726	6	6	3.6	C
	2801.1g	35 690	6	4	3.7	C
	4030.8g	24 802	6	8	0.19	C
	4033.1g	24 788	6	6	0.18	C
	4034.5g	24 779	6	4	0.18	C
N	1199.6g	83 365	4	6	5.5	D
	1492.6	86 221	6	4	5.3	D
	4223.1	107 037	6	6	0.051	D
	4935.1	106 478	4	2	0.0158	B
	7468.3	96 751	6	4	0.161	C
Na	5890.0g	16 973	2	4	0.622	A
	5895.9g	16 956	2	2	0.618	A
	5682.6	34 549	2	4	0.103	C
	8183.3	29 173	2	4	0.453	C
Ne	735.90g	135 889	1	3	6.11	B
	743.72g	134 459	1	3	0.486	B
	5852.5	152 971	3	1	0.682	B
	6402.2	149 657	5	7	0.514	B
	6074.3	150 917	3	1	0.603	B
Ni	3101.6	33 112	5	7	0.72	C
	3134.1	33 611	3	5	0.71	C
	3369.6g	29 669	9	7	0.17	C
	3414.8	29 481	7	9	0.55	C
	3524.5	28 569	7	5	1.0	C
	3619.4	31 031	5	7	0.73	C
O	1302.2g	76 795	5	3	3.3	C
	5436.9	105 019	7	5	0.0142	C
	6653.8	130 943	3	1	0.600	B
	7156.7	116 631	5	5	0.473	B
	7771.9	86 631	5	7	0.340	B
P	1775.0g	56 340	4	6	2.17	C
	1782.9g	56 090	4	4	2.14	C
	2136.2	58 174	6	4	2.83	C
	2535.6	58 174	4	4	0.95	C
Pb	2802.0g	46 329	5	7	1.6	D
	2833.1g	35 287	1	3	0.58	D
	3683.5g	34 960	3	1	1.5	D
	4057.8g	35 287	5	3	0.89	D
Rb	4201.8g	23 793	2	4	0.018	C
	4215.5g	23 715	2	2	0.015	C
	7800.3g	12 817	2	4	0.370	B
	7947.6g	12 579	2	2	0.340	B
S	1474.0g	67 843	5	7	1.6	D
	1666.7	69 238	5	5	6.3	C
	1807.3g	55 331	5	3	3.8	C
	4694.1	73 921	5	7	0.0067	D
Sc	3907.5g	25 585	4	6	1.28	C
	3911.8g	25 725	6	8	1.37	C
	4020.4g	24 866	4	4	1.65	C
	4023.7g	25 014	6	6	1.44	C
Si	2506.9g	39 955	3	5	0.466	C
	2516.1g	39 955	5	5	1.21	C
	2881.6	40 992	5	3	1.89	C
	5006.1	60 962	3	5	0.028	D
	5948.5	57 798	3	5	0.022	D
Sn	2840.0g	38 629	5	5	1.7	D
	3034.1g	34 641	3	1	2.0	D
	3175.1g	34 914	5	3	1.0	D
	3262.3	39 257	5	3	2.7	D
Sr	2428.1g	41 172	1	3	0.17	C
	4607.3g	21 698	1	3	2.01	B
Ti	3642.7g	27 615	7	9	0.67	C
	3653.5g	27 750	9	11	0.66	C
	3998.6g	25 388	9	9	0.39	C
	4981.7	26 911	11	13	0.59	C
	5210.4g	19 574	9	9	0.034	C
Tl	2767.9g	36 118	2	4	1.26	C
	3519.2g	36 200	4	6	1.24	C
	3775.7g	26 478	2	2	0.625	B
	5350.5g	26 478	4	2	0.705	B
U	3566.6g	28 650	11	11	0.24	B
	3571.6	38 338	17	15	0.13	C
	3584.9g	27 887	13	15	0.18	B
V	3183.4g	31 541	6	8	1.3	D
	4111.8	26 738	10	10	0.91	D
	4379.2	25 254	10	12	1.2	D
	4384.7	25 112	8	10	0.97	D
Xe	1192.0g	83 890	1	3	6.2	C
	1295.6g	77 186	1	3	2.5	C
	1469.6g	68 046	1	3	2.8	B
	4671.2	88 470	5	7	0.010	D
	7119.6	92 445	7	9	0.066	D
Zn	2138.6g	46 745	1	3	7.09	B
	3302.6	62 772	3	5	1.2	B
	3345.0	62 777	5	7	1.7	B
	6362.3	62 459	3	5	0.474	C

[a]A "g" next to the λ value indicates that the lower level of the transition belongs to the ground term; i.e., the line is a resonance line.
[b]Accuracy ratings pertain to A_{ki} values: AA, uncertainty within 1%; A, within 3%; B, within 10%; C, within 25%; D, within 50%.
[c]The A_{ki}, λ, g_i, and g_k values are averages for the entire multiplet.

quencies are interrelated through λ_{vac}, wavelengths measured in air have to be corrected before conversion to σ, etc.; this can be done by an iterative process involving the refractive index $n(\sigma)$ of standard (dry) air at 101325 Pa and 288 K:

$$\sigma = 1/n(\sigma)\lambda_{air},$$

$$n(\sigma) = 1 + 6432.8\times10^{-8} + \frac{2\,949\,810}{146\times10^8-\sigma^2} + \frac{25\,540}{41\times10^8-\sigma^2},$$

with σ in cm^{-1}. Wavelengths for prominent lines of selected elements are given in Table 5.02.C.1. For comprehensive tabulations of wavelengths of atomic lines and atomic energy levels, see Refs. 2–9.

D. Wave number σ

Customary unit cm^{-1} = kayser (K); in addition, millikayser (mK) [1 mK≡29.979 MHz] and kilokayser (kK) are sometimes used.

E. Energy equivalents

Wave number associated with 1 eV:

$\sigma_0 = 8065.5\ cm^{-1}$.

Wavelength associated with 1 eV:

$\lambda_0 = 12\,399$ Å.

Temperature associated with 1 eV:

$T_0 = 11\,605$ K.

Frequency associated with 1 eV:

$\nu_0 = 2.4180\times10^{14}$ Hz.

Energy of 1 eV in joules:

$E_0 = 1.6022\times10^{-19}$ J.

Also used is the atomic unit of energy = 2 Ry:

$$1\ \text{Ry} = R_\infty hc = e^2/8\pi\epsilon_0 a_0 = 2.180\times10^{18}\ \text{J} = 13.606\ \text{eV}.$$

(R_∞ = Rydberg constant = $1.09737\times10^7\ m^{-1}$).

TABLE 5.03.A. Other recommended secondary standards[7]

	Source		
	^{86}Kr	^{198}Hg	^{114}Cd
Air wavelengths (Å)	6456.2876	5790.6626	6438.4685
	6421.0257	5769.5982	5085.8203
	5649.5606	5460.7530	4799.9104
	4502.3533	4358.3374	4678.1487

5.03. WAVELENGTH STANDARDS

Primary standard of length (defined at the 10th General Conference on Weights and Measures, 1960): ^{86}Kr line at $\lambda_{vac} = 6057.80210$ Å ($2p_{10}$-$5d_5$ transition in Paschen notation) [$\lambda_{air} = 6056.12525$ Å, in standard (dry) air at 101 325 Pa and 288 K].

Secondary standards, realized by methane- or iodine-stabilized He-Ne lasers: CH_4, $P(7)$, band ν_3: $\lambda_{vac} = 33\,922.3140$ Å; $^{127}I_2$, $R(127)$, band 11-5, component i: $\lambda_{vac} = 6329.91399$ Å ($\lambda_{air} = 6328.16414$ Å).

Other recommended secondary standards are given in Table 5.03.A.

In addition, numerous lines of the Fe I, Fe II, Th I, Th II, Hg I, and noble gas spectra (see Refs. 5–9) are used as working standards.

5.04. SELECTION RULES FOR DISCRETE TRANSITIONS

For the selection rules for discrete transitions, see Table 5.04.A.

5.05. SPECTRAL LINE INTENSITIES, TRANSITION PROBABILITIES, *f* VALUES, AND LINE STRENGTHS

A. Emission intensities

The total power ϵ radiated in a spectral line of frequency ν per unit source volume and per unit solid angle is

$$\epsilon_{line} = (4\pi)^{-1}h\nu A_{ki}N_k,$$

TABLE 5.04.A. Selection rules for discrete transitions

	Electric dipole ("allowed")	Magnetic dipole ("forbidden")	Electric quadrupole ("forbidden")
Rigorous rules	1. $\Delta J = 0, \pm1$ (except 0↮0) 2. $\Delta M = 0, \pm1$ (except 0↮0 when $\Delta J = 0$) 3. Parity change	$\Delta J = 0, \pm1$ (except 0↮0) $\Delta M = 0, \pm1$ (except 0↮0 when $\Delta J = 0$) No parity change	$\Delta J = 0, \pm1, \pm2$ (except 0↮0, 1/2↮1/2, 0↮1) $\Delta M = 0, \pm1, \pm2$ No parity change
With negligible configuration interaction	4. One electron jumping, with $\Delta l = \pm1$, Δn arbitrary	No change in electron configuration; i.e., for all electrons, $\Delta l = 0$, $\Delta n = 0$	*No* change in electron configuration; *or* one electron jumping, with $\Delta l = 0, \pm2$, Δn arbitrary
For *LS* coupling only	5. $\Delta S = 0$ 6. $\Delta L = 0, \pm1$ (except 0↮0)	$\Delta S = 0$ $\Delta L = 0$ $\Delta J = \pm1$	$\Delta S = 0$ $\Delta L = 0, \pm1, \pm2$ (except 0↮0, 0↮1)

where A_{ki} is the atomic transition probability and N_k the number per unit volume (number density) of excited atoms in the upper (initial) level k. For a homogeneous light source of length l and for the optically thin case, where all radiation escapes, the total emitted line intensity (SI quantity: radiance) is

$$I_{\text{line}} = \epsilon_{\text{line}} l = \int_0^{+\infty} I(\lambda) d\lambda$$
$$= (4\pi)^{-1}(hc/\lambda_0) A_{ki} N_k l,$$

where $I(\lambda)$ is the specific intensity at wavelength λ, and λ_0 the wavelength at line center.

B. Absorption

In absorption, the reduced absorption

$$W(\lambda) = [I(\lambda) - I'(\lambda)]/I(\lambda)$$

is used, where $I(\lambda)$ is the incident intensity at wavelength λ, e.g., from a continuum background source, and $I'(\lambda)$ the intensity after passage through the absorbing medium. The reduced line intensity from a homogeneous and optically thin absorbing medium of length l follows as

$$W_{ik} = \int_0^{+\infty} W(\lambda) d\lambda = \frac{e^2}{4\epsilon_0 m_e c^2} \lambda_0^2 N_i f_{ik} l,$$

where f_{ik} is the atomic (absorption) oscillator strength (dimensionless).

C. Line strength

A_{ki} and f_{ik} are thus the principal atomic quantities related to line intensities. Furthermore, in theoretical work, the *line strength S* is widely used:

$$S = S(i,k) = S(k,i) = |R_{ik}|^2, \quad R_{ik} = \langle \psi_k | P | \psi_i \rangle,$$

where ψ_i and ψ_k are the initial- and final-state wave functions (Sec. 1.07.D) and R_{ik} is the *transition matrix element* of the appropriate multipole operator P (R_{ik} involves an integration over spatial and spin coordinates of all N electrons of the atom or ion). These operators for the most common types of radiation are:

Electric dipole:

$$P = -|e| \sum_{m=1}^{N} \mathbf{r}_m,$$

Electric quadrupole:

$$P = -|e| \sum_{m=1}^{N} \left[\mathbf{r}_m \mathbf{r}_m - \tfrac{1}{3} r_m^2 (\hat{i}\hat{i} + \hat{j}\hat{j} + \hat{k}\hat{k}) \right],$$

Magnetic dipole:

$$P = -\frac{|e|}{2m_e} \sum_{m=1}^{N} (\mathbf{l}_m + 2\mathbf{s}_m),$$

where the $\mathbf{r}_m$ are the position vectors of the electrons, $\mathbf{l}_m$ and $\mathbf{s}_m$ their orbital and spin angular momentum vectors, and $\hat{i}, \hat{j}$, and $\hat{k}$ the unit vectors in Cartesian coordinates. S in normally given in atomic units (for electric dipole transitions: $a_0^2 e^2 = 7.188 \times 10^{-59}$ m^2 C^2).

TABLE 5.05.D.1. Conversions between S and A_{ki} for forbidden transitions

	SI units [a]	Numerically, in customary units [b]
Electric quadrupole	$A_{ki} = \frac{8\pi^5}{5h\epsilon_0 \lambda^5 g_k} S$	$A_{ki} = \frac{1.680 \times 10^{18}}{g_k \lambda^5} S$
Magnetic dipole	$A_{ki} = \frac{16\pi^3 \mu_0}{3h\lambda^3 g_k} S$	$A_{ki} = \frac{2.697 \times 10^{13}}{g_k \lambda^3} S$

[a] A in s^{-1}, λ in m. Electric quadrupole: S in m^4 C^2. Magnetic dipole: S in J^2 T^{-2}.
[b] A in s^{-1}, λ in Å. S in atomic units: $a_0^4 e^2 = 2.013 \times 10^{-79}$ m^4 C^2 (electric quadrupole), $e^2 h^2 / 16\pi^2 m_e^2 = \mu_B^2 = 8.601 \times 10^{-47}$ J^2 T^{-2} (magnetic dipole). μ_B is the Bohr magneton.

D. Relationships between A, f, and S

Relationships between A, f, and S for electric dipole (allowed) transitions in SI units (A in s^{-1}, λ in m, S in m^2 C^2):

$$A_{ki} = \frac{2\pi e^2}{m_e c \epsilon_0 \lambda^2} \frac{g_i}{g_k} f_{ik} = \frac{16\pi^3}{3h\epsilon_0 \lambda^3 g_k} S.$$

Numerically, in customary units (A in s^{-1}, λ in Å, S in atomic units),

$$A_{ki} = \frac{6.6702 \times 10^{15}}{\lambda^2} \frac{g_i}{g_k} f_{ik} = \frac{2.0261 \times 10^{18}}{\lambda^3 g_k} S,$$

and for S and ΔE in atomic units,

$$f_{ik} = \tfrac{2}{3}(\Delta E / g_i) S.$$

g_i and g_k are the statistical weights, which are obtained from the appropriate angular momentum quantum numbers. Thus for the lower (upper) level of a spectral line,

$$g_{i(k)} = 2J_{i(k)} + 1,$$

and for the lower (upper) term of a multiplet,

$$\bar{g}_{i(k)} = \sum_{J_{i(k)}} (2J_{i(k)} + 1)$$
$$= (2L_{i(k)} + 1)(2S_{i(k)} + 1).$$

For forbidden transitions, i.e., electric quadrupole, magnetic dipole, etc., f is not in use.

[Numerical example: For the $2p\,^1P_1^o$-$3d\,^1D_2$ (allowed) transition in He I at 6678.15 Å: $g_i = 3$; $g_k = 5$; $A_{ki} = 6.38 \times 10^7$ s^{-1}; $f_{ik} = 0.711$; $S = 46.9 a_0^2 e^2$.]

Conversions between S and A_{ki} for forbidden transitions are given in Table 5.05.D.1.

E. Relationships between line and multiplet values

Relations between total strengths and f values of multiplets (M) and the corresponding quantities for lines within multiplets of allowed transitions:

$$(1)\ S_M = \sum S_{\text{line}}.$$

$$(2)\ f_M = (\bar{\lambda} \bar{g}_i)^{-1} \sum_{J_k, J_i} g_i \lambda(J_i, J_k) f(J_i, J_k).$$

$\bar{\lambda}$ is the weighted ("multiplet") wavelength:

$$\bar{\lambda} = hc/n\,\overline{\Delta E},$$

where

$$\overline{\Delta E} = \overline{E_k} - \overline{E_i} = (\bar{g}_k)^{-1}\sum_{J_k} g_k E_k - (\bar{g}_i)^{-1}\sum_{J_i} g_i E_i,$$

and n is the refractive index of standard air (Sec. 5.02.C).

F. Tabulations

A values for strong lines of selected elements are given in Table 5.02.C.1. For comprehensive numerical tables of *A*, *f*, and *S*, including forbidden lines, see Refs. 10–13.

If multiplet strengths S_M are known, and individual line strengths are needed, these may be obtained for *LS* coupling from general tables given, e.g., in Ref. 14.

5.06. ATOMIC LIFETIMES

The radiative lifetime τ_k of an atomic level *k* is related to the sum of transition probabilities to all levels *i* lower in energy than *k*:

$$\tau_k = \left(\sum_i A_{ki}\right)^{-1}.$$

The "branching" ratio of a particular transition, say to state i', is defined as

$$A_{ki'}/\sum_i A_{ki} = A_{ki'}\tau_k.$$

Relation between τ_k and (absorption) oscillator strengths of transitions to levels lower in energy than *k*:

$$\tau_k = \frac{m_e c \epsilon_0 g_k}{2\pi e^2}\left(\sum_{i,\, E(i)<E(k)} \frac{g_i f_{ik}}{[\lambda(J_i,J_k)]^2}\right)^{-1}.$$

If only one branch (i') exists or is *dominant*, one obtains $A_{ki'}\tau_k \simeq 1$, and

$$\tau_k = 1/A_{ki'}.$$

5.07. HYDROGENIC (ONE-ELECTRON) SPECIES

A. Energy levels

Nonrelativistic *energy levels* with respect to the ground level of a hydrogenic atom ($1s\,^2S_{1/2}$) of nuclear charge *Z* ($Z=1$ for H, 2 for He^+,...) and nuclear mass M_Z:

$$E_{Z,n} = \frac{m_e e^4 Z^2(1-1/n^2)}{8h^2\epsilon_0^2(1+m_e/M_Z)}.$$

The factor $1+m_e/M_Z$ takes into account the reduced mass of the electron with respect to the nucleus.

(1) In terms of the Rydberg constant R_∞ $(=m_e e^4/8h^3 c\epsilon_0^2)$:

$$E_{Z,n} = \frac{R_\infty hcZ^2(1-1/n^2)}{1+m_e/M_Z}.$$

Thus if $R_Z \equiv R_\infty/(1+m_e/M_Z)$,

$$E_{Z,n} = R_Z hcZ^2(1-1/n^2).$$

$E_{Z,n}$ is obtained in wave-number units by omitting the factor hc. In customary units: $R_\infty = 109\,737\ \text{cm}^{-1}$,

TABLE 5.07.B.1. Spectral series of hydrogen

Transition	Customary name[a]	λ (Å)	g_i[b]	g_k	A_{ki} ($10^8\ s^{-1}$)
1-2	(L_α)	1 215.67	2	8	4.699
1-3	(L_β)	1 025.73	2	18	5.575(−1)[c]
1-4	(L_γ)	972.537	2	32	1.278(−1)
1-5	(L_δ)	949.743	2	50	4.125(−2)
1-6	(L_ϵ)	937.803	2	72	1.644(−2)
2-3	(H_α)	6 562.80	8	18	4.410(−1)
2-4	(H_β)	4 861.32	8	32	8.419(−2)
2-5	(H_γ)	4 340.46	8	50	2.530(−2)
2-6	(H_δ)	4 101.73	8	72	9.732(−3)
2-7	(H_ϵ)	3 970.07	8	98	4.389(−3)
3-4	(P_α)	18 751.0	18	32	8.986(−2)
3-5	(P_β)	12 818.1	18	50	2.201(−2)
3-6	(P_γ)	10 938.1	18	72	7.783(−3)
3-7	(P_δ)	10 049.4	18	98	3.358(−3)
3-8	(P_ϵ)	9 545.97	18	128	1.651(−3)

[a] L_α is often called Lyman α, H_α = Balmer α, P_α = Paschen α, etc.
[b] For transitions in hydrogen, $g_{i(k)} = 2(n_{i(k)})^2$, where $n_{i(k)}$ is the principal quantum number of the lower (upper) shell.
[c] The number in parentheses indicates the power of 10 by which the value has to be multiplied.

R_H $(Z=1) = 109\,678\ \text{cm}^{-1}$.

(2) Relativistic and quantum electrodynamic effects cause small shifts and splittings of the levels.[15] For example, the wave number of the transition between the hyperfine components of the ground state of hydrogen is $\sigma = 0.047380\ \text{cm}^{-1}$ (≡1420.406 MHz). The hyperfine splitting is due to the interaction of the nuclear magnetic moment with the magnetic field produced by the electron.

B. Transitions

(1) Nonrelativistic *energy* of a hydrogenic transition:

$$(\Delta E)_Z = (E_k - E_i)_Z = R_Z hcZ^2(1/n_i^2 - 1/n_k^2).$$

(The corresponding wave number is obtained by omitting the factor *hc*.) Isotope shifts arise from differences in the masses M_Z of the various isotopes of a given ion of nuclear charge *Z*.

(2) *Hydrogenic Z scaling.* The spectroscopic quantities for a hydrogenic ion of nuclear charge *Z* relate to the equivalent quantities in hydrogen ($Z=1$) as follows (neglecting small differences in the values of R_Z):

$$(\Delta E)_Z = Z^2(\Delta E)_H, \quad (\lambda_{vac})_Z = Z^{-2}(\lambda_{vac})_H,$$
$$S_Z = Z^{-2}S_H, \quad f_Z = f_H, \quad A_Z = Z^4 A_H.$$

For large values of *Z*, roughly $Z>20$, relativistic corrections become noticeable and must be taken into account.

(3) *f-value trends. f* values for high series members (large n' values) of hydrogenic ions decrease according to

$$f(n,l \to n', l\pm 1) \propto (n')^{-3}.$$

For the spectral series of hydrogen see Table 5.07.B.1.

5.08. ATOMS AND IONS WITH TWO OR MORE ELECTRONS

A. Systematic trends and regularities

(1) *Isoelectronic sequence:* comprises all atomic species with the same number of electrons but different Z (e.g., Li sequence = Li, Be^+, B^{2+}, C^{3+}, N^{4+},...).

Atomic quantities for a given state or transition can be expressed as power series expansions in Z^{-1}:

$$Z^{-2}E = \epsilon_0 + \epsilon_1 Z^{-1} + \epsilon_2 Z^{-2} + \cdots,$$

$$Z^2 S = S_0 + S_1 Z^{-1} + S_2 Z^{-2} + \cdots,$$

$$f = f_0 + f_1 Z^{-1} + f_2 Z^{-2} + \cdots,$$

where ϵ_0, f_0, and S_0 are hydrogenic quantities. For transitions in which n does not change ($n_i = n_k$), $f_0 = 0$, since states i and k are degenerate.

(2) *Spectral series* (appearing in simple atomic structures such as alkalis, alkaline earths): group of spectral lines from a given lower state (n,l) to all upper states with increasing $n' = n, n+1, n+2, \ldots$ and constant orbital angular momentum quantum number (either $l+1$ or $l-1$):

$$(n,l)^{a\ \ 2S+1}L_J \rightarrow \sum_{n'} (n,l)^{a-1}(n', l \pm 1)\ ^{2S'+1}L'_{J'}.$$

Line positions and f values will generally show a regular pattern with increasing n', since the final orbital quantum number $l \pm 1$ of the electron undergoing the transition and the angular structure $^{2S'+1}L'_{J'}$ are the same for all n'.

(3) *Homologous atoms* have similar valence structures but different numbers of closed shells of electrons. For example, atoms with a single valence electron outside closed shells, i.e., in the spectroscopic sense the "alkalis":

Li: $1s^2$ + valence electron,

Na: $1s^2 2s^2 2p^6$ + valence electron,

Cu: $1s^2 2s^2 2p^6 3s^2 3p^6 3d^{10}$ + valence electron,

⋮

For equivalent transitions, f values vary gradually within a group of homologous atoms. Transitions to be compared in the case of the "alkalis" are

$$(nl - n'l')_{\rm Li} \rightarrow [(n+1)l - (n'+1)l']_{\rm Na} \rightarrow [(n+2)l - (n'+2)l']_{\rm Cu} \rightarrow \cdots.$$

Complex atomic structures, as well as cases of strong cancellation in the integrand of the transition integral (see below), generally do not adhere to this regular behavior.

B. Some effects peculiar to complex spectra

(1) *Quantum defect.* For atoms or ions of more than one electron, the effective principal quantum number n^* corresponding to a particular energy level E is used:

$$n^* = (Z - N + 1)\sqrt{\frac{R_Z hc}{I - E}}.$$

N is the total number of electrons, I the appropriate ionization limit. Quantum defect: $\mu = n - n^*$ (n is the principal quantum number of the valence electron).

Within a spectral series [Sec. 5.08.A (2)], transition energies are

$$\Delta E \propto 1/(n^*)^2 - 1/(n'^*)^2.$$

Multichannel quantum defect theory (MQDT) exploits regularities of quantum defects within spectral series to predict the positions and relative intensities of transitions to high-lying series members (large values of n').[16]

(2) *Configuration interaction.* Because of interactions among the electrons of an atom or ion, the electronic wave function often cannot be accurately described by a single configuration. Since both the energy levels and the transition probabilities can be significantly affected by configuration interaction, many recent atomic structure calculations have been carried out in a multiconfigurational framework, such as the "multiconfiguration self-consistent field" (MCSCF) and the "superposition of configurations" (SOC) treatments.[17] For example, the ground term of a Be-like ion ($2s^2\ ^1S$) is better described as follows:

$$a_1 2s^2\ ^1S + a_2 2p^2\ ^1S + \text{other configurations of even parity which form a } ^1S \text{ term.}$$

The a_i are mixing coefficients ($\Sigma_i |a_i|^2 = 1$). The squares of the mixing coefficients, $|a_i|^2$, multiplied by 100 are known as percentages and are included in recent energy level tables.[2,4]

(3) *Cancellation effects.* In calculations of transition matrix elements, cases occur where the positive and negative contributions to the transition integrand are very nearly equal, so that the integral is a small number and very sensitive to the numerical integration method used.

5.09. SPECTRAL LINE SHAPES, WIDTHS, AND SHIFTS

Principal causes of spectral line broadening:

A. Doppler broadening

Doppler broadening is due to thermal motion of emitters. For Maxwellian velocity distribution, line shape is *Gaussian*; the half half-width, i.e., half width at half maximum intensity (HWHM), is

$$\Delta\lambda^D_{1/2} = 3.58 \times 10^{-7} \lambda (T/M)^{1/2}.$$

T is the temperature of emitters, M the atomic weight in atomic mass units (u or amu).

B. Pressure broadening

Pressure broadening is due to collisions between emitters and surrounding particles. Shapes are often approximately Lorentzian, i.e., $I(\lambda) \propto [1 + (\Delta\lambda/\Delta\lambda_{1/2})^2]^{-1}$. In the following formulas, all HWHM's and wavelengths are expressed in Å and the particle densities N in cm^{-3}.

(1) *Resonance broadening* (self-broadening) occurs only between identical species and is confined to lines with upper or lower states combining by dipole transition (resonance line) to the ground state. The HWHM may be estimated as

$$\Delta\lambda^R_{1/2} \simeq 4.3 \times 10^{-30} (g_i/g_k)^{1/2} \lambda^2 \lambda_r f_r N_i.$$

λ is the wavelength of the observed line. f_r and λ_r are the oscillator strength and wavelength of the resonance line; g_k and g_i are the statistical weights of its upper and lower state. N_i is the ground state number density.

TABLE 5.09.B.1. Values of Stark-broadening parameter $\alpha_{1/2}$ of H_β line of hydrogen (4861 Å) for various temperatures and electron densities

	N_e (cm^{-3})			
T (K)	10^{15}	10^{16}	10^{17}	10^{18}
5 000	0.0787	0.0808	0.0765	...
10 000	0.0803	0.0840	0.0851	0.0781
20 000	0.0815	0.0860	0.0902	0.0896
30 000	0.0814	0.0860	0.0919	0.0946

For the $2p\,^1P_1^\circ$-$3d\,^1D_2$ transition in neutral He [$\lambda = 6678.15$ Å; λ_r ($1s^2\,^1S_0$-$1s2p\,^1P_1^\circ$) $= 584.334$ Å; $g_i = 1$; $g_k = 3$; $f_r = 0.2762$] at $N_i = 1\times10^{18}$ cm^{-3}: $\Delta\lambda_{1/2}^R = 0.018$ Å.

(2) *Van der Waals broadening* arises from the dipole interaction of an excited atom with the induced dipole of a ground state atom. (In the case of foreign gas broadening, both the perturber and the radiator may be in their respective ground states.) An approximate formula for the HWHM, strictly applicable to hydrogen and similar atomic structures only, is

$$\Delta\lambda_{1/2}^W \simeq 1.5\times10^{-16}\lambda^2 C_6^{2/5}(T/\mu)^{3/10}N,$$

where μ is the atom-perturber reduced mass in amu, N the perturber density, and C_6 the interaction constant. C_6 may be roughly estimated as follows: $C_6 = C_k - C_i$, with $C_{i(k)} = 9.8\times10^{-10}\,\alpha_P R_{i(k)}^2$ (α_P in cm^3, R^2 in a_0^2). Mean atomic polarizability $\alpha_P \approx 6.7\times10^{-25}\,(3I_H/4E^*)^2$ cm^3, where I_H is the ionization energy of hydrogen and E^* the energy of the first excited level of the perturber atom. $R_{i(k)}^2 \approx 2.5[I_H/(I - E_{i(k)})]^2$, where I is the ionization energy of the radiator. Van der Waals broadened lines are red shifted by about two-thirds the size of the HWHM.

For the $2p\,^1P_1^\circ$-$3d\,^1D_2$ transition in neutral He, and with He as perturber: $\lambda = 6678.15$ Å; $I = 198\,311$ cm^{-1}; $E^* = E_i = 171\,135$ cm^{-1}; $E_k = 186\,105$ cm^{-1}; $\mu = 2$. At $T = 15\,000$ K and $N = 1\times10^{18}$ cm^{-3}: $\Delta\lambda_{1/2}^W = 0.022$ Å.

(3) *Stark broadening* due to charged perturbers, i.e., ions and electrons, usually dominates resonance and van der Waals broadening in discharges and plasmas. The HWHM for hydrogen lines is

$$\Delta\lambda_{1/2}^{S,H} = 1.25\times10^{-9}\alpha_{1/2}N_e^{2/3},$$

where N_e is the electron density. The half-width parameter $\alpha_{1/2}$ for the widely used H_β line at 4861 Å is tabulated in Table 5.09.B.1 for some typical temperatures and electron densities.[18] This reference also contains $\alpha_{1/2}$ parameters for other hydrogen lines, as well as Stark width and shift data for numerous lines of the other elements, i.e., neutral atoms and singly charged ions (in the latter, Stark widths and shifts depend linearly on N_e). Other tabulations of complete hydrogen Stark profiles exist.[19]

5.10. SPECTRAL CONTINUUM RADIATION

A. Hydrogenic species

Precise quantum-mechanical calculations exist only for hydrogenic species. The total power ϵ_{cont} radiated (per unit source volume and per unit solid angle, and expressed in SI units) in the wavelength interval $\Delta\lambda$ is the sum of radiation due to the recombination of a free electron with a bare ion (free-bound transitions) and bremsstrahlung (free-free transitions):

$$\epsilon_{cont} = \frac{e^6}{2\pi\epsilon_0^3(6\pi m_e)^{3/2}}N_e N_Z Z^2 \times \frac{1}{(kT)^{1/2}}\exp\left(-\frac{hc}{\lambda kT}\right)\frac{\Delta\lambda}{\lambda^2} \times\left\{\frac{2Z^2 I_H}{kT}\sum_{n>(Z^2 I_H\lambda/hc)^{1/2}}^{n'}\frac{\gamma_{fb}}{n^3}\exp\left(\frac{Z^2 I_H}{n^2kT}\right) + \bar{\gamma}_{fb}\left[\exp\left(\frac{Z^2 I_H}{(n'+1)^2kT}\right)-1\right]+\gamma_{ff}\right\},$$

where N_e is the electron density, N_Z the number density of hydrogenic (bare) ions of nuclear charge Z, I_H the ionization energy of hydrogen, n' the principal quantum number of the lowest level for which spacings to adjacent levels are so close that they approach a continuum and summation over n may be replaced by an integral. (Choice of n' is rather arbitrary—n' as low as 6 is found in the literature.) γ_{fb} and γ_{ff} are the Gaunt factors, which are generally close to unity.[20] (For the higher free-bound continua, starting with $n'+1$, an average Gaunt factor $\bar{\gamma}_{fb}$ is used.) For neutral hydrogen, the recombination continuum forming H^- becomes important, too.[21]

In the equation above, the value of the constant factor is 6.065×10^{-55} W m^4 J$^{1/2}$ sr^{-1}. [Numerical example: For atomic hydrogen ($Z = 1$), the quantity ϵ_{cont} has the value 2.9 W m^{-3} sr^{-1} under the following conditions: $\lambda = 3\times10^{-7}$ m; $\Delta\lambda = 1\times10^{-10}$ m; N_e ($= N_{Z=1}$) $= 1\times10^{21}$ m^{-3}; $T = 12\,000$ K. The lower limit of the summation index n is 2; the upper limit n' has been taken to be 10. All Gaunt factors—γ_{fb} ($n = 2,3,\ldots,10$), $\bar{\gamma}_{fb}$, and γ_{ff}—have been assumed to be unity.]

B. Many-electron systems

For many-electron systems, only approximate treatments exist, based on the quantum-defect method (for results of calculations for noble gases, see, e.g., Ref. 22). Modifications of the continuum by autoionization processes must also be considered.[23]

By analogy with the line intensities (Sec. 5.05.A), the continuum intensity emitted by a homogeneous source of length l is $I_{cont} = \epsilon_{cont} l$.

C. Continuity across series limit

Near the ionization limit ($n'^* \to \infty$), the f values for bound-bound transitions of a spectral series [Sec. 5.08.A(2)] should make a smooth connection, across the limit, to the differential oscillator strength distribution $df/d\epsilon$:[23]

$$\text{(bound)}\ \frac{(n'^*)^3 f}{2(Z-N+1)^2} \longleftarrow \big| \longrightarrow \frac{df}{d\epsilon}\ \text{(continuum)}. \quad (\big| = \text{ionization limit } I)$$

$\epsilon = (E-I)/R_Z hc$ for a continuum state ($E>I$). (For E, I in wave-number units, omit the factor hc.) In the case of hydrogenic species, $n'^* \equiv n'$.

5.11. REFERENCES

[1]E. U. Condon and H. Odabasi, *Atomic Structure* (Cambridge University, Cambridge, England, 1980).

[2]W. C. Martin, R. Zalubas, and L. Hagan, *Atomic Energy Levels—The Rare-Earth Elements,* Natl. Stand. Ref. Data Ser. Natl. Bur. Stand. **60** (1978).

[3]C. E. Moore, *Atomic Energy Levels (As Derived from the Analyses of Optical Spectra),* Natl. Stand. Ref. Data Ser. Natl. Bur. Stand. **35** (1971) [reprinted from Natl. Bur. Stand. (U.S.) Circ. **467**, Vols. I (1949) (H through V), II (1952) (Cr through Nb), and III (1958) (Mo through La, Hf through Ac)].

[4]New, separate compilations for He, Na, Mg, Al, K, Ca, Sc, Ti, V, Cr, Mn, Fe, and Ni are in recent issues of J. Phys. Chem. Ref. Data, and for H, D, T, C, N, O, and Si in Natl. Stand. Ref. Data Ser. Natl. Bur. Stand. **3**.

[5]J. Reader and C. H. Corliss, in *Handbook of Chemistry & Physics,* 62nd ed., edited by R. C. Weast and M. J. Astle (Chemical Rubber, Boca Raton, FL, 1981), pp. E205–E334; same tables, with finding list, in Natl. Stand. Ref. Data Ser. Natl. Bur. Stand. **68** (1980), Pt. I (45 000 lines, most elements included).

[6]R. L. Kelly and L. J. Palumbo, *Atomic and Ionic Emission Lines Below 2000 Angstroms—Hydrogen through Krypton,* Naval Research Laboratory Report 7599, 1973; R. L. Kelly, *Atomic Emission Lines in the Near Ultraviolet—Hydrogen through Krypton,* NASA Technical Memorandum 80268, 1979.

[7]V. Kaufman and B. Edlén, J. Phys. Chem. Ref. Data **3**, 825 (1974) (reference wavelengths).

[8]C. E. Moore, *A Multiplet Table of Astrophysical Interest, Revised Edition,* Natl. Stand. Ref. Data Ser. Natl. Bur. Stand. **40** (1972) (comprehensive).

[9]C. E. Moore, *An Ultraviolet Multiplet Table,* Natl. Bur. Stand. (U.S.) Circ. **488**, Secs. 1 (1950) (H through V), 2 (1952) (Cr through Nb), 3 (1962) (Mo through La, Hf through Ra), 4 (1962) (finding list—H through Nb), and 5 (1962) (finding list—Mo through La, Hf through Ra).

[10]W. L. Wiese and G. A. Martin, in *Handbook of Chemistry & Physics,* 62nd ed., edited by R. C. Weast and M. J. Astle (Chemical Rubber, Boca Raton, FL, 1981), pp. E335–E370; Natl. Stand. Ref. Data Ser. Natl. Bur. Stand. **68** (1980), Pt. II (5000 lines, most elements included).

[11]W. L. Wiese, M. W. Smith, and B. M. Glennon, *Atomic Transition Probabilities—Vol. I, Hydrogen through Neon,* Natl. Stand. Ref. Data Ser. Natl. Bur. Stand. **4** (1966).

[12]W. L. Wiese, M. W. Smith, and B. M. Miles, *Atomic Transition Probabilities—Vol. II, Sodium through Calcium,* Natl. Stand. Ref. Data Ser. Natl. Bur. Stand. **22** (1969).

[13]G. A. Martin, W. L. Wiese, J. R. Fuhr, and S. M. Younger, *Atomic Transition Probabilities—Vol. III, Scandium through Nickel,* Natl. Stand. Ref. Data Ser. Natl. Bur. Stand. (to be published).

[14]C. W. Allen, *Astrophysical Quantities,* 3rd ed. (Athlone, London, 1973).

[15]H. A. Bethe and E. E. Salpeter, *Quantum Mechanics of One- and Two-Electron Atoms* (Springer, Berlin, 1957).

[16]K. T. Lu and U. Fano, Phys. Rev. A **2**, 81 (1970).

[17]A. W. Weiss, Adv. At. Mol. Phys. **9**, 1 (1973).

[18]H. R. Griem, *Spectral Line Broadening by Plasmas* (Academic, New York, 1974).

[19]C. R. Vidal, J. Cooper, and E. W. Smith, Astrophys. J. Suppl. Ser. **25**, 37 (1973).

[20]W. J. Karzas and R. Latter, Astrophys. J. Suppl. Ser. **6**, 167 (1961).

[21]J. R. Roberts and P. A. Voigt, J. Res. Natl. Bur. Stand. Sect. A **75**, 291 (1971).

[22]D. Schlueter, Z. Astrophys. **61**, 67 (1965).

[23]U. Fano and J. W. Cooper, Phys. Rev. **137**, A1364 (1965).

6.00. Biological physics

HANS FRAUENFELDER AND MICHAEL C. MARDEN

University of Illinois at Urbana-Champaign

CONTENTS

6.01. INTRODUCTION: HIERARCHY OF LIVING SYSTEMS

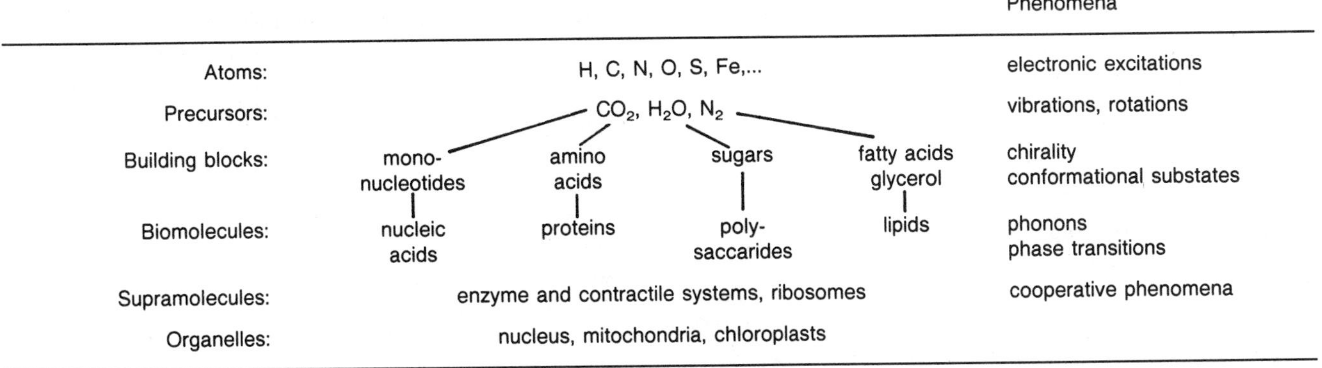

					Phenomena
Atoms:		H, C, N, O, S, Fe,...			electronic excitations
Precursors:		CO_2, H_2O, N_2			vibrations, rotations
Building blocks:	mono-nucleotides	amino acids	sugars	fatty acids glycerol	chirality conformational substates
Biomolecules:	nucleic acids	proteins	poly-saccarides	lipids	phonons phase transitions
Supramolecules:		enzyme and contractile systems, ribosomes			cooperative phenomena
Organelles:		nucleus, mitochondria, chloroplasts			

Organelles, in turn, are the components of cells, tissues, organs, and finally organisms. Biomolecular physics is concerned mainly with the properties of biomolecules, although knowledge of their building blocks and functions is essential.

Nucleic acids and proteins are the most important biomolecules. While nucleic acids store and transmit the information necessary for life, proteins form the machinery that executes essentially all functions involved in living.

6.02. DNA, RNA, AND THE GENETIC CODE

The entire genetic information of an organism (some 5×10^9 bits in humans), carried by deoxyribonucleic acid (DNA), is contained in every cell. The information is stored by using four nucleotides, which form complementary strands in a double helix structure. The nucleotides are Adenine (A), Cytosine (C), Guanine (G), and Thymine (T). Various ribonucleic acids (RNA) are involved in the execution of the commands stored in the DNA. In RNA, Thymine is replaced by Uracil (U). The pairs A-T (A-U in RNA) and C-G are energetically favored (Figure 6.02.A).

In DNA a sequence of base pairs (A-T, C-U) is formed along a helical sugar-phosphate backbone (Figures 6.02.B and 6.02.C). The sequence, as executed by the RNA, carries the genetic information, each bit being a nucleotide triplet which defines a specific amino acid (the building blocks of the proteins). The triplet UGG, for example, codes for tryptophan=TRP (Table 6.02.D). The DNA content and number of nucleotide pairs for various organisms is shown in Table 6.02.E.

FIGURE 6.02.A. Nucleotide pairs, the building blocks of DNA. All distances in pm. For RNA, replace T with U by substituting H for CH_3. From Ref. 1.

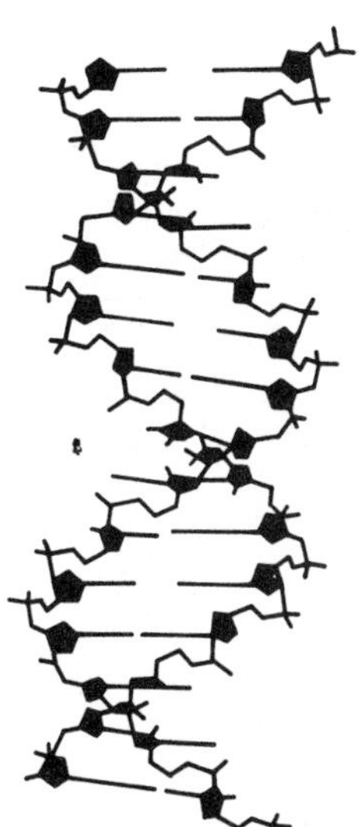

FIGURE 6.02.B. Double helical structure of DNA. From Ref. 2.

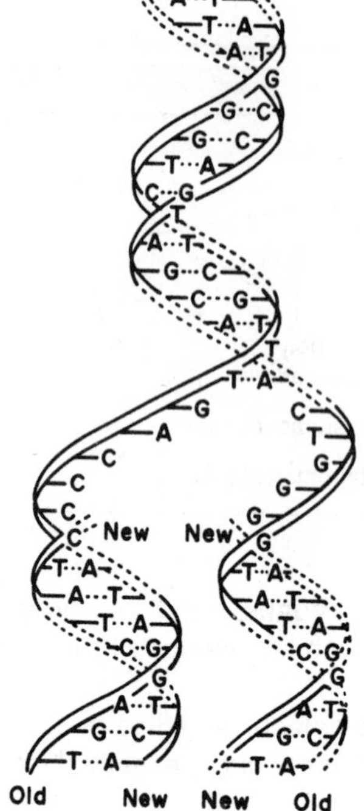

FIGURE 6.02.C. DNA replication. From Ref. 3.

TABLE 6.02.D. Genetic code

First	Second position U	C	A	G	Third
	PHE	SER	TYR	CYS	U
	PHE	SER	TYR	CYS	C
U	LEU	SER	Stop	Stop	C
	LEU	SER	Stop	TRP	G
	LEU	PRO	HIS	ARG	U
	LEU	PRO	HIS	ARG	C
C	LEU	PRO	GLN	ARG	A
	LEU	PRO	GLN	ARG	G
	ILE	THR	ASN	SER	U
	ILE	THR	ASN	SER	C
A	ILE	THR	LYS	ARG	A
	MET	THR	LYS	ARG	G
	VAL	ALA	ASP	GLY	U
	VAL	ALA	ASP	GLY	C
G	VAL	ALA	GLU	GLY	A
	VAL	ALA	GLU	GLY	G

TABLE 6.02.E. DNA content

Species	DNA pg/cell	No. of pairs
Mammals	6	5.5G
Amphibia	7	6.5G
Fishes	2	2.0G
Reptiles	5	4.5G
Birds	2	2.0G
Crustaceans	3	2.8G
Mollusks	1.2	1.1G
Sponges	0.1	0.1G
Plants	2.5	2.3G
Fungi	20m	20M
Bacteria	2–6m	2M
Bacteriophage T4	0.24m	170k
Bacteriophage	0.08m	50k

6.03. PROTEINS

Proteins perform all essential functions of life; they store and transfer energy, matter, and charge, are involved in the transfer and storage of information, perform catalysis, control all reactions, and are the construction elements of the organism (Table 6.03.A). Proteins are

TABLE 6.03.A. Classification of proteins by biological function. From Ref. 3.

Type and examples	Occurrence or function
Enzymes	
Hexokinase	Phosphorylates glucose
Lactate dehydrogenase	Dehydrogenates lactate
Cytochrome *c*	Transfers electrons
DNA polymerase	Replicates and repairs DNA
Storage proteins	
Ovalbumin	Egg-white protein
Casein	A milk protein
Ferritin	Iron storage in spleen
Gliadin	Seed protein of wheat
Zein	Seed protein of corn
Transport proteins	
Hemoglobin	Transports O_2 in blood of vertebrates
Hemocyanin	Transports O_2 in blood of some invertebrates
Myoglobin	Transports O_2 in muscle cells
Serum albumin	Transports fatty acids in blood
β_1-lipoprotein	Transports lipids in blood
Iron-binding globulin	Transports iron in blood
Ceruloplasmin	Transports copper in blood
Contractile proteins	
Myosin	Thick filaments in myofibril
Actin	Thin filaments in myofibril
Dynein	Cilia and flagella
Protective proteins in vertebrate blood	
Antibodies	Form complexes with foreign proteins
Complement	Complexes with some antigen-antibody systems
Fibrinogen	Precursor of fibrin in blood clotting
Thrombin	Component of clotting mechanism
Toxins	
Clostridium botulinum toxin	Causes bacterial food poisoning
Diphtheria toxin	Bacterial toxin
Snake venoms	Enzymes that hydrolyze phosphoglycerides
Ricin	Toxic protein of castor bean
Gossypin	Toxic protein of cottonseed
Hormones	
Insulin	Regulates glucose metabolism
Adrenocorticotrophic hormone	Regulates corticosteroid synthesis
Growth hormone	Stimulates growth of bones
Structural proteins	
Viral-coat proteins	Sheath around nucleic acid
Glycoproteins	Cell coats and walls
α-keratin	Skin, feathers, nails, hoofs
Sclerotin	Exoskeletons of insects
Fibroin	Silk of cocoons, spider webs
Collagen	Fibrous connective tissue (tendons, bone cartilage)
Elastin	Elastic connective tissue (ligaments)
Mucoproteins	Mucous secretions, synovial fluid

formed from 20 amino acids whose main properties are listed in Tables 6.03.B and 6.03.C. All amino acids have the same backbone, but differ in their sidechains (Figure 6.03.D). Proteins are soluble only in a narrow range of solvents. Typically, they are stable in certain solvents below temperatures of 340 K and pressures of 2 MPa. Some useful solvents for studying proteins at high viscosity or low temperature are shown in Table 6.03.E.

The backbones of amino acids can link covalently, as indicated in Figure 6.03.F, and thus form the primary protein sequence (the arrangement is determined by the DNA). The primary sequence of myoglobin is given as an example in Table 6.03.G. (For details and further information on sequences and structures of a large number of proteins see Ref. 8.) For a given protein, various species will differ in many of the amino acids; however, certain positions are invariant. For example, in cytochrome *c*, species can differ in their sequence from 1 (man versus rhesus monkey) to over 50 amino acids, but about 20 of the 125 (length varies slightly) positions are independent of the species. Properties and function of a protein are determined by the primary sequence. In the proper solvent, the linear chain will spontaneously fold into the secondary and tertiary structure (Figure 6.03.H). The two most common secondary structures are the alpha helix and the beta sheet. The picture of myoglobin shown in Figure 6.03.H is somewhat misleading: the actual protein does not have large holes, but is a rather tightly constructed system.

A protein is not a rigid, unique structure, but can exist in a large number of conformational substates. These have the same coarse structure and perform the same biological function, but differ in their detailed structure. An indication of some features of conformational substates is obtained by extraction of the mean-square displacements $\langle x^2 \rangle$ for all nonhydrogen atoms in a protein.[9] An example is given in Figure 6.03.I.

Since proteins require some medium (solvent or membrane), direct measurements of physical properties are difficult. One indirect technique is the determination

TABLE 6.03.B. Properties of the amino acids, the protein building blocks. The length (L) and volume are for the sidechain only. The molecular weight is for the entire amino acid—subtract 17.9 to obtain molecular weight of residue. The polarity indicates whether the amino acid is nonpolar (NP) or polar with a net positive, negative, or neutral charge at pH = 6. The relative frequency of occurrence is for E. Coli.

Amino acid	Symbol	Molecular weight (amu)	L (nm)	Volume (nm^3)	Polarity	Relative frequency	Sidechain (X = benzene)
Alanine	ALA A	89	0.28	0.0322	NP	100	—C
Arginine	ARG R	174	0.88	0.1257	+	41	—C—C—C—N—C(=N)—N
Asparagine	ASN N	132	0.51	0.0654	0	76	—C—C(=O)—N
Aspartic acid	ASP D	133	0.50	0.0584	−		—C—C(=O)—O
Cysteine	CYS C	121	0.43	0.0579	0	14	—C—S
Glutamine	GLN Q	146	0.64	0.0925	0	83	—C—C—C(=O)—N
Glutamic acid	GLU E	147	0.63	0.0855	−		—C—C—C(=O)—O
Glycine	GLY G	75	0.15	0.0051	0	60	—H
Histidine	HIS H	155	0.65	0.0890	+	5	—C—C=C, N, N, C (ring)
Isoleucine	ILE I	131	0.53	0.1134	NP	34	—C(—C)—C—C
Leucine	LEU L	131	0.53	0.1134	NP	60	—C—C(—C)—C
Lysine	LYS K	146	0.77	0.1210	+	54	—C—C—C—C—N
Methionine	MET M	149	0.69	0.1121	NP	29	—C—C—S—C
Phenylalanine	PHE F	165	0.69	0.1366	NP	25	—C—X
Proline	PRO P	115			NP	35	C—C—C—N—C (ring)
Serine	SER S	105	0.38	0.0360	0	46	—C—O
Threonine	THR T	119	0.40	0.0631	0	35	—C(—O)—C
Tryptophan	TRP W	204	0.81	0.1755	NP	8	—C—C=C, N (indole ring)
Tyrosine	TYR Y	181	0.77	0.1388	0	17	—C—X—O
Valine	VAL V	117	0.40	0.0863	NP	46	—C(—C)—C

TABLE 6.03.C. Properties of amino acids. An asterisk indicates amino acids not produced by the human body. The definition of pK is given in Sec. 6.07.B; pK_1 characterizes the carboxyl group, pK_2 the ammonium ion, and pK_R the sidechain. For CYS, the lower pK_R refers to the sulphydryl adjacent to the NH_3, the higher to the sulphydryl adjacent to the amine NH_2. pI is the pH of an aqueous solution such that the net charge on the molecule is zero. Cost is the energy of synthesis in units of ATP-ADP equivalents, 30.6 kJ mol^{-1}.[4] The molar polarizability P is calculated for the sidechains.[5] The volume per mole includes any bound water (which has a negative contribution to the compressibility β). Φ is the hydrophobicity. Refer to C. Tanford in reference text list.

Amino acid	pK_1 (COOH)	pK_2 (NH_3^+)	pK_R	pI	Cost (ATP)	P (10^6 m^{-3}/mol)	V (10^{-6} m^3/mol)	$-\beta$ (10^{-12} kg^{-1} ms^2)	Φ (kJ/mol)
ALA	2.35	9.87		6.11	20	18.6	60.5	423	3.1
ARG*	2.01	9.04	12.48	10.76	44	42.2	127.3	209	3.1
ASN					22	28.4			−0.04
ASP	2.10	9.82	3.86	2.98	21	24.9	73.8	449	2.3
CYS	1.04	10.50	8, 10.3	5.02	19	26.4	73.4	447	2.8
GLN					31	36.1			−0.4
GLU	2.10	9.47	4.07	3.08	30	29.5	85.9	421	2.4
GLY	2.35	9.78		6.06	12	14.0	43.2	625	0
HIS*	1.77	9.18	6.10	7.64	42	39.5	98.8	322	6.0
ILE*					55	32.6			12.8
LEU*	2.36	9.60			47	32.6	107.7	295	9.7
LYS*	2.18	8.95	10.53	9.47	50	36.0			6.5
MET*					44	35.9	105.4	296	5.6
PHE*	1.8	9.1			65	37.8	121.5	284	11.4
PRO					39	26.9	82.8	281	11.2
SER	2.21	9.15			18	20.2			0.2
THR*	2.63	10.43			31	24.8			12.3
TRP*					78	54.1	143.9	210	12.9
TYR	2.20	9.11	10.07	5.63	62	40.3			12.3
VAL*	2.3	9.6			39	27.9	90.8	337	7.3

of the dielectric coefficient of a solvent with (D) and without (D_0) the protein: $D = D_0 + (\partial D/\partial c)c$, where c is the protein concentration and $\partial D/\partial c$ is called the dielectric increment. The frequency dependence of the dielectric coefficient yields the relaxation time t_r, since dipoles will not align for frequencies greater than $\nu = 1/2\pi t_r$. The dipole moment is calculated from the limiting values (high and low frequency) of the dielectric coefficient. Values for some proteins are given in Table 6.03.K.

Measurements have also been made of the density ρ and sound velocity c versus protein concentration. The protein compressibility can be extracted from these values: $\beta = 1/\rho c^2$. The interaction with the bound water has a negative contribution to the compressibility. This effect causes an overall negative compressibility for the amino acids and fibular proteins. Values for the globular proteins are less influenced. Published values vary by over 50%.

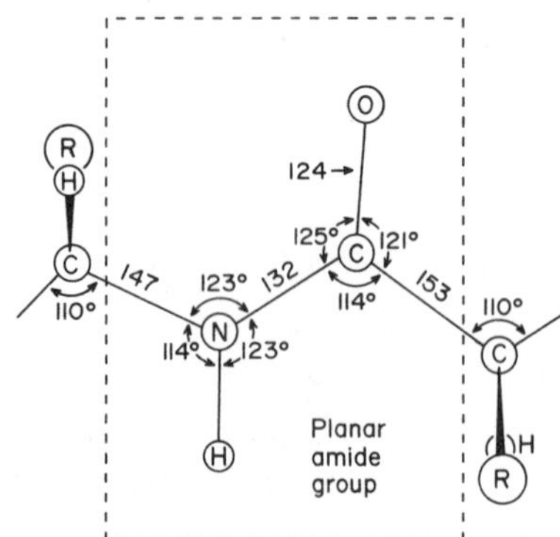

FIGURE 6.03.D. Backbone structure of amino acids. Bond lengths are in pm. After Ref. 6.

TABLE 6.03.E. Solvent properties. Mixtures of water with the various cosolvents can be used to obtain other values of these physical properties.[7] Freezing points below 200 K can be obtained from mixtures of about 60–90% of solvents 2–5 with water. Values are at 298 K for the density ρ, index of refraction n, thermal expansion α [$= (1/V)(\partial V/\partial T)$], compressibility β [$= -(1/V)(\partial V/\partial P)$], dielectric coefficient ϵ, and viscosity η (P = poise = g/cm s).

Solvent	Molecular weight (amu)	bp (K)	fp (K)	ρ (g/cm^3)	n	α (10^{-6} K^{-1})	β (10^{-12} kg^{-1} ms^2)	ϵ	η (P)
1 Water	18.02	373	273	0.997	1.333	257	460	78.5	10m
2 Methanol	32.04	338	179	0.787	1.329		1260	32.6	5.5m
3 Ethanol	46.07	352	156	0.785	1.361	750	1150	24.3	12m
4 Glycerol	92.11	563	291	1.261	1.475	490	210	42.5	14.9
5 E glycol	62.07	471	260	1.11	1.432		370	37.7	199m
6 N_2 (gas)	28.01	77	63	0.0012	1.0003			1.0	175μ

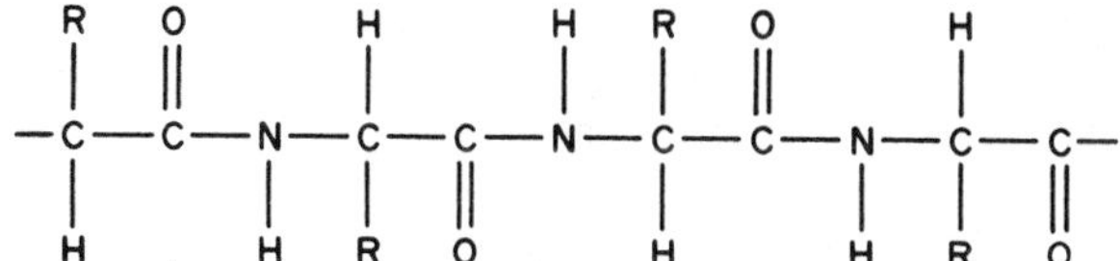

FIGURE 6.03.F. Schematic representation of the primary structure of proteins. R denotes the residues that distinguish the 20 different amino acids, as indicated in Table 6.03.B.

TABLE 6.03.G. Amino acid sequence of sperm whale myoglobin

1	VLSEG	EWQLV	LHVWA	KVEAD	VAGHG	QDILI
31	RLFKS	HPETL	EKFDR	FKHLK	TEAEM	KASED
61	LKKHG	VTVLT	ALGAI	LKKKG	HHEAE	LKPLA
91	QSHAT	KHKIP	IKYLE	FISEA	IIHVL	HSRHP
121	GNFGS	DAQGA	MNKAL	ELFRK	DIAAK	YKELG
151	YQG					

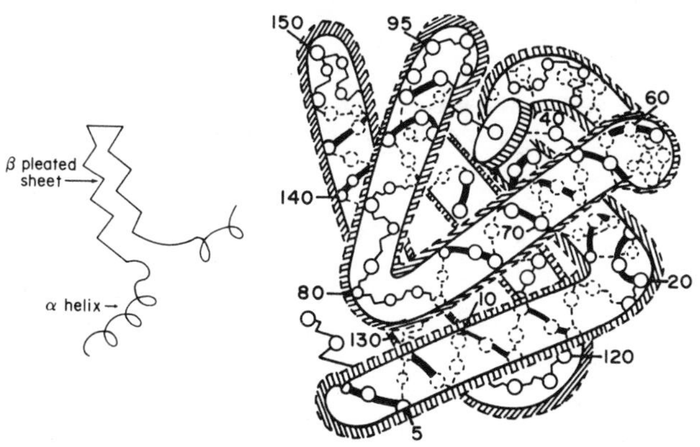

FIGURE 6.03.H. Secondary and tertiary structure. At left the secondary structure is shown schematically. The two most important secondary elements, the alpha helix and the beta pleated sheet, are sketched. At right the backbone of myoglobin in the globular tertiary structure is shown. Also indicated through the shaded outside layer is the amplitude of the displacement along the chain. The numbers denote the residues given also in Table 6.03.G. From Ref. 9.

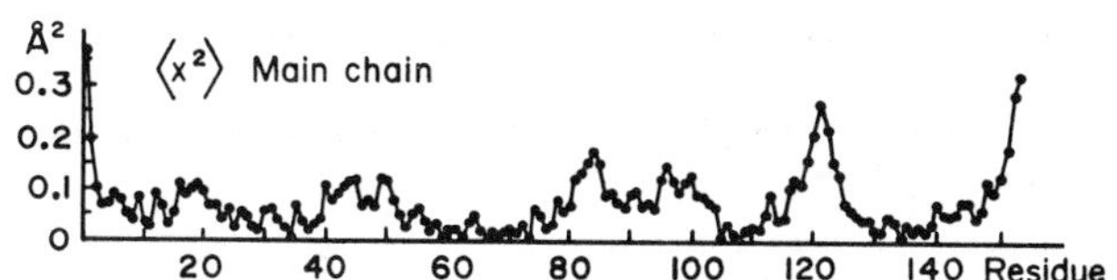

FIGURE 6.03.I. Example of the mobility of the proteins: the mean-square displacements as derived from x-ray diffraction are plotted along the residue number of myoglobin. This plot describes the shaded area in Figure 6.03.H in detail. From Ref. 9.

The protein environment does affect the physical properties. There is approximately 0.3 g of water per gram of dry protein. Proteins have a water layer about 1–2 molecules thick. The exterior of the protein consists mainly of hydrophilic amino acids whose ionic state will depend on the solvent pH. For example, BSA (bovine serum albumin) has a net charge of -19 at $pH=7.4$, but is neutral at $pH=pI=4.7$. Two units of charge separated by a typical protein size of 3 nm produce a dipole moment of 300 D (debyes: $1\,D=3.336\times10^{-30}$ C m). Thus large dipoles and many ionic interactions on the surface are possible. The interior of the protein is mainly composed of hydrophobic amino acids. The dielectric coefficient of the protein interior is estimated to be about 2, which would shield charge interactions much less than water ($\epsilon=80$).

6.04. MEMBRANES

Membranes, the interfaces between cells and tissues, govern the transport and gradients of various small molecules. The building blocks of membranes are the lipids. Lipids consist of a polar, hydrophilic head and a hydrocarbon, hydrophobic tail. They are insoluble in water, but highly soluble in organic solvents. In aqueous solution the lipids will arrange to minimize their total free energy. Three typical structures are shown in Figure 6.04.A.

Since membranes form the boundary between biological systems, one important characterization is their

TABLE 6.03.J. Properties of chemical bonds. L is the bond length and P the polarizability per molecule.

Bond	Energy (kJ/mol)	L (pm)	P (10^{-18} m^3)
H—H	450	74	0.79
C—C	350	154	0.475
C═C	680	134	1.59
C≡C	960	121	2.31
C—H	410	110	0.655
C—N	340	147	0.598
C═N	660	129	1.86
C≡N	1000	116	
C—O	380	142	0.559
C═O	730	121	1.31
C≡O	1100	113	
O—H	460	96	0.733
N—H	390	100	0.721

Interaction	Energy (kJ/mol)
Covalent	200–1000
Electrostatic	10–30
Hydrogen bond	5–20
Van der Waals	1–10
Hydrophobic (not a true force)	0–20

TABLE 6.03.K. Properties of biomolecules. TMV is tobacco mosaic virus, BSA bovine serum albumin. The length L refers to the maximum dimension. % helix indicates the fraction of the protein that is built from helical secondary structures. Compressibility (β) measurements are at a frequency of 7 MHz. The other entries are the dipole moment μ, sedimentation coefficient s, diffusivity D, and log of rotational rate k_r (in s^{-1}). See Ref. 10.

Molecule	Molecular weight (amu)	L (nm)	% helix	ρ (g/cm^3)	β (10^{-12} kg^{-1} m s^2)	pI	μ (D)	s (10^{-13} s)	D (10^{-6} cm^2/s)	logk_r
Glycine	75	0.15			−625	6.1	16		9.5	
Sucrose	342	0.7		1.58					4.6	
Insulin	5.7k		31	1.34		5.4	360	3.58	0.75	
Lysozyme	14.4k	4.5	29	1.38	44	11.1		1.91	1.19	
Myoglobin	17k	4.5	77	1.36	43	7.0	170	2.04	1.13	
Hemoglobin	65k	6.8	72	1.33		6.8	480	4.13	0.69	6.8
BSA	65k		50	1.35				4.6	0.59	
Myosin	470k	160.0		1.37	−180			6.43	0.12	0.85
TMV	40M	300.0		1.37				185	0.053	2.7

permeability (Figure 6.04.B), which is usually small, except for water. The interior of the membranes can exist in an ordered and rigid, or in a semiliquid, state. The transition occurs at a temperature that depends on the length and degree of saturation of the fatty acid chain. Membranes contain proteins (Figure 6.04.C); the content varies from 20 to 75%.

FIGURE 6.04.A. Three typical arrangements of fatty acids: miscelle bilayer membrane, and liposome (closed bilayer). From Ref. 2.

6.05. ENERGY

Living systems require energy for synthesis of molecules, molecular transport, and muscle contraction (mechanical work). Organisms derive their energy directly from sunlight (photosynthesis) or indirectly by the oxidation of energy-rich materials. The basic unit of energy currency is ATP (adenine triphosphate). The reaction ATP-ADP, releasing 30.6 kJ/mol, is the immediate energy source for most reactions. (The ATP level does not act as an energy reservoir. A certain ATP/ADP ratio is maintained, and the ATP-ADP cycle acts as an interface that channels energy from bulk sources to the specific place where it is needed.) Over 10^{19} kJ/yr (10^{15} W) of solar energy are used by plants to produce the fuels used by other organisms. Some typical values of energy use and storage by humans are shown in Table 6.05.A. As machines, muscles are efficient (25%); the biosphere makes excellent use of both materials and energy.

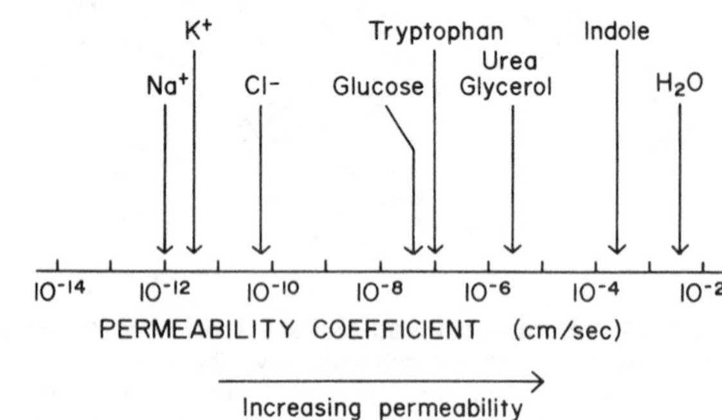

FIGURE 6.04.B. Permeability of a typical membrane for various substances. From Ref. 2.

6.06. TIME SCALES

The times involved in biological events range from about 10^{-15} s in elementary relaxation steps[11] to 10^{14} s, characteristic of the mutation rate of a particular protein.

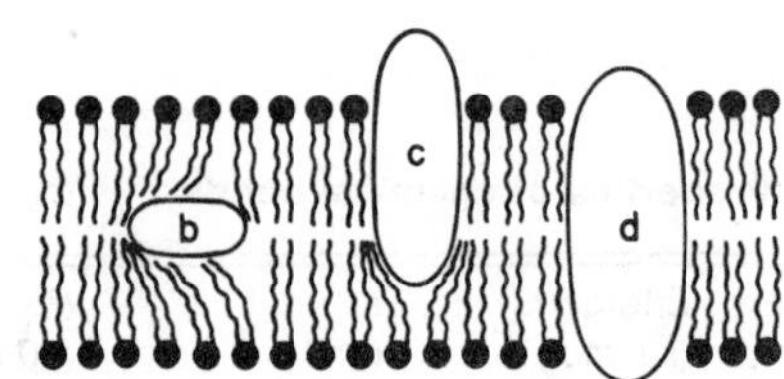

FIGURE 6.04.C. Various arrangements of proteins embedded in a membrane. From Ref. 2.

TABLE 6.05.A. Typical energy values for a 70-kg human

Use		Storage	
Input	10 MJ/day = 100 W	Tryacylglycerols	420 MJ at 38 MJ/kg
Rest	85 W	Protein	100 MJ at 17 MJ/kg
Steady work	85 + (30–40) W	Glycogen	2.5 MJ at 6.3 MJ/kg
Maximum exertion	4000 W	Glucose	170 kJ

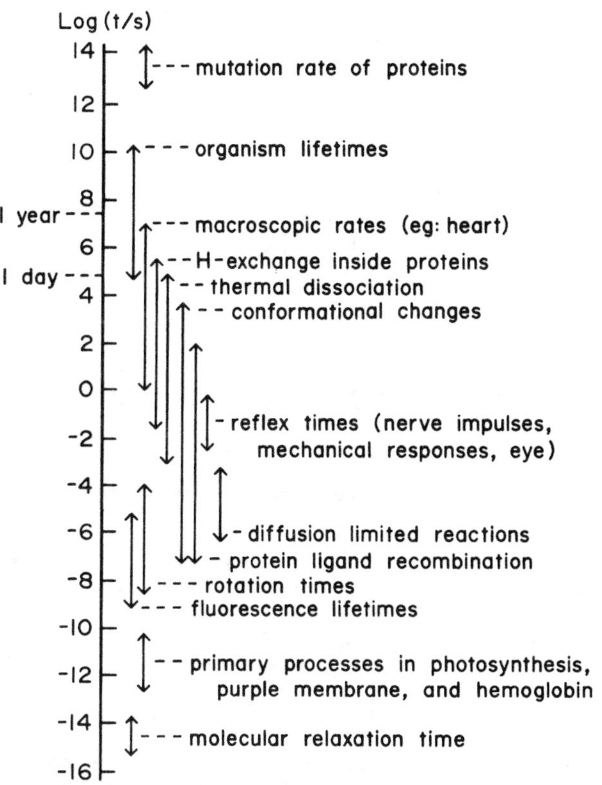

FIGURE 6.06.A. Time scale of biological events

Figure 6.06.A provides a survey of the times of important biological phenomena.

6.07. SELECTED FORMULAS AND SPECTRA

Nearly all relations used in biological physics are covered in other sections of the handbook. We list here only a few formulas that are useful in describing biomolecular phenomena. In Figs. 6.06.B–6.06.D, we add some typical spectra.

A. Reaction rate theory

The rate coefficient k for transition of a system over a barrier of height $G^{\neq}$ at temperature T is given by transition state theory as

$$k = \nu \exp(-G^{\neq}/RT). \tag{1}$$

Here $G^{\neq} = H^{\neq} - TS^{\neq}$ is the activation Gibbs energy, $H^{\neq}$ the activation enthalpy, $S^{\neq}$ the activation entropy, and R the gas constant ($= 8.31\ \mathrm{J\ mol^{-1}\ K^{-1}}$). The preexponential is an attempt frequency. Equation (1) applies best to gas phase reactions.[12] For reactions within proteins, a modified Kramers relation is more appropriate; the rate coefficient for a wide range of solvents is given by

$$k = (A/\eta^{\kappa}) \exp(-G^{*}/RT). \tag{2}$$

Here A is a preexponential factor connected to the shape of the barrier, η the solvent viscosity, κ a parameter that describes the solvent-protein coupling ($0 \leqslant \kappa \leqslant 1$), and G^{*} the activation Gibbs energy evaluated at constant viscosity.[13]

For a diffusion-controlled reaction, the rate coefficient is given by

$$k_D = 7.57 \times 10^{21} D r_i c\ \mathrm{s^{-1}}. \tag{3}$$

D is the sum of the diffusivities of the reactants (cm^2/s), r_i the interaction distance (cm), and c the concentration (M). When Stokes' law applies, the viscous drag force on a spherical molecule of radius b and velocity v is $F = 6\pi\eta b v$, the diffusivity $D = k_B T/6\pi\eta b$. This relationship may not be valid for small molecules. For a random walk, the mean-square displacement $\langle r^2 \rangle = 6Dt$. Typical values for the diffusion of dioxygen in water at 298 K are

$$D = 2 \times 10^{-5}\ \mathrm{cm^2/s}, \quad k_D = 10^{10}\ \mathrm{s^{-1}}$$

when $c = 1$ M and $r_i = 0.6$ nm.

B. Amino acid charges

The charges on amino acids are crucial for the function of proteins. Amino acids with neutral sidechains can assume three charge states, A^{+}, A^{0}, and A^{-}. For the

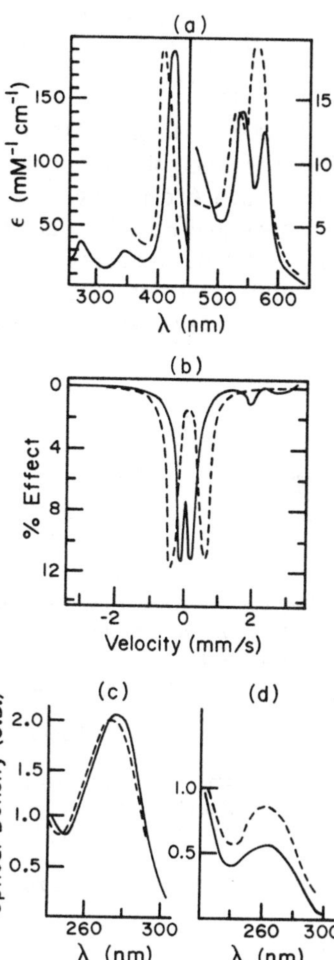

FIGURE 6.06.B. Biomolecular spectra. (a) Visible absorption spectra of protoheme-CO (dashed line) and myoglobin-CO. The optical density for a given sample is defined by $I = I_0 \times 10^{-\mathrm{OD}}$, where I_0 is the intensity of the incident, I the intensity of the transmitted beam. (b) Mössbauer spectra of protoheme-CO (dashed line) and myoglobin-CO. (c) UV spectra of a protein, serum globulin, and its constituent amino acids (dashed line). (d) UV spectra of DNA and the sum of its constituents (dashed line). The extinction coefficient ϵ is defined by $\mathrm{OD} = \epsilon c l$, where c is the concentration (moles/l) and l is the optical path in cm. For further information in regards to (c) and (d), see Setlow and Pollard in reference text list.

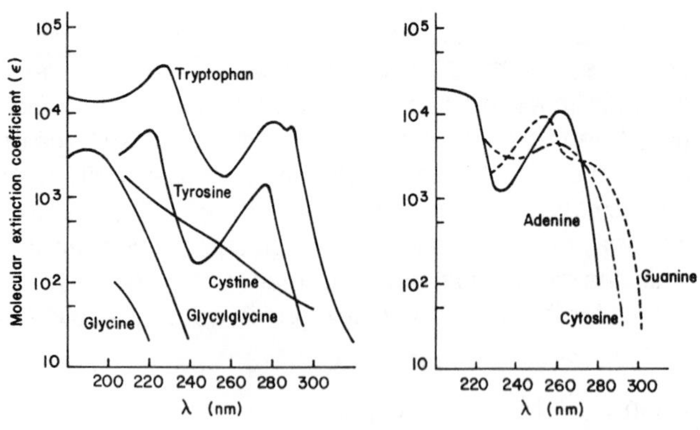

FIGURE 6.06.C. UV spectra of some amino and nucleic acids

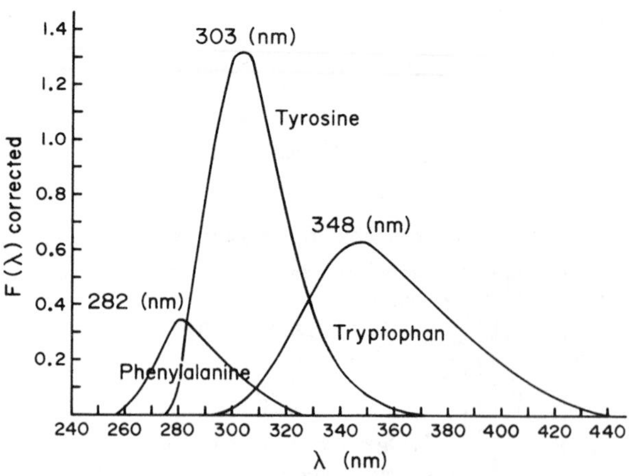

FIGURE 6.06.D. Fluorescence spectra of some amino acids

reactions $A^+ \rightleftarrows A^0 + H^+$ and $A^0 \rightleftarrows A^- + H^+$, two equilibrium coefficients K_1 and K_2 can be defined by $K_1 = [H^+][A^0]/[A^+]$ and $K_2 = [H^+][A^-]/[A^0]$, where brackets denote concentrations. With $\log[H^+] = pH$ and $\log K_1 = pK_1$, the definition of K_1 can be written

$$pH = pK_1 + \log\{[A^0]/[A^+]\}. \quad (4)$$

A similar expression holds for pK_2. At low pH ($pH < pK_1$) the amino acid will be predominantly positive, at $pH > pK_2$ predominantly negative. At $pH = pK_1$, the probabilities of finding A^+ or A^0 are equal. For amino acids with charged sidechains pK_R similarly indicates the pH at which charged and neutral sidechains have equal probabilities. Values for pK_1, pK_2, and pK_R are given in Table 6.03.C.

6.08. REFERENCES

[1]L. Pauling and R. B. Corey, Arch. Biochem. Biophys. **65**, 164 (1956).

[2]L. Stryer, *Biochemistry,* 2nd ed. (Freeman, San Francisco, 1981).

[3]A. L. Lehninger, *Biochemistry* (Worth, New York, 1987).

[4]D. E. Atkinson, *Cellular Energy Metabolism and Its Regulation* (Academic, New York, 1977).

[5]R. Gabler, *Electrical Interactions in Molecular Biophysics* (Academic, New York, 1978).

[6]R. B. Corey and L. Pauling, Rend. 1st. Lomb. Sci. Lett. Cl. Sci. Mat. Nat. **89**, 10 (1955).

[7]P. Douzou, *Cryobiochemistry* (Academic, London, 1977).

[8]*Atlas of Protein Sequence and Structure,* edited by M. O. Dayhoff (National Biomedical Research Foundation, Silver Springs, MD, 1972–1977).

[9]H. Frauenfelder, G. A. Petsko, and D. Tsernoglou, Nature **280**, 558 (1979).

[10]K. Gekko and H. Noguchi, J. Phys. Chem. **83**, 2706 (1979); *Biophysical Science: A Study Program,* edited by J. L. Oncely (Wiley, New York, 1979); R. Pethig, *Dielectric and Electronic Properties of Biological Materials* (Wiley, New York, 1979).

[11]J. A. McCammon and M. Karplus, Ann. Rev. Phys. Chem. **31**, 29 (1980).

[12]S. Glasstone, K. J. Laidler, and H. Eyring, *The Theory of Rate Processes* (McGraw-Hill, New York, 1941).

[13]H. A. Kramers, Physica (Utrecht) **7**, 284 (1940); J. L. Skinner and P. Wolynes, J. Chem. Phys. **69**, 2143 (1978); R. S. Larson and M. D. Kostin, *ibid.* **69**, 4821 (1978); D. Beece, L. Eisenstein, H. Frauenfelder, D. Good, M. C. Marden, L. Reinisch, A. H. Reynolds, L. B. Sorensen, and K. T. Yue, Biochemistry **19**, 5147 (1980).

The following texts treat various aspects of biological physics in easily accessible form:

C. R. Cantor and P. R. Schimmel, *Biophysical Chemistry* (Freeman, San Francisco, 1980).

R. E. Dickerson and I. Geis, *The Structure and Action of Proteins* (Benjamin, Menlo Park, CA, 1969).

A. L. Lehninger, *Biochemistry* (Worth, New York, 1975).

G. E. Schulz and R. H. Schirmer, *Principles of Protein Structure* (Springer, New York, 1979).

R. B. Setlow and E. C. Pollard, *Molecular Biophysics* (Addison-Wesley, Reading, MA, 1962).

L. Stryer, *Biochemistry* (Freeman, San Francisco, 1981).

C. Tanford, *Physical Chemistry of Macromolecules* (Wiley, New York, 1961).

Additional data can be found in:

Handbook of Biochemistry & Molecular Biology, 3rd ed., edited by G. D. Fasman (Chemical Rubber, Cleveland, 1976).

Handbook of Chemistry & Physics, 68th ed., edited by Robert C. Weast (Chemical Rubber, Cleveland, 1987).

J. D. Watson, *The Molecular Biology of the Gene* (Benjamin, Menlo Park, CA, 1976).

7.00. Cryogenics

Russell J. Donnelly

University of Oregon

CONTENTS

7.01. INTRODUCTION

The past century has seen steady progress in the achievement of ever lower temperatures by physicists, chemists, and engineers, and the gradual commercial development of successively lower temperatures has been one area where pure research has produced a steady flow of new technology into the marketplace.

The meaning of "low-temperature physics" and "cryogenics" has therefore evolved steadily to include an ever wider range of phenomena. In the 1940s and 1950s low-temperature physics meant investigations with liquid nitrogen, liquid hydrogen, and liquid helium, and in particular the remarkable macroscopic quantum phenomena exhibited by superconductors and by the superfluid phase of liquid ^{4}He. There are many references for these subjects, but a good starting point for both would be *Progress in Low Temperature Physics,* originally edited by C. J. Gorter, and now by D. F. Brewer.[11] For superconductivity, there is the two-volume series *Superconductivity,* edited by Parks,[13] and for superfluidity, the two-volume series *The Physics of Liquid and Solid Helium,* edited by Benneman and Ketterson.[12]

There is often confusion about the designation of the various helium liquids: He I is the normal liquid phase of ^{4}He, He II is the phase of liquid ^{4}He below the λ temperature exhibiting superfluidity. Solutions of liquid ^{3}He and ^{4}He can form a whole series of "custom-made" superfluids, and offer an interesting important new field of research. Liquid ^{3}He is not only important as a Fermi liquid, but also exhibits superfluidity in the millidegree range of temperatures. The reader should note that experiments are now conducted routinely at such temperatures—a full factor of 1000 times colder than the frontiers of research in the first half of this century.

7.02. PROPERTIES OF SUPERCONDUCTORS

A. Table: Superconducting elements

Compiled by S. L. Wipf, LANL, from: B. W. Roberts, J. Phys. Chem. Ref. Data **5**, 581 (1976) (Reprint 84); "Properties of Selected Superconductive Materials, 1978 Supplement," Natl. Bur. Stand. (U.S.) Tech. Note **983** (1978); and (for Am, Pa, and Sc) J. L. Smith (private communication).

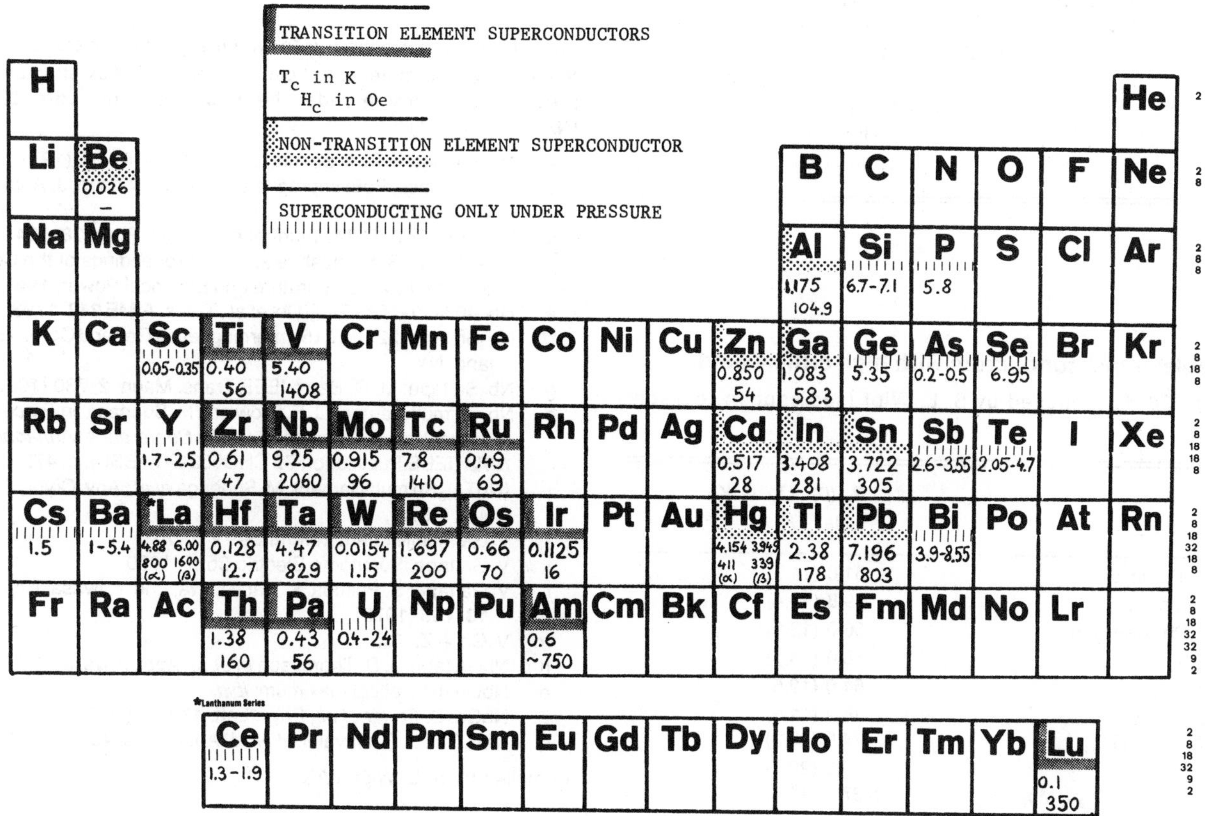

B. Table: Superconductors with high critical temperature

Ternary and higher alloys derived from binaries by addition or substitution are listed only if their critical temperature T_c surpasses that of the binary. The highest reported T_c is listed in parentheses; usually T_c is sensitive to the metallurgical preparation. Compiled by S. L. Wipf (see caption to Table 7.02.A).

T_c (K)	Binary	Ternary and higher
> 20	Nb_3 Ge (23)	$Nb_{3.16}Al_{0.64}Ge_{0.2}$ (20.7)
	Nb_3 Ga (20.7)	$NbAl_{0.24}Ga_{0.05}Ge_{0.05}$ (20.4)
		$Nb_3Ge_{0.9}Si_{0.1}$ (20.3)
18–20	Nb_3Al (18.7)	$Nb_3Al_{0.8}Ge_{0.2}$ (19.7)
	Nb_3Sn (18.3)	$Nb_3Al_{0.95}Be_{0.05}$ (19.6)
		$Nb_{0.75}Al_{0.22}Si_{0.3}$ (19.2)
		$Nb_3Al_{1-y}B_y$ (< 19.1)
		$Nb_3Al_{0.5}Ga_{0.5}$ (19)
		$V_3Si_{0.75}Ga_{0.25}$ (18.6)
16–18	V_3Si (16.9)	$NbN_{1-0.6}N_{1-0.6}C_{0-0.4}Ti_{0-0.4}$ (< 18)
	NbN (16.5)	$NbN_{0.65}C_{0.35}$ (17.8)
		$Nb_{0.66}N_{0.85}Ti_{0.34}$ (17.6)
		$Nb_{1-x}N_{0.75}C_{0.25}Hf_x$ (< 17.6)
		$Nb_{0.998}NO_{0.002}$ (17.3)
		$C_{1.55}Th_{0.3}Y_{0.7}$ (17.1)
		$V_3Si_{1-x}B_x$ (< 17)
		$Cu_{0-0.85}H_{0.7}Pd_{1-0.15}$ implant (< 16.6)
14–16	Tc_3Mo (15.8)	$Ag_{0-0.4}H_xPd_{1-0.15}$ implant (< 15.6)
	V_3Ga (15.1)	$C_{1.55}W_{0.1}Y_{0.9}$ (14.8)
	MoC (14.3)	$PbMo_6S_8$ (14.7)
	$Mo_{0.51}Re_{0.43}$ (14.0)	$Al_{0.5}SnMo_{5.6}S_6$ (14.4)
		$SnMo_3S_4$ (14.2)
		$C_{0.75}Hf_{0.05}Mo_{0.95}$ (14.2)
		$C_{1.45}La_{0.5}Th_{0.5}$ (14.2)

C. Table: Superconductors with high critical field

$B_{c2}(0) > 30$ T. Compiled by S. L. Wipf (see caption to Table 7.02.A).

	$B_{c2}(0)$ tesla	(T_c in K) K
$Gd_{0.2}PbMo_6S_8$	61.0	(14.3)
$PbMo_{5.1}S_6$	59.8	(14.4)
$La_{0.2}Pb_{0.8}Mo_{6.35}S_8$	56.0	(13.2)
$Al_{0.5}Sn\,Mo_5S_6$	56.0	(14.4)
$PbMo_{6.35}S_8$	54.0	(12.6)
$La\,Mo_6S_8$	46.3	(11.3)
$Nb_3Al_{0.57}Ge_{0.23}$	44.0	(20.1)
Nb_3Ge	39.0	(20.7)
$Eu_{0.1-0.8}Sn_{1.1-0.24}Mo_{6.35}S_8$	≲37.0	(11.5)
$Eu_{0.2}La_{0.8}Mo_8Se_8$	35.0	(11.4)
$SnMo_5S_6$	34.4	(13.4)
Nb_3Ga	34.1	(20.2)
Nb_3Al	32.5	(18.7)
$Nb_3Al_{0.5}Ga_{0.5}$	31.0	(19.0)

D. Figure: Highest flux pinning forces $j_c \times B$ of various superconductors at 4.2 K

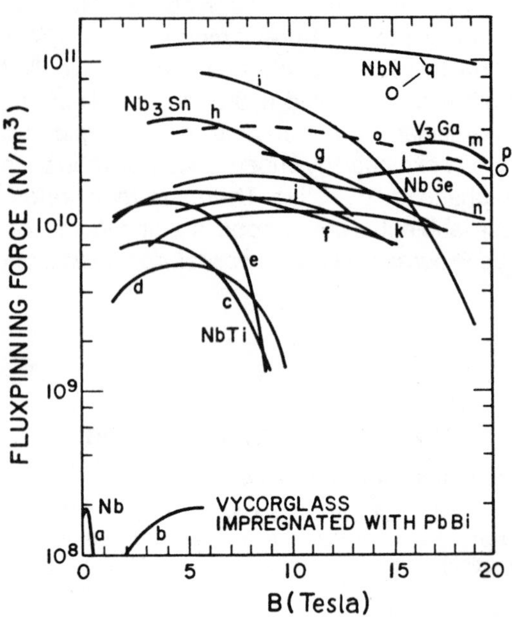

The critical current density j_c in type-II superconductors is given by the ability of the material to "pin" the magnetic flux structure, thus preventing its motion under the influence of the Lorentz force ($N/m^3 = A{\cdot}m^{-2}{\cdot}T$).

- a Nb, irradiated: S. T. Sekula, J. Appl. Phys. **42**, 16 (1971).
- b Porous glass, PbBi impregnated: J. H. P. Watson, J. Appl. Phys. **42**, 46 (1971).
- c Nb–78% Ti: A. D. McInturff *et al.*, J. Appl. Phys. **38**, 524 (1967).
- d Nb–61% Ti: R. Hampshire *et al.*, in Proceedings of the Conference on Low Temperature and Electrical Power, 1969.
- e Nb–40% Zr–10% Ti: T. Doi *et al.*, Trans. AIME **242**, 1793 (1968).
- f Nb_3Sn tape: 22 CYO 15, Intermagnetics General Corp., Guilderland, NY.
- g Nb_3Sn tape: M. G. Benz, IEEE Trans. Magn. **2**, 760 (1966).
- h Nb_3Sn multifilament: J. E. Crow and M. Suenaga, in *Proceedings of the Applied Superconductivity Conference, Annapolis, 1972* (IEEE Publication 72 CHO682-5-TABSC), p. 472.
- i $(NbTa)_3Sn$ multifilament: M. Suenaga *et al.*, Adv. Cryog. Eng. **26**, 442 (1979).
- j V_3Ga multifilament: reference for curve h.
- k V_3Ga tape: Sumitomo Electric Corp., 1970.
- l V_3Ga tape: S. Fukuda, K. Tachikawa, and Y. Iwasa, Cryogenics **13**, 153 (1973).
- m V_3Ga + Zr: *ibid.*
- n NbGe tape: J. D. Thompson *et al.*, J. Appl. Phys. **50**, 977 (1979).
- o NbGe theoretical maximum: *ibid.*
- p NbGe: S. Foner *et al.*, Phys. Lett. **47A**, 485 (1974).
- q NbN film: J. R. Gavaler *et al.*, J. Appl. Phys. **42**, 54 (1971).

Compiled by S. L. Wipf, LANL.

E. Table: Superconducting material parameters

The values in (a) are for pure, clean elements (i.e., long mean-free path); those in (b) are only representative because the parameters for alloys and compounds are sample dependent. Formulas for estimating the parameters for a particular sample are listed in T. P. Orlando *et al.*, Phys. Rev. B **19**, 4545 (1979); for more thorough listings of superconducting materials and parameters see B. W. Roberts, J. Phys. Chem. Ref. Data **5**, 823 (1976). T_c is the transition temperature. The penetration depth is given at zero temperature, λ_0, in (a), and as the Ginzburg-Landau coefficient $\lambda(T)=\lambda_{GL}(0)(1-T/T_c)^{-1/2}$ in (b). The BCS coherence length is ξ_0 in (a), and the coefficient of the Ginzburg-Landau coherence length $\xi_{GL}=\xi_{GL}^{(0)}(1-T/T_c)^{-1/2}$ is given in (b). Δ_0 is the energy gap. The critical field at zero temperature beyond which the material is not superconducting is given as H_{c0} in (a) and as H_{c20} in (b). λ_{ep} is the electron-phonon coupling constant, Θ_D the Debye temperature, and γ the electronic part of the heat capacity. Compiled by M. R. Beasley and T. P. Orlando, Stanford University.

(a) Type I

Material	T_c (K)	λ_0 (nm)	ξ_0 (nm)	Δ_0 (meV)	H_{c0} (kOe)	$2\Delta/k_BT_c$	λ_{ep}	Θ_D (K)	γ (mJ cm^{-3} K^{-2})
Al	1.18	50	1600	0.18	0.105	3.53	0.38	420	0.14
In	3.41	64	364	0.54	0.282	3.68	0.81	109	0.11
Sn	3.72	51	230	0.59	0.305	3.7	0.72	195	0.11
Ta	4.47			0.71	0.831	3.66	0.69	258	0.57
V	5.40			0.81	1.408	3.5	0.60	383	1.18
Pb	7.20	39	87	1.35	0.803	4.7	1.6	96	0.17
Nb	9.25	44	38	1.5	1.98	3.89	0.82	277	0.72

(b) Type II

Material	T_c (K)	$\lambda_{GL}^{(0)}$ (nm)	$\xi_{GL}^{(0)}$ (nm)	Δ_0 (meV)	H_{c20} (kOe)	$2\Delta/k_BT_c$	λ_{ep}	Θ_D (K)	γ (mJ cm^{-3} K^{-2})
Pb-In	7	150	30	1.2					
Pb-Bi	8.3	200	20	1.7		4.9	2		
Nb-Ti	9.5	300	4	1.5	130				
Nb-Zr	10.7			1.9		4.13	1.3	253	
Nb-N	16	200	5		150		0.9		0.21
V_3Ga	15.3	90	2–3		230		1.1	310	2.9
V_3Si	16.3	60	3	2.3	200	3.8	0.9	530	2.2
Nb_3Sn	18.0	65	3	3.4	230	4.3	1.7	290	1.2
Nb_3Ge	23.2		3	3.7	380	4.2	1.7	378	0.7

F. Table: Properties of superconducting cavities

The Q of a microwave cavity is related to the surface resistance of the conducting walls R_S by $QR_S=\Gamma$. Γ typically ranges from 300 to 1000 Ω depending on the shape of the cavity and the mode which is excited. The surface resistance of superconducting metals can be written $R_S=R_{BCS}+R_0$, where R_{BCS} is due to thermal excitations inherent in the superconductor, and R_0 is a residual, temperature-independent loss due to dirt, stray magnetic field, etc. R_{BCS} varies with the microwave frequency ω approximately as $R_{BCS}\propto\omega^{1.7}$, while residual losses have no clear frequency dependence. The maximum surface field attainable B_C^{rf} is apparently constant with frequency and compares with the thermodynamic critical field B_C and the extrapolated Ginzburg-Landau field for first flux penetration $B_{C1}(0)$. This table lists properties for the three materials most commonly applied to superconducting cavities. Compiled by G. J. Dick, Cal. Tech.

Material	R_{BCS}(1.5 K)[a] (Ω)	R_{BCS}(4.2 K)[a] (Ω)	R_0(typical)[b] (Ω)	R_0(best) (Ω) [Q_0(best)]	$B_{C1}(0)$ (mT)	$B_C(0)$ (mT)	B_C^{rf}(best) (mT)
Pb	7.0×10^{-8}	4×10^{-5}	3×10^{-7}	2×10^{-8} [$Q_0=4\times10^{10}$]	...	80	80
Nb	2.6×10^{-8}	2×10^{-5}	1×10^{-7}	1×10^{-9} [$Q_0=5\times10^{11}$]	185	200	159
Nb_3Sn	5.0×10^{-13}	5×10^{-7}	1×10^{-6}	1×10^{-7} [$Q_0=6\times10^{9}$]	50	320	101

[a]Thermal surface resistance R_{BCS} is given for a frequency of 10 GHz.
[b]Typical values for R_0 in the literature lie within a factor of 3 of the value given.

Tables G and H are representative of materials of interest in the rapidly changing field of high temperature superconductivity. They should be regarded as representative of the situation in 1988 and expected to change.

Table G. Oxide superconductors with high critical temperatures

COMPILED BY I. Bozovic and M. Beasley, Stanford University.

	T_c (K)	Reference
$BaPb_{0.75}B_{0.25}O_3$	13	1
$Ba_{0.6}K_{0.4}BiO_3$	30	2,3
$La_{1.85}Ba_{0.15}CuO_4$	36	4,5
$YBa_2Cu_3O_7$	94	6,7
$Y_2Ba_4Cu_8O_{16}$	81	8,9
$Bi_2Sr_2CaCu_2O_8$	84	10
$Tl_2Ba_2CaCu_2O_8$	108	11,12
$Tl_2Ba_2Ca_2Cu_3O_{10}$	125	

Note that the oxygen stoichiometry is only approximate; also, the superconducting transitions are typically few degrees wide.

References (for table G)

[1]A. W. Sleight *et al.*, Solid State Commun. **17**, 27 (1975).
[2]L. F. Matheiss *et al.*, Phys. Rev. B **37**, 3745 (1988).
[3]R. J. Cava *et al.*, Nature (London) **332**, 814 (1988).
[4]J. G. Bednorz and K. A. Muller, Z. Phys. B. **64**, 189 (1986).
[5]In isostructural $La_{1.85}Sr_{0.15}CuO_4$ one finds $T_c = 36$–40, see R. J. Cava *et al.*, Phys. Rev. Lett. **58**, 908 (1987).
[6]M. K. Wu *et al.*, Phys. Rev. Lett. **58**, 908 (1987).
[7]In isostructural $RBa_2Cu_3O_7$ one has [see J. M. Tarascon *et al.*, Phys. Rev. B **36**, 226 (1987)], for

R =	Nd	Sm	Eu	Gd	Dy	Ho	Er	Tm	Yb	Lu
T_c =	95	93.5	95	94	93	93	92.5	92.5	87	89.5

[8]A. F. Marshal *et al.*, Phys. Rev. B **37**, 9353 (1988).
[9]K. Char *et al.*, Phys. Rev. B **38**, 834 (1988).
[10]H. Maeda *et al.*, Jap. J. Appl. Phys. Lett. **27**, pt 2, 209 (1988).
[11]Z. Z. Sheng and A. M. Hermann, Nature (London) **332**, 55 (1988).
[12]S. S. Parkin *et al.*, Phys. Rev. Lett. **60**, 2539 (1988).

Table H. Material parameters of high-T_c superconductors $La_{1.85}Sr_{0.15}CuO_4$ and $YBa_2Cu_3O_7$

Compiled by I. Bozovic and M. Beasley, Stanford University.

(a) **Normal state**. Here n is the free-carrier density (determined from Hall effect data assuming $R_H = 1/ne$); ω_p is the *bare* plasma frequency (from optical reflectance and electron energy loss spectroscopy); $\rho_\parallel(T)$ is the resistivity for current parallel to the crystal *ab* planes (i.e. to the CuO_2 layers) and $d\rho_\parallel/dT$ is the slope of the linear part of $\rho_\parallel(T)$; γ is the Sommerfeld constant (calculated from critical field and resistivity data using Gorkov's theory); θ_D is the Debye temperature (from specific heat measurements and phonon spectroscopy).

		$La_{1.85}Sr_{0.15}CuO_4$	$YBa_2Cu_3O_7$	$Bi_2Sr_2CaCu_2O_8$
$n(T = 300\ K)$	$[10^{21}\ cm^3]$	3	7	3
$h\omega_p\ (T = 300\ K)$	[eV]	2.6	3.1	2.9
$\rho_\parallel\ (T = 300\ K)$	$[\mu\Omega\ cm]$	1200	250	130
$d\rho_\parallel/dT$	$[\mu\Omega\ cm\ K^{-1}]$	3	0.8	0.45
γ	$[\mu J\ cm^{-3}\ K^{-2}]$	95	100	
θ_D	[K]	440	440	

(b) **Superconducting state**. Here T_c is the critical temperature; $j_c^\parallel$ and $j_c^\perp$ are the critical currents parallel and perpendicular to *ab* planes, respectively, for single crystal and epitaxial thin film samples. H_{c2} is the (extrapolated) zero-temperature critical field; λ_{ab} is the penetration depth (from magnetization measurements and from muon-spin-relaxation data), for **H**‖**c**; ξ is the coherence length (from critical field measurements), and $\Delta C_p/T_c$ is the jump of the specific heat at the superconducting transition.

		$La_{1.85}Sr_{0.15}CuO_4$	$YBa_2Cu_3O_7$	$Bi_2Sr_2CaCu_2O_8$
T_c [K]		36	94	84
$j_c^\parallel$ [A/cm²]	at $T = 4$ K	$> 10^4$	1.2×10^7	5×10^6
	at $T = 77$ K	/	1.8×10^6	2×10^5
$j_c^\perp$ [A/cm²]	at $T = 4$ K		4.2×10^5	
	at $T = 77$ K	/	2×10^4	
H_{c2} [T]	for **H**‖**c**	> 45	> 30	20
	for **H**⊥**c**	> 200	> 220	400
λ_{ab} [nm]		200	140	
ξ_{ab} [nm]		4	1.5–3	2–4
ξ_c [nm]		0.6	0.2–0.6	0.2
$\Delta C_p/T_c$ [mJ cm⁻³ K⁻²]		0.3	0.48	

7.03. PROPERTIES OF LIQUID ^{3}He, ^{4}He, AND MIXTURES

A. Table: Properties of liquid He I at vapor pressure

Except as noted, all parameters are from C. F. Barenghi, P. G. J. Lucas, and R. J. Donnelly, "Cubic spline fits to thermodynamic and transport parameters of liquid ^{4}He above the λ transition," J. Low Temp. Phys. **44** (5/6), 491 (1981). Compiled by C. F. Barenghi and R. J. Donnelly.

T (K)	Viscosity η (μP)	Density ρ (g/cm^3)	Thermal conductivity $k\times10^{-4}$ (W/cm K)	Specific heat C_s (J/mol K)	Expansion coefficient β_s (1/K)	Thermal diffusivity $D_T\times10^{-4}$ (cm^2/s)	Kinematic viscosity $\nu\times10^{-6}$ (cm^2/s)	Prandtl No.	Vapor pressure[a] P (μ Hg at 0°C, standard, gravity)	Entropy[b] S (J/g K)	Sound velocity[c] u (m/s)
2.18	25.31	0.146 09	1.843	22.389	0.0005	2.255	173.2	0.7681	38 550		
2.2	26.30	0.146 07	1.514	16.166	0.0110	2.564	180.1	0.7022	40 465	1.61	
2.4	31.15	0.145 36	1.458	9.530	0.0327	4.196	214.3	0.5017	63 304	1.85	
2.6	33.43	0.144 24	1.559	9.088	0.0436	4.722	231.8	0.4909	93 733	2.02	222
2.8	34.74	0.142 85	1.653	9.358	0.0535	4.886	243.2	0.4977	132 952	2.19	221
3.0	35.54	0.141 19	1.738	9.951	0.0636	4.855	251.7	0.5184	182 073	2.36	218
3.2	35.88	0.139 26	1.812	10.807	0.0742	4.688	257.7	0.5496	242 266	2.53	214
3.4	35.81	0.137 05	1.872	11.859	0.0858	4.443	261.3	0.5881	314 697	2.71	208
3.6	35.37	0.134 55	1.919	13.048	0.0985	4.167	262.9	0.6309	400 471	2.89	204
3.8	34.62	0.131 74	1.955	14.385	0.1126	3.877	262.8	0.6778	500 688	3.06	197
4.0	33.59	0.128 61	1.980	15.991	0.1283	3.560	261.2	0.7337	616 537	3.26	190
4.2	33.34	0.125 13	1.998	17.323	0.1459	3.231	258.5	0.7999	749 328	3.46	180

[a] *The 1958 ^{4}He Scale of Temperatures,* Natl. Bur. Stand. (U.S.) Monogr. **10** (1960).
[b] R. W. Hill and O. V. Lounasmaa, Philos. Mag. **2**, 143 (1957).
[c] J. H. Vignos and H. A. Fairbank, Phys. Rev. **147**, 185 (1966); K. R. Atkins and R. A. Staior, Can. J. Phys. **31**, 1156 (1953).

B. Table: Properties of liquid He II at vapor pressure

All tabulations are at saturated vapor pressure (SVP). Compiled by C. F. Barenghi and R. J. Donnelly.

T (K)	Viscosity[a] η (μP)	Density[b] ρ (g/cm^3)	Normal-fluid density ratio[c] ρ_n/ρ	Specific heat[d] C_{sat} (J·g^{-1}·K^{-1})	First-sound speed[e] u_1 (m/s)	Entropy[f] S (J·g^{-1}·K^{-1})	Second-sound velocity[g] u_2 (m/s)	Expansion coefficient[h] $\alpha\times10^{-6}$ (1/K)	Vapor pressure[i] (μ Hg at 0°C, standard gravity)
0.2		0.145 12		1.63×10^{-4}	238	5.52×10^{-5}		9.023	
0.4		0.145 12		1.27×10^{-3}	238	4.32×10^{-4}		67.28	
0.6		0.145 11	4.27×10^{-5}	4.60×10^{-3}	238	1.43×10^{-3}	83	207.7	
0.8	162	0.145 11	9.66×10^{-4}	2.24×10^{-2}	238	4.50×10^{-3}	29	373.3	0.281
1	35.1	0.145 09	7.52×10^{-3}	0.101	238	0.0162	18.9	279.8	11.4
1.2	17.60	0.145 09	2.92×10^{-2}	0.322	237	0.0514	18.4	− 330.4	120
1.4	14.31	0.145 12	7.54×10^{-2}	0.780	236	0.132	19.7	− 1 622.0	625
1.6	12.71	0.145 20	0.17	1.62	234	0.285	20.3	− 3 652.0	2 155
1.8	12.68	0.145 34	0.32	2.98	231	0.542	19.9	− 6 699.0	5 690
2.0	13.78	0.145 61	0.56	5.27	226	0.947	16.4	− 12 140.0	12 466
2.1	16.40	0.145 82	0.74	7.38	222	1.23	12.2	− 17 990.0	23 767
									31 428

[a] A. D. B. Woods and A. C. Hollis Hallet, Can. J. Phys. **41**, 596 (1963); J. M. Goodwin, Ph.D. thesis, University of Washington, 1968. Near T_λ Ahlers recommends

$$1-\eta/\eta_\lambda=5.19[(T_\lambda-T)/T_\lambda]^{0.85}\quad \text{for } 10^{-5}<(T_\lambda-T)/T_\lambda<10^{-1.5}\quad (n_\lambda=24.7\pm0.3\ \mu\text{P})$$

[G. Ahlers in *The Physics of Liquid and Solid Helium,* edited by K. H. Bennemann and J. B. Ketterson (Wiley, New York, 1976), p. 178].

[b] C. T. Van Degrift, Ph.D. thesis, University of California at Irvine, 1974. For values under different pressures, see J. S. Brooks and R. J. Donnelly, J. Phys. Chem. Ref. Data **6**, 51 (1977); J. Maynard, Phys. Rev. B **14**, 3868 (1976).

[c] A. D. B. Woods and A. C. Hollis Hallet, Can. J. Phys. **41**, 596 (1963). Near T_λ for the superfluid density ratio use the fit

$$\rho_s/\rho=2.40[(T_\lambda-T)/T_\lambda]^{0.666\pm0.006}\quad \text{for } 3\times10^{-5}<(T_\lambda/T)/T_\lambda<2.3\times10^{-2}$$

[G. Ahlers (Ref. a)]. For values under pressure use J. S. Brooks and R. J. Donnelly (Ref. b) and J. Maynard (Ref. b).

[d] D. S. Greywall, Phys. Rev. B **18**, 2127 (1978); **21**, 1329(E) (1979); R. J. Donnelly, J. A. Donnelly, and R. N. Hills, J. Low Temp. Phys. **44** (5/6), 471 (1981).

[e] J. S. Brooks and R. J. Donnelly (Ref. b) and J. Maynard (Ref. b). Near T_λ see M. Barmatz and I. Rudnick, Phys. Rev. **170**, 224 (1968).

[f] R. J. Donnelly and J. A. Donnelly, J. Low Temp. Phys. (to be published).

[g] V. P. Peshkov, Sov. Phys. JETP **11**, 580 (1960). For values under different pressures see J. S. Brooks and R. J. Donnelly (Ref. b) and J. Maynard (Ref. b).

[h] C. T. Van Degrift (1974). For values under different pressures see J. S. Brooks and R. J. Donnelly (Ref. b) and J. Maynard (Ref. b).

[i] *The 1958 ^{4}He Scale of Temperatures,* Natl. Bur. Stand. (U.S.) Monogr. **10** (1960).

C. Table: Properties of liquid ^3He at vapor pressure

Compiled by C. F. Barenghi and R. J. Donnelly.

T (K)	Entropy[a] S (J/mol K)	Viscosity[b] η (μP)	Sound speed[c] u (m/s)	Expansion coefficient[d] $\alpha \times 10^{-3}$ (1/K)	Specific heat[e] C_{sat} (J/mol K)	Molar volume[f] V (cm³)	Vapor pressure[g] p (mm Hg at 0°C)
0.2	3.703		183.6	− 12.5	2.728	36.78	1.21×10^{-3}
0.4	5.731	4.10	183.1	− 5.2	3.146	36.72	28.1×10^{-3}
0.6	7.071	3.05	181.3	4.9	3.477	36.72	545×10^{-3}
0.8	8.117	2.63	180.1	12.9	3.812	36.78	2.892
1.0	9.010	2.34	177.7	20.7	4.222	36.91	8.842
1.2	9.824	2.14	174.8	30.3	4.753	37.09	20.16
1.4	10.605	2.02	171.5	41.55	5.414	37.35	38.52
1.6	11.378	1.95	167.8	55.65	6.201	37.72	65.47
1.8	12.159	1.91	163.7	75.10	7.079	38.20	102.5
2.0	12.951	1.88	158.9	94.57	7.983	38.85	151.1
2.2		1.83	153.1	116.84		39.68	212.7
2.4		1.79	146.0	146.93		40.73	288.6
2.6		1.74	137.4	191.88		42.12	380.4
2.8		1.69	126.9	259.33		44.05	489.5
3.0		1.66	114.4	355.24		46.82	617.9
3.2		1.61	99.4	480.86		50.88	767.7

[a]T. R. Roberts, R. H. Sherman, S. G. Sydoriak, and F. G. Brickwedde, Prog. Low Temp. Phys. **4**, 19480 (1964).
[b]K. N. Zinoveva, Sov. Phys. JETP **34**, 421 (1958).
[c]E. R. Grilly and E. F. Hammel, Prog. Low Temp. Phys. **3**, 19144 (1961); K. R. Atkins and M. Flicker, Phys. Rev. **116**, 1063 (1959).
[d]E. C. Kerr and R. D. Taylor, Ann. Phys. (N.Y.) **20**, 450 (1962); C. Boghosian, H. Meyer, and J. E. River, Phys. Rev. **146**, 110 (1966).
[e]T. R. Roberts, R. H. Sherman, S. G. Sydoriak, and F. G. Brickwedde (Ref. a); D. F. Brewer, J. G. Daunt, and A. K. Sreedhar, Phys. Rev. **115**, 836 (1959).
[f]E. C. Kerr and R. D. Taylor (Ref. d).
[g]R. H. Sherman, S. G. Sydoriak, and T. R. Roberts, J. Res. Natl. Bur. Stand. **68A**, 579 (1964).

D. Table: Thermal conductivity of pure ^3He at saturated vapor pressure and at 0.12 atm

From J. C. Wheatley, Prog. Low Temp. Phys. **4**, Chap. 3 (1979).

Saturated vapor pressure				0.12 atm	
T (mK)	κ (ergs/s cm K)	T (mK)	κ (ergs/s cm K)	T (mK)	κ (ergs/s cm K)
2.98	10 630	2.85	11 010	60	893
3.38	9 670	3.32	9 540	65	835
4.12	8 160	3.62	9 340	70	787
4.82	7 160	3.88	8 660	75	751
5.28	6 590	4.36	7 770	80	722
5.81	6 090	5.25	6 600	90	682
6.69	5 430	5.77	6 020	100	655
7.25	4 900	5.95	5 840	120	624
7.95	4 420	6.93	5 060	140	605
9.75	3 670	7.27	4 880	160	594
10.30	3 480	8.31	4 270	180	588
10.81	3 380	8.51	4 200	200	585
12.06	3 085	9.05	3 920	250	585
13.31	2 820	11.22	3 260	300	595
14.67	2 540	11.93	3 090	350	612
15.46	2 390	13.48	2 630	400	630
16.01	2 400	14.99	2 500	450	650
16.71	2 330	16.61	2 360	500	669
17.33	2 260	17.84	2 160		
18.72	2 060	20.6	1 890		
21.8	1 836	22.8	1 780		
25.5	1 659	26.5	1 580		
30.7	1 348				

E. Table: Parameters derived from the specific heat of normal ^{3}He

$\gamma = C/RT$ where C is the heat capacity and R the gas constant. m_3^*/m_3 is the ratio of the effective mass to the physical mass of ^{3}He and F_1^s is a Landau parameter. From D. S. Greywall, Phys. Rev. B **33**, 7520 (1988).

P (bars)	V (cm^3)	γ (K^{-1})	m_3^*/m^3	F_1^s	P (bars)	V (cm^3)	γ (K^{-1})	m_3^*/m^3	F_1^s
0	36.84	2.78	2.80	5.39	18	28.18	3.77	4.53	10.60
1	35.74	2.85	2.93	5.78	19	27.96	3.82	4.61	10.84
2	34.78	2.92	3.05	6.14	20	27.75	3.87	4.70	11.09
3	33.95	2.98	3.16	6.49	21	27.55	3.92	4.78	11.34
4	33.23	3.04	3.27	6.82	22	27.36	3.97	4.86	11.58
5	32.59	3.10	3.38	7.14	23	27.18	4.02	4.94	11.83
6	32.03	3.16	3.48	7.45	24	27.01	4.06	5.02	12.07
7	31.54	3.21	3.58	7.75	25	26.85	4.11	5.10	12.31
8	31.10	3.27	3.68	8.03	26	26.70	4.16	5.18	12.55
9	30.71	3.32	3.77	8.31	27	26.56	4.21	5.26	12.79
10	30.35	3.37	3.86	8.57	28	26.42	4.26	5.34	13.03
11	30.02	3.42	3.95	8.84	29	26.29	4.31	5.42	13.26
12	29.71	3.48	4.03	9.09	30	26.17	4.36	5.50	13.50
13	29.42	3.53	4.12	9.35	31	26.04	4.40	5.58	13.73
14	29.15	3.58	4.20	9.60	32	25.90	4.45	5.66	13.97
15	28.89	3.62	4.28	9.85	33	25.75	4.50	5.74	14.21
16	28.64	3.67	4.37	10.10	34	25.58	4.54	5.82	14.46
17	28.41	3.72	4.45	10.35	34.39	25.50	4.56	5.85	14.56

F. Tables: Parameters for dilute solutions of ^{3}He in ^{4}He

These are most useful parameters for discussing the physics of dilute solutions of ^{3}He in ^{4}He. Subscripts F refer to the Fermi temperature, momentum, velocity, etc. x is the concentration as defined in (b) in terms of the number densities of ^{3}He and ^{4}He atoms. $V(q)$ is the Bardeen-Baym-Pines interaction potential between ^{3}He particles in dilute solutions where s is the velocity of first sound and q the momentum exchanged in a collision between quasiparticles. m^* is the effective mass of ^{3}He at concentration x, $m = 2.34m_3$ the effective mass of a single ^{3}He atom in ^{4}He. D is the spin diffusion constant, X the magnetic susceptibility, and Z_0 a Landau parameter; K is the thermal conductivity, F_1 a Landau parameter measuring the strength of interactions between quasiparticles on the Fermi surface, and τ_D, τ_κ, and τ_η are relaxation times associated with diffusion, conduction, and viscosity. From J. C. Wheatley, Prog. Low Temp. Phys. **4**, Chap. 3 (1970). See also C. Ebner and D. O. Edwards, Phys. Rep. **2C**, 77 (1970), and G. Baym, in *The Physics of Liquid and Solid Helium,* edited by K. H. Bennemann and J. B. Ketterson (Wiley-Interscience, New York, 1980), Pt. 2, Chap. 9 (1978).

(a) Numerical results for dilute solutions of ^{3}He in ^{4}He at zero temperature and pressure

^{4}He atomic mass m_4	6.646×10^{-24} g
^{3}He atomic mass m_3	5.008×10^{-24} g
^{3}He quasiparticle mass $2.34m_3$	11.73×10^{-24} g
Volume per atom, pure ^{4}He, $\omega_{4.0}$	45.8 Å^3
Volume per atom, pure ^{3}He, $\omega_{3.0}$	61.2 Å^3
Fractional excess volume of ^{3}He, α_0	0.29 ± 0.01
Solubility of ^{3}He in ^{4}He	0.0640 ± 0.0007
$T_F = p_F^2/2mk$	$2.58x^{2/3}$ K
^{4}He interaction energy m_4s^2/k	27.4 K
Binding energy, ^{3}He to ^{3}He, L_{03}/k	2.47 K
Binding energy, ^{4}He to ^{4}He, L_{04}/k	7.17 K
Excess binding energy at $x=0$, ^{3}He to ^{4}He, $(E_{03}-L_{03})/k$	0.312 ± 0.007 K
Effective interaction $V(q)$	$-0.064(m_4s^2/N_4)(1-1.15y-4.16y^2+6.21y^3-2.32y^4)$, $y=\hbar q/2(0.318)$ Å^{-1}

(b) Measured and derived parameters for two dilute solutions of ^{3}He in ^{4}He and for pure ^{3}He at low pressure

	1.3%	5.0%	Pure (low pressure)
$T_F \equiv p_F^2/2m^*k$	0.141 ± 0.002 K	0.331 ± 0.005 K	1.62 K
$x \equiv N_3/(N_3+N_4)$	0.0132 ± 0.0001	0.0502 ± 0.0003	1
m^*/m_3	$2.38 \pm 0.04(0.02)$	$2.46 \pm 0.04(0.02)$	3.0
$F_1 = 3(m^*/m-1)$	$F_{1,5.0\%} - F_{1,1.3\%} = 0.11 \pm 0.04$		6.0
$k_F \equiv p_F/\hbar$	2.04×10^7/cm	3.17×10^7/cm	7.88×10^7/cm
$V_F \equiv p_F/m^*$	1.80×10^3 cm/s	2.71×10^3 cm/s	5.38×10^3 cm/s
$1+Z_0/4 \equiv (\chi_{ideal}/\chi)_{0\ K}$	1.09 ± 0.03	1.08 ± 0.03	0.34
Z_0	$+0.35 \pm 0.1$	$+0.34 \pm 0.1$	-2.66
DT^2	$[17.2 \pm 1.7(0.5)]\times10^{-6}$ cm^2 K^2/s	$[90 \pm 9(4)]\times10^{-6}$ cm^2 K^2/s	$(1.4 \pm 0.14)\times10^{-6}$ cm^2/K^2 s
$\tau_D T^2 \equiv 3DT^2/v_F^2(1+Z_0/4)$	2.0×10^{-11} s K^2	4.5×10^{-11} s K^2	6.3×10^{-13} s K^2
κT	11 ± 1.1 ergs/s cm	24 ± 2.4 ergs/s cm	33 ± 3.3 ergs/s cm
$\tau_\kappa T^2$	1.8×10^{-11} s K^2	1.0×10^{-11} s K^2	1.2×10^{-12} s K^2
$\tau_\eta T^2$	2.0×10^{-11} s K^2	2.4×10^{-11} s K^2	2.0×10^{-12} s K^2

G. Table: Averaged specific heat of two dilute solutions of ^3He in ^4He

Averaged specific heat of dilute solutions of nominal concentrations $(1.32 \pm 0.01)\%$ and $(5.02 \pm 0.03)\%$ ^3He, where n is the number of moles of *solution.* After J. C. Wheatley, Prog. Low Temp. Phys. **4**, Chap. 3 (1970).

1.3%		5.0%			
T (mK)	C/nRT (K^{-1})	T (mK)	C/nRT (K^{-1})	T (mK)	C/nRT (K^{-1})
7.31	0.4645	3.11	0.6867	50.19	0.6863
9.41	0.4712	3.89	0.7541	53.20	0.6704
10.42	0.4523	4.70	0.7778	57.93	0.6591
11.87	0.4469	5.47	0.8205	64.33	0.6363
14.21	0.4479	6.23	0.7649	69.45	0.6237
17.62	0.4374	7.06	0.7758	73.12	0.6138
18.19	0.4378	7.95	0.7463	76.89	0.5939
22.38	0.4226	8.98	0.7616	81.61	0.5840
27.73	0.3969	9.83	0.7458	87.40	0.5675
30.04	0.3879	10.98	0.7693	92.08	0.5400
34.77	0.3618	12.16	0.7451	97.22	0.5375
41.97	0.3314	13.49	0.7481	106.0	0.5012
44.36	0.3210	14.85	0.7475	114.2	0.4961
51.33	0.2911	16.51	0.7421	126.1	0.4664
55.53	0.2795	18.98	0.7411	135.9	0.4457
59.43	0.2642	21.23	0.7446	151.4	0.4088
68.99	0.2363	23.46	0.7423	165.3	0.3870
83.63	0.2054	26.72	0.7327	184.1	0.3524
98.28	0.1798	29.67	0.7301	200.9	0.3359
105.4	0.1674	34.30	0.7192	228.1	0.3029
116.2	0.1558	35.92	0.7226		
135.0	0.1353	39.68	0.7086		
145.2	0.1274	42.21	0.7142		
180.5	0.1073	44.05	0.6929		

H. Table: Thermal conductivity of two dilute solutions of ^3He in ^4He

Thermal conductivity of dilute solutions of nominal concentrations 1.3% and 5.0% ^3He at saturated vapor pressure. After J. C. Wheatley, Prog. Low Temp. Phys. **4**, Chap. 3 (1970).

1.3%		5.0%			
T (mK)	κ (10^3 ergs/s cm K)	T (mK)	κ (10^3 ergs/s cm K)	T (mK)	κ (10^3 ergs/s cm K)
4.9	2.30	3.0	8.48	18.3	1.95
6.1	1.89	4.0	5.63	19.8	1.94
7.3	1.64	4.9	5.07	20.8	1.98
8.5	1.45	5.3	4.65	21.5	1.93
9.3	1.36	5.5	4.38	21.6	1.97
9.6	1.32	5.7	4.25	23.0	1.96
10.7	1.26	6.7	3.63	26.3	2.04
11.7	1.22	7.2	3.42	27.3	2.06
11.8	1.20	7.4	3.30	29.8	2.15
12.6	1.16	8.0	3.14		
13.2	1.17	8.3	3.04		
13.8	1.15	8.7	2.84		
14.7	1.16	9.6	2.70		
15.7	1.16	10.3	2.62		
17.2	1.20	10.4	2.49		
18.0	1.24	10.9	2.46		
18.2	1.24	11.6	2.37		
18.9	1.26	12.4	2.28		
19.6	1.30	12.7	2.26		
20.8	1.36	12.9	2.23		
21.3	1.40	13.2	2.24		
22.4	1.50	14.1	2.13		
23.4	1.62	14.6	2.10		
25.7	1.76	14.9	2.06		
27.2	1.96	16.2	2.02		
28.0	2.10	16.8	2.00		
29.2	2.28	17.3	1.98		

I. ^3He thermometry at millikelvin temperatures. [Taken from Dennis S. Greywall, Phys. Rev. B **33**, 7520 (1986)].

The ^3He phase diagram is shown in Figure 7.03.I.1. The transition between normal and superfluid ^3He is described by Eq. (1) and Table 7.03.I.2. It begins at $T_c(0) = 0.929$ mK and meets the melting curve at $T_A = 2.491$ mK and $P_A = 34.338$ bars. The A–B transition begins at the polycritical point, $T_{pcp} = 2.273$ mK, $P_{pcp} = 21.22$ bars and extends to the melting curve as described by Eq. (2) and Table 7.03.I.3, meeting the melting curve at $T_{A-B} = 1.932$ mK and 34.358 bars. The melting curve itself is described over a large range by Eq. (3) and Table 7.03.I.4.

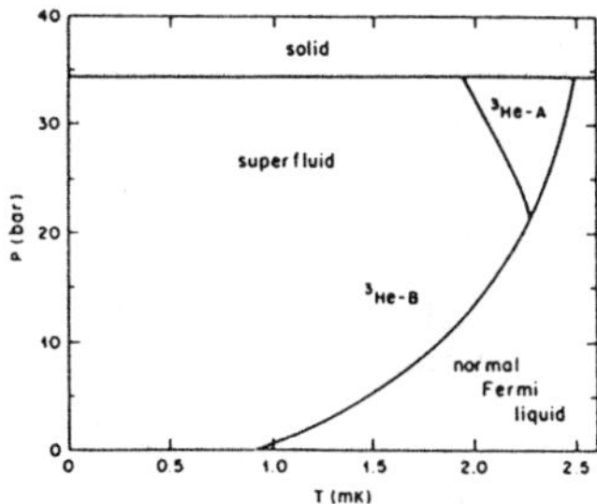

FIGURE 7.03.I.1. ^3He phase diagram determined by Eqs. (1), (2), and (3).

The three equations and corresponding tables of representative values follow:

$$T_c = \sum_{i=0}^{5} a_i p^i, \tag{1}$$

with

$a_0 = 0.929\,383\,75$, $a_1 = 0.138\,671\,88$,

$a_2 = -0.693\,021\,85 \times 10^{-2}$, $a_3 = 0.256\,851\,69 \times 10^{-3}$,

$a_4 = -0.572\,486\,44 \times 10^{-5}$, $a_5 = 0.530\,109\,18 \times 10^{-7}$,

where T_c is in mK and P is in bars.

TABLE 7.03.I.2. Pressure–temperature coordinates for the transition between normal and superfluid ³He. The smoothed values were computed using Eq. (1).

P (bars)	T_c (mK)	P (bars)	T_c (mK)
0	0.929	18	2.177
1	1.061	19	2.209
2	1.181	20	2.239
3	1.290	21	2.267
4	1.388	22	2.293
5	1.478	23	2.317
6	1.560	24	2.339
7	1.636	25	2.360
8	1.705	26	2.378
9	1.769	27	2.395
10	1.828	28	2.411
11	1.883	29	2.425
12	1.934	30	2.438
13	1.981	31	2.451
14	2.026	32	2.463
15	2.067	33	2.474
16	2.106	34	2.486
17	2.143	34.338	2.491

$$T_{A\text{-}B} = T_{\text{PCP}} + \sum_{i=1}^{5} a_i (P - P_{\text{PCP}})^i , \tag{2}$$

with $P_{\text{PCP}} = 21.22$ bars, $T_{\text{PCP}} = 2.273$ mK,

and

$a_1 = -0.103\,226\,23 \times 10^{-1}$, $a_2 = -0.536\,331\,81 \times 10^{-2}$, $a_3 = 0.834\,370\,32 \times 10^{-3}$, $a_4 = -0.617\,097\,83 \times 10^{-4}$, $a_5 = 0.170\,389\,92 \times 10^{-5}$.

TABLE 7.03.I.3. Pressure–temperature coordinates for the A–B transition in superfluid ³He. The smoothed values were computed using Eqs. (15).

P (bars)	$T_{A\text{-}B}$ (mK)
21.22	2.273
22	2.262
23	2.242
24	2.217
25	2.191
26	2.164
27	2.137
28	2.111
29	2.083
30	2.056
31	2.027
32	1.998
33	1.969
34	1.941
34.358	1.932

$$P - P_A = \sum_{n=-3}^{5} a_n T^n , \tag{3}$$

with

$a_{-3} = -0.196\,529\,70 \times 10^{-1}$, $a_{-2} = 0.618\,802\,68 \times 10^{-1}$, $a_{-1} = -0.788\,030\,55 \times 10^{-1}$, $a_0 = 0.130\,506\,00$, $a_1 = -0.435\,193\,81 \times 10^{-1}$, $a_2 = 0.137\,527\,91 \times 10^{-3}$, $a_3 = -0.171\,804\,36 \times 10^{-6}$, $a_4 = -0.220\,939\,06 \times 10^{-9}$, $a_5 = 0.854\,502\,45 \times 10^{-12}$.

TABLE 7.03.I.4. ³He-melting curve coordinates determined using Eq. (3).

T (mK)	$P-P_A$ (bars)	dp/dT (bars K^{-1})	T (mK)	$P-P_A$ (bars)	dP/dT (bars K^{-1})	T (mK)	$P-P_A$ (bars)	dP/dT (bars K^{-1})
0.931 (T_c)	0.052 52	−27.2	4.0	−0.057 52	−39.2	80.0	−2.566 06	−25.1
1.0	0.050 55	−29.2	5.0	−0.097 12	−39.9	85.0	−2.689 17	−24.2
1.1	0.047 54	−30.8	6.0	−0.137 20	−40.2	90.0	−2.807 83	−23.3
1.2	0.044 41	−31.7	7.0	−0.177 50	−40.3	95.0	−2.922 15	−22.4
1.3	0.041 21	−32.2	8.0	−0.217 86	−40.3	100.0	−3.032 29	−21.6
1.4	0.037 97	−32.7	9.0	−0.258 18	−40.3	110.0	−3.240 51	−20.0
1.5	0.034 68	−33.1	10.0	−0.298 39	−40.2	120.0	−3.433 50	−18.6
1.6	0.031 35	−33.5	11.0	−0.338 47	−40.0	130.0	−3.612 23	−17.2
1.7	0.027 98	−33.9	12.0	−0.378 37	−39.8	140.0	−3.777 57	−15.9
1.8	0.024 56	−34.3	14.0	−0.457 61	−39.4	150.0	−3.930 35	−14.7
1.9	0.021 11	−34.7	16.0	−0.536 00	−39.0	160.0	−4.071 28	−13.5
1.932 ($T_{A\text{-}B}$)	0.020 00	−34.8	18.0	−0.613 50	−38.5	170.0	−4.200 97	−12.4
2.0	0.017 63	−35.1	20.0	−0.690 07	−38.0	180.0	−4.319 95	−11.4
2.1	0.014 10	−35.4	25.0	−0.877 34	−36.9	190.0	−4.428 59	−10.4
2.2	0.010 55	−35.7	30.0	−1.058 66	−35.7	200.0	−4.527 14	−9.4
2.3	0.006 96	−36.1	35.0	−1.234 06	−34.5	210.0	−4.615 74	−8.4
2.4	0.003 33	−36.4	40.0	−1.403 63	−33.3	220.0	−4.694 32	−7.4
2.491 (T_A)	0.000 00	−36.6	45.0	−1.567 50	−32.2	230.0	−4.762 70	−6.3
2.5	−0.000 32	−36.6	50.0	−1.725 79	−31.1	240.0	−4.820 50	−5.2
2.6	−0.003 99	−36.9	55.0	−1.878 63	−30.0	250.0	−4.867 17	−4.1
2.7	−0.007 69	−37.1	60.0	−2.026 16	−29.0			
2.8	−0.011 42	−37.4	65.0	−2.168 53	−28.0			
2.9	−0.015 17	−37.6	70.0	−2.305 88	−27.0			
3.0	−0.018 94	−37.8	75.0	−2.438 34	−26.0			

7.04. PROPERTIES OF CRYOGENIC GASES

A. Table: Densities of gases (g/cm³)

(sat) refers to the saturated vapor pressure. Data from Ref. 6 and (for hydrogen) Ref. 1.

T (K)	Air (1 atm)	Argon (1 atm)	CO (sat)	Helium (1 atm)	Hydrogen (1 atm)	Methane (sat)	Nitrogen (sat)	Neon (1 atm)	Oxygen (1 atm)
10				0.005 0					
20				0.002 5					
40				0.001 2	0.000 6			0.0063	
60				0.000 8	0.000 4			0.0041	
80			0.0038	0.000 6	0.000 3		0.006	0.0031	
100	0.00359	0.004 97	0.0219	0.000 48	0.000 25	0.000 67	0.032	0.0025	
120	0.00297	0.004 12	0.0782	0.000 40	0.000 2	0.003 4	0.120	0.0020	
140	0.00253	0.003 50		0.000 34	0.000 18	0.011		0.0018	
160	0.00222	0.003 05			0.000 15	0.025		0.0015	0.0260
180	0.00197	0.002 70			0.000 14	0.050		0.0014	0.0225
200	0.00177	0.002 43		0.000 24	0.000 12			0.0012	0.0200
220	0.00161	0.002 22			0.000 11			0.0011	0.0181
240	0.00148	0.002 04			0.000 10			0.0010	0.0165
260	0.00136	0.001 88			0.000 10			0.0009	0.0151
280	0.00126	0.001 75			0.000 09			0.0009	0.0140
300	0.00118	0.001 63		0.000 16	0.000 08			0.0008	0.0131

B. Table: Thermal conductivities of gases at 1 atm (mW/cm K)

Data from Ref. 6 and (for neon to 220 K) Ref. 3.

T						Hydrogen					
(K)	Air	Argon	CO	Fluorine	Helium	Normal	Para	Methane	Nitrogen	Neon	Oxygen
10					0.17	0.074	0.074				
20					0.27	0.155	0.155				
40					0.41	0.298	0.298				
60					0.57	0.422	0.429			0.1584	
80			0.069		0.64	0.542	0.578		0.08	0.1893	
100	0.093		0.088	0.086	0.74	0.664	0.754		0.10	0.2169	
120	0.105		0.105	0.105	0.84	0.790	0.941	0.100	0.12	0.2416	0.110
140	0.128	0.095	0.123	0.124	0.93	0.918	1.105	0.125	0.14	0.2645	0.130
160	0.146	0.106	0.141	0.144	1.02	1.043	1.246	0.150	0.16	0.2859	0.150
180	0.164	0.118	0.167	0.163	1.08	1.166	1.361	0.175	0.18	0.3059	0.165
200	0.181		0.175	0.183	1.17	1.282	1.455	0.201	0.19	0.3686	0.180
220	0.198		0.191	0.201	1.24	1.398	1.542	0.227	0.21	0.396	0.195
240	0.214		0.207	0.218	1.32	1.507	1.621	0.253	0.22	0.422	0.215
260	0.230		0.222	0.237	1.38	1.613	1.702	0.278	0.23	0.447	0.230
280	0.247		0.238	0.253	1.47	1.717	1.784	0.305	0.24	0.469	0.248
300	0.265		0.251	0.269	1.53	1.816	1.865			0.492	0.266

C. Table: Viscosities of gases at 1 atm (μP)

Data from Ref. 6.

T (K)	Air	CO	Fluorine (10^2)	Helium	Hydrogen	Methane	Nitrogen	Neon	Oxygen
10				23.5	5				
20				36	11			41	
40				56	21			72	
60				72	28			98	
80		53		88	36			124	
100	70	67	85	100	42	40	70	146	78
120	83	79	102	114	48	48	82	165	93
140	97	92	118	126	53	55	95	184	107
160	109	104	135	137	58	63	107	202	121
180	122	116	152	148	63	70	119	219	135
200	133	127	168	158	68	78	130	237	147
220	144	138	184	168	73	85	141	253	160
240	155	148	200	178	77	92	151	270	172
260	165	159	215	188	82	98	160	287	184
280	176	169	229	197	86	105	170	302	195
300	185	179	241		89	112	179	318	206

7.05. PROPERTIES OF CRYOGENIC LIQUIDS

In the following tables, (sat) refers to the saturated vapor pressure.

A. Table: Properties of cryogenic fluids

Except as noted, all data are from Ref. 1. For more detailed, and other information concerning cryogenic fluids, see the cited sources.

Substance	Triple point [a] (K, mm Hg)	Critical temp. (K)	Critical pressure (atm)	Critical density (g/cm^3)	Freezing temp. at 1 atm (K)	Boiling temp. at 1 atm (NBT) (K)	Liquid density [b] at NBT (g/cm^3)	Heat of vapor [b] at NBT (J/cm^3)
Air		132.4	37.25	0.238		78.8		
Argon	83.8, 516	150.86	48.34	0.5358	83.9	87.28	1.4	230
CO [a]	68.09, 115.3	132.91	34.53	0.3010	68.09	81.61		
Fluorine		144.3[c]	51.5[c]	0.5740[c]	53.5[a]	85.21[a]		
Helium-4 [d]		5.20	2.26	0.0693		4.21	0.13	2.6
Helium-3 [d]		3.324	1.15	0.0413		3.19		
Hydrogen								
Normal	13.96, 54.0	33.18	12.98	0.0301	14.0	20.39	0.07	31
Para	13.81, 52.8	32.98	12.76	0.0314	13.8	20.27	0.07	31
Methane	88.7, 75.24	190.66	45.80	0.1618	90.72	111.67		
Neon	24.57, 323.5	44.40	26.19	0.4829	24.54	27.09	1.2	108
Nitrogen	63.14, 96.37	126.26	33.56	0.3146	63.17	77.36	0.81	160
Oxygen	54.35, 1.13	154.77	50.14	0.4266	54.4 [c]	90.18	1.14	240
Krypton [c]		209.4	54.3	0.9189	115.8	119.8		
Xenon [c]		289.7	57.6	1.1127	161.3	165.0		

[a] Reference 6.
[b] Reference 8.
[c] Reference 7.
[d] Reference 10.

B. Table: Vapor pressure data for a number of common gases

Gas	Data temp. range (K)	10^{-13}	10^{-12}	10^{-11}	10^{-10}	10^{-9}	10^{-8}	10^{-7}	10^{-6}	10^{-5}	10^{-4}	10^{-3}	10^{-2}	10^{-1}	1	10^{1}	10^{2}	10^{3}
		Temperatures (K) for vapor pressure (Torr)																
He	0.9–5.2													0.98	1.268	1.738	\ 2.634	4.518
H_2	14–21	2.67	2.83	3.01	3.21	3.45	3.71	4.03	4.40	4.84	5.38	6.05	6.90	8.03	9.55	11.70⊙	15.10	21.4
Ne	15–45	5.50	5.79	6.11	6.47	6.88	7.34	7.87	8.48	9.19	10.05	11.05	12.30	13.85	15.80	18.45	22.10⊙	27.5
CH_4	48–112	24.0	25.3	26.7	28.2	30.0	32.0	34.2	36.9	39.9	43.5	47.7	52.9	59.2	67.3	77.7⊙	91.7	115.0
F_2	54–89													\	(55.2)⊙	59.5	70.5	87.5
N_2	54–128	18.1	19.0	20.0	21.1	22.3	23.7	25.2	27.0	29.0	31.4	34.1	37.5	41.5	47.0	54.0⊙	63.4	80.0
CO	56–133	20.5	21.5	22.6	23.8	25.2	26.7	28.4	30.3	32.5	35.0	38.0	41.5	45.8	51.1	57.9\	67.3⊙	84.1
O_2	57–154	21.8	22.8\	24.0	25.2	26.6	28.2	29.9	31.9	34.1	36.7	39.8	43.3\	48.1	54.1⊙	62.7	74.5	92.8
Kr	63–121	27.9	29.4	30.9	32.7	34.6	36.8	39.3	42.2	45.5	49.4	53.9	59.4	66.3	74.8	85.9	101.0⊙	123.5
NO	73–180	37.7	39.4	41.3	43.4	45.6	48.1	50.9	54.0	57.6	61.6	66.3	71.7	78.1	85.7	95.0	106.5⊙	123.5
Ar	82–88	20.3	21.3	22.5	23.7	25.2	26.8	28.6	30.6	33.1	35.9	39.2	43.2	48.2	54.4	62.5	73.4⊙	89.9
N_2O	103–186	55.8	58.3	61.1	64.2	67.6	71.3	75.5	80.3	85.7	91.9	99.0	107.5	117.5	129.5	144.0	162.5⊙	189.5
CO_2	107–196	59.5	62.2	65.2	68.4	72.1	76.1	80.6	85.7	91.5	98.1	106.0	114.5	125.0	137.5	153.5	173.0	198.0⊙
Xe	110–166	38.5	40.5	42.7	45.1	47.7	50.8	54.2	58.2	62.7	68.1	74.4	82.1	91.5	103.5	118.5	139.5⊙	170.0
HBr	120–205	51.8	54.3	57.1	60.2	63.7	67.6	72.1	77.1	82.9\	89.6	97.5	107.0\	118.5	132.5	151.0	175.0⊙	209.0
HCl	132–195	49.7	52.1	54.6	57.5	60.6	64.1	68.1	72.5	77.6	83.4	90.1	98.1\	108.5	121.0	137.0	158.5⊙	193.0
NH_3	145–240	70.9	74.1	77.6	81.5	85.8	90.6	95.9	102.0	108.5	116.5	125.5	136.0	148.0	163.0	181.0⊙	206.0	245.0
H_2S	153–213	57.1	59.8	62.7	65.9	69.5	73.5	78.0	83.1	89.0	95.7	103.5\	113.5	124.5\	138.5	156.5	180.5⊙	218.0
COS	162–224													(124.5)⊙	139.5	159.5	187.0	229.0
Cl_2	162–420	66.1	69.1	72.4	76.0	80.0	84.4	89.4	95.1	101.5	109.0	117.5	127.5	140.0	155.0⊙	173.0	201.0	245.0
H_2O	175–380	113.0	118.5	124.0	130.0	137.0	144.5	153.0	162.0	173.0	185.0	198.5	215.0	233.0	256.0⊙	284.0	325.0	381.0
SO_2	178–263	78.9	82.4	86.3	90.4	95.1	100.0	106.0	112.5	119.5	128.0	137.5	148.5	161.5	177.0	195.5⊙	225.0	269.0
CS_2	194–319												(160.0)⊙	177.5	199.5	228.0	269.0	329.0
HF	240–290														(179.0)⊙	207.0	45.0	301.0
Br_2	253–331	102.0	106.5	111.0	116.5	122.0	128.5	135.5	143.5	152.5	163.0	174.5	188.5	204.0	224.0	248.0⊙	282.0	339.0
I_2	298–456	141.5	147.5	154.0	161.5	169.5	178.5	188.5	199.5	212.0	226.0	243.0	262.0	285.0	312.0	345.0⊙	389.0	471.0

⊙ Melting point.
\ Transition point.

C. Table: Thermal conductivity of various cryogenic liquids (mW/cm K)

Data from Ref. 6.

T (K)	Hydrogen	*T* (K)	Nitrogen (sat)	*T* (K)	CO	Methane
16	1.08	70	1.498	80	1.42	
18	1.13	72	1.472	90	1.25	
20	1.18	76	1.417	100	1.08	
22	1.22	78	1.390	110	0.90	1.98
24	1.27	80	1.362	120		1.80
26	1.32	82	1.335	130		1.64
28	1.36	84	1.308	140		1.48
30	1.41	86	1.280	150		1.32

D. Table: Dielectric constants of various cryogenic liquids

Data from Ref. 6. For exact condition of measurement consult original references.

T (K)	Hydrogen	*T* (K)	Nitrogen (1 atm)	Oxygen	*T* (K)	Argon	*T* (K)	Methane (sat)
		55		1.570				
14.10	1.2533	64	1.472	1.544	82.5	1.5365	94.0	1.714
14.29	1.2505	66	1.466	1.540	83.0	1.5349	96.0	1.711
14.56	1.2497	68	1.460	1.535	84.0	1.5312	98.0	1.708
14.69	1.2512	70	1.454	1.530	85.0	1.5278	100.0	1.704
14.71	1.2492	72	1.448	1.526	86.0	1.5240	102.0	1.701
15.47	1.2484	74	1.443	1.521	87.0	1.5204	104.0	1.698
16.73	1.2455	76	1.438	1.516	87.5	1.5188	106.0	1.695
17.73	1.2408	78	1.433	1.512	88.0	1.5178	108.0	1.690
19.11	1.2356	80		1.507	88.5	1.5172	110.0	1.682
20.40	1.2315	90		1.483			112.0	1.675

E. Table: Surface tension of various cryogenic liquids (dyn/cm)

Data from Ref. 6.

T (K)	^{4}He	*T* (K)	Hydrogen	*T* (K)	Neon	*T* (K)	Nitrogen	CO	*T* (K)	Fluorine	*T* (K)	Argon	*T* (K)	Methane
									66	18.9				
3.0	0.215	15	2.83	25	5.55	70	10.5	12.1	68	18.3	84	11.45	95	17.6
3.5	0.165	16	2.66	26	5.20	75	9.3	10.9	70	17.7	85	11.30	100	16.5
4.0	0.115	17	2.49	27	4.85	80	8.2	9.8	72	17.1	86	11.15	105	15.4
4.5	0.065	18	2.32	28	4.50	85	7.2	8.7	74	16.6	87	11.00	110	14.3
5.0	0.012	19	2.15	28.4	4.38	90	6.1	7.7	76	16.0	88	11.85	114	13.5
		20	1.97						78	15.4	89	11.69		
									80	14.8	90	11.54		

F. Table: Viscosity of various cryogenic liquids (μP)

Data from Ref. 6.

T (K)	Hydrogen	*T* (K)	Nitrogen (sat)	Oxygen (sat)	*T* (K)	CO (1 atm)	Fluorine (1 atm)	*T* (K)	Argon (sat)	Methane (1 atm)
15	217	60		5800	72	2430	3660	85	2720	
16	197	70	2200	3580	73	2320	3525	90	2300	2100
17	178	80	1410	2500	74	2220	3410	95	1970	1770
18	161	90	1040	1890	75	2130	3300	100	1720	1530
19	147	100	850	1520	76	2050	3200	105	1540	1360
20	134	110	760	1280	77	1970	3100	110	1410	1240
21	126	120		1130	78	1900	3010	115	1270	
		130		1070	79	1830	2920	120	1150	
		140		1010	80	1770	2830	125	1020	

G. Table: Latent heats (J/g)

Data from Ref. 6 and (for hydrogen) Ref. 1.

T (K)	Air	CO	Fluorine	Helium	Hydrogen	Nitrogen	Neon	Oxygen
T_c	132.4	132.91		5.20	33.19	126.1	44.74	154.77
2.2				22.8				
3				23.7				
4				21.9				
5				12.0				
15					458			
20					445			
25					410		89.5	
30					295		80.9	
32					185			
35							70.9	
40							53.0	
70						207		231
75						202		227
80	204					195		223
85	198					188		218
90	192	200	161			180		213
100	178	182	149			161		
110	154	159	134			137		
120	119	130	110			95		
130	56	72	95.9					

7.06. CRYOGENIC DESIGN INFORMATION

A. Table: Electrical resistivities of ideally pure elements ($\mu\Omega$ cm)

Adapted from Ref. 9.

T (K)	Ag	Al	Au	Cu	Fe	Na	Ni	Pb	Pt	W
295	1.61	2.74	2.20	1.70	9.8	4.75	7.0_4	21.0	10.42	5.3_3
273	1.47	2.50	2.01	1.55	8.7	4.29	6.2	19.3	9.59	4.8_2
250	1.34	2.24	1.83	1.40	7.5_5	3.82	5.4	17.6	8.70	4.3_2
200	1.04	1.65	1.44	1.06	5.3	2.88	3.7	13.9	6.76	3.2_2
150	0.73_5	1.0_6	1.04	0.70_5	3.1_5	2.00	2.2_4	10.2	4.78	2.1_1
100	0.42	0.47	0.63	0.35	1.2_4	1.15	1.0_0	6.5	2.74	1.02
80	0.29	0.25	0.46	0.21_5	0.64	0.81	0.55	5.0	1.91	0.60
50	0.11	0.05	0.20	0.050	0.13_5	0.32	0.15	2.7_5	0.72	0.15
25	0.010	...	0.027	0.0025	0.012_5	0.039	0.017	1.0	0.084	0.011_5
15	0.0011	...	0.0037	0.0001_7	0.003_4	0.005	0.004_5	0.25	0.0116	0.002_4

B. Table: Electrical resistivity of alloys ($\mu\Omega$ cm)

Adapted from Ref. 9.

Alloy	Physical	Temperature (K)			
		295	90	77	4.2
Brass (70 Cu, 30 Zn)	strained	7.2	5.0	...	4.3
	annealed	6.6	4.2	...	3.6
Constantan	as received	52.5	45	...	44
Cupro-nickel (80 Cu, 20 Ni)	as received	26	...	24	23
Evanohm	as received	134	133	...	133
German silver	as received	30	27.5	...	26
Gold + 2.1 Co	as received	≈13	12.3	...	12.0
Manganin	as received	48	46	45	43
Monel	annealed	50	...	32	30
Niobium + 25 Zr	as received	40	28	...	25 (11 K)

C. Table: Thermal conductivity (mW cm^{-1} deg^{-1}) for some technical alloys, glasses, and plastics

Adapted from Ref. 9.

	Temperature (K)								
Substance	0.1	0.4	1	4	10	40	80	150	300
Al 5083 (≈4.5 Mg, 0.7 Mn)	...	...	7	30	82	340	560	800	1200
Al 6063 (≈0.6 Mg, 0.4 Si)	...	...	85	350	870	2700	2300	2000	2000
Brass (70 Cu, 30 Zn)	0.6	2.5	7	30	100	375	650	850	1200
Brass (70 Cu, 27 Zn, 3 Pb)	...	...	...	30	90	350	550	900	1100
Cu + 30 Ni	0.06	0.3	0.9	5	20	~120	~200	~250	300
Constantan	0.06	0.23	1	8	35	140	180	200	230
German silver	...	...	...	7	28	130	170	180	220
Gold + 2.1 Co	...	...	2	10	40	135	200	...	...
Inconel	...	...	0.5	4	15	70	100	125	140
Manganin	...	...	0.6	5	21	75	130	160	220
Silicon bronze (96 Cu, 3 Si, 1 Mn)	...	...	...	...	15	69	140	...	250
Soft solder (60 Sn, 40 Pb)	...	...	...	160	425	525	525	...	~500
Stainless steels (18/8)	...	...	...	2.5	7	46	80	110	150
Wood's metal	...	...	...	4.0	120	200	230	...	...
Nylon	...	0.006	0.02_5	0.12_5	0.3_9	...	...	...	...
Perspex	...	...	0.21	0.56	0.62	...	...	...	...
Plexiglas	...	...	0.21	0.50	...	...	...	...	...
Polystyrene	...	...	0.12	0.26	...	...	...	...	...
Pyrex	0.003	0.03	0.12	...	...	...	4.8	7.6	11
Soft glass	...	...	0.15	1.15	1.9	2.6	4.6	...	...
Teflon	0.0002	0.004	0.04	0.4_5	0.9_5	1.9_6	2.3	...	...
Vitreous silica	...	...	...	1	1.2	2.5	4.8	8.0	14

D. Table: Linear thermal contractions relative to 293 K

Units are $10^4(L_{293} - L_T)/L_{293}$. Adapted from Ref. 9.

	Temperature (K)								
Substance	0	20	40	60	80	100	150	200	250
Aluminum	41.5	41.5	41.3	40.5	39.1	37.0	29.6	20.1	9.7
Chromium	9.8	9.8	9.8	9.7	9.5	9.0	7.4	5.05	2.4
Copper	32.6	32.6	32.4	31.6	30.2	28.3	22.2	14.9	7.1
Germanium	9.2	9.2	9.2	9.2	9.0_5	8.7	7.1	4.9	2.4
Iron	19.8	19.8	19.7	19.5	18.9	18.1	14.8	10.2	4.9
Lead	70.8	70.0	66.7	62.4	57.7	52.8	39.9	26.3	12.4
Nickel	22.4	22.4	22.3	21.9	21.1	20.1	16.2	11.1	5.0
Silver	41.3	41.2	40.5	38.9	36.6	33.9	25.9	17.3	8.2
Titanium	15.1	15.1	15.0	14.8	14.2	13.4	10.7	7.3	3.5
Tungsten	8.6	8.6	8.5	8.3	8.0	7.5	5.9	4.0	1.9
Brass (65 Cu, 35 Zn)	38.4	38.3	38.0	36.8	35.0	32.6	25.3	16.9	8.0
Constantan	...	...	26.4	25.8	24.7	23.2	18.3	12.4	5.8_5
German silver	37.6	37.6	37.3	36.2	34.5	32.3	25.4	17.0	8.1
Invar[a]	4.5	4.6	4.8	4.9	4.8	4.5	3.0	2.0	1.0
304, 316 stainless steel	...	29.7	29.6	29.0	27.8	26.0	20.3	13.8	6.6
Pyrex	5.6	5.6	5.7	5.6	5.4	5.0	3.9_5	2.7	0.8
Vitreous silica	− 0.7	− 0.65	− 0.5	− 0.3	-0.1_5	0.0	0.1_5	0.2_5	0.1_5
Araldite	106	105	102	98	94	88	71	50_5	25.0
Nylon	139	138	135	131	125	117	95	67	34.0
Polystyrene	155	152	147	139	131	121	93	63	30.0
Teflon	214	211	206	200	193	185	160	124	75.0

[a]The expansion of Invar or NiFe alloys containing ≈36% Ni is very sensitive to composition and heat treatment.

E. Cryogenic design: Sources of further information

For data on thermal conductivity:

R. L. Powell and W. A. Blanpied, *Thermal Conductivity of Metals and Alloys at Low Temperatures*, Natl. Bur. Stand. (U.S.) Circ. **556** (1966).

V. L. Johnson, *Properties of Materials at Low Temperatures* (Pergamon, New York, 1961).

R. W. Powell, C. Y. Ho, and P. E. Liley, *Thermal Conductivity of Selected Materials*, Natl. Stand. Ref. Data Ser. Nat. Bur. Stand. **58** (1966).

For data on specific heats and linear thermal contraction of many materials:

R. J. Corruccini and J. J. Gruewek, *Specific Heat and Enthalpies of Technical Solids at Low Temperatures*, Natl. Bur. Stand. (U.S.) Monogr. **21** (1960).

R. J. Corruccini and J. J. Gruewek, *Thermal Expansion of Technical Solids at Low Temperatures*, Natl. Bur. Stand. (U.S.) Monogr. **29** (1961).

Smithsonian Physical Tables, 9th ed. (Smithsonian, Washington, DC, 1954).

W. B. Pearson, *Handbook of Lattice Spacings of Metals and Alloys* (Pergamon, London, 1958).

For principles of dilution and magnetic refrigeration:

Reference 8, Chap. 6.
Reference 9, Chaps. 5 and 9.
Reference 5, Chaps. 3, 5, and 6.

For Kapitza resistivity and thermal conductivity of some of the above materials below 1.0 K:

Reference 5, p. 266.

For thermal conductivity of copper samples:

Reference 9, p. 356.

For fluid properties:

Interactive Fortran IV Computer Programs for the Thermodynamic and Transport Properties of Selected Cryogens [Fluids Pack], R. D. McCarty, NBS Technical Note 1025 (1980).

A Computer Program for the Prediction of Viscosity and Thermal Conductivity in Hydrocarbon Mixtures, J. F. Ely and H. J. M. Hanley, NBS Technical Note 1039 (1981).

The Thermodynamic Properties of Helium II from 0 K to the Lambda Transitions, R. D. McCarty, NBS Technical Note 1029 (1980).

Thermophysical Properties of Helium-4 from 2 to 1500 K with Pressures to 1000 Atmospheres, R. D. McCarty, NBS Technical Note 631 (1972).

Materials at Low Temperatures, edited by R. P. Reed and A. F. Clark, American Society for Metals, Metals Park, Ohio (1983).

For properties of materials at low temperatures:

A Compilation and Evaluation of Mechanical, Thermal, and Electrical Properties of Selected Polymers, R. E. Schramm, A. F. Clark, and R. P. Reed, NBS Monograph 132 (1973).

Thermal Conductivity of Solids at Room Temperature and Below, G. E. Childs, L. J. Ericks, and R. L. Powell, NBS Monograph 131 (1973).

Materials at Low Temperatures, edited by R. P. Reed and A. F. Clark, American Society for Metals, Metals Park, Ohio (1983).

7.07 REFERENCES

[1]ASHRAE *Thermodynamic Properties of Refrigerants* (ASHRAE, New York, 1969).
[2]R. E. Honig and H. O. Hook, RCA Rev. **21**, 360 (1960).
[3]R. R. Conte, *Elements de cryogenie* (Masson, Paris, 1970).
[4]*American Institute of Physics Handbook*, 3rd ed., edited by Dwight E. Gray (McGraw-Hill, New York, 1972).
[5]O. V. Lounasmaa, *Experimental Principles and Methods Below 1K* (Academic, London, 1974).
[6]National Bureau of Standards, *Properties of Materials at Low Temperature* (Pergamon, New York, 1959).
[7]R. C. Reid, J. M. Prausnitz, and T. K. Sherwood, *The Properties of Gases and Liquids* (McGraw-Hill, New York, 1977).
[8]A. C. Rose-Innes, *Low Temperature Laboratory Techniques* (English Universities, Liverpool, 1973).
[9]G. K. White, *Experimental Techniques in Low Temperature Physics*, 2nd ed. (Oxford, London, 1968).
[10]W. E. Keller, *Helium-3 and Helium-4* (Plenum, New York, 1969).
[11]*Progress in Low Temperature Physics*, edited by C.J. Gorter (Vols. 1–6) and D. F. Brewer (Vols. 7–) (North-Holland, Amsterdam, 1955–).
[12]*The Physics of Liquids and Solid Helium*, edited by K. H. Benneman and J. B. Ketterson (Wiley-Interscience, New York, 1976, 1978).
[13]*Superconductivity*, edited by R. D. Parks (Dekker, New York, 1969).
[14]J. Wilks, *The Properties of Liquid and Solid Helium* (Clarendon, Oxford, 1967).

8.00. Crystallography

GEORGE A. JEFFREY

University of Pittsburgh

CONTENTS

8.01. HISTORICAL SKETCH

Prior to the discovery of x-ray diffraction by crystals by Friedrich, Knipping, and Laue,[1] crystallography was concerned with the external morphology of crystals. Groth's five-volume *Chemische Krystallographie* published between 1906 and 1919 contains the results of very accurate measurements of interfacial angles of crystals of 7350 chemical compounds. Lack of knowledge about the atomic structure of crystals did not prevent the classical crystallographers from obtaining significant insight into the symmetry properties of crystals. The concepts of unit cell (Haüy, 1800), lattices (Bravais, 1840), and space groups (Barlow, 1883; Federov, 1890; Schoenflies, 1891) were well developed, albeit hypothetical, until W. H. and W. L. Bragg used Laue's discovery of x-ray diffraction by crystals for the experimental determination of crystal structures.[2] With the Braggs, the frontier of crystallography moved out of mineralogy into physics. There it remained, generally, with some excursions into physical metallurgy and material sciences, until the early 1950s, when the availability of the general-purpose digital computer made crystal structure determination an important component of chemistry. This development into chemistry was pioneered primarily by Linus Pauling and his colleagues at the California Institute of Technology and by Monteath Robertson at the Royal Institution in London. An excellent example is Lipscomb's work on the boron hydrides,[3] following the first crystal structure of these compounds by Kasper, Lucht, and Harker.[4] By 1960, the work of Watson and Crick,[5] Hodgkin,[6] Perutz,[7] and Kendrew[8] had sufficient impact on the biological sciences that crystallography in the 1970s began to play the same crucial role in the science of molecular biology as it had for chemistry some twenty years earlier. Crystallography continues to be a valuable discipline for studying the atomic and electronic structures of minerals and for both crystal structural and texture studies in metallurgy and the material sciences.

The development of the "direct methods" for solving the crystal structure phase-problem, recognized by the Nobel Prize in 1986 to Karle and Hauptman, has made crystal structure analysis an indispensable tool for determining the configuration and conformation of molecules with molecular weights of less than 500 daltons. More than 60,000 organic and organo-metallic crystal structures are now contained in the Cambridge Crystallographic Data Base. A similar data base is being developed for inorganic crystal structures.

X-ray crystal structure analysis is the only source of the complete structure determination of macromolecules such as proteins, nucleic acids, and polysaccharides. Charge density studies have begun to provide information concerning the electronic distribution which relates directly to the chemical reactivity of molecules.

The study of crystals by x rays has been the major activity of crystallographers since the Laue experiment in the summer of 1912. The use of the electrons[9] and neutrons[10] have developed as complementary diffraction methodologies for studies where they provide particular advantages. In the case of electron diffraction, these advantages are in the study of very small crystals and of crystal surfaces and for revealing the systematics of solid-state defect structures. In the case of neutrons, the advantages are for locating light atoms, especially hydrogen, with an accuracy comparable to that of heavier atoms, and for studying magnetic spin structures.

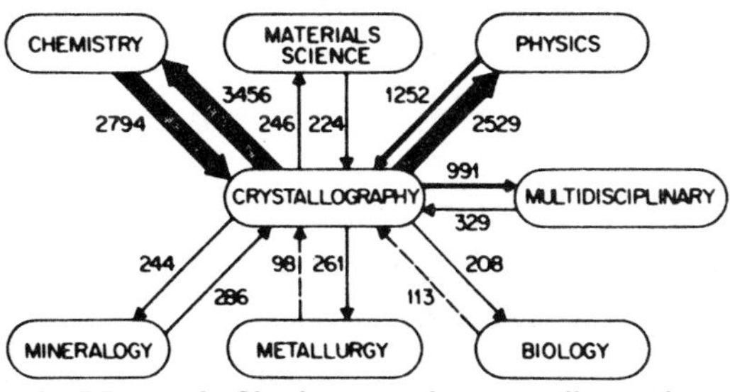

FIGURE 8.01.A. Citation map for crystallography and other disciplines with the number of citations between these shown

The new tools for crystallographers in the 1980s are the synchrotron x-ray sources and spallation neutrons. It will be interesting to see which of the many facets of crystallography benefit most from these powerful pulsed-beam sources. The interplay between crystallography and the other disciplines at the end of the 1970s is shown by the "citation map" in Figure 8.01.A.[11]

8.02. CRYSTAL DATA AND SYMMETRY

A. Crystal system, space group, lattice constants, and structure type

The seven *crystal systems* are based on the principal axial symmetry of the 32 *point-group symmetries* of the crystal lattice, as shown in Table 8.02.A.1. The lattice divides the crystal structure into *unit cells,* which are defined by three concurrent cell edges, **a**, **b**, **c**, which, by convention, form a right-handed system. The corresponding scalar *lattice constants* are $a, b, c, \alpha, \beta, \gamma$. Also commonly used in crystallography is the reciprocal lattice[12] $\mathbf{a}^*, \mathbf{b}^*, \mathbf{c}^*$, where

$$\mathbf{a}\cdot\mathbf{a}^* = \mathbf{b}\cdot\mathbf{b}^* = \mathbf{c}\cdot\mathbf{c}^* = 1,$$

$$\mathbf{a}\cdot\mathbf{b}^* = \mathbf{a}\cdot\mathbf{c}^* = \mathbf{b}\cdot\mathbf{a}^* = \mathbf{b}\cdot\mathbf{c}^* = \mathbf{c}\cdot\mathbf{a}^* = \mathbf{c}\cdot\mathbf{b}^* = 0.$$

The corresponding scalar reciprocal-lattice constants are $a^*, b^*, c^*, \alpha^*, \beta^*, \gamma^*$. The point-group symmetry of the lattice entails the specialization of the lattice constants shown in Table 8.02.A.1. The lattices are defined by $a, b, c, \alpha, \beta, \gamma$ for triclinic crystals; a, b, c, β for monoclinic; a, b, c for orthorhombic; a, c for tetragonal and hexagonal; a, α for rhombohedral; and a for cubic crystals. In the monoclinic system, the unique symmetry axis, by convention, is **b**. In the tetragonal and hexagonal systems, it is **c**. The *Laue symmetry,* of which there are 11 types, describes the point-group symmetry of the corresponding diffraction spectra.

Table 8.02.A.2 gives the crystallographic data for selected elements, intermetallic phases, inorganic and organic compounds. Complete compilations of crystal data are available elsewhere.[13–19]

The *Hermann-Mauguin (HM) point- and space-group symbols* are used by crystallographers. The

TABLE 8.02.A.1. Point-group symmetries (class), Laue symmetries, and lattice constants for the seven crystal systems

System	Point-group symmetry, crystal class	Laue symmetry	Symmetry directions	Lattice constraints	Lattice constants
Triclinic (Tric)	1, $\bar{1}$	$\bar{1}$	None	None	a,b,c α,β,γ
Monoclinic (M)	2, $\bar{2}$, $1/m$	$2/m$	b	$\alpha=\gamma=90°$	a,b,c,β
Orthorhombic (O)	222 $mm2$ mmm	mmm	a,b,c	$\alpha=\beta=\gamma=90°$	a,b,c
Tetragonal (Tetr)	4, $\bar{4}$, $4/m$ 422, $4mm$ $\bar{4}2m$, $4/mmm$	$4/m$ $4/mmm$	c,a,b [110]	$a=b$ $\alpha=\beta=\gamma=90°$	a,c
Hexagonal (H)	6, $\bar{6}$, $6/m$ 622, $6mm$, $\bar{6}m2$, $6/mmm$	$6/m$ $6/mmm$	c,a,b [110]	$a=b$ $\alpha=\beta=90°$ $\gamma=120°$	a,c
Rhombohedral (R) or Trigonal (Trig)	3, $\bar{3}$ 32, $3m$ $\bar{3}m$	$\bar{3}$ $\bar{3}m$	c [110]	$a=b=c$ $\alpha=\beta=\gamma$ $a=b\neq c$ $\alpha=\beta=90°$, $\gamma=120°$	a,γ a,c
Cubic (C)	23, $m3$ 432, $\bar{4}3m$ $m3m$	$m3$ $m3m$	a,b,c [110], [111]	$a=b=c$ $\alpha=\beta=\gamma=90°$	a

space-group symbol starts with a lattice descriptor: P is primitive; A, B, C are (100), (010), (001) face centered; I is body centered; F is all faces centered. R is a rhombohedral lattice. It is followed by the symmetry operator in the direction of the principal symmetry axis, followed by the other symmetry-independent axes. The space-group symbol, given in Table 8.02.A.2, is that used in the *International Tables for X-ray Crystallography,* Vol. I.[20] It is the *short symbol;* containing only those symmetry elements necessary to generate the full symmetry of the space group. The order of precedence in use of symbols is glide and mirror planes, screw axes, axes. In a few examples, a redundant symmetry element is introduced to indicate a distinction in the orientation of a symmetry element, as in $P\bar{6}m2$ and $P\bar{6}2m$. Further details are given in Int. Tables, Vol. I, p. 29. The position of the symbol denotes the direction of the symmetry element with respect to the axes **a**, **b**, **c**. A list of the diffraction symmetry for each space group is given in Int. Tables, Vol. 1, pp. 349–352.

TABLE 8.02.A.3. Conversion from Schoenflies to Hermann-Mauguin point-group notations. In crystals, $n=1, 2, 3, 4, 6$.

Schoenflies	HM	Schoenflies	HM
C_n	n	$S_2\equiv C_i$	$\bar{1}$
$C_S\equiv C_1^h(\equiv C_1^v\equiv S_1)$	m	S_4	$\bar{4}$
C_n^h	n/m	$S_6\equiv C_3^i$	$\bar{3}$
C_2^v	mm	D_2	222
C_3^v	$3m$	D_3	32
C_4^v	$4mm$	D_4	42
C_6^v	$6mm$	D_6	62
D_2^d	$\bar{4}2m$	D_2^h	mmm
D_3^d	$\bar{3}m$	D_3^h	$\bar{6}m$ ($\equiv 3/mm$)
T	23	D_4^h	$4/mmm$
T_h	$m3$	D_6^h	$6/mmm$
T_d	$\bar{4}3m$	O	43
		O_h	$m3m$

n-fold rotation axes are denoted by 1, 2, 3, 4, and 6, n-fold rotation-inversion axes by $\bar{1}, \bar{2}, \bar{3}, \bar{4}$, and $\bar{6}$. 1 implies absence of symmetry, $\bar{1}$ a center of symmetry; $\bar{2}$ is a mirror plane, also denoted by the symbol m. The symbol n/m denotes a mirror plane perpendicular to an n-fold rotation axis. Screw axes are denoted by $2_1, 3_1, 3_2, 4_1, 4_2, 4_3, 6_1, 6_2, 6_3, 6_4$, and 6_5, where the subscript denotes the fractional translation in the direction of the screw axis. Glide planes are denoted by a, b, c, n, and d, indicating the translation direction.

There are 230 space groups, of which 50 can be uniquely identified from their diffraction symmetry, as can nine enantiomorphous pairs and two special pairs, leaving 158 space groups which cannot be determined by inspection of the diffraction patterns.

Most space groups rarely occur, and a significant number (≈ 18) have never been observed.[21] Sixty-five percent of the organic compounds occur in six space groups,[22] $P\bar{1}$, $P2_1$, $P2_1/c$, $C2/c$, $P2_12_12_1$, and $Pbca$, of which only $P2_1$ and $P2_12_12_1$ occur in natural products.

The older *Schoenflies point-group notation* is still used by spectroscopists, inorganic and theoretical chemists, and some solid-state physicists. It is based on rotation axes and combinations of rotation axes and mirror planes, known as rotation-reflexion, or alternating axes. Table 8.02.A.3 gives the conversion from Schoenflies to Hermann-Mauguin notation. The Schoenflies space-group notation is obsolete.

The lattice constants, in Å, define the unit cell with volume V in Å^3. This is related to the calculated crystal densities, by

$$Vd_c = ZM/N_A,$$

where d_c is the calculated density of the crystal in Mg/m^3; Z is the number of asymmetric formula units in

TABLE 8.02.A.2. Crystallographic data for selected crystalline materials

Formula or name	Crystal system	HM space group	Z	Lattice constants a, b, c (Å), α, β, γ (deg)	Structure type and comments
				Elements	
As, α	R	$R\bar{3}m$	2	4.1320(2), 54.12(1)	*A*7 There are 5 other allotropes
Au	C	$Fm3m$	4	4.0782(2)	*A*1 Face-centered cubic, cubic close packed (ccp)
B, rhomb-12	R	$R\bar{3}m$	12	5.057, 58.06°	Structure contains nearly regular icosahedra; 16 other allotropes have been reported
C, diamond	C	$Fd3m$	8	3.56688(15)	*A*4
C, hexag. diam.	H	$P6_3/mmc$	4	2.52, 4.12	Found in meteorites, and made synthetically
C, graphite	H	$P6_3mc$	4	2.4612(2), 6.7090(12)	*A*9
Fe, α	C	$Im3m$	2	2.8664(2)	*A*2 Body-centered cubic
γ	C	$Fm3m$	4	3.6467(2)	*A*1
δ	C	$Im3m$	2	2.9315(2)	*A*2
Hg, α	R	$R\bar{3}m$	1	2.982, 77°, 75°	*A*10 There are 2 other allotropes
Mg	H	$P6_3/mmc$	2	3.2029	*A*3 Hexagonal close packed (hcp)
Mn, α	C	$I\bar{4}3m$	58	8.9129(6)	*A*12 There are 3 other allotropes
S_6	R	$R\bar{3}$	3	4.280, 115.2°	More than 50 allotropes of S have been described; many are doubtful
S_8	O	$Fddd$	128	10.46(1), 12.879(2), 24.478(5)	
Se, α	H	$P3_121$ $P3_221$	3	4.3655(10), 4.9576(24)	*A*8 There are 5 other allotropes
Sn, α	C	$Fd3m$	8	6.4892	*A*4
β	Tetr	$I4_1/amd$	4	5.8316(2), 3.1815(2)	*A*5
U, α	O	$Cmca$	4	2.838(4), 5.868(2), 4.956(1)	*A*20
β	Tetr	$P4_2/mmm$	30	10.759, 5.656	
γ	C	$Im3m$	2	3.524	*A*2
				Intermetallic phases	
βCuZn	C	$Im3m$	2	2.9907 (871°C)	*A*2 } Hume Rothery phases
β'CuZn	C	$Pm3m$	1	2.9539 (47% Zn, 20°C)	*B*2 } Hume Rothery phases
γ CuZn	C	$I\bar{4}3m$	52	8.852 (63% Zn, 20°C)	$D8_2$ } Hume Rothery phases
$MgCu_2$	C	$Fd3m$	8	7.034	*C*15 } Laves phases
$MgNi_2$	H	$P6_3/mmc$	8	4.805, 15.77	*C*36 } Laves phases
$MgZn_2$	H	$P6_3mmc$	4	5.17, 8.50	*C*14 } Laves phases
Mg_2Cu	O	$Fddd$	16	9.05, 18.21, 5.273	
AuCu, I	Tetr	$P4/mmm$	2	3.9512, 3.6798 (350°C)	$L1_o$
$AuCu_3$	C	$Pm3m$	1	3.7432 (20°C)	$L1_2$ Zintl phase
$AuMg_3$	H	$P6_3/mmc$	2	4.63, 8.44	DO_{18}
$CaCu_5$	H	$P6/mmm$	1	5.082, 4.078	$D2_d$
$AlFe_3$	C	$Fm3m$	4	5.780	DO_3
$AlCu_2Mn$	C	$Fm3m$	4	5.937	$L2_1$ Heusler alloys
Cu_5Zn_8	C	$I\bar{4}3m$	52	8.854 (65% Zn)	$D8_2$
Cr_3Ge	C	$Pm3n$	2	4.614	*A*15
Cr_5Si_3	Tetr	$I4/mcm$	4	9.170, 4.636	$D8_m$
$NaZn_{13}$	C	$Fm3c$	8	12.2836	$D2_3$
$Al_{12}W$	C	$Im3$	2	7.5803	
$Cr_{23}C_6$	C	$Fm3m$	4	10.638	$D8_4$
Fe_3C	O	$Pnma$	4	5.0787, 6.7297, 4.5144	DO_{11} Cementite
				Solid-state physics	
Fe_3O_4	C	$Fd3m$	8	8.3940	$H1_1$ Ferrimagnet, Magnetite
	O	$Imcm$	4	5.945, 8.388, 5.912 (−195°C)	
MnO	C	$Fm3m$	4	4.445	*B*1 Antiferromagnet, Magnanosite
MnF_2	Tetr	$P4_2/mnm$	2	4.8734, 3.3099	*C*4 Rutile structure, antiferromagnet
$CsNiCl_3$	H	$P6_3/mmc$	2	6.236, 5.225	$BaNiO_3$ structure
$CrBr_3$	R	$R\bar{3}$	2	7.05, 52°36′	Ferromagnet
$MnAu_2$	Tetr	$I4/mmm$	2	3.197, 7.871	$C11_b$ Helimagnet
αFe_2O_3	R	$R\bar{3}c$	2	5.4271, 55.26°	$D5_1$ Weak ferromagnet
$KMnF_3$	C	$Pm3m$	1	4.190	Weak ferromagnet
$YFeO_3$	C	$Pm3m$	1	3.785	Weak ferromagnet
$CuCl_2{\cdot}2H_2O$	O	$Pbmn$	2	7.38, 8.04, 3.72	$E2_1$ 4-sublattice antiferromagnet
$\alpha CoSO_4$	O	$Cm3m$	4	5.198, 7.871, 6.522	4-sublattice antiferromagnet

TABLE 8.02.A.2–*Continued*

Formula or name	Crystal system	HM space group	Z	Lattice constants a, b, c (Å), α, β, γ (deg)	Structure type and comments
$BaTiO_3$	C	*Pm3m*	1	4.0118 (201°C)	
	Tetr	*P4mm*	1	3.9947, 4.0336 (25°C)	Ferroelectric
	O	*Amm2*	2	3.990, 5.669, 5.682 (5°C)	
$SrTiO_3$	C	*Pm3m*	1	3.9051	Ferroelectric
KH_2PO_4	Tetr	$I\bar{4}2d$	4	7.453, 6.959	Ferroelectric
$PbNb_2O_6$	Tetr	*P4/?b?*	5	12.46, 3.907 (600°C)	Paraelectric
$BaMnO_4$	O	*Pbnm*	4	7.304, 9.065, 5.472	Barite-type structure
Cr_2O_3	R	$R\bar{3}c$	2	5.35, 55.0°	$D5_1$ Magnetoelectric
$NaKC_4H_4O_6{\cdot}4H_2O$	O	$P2_12_12$	4	11.93, 14.30, 6.17	Rochelle salt, anomalous dielectric
$YBa_2Cu_3O_{6.9}$	O	*Pmmm*	1	3.8218, 3.8913, 11.677	High T_c superconductors
$Y_2BaCu_5O_5$	O	*Pnma*	4	12.176, 5.655, 7.130	″
$YBa_2Cu_3O_{7-x}$	T	*P4/mmm*	1	3.8683, 11.708	″
$Y_2BaCu_3O_{6-x}$	T	$P\bar{4}m2$	1	3.859, 11.71	″
$La_{1.85}Ba_{0.15}CuO_4$	T	*I4/mnm*	1	3.7817, 13.2487	″
				Inorganics (see also **Minerals**)	
CsCl	C	*Pm3m*	1	4.123	*B2*
H_2O I	H	$P6_3/mmc$	4	4.48, 7.31	Normal ice
Ic	C	*Fd3m*	8	6.35	Low-temperature phase
II	R	$R\bar{3}$	12	7.78, 113.1°	Ices II–VII are high-pressure phases
III	Tetr	$P4_12_12$	12	6.73, 6.83	
IV					
V	M	*A2/a*	28	9.22, 7.54, 10.35, 109.2°	
VI	Tetr	$P4_2/mmc$	10	6.27, 5.79	Self-clathrate
VII	C	*Pn3m*	2	3.41	Self-clathrate
VIII	C		32	9.70 (−30°)	
H_2O_2	Tetr	$P4_22_1$	4	4.06, 8.00	Below −0.9°C
N_2O_2	M	$P2_1/a$	2	6.68, 3.96, 6.55, 127°54′	At −175°C
$6X,2Y,46H_2O$	C	*Pm3n*	1	∼12.0	Type I, gas hydrate, X is a molecule not larger than CH_3Cl
$8X,16Y,136H_2O$	C	*Fd3m*	1	∼17.3	Type II, gas hydrate, Y is a molecule not larger than CCl_4
Fe_2Ti_4O	C	*Fd3m*	16	11.275	$E9_3$
$Na_2O{\cdot}11Al_2O_3$	H	$P6_1/mmc$	2	5.595, 22.49	*β-Alumina*
$Na_2SO_4{\cdot}10H_2O$	M	$P2_1/c$	4	11.51, 10.38, 12.83, 107°45′	Glaubers salt
$KAl(SO_4)_2{\cdot}12H_2O$	C	*Pa3*	4	12.158	Typical of *alums*
				Minerals	
NaCl	C	*Fm3m*	4	5.64056 (26°C)	*B*1 Halite
ZnS	H	$P6_3mc$	2	3.076, 5.048	*B*3 Wurtzite
ZnS	C	$F\bar{4}3m$	4	5.4093 (26°C)	*B*4 Zinc blende
NiAs	H	$P6_3/mmc$	2	3.602, 5.009	$B8_1$ Niccolite
CaF_2	C	*Fm3m*	4	5.463	*C*1 Fluorite
FeS_2	C	*Pa3*	4	5.40667 (26°C)	*C*2 Pyrite
FeS_2	O	*Pnnm*	2	4.436, 5.414, 3.381	*C*18 Marcasite
SiO_2	Trig	$P3_121$	3	4.9027, 5.3934	α Quartz
	H	$P6_322$	3	5.45, 4.99	β Quartz
	H	$P6_3/mmc$		8.22, 5.03	β Tridymite
	Tetr	$P4_12_12$	4	4.97, 6.92	α Cristobalite
	C	*Fd3m*	8	7.05	β Cristobalite
TiO_2	Tetr	$P4_2/mmn$	2	4.5937, 2.9581	*C*4 Rutile
TiO_2	Tetr	*I4/amd*	4	3.785, 9.514	Anatase
TiO_2	O	*Pbca*	8	9.184, 5.447, 5.145	Brookite
Cu_2O	O	*Pn3m*	2	4.2696 (26°C)	*C*3 Cuprite
Al_2O_3	R	$R\bar{3}c$	2	(4.758, 12.991)	$D5_1$ Corundun
$BaSO_4$	O	*Pnma*	4	8.8701, 5.4534, 7.1507	Barite
$CaCO_3$	R	$R\bar{3}c$	2	6.75, 46°5′	Calcite
$CaCO_3$	O	*Pcmm*	4	5.72, 7.94, 4.94	Aragonite
$CuFeS_2$	Tetr	$I\bar{4}2d$	4	5.24, 10.30	$E1_1$ Chalcopyrite
Al_2MgO_4	C	*Fd3m*	8	8.0800 (26°C)	$H1_1$ Spinel
Fe_2SiO_4	O	*Pbnm*		4.820, 10.485, 6.093	Olivine
$MgAl_2O_4$	C	*Fd3m*	8	8.083	Spinel
$ZrSiO_4$	Tetr	$I4_1/amd$	4	6.607, 5.982	Zircon
$CaSO_4$	O	*Amma*	4	6.991, 6.996, 6.238	Anhydrite
$CaSO_4{\cdot}2H_2O$	M	*C2/c*	4	6.28, 5.67, 15.15, 113°51′	Gypsum

TABLE 8.02.A.2-*Continued*

Formula or name	Crystal system	HM space group	Z	Lattice constants a, b, c (Å), α, β, γ (deg)	Structure type and comments
$CaWO_4$	Tetr	$I4_1/a$	4	5.243, 11.376	Scheelite
$FeWO_4$	M	$P2/c$	2	2.730, 5.703, 4.952, 90°05′	Wolframite
$Ca_5F(PO_4)_3$	H	$P6_3m$	2	9.3684, 6.8841	Fluorapatite
$Al_2Mg_3(SiO_4)_3$	C	$Ia3d$	8	11.459	A synthetic garnet
$Be_3Al_2(SiO_3)_6$	H	$P6/mcc$	4	9.206, 9.025	Beryl
$[Al(F,OH)]_2SiO_4$	O	$Pbnm$	4	4.6499, 8.7968, 8.3909	Topaz
				Organics	
C_2H_6	H	$P6/mmc$	2	8.19, 4.46	Ethane
C_2H_4	O	$Pnnm$	2	4.87, 6.46, 4.15	Ethylene
C_8H_8	R	$R\bar{3}$	1	5.340, 72°26′	Cubane
CH_3OH	M	$P2_1/m$	2	4.53, 4.91, 4.69, 90.0°	α Methanol
CH_3OH	O	$Cmcm$	4	4.67, 7.24, 6.43	β Methanol
C_6H_6	O	$Pbca$	4	7.034, 7.460, 9.666	Benzene −3°C
$C_{10}H_8$	M	$P2_1/c$	2	8.080, 6.003, 8.235	Naphthalene
$C_{14}H_{10}$	M	$P2_1/c$	2	9.439, 6.036, 8.561	Anthracene
$C_{10}H_{16}$	C	$F\bar{4}3m$ (or $Fm3m$)	4	9.426	Adamantine
	Tetr	$P\bar{4}2_1c$	2	6.60, 8.81 (−65°)	
CH_2N_2O	Tetr	$P\bar{4}2_1m$	2	5.670, 4.726	Urea
C_2H_5NO	R	$R3c$		13.49, 91.6°	Acetamide
$C_5H_{12}O_4$	Tetr	$I\bar{4}$	2	6.10, 8.73	Pentaerythritol
$C_6H_{16}N_2$	O	$Pbca$	4	6.94, 5.77, 19.22	Hexamethylenediamine
$C_2H_2O_4$	M	$P2_1/c$	2	5.30, 6.09, 5.51, 115°30′	Oxalic acid
$C_2H_2O_4$	M	$P2_1/n$	2	6.119, 3.604, 12.051, 106°12′	Oxalic acid dihydrate
$C_4H_6O_6$	M	$P2_1$	2	7.72, 6.00, 6.20, 100°10′	*D*-tartaric acid
$C_2H_4NO_2$	M	$P2_1/n$	4	5.10, 11.96, 5.45	α Glycine
$C_{10}H_{14}$	M	$P2_1/a$	2	11.57, 5.77, 7.03, 113.3°	1,2,4,5-Tetramethylbenzene
$C_{12}H_{18}$	Tric	$P\bar{1}$	1	8.92, 8.86, 5.30, 44°27′, 116°43′, 119°34′	Hexamethylbenzene
$C_{25}H_{20}$	Tetr	$P\bar{4}2_1c$	2	10.87, 7.23	Tetraphenylbenzene
$C_{20}H_{12}$	M	$P2_1/a$	2	16.10, 4.695, 10.15, 110°8′	Coronene
$C_6H_{12}O_6$	O	$P2_12_12_1$	4	10.36, 14.84, 4.93	α Glucose
$C_6H_8O_6$	M	$P2_1$	4	16.95, 6.32, 6.38, 102.5°	Ascorbic acid
$C_{16}H_{17}N_2O_4SNa$	M	$P2_1$	2	8.48, 6.33, 16.63, 94.2°	Penicillin
				Organic macromolecules	
Cellulose I	M	$P2_1$	8	16.34, 15.72, 10.38, 97°	
Cellulose II	M	$P2_1$	2	8.01, 9.04, 10.36, 117.1°	Fortisan
α-chitin	O	$P2_12_12_1$	4	4.74, 18.86, 10.32	Lobster tendon
β-chitin	M	$P2_1$	2	4.85, 9.26, 10.38, 97.5°	
DNA, A-form	M	$C2$	4	22.1, 40.4, 28.1, 97.1°	SNa, salt, fiber axis *c*
Transfer RNA, yeast phenylalanine	O	$P2_122_1$	4	330, 560, 161	Molec. wt. 23 788
Myoglobin	M	$P2_1$	2	64.6, 31.3, 34.8, 105.5°	Sperm whale
Haemoglobin	M	$C2_1$	2	108.95, 63.51, 54.92, 110°53′	Horse
Satellite tobacco necrosis virus	M	$C2$	4	318.4, 305.0, 185.3, 94°37′	Molec. wt. 1.7×10^6
DNA, β-form	O		4	22.7, 31.2, 33.7	Salt
Polyoma virus	C	$I23$		490	Molec. wt. 3.6×10^6
Southern bean mosaic virus	R	$R32$	32	757.5, 63.6°	Molec. wt. asymmetric unit, 9×10^6

the unit cell, usually an integral; M is the molecular weight of the formula unit in daltons; and N_A is Avogadro's constant, 6.0226×10^{23}.

Experimental crystal densities d_m are generally measured by flotation in mixtures of suitable liquids of known densities. Suggested liquids and their densities at 25°C are given in Int. Tables, Vol. III, p. 19.

B. Reduced cells

In the case of a triclinic crystal, the selection of the lattice translations is arbitrary. (It is also arbitrary for the other crystal systems if the unit cells are defined independently of symmetry, which is rarely necessary.)

For a unit cell represented by

$$\mathbf{P} = \begin{pmatrix} \mathbf{a\cdot a} & \mathbf{b\cdot b} & \mathbf{c\cdot c} \\ \mathbf{b\cdot c} & \mathbf{a\cdot c} & \mathbf{a\cdot b} \end{pmatrix}$$

the *positive reduced cell* (type I) has all angles $<90°$:

$$\mathbf{a\cdot a} \leqslant \mathbf{b\cdot b} \leqslant \mathbf{c\cdot c},$$

$$\mathbf{b\cdot c} \leqslant \tfrac{1}{2}\,\mathbf{b\cdot b}, \quad \mathbf{a\cdot c} \leqslant \tfrac{1}{2}\,\mathbf{a\cdot a}, \quad \mathbf{a\cdot b} \leqslant \tfrac{1}{2}\,\mathbf{a\cdot a}.$$

The *negative reduced cell* (type II) has all angles $\geqslant 90°$:

$$\mathbf{a\cdot a} \leqslant \mathbf{b\cdot b} \leqslant \mathbf{c\cdot c}, \quad |\mathbf{b\cdot c}| \leqslant \tfrac{1}{2}\,\mathbf{b\cdot b}, \quad |\mathbf{a\cdot c}| \leqslant \tfrac{1}{2}\,\mathbf{a\cdot a},$$

$$|\mathbf{a\cdot b}| \leqslant \tfrac{1}{2}\,\mathbf{a\cdot a}, \quad |\mathbf{b\cdot c}| + |\mathbf{a\cdot c}| + |\mathbf{a\cdot b}| \leqslant \tfrac{1}{2}(\mathbf{a\cdot a} + \mathbf{b\cdot b}).$$

Other special conditions and the transformations from an

unreduced to a reduced cell are given in Int. Tables (1969 edition), Vol. I, pp. 530–535.

C. Physical properties of crystals

The properties of the crystal classes according to the type of piezoelectric moment and optical activity are given in Int. Tables, Vol. 1, p. 42. Of the 32 crystal classes, 21 are noncentrosymmetrical. Optical activity occurs in 15 classes. Pyroelectricity can theoretically occur only when there is a unique polar axis in the crystal, but owing to the accompanying piezoelectricity of the measurement, it can occur in all noncentrosymmetrical crystals. Piezoelectricity can occur in all noncentrosymmetrical crystals except those in class 432, where the moduli are all zero owing to the high symmetry.

The relations between electrical, magnetic, thermal, optical, and elastic stress and strain variables and their scalar and tensorial properties are described in detail in Sec. 9a-2 of the *American Institute of Physics Handbook,* 3rd ed. (McGraw-Hill, New York, 1972).

8.03. CRYSTAL DIFFRACTION

A. Conditions for diffraction

Diffraction occurs if the wavelength λ of the incident radiation on a crystal is comparable with the periodicity of the atomic structure expressed by the lattice constants. Thus crystals diffract x rays, electrons, and neutrons with wavelengths from 0.1 to 10 Å. Crystallography uses the *angstrom,* Å, as the unit of length. The conversion factor to the *kx unit,* 10^{-10} m, is kx = 1.002 077 6(54). X-ray wavelengths calculated from excitation potential from $\lambda = 12.398\,10$ keV are given in Int. Tables, Vol. IV, pp. 5–43.

The condition for the occurrence of a diffraction spectrum, *hkl*, is given by

$$\mathbf{d}^*_{hkl} = \frac{1}{d_{hkl}} = \frac{2\sin\theta}{\lambda} = \mathbf{h}a^* + \mathbf{k}b^* + \mathbf{l}c,$$

where $\mathbf{d}^*_{hkl}$ is the reciprocal-lattice vector, or *scattering vector,* of the spectrum *hkl* (also referred to as ρ_{hkl}[27] and σ_{hkl}[28]). d_{hkl} is the spacing of the *hkl* crystal planes, θ is the Bragg reflection angle, and 2θ is the angle between incident beam $\mathbf{s}_0$ and diffracted beam $\mathbf{s}$. The direction of the diffracted beam $\mathbf{s}$ relative to that of the incident beam $\mathbf{s}_0$ and the reciprocal-lattice vector $\mathbf{d}^*_{hkl}$ is given by $\mathbf{s} - \mathbf{s}_0 = \mathbf{d}^*_{hkl}$. This is conveniently represented graphically by the Ewald construction[23] shown in Figure 8.03.A.1.

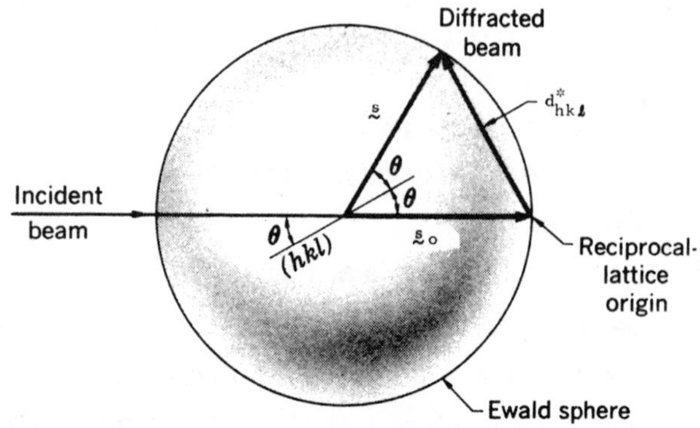

FIGURE 8.03.A.1. Ewald reciprocal-lattice condition for x-ray diffraction with monochromatic radiation

For a particular wavelength λ, the number of diffraction spectra that can be observed is limited to those lying within the *sphere of diffraction* of radius $2\lambda^{-1}$. The number of diffraction spectra is therefore $32\pi V_p/3\lambda^3$, where V_p is the volume of the primitive unit cell.

Diffraction will occur for a stationary crystal, when the value of λ varies. This is the *Laue method.* A more common procedure is to use *monochromatic radiation,* with fixed λ, and oscillate, rotate, or precess the crystal. In these moving-crystal methods, diffraction occurs when the reciprocal-lattice point d^*_{hkl} passes through the Ewald sphere. Both the reciprocal-lattice point and the surface of the sphere are finite, the former because of the mosaic structure and size of the diffracting crystal, the latter because of the finite width of monochromatic K wavelengths used ($\Delta\lambda/\lambda \approx 10^{-4}$), the substructure of the radiation, e.g., $\lambda_{\mathrm{Cu}K\alpha_1}$, $\lambda_{\mathrm{Cu}K\alpha_2}$, and the finite size of the focal spot of the x-ray tube or monochromator.

The most useful theory for relating the intensities of the x-ray diffracted spectra to the atomic coordinates and thermal parameters of a crystal structure is the *kinematical theory.* This classical theory is based on a model in which the radiation is represented as a wave which transverses the whole crystal. The amplitude phase differences between the radiation scattered at different points in the crystal depend only on the difference in length of the paths of the incident and diffracted waves to and from those points. This theory applies only to perfectly *mosaic crystals.* Departures from ideality due to the crystal being too perfect, or insufficiently mosaic, are treated as corrections. For diffraction by *perfect crystals,* the *dynamical theory*[24] is more appropriate. Perfection is as hard to come by in crystals as in humans,[25] and the dynamical theory is applicable only to a limited class of crystals.

In the kinematical theory, the intensity of the diffracted spectra depends on the wavelength of the incident radiation, the crystal structure and the nature of the diffracting specimen, the direction of diffraction, and the experimental conditions of the measurement. Intensity measurements are generally made with monochromatic radiation produced by means of appropriate filters, balanced filters, or crystal monochromators. The properties of filters and of various monochromating crystals are given in Int. Tables, Vol. III, pp. 73–87. The most commonly used monochromator in single-crystal structure analysis is the graphite monochromator manufactured by Union Carbide Corp., Carbon Products Division, Parma, OH.

For a small crystal, volume v, completely bathed in the incident x-ray beam,

$$P_{hkl} = Q_{hkl}v = I_{hkl}\omega/I_0,$$

where P_{hkl} is the integrating diffracting power of the crystal, with the dimensions of area, Q_{hkl} is the integrated diffracting power per unit volume, I_0 and I_{hkl} are the energy per unit area per unit time of the incident and diffracted beams, and ω is the constant angular velocity with which the reciprocal-lattice point passes through the *Ewald sphere.*

For extended faces of crystals or powder specimens which are larger than the incident beam, a more useful expression is

$$I'_{hkl} = \int_{\theta-\epsilon}^{\theta+\epsilon} R_{hkl}(\theta)d\theta,$$

where R_{hkl} is the ratio of the power (energy per second) of the diffracted beam to that of the incident beam as the crystal moves through the diffracting position, starting at $\theta - \epsilon$ and finishing at $\theta + \epsilon$, beyond which no diffraction occurs. I' is dimensionless.

For x-ray diffraction by a small crystal, volume v, entirely in the x-ray incident beam,

$$I_{hkl} = \frac{I_0}{\omega} = \frac{1+\cos^2 2\theta}{2\sin 2\theta}\left(\frac{e^2}{4\pi\epsilon_0 mc^2}\right)^2 N^2\lambda^3 F_{hkl}^2 v,$$

where N^2 is the number of unit cells per unit volume of the crystal. The θ-dependent factors are the Lorentz factor $1/\sin 2\theta$ and the polarization factor $(1+\cos^2 2\theta)/2$. F_{hkl} is the structure factor for the diffracted spectrum hkl. The classical radius of an electron is $e^2/4\pi\epsilon_0 mc^2$. For a particular wavelength and crystal, the quantity

$$(4\pi\epsilon_0)^{-1}e^4m^{-2}c^{-4}\lambda^3N^2v$$

is a constant. It is generally not calculated, since for a given crystal and radiation, the absolute values of I_{hkl} are rarely measured experimentally. Note, however, that the diffracting power is proportional to λ^3; therefore Cu$K\alpha$ radiation diffracts ten times more efficiently than Mo$K\alpha$. Very small crystals with very large unit cells, e.g., proteins, diffract more poorly than large crystals with small unit cells, because of N^2. Because of m^{-2}, the scattering of x rays by the protons is negligible.

Neutron diffraction uses wavelengths similar to those of x rays. The scattering is by the atomic nuclei, and the constant term in the intensity expression is $m^2\lambda^3h^{-2}N^2$. The kinematic theory can be used in the same way as for x-ray diffraction except that the departures from the ideal theory are greater.

In *electron diffraction,* the wavelengths provided in Int. Tables, Vol. IV, p. 174 depend on the accelerating voltage. They are an order of magnitude shorter. The scattering involves both atomic nuclei and electrons.

For qualitative (rough) crystal structure determination, the kinematic theory can be used, or a compromise between the kinematical and two-beam dynamical theory, wherein $I_{hkl} \propto |F_{hkl}|^\alpha$, where $1 < \alpha < 2$.[26] For quantitative crystal structure analyses, n-beam dynamical theory must be used.[27] Because of the dynamical scattering effects, these methods are applicable only to very thin crystals of light-atom compounds with carefully controlled morphologies.[28]

The principal formulas for x-ray diffraction experiments are given in Int. Tables, Vol. II, p. 314.

The *polarization factor* $(1+\cos^2 2\theta)/2$ is for an unattenuated or attenuated, i.e., filtered, x-ray beam. If a crystal monochromator is used, this factor becomes, for an ideally imperfect monochromating crystal,

$$\frac{\cos^2 2\theta_m + \cos^2 2\theta}{1+\cos^2 2\theta_m}.$$

For a perfect monochromating crystal, it is

$$\frac{\cos 2\theta_m + \cos^2 2\theta}{1+\cos^2 2\theta_m},$$

where θ_m is the Bragg angle at the monochromator. These differences are not significant for Mo$K\alpha$ radiation ($<5\%$ for $0 \leqslant 2\theta \leqslant 90°$), but may be significant for very accurate measurements with Cu$K\alpha$ radiation.

The *Lorentz factor* $1/\sin 2\theta$ depends on the angular velocity with which the reciprocal-lattice point hkl passes through the Ewald reflecting sphere. Its form varies with the geometry of the instrument which moves the crystal and records the diffraction spectra. The Lorentz factors

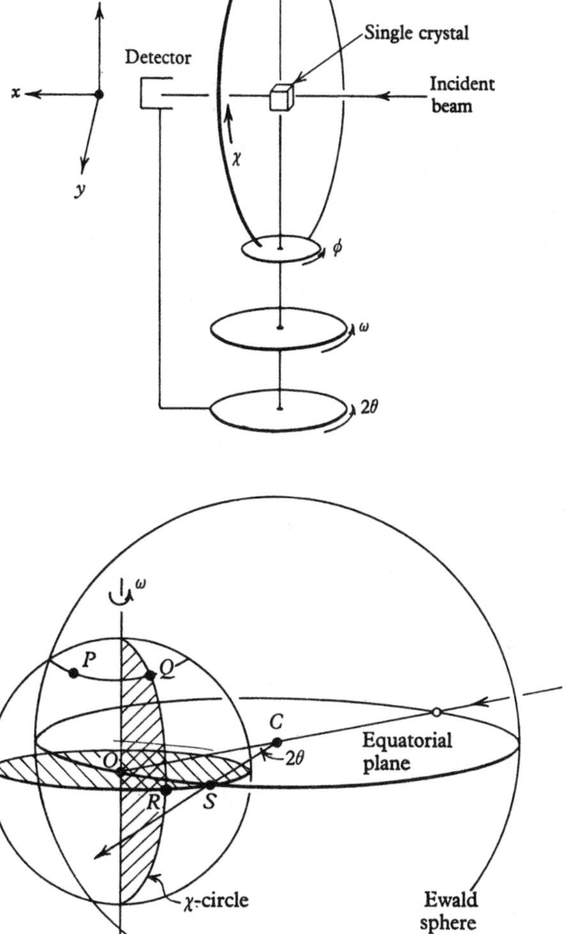

FIGURE 8.03.B.1. Geometry of four-circle diffractometer. Reciprocal-lattice point P can be moved to reflecting position S on the Ewald sphere as follows:

$$P \xrightarrow{\omega} Q \xrightarrow{\chi} R \xrightarrow{\omega} S.$$

TABLE 8.03.C.1. Mass attenuation coefficients (μ/σ in cm^2 g^{-1}) of the atoms for $\left.\begin{matrix}\text{Cu}K\alpha\\ \text{Mo}K\alpha\end{matrix}\right\}$ radiation

IA	IIA	IIIB	IVB	VB	VIB	VIIB	VIII	VIII	VIII	IB	IIB	IIIA	IVA	VA	VIA	VIIA	VIIIA
H 0.391 0.173																	**He** 0.284 0.202
Li 0.477 0.197	**Be** 1.001 0.245											**B** 2.142 0.345	**C** 4.219 0.515	**N** 7.142 0.790	**O** 11.01 1.147	**F** 15.95 1.584	**Ne** 22.13 2.209
Na 30.30 2.939	**Mg** 40.88 3.979											**Al** 50.23 5.041	**Si** 65.32 6.23	**P** 77.28 7.87	**S** 92.53 9.63	**Cl** 109.2 11.66	**Ar** 119.5 12.62
K 148.4 16.20	**Ca** 171.6 19.00	**Sc** 186.0 21.04	**Ti** 202.4 23.25	**V** 222.6 25.24	**Cr** 252.1 29.25	**Mn** 272.5 11.86	**Fe** 104.4 17.74	**Co** 338.6 41.0	**Ni** 48.8 47.2	**Cu** 51.5 49.3	**Zn** 59.5 55.5	**Ga** 62.1 56.9	**Ge** 67.9 60.5	**As** 75.7 66.0	**Se** 82.9 68.8	**Br** 90.3 74.7	**Kr** 97.0 79.1
Rb 106.3 83.0	**Sr** 115.3 88.0	**Y** 127.1 97.6	**Zr** 136.8 16.1	**Nb** 168.8 16.97	**Mo** 158.3 18.44	**Tc** 167.7 19.78	**Ru** 180.8 21.11	**Rh** 194.1 23.05	**Pd** 205.0 24.4	**Ag** 218.1 26.38	**Cd** 229.3 27.73	**In** 242.1 29.11	**Sn** 253.3 31.18	**Sb** 266.5 33.01	**Te** 273.4 33.92	**I** 291.7 36.33	**Xe** 309.8 38.31
Cs 325.6 29.5	**Ba** 336.1 31.0	**La** 353.5 45.4	**Hf** 157.7 59.7	**Ta** 161.5 89.5	**W** 170.5 95.8	**Re** 178.1 98.7	**Os** 181.8 100.2	**Ir** 192.9 101.4	**Pt** 198.2 108.6	**Au** 207.8 111.3	**Hg** 216.2 114.7	**Te** 222.2 119.4	**Pb** 232.1 122.8	**Bi** 242.9 125.9	**Po**	**At**	**Rn** 263.7 117.2
Fr	**Ra**	**Ac**															
				Ce 378.8 48.6	**Pr** 402.2 50.8	**Nd** 417.9 53.3	**Pm** 441.1 55.5	**Sm** 453.5 58.0	**Eu** 417.9 61.2	**Gd** 426.7 62.8	**Tb** 321.9 66.8	**Dy** 336.6 68.9	**Ho** 128.4 72.1	**Er** 134.1 75.4	**Tm** 140.2 79.0	**Yb** 144.7 80.2	**Lu** 152.0 84.2
				Th 306.8 99.5	**Pa**	**U** 305.7 96.7	**Np**	**Pu** 352.9 48.8	**Am**	**Cm**	**Bk**	**Cf**	**Es**	**Fm**	**Md**	**No**	**Lw**

for different methods of recording diffraction spectra are given in Int. Tables, Vol. II, pp. 266–267.

Polarization, Lorentz, and other angle factors are usually combined in the computer programs which reduce intensities to structure amplitudes. Tabulated values are given in Int. Tables, Vol. II, pp. 268–273.

B. Single-crystal diffractometer[29,30]

The standard x-ray or neutron diffractometer used in crystallography is a four-angle instrument. The angles θ, Φ, χ, and ω are defined in Figure 8.03.B.1.

The *diffraction plane* is defined by the source (focal spot), the crystal, and the detector. The angle subtended by the source and the detector at the crystal is $180° - 2\theta$, and the bisector of this angle is the *diffraction vector.* Rotation about the diffraction vector is denoted by ψ.

The *orientation matrix* **U** is such that

$$\mathbf{A}^* = \mathbf{U}\mathbf{A}_G$$

when

$$A^* = \begin{pmatrix}\mathbf{a}^*\\ \mathbf{b}^*\\ \mathbf{c}^*\end{pmatrix}$$

with metric

$$G^{-1} = \begin{pmatrix} a^*a^* & a^*b^* & a^*c^* \\ a^*b^* & b^*b^* & b^*c^* \\ a^*c^* & b^*c^* & c^*c^* \end{pmatrix};$$

$$A_g = A_D$$

when (Φ,χ,ω) are zero, where A_g refers to the orientation of the crystal on its goniometer head and A_D is the orientation of the diffractometer axes.

The search for crystal diffraction spectra, determination of the orientation matrix, and indexing and determination of the lattice parameters is a computer software component supplied with or developed for the particular computer-controlled diffractometer. The formulas for calculating setting angles and determining the orientation matrix are given in Int. Tables, Vol. IV, pp. 278–884.

C. Absorption

The diffracted intensity I_{hkl} is reduced by a transmission factor A_{hkl} which is less than unity (or absorption factor A^*_{hkl}):

$$A_{hkl} = \frac{1}{V}\int \exp[-\mu(r_\alpha + r_\beta)]dV,$$

where μ is the linear absorption coefficient in cm^{-1}, and r_α and r_β are the path lengths in the crystal transversed by the incident and diffracted beams to and from the element of volume dV. The *linear absorption coefficient* is independent of the physical state of the material and is calculated from the *mass absorption coefficient* μ/ρ, where ρ is the density. To a good approximation, mass absorption coefficients are additive properties given by

$$\mu/\rho = \sum_i g_i(\mu/\rho)_i,$$

where g_i is the mass fraction contributed by the element i whose mass absorption (attenuation) coefficient is $(\mu/\rho)_i$. For a material of known formula weight, the most convenient formula is

$$\mu = \frac{Z}{V}\sum_i (\mu/\rho)_i,$$

where the summation is over one formula weight in the unit cell. Values of the x-ray mass atomic absorption coefficient (in cm^2 g^{-1}), for the most commonly used radiations, Cu$K\alpha$ and Mo$K\alpha$, are given in Table 8.03.C.1. Those for other wavelengths are given in Int. Tables, Vol. IV, pp. 61–66.

D. X-ray absorption corrections

X-ray absorption corrections are very important in accurate x-ray crystal structure analysis. The quantity A_{hkl} is calculated by analytical procedures,[31,32] requiring as input μ, (hkl), and Δ_{hkl}, where Δ_{hkl} is the perpendicular distance of the *hkl* face to an origin *within* the crystal; *hkl* need not be integral, so that noncrystallographic faces can be defined. The determination of Δ_{hkl} requires careful measurements of the crystal using an optical goniometer or the optics of a diffractometer. If the faces can be indexed, the calculated face normals can be refined by least-squares fit to the known axial ratios.

E. Extinction

The intensity formulas given above are based on the kinematic theory of x-ray diffraction by a crystal. This assumes that the crystal is ideally mosaic, i.e., that the crystal lattice is not coherent over large regions and in consequence the diffracted amplitudes within each mosaic block are small enough that interaction between the incident and diffracted waves can be neglected. If this assumption does not hold, the dynamical theory of crystal diffraction may be more appropriate.

In practice, crystals which are perfect enough for application of the dynamical theory are rare. Since the dynamical theory gives smaller values for I_{hkl} than the kinematic theory $(I_{hkl} \approx |F_{hkl}|)$, deviations from the *ideally mosaic model* are therefore treated as *extinction corrections* to the kinematic theory.[33]

If the mosaic blocks are too large, interference occurs between the diffracted and incident beams within each block. This is known as *primary extinction.* If the mosaic blocks are small, but so well aligned that several blocks are simultaneously in the reflecting position for the same incident x-ray beam, then lower blocks receive incident beams of lower intensity. This is referred to as *secondary extinction.*

Electrons with the wavelengths used in diffraction experiments are very strongly absorbed, and diffraction by transmission is only possible with thin crystals (less than $\approx$500 Å). The small penetration of low-energy electrons is used to study the surfaces of crystals by the LEED (low-energy electron diffraction) technique.[34,35]

Neutron absorption coefficients[10] are generally negligible (i.e., $\mu/\rho < 0.5$), with the exception of elements such as lithium (3.5), boron (24), cadmium (14), samarium (47), europium (6), and gadolinium (73), where the neutron capture resonance energies are in the region of the wavelengths used for crystal diffraction.

Extinction is generally of lesser importance than absorption in x-ray diffraction, but is very important in neutron and electron diffraction; so much so in electron diffraction that the dynamical theory generally has to be applied.

Satisfactory treatments of extinction in x-ray and neutron diffraction are available.[36–39] These utilize the transmission factors A_{hkl} calculated for the absorption corrections, and permit the determination of anisotropic extinction corrections by including an extinction parameter tensor $\mathbf{g}_{ij}$ in the least-squares refinement of the atomic and thermal parameters.

F. Multiple reflections

When a crystal is oriented with respect to the beam so that several reciprocal-lattice points lie on the Ewald sphere simultaneously,

$$\mathbf{d}_i^* = \frac{\mathbf{s}_i - \mathbf{s}_0}{\lambda} \quad \text{and} \quad \mathbf{d}_i^* - \mathbf{d}_j^* = \frac{\mathbf{s}_i - \mathbf{s}_j}{\lambda}.$$

Thus the diffracted beam corresponding to $\mathbf{d}_j^*$ can act as an incident beam for $\mathbf{d}_i^*$. This is know as the *Renninger effect.*[40,41]

The intensity of the multiple-diffracted beam is small compared with the single-diffracted beam. However, when one of the intensities is large, this can introduce significant errors.[42,43] It can also lead to space-group ambiguities in the identification of systematically absent reflections. The occurrence of multiple reflections is increased with shorter values of λ (i.e., larger Ewald spheres) and larger unit cells (closer density of reciprocal-lattice points). Since the occurrence is greater when $\mathbf{s}_0$ is in the same plane as a well-populated layer of reciprocal-lattice points, it is customary to offset the crystal axis from the rotation axis by a few degrees to reduce this condition. The importance of this effect for a particular reflection is examined by recording the diffracted intensity while rotating the crystal about the diffraction vector $\mathbf{d}_{hkl}^*$.

G. Diffraction by perfect crystals

Since the diffracting power depends on the *crystal perfection,* it can be used to investigate faults in crystals, especially for investigating imperfections in nearly perfect crystals. This is the basis of the Berg-Barrett method.[44–46] The image of the face of a crystal can be obtained by illuminating it with an x-ray beam placed about 30 cm away, and recording on a film parallel to the face and as close as possible. The resolving power, of the order of microns, can reveal dislocations in the crystal surface. Thin crystal plates can also be examined by transmission in this way. This method is most sensitive for nearly perfect crystals, and ceases to be useful for the mosaic crystals, where the kinematic theory of diffraction is a good approximation.

H. "Borrmann" or "anomalous transmission" effect[47]

Transmitted and diffracted x-ray beam intensities are observed under circumstances where the absorption is such that no penetration of the x-ray beam would be expected. This occurs in x-ray transmission through a thick and highly perfect crystal, which is in the symmetrical Laue reflection orientation. In dynamical theory, if the nodes of the standing waves generated in the crystal correspond to the main absorption centers, i.e., the heavier atoms, energy is transmitted unabsorbed. This phenomenon is rarely observed in other than large very perfect crystals, of semiconductors for example. The transmitted beams are highly monochromatic, parallel, and polarized, but with insufficient intensity to be used as monochromated sources except perhaps with synchrotron radiation. They are used in *x-ray topography.* If the transmitted and reflected beams are recorded with a fine-grain emulsion, they provide images of the dislocations and other departures from lattice perfection in the crystal.

I. Kossel and Kikuchi lines

When a perfect crystal is the target of the x-ray tube, the absorption of the characteristic diffraction spectra from the source within the crystal produces deficiency lines on the scattering pattern, which are known as Kossel lines.[48] Similar patterns can be produced by divergent beam diffraction when a divergent monochromatic x-ray beam is transmitted through a thin slice of perfect crystal. Measurement of the point of intersection of these lines, which are conic sections, can provide very accurate lattice parameters.[49,50] These patterns are also characteristic of the crystal perfection and have been used to distinguish between type-I and type-II diamonds.[51] A quantum-mechanical interpretation has been provided for this phenomenon.[52]

The same phenomenon, when it occurs in electron diffraction, is referred to as Kikuchi lines.[53] Owing to the shorter waves, the lines appear straight rather than as curved conic sections.

X-ray diffraction from thin wedge-shaped perfect crystals produces interference patterns known as *pendellösung fringes.*[54] This phenomenon has been explored experimentally[55] and is used in x-ray topography. The dynamical theory has been developed by Kato.[56] It can be used for very accurate measurements of structure amplitudes in special cases.[57]

J. Powder diffractometry

The diffraction spectra from a crystalline powder lie on the surface of cones with semivertical angles of $2\theta_{hkl}$ with respect to the direct beam. When the powder consists of suitably small crystals, the diffraction spectra form uniform *powder lines,* which are the intersection of the diffraction cones with the recording device. The *powder diffraction pattern* is a one-dimensional record of I_{hkl} vs 2θ, for all reciprocal-lattice points within the sphere of diffraction. It takes the form of a record of 2θ or d, and I/I_0 for each symmetry-independent *hkl* diffraction spectra. It is characteristic of the crystal, and can be used as a *fingerprint* for identification purposes. In favorable cases, the presence of impurities of 1–2% can be detected. A *powder diffraction* file containing the diffraction pattern of 35 000 crystalline materials[58] is available for identification purposes.

When very small quantities of material are available, film methods using a 57.3-mm-radius powder camera are commonly used. Larger quantities permit the use of powder diffractometers with slabs of powder, making use of the *Bragg-Brentano parafocusing* instrumentation. A very effective film instrument for high resolution of powder lines is the Guinier focusing powder camera. Back-reflection powder patterns with large-radius evacuated (or helium-filled) cameras are used for high-precision lattice parameter measurements of high-symmetry crystals. For cubic crystals systematic errors can be reduced[59] by extrapolation of a to $\theta = 90°$ by plotting the lattice parameters versus

$$\frac{1}{2}\left(\frac{\cos^2\theta}{\theta} + \frac{\cos^2\theta}{\sin\theta}\right).$$

This function is tabulated in Int. Tables, Vol. II, pp. 228–229. For noncubic crystals, an analytical least-squares refinement of the lattice parameters is preferable.[60] If the crystals are too small, there is line broadening which is additional to that characteristic of the collimation and specimen. The increase in width of the diffraction line $\Delta 2\theta \approx \lambda / t \cos\theta$, where t is the linear dimension of the crystal.

When the crystalline powder is sufficiently coarse, discrete diffraction spots appear on the powder line. For metals and alloys, this provides a method of studying *preferred orientation* in the polycrystalline *grain texture*. Preferred orientation is also common in compressed powders. This can seriously interfere with qualitative and quantitative analysis by means of powder diffractometry. Under special circumstances, information concerning the structure of the grain boundaries can be obtained from "extra" weak or diffuse spectra observed on x-ray or electron diffraction photographs.[61]

K. Powder diffraction profile refinement: Rietveld method

The whole-pattern-fitting method of interpreting powder diffraction was first introduced by Rietveld[62,63] for neutron powder diffraction patterns, for which the theory is mathematically simpler. Several hundred crystal structure analyses have been successfully determined from neutron powder data.[64] More recently, the theory was adapted to x-ray patterns,[65,66] and a number of structures have been determined. The method can also be applied to mixtures of several phases.[67] The method has been combined with a well-known single-crystal atomic parameter refinement method.[68]

The powder pattern intensity is measured stepwise across the whole pattern. The scan is usually $\approx 0.05°$ in 2θ. A set of data will therefore contain about 2000 separate I_i values. The method aims to minimize by least squares the residual

$$R = \sum_i \omega_i \left(I_i(\text{obs}) - \frac{1}{k} I_i(\text{calc}) \right)^2,$$

where ω_i is a weight assigned to the ith datum point, usually derived from the counting statistics, and k is a scale factor parameter. The success of the method depends upon how well $I_i(\text{calc})$ can be derived from an expression such as

$$I_i(\text{calc}) = I_{ib} + \sum_{\mathbf{s}} G(\theta_i - \theta_{\mathbf{s}}) m_{\mathbf{s}} T_{\mathbf{s}} (\text{Lp})_i |F(\mathbf{s})|,$$

where I_{ib} is the background intensity, $G(\theta_i - \theta_{\mathbf{s}})$ a convo-

Table 8.04.A.1. Neutron scattering factors (cross sections) of the atoms, in 10^{-13} cm, and, in [ ], mass absorption coefficients (μ/ρ cm^2 gm^{-1}) for $\lambda = 1.08$ Å.

IA	IIA	IIIB	IVB	VB	VIB	VIIB		VIII		IB	IIB	IIIA	IVA	VA	VIA	VIIA	VIIIA
H -3.7409 [25.3]	D 6.674																He 3.26
Li -2.03 [3.5]	Be 7.79 [0.0003]											3 5.35 [24]	C 6.6484 [0.0001]	N 9.3 [0.048]	O 5.805 [0.0000]	F 5.665 [0.0002]	Ne 4.55 [0.006]
Na 3.63 [0.007]	Mg 5.375 [0.001]											Al 3.449 [0.003]	Si 4.149 [0.002]	P 5.13 [0.002]	S 2.847 [0.0055]	Cl 9.579 [0.33]	Ar 1.884 [0.006]
K 3.67 [0.018]	Ca 4.90 [0.0037]	Sc 12.3 [0.25]	Ti -3.3438 [0.044]	V -0.380 [0.033]	Cr 3.635 [0.021]	Mn -3.73 [0.083]	Fe 9.54 [0.015]	Co 2.53 [0.21]	Ni 10.3 [0.028]	Cu 7.772 [0.021]	Zn 5.680 [0.0055]	Ca 7.29 [0.015]	Ge 8.193 [0.011]	As 6.58 [0.020]	Se 7.97 [0.056]	Br 6.79 [0.029]	Kr 7.85 [0.13]
Rb 7.22 [0.003]	Sr 7.02 [0.005]	Y 7.75 [0.006]	Zr 7.16 [0.0006]	Nb 7.054 [0.004]	Mo 6.95 [0.009]	Tc 6.8	Ru 7.21 [0.009]	Rh 5.93 [0.53]	Pd 5.91 [0.023]	Ag [0.20]	Cd 0.50 [14]	In 4.06 [0.6]	Sn 6.228 [0.002]	Sb 5.64 [0.016]	Te 5.80 [0.013]	I 5.28 [0.018]	Xe 4.89 [0.083]
Cs 5.42 [0.077]	Ba 5.25 [0.0027]	La 8.27 [0.023]	Hf 7.7 [0.20]	Ta 6.91 [0.044]	W 4.77 [0.036]	Re 9.2 [0.16]	Os 10.7 [0.028]	Ir 10.6 [0.80]	Pt 9.5	Au 7.63 [0.17]	Hg 12.66 [0.63]	Tl 8.79 [0.006]	Pb 9.401 [0.0003]	Bi 8.5233 [0.0000]	Po	At [0.79]	Rn
				Ca 4.84 [0.0021]	Pr 4.45 [0.029]	Hd 7.69 [0.11]	Pm 12.6	Sm -5.0 [47]	Lu 6.0 [6]	Cd 9.5 [73]	Tb 7.38 [0.09]	Dy 16.9 [2.0]	Ho 8.08 [0.15]	Er 8.03 [0.36]	Tm 7.05 [0.25]	Ib 12.4 [0.076]	Lu 7.3 [0.22]
				Th 9.84 [0.01]	Pa 9.1	U 8.42 [0.005]	Np 10.6	Pu 7.7	Am 8.3								

lution, in analytical form, of the instrumental profile function and the intrinsic diffraction profile, both of which are dependent on θ, m_s the multiplicity of the diffraction spectrum, T_s a preferred orientation function, $(Lp)_i$ the Lorentz-polarization factor at θ_i, and $F(\mathbf{s})$ the structure factor corresponding to diffraction vector, **s**.

8.04. STRUCTURE FACTOR

The structure factor $F(\mathbf{s})$, F_{hkl}, or $F_{\mathbf{h}}$ is a dimensionless quantity. It depends upon the scattering properties of the matter in the crystal. For Daltonian convenience, it is separated into the scattering power of the atoms and their geometrical arrangement with respect to the crystal lattice. The former are referred to as *atomic scattering factors,* the latter as *geometrical structure factors.*

A. Atomic scattering factors $f_i(\mathbf{s})$

For x rays,

$$f_i^x(\mathbf{s}) = \int_{\text{atom}} \psi_f^*(\mathbf{r})\exp(i\mathbf{s}\cdot\mathbf{r})\psi_i(\mathbf{r})dr,$$

where ψ_f^* and ψ_i refer to the wave functions for the initial and final states. For coherent scattering, these states are the same. $|s| = 4\pi\lambda^{-1}\sin\theta$. Except in special problems, it is assumed that atoms are spherical; then

$$f(\mathbf{s}) = \int \frac{r^2 P(r)^2 \sin(sr)}{sr}\, dr,$$

where $P(r)$ is the radial component of $\psi(r)$. Numerical values for the x-ray scattering factors for free atoms and chemically significant ions as a function of $(\sin\theta)/\lambda$ are given in Int. Table, Vol. IV, pp. 71–98. An analytical expression convenient for computer generation of atomic scattering factors is

$$f\left(\frac{\sin\theta}{\lambda}\right) = \sum_{i=1}^{4} a_i \exp\left(-b_i \frac{\sin^2\theta}{\lambda}\right) + c,$$

the coefficients a_i, b_i, and c of which are given in Int. Tables, Vol. IV, pp. 99–102. These expressions reproduce those from relativistic Hartree-Fock or relativistic Dirac Slater calculations within a mean error generally less than 0.010 and seldom greater than 0.020. Scattering factors for spherically bonded hydrogen atoms are given in Int. Tables, Vol. IV, p. 102.

For electrons,

$$f_i^e(\mathbf{s}) = 2\frac{m^2}{h^2}\frac{Z_i - f_i^x(\mathbf{s})}{s^2},$$

where Z_i is the nuclear charge. If θ is small,

$$f_i^e(s) = 0.023\,93[Z_i - f_i^x(s)]\lambda^2/\theta^2.$$

Because of the denominator, electron scattering falls off much more rapidly with scattering angle than does x-ray scattering. Tables of electron scattering factors, in Å, for neutral atoms are given in Int. Tables, Vol. IV, pp. 155–174.

For neutrons,[69] the wavelengths used for diffraction, ≈ 1 Å, are much larger than nuclear dimensions. As a result, nuclei act as point scatterers:

$$f_i^n(\mathbf{s}) = \frac{m}{2\pi\hbar^2} a(\tfrac{4}{3}\pi r_0^3),$$

where a is the Fermi pseudopotential, which is zero outside the radius r_0 ($\approx 10^{-3}$ Å). $f_n(s)$ is independent of scattering angles and is denoted by b in 10^{12} cm, the nuclear

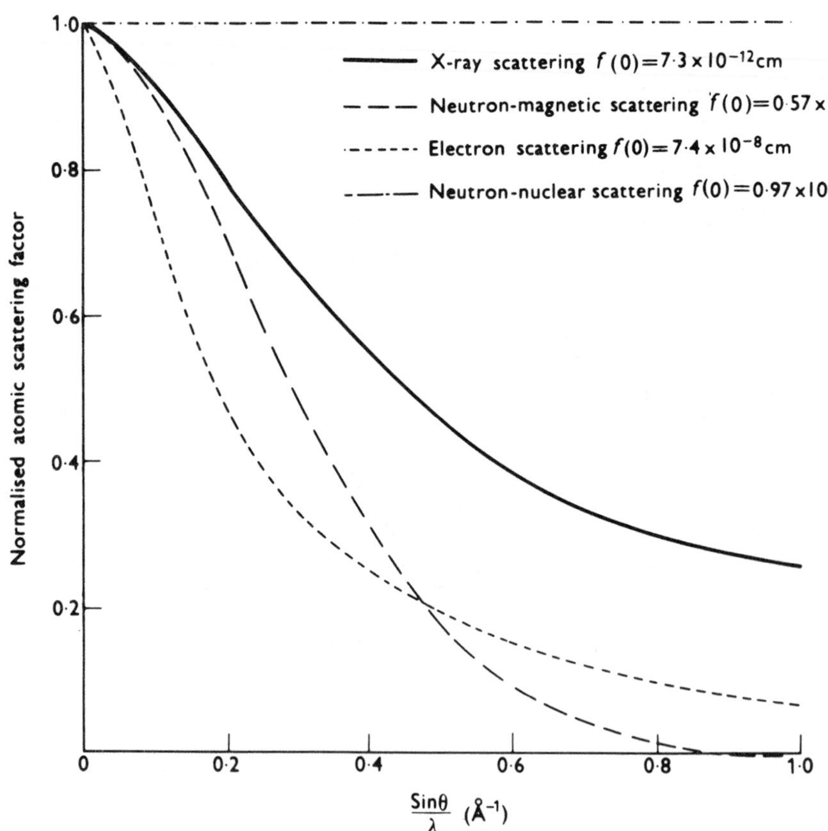

FIGURE 8.04.A.2. Relative dependence of scattering factors with scattering angle for the types of radiation used in crystallography for an iron atom

scattering length, which is a constant for each isotopic species of the elements.

The nuclear neutron scattering factors[70] are shown in Table 8.04.A.1. Nuclei that have magnetic moments give additional neutron scattering. The neutron magnetic scattering factor b_m is given by

$$b_m = (e^2\gamma/mc^2)Sf,$$

where γ is the neutron magnetic moment in nuclear magnetons, S the electronic-spin quantum number, and f the atomic scattering factor of the unpaired nuclear electron (dependent on $\sin\theta/\lambda$, $f=1$ for $\theta=0°$). Magnetic scattering is of the same order as nuclear scattering.

These atomic scattering factors for x rays, electrons, and neutrons are for isolated atoms *at rest.* The effects of thermal motion on the scattering factors are discussed separately. The relative dependence of the scattering factors with scattering angle for the types of radiation used in crystallography is illustrated in Figure 8.04.A.2 for an iron atom.

B. Dispersion corrections for x-ray atomic scattering factors

The x-ray atomic scattering factors take account of the spatial distribution of electrons in the atom, but they are calculated on the assumption that the electronic binding energy is so small compared with the energy of the x-ray photon that the scattering power of each electron is like that of a free electron. When this is true, $f_{hkl} = f_{\bar{h}\bar{k}\bar{l}}$, $F(\mathbf{s}) = F(-\mathbf{s})$, and $F_{hkl} = F_{\bar{h}\bar{k}\bar{l}}$. This is known as *Friedel's law.* When the incident x-ray wavelength approaches a characteristic resonance frequency of an electronic transition in the atom, the scattering power of a bound electron will be greater than or less than that of a free electron, and the phase of the scattered wave will be different. These effects are taken into account by representing the atomic scattering factor f as a complex number:

$$f = f_0 + \Delta f' + i\Delta f'',$$

where $\Delta f'$ and $\Delta f''$ are the real and imaginary dispersion corrections.

The dispersion corrections depend on the x-ray wavelength λ and the diffraction angle θ. They are less sensitive functions of θ than is f_0, because the tightly bound electrons responsible for these effects are concentrated in a small volume near the atomic nucleus. Values of f' generally increase, and those of f'' decrease, by about 10% in going from $(\sin\theta)/\lambda = 0$ to 1.0 (see Int. Tables, Vol. III, pp. 214–216).

Values for $\Delta f'$ and $\Delta f''$ for Cu$K\alpha$ and Mo$K\alpha$ are given in Table 8.04.B.1. The values for other radiations are given in Int. Tables, Vol. IV, pp. 149–150. It is difficult to assess the accuracy of these values owing to approximations in the theory and the absence of any accurate experimental measurements. The use of tunable synchrotron x radiation has made possible some experimental measurments of $\Delta f'$ and $\Delta f''$ close to the L-shell absorption edges of particular elements.[71]

C. Geometrical structure factor

The structure factor

$$F(\mathbf{s}) = \sum_{i=1}^{n} f_i \exp(2\pi i \mathbf{r}_j \cdot \mathbf{s}),$$

TABLE 8.04.B.1 Real and imaginary dispersion corrections for x-ray atomic scattering factors for Cu$K\alpha$ and Mo$K\alpha$ radiation: $\begin{matrix} & \text{Cu}K\alpha & \text{Mo}K\alpha \\ \Delta f' & & \\ \Delta f'' & & \end{matrix}$

IA	IIA	IIB	IVB	VB	VIB	VIIB	VIII			IB	IIB	IIIA	IVA	VA	VIA	VIIA	VIIIA
H																	**He**
Li 0.001 0.000 0.000 0.000	**Be** 0.003 0.000 0.001 0.000											**B** 0.008 0.000 0.004 0.001	**C** 0.017 0.002 0.009 0.002	**N** 0.029 0.004 0.018 0.003	**O** 0.407 0.008 0.032 0.006	**F** 0.069 0.014 0.053 0.010	**Ne** 0.097 0.021 0.083 0.106
Na 0.129 0.030 0.124 0.025	**Mg** 0.165 0.042 0.177 0.036											**Al** 0.204 0.056 0.246 0.052	**Si** 0.244 0.072 0.330 0.071	**P** 0.283 0.090 0.434 0.095	**S** 0.319 0.110 0.557 0.124	**Cl** 0.348 0.132 0.702 0.159	**Ar** 0.366 0.155 0.872 0.201
K 0.365 0.179 1.066 0.250	**Ca** 0.341 0.203 1.286 0.306	**Sc** 0.285 0.226 1.533 0.372	**Ti** 0.189 0.248 1.807 0.446	**V** 0.035 0.267 2.110 0.530	**Cr** −0.198 0.284 2.443 0.624	**Mn** −0.568 0.295 2.808 0.729	**Fe** −1.179 0.301 3.204 0.845	**Co** −2.464 0.299 3.608 0.973	**Ni** −2.956 0.285 0.506 1.113	**Cu** −2.019 0.263 0.589 1.266	**Zn** −1.612 0.222 0.678 1.431	**GA** −1.354 0.163 0.777 1.609	**Ge** −1.163 0.081 0.886 1.801	**As** −1.011 −0.030 1.006 2.007	**Se** −0.879 −0.178 1.139 2.223	**Br** −0.767 −0.374 1.283 2.456	**Kr** −0.665 −0.652 1.439 2.713
Rb −0.574 1.608	**Sr** −0.465 1.820	**Y** −0.386 2.025	**Zr** −0.314 2.245	**Nb** −0.248 2.482	**Mo** −0.191 2.735	**Tc** −0.145 3.005	**Ru** −0.105 3.296	**Rh** −0.077 3.605	**Pd** −0.059 3.934	**Ag** −0.060 4.282	**Cd** −0.079 4.653	**In** −0.126 5.045	**Sn** −0.194 5.459	**Sb** −0.287 5.894	**Te** −0.418 6.352	**I** −0.579 6.835	**Xe** −0.783 7.348
Cs −1.022 7.904	**Ba** −1.334 8.460	**La** −1.716 9.036	**Hf** −6.715 4.977	**Ta** −6.351 5.271	**W** −6.048 5.577	**Re** −5.790 5.891	**Os** −5.581 6.221	**Ir** −5.391 6.566	**Pt** −5.233 6.925	**Au** −5.096 7.297	**Hg** −4.990 7.686	**Tl** −4.883 8.089	**Pb** −4.818 8.505	**Bi** −4.776 8.930	**Po** −4.756 9.383	**At** −4.772 9.842	**Rn** −4.787 10.317
Fr −4.833 10.803	**Ra** −4.898 11.296	**Ac** −4.994 11.799															
				Ce −2.170 9.648	**Pr** −2.939 10.535	**Nd** −3.431 10.933	**Pm** −4.357 11.614	**Sm** −5.696 12.320	**Eu** −7.718 11.276	**Gd** −9.242 11.946	**Tb** −9.498 9.242	**Dy** −10.423 9.748	**Ho** −12.255 3.704	**Er** −9.733 3.937	**Tm** −8.488 4.181	**Yb** −7.701 4.432	**Lu** −7.133 4.693
				Th −5.091 12.330	**Pa** −5.216 12.868	**U** −5.359 13.409	**Np** −5.529 13.969	**Pu** −5.712 14.536	**Am** −5.930 15.087	**Cm** −6.176 15.634	**Bk** −6.498 16.317	**Cf** −6.798 16.930	**Es**	**Fm**	**Md**	**No**	**Lw**

or

$$F_{hkl} = \sum_{i=1}^{n} f_i A + \sum_{i=1}^{n} f_i B,$$

where

$$A = \cos 2\pi \mathbf{r}_j \cdot \mathbf{s} = \cos 2\pi(hx_i + ky_i + lz_i),$$

$$B = i \sin 2\pi \mathbf{r}_j \cdot \mathbf{s} = i \sin 2\pi(hx_i + ky_i + lz_i),$$

where x, y, and z are the *fractional atomic coordinates* x_i/a, y_i/b, and z_i/c.

The geometrical structure factors A and B can be simplified by summing over the symmetry-equivalent atomic positions characteristic of the space group. The trigonometric expressions for A, B, and α appropriate for each space group are given in Int. Tables, Vol. I, pp. 353–525. These simplified expressions are sometimes advantageous for computer-programming structure factor calculations for highly symmetrical space groups.

The *phase angle* α_{hkl} is given by

$$\alpha_{hkl} = \tan^{-1}(B/A).$$

Then

$$F_{hkl} = \sum_{i=1}^{n} f_i \cos 2\pi(hx_i + ky_i + lz_i - \alpha_{hkl}).$$

The phase angle plays a key role in crystal structure determination, because the experimentally measured quantity $I_{hkl} \approx F_{hkl}^2$, and therefore α_{hkl} cannot be measured experimentally for a particular diffracted spectrum, except under the special conditions of anomalous scattering. The quantity that can be measured experimentally by normal diffraction methods, $|F_{hkl}|$, is referred to as the *structure amplitude.*

The structure factor expression including the dispersion effect is[72]

$$F(\mathrm{s}) = \sum_{i=1}^{n} f_i A + \sum_{i'=1}^{n} (\Delta f_{i'}' A' - \Delta f_{i'}'' B')$$

$$+ \sum_{i=1}^{n} f_i B + \sum_{i=1}^{n'} (\Delta f_{i'}' B' + \Delta f_{i'}'' B'),$$

$$F(\bar{\mathrm{s}}) = \sum_{i=1}^{n} f_i A + \sum_{i'=1}^{n'} (\Delta f_{i'}' A' + \Delta f_{i'}'' B')$$

$$- \sum_{i=1}^{n} f_i B - \sum_{i'=1}^{n'} (\Delta f_{i'}' A' - \Delta f_{i'}'' B'),$$

where A and B are summed over *all* atoms, A' and B' over only those atoms with significant values of the dispersion corrections $\Delta f'$ and $\Delta f''$.

D. Unitary and normalized structure factors

Unitary and normalized structure factors are used particularly in phase determination. The *unitary structure factor*[73]

$$U_{hkl} = F_{hkl} \Big/ \left(\sum_{i}^{n} f_i \right)^{-1}.$$

The *normalized structure factor (E value)*

$$E_{hkl} = F_{hkl} \Big/ \left(\epsilon \sum_{i}^{n} f_i^2 \right)^{-1/2}.$$

ϵ is an integral multiple to account for the multiplicity of certain classes of *hkl*. It is the integral multiple that occurs in the reduced geometrical structure for the class of *hkl* (see Int. Table, Vol. I, pp. 353–525).

8.05. THERMAL MOTION[74,75]

The thermal motion of the atoms in a crystal reduces the *Bragg intensities* of the x-ray, neutron, or electron diffracted spectra relative to that of an atomic structure *at rest.* This is accounted for by means of atomic temperature factors, which are included with the atomic scattering factors in the calculation of the structure factors. These atomic thermal parameters are included with the atomic coordinates in the variables which are refined by least-squares methods and are an integral part of the results of a modern crystal structure analysis. (An alternative approach[76] is to determine the molecular rigid-body motion parameters directly by constrained least-squares refinement of the diffraction data without determing individual atomic vibration parameters.) The structure factor expression which includes thermal motion in matrix-vector notation is then

$$F(\mathbf{h}) = \sum_j f_i(\mathbf{h}) \exp(2\pi i\, \mathbf{h}^t \mathbf{x}_j - 2\pi^2 h^t \mathbf{Q}^t \mathbf{U} \mathbf{Q} \mathbf{h}),$$

where $\mathbf{h}$ is the Miller index triplet vector and h^t its transpose, $\mathbf{x}_j$ is the position parameter vector,

$$\mathbf{Q}_{ij} = \mathbf{a}_i^* \quad \text{for } i=j$$
$$= 0 \quad \text{otherwise,}$$

with Q^t its transpose, where $\mathbf{a}_i^*, i = 1, 2, 3$, are the reciprocal-lattice lengths, and $\mathbf{U}$ is the symmetric 3×3 matrix describing the thermal motion.

In addition to the dynamic time-dependent motion of an atom due to vibrational displacements relating to molecular vibrations or lattice mode librations, there may be static lattice-dependent effects due to disordering of the atoms. Unless these are specifically separated by appropriate experimental or interpretative methods, the atomic parameters, i.e., coordinates and thermal parameters, describe the crystal structure "averaged" over the *space* of the crystal lattice and the *time* of the measurement of the diffracted spectra.

The commonly used method for including the effect of thermal motion on the atomic scattering factor is to assume that the atom vibrates anisotropically in a harmonic potential field, and can therefore be represented by a Gaussian smearing function.[77] These vibrations can then be described by a symmetric tensor $\mathbf{U}$ with six independent components such that the mean-square amplitude of vibration $\langle u^{-2} \rangle$ in the direction of the unit vector with components l_i is

$$\langle u^2 \rangle = \sum_{i=1}^{3} \sum_{j=1}^{3} U_{ij} l_i l_j .$$

The smearing function to be applied to each atomic scattering factor f_i^{hkl} is then

$$\exp[-2\pi^2(h^2a^{*2}U_{11} + k^2b^{*2}U_{22} + l^2c^{*2}U_{33} + 2hka^*b^*U_{12} + 2hla^*c^*U_{13} + 2klb^*c^*U_{23})].$$

This is also expressed as

$$\exp[-(\beta_{11}h^2 + \beta_{22}k^2 + \beta_{33}l^2 + 2\beta_{12}hk + 2\beta_{13}hl + 2\beta_{23}kl)].$$

The reporting of U_{ij} rather than the dimensionless β_{ij} values has the advantage that it reveals the anisotropy of the atomic thermal motion more directly. The diagonal elements are equal in magnitude to the mean-square displacements in Å^2 along the reciprocal axes. The exponential factor in the equation for $F(h)$ is the characteristic function or Fourier transform of a Gaussian probability density function. The density function is

$$P(\underline{u}) = \frac{\det(\underline{P}_j)^{1/2}}{(2\pi)^{3/2}} \exp[-\tfrac{1}{2}(u - \underline{x}_j)^t P_j (u - \underline{x}_j)],$$

where P is the matrix inverse to Q^tUQ. Properties of this density function are given in Int. Tables, Vol. IV, Sec. 5.

Graphical representations of the atomic parameters, which include the thermal motion, are obtained by plotting these probability functions using the program ORTEP.[78] In these representations, an example of which is shown in Figure 8.05.A, the atoms appear as ellipsoids. The distances to the ellipsoidal surface along the principal axes of the ellipsoids are equal to the root-mean-square displacements in those directions. The ellipsoid may contain a chosen fraction of the total probability function. For room-temperature measurements, a com-

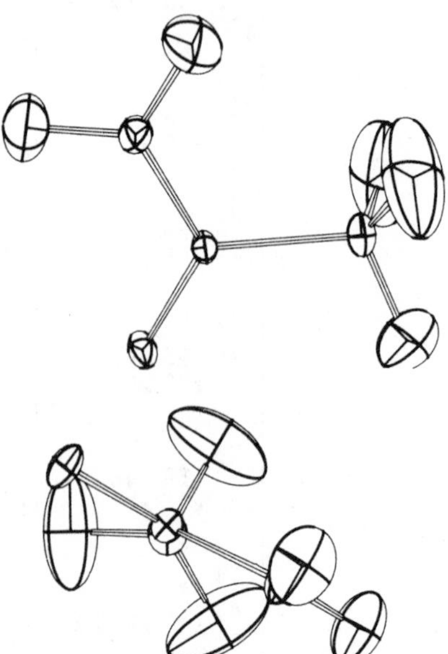

FIGURE 8.05.A. ORTEP representation, at 75% probability, of the nuclear thermal motion in acetamide,

$H_2N-\underset{\underset{O}{\|}}{C}-CH_3$,

at 23 K, viewed in the plane of the molecule and down the C—C bond, from a neutron diffraction analysis of the rhombohedral crystal form: G. A. Jeffrey, J. R. Ruble, R. K. McMullan, D. J. DeFrees, J. S. Binkley, and J. A. Pople, Acta Crystallogr. Sect. B **36**, 2292 (1980).

monly used level is 50%, corresponding to a scale of 1.5. For low-temperature work, 74% may be used, corresponding to a scale of 2.0.

If the motion is isotropic, as in cubic crystals, the isotropic temperature factor $B = 8\pi^2\langle u^2_{iso}\rangle$, where $\langle u^2_{iso}\rangle$ is the mean-square displacement of the atom from its equilibrium position. For cubic crystals,

$$\langle u^2_{iso}\rangle = \frac{3\hbar^2}{4\pi^2 m\Theta}\left(\frac{\Phi(\Theta/T)}{\Theta/T} + \frac{1}{4}\right),$$

where Θ_D is the *Debye characteristic temperature,* defined by

$$\hbar\nu_m = k_B\Theta_D,$$

where ν_m is the maximum (Debye cutoff) frequency of the elastic vibrations of the crystal, k_B Boltzmann's constant, and T the absolute temperature. The Debye integral function $\Phi(x)$ is defined by

$$\Phi(x) = \frac{1}{s}\int_0^x \frac{y}{e^y - 1}\,dy.$$

At high temperatures, $T > \Theta_D$ or $x < 1$, then

$$\langle u^2_{iso}\rangle = 3\hbar^2 T/4\pi m k_B \Theta_D^2,$$

so that $\langle u^2\rangle$ is proportional to the absolute temperature T. At low temperatures, $T < \Theta_D$ or $1/x < 1$, zero-point motion is dominant. $\Phi(x) = 0$, and

$$\langle u^2_{iso}\rangle = 3\hbar^2/4\pi m k_B \Theta_D,$$

which is independent of T. Values of the Debye characteristic temperatures and vibration amplitudes for some cubic elemental crystal structures are given in Int. Tables, Vol. III, pp. 234–244.

Even when an atom is undergoing anisotropic thermal motion, it is useful to have a single parameter to represent the degree of thermal motion. Such a parameter is the *equivalent isotropic temperature* B_{eq}. In general,

$$B_{eq} = \tfrac{8}{3}\pi^2\hbar(Q^t U Q)$$

expressed in a Cartesian axis system. For an orthogonal axis system,

$$B_{eq} = \tfrac{8}{3}\pi^2(U_{11} + U_{22} + U_{33}).$$

For cases where the Gaussian probability density is an inadequate model, more elaborate models are needed. Gram-Charlier series expansion, cumulant expansion of the Gaussian model,[79] and models based on curvilinear density functions are described in Int. Tables, Vol. IV, Sec. 5.

8.06. DIFFRACTING DENSITY FUNCTION

For x rays, the diffracting density function is the electron density function in the unit cell; for electrons, it is the nuclear and electron potential; for neutrons, it is the nuclear scattering distribution. In all cases,

$$\rho(\mathbf{r}) = \int_v F(\mathbf{s})\exp(-2\pi i\mathbf{s}\cdot\mathbf{r})dr.$$

This is the Fourier transform of the structure factor expression

$$F(s) = V\int_v \rho(r)\exp 2\pi i\mathbf{r}\cdot\mathbf{s}\,ds.$$

In a crystal, the electron, or scattering density, distribution ρ_{xyz} is a Fourier synthesis of waves, the amplitudes and phases of which are given by the structure factors F_{hkl}[80]:

$$\rho_{xyz} = \frac{1}{V}\int_{-\infty}^{\infty}\int_{-\infty}^{\infty}\int_{-\infty}^{\infty} F_{hkl} \times\exp[-2\pi i(hx + ky + lz)].$$

Separating out the measurable structure amplitude $|F_{hkl}|$ from the unknown phase angle α_{hkl},

$$\rho_{xyz} = \frac{1}{V}\sum_{-\infty}^{\infty}\sum_{-\infty}^{\infty}\sum_{-\infty}^{\infty}|F_{hkl}| \times\cos[2\pi(hx + ky + lz - \alpha_{hkl})].$$

In some cases, it is desirable to remove from the electron density map a known part of the crystal structure. A *difference Fourier synthesis* is then calculated:

$$\rho'_{xyz} = \frac{1}{V}\sum_{-\infty}^{\infty}\sum_{-\infty}^{\infty}\sum_{-\infty}^{\infty}|\Delta F_{hkl}| \times\cos[2\pi(hx + ky + lz - \alpha'_{hkl})],$$

where $|\Delta F_{hkl}|$ is the difference between the observed structure amplitudes and those calculated for the known atomic positions, and α'_{hkl} is calculated from the known atomic positions. This method is commonly used for completing the analysis of a structure containing heavy atoms, or for determining hydrogen atom positions when the nonhydrogen atoms have been located. It is more commonly used these days in protein crystallography.

Normalized structure factors E_{hkl} are frequently used instead of F_{hkl} in Fourier syntheses, since these give sharper and better-resolved atomic peaks. The three-dimensional plot of these Fourier syntheses is referred to as an E map.

8.07. PHASE PROBLEM

Since $I_{hkl} = kF^2_{hkl}$, only the structure amplitudes $|F_{hkl}|$ can be measured experimentally. The determination of the phases α_{hkl} is therefore the central problem in crystal structure analysis.

A. Phase-solving methods

There are two basic approaches. One makes use of the *Patterson function;* the other, called the *direct method,* uses the statistical properties of the structure factors which arise from the fact that the electron density distribution in a crystal is always positive. (Although the neutron scattering density in a crystal can be negative, owing to the negative scattering cross section of deuterium and other nuclei, the direct method has been applied successfully to neutron crystal structure analysis.)

Phase-solving methods based on these alternative approaches are given in Table 8.07.A.1.

B. Patterson synthesis

The Patterson Fourier synthesis[81] $P(UVW)$ can be calculated from experimental data alone, since

$$P(\mathbf{U}) = P(UVW) = \frac{1}{V}\sum_{\infty}^{\infty}\sum_{\infty}^{\infty}\sum_{\infty}^{\infty} F_{hkl}^2 \times \cos[2\pi(hU + kV + lW)].$$

Since

$$F_{\mathbf{s}}^2 = \sum_{i=1}^{N}\sum_{j=1}^{N} f_i f_j \cos[2\pi(\mathbf{r}_i - \mathbf{r}_j)\cdot\mathbf{s}],$$

this synthesis gives a vectorial pattern of the interatomic distances in the unit cell. Interpretation of the "Patterson" is complex except for very simple structures, since n atoms in the cell will give rise to $n(n-1)$ peaks on the Patterson Fourier synthesis. Nevertheless, this synthesis is the basis of several very important methods of phase determination.

Since the overlap of the interatomic vector peaks is a serious obstacle to interpretation of the Patterson synthesis, a function commonly used is the *sharpened Patterson synthesis, with origin peak removed.* This is obtained by using normalized structure amplitudes $E_{hkl}^2 - 1$ as the Fourier series coefficients.

C. Direct methods

The "direct methods" are now the most powerful methods of phase determination. Methods such as MULTAN, SHELX, MITHRIL are used for the majority of crystal structure analyses. Only in exceptional cases and for structures having more than 1000 atomic positional parameters do they fail. Programs of this type are incorporated within the x-ray diffractometer computer software.

The direct methods of phase determination depend upon the non-negative electron density inequality[107]

$$\begin{vmatrix} F_0 & F_{\bar{\mathbf{h}}_1} & F_{\bar{\mathbf{h}}_{n-1}} \\ F_{\mathbf{h}_1} & F_0 & F_{\mathbf{h}_1\bar{\mathbf{h}}_{n-1}} \\ F_{\mathbf{h}_{n-1}} & F_{\mathbf{h}_{n-1}\bar{\mathbf{h}}_1} & F_0 \end{vmatrix} \geqslant 0,$$

where $\mathbf{h}_n$ refer to different hkl's.

TABLE 8.07.A.1. Phase-solving methods used in crystal structure analysis

Based on use of Patterson synthesis	Based on use of structure factor statistics
Heavy-atom method[82]	Inequalities method[92]
Deconvolution methods by superposition, image seeking, or vector search[83–85]	Sayre equation[93]
	Symbolic addition[94–96]
Rotation function[86–88]	Tangent formula[97–100]
Multiple isomorphous replacement[89–91]	Multisolution methods[101]
Anomalous scattering	Structure invariants and semi-invariants[102–104]
	Magic integers[105]
	Anomalous scattering[106]

8.08. CRYSTAL STRUCTURE REFINEMENT: METHOD OF LEAST SQUARES

In crystal structure analysis, the number m of observations $|F_{hkl}|$ generally exceeds the number n of unknown parameters (approximately nine times the number of atoms in the asymmetric unit). The method of least squares is therefore commonly used for parameter refinement.[108] The functions most frequently minimized are

$$R_1 = \sum \omega_{hkl}(|F_{hkl}^{\rm obs}| - k|F_{hkl}^{\rm calc}|)^2,$$

$$R_2 = \sum \omega'_{hkl}(|F_{hkl}^{\rm obs}| - k|F_{hkl}^{\rm calc}|)^2,$$

where the sums are over the sets of crystallographically independent hkl reflections. k is a scaling factor between $|F_{hkl}^{\rm obs}|$ and $|F_{hkl}^{\rm calc}|$. The weighting factors

$$\omega_{hkl} = 1/\sigma_{hkl}^2,$$

where σ is the standard deviation of $|F_{hkl}^{\rm obs}|$ or $|F_{hkl}^{\rm calc}|^2$. The n simultaneous linear equations in m unknowns are then

$$\sum_{i-1}^{n} \Delta u_i\left(\sum \omega_{hkl}\frac{\delta F_{hkl}^{\rm calc}}{\delta u_j}\frac{\delta F_{hkl}^{\rm calc}}{\delta u_i}\right) = -\sum \omega_{hkl}\Delta_{hkl}\frac{\delta F_{hkl}^{\rm calc}}{\delta u_j} \quad (j = 1,2,\ldots,m),$$

where $\Delta_{hkl} = |F_{hkl}^{\rm obs}| - |F_{hkl}^{\rm calc}|$. For R_1,

$$\frac{\delta\Delta_{hkl}}{\delta u_j} = -\frac{\delta|F_{kl}^{\rm calc}|}{\delta u_j}.$$

For R_2,

$$\frac{\delta\Delta_{hkl}}{\delta u_j} = -2\,|F_{kl}^{\rm obs}|\,\frac{\delta|F_{hkl}^{\rm calc}|}{\delta u_j}.$$

In matrix form, $\mathbf{Nx} = \mathbf{e}$, where $\mathbf{x}$ and $\mathbf{e}$ are column vectors of order n and m, respectively; $\mathbf{N}$ is a matrix of order mn. The elements of $\mathbf{N}$ are

$$N_{jk} = \sum_{HKL} \omega_{hkl}\frac{\delta|F_{hkl}^{\rm calc}|}{\delta u_j}\frac{\delta|F_{hkl}^{\rm calc}|}{\delta u_i}.$$

The elements of $\mathbf{e}$ are

$$E_j = \sum \omega_{hkl}\frac{\delta|F_{hkl}^{\rm calc}|}{\delta u_j}\Delta_{hkl},$$

where

$$\Delta_{hkl} = |F_{hkl}^{\rm obs}| - |F_{hkl}^{\rm calc}|, \quad \mathbf{x} = \Delta u_i.$$

The normal set of equations is

$$\tilde{\mathbf{N}}\mathbf{N}\mathbf{x} = \tilde{\mathbf{N}}\mathbf{e},$$

where $\tilde{\mathbf{N}}$ is the transpose of $\mathbf{N}$.

The solution vector for the x's is

$$x = (\tilde{\mathbf{N}}\mathbf{N})^{-1}\tilde{\mathbf{N}}\mathbf{e}.$$

The variance of the derived parameters u_i is given by

$$\sigma^2(u_i) = \frac{\mathbf{M}_{ii}^{-1}\Sigma\omega\Delta u_i^2}{m-n},$$

where $\mathbf{M}_{ii}^{-1}$ is the inverse of the normal equation matrix.

The covariance between derived parameters u_i and u_j is given by

$$\sigma(u_iu_j) = \frac{\mathbf{M}_{ij}^{-1}\Sigma\omega\Delta u_i^2}{m-n}.$$

For a parameter x_i such as an atomic coordinate or thermal parameter, its variance

$$\sigma(x_i)^2 = \mathbf{M}_{ii}^{-1}\frac{\Sigma\omega(|F_O| - |F_C|)^2}{m-n}.$$

The goodness or error of fit,

$$S = \frac{\Sigma\omega(|F_O| - |F_c|)^2}{m-n},$$

is a more meaningful measure. It should be unity if the model is complete and correct and the weights are properly assigned.

The choice of the weights ω is important. Since the individual values of ω_{hkl} are not determined experimentally (it would require repeated measurement of each intensity), a weighting scheme is used. The simplest, and least useful, are unit weights. If the weights are inversely proportional to the overall atomic scattering factors, including the thermal motion, the atomic positions obtained by the least-squares minimization are the same as those given by the zero gradients of an unweighted difference Fourier synthesis.[109]

With the advent of automatic diffractometers, it is usual to use the σ_c from counting statistics, combined with some additional terms that correct for systematic trends noted in the values of ΔF_h. Such expressions generally take the form suggested by Cruickshank,[110]

$$\omega^{-1/2} = (a + b|F_\mathbf{h}| + c|F_\mathbf{h}|^2)^{-1},$$

where $a = \sigma_c^2$ from counting statistics, and b and c are empirically determined constants. Values of $F_h < 2\sigma$ or 3σ are considered to be unobserved. However, the rejection of low-intensity data leads to an underestimate of the scale and thermal parameters.[111] The information content of the unobserved reflections and how best to weight them has been discussed.[112] The consistency of the weights as a function of 2θ, I_{hnk}, or time, should be examined for systematic trends, as described in Int. Tables, Vol. IV, pp. 293–294.

8.09. REFERENCES

[1]W. Friedrich, P. Knipping, and M. Laue, Sitzungsber. Bayer. Akad. Wiss. **1912**, 303 [reprinted in Naturawissenschaften **39**, 361 (1952)].

[2]W. L. Bragg, Proc. R. Soc. London **A89**, 248 (1913); **A89**, 468 (1913).

[3]W. N. Lipscomb, *Boron Hydrides* (Benjamin, New York, 1963).

[4]J. S. Kasper, C. M. Lucht, and D. Harker, Acta Crystallogr. **3**, 436 (1950).

[5]J. D. Watson and F. H. C. Crick, Nature **171**, 737 (1953); **171**, 964 (1953).

[6]D. C. Hodgkin, J. Kamper, J. Lindsey, M. MacKay, J. Pickworth, J. H. Robertson, C. B. Shoemaker, J. G. White, R. J. Prosen, and K. N. Trueblood, Proc. R. Soc. London **A242**, 228 (1957).

[7]M. F. Perutz, M. G. Rossmann, A. F. Cullis, H. Muirhead, G. Will, and A. C. T. North, Nature **185**, 416 (1960).

[8]J. C. Kendrew, H. C. Watson, B. E. Strandberg, R. E. Dickerson, D. C. Phillips, and V. C. Shore, Nature **190**, 666 (1961).

[9]J. M. Cowley, *Diffraction Physics* (North-Holland, Amsterdam, 1975).

[10]G. E. Bacon, *Neutron Diffraction,* 3rd ed. (Clarendon, Oxford, 1975).

[11]D. T. Hawkins, Acta Crystallogr. Sect. A **36**, 475 (1980).

[12]P. P. Ewald, Z. Krist. **56**, 129 (1921).

[13]J. Donohue, *The Structure of the Elements* (Wiley, New York, 1974).

[14]W. B. Pearson, *A Handbook of Lattice Spacings and Structures of Metals and Alloys* (Pergamon, New York, 1958), Vols. 1 and 2.

[15]J. D. H. Donnay and Helen M. Ondik, *Crystal Data Determinative Tables,* 3rd ed. (U.S. Dept. of Commerce, National Bureau of Standards, Washington, DC, 1973), Vols. 1 and 2.

[16]*Structure Berichte* (Johnson, New York, 1966), Vols. 1–7.

[17]R. W. G. Wyckoff, *Crystal Structures* (Interscience, New York, 1963–1971), Vols. 1–6.

[18]*Structure Reports* (Oosthoek, Utrecht, 1940–1975), Vols. 8–41.

[19]*Molecular Structures and Dimensions, Guide to the Literature, 1935–1976, Organic and Organo-metallic Crystal Structures,* and *Molecular Structures and Dimensions,* Vols. 1–11 (1935–1979), edited by O. Kennard *et al.* (Reidel, Hingham, MA).

[20]The definitive reference sources for this chapter are Vols. I–IV of the *International Tables for X-ray Crystallography,* published for the International Union of Crystallography (Kynoch, Birmingham, England, 1959-1974). These are abbreviated herein as Int. Tables, Vol. xx.

[21]A.L. Mackay, Acta Crystallogr. **22**, 329 (1967).

[22]W. Nowacki, T. Matsumoto, and A. Edenharter, Acta Crystallogr. **22**, 935 (1967).

[23]P. P. Ewald, Phys. Z. **14**, 465 (1913).

[24]L. V. Azaroff, R. Kaplow, N. Kato, R. J. Weiss, A. J. C. Wilson, and R. A. Young, *X-ray Diffraction* (McGraw-Hill, New York, 1974).

[25]Extinction effects in x-ray crystal structure analysis are recognized by $|F_{\rm obs}^{hkl}| \ll |F_{\rm calc}^{hkl}|$ for small values of h,k,l. These data are omitted from the least squares refinement calculations.

[26]B. K. Vainshtein, *Structure Analysis by Electron Diffraction* (Pergamon, Oxford, 1964).

[27]P. S. Turner and J. M. Cowley, Acta Crystallogr. Sect. A **25**, 475 (1969).

[28]P. Goodman, Acta Crystallogr. Sect. A **32**, 793 (1976).

[29]U. W. Arndt and B. T. M. Willis, *Single Crystal Diffractometry* (Cambridge University, London, 1966).

[30]W. R. Busing and H. A. Levy, Acta Crystallogr. **22**, 457 (1967).

[31]W. R. Busing and H. A. Levy, Acta Crystallogr. **10**, 180 (1957).

[32]J. de Meulenaer and H. Tompa, Acta Crystallogr. **19**, 1014 (1965).

[33]W. H. Zachariasen, Acta Crystallogr. **23**, 558 (1967).

[34]R. M. Stern, Trans. Am. Cryst. Assoc. **4**, 14 (1968).

[35]G. A. Somorjai and L. L. Kesmodel, Trans. Am. Cryst. Assoc. **13**, 67 (1977).

[36]W. C. Hamilton, Acta Crystallogr. Sect. A **25**, 194 (1969).

[37]P. Coppens and W. C. Hamilton, Acta Crystallogr. Sect. A **26**, 71 (1970).

[38]P. J. Becker and P. Coppens, Acta Crystallogr. Sect. A **30**, 129 (1974); **31**, 417 (1975).

[39]F. R. Thornley and R. J. Nelmes, Acta Crystallogr. Sect. A **30**, 748 (1974).

[40]O. Berg, Veroeff. Siemens Konzern **5**, 89 (1926).

[41]M. Renninger, Z. Krist. **97**, 107 (1937).

[42]W. H. Zachariasen, Acta Crystallogr. **18**, 705 (1965).

[43]R. D. Burbank, Acta Crystallogr. **19**, 957 (1965).

[44]C. S. Barrett, *Structure of Metals* (McGraw-Hill, New York, 1952).

[45]A. P. L. Turner, T. Vreeland, Jr., and D. P. Pope, Acta Crystallogr. Sect. A **24**, 452 (1968).

[46]C. S. Barrett and T. B. Massalski, *Structure of Metals* (McGraw-Hill, New York, 1966).
[47]B. Borie, Acta Crystallogr. **21**, 470 (1966).
[48]W. Kossel, V. Loeck, and H. Voges, Z. Phys. **94**, 139 (1935).
[49]W. Kossel and H. Voges, Ann. Phys. (Paris) **23**, 677 (1935).
[50]B. J. Isherwood and C. A. Wallace, Acta Crystallogr. Sect. A **27**, 119 (1971).
[51]K. Lonsdale, Nature **151**, 52 (1943); **153**, 22 (1944).
[52]M. Kohler, Berl. Sitzunsber. **1935**, 334.
[53]S. Kikuchi, Proc. Jpn. Acad. Sci. **4**, 271 (1928); **4**, 275 (1928); **4**, 354 (1928); **4**, 475 (1928).
[54]P. P. Ewald, Ann. Phys. (Paris) **54**, 519 (1917).
[55]N. Kato and A. R. Lang, Acta Crystallogr. **12**, 787 (1959).
[56]N. Kato, Acta Crystallogr. **14**, 526 (1961); J. Appl. Phys. **39**, 2225 (1968); **39**, 2231 (1968).
[57]M. Hart and A. D. Milne, Acta Crystallogr. Sect. A **26**, 223 (1970).
[58]JCPDS (Joint Committee on Powder Diffraction Standards, ASTM, Powder Diffraction File, 1601 Park Lane, Swarthmore, PA 19081).
[59]J. B. Nelson and D. P. Riley, Proc. Phys. Soc. London **57**, 160 (1954).
[60]M. U. Cohen, Rev. Sci. Instrum. **6**, 68 (1935).
[61]S. L. Sass, J. Appl. Crystallogr. **13**, 109 (1980).
[62]H. M. Reitveld, Acta Crystallogr. **22**, 151 (1967).
[63]H. M. Reitveld, J. Appl. Cryst. **2**, 65 (1969).
[64]A. K. Cheetham and J. C. Taylor, J. Solid State Chem. **21**, 253 (1977).
[65]W. Parrish, T. C. Huang, and G. L. Ayers, Trans. Am. Cryst. Assoc. **12**, 55 (1976).
[66]R. A. Young, P. E. Mackie, and R. D. Van Dreche, J. Appl. Phys. **10**, 262 (1977).
[67]P.-E. Werner, S. Salome, G. Malmros, and J. O. Thomas, J. Appl. Crystallogr. **12**, 107 (1979).
[68]G. S. Pawley, J. Appl. Crystallogr. **13**, 630 (1980).
[69]C. G. Shull, Trans. Am. Cryst. Assoc. **3**, 1 (1967).
[70]L. Koester, *Neutron Physics,* Vol. 80 of *Springer Tracts in Modern Physics* (Springer, New York, 1977).
[71]J. C. Phillips, D. H. Templeton, L. K. Templeton, and K. O. Hodgson, Science **201**, 257 (1978).
[72]C. H. Dauben and D. H. Templeton, Acta Crystallogr. **8**, 841 (1955).
[73]A. L. Patterson, Phys. Rev. **46**, 372 (1934).
[74]B. T. M. Willis and A. W. Pryor, *Thermal Vibrations in Crystallography* (Cambridge University, London, 1975).
[75]Int. Tables, Vol. IV, pp. 314–319.
[76]G. S. Pawley, Acta Crystallogr. **20**, 631 (1966).
[77]D. W. J. Cruickshank, Acta Crystallogr. **9**, 747 (1956).
[78]C. K. Johnson, ORTEP II (Oak Ridge National Laboratory, Oak Ridge, TN, 1976), Report ORNL-5138.
[79]C. K. Johnson, Acta Crystallogr. Sect. A **25**, 187 (1969).
[80]W. H. Bragg, Philos. Trans. R. Soc. London **A216**, 254 (1915).
[81]A. L. Patterson, Z. Krist. **90**, 517 (1935).
[82]J. M. Robertson, J. Chem. Soc. **1936**, 1195.
[83]D. M. Wrinch, Philos. Mag. J. Sci. **27**, 98 (1939).
[84]M. J. Buerger, *Vector Space, and Its Application in Crystal Structure Investigation* (Wiley, New York, 1959).
[85]C. E. Nordman, in *Computing in Crystallography,* edited by R. Diamond, S. Rameseshan, and K. Venkatesan (Indian Academy of Science, Bangalore, 1980).
[86]M. G. Rossman and D. M. Blow, Acta Crystallogr. **15**, 24 (1962).
[87]M. G. Rossman, *The Molecular Replacement Method,* Vol. 13 of *International Science Review Series* (Gordon and Breach, New York, 1972).
[88]P. Tollin and W. Cochran, Acta Crystallogr. **17**, 1322 (1964).
[89]D. Harker, Acta Crystallogr. **9**, 1 (1956).
[90]C. Bokhoven, J. C. Schoone, and J. M. Bijvoet, Acta Crystallogr. **4**, 245 (1951).
[91]Y. Okaya, Y. Saito, and R. Pepinsky, Phys. Rev. **98**, 1857 (1955).
[92]D. Harker and J. S. Kasper, Acta Crystallogr. **1**, 70 (1948).
[93]D. Sayre, Acta Crystallogr. **5**, 60 (1952).
[94]H. A. Hauptman and J. Karle, ACA Monogr. **3** (1953).
[95]W. H. Zachariasen, Acta Crytallogr. **5**, 68 (1952).
[96]I. L. Karle and J. Karle, Acta Crystallogr. **16**, 969 (1963); **17**, 835 (1964).
[97]J. Karle and H. Hauptman, Acta Crystallogr. **9**, 635 (1956).
[98]J. Karle and I. L. Karle, Acta Crystallogr. **21**, 849 (1966).
[99]G. Germain, P. Main, and M. M. Wolfson, Acta Crystallogr. Sect. A **27**, 368 (1971); P. Main, I. Lessinger, M. M. Woolfson, G. Germain, and J. P. Declercq, MULTAN-77 (University of York, UK, and Louvain, Belgium, 1977).
[100]G. M. Sheldrick, SHELX, Program for Crystal Structure Determination (Cambridge University, UK).
[101]P. Main, in *Computing in Crystallography,* edited by R. Diamond, S. Rameseshan, and K. Venkatesan (Indian Academy of Science, Bangalore, 1980).
[102]J. Karle and H. Hauptman, Acta Crystallogr. **14**, 217 (1961).
[103]H. A. Hauptman, *Crystal Structure Determination; the Role of the Cosine Semiinvariants* (Plenum, New York, 1972).
[104]H. Schenk, in *Computing in Crystallography,* edited by R. Diamond, S. Rameseshan, and K. Venkatesan (Indian Academy of Science, Bangalore, 1980).
[105]P. S. White and M. M. Woolfson, Acta Crystallogr. Sect. A **31**, 53 (1975).
[106]J. Karle, Acta Crystallogr. Sect. A **41**, 387 (1985); A **42**, 246 (1986).
[107]J. Karle and H. Hauptman, Acta Crystallogr. **3**, 181 (1950).
[108]E. W. Hughes, J. Am. Chem. Soc. **63**, 1737 (1941).
[109]W. Cochran, Acta Crystallogr. **1**, 138 (1948).
[110]D. W. J. Cruickshank, in *Computing Methods and the Phase Problem,* edited by R. Pepinsky, J. M. Robertson, and J. C. Speakman (Pergamon, New York, 1961).
[111]F. L. Hirshfeld and D. Rabinovich, Acta Crystallogr. Sect. A **29**, 510 (1973).
[112]L. Arnberg, S. Hovmoller, and S. Westman, Acta Crystallogr. Sect. A **35**, 497 (1979).

9.00. Elementary particles

THOMAS G. TRIPPE

Lawrence Berkeley Laboratory

CONTENTS

This chapter is an update of the original version by Robert L. Kelly. Most of the contents of this chapter are adapted from *Review of Particle Properties,* Particle Data Group, Phys. Lett. **204B** (1988).

9.01. TABLE: STABLE PARTICLES

Stable particles under strong decay, i.e., which decay only via the weak or electromagnetic interactions. Quantities which are not well established are enclosed in parentheses. Uncertainties are implied by the number of significant figures given. Only large uncertainties are shown.

Particle	$I^G[J^P]C$	Mass (MeV)	Mean life (s)	[Branching ratios (%)] Principal observed decays	Baryon magnetic moments ($e\hbar/2m_pc$)
			Gauge bosons		
γ	$0,1[1^-]-$	$<3\times10^{-33}$	stable		
$W^\pm$	$J=1$	$81\,000\pm1300$	$>1.0\times10^{-25}$	$e^\pm\nu[10\pm3]\mu^\pm\nu[12\pm7]\tau^\pm\nu[10\pm4]$	
Z		$92\,400\pm1800$	$>1.2\times10^{-25}$	$e^+e^-[5\pm2]\mu^+\mu^-$[seen]	
			Leptons		
ν_e	$[\frac{1}{2}]$	$<1.8\times10^{-5}$	(stable)		
$e^\mp$	$[\frac{1}{2}]$	0.510 999 1	$>6\times10^{29}$		
ν_μ	$[\frac{1}{2}]$	<0.25	(stable)		
$\mu^\mp$	$[\frac{1}{2}]$	105.6595	2.1971×10^{-6}	$e^\mp\nu\nu[99]e^\mp\nu\nu\gamma[1]$	
ν_τ	$[\frac{1}{2}]$	<35			
$\tau^\mp$	$[(\frac{1}{2})]$	1784	3.0×10^{-13}	$\mu^\mp\nu\nu[18]e^\mp\nu\nu[18]$ charged hadrons + neutrals [64]	
			Light mesons		
$\pi^\pm$	$1^-[0^-]$	139.5675	2.603×10^{-8}	$\mu^\pm\nu[100]$	
π^0	$1^-[0^-]+$	134.973	8×10^{-17}	$\gamma\gamma[98.80]$ $\gamma e^+e^-[1.20]$	
η	$0^+[0^-]+$	549	6×10^{-19}	$\gamma\gamma[39]$ $3\pi^0[32]$ $\pi^+\pi^-\pi^0[24]$ $\pi^+\pi^-\gamma[5]$	
			Strange mesons		
$K^\pm$	$\frac{1}{2}[0^-]$	493.65	1.237×10^{-8}	$\mu^\pm\nu[63.5]$ $\mu^\pm\nu\pi^0[3.2]$ $e^\pm\nu\pi^0[4.8]$ $\pi^\pm\pi^0[21.2]\pi^\pm\pi^+\pi^-[5.59]\pi^\pm\pi^0\pi^0[1.73]$	
K_S^0	$\frac{1}{2}[0^-]-$	497.67	8.92×10^{-11}	$\pi^+\pi^-[68.6]\pi^0\pi^0[31.4]$	
K_L^0	$\frac{1}{2}[0^-]+$	$m_{K_L^0}-m_{K_S^0}=3.52\times10^{-12}$	5.18×10^{-8}	$\pi^\pm e^\mp\nu[38.6]\pi^\pm\mu^\mp\nu[27.0]$ $3\pi^0[22]$ $\pi^+\pi^-\pi^0[12.4]$	
			Charmed mesons		
$D^\pm$	$\frac{1}{2}[0^-]$	1869	10.7×10^{-13}	$D^+\to e^+X[19]$ $K^-X[16]$ $K^+X[7\pm3]$ $\bar{K}^0X+K^0X[48\pm15]$	
$D^0/\bar{D}^0$	$\frac{1}{2}[0^-]$	1863	3.5×10^{-13} $+3.5/-1.7$	$D^0\to e^+X[8]$ $K^-X[43]$ $K^+X[6\pm3]$ $\bar{K}^0X+K^0X[33\pm10]$ $\pi^+\pi^+\pi^-\pi^-[1]$ $\pi^+\pi^-\pi^0[1]$	
$D_s^\pm$	$(0[0^-])$	1969	4.3×10^{-13}	$D_s^+\to\phi\pi^+[8\pm5]$ $\phi\pi^+\pi^+\pi^-[4\pm3]$ $\bar{K}^*(892)^0K^+[8\pm5]$	
$D_s^{*\pm}$		2112	$>3\times10^{-23}$	$D_s^{*+}\to D_s^+\gamma$[dominant]	
			Bottom mesons		
$B^\pm$	$(\frac{1}{2}[0^-])$	5278	1.3×10^{-12} (both)	$B\to e^\pm\nu$ hadrons [12] $\mu^\pm\nu$ hadrons [11] $D^0/\bar{D}^0X[39\pm6]$ $K^\pm X[85\pm11]$	
$B^0/\bar{B}^0$	$(\frac{1}{2}[0^-])$	5279		$K^0/\bar{K}^0X[63\pm8]$ $pX[>2]\Lambda X[>1]J/\Psi(1S)X[1]$ $D^*(2010)^\pm X[22\pm7]$ $D^\pm X[17\pm6]$	
			Nonstrange baryons		
p	$\frac{1}{2}[\frac{1}{2}^+]$	938.2723	$>3\times10^{38}$		2.792 8474
n	$\frac{1}{2}[\frac{1}{2}^+]$	939.5656	896	$pe^-\nu[100]$	−1.913 043
			Strangeness −1 baryons		
Λ	$0[\frac{1}{2}^+]$	1115.6	2.63×10^{-10}	$p\pi^-[64]$ $n\pi^0[36]$	−0.61
Σ^+	$1[\frac{1}{2}^+]$	1189.4	7.99×10^{-9}	$p\pi^0[51.6]$ $n\pi^+[48.3]$	2.4
Σ^0	$1[(\frac{1}{2}^+)]$	1192.6	7×10^{-20}	$\Lambda\gamma[100]$	
Σ^-	$1[\frac{1}{2}^+]$	1197.4	1.48×10^{-10}	$n\pi^-[99.85]$	−1.16
			Strangeness −2 baryons		
Ξ^0	$\frac{1}{2}[\frac{1}{2}^{(+)}]$	1315	2.9×10^{-10}	$\Lambda\pi^0[100]$	−1.25
Ξ^-	$\frac{1}{2}[\frac{1}{2}^{(+)}]$	1321.3	1.64×10^{-10}	$\Lambda\pi^-[100]$	−0.7
			Strangeness −3 baryon		
Ω^-	$0[(\frac{3}{2}^+)]$	1672.4	8.2×10^{-11}	$\Lambda K^-[68]$ $\Xi^0\pi^-[24]$ $\Xi^-\pi^0[8.6]$	
			Charmed baryons		
Λ_c^+	$0[(\frac{1}{2})^+]$	2284	1.8×10^{-13}	$p\bar{K}^0[2\pm1]$ $pK^-\pi^+[3\pm1]$ $p\bar{K}^0\pi^+\pi^-[7\pm4]$ $\Lambda X[27\pm9]$ $\Lambda\pi^+\pi^+\pi^-[2\pm1]$ $\Sigma^\pm X[10\pm5]$ $\Sigma^+\pi^+\pi^-[10\pm8]$ $e^+X[4\pm2]$ $pe^+X[2\pm1]$ $\Lambda e^+X[1\pm1]$	
Ξ_c^+		2460 ± 19	4×10^{-13} $+2/-1$	$\Lambda K^-\pi^+\pi^+$	

9.02. TABLE: MESONS AND MESON RESONANCES

Approximate masses in MeV are indicated in parentheses. Assignments to $q\bar{q}$ states are given in the first column; L = quark orbital angular momentum, S = quark spin. Only the lowest mass entries are shown; many higher mass radial excitations have been observed. The $q\bar{q}$ composition is shown in terms of the known quarks: u(up), d(down), s(strange), c(charmed), b(bottom). Note that only the states in the $u\bar{u}$, $d\bar{d}$, $s\bar{s}$, $c\bar{c}$, and $b\bar{b}$ columns and the neutral states in the $I=1$ column are eigenstates of charge conjugation C.

		Light quark nonets			Heavy quark states				
$^{2S+1}L_J$	J^{PC}	Nonstrange $u\bar{d},u\bar{u},d\bar{d}$ $I=1$	Nonstrange $u\bar{u},d\bar{d},s\bar{s}$ $I=0$	Strange $\bar{s}u,\bar{s}d$ $I=1/2$	Charmonium $c\bar{c}$ $I=0$	Bottomonium $b\bar{b}$ $I=0$	Charm $c\bar{u},c\bar{d}$ $I=1/2$	Charm-Strange $c\bar{s}$ $I=0$	Bottom $\bar{b}u,\bar{b}d$ $I=1/2$
$^1S_0$	0^{-+}	$\pi(137)$	$\eta(549),\eta'(958)$	$K(496)$	$\eta_c(2980)$		$D(1867)$	$D_s(1972)$	$B(5273)$
$^3S_1$	1^{--}	$\rho(770)$	$\phi(1020),\omega(782)$	$K^*(892)$	$J/\psi(3097)$	$\Upsilon(9460)$	$D^*(2010)$	$D_s^*(2113)$	
$^1P_1$	1^{+-}	$b_1(1235)$	$h_1(1170)$	$K_{1B}{}^{\dagger}$					
$^3P_0$	0^{++}	$a_0(980)$	$f_0(975)\ f_0(1400)$	$K_0^*(1430)$	$\chi_{c0}(3415)$	$\chi_{b0}(9860)$			
$^3P_1$	1^{++}	$a_1(1260)$	$f_1(1285),f_1(1420)$	$K_{1A}{}^{\dagger}$	$\chi_{c1}(3510)$	$\chi_{b1}(9890)$			
$^3P_2$	2^{++}	$a_2(1320)$	$f_2'(1525),f_2(1270)$	$K_2^*(1430)$	$\chi_{c2}(3555)$	$\chi_{b2}(9915)$			
$^1D_2$	2^{-+}	$\pi_2(1670)$							
$^3D_1$	1^{--}	$\rho(1700)$?			$\psi(3770)$				
$^3D_2$	2^{--}			$K_2(1770)$					
$^3D_3$	3^{--}	$\rho_3(1690)$	$\omega_3(1670)$	$K_3^*(1780)$					
$^3F_4$	4^{++}		$f_4(2050)$?	$K_4^*(2075)$					

† K_{1B} and K_{1A} are mixtures of the states $K_1(1270)$ and $K_1(1400)$.

9.03. TABLE: BARYONS AND BARYON RESONANCES

Approximate masses in MeV are followed by J^P. Level assignments in the nonrelativistic harmonic-oscillator quark model are given in the first column; these are rather unambiguous for $N=0$ and $N=1$—the $[56,0^+]$ and $[70,1^-]$SU(6) supermultiplets for noncharmed baryons—but may be unreliable for higher N.

							Charmed baryons		
Level	N $S=0$ $I=\frac{1}{2}$	Δ $S=0$ $I=\frac{3}{2}$	Λ $S=-1$ $I=0$	Σ $S=-1$ $I=1$	Ξ $S=-2$ $I=\frac{1}{2}$	Ω $S=-3$ $I=0$	Λ_c $S=0$ $I=0$	Σ_c $S=0$ $I=1$	Ξ_c $S=-1$ $I=\frac{1}{2}$
0	$939\frac{1}{2}^+$	$1232\frac{3}{2}^+$	$1116\frac{1}{2}^+$	$1193\frac{1}{2}^+$ $1385\frac{3}{2}^+$	$1318\frac{1}{2}^+$ $1530\frac{3}{2}^+$	$1672\frac{3}{2}^+$	$2285\frac{1}{2}^+$	$2455\frac{1}{2}^+$	$2460\frac{1}{2}^+$
1	$1535\frac{1}{2}^-$ $1650\frac{1}{2}^-$ $1520\frac{3}{2}^-$ $1700\frac{3}{2}^-$ $1675\frac{5}{2}^-$	$1620\frac{1}{2}^-$ $1700\frac{3}{2}^-$	$1405\frac{1}{2}^-$ $1670\frac{1}{2}^-$ $1800\frac{1}{2}^-$ $1520\frac{3}{2}^-$ $1690\frac{3}{2}^-$ $1830\frac{5}{2}^-$	$1750\frac{1}{2}^-$ $1670\frac{3}{2}^-$ $1940\frac{3}{2}^-$ $1775\frac{5}{2}^-$	$1820\frac{3}{2}^-$				
2	$1440\frac{1}{2}^+$ $1710\frac{1}{2}^+$ $1720\frac{3}{2}^+$ $1680\frac{5}{2}^+$	$1910\frac{1}{2}^+$ $1905\frac{5}{2}^+$ $1950\frac{7}{2}^+$	$1600\frac{1}{2}^+$ $1810\frac{1}{2}^+$ $1890\frac{3}{2}^+$ $1820\frac{5}{2}^+$ $2110\frac{5}{2}^+$	$1660\frac{1}{2}^+$ $1915\frac{5}{2}^+$ $2030\frac{7}{2}^+$	$2030\frac{5}{2}^+$				
>2	$2190\frac{7}{2}^-$ $2250\frac{9}{2}^-$ $2220\frac{9}{2}^+$ $2600\frac{11}{2}^-$	$1930\frac{5}{2}^-$ $2420\frac{11}{2}^+$	$2100\frac{7}{2}^-$ $2350\frac{9}{2}^+$						

9.04. RELATIVISTIC KINEMATICS OF REACTIONS AND DECAYS

Boost of a 4-vector $p=(E,\mathbf{p})$ by velocity β in the z direction:

$$\begin{pmatrix}E'\\p'_z\end{pmatrix}=\begin{pmatrix}\gamma & \gamma\beta\\ \gamma\beta & \gamma\end{pmatrix}\begin{pmatrix}E\\p_z\end{pmatrix},$$

$$(p'_x,p'_y)=(p_x,p_y),\quad \gamma\equiv(1-\beta^2)^{-1/2}.$$

Survival probability of a particle of lifetime τ_0 moving with velocity β for time $>t$: $\exp(-t/\gamma\tau_0)$; for path length $>x$: $\exp(-x/\gamma\beta\tau_0)$.

Kinematic function λ:

$$\lambda(x,y,z)=x^4+y^4+z^4-2x^2y^2-2x^2z^2-2y^2z^2$$
$$=[x^2-(y+z)^2][x^2-(y-z)^2].$$

Two-body decay $m\rightarrow m_1+m_2$ in rest system of m.

$$E_1=(m^2+m_1^2-m_2^2)/2m,$$
$$E_2=(m^2+m_2^2-m_1^2)/2m,$$
$$|\mathbf{p}_1|=|\mathbf{p}_2|=\lambda^{1/2}(m^2,m_1^2,m_2^2)/2m.$$

Three-body decay $m\rightarrow m_1+m_2+m_3$. The two-body invariant masses are related by

$$m_{12}^2+m_{23}^2+m_{31}^2=m^2+m_1^2+m_2^2+m_3^2,$$

where $m_{ij}^2\equiv(p_i+p_j)^2$. In a rectangular Dalitz plot events are scatter plotted versus two invariant mass-squared's. The phase space density is uniform over the area of the plot. For m_{12}^2 vs m_{13}^2,

$$(m_1+m_2)^2<m_{12}^2<(m-m_3)^2,$$

and the limits on m_{13}^2 for fixed m_{12}^2 are

$$(E_1+E_3)^2-(|\mathbf{p}_1|+|\mathbf{p}_3|)^2$$
$$<m_{13}^2<(E_1+E_3)^2-(|\mathbf{p}_1|-|\mathbf{p}_3|)^2,$$

where $E_1E_3\mathbf{p}_1\mathbf{p}_3$ are the energies and momenta of 1 and 3 in the (12) rest frame:

$$E_1=(m_{12}^2+m_1^2-m_2^2)/2m_{12},$$
$$E_3=(m^2-m_{12}^2-m_3^2)/2m_{12}.$$

Two-body reactions $p_1=p_2\rightarrow p_3+p_4$. Mandelstam variables:

$$s=(p_1+p_2)^2=(p_3+p_4)^2,$$
$$t=(p_1-p_3)^2=(p_2-p_4)^2,$$
$$u=(p_1-p_4)^2=(p_2-p_3)^2;$$
$$s+t+u=m_1^2+m_2^2+m_3^2+m_4^2.$$

In the c.m. frame $p_i=(E_i,\mathbf{p}_i)$ and $\cos\theta=\hat{p}_1\cdot\hat{p}_3$, where

$$s^{1/2}=m_1^2+m_2^2+2m_{\text{target}}E_{\text{lab}},$$
$$|\mathbf{p}_1|=|\mathbf{p}_2|=m_{\text{target}}p_{\text{lab}}/s^{1/2},$$
$$\left.\begin{aligned}E_i&=(s+m_i^2-m_j^2)/2s^{1/2}\\|\mathbf{p}_i|&=\lambda^{1/2}(s,m_i^2,m_j^2)/2s^{1/2}\end{aligned}\right\}\quad (ij)=\text{initial or final pair of particles}$$
$$t=m_1^2+m_3^2-2E_1E_3+2|\mathbf{p}_1||\mathbf{p}_3|\cos\theta,$$
$$u=m_1^2+m_4^2-2E_1E_4-2|\mathbf{p}_1||\mathbf{p}_3|\cos\theta.$$

The lab (m_2 at rest) scattering angle Θ is given by

$$\tan\Theta=\frac{\sin\theta}{\gamma(\cos\theta+\beta/\beta_3)},$$
$$\gamma=(E_{\text{lab}}+m_{\text{target}})/s^{1/2},$$
$$\gamma\beta=p_{\text{lab}}/s^{1/2},\ \beta_3=|\mathbf{p}_3|/E_3.$$

Lepton-inclusive reactions $l+2\rightarrow l'+X$. Common variables used for these reactions are

$$E_l,E_{l'}=\text{lab energies of } l \text{ and } l',$$
$$q=p_l-p_{l'},\ Q^2=-q^2,$$
$$\nu=p_2\cdot q/m_2=E_l-E_{l'},$$
$$x=Q^2/2m_2\nu,$$
$$y=m_2\nu/p_l\cdot p_2=\nu/E_l.$$

For massless leptons,

$$E_{l'}\frac{d\sigma}{d^3p_{l'}}=\frac{1}{2\pi M\nu}\frac{d\sigma}{dx\,dy}=\frac{E_l}{\pi}\frac{d\sigma}{dQ^2d\nu}=\frac{x}{\pi y}\frac{d\sigma}{dx\,dQ^2}.$$

Hadronic-inclusive reaction $h+2\rightarrow 3+X$. Common variables used for these reactions are

$p_\parallel,p_\perp=$ c.m. momentum components of $\mathbf{p}_3$ parallel and perpendicular to $\mathbf{p}_h$,

$E=$ c.m. energy of particle 3,

$\theta=$ c.m. scattering angle of particle 3,

$$m_\perp=(m_3^2+p_\perp^2)^{1/2},$$
$$y=\text{rapidity}=\tfrac{1}{2}\ln\left(\frac{E+p_\parallel}{E-p_\parallel}\right)$$
$$=\ln\left(\frac{E+p_\parallel}{m_\perp}\right)\underset{m_3\rightarrow 0}{\rightarrow}-\ln\tan\left(\frac{\theta}{2}\right),$$
$$E=m_\perp\cosh y,$$
$$p_\parallel=m_\perp\sinh y,$$
$$E\frac{d\sigma}{d^3\mathbf{p}_3}=\frac{1}{\pi}\frac{d\sigma}{dy\,dp_\perp^2}.$$

9.05. PARTICLE DETECTORS, ABSORBERS, AND RANGES

A. Particle detectors*

In this section we give various parameters for common detectors. The quoted numbers are usually based on some typical apparatus, and obviously should be regarded as rough approximations, valid only for preliminary design when applied to other cases. A more detailed introduction to detectors can be found in *Experimental Techniques in High Energy Physics*, edited by T. Ferbel (Addison-Wesley, Menlo Park, CA, 1987).

* Updated April 1988 by D. Anderson, G. Hall, and R. Wigmans.

1. Scintillators

The photon yield in the frequency range of practical photomultiplier tubes is $\cong 1\gamma$ per 100 eV of charged particle ionization energy loss in plastic scintillator[1] and as below in other materials.

Properties of three scintillators[1–6]

	BaF_2	BGO	NaI(Tl)
Density (g/cm^3)	4.9	7.1	3.7
Radiation length (cm)	2.1	1.1	2.6
dE/dx (avg. for MIP) (MeV/cm)	6.6	9.0	4.8
Peak emission (nm)	220 310	480	410
Decay constant (ns)	0.6 620	300	250
Index of refraction	1.56	2.15	1.85
Light yield (photons/MeV)	2×10^3 6.5×10^3	2.8×10^3	4×10^4
Hygroscopic	slightly	no	very

2. Čerenkov[7]

The half-angle θ_c of the Čerenkov cone aperture in terms of the velocity β and the index of refraction n is

$$\theta_c = \arc\cos\left(\frac{1}{\beta n}\right) \cong \left[2\left(1-\frac{1}{\beta n}\right)\right]^{1/2}.$$

The threshold velocity is $\beta_t = 1/n$; $\gamma_t = 1/(1-\beta_t^2)^{1/2}$. Therefore, $\beta_t\gamma_t = 1/(2\delta+\delta^2)^{1/2}$, where $\delta = n-1$. Values of δ for various commonly used gases are given as a function of pressure and wavelength in Ref. 8; for values at atmospheric pressure, see Table 9.05.I, Atomic and Nuclear Properties of Materials.

The number of photons N per cm of path length is given by:

$$N = \frac{\alpha}{c}\int\left(1-\frac{1}{\beta^2 n^2}\right)2\pi\,d\nu = \frac{\alpha}{c}\beta_t^2\int\left(\frac{1}{\beta_t^2}-\frac{1}{\beta^2}\right)2\pi\,d\nu$$

$$\cong 500\sin^2\theta_c/\text{cm (visible spectrum)}.$$

3. Photon collection

In addition to the photon yield, one should take into account the light collection efficiency ($\lesssim 10\%$ for typical 1-cm-thick scintillator), the attenuation length ($\cong 1$ to 4 m for typical scintillators[9]), and the quantum efficiency of the photomultiplier cathode ($\lesssim 25\%$).

4. Typical detector characteristics

Detector Type	Accuracy (rms)	Resolution Time	Dead Time
Bubble chamber	$\cong \pm 10$ to $\cong \pm 150\ \mu m$	$\cong$ 1 ms	$\cong 1/20$ s[a]
Streamer chamber	$\pm 300\ \mu m$	$\cong$ 2 μs	$\cong$ 100 ms
Proportional chamber	$\geqslant \pm 300\ \mu m$[b,c]	$\cong$ 50 ns	$\cong$ 200 ns
Drift chamber	± 50 to 300 μm	$\cong$ 2 ns[d]	$\cong$ 100 ns
Scintillator	—	$\cong$ 150 ps	$\cong$ 10 ns
Emulsion	$\pm 1\ \mu m$	—	—
Silicon strip	$\pm 2.5\ \mu m$	e	e

[a] Multiple pulsing time.
[b] 300 μm is for 1 mm pitch.
[c] Delay line cathode readout can give $\pm 150\ \mu m$ parallel to anode wire.
[d] For two chambers.
[e] Limited at present by noise and readout time of attached electronics.

5. Electromagnetic shower detectors

We give below typical energy resolutions (FWHM) for an incident electron in the 1 GeV range; E is in GeV. For a fixed number of radiation lengths, FWHM in the last three detectors would be expected to be proportional to $\sqrt{t}$ for t (= plate thickness)$\geqslant 0.2$ radiation lengths.[10]

For all detectors, operational resolution may be up to 50% worse due to dead areas, non-normally incident tracks, and other effects.

NaI (20 rad. lengths)[11]: $\frac{2\%}{E^{1/4}}$

Lead glass (14 rad. lengths)[12]: $\frac{10-12\%}{\sqrt{E}}$

Lead-liquid argon (15.75 rad. lengths)[10]: $\frac{16\%}{\sqrt{E}}$

(42 cells: 1.1 mm lead, 2 mm liquid argon, 2.3 mm lead-G10, 2 mm liquid argon)

Lead-scintillator sandwich (12.5 rad. lengths)[13]: $\frac{17\%}{\sqrt{E}}$

(66 cells: 1 mm lead, 5 mm scintillator)

Proportional wire shower chamber (17 rad. lengths)[14]: $\frac{40\%}{\sqrt{E}}$

(36 cells: 0.474 rad. length type-metal + Al, 9.5 mm 80% Ar – 20% CH_4 gas)

6. Hadronic shower detectors[15]

The performance of hadron calorimeters is crucially influenced by the relative response to the e.m. and non-e.m. shower components (e/h ratio). Ideally, this ratio should be 1 (compensation), thus eliminating the effects of the large non-Gaussian fluctuations in the π^0 content of hadron showers. A noncompensated calorimeter has the following problems:

a) A non-Gaussian signal distribution for monoenergetic hadrons;

b) A nonlinear response to hadrons;

c) an energy resolution σ/E that does not scale as $E^{-1/2}$, but rather as $c_1E^{-1/2}+c_2$, where c_2 is determined by the e/h value ($c_2 \sim 0$ for $e/h = 1$).

These effects may severely deteriorate the performance, particularly at high energies ($E \gtrsim 100$ GeV): a 20% deviation from linearity over one order of magnitude in energy, considerable non-Gaussian tails, and a constant terms $c_2 = 5$–7% in the energy resolution were observed for calorimeters with $e/h \sim 1.3$ and 0.8.[16]

Fully sensitive detectors *cannot* be made compensating, unless made out of hydrogen. In all more practical cases $e/h \gtrsim 1.4$, since the undetectable binding energy lost when protons and neutrons are released from their nuclear environment accounts for 30–40% of the energy carried by the non-e.m. shower component.

Compensation *can* be achieved in sampling calorimeters with hydrogenous active layers sandwiched between passive layers, making use of the fact that the calorimeter response to the various shower components

(π^0's, minimum-ionizing particles, nonrelativistic protons and neutrons) may be very different in that case. The relative contribution of neutrons (through elastic *np* scattering) to the non-e.m. calorimeter signal, which is $\sim 30\%$ for a compensating calorimeter, can be tuned through the sampling fraction. Efficient neutron detection as provided by hydrogeneous active layers also considerably reduces the contribution of fluctuations in nuclear binding-energy losses to the energy resolution.

Compensation has been experimentally demonstrated in calorimeters using 2.5 mm plastic scintillator active layers and ^{238}U(3 mm)[17] or Pb(10 mm)[18] passive layers, with a total energy resolution σ/E of 0.34 $E^{-1/2}$ or 0.44 $E^{-1/2}$, respectively (E in GeV). The former has also been shown linear to within 2% over three orders of magnitude in energy with Gaussian signal distributions.

7. dE/dx resolution in argon

Particle identification (relativistic, $Q=1$ incident particles) by dE/dx is dependent on the width of the distribution:

Multiple-sample Ar gas counters (no lead)[19]:

$$\frac{\text{FWHM}\left(\left.\dfrac{dE}{dx}\right|_{\text{most probable}}\right)}{\left.\dfrac{dE}{dx}\right|_{\text{most probable}}} = 0.96N^{-0.46}(tp)^{-0.32}$$

$N=$ no. of samples, $t=$ thickness per sample (cm), $p=$ pressure (atm); most commonly used chamber gases (except Xe) give approximately the same resolution.

8. Free electron drift velocities in liquid ionization chambers[20–23]

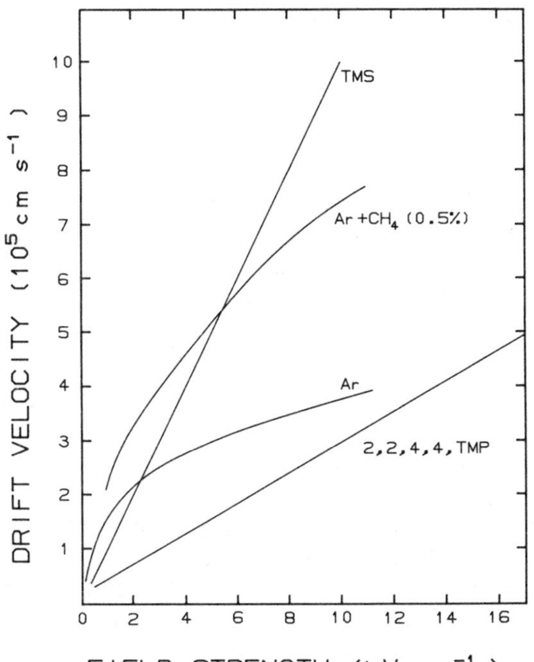

9. An approximation for the calculation of particle momenta in a uniform magnetic field[24]

The path of motion of a charged particle of momentum p, in GeV/c, is a helix of constant radius of curvature and constant pitch angle λ, with the axis of the helix along $\vec{H}$ and:

$$p\cos\lambda = 0.299\,79\; ZH/k\,,$$

where the field strength H is in tesla, the charge Z is in units of the electronic charge, and the curvature k, equal to 1/radius of curvature, is measured in a plane normal to the field, in m^{-1}.

The distribution of measurements of curvature about its true value is approximately Gaussian. The curvature error for a large number of uniformly spaced points measured on the trajectory of a charged particle in a uniform magnetic field can be approximated by the following expression:

$$(\delta k)^2 = (\delta k_{\text{res}})^2 + (\delta k_{\text{ms}})^2\,,$$

where

δk = curvature error,
δk_{res} = curvature error due to finite measurement resolution,
δk_{ms} = curvature error due to multiple scattering.

For a charged particle measured many times along its path ($\geqslant$10 measurements) in a uniform medium,

$$\delta k_{\text{res}} = \frac{\epsilon}{L'^2}\sqrt{\frac{720}{N+5}}\,,$$

where

N = number of points (uniformly spaced), measured along track,
L' = the projected length of the track onto the bending plane,
ϵ = measurement error for each point, perpendicular to the trajectory.

The contribution due to multiple Coulomb scattering is approximately:

$$\delta k_{\text{ms}} \approx \frac{(0.016)(\text{GeV}/c)Z}{L'^2 p\beta}\sqrt{\frac{L}{L_R}}$$

where

p = momentum (GeV/c),
Z = charge of incident particle in units of e,
L_R = radiation length of the scattering medium,
β = the kinematic variable v/c,
L = the total track length.

More accurate approximations for multiple scattering may be found in the section on Passage of Particles Through Matter (9.05.C) following. The contribution to the curvature error is given approximately by $\delta k_{\text{ms}} \simeq \delta s^{\text{rms}}_{\text{plane}}/L'^2$, where $s^{\text{rms}}_{\text{plane}}$ is defined there.

10. Proportional chamber wire instability

The limit on the voltage V for a wire tension T, due to mechanical effects when the electrostatic repulsion of adjacent wires exceeds the restoring force of wire tension, is given by (MSKA)[25]

$$V \leqslant \frac{s}{\ell C}\sqrt{4\pi\epsilon_0 T},$$

where s, ℓ, and C are the wire spacing, length, and capacitance per unit length. An approximation to C for chamber half-gap t and wire diameter d (good for $s \lesssim t$) gives[26]

$$V \lesssim 59T^{1/2}\left[\frac{t}{\ell} + \frac{s}{\pi\ell}\ln\left(\frac{s}{\pi d}\right)\right],$$

where V is in kV, and T is in grams-weight equivalent.

11. Proportional and drift chamber potentials

The potential distributions and fields in a proportional or drift chamber can usually be calculated with good accuracy from the exact formula for the potential around an array of parallel line charges q (coul/m) along z and located at $y = 0$, $x = 0$, $\pm s$, $\pm 2s$,...,

$$V(x,y) = -\frac{q}{4\pi\epsilon_0}\ln\left\{4\left[\sin^2\left(\frac{\pi x}{s}\right) + \sinh^2\left(\frac{\pi y}{s}\right)\right]\right\}.$$

Errors from the presence of cathodes, mechanical defects, TPC-type edge effects, etc., are usually small and are beyond the scope of this review.

12. Silicon strip detectors and photodiodes

These are silicon diodes operated with a reverse bias voltage V (typically 30–300 volts) sufficient to deplete the sensitive volume of most mobile charge carriers (electrons and holes). The active (depletion layer) thickness t (cm) is given in a simple model by

$$t = \sqrt{\frac{2\epsilon V}{ne}} = \sqrt{2\rho\mu\epsilon V},$$

where

$n =$ number of impurity centers/cm^3,

$e =$ electron charge,

$\epsilon =$ dielectric constant $\cong 1$ pF/cm $\cong 11.9\epsilon_0$,

$\rho =$ resistivity $\cong 1$–20 kΩ-cm,

$\mu =$ majority charge carrier mobility

$\cong$ 1300–1500 cm^2/volt-sec (electrons)

$\cong$ 450–600 cm^2/volt-sec (holes).

The capacitance of the diode is ϵ/t per unit area, but in the case of microstrips this is usually dominated by the interstrip capacitance of ~ 1 pF per cm of strip length. A minimum-ionizing particle has a Landau energy-loss distribution with average energy loss 39 keV/100 μm, most probable energy loss 26 keV in 100 μm (which scales within $\sim \pm 10\%$ from ~ 20 to ~ 300 μm), and full width at half-maximum of roughly $0.1t/\beta^2$ keV, where t is the detector thickness in microns and $\beta = v_{\rm inc}/c$. The width is usually increased further by electronic noise ($\sigma \sim 1$–10 keV) and for thin layers by a Gaussian contribution due to atomic effects [$\sigma \sim (0.3$–$0.4)\sqrt{t}$ keV]. The average energy required to produce an electron-hole pair is 3.6 eV, from which one can estimate total charge of either sign released. Silicon detectors can tolerate integrated charged-particle fluxes of up to $\sim 10^{10}$–10^{14}/cm^2 and still operate as efficient detectors.

Typical photodiodes are sensitive (quantum efficiencies greater than $\sim 10\%$) to wavelengths from ~ 200 nm to 1100 nm.

B. Cosmic ray fluxes

The fluxes of particles of different types depend at the $\sim 10\%$ level on the latitude, their energy, and the conditions of measurement. Some typical sea-level values[27] for charged particles are given below:

I_v flux per unit solid angle per unit horizontal area about vertical direction $\equiv j(\theta = 0,\phi)$ [$\theta =$ zenith angle, $\phi =$ azimuthal angle];

J_1 total flux crossing unit horizontal area from above

$$\equiv \int_{\theta \leqslant \pi/2} j(\theta,\phi)\cos\theta\, d\Omega\ [d\Omega = \sin\theta\, d\theta\, d\phi];$$

J_2 total flux from above (impinging on a sphere of unit cross-sectional area)

$$\equiv \int_{\theta \leqslant \pi/2} j(\theta,\phi)d\Omega.$$

	Total Intensity	Hard Component	Soft Component	
I_v	1.1×10^2	0.8×10^2	0.3×10^2	m^{-2} sec^{-1} sterad^{-1}
J_1	1.8×10^2	1.3×10^2	0.5×10^2	m^{-2} sec^{-1}
J_2	2.4×10^2	1.7×10^2	0.7×10^2	m^{-2} sec^{-1}

Very approximately, about 75% of all particles at sea level are penetrating, and are muons (the dominant portion of the hard component at sea level). The sea-level vertical flux ratio for protons to muons (both charges together) is about $3\frac{1}{2}$% at 1 GeV/c, decreasing to about $\frac{1}{2}$% at 10 GeV/c.

The muon flux at sea level has a mean energy of 2 GeV and a differential spectrum falling as E^{-2}, steepening smoothly to $E^{-3.6}$ above a few TeV. The angular distribution is $\cos^2\theta$, changing to $\sec\theta$ at energies above a TeV, where θ is the zenith angle at production. The $+\ -$ charge ratio is 1.25 — 1.30. The mean energy of muons originating in the atmosphere is roughly 300 GeV at slant depths $\gtrsim$ a few hundred meters. Beyond slant depths of ~ 10 km water-equivalent, the muons are due primarily to in-the-earth neutrino interactions (roughly 1/8 interaction ton^{-1} year^{-1} for $E_\nu > 300$ MeV, $\sim$constant throughout the earth).[28] Muons from this source arrive with a mean energy of 20 GeV, and have a flux of 2×10^{-9} m^{-2} sec^{-1} sterad^{-1} in the vertical direction and about twice that in the horizontal,[29] down at least as far as the deepest mines.

C. Passage of particles through matter*

1. Energy loss rates for heavy charged projectiles

A heavy projectile (much more massive than an electron) of charge $Z_{\rm inc}e$, incident at speed βc ($\beta \gg 1/137$) through a slowing medium, dissipates energy princi-

* Revised April 1988 by J. J. Eastman.

pally via interactions with the electrons of the medium. The mean rate of such energy loss per unit path length x, called the stopping power, is given by the Bethe–Bloch equation[30]:

$$\left(\frac{dE}{dx}\right)_{inc} = \frac{D Z_{med}\rho_{med}}{A_{med}} \left(\frac{Z_{inc}}{\beta}\right)^2 \times\left[\ln\left(\frac{2m_e\gamma^2\beta^2c^2}{I}\right) - \beta^2 - \frac{\delta}{2} - \frac{C}{Z_{med}}\right]\{1+\nu\},$$

where $D = 4\pi N_A r_e^2 m_e c^2 = 0.3071$ MeV cm^2/g (see Precise Physical Constants Table). A mean range and energy loss figure appears in Section 9.05.D.

Here Z_{med} and A_{med} are the charge and mass numbers of the medium and ρ_{med} is the mass density of the medium; I, δ, C, and ν are phenomenological functions. Frequently, the values of δ, C, and ν are negligibly small; the parameter I characterizes the binding of the electrons of the medium. As a rule of thumb, we may estimate I for an idealized medium as $I \cong 16\ (Z_{med})^{0.9}$ eV when $Z_{med} > 1$. For realistic media the value of I will vary at the 10% level from this estimate. Variations of this order occur due to atomic effects such as completion of a shell, also due to chemical binding, and even due to the phase of the substance. Hydrogen, perhaps the most sensitive, has I of about 15 eV in the atomic mode, rising to about 19.2 eV as H_2 gas and to 21.8 eV as H_2 liquid.[31] For many substances, the transition from gas to solid is accompanied by a 20–30% increase in I.[31] We may *approximately* treat media which are chemical mixtures or compounds by computing

$$\frac{dE}{dx} \cong \sum_{n=1}^{N}\left(\frac{dE}{dx}\right)_n,$$

with $(dE/dx)_n$ appropriate to the nth chemical constituent (using $\rho_{med}^{(n)}$ as the partial density in the formula for dE/dx).[32] For many chemical compounds, small corrections to this addivity rule may be found in Ref. 31.

The function δ represents the density effect upon the energy loss rate; it is non-negligible only for highly relativistic projectiles in denser media.[33] For ultrarelativistic projectiles, δ approaches $2\ln\gamma +$ constant, where the value of the constant depends upon the density of the medium as well as its chemical composition.

The function C represents shell corrections to the energy loss rate.[30] These effects are non-negligible only for projectiles with speeds not much faster than the speeds of the fastest electrons bound in the medium.

The function ν represents corrections due to higher order electrodynamics.[34] These effects become important when $|Z_{inc}/\beta|$ is comparable to 137. For relavistic unit-charge projectiles, $|\nu|$ is of the order of 1%; positively charged projectiles lose energy more rapidly than do their charge conjugates.[34,35]

For nonrelativistic projectiles, our formulas above are inapplicable. At the very slowest speeds, total energy loss rates are believed to be proportional to β, rising through a peak at projectile speeds comparable to atomic speeds (β on the order of αc), after having passed through a smaller peak (due to elastic Coulomb collisions with the *nuclei* of the slowing medium[36]) at intermediate speeds. For example, for protons in Si, $dE/dx = 61.23\beta$ GeV/(gm cm^{-2}) for $\beta < 0.005$; the peak occurs at $\beta = 0.0126$ where $dE/dx = 522$ MeV/(gm cm^{-2}). In some cases, energy loss rates depend significantly upon the relation of the projectile trajectory to the crystalline structure of the slowing medium.[37]

For relativistic projectiles, $(dE/dx)_{inc}$ falls rapidly with increasing β until reaching a minimum around $\beta = 0.96$ (almost independent of medium), followed by a slow rise. Because of the density effect, the quantity in square brackets approaches $\ln\gamma +$ constant for large γ.

The quantity $(dE/dx)_{inc}\,\delta x$ is the *mean* total energy loss via interactions with electrons of the medium in a layer of thickness δx. For any finite δx, Poisson fluctuations can cause the actual energy loss to deviate from the mean. For thin layers, the distribution is broad and skewed, being peaked below $(dE/dx)\delta x$, and having a long tail toward large energy losses.[38] Only for a very thick layer $[(dE/dx)\delta x \gg 2m_e\beta^2\gamma^2c^2]$ will the distribution of energy losses become nearly Gaussian. The large fluctuations of the total energy loss rate from the mean are due to a small number of collisions involving large energy transfers. The fluctuations are greatly reduced for the so-called restricted energy loss rate, described in Section 9.05.C.4.

2. Ionization yields

Physicists frequently relate total energy loss to the number of ion pairs produced near the projectile's track. This relation becomes complicated for relativistic projectiles due to the wandering of energetic knock-on electrons whose ranges exceed the dimensions of the fiducial volume. For a qualitative appraisal of the nonlocality of energy deposition by such modestly energetic knock-on electrons in various media, see Ref. 39. Furthermore, the mean local energy dissipation per local ion pair produced, W, while essentially constant for relativistic projectiles, increases at slow projectile speeds.[40] The numerical value of W for gases can be surprisingly sensitive to trace amounts of various contaminants.[40] Of course, in addition to the preceding effects, practical ionization yields may be greatly influenced by subsequent recombinations and other factors.[41]

3. Energetic knock-on electrons

For a relativistic point-charge projectile, the production of high energy (kinetic energy $T \gg I$) electrons is given by[42]

$$\frac{d^2N}{dTdx} = \tfrac{1}{2} D\left(\frac{Z_{med}}{A_{med}}\right)\left(\frac{Z_{inc}}{\beta}\right)^2 \rho_{med}\frac{1}{T^2}F,$$

for $I \ll T \leqslant T_{max}$, where

$$T_{max} = \frac{2m_e\beta^2\gamma^2c^2}{1 + 2\gamma\dfrac{m_e}{M_{inc}} + \left(\dfrac{m_e}{M_{inc}}\right)^2},$$

M_{inc} is the mass of the incident projectile, and all other quantities except F are as in Sec. 9.05.C.1. F ($\cong 1$ for $T \ll T_{max}$) is a factor dependent upon the spin of the projectile.

For spin-0 projectiles,

$$F = 1 - \beta^2 \frac{T}{T_{max}};$$

for spin-1/2 projectiles,

$$F = 1 - \beta^2 \frac{T}{T_{max}} + \frac{1}{2}\left(\frac{T}{T_{inc} + M_{inc}c^2}\right)^2,$$

where T_{inc} is the kinetic energy of the projectile; for electrons incident,

$$F = \beta^2 T^2 \left[\frac{T_{inc}}{T(T_{inc} - T)} - \frac{1}{T_{inc}}\right]^2;$$

and for positrons incident,

$$F = \beta^2 \left[1 - \frac{T}{T_{inc}} + \left(\frac{T}{T_{inc}}\right)^2\right]^2.$$

For incident electrons, the indistinguishability of projectile and target means that the range of T is only up to $T_{inc}/2$. For additional formulas see Ref. 43. Our formula is inaccurate for T close to I; for $2I \lesssim T \lesssim 10I$, the $1/T^2$ dependence above becomes $\cong T^{-\eta}$ with $3 \lesssim \eta \lesssim 5$.[44]

4. Rates of restricted energy loss for relativistic charged projectiles

The variability of energy loss for heavy projectiles is due primarily to the variability in the production of energetic knock-on electrons. Bremsstrahlung and pair-production processes make this variability even greater for electrons than for heavy particles as projectiles (see, e.g., Plot 9.05.G, Fractional Energy Loss for Electrons and Positrons in Lead). If an instrument, such as a bubble chamber, is capable of isolating these high-energy-loss interactions, then it is appropriate to consider the rate of energy loss excluding them, i.e., a restricted energy loss rate. The mean energy loss rate via all collisions which have energy transfer T such that $T \leqslant E_{max} \ll T_{max}$ is[1]

$$\left(\frac{dE}{dx}\right)_{\leqslant E_{max}} = \tfrac{1}{2} D \frac{Z_{med}\rho_{med}}{A_{med}} \left(\frac{Z_{inc}}{\beta}\right)^2 \times \left[\ln\left(\frac{E_{max} T_{max}}{I^2}\right) - \beta^2 - \delta - \frac{2C}{Z_{med}}\right].$$

Notice the overall factor of 1/2. See Sec. 9.05.C.1 above for definitions of the quantities in this equation.

The density effect causes the restricted energy loss rate to approach a constant, the Fermi plateau value, for the fastest projectiles.

5. Multiple scattering through small angles

As a charged particle traverses a medium it is deflected by many small-angle elastic scatterings. The bulk of this deflection is due to elastic Coulomb scattering from the nuclei within the medium, hence the usual identification as multiple Coulomb scattering (note, however, that strong interactions do contribute to the total multiple scattering for hadronic projectiles). For both Coulomb and strong interactions, the Central Limit Theorem provides little useful guidance in establishing the precise nature of the distribution of the total deflections resulting from multiple scattering. The true distribution is roughly Gaussian only for small deflection angles, while it shows much greater probability for large-angle scatterings ($\gtrsim$ a few θ_0, see below, depending on absorber) than the Gaussian would suggest. These tails on the distribution (a few per cent of peak height in the region where the Gaussian part becomes negligible) are more pronounced for hadrons than for muons as projectiles. The large-angle behavior of these distributions is best estimated by computing the exact distribution for the vectorial sum of the largest deflections based upon the true elastic scattering cross section of the projectile against the medium,[45] or, when applicable, by interpolation from tabular data.[46] An easier alternative which may suffice for noncritical applications would be to use a Gaussian approximation with the following width[47]:

$$\theta_0 = \frac{14.1\ \mathrm{MeV}/c}{p\beta} Z_{inc} \sqrt{L/L_R} \times [1 + \tfrac{1}{9}\log_{10}(L/L_R)]\ (\text{radians}),$$

where p, β, and Z_{inc} are the momentum (in MeV/c), velocity, and charge number of the incident particle, and L/L_R is the thickness, in radiation lengths, of the scattering medium. L_R for certain materials is given in Table 9.05.I, Atomic and Nuclear Properties of Materials. See also Sec. 9.05.C.8 below. The angle θ_0 is a fit to Moliere[45] theory, accurate to about 5% for $10^{-3} < L/L_R < 10$ except for very light elements or low velocity where the error is about 10 to 20%. In this Gaussian approximation, θ_0 has the meaning

$$\theta_0 = \theta_{plane}^{rms} = \frac{1}{\sqrt{2}} \theta_{space}^{rms}.$$

The nonprojected (space) and projected (plane) angular distributions are given approximately[45] by the Gaussian forms:

$$\frac{1}{2\pi\theta_0^2} \exp\left(-\frac{\theta_{space}^2}{2\theta_0^2}\right) d\Omega,$$

$$\frac{1}{\sqrt{2\pi}\theta_0} \exp\left(-\frac{\theta_{plane}^2}{2\theta_0^2}\right) d\theta_{plane},$$

where θ is the deflection angle. In this approximation, $\theta_{space}^2 \cong (\theta_{plane,x}^2 + \theta_{plane,y}^2)$, where the x and y axis are orthogonal to the direction of motion, and $d\Omega \cong d\theta_{plane,x} d\theta_{plane,y}$. Deflections into $\theta_{plane,x}$ and $\theta_{plane,y}$ are independent and identically distributed.

Other quantities are sometimes used to describe the amount of multiple Coulomb scattering: the auxiliary quantities $\psi_{\rm plane}$, $y_{\rm plane}$, and $s_{\rm plane}$ (see the figure) obey

$$\psi^{\rm rms}_{\rm plane} = \frac{1}{\sqrt{3}}\,\theta^{\rm rms}_{\rm plane} = \frac{1}{\sqrt{3}}\theta_0\,,$$

$$y^{\rm rms}_{\rm plane} = \frac{1}{\sqrt{3}}\,L\,\theta^{\rm rms}_{\rm plane} = \frac{1}{\sqrt{3}}\,L\theta_0\,,$$

and

$$s^{\rm rms}_{\rm plane} = \frac{1}{4\sqrt{3}}\,L\,\theta^{\rm rms}_{\rm plane} = \frac{1}{4\sqrt{3}}\,L\theta_0\,.$$

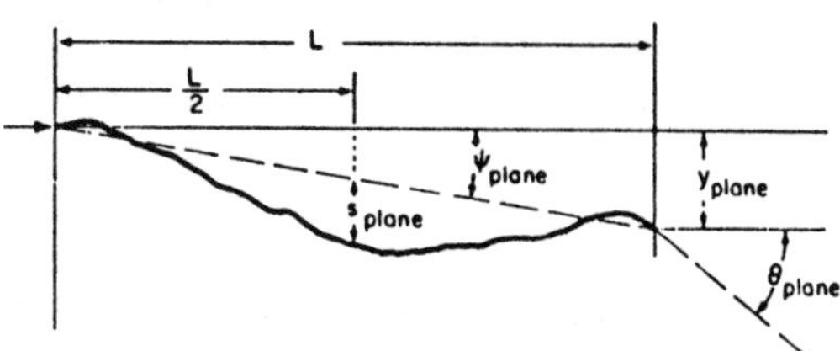

All the quantitative estimates in this section apply only in the limit of small $\theta^{\rm rms}_{\rm plane}$ and in the absence of large-angle scatters. The random variables s, ψ, y, and θ in a given plane are distributed in a correlated fashion. Obviously, $y \cong L\psi$. In addition, y and θ have correlation coefficient $\rho_{y\theta} = \sqrt{3}/2 \cong 0.87$. For Monte Carlo generation of a joint $(y_{\rm plane}, \theta_{\rm plane})$ distribution or for other calculations, it may be most convenient to work with independent Gaussian random variables (z_1, z_2) with mean zero and variance one and subsequently set

$$y_{\rm plane} = z_1 L\theta_0(1-\rho^2_{y\theta})^{1/2}/\sqrt{3} + z_2\,\rho_{y\theta}L\theta_0/\sqrt{3}$$

$$= z_1 L\theta_0/\sqrt{12} + z_2 L\theta_0/2\,;$$

$$\theta_{\rm plane} = z_2\theta_0\,.$$

Note that the second term for $y_{\rm plane}$ equals $L\theta_{\rm plane}/2$ and represents the displacement that would have occurred had the deflection $\theta_{\rm plane}$ all occurred at the single point $L/2$.

6. Muon energy loss at high energy

At muon energies above a few hundred GeV, radiative processes (bremsstrahlung, direct pair production, and photonuclear interactions) dominate over ionization as sources of energy loss. The figure below shows the total average energy loss of a muon per g/cm^2 in various materials plotted against the muon energy. In the case of iron, the contribution of each process is also shown. Average dE/dx values for many media for muons in the energy range 1–10000 GeV are tabulated in Ref. 48. The radiative processes are characterized by small cross sections, hard spectra, large energy fluctuations, and the associated generation of electromagnetic and (in the case of photonuclear interactions) hadronic showers. As a consequence, the treatment of energy loss as a uniform and continuous process at these energies is inadequate for many purposes. Detailed calculations of the differential cross sections for these processes with respect to the energy loss fraction v are available.[49–52] In addition, Ref. 53 provides useful parametrizations of the differential cross sections.

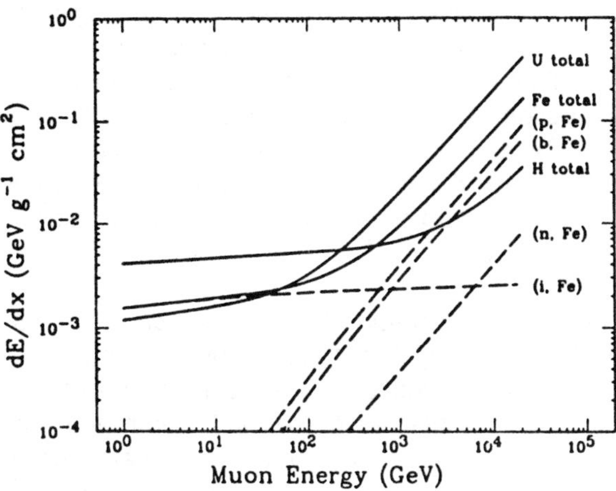

The average energy loss of a muon per g/cm^2 of hydrogen, iron, and uranium as a function of muon energy. Contributions of several processes to dE/dx in iron are also shown: (p) direct e^+e^- pair production, (b) bremsstrahlung, (n) photonuclear interactions, and (i) ionization.

The total cross section for muon bremsstrahlung is not large—the probability for a 1 TeV muon to bremsstrahlung in 1 m of iron is 5%—but the probability density $v\,d\sigma/dv$ for producing a photon of energy vE_μ is nearly flat across a wide range of values of v. This means that bremsstrahlung contributes a tail of catastrophic loss to the muon energy loss distribution. Petrukhin and Shestakov[49] provide an expression for the differential cross section for muon bremsstrahlung, taking into account the effects of nuclear and atomic form factors. The form factor Z-dependence is nontrivial; the average dE/dx from bremsstrahlung departs from Z^2/A scaling by 10–15% as one extrapolates from iron to uranium.

At energies above $\sim$500 GeV in iron, and $\sim$150 GeV in uranium, direct e^+e^- pair production becomes the single most important source of muon energy loss. Unlike the bremsstrahlung case, the pair energy probability peaks at low energies, making high-energy pairs relatively unlikely. Kel'ner and Kotov[51] give an expression for the differential cross section for e^+e^- production by muons which includes screening. See Ref. 50 for an evaluation of various theoretical treatments and approximations. For muon energies above $\sim$100 GeV, $\mu^+\mu^-$ pair production is also possible. Such $\mu^+\mu^-$ production by muons is a potentially important process that can lead to misassignment of the sign of the incident muon, but this mechanism contributes less than 0.01% to the total energy loss.[48]

Photonuclear interactions account for about 5% of the total energy loss of high-energy muons in iron, and about 2% in uranium. The losses are concentrated in rare, relatively hard events. Bezrukov and Bugaev[52] derive an expression for the differential cross section that includes nucleon shadowing effects.

These energy-loss processes will have a significant impact on the design of muon detectors operating in the multihundred-GeV regime. Energy fluctuations can exceed nominal detector resolutions,[54] necessitating the reconstruction of lost energy. Electromagnetic and hadronic showers in detector materials can obscure muon tracks in detector planes and reduce the tracking efficiency.[55] Unresolved shower particles can also degrade position measurements.[55]

7. Longitudinal distribution of electromagnetic showers

A photon energy $E \geqslant 0.1$ GeV converting in a semi-infinite medium produces an electromagnetic cascade whose intensity initially increases with depth and then falls off. The average number of $e^{\pm}$ with kinetic energy above 1.5 MeV, crossing a plane at a depth of L radiation lengths from the beginning of the medium, in a material of atomic number Z, calculated using the Monte Carlo program EGS,[56] can be fit by the empirical formula[57]

$$N = N_0 L^a e^{-bL},$$

where

$$N_0 = 5.51\, E(\text{GeV})\sqrt{Z} b^{a+1}/\Gamma(a+1)$$

and $b = 0.634 - 0.0021\ Z$. For $Z \geqslant 26$, $a = 2.0 - Z/340 + (0.664 - Z/340)\ln E$. For $Z = 13, a = 1.77 - 0.52 \ln E$. The maximum intensity, $N_{\max}$, occurs at the depth $L = a/b$. The maximum error of the fit occurs in the vicinity of this depth and is less than $0.15\ N_{\max}$. The integral of the tail

$$\int_{1.5a/b}^{\infty} N\, dL$$

is fit to better than 2.5%. The total longitudinally projected $e^{\pm}$ path length,

$$\int_0^{\infty} N\, dL = 5.51\, E\sqrt{Z},$$

is less than the total $e^{\pm}$ path length due primarily to multiple Coulomb scattering.

8. Radiation length

For the passage of electromagnetically interacting particles through a medium it is convenient to measure thickness in terms of radiation length.[58] For most electromagnetic processes (bremsstrahlung, Coulomb scattering, showering, pair production, etc.), over large energy intervals, some or all of the dependence upon the medium is contained in the radiation length.

The radiation length may be defined as the distance L_R over which a high energy electron ($\gtrsim 1$ GeV for most materials) loses all but a fraction $1/e$ of its energy to bremsstrahlung, on average. For a homogeneous monoatomic medium, $Z \geqslant 5$,

$$\frac{1}{L_R} = \frac{4\alpha r_e^2 N_A Z^2}{A}\left[\ln\left(\frac{184.15}{Z^{1/3}}\right) + \frac{1}{Z}\ln\left(\frac{1194}{Z^{2/3}}\right) - 1.202\alpha^2 Z^2 + 1.0369\alpha^4 Z^4 - \frac{1.008\alpha^6 Z^6}{1+\alpha^2 Z^2}\right]$$

$$= \frac{Z^2[\,]}{716.405A},$$

where α, r_e, and N_A are found in Precise Physical Constants, Table 1.01, and Z and A are the atomic number and weight of the medium. If r_e is expressed in cm, L_r will have the conventional units of g/cm^2. For $Z < 5$, a more complex numerical calculation is required. Radiation lengths for many substances are tabulated in Table 9.05.I, Atomic and Nuclear Properties of Materials. For media which are chemical mixtures or compounds,

$$\frac{1}{L_R} \cong \sum_i \frac{f_i}{L_R^i},$$

where f_i is the fraction by mass of atoms of type i, radiation length L_R^i. Chemical binding can lower L_R from this, typically by a few per cent.

For electrons of energy below about one GeV, the average fractional energy loss per unit length decreases as the energy decreases (see Plot 9.05.G, Fractional Energy Loss for Electrons and Positrons in Lead). With distances measured in units of L_R, dependence of the bremsstrahlung fractional energy loss upon Z of the medium in the low energy region ($\gtrsim 10$ MeV) is of order a few percent or less.

For photons of infinite energy, the total pair-production cross section is

$$\sigma = \frac{7}{9}(A/L_R N_A).$$

This is accurate to within a few per cent down to ~ 1 GeV for most materials. For energies below about 1 GeV, the cross section varies in a manner which may be determined from Plots 9.05.E and 9.05.F of the photon attenuation length. See also Plots 9.05.H, Contributions to Photon Cross Section in Lead.

9. Electron practical range

The electron "practical range"—a common measure of straight-line penetration distance—is shorter than the total path length because of multiple Coulomb scattering, which becomes increasingly important as the electron slows down. For example, for a fast electron the rms projected angle due to multiple Coulomb scattering reaches 1 radian by the time the electron has slowed to 0.4 MeV in hydrogen, 1.5 MeV in carbon, 9 MeV in copper, and 24 MeV in lead. Electrons which have energy less than 0.2 MeV in Ar, 1.5 MeV in Cu, 3.5 MeV in Sn, and 5 MeV in Pb are likely to deposit 10% of their energy *behind* their starting plane. The practical range, R_p, is defined as that absorber thickness obtained by extrapolating to zero the linearly decreasing part of the curve of penetration probability vs. absorber thickness. Data for Al in the T range up to about 10 MeV are available, and fit (to $\sim \pm 10\%$) $R_p = AT[1 - B/(1 + CT)]$ mg cm^{-2}, a from suggested in Ref. 59, with $A = 0.55$ mg cm^{-2} keV^{-1}, $B = 0.9841$, and $C = 0.0030$ keV^{-1}. At this penetration depth, 90–95% of the incident electrons have stopped. Data for other elements are sketchy, but suggest that higher-$Z(\lesssim 50)$ elements have $1 \lesssim R_p/R_p(\mathrm{Al}) \lesssim 1.4$ below $\sim$10 keV, and $0.6 \lesssim R_p/R_p(\mathrm{Al}) \lesssim 1$ above $\sim$100 keV. The "critical energy" (above which the energy loss due to bremsstrahlung exceeds that due to ionization, and showering becomes important) is 400 MeV for hydrogen, 100 MeV for carbon, 25 MeV for copper, and 10 MeV for lead. The mean positron range may differ from the mean electron range by several percent. See Refs. 60 and 61. Electron energy deposition and penetration probability vs. range are discussed in Refs. 39, 62, and 63.

D. Plots: Mean range and energy loss in lead, copper, aluminum, and carbon

Mean range and energy loss due to ionization for the indicated particles in Pb, with scaling to Cu, Al, and C indicated, using Bethe–Bloch equation (Sec. 9.05.C.1 above) with corrections. Calculated by M. J. Berger.

Scaling law for particles of other mass or charge (except electrons). For a given medium, the range R_b of any beam particle with mass M_b, charge z_b, and momentum p_b is given in terms of the range R_a of any other particle with mass M_a, charge z_a, and momentum $p_a = p_b M_a/M_b$ (i.e., having the same velocity) by

$$R_b(M_b, z_b, p_b) = \frac{M_b/M_a}{z_b^2/z_a^2} R_a(M_a, z_a, p_a = p_b M_a/M_b) .$$

For ranges of various particles in liquid hydrogen, see *Review of Particle Properties*, Phys. Lett. **204B** (1988).

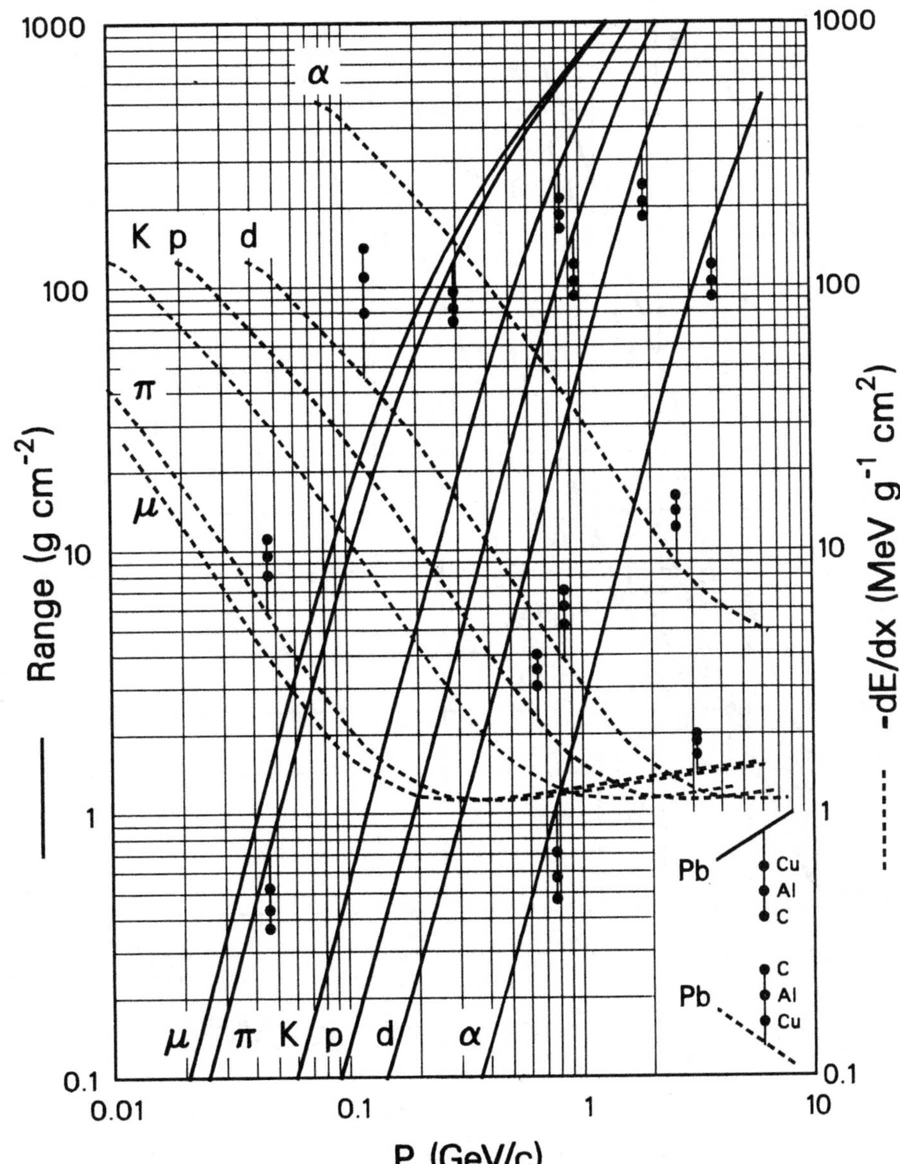

E. Plots: Photon attenuation length

The photon mass attenuation length $\lambda = 1/(\mu/\rho)$ (also known as mfp, mean free path) for various absorbers as a function of photon energy, where μ is the mass attenuation coefficient. For a homogeneous medium of density ρ, the intensity I remaining after traversal of thickness t is given by the expression $I = I_0 \exp(-t\rho/\lambda)$. The accuracy is a few percent. Interpolation to other Z should be done in the cross section $\sigma = A/\lambda N_A$ cm^2/atom, where A is the atomic weight of the absorber material in grams and N_A is the Avogadro number. For a chemical compound or mixture, use $(1/\lambda)_{\text{eff}} \cong \Sigma w_i (1/\lambda)_i$, accurate to a few percent, where w_i is the proportion by weight of the ith constituent. See Plots 9.05.F for high energy range. The processes responsible for attenuation are given in Plots. 9.05.H, Contributions to Photon Cross Section in Lead. Not all of these processes necessarily result in detectable attenuation. For example, coherent Rayleigh scattering off an atom may occur at such low momentum transfer that the change in energy and momentum of the photon may not be significant. From Hubbell, Gimm, and Øverbø, J. Phys. Chem. Ref. Data **9**, 1023 (1980). See also J. H. Hubbell, Int. J. of Applied Rad. and Isotopes **33**, 1269 (1982). Data courtesy J. H. Hubbell.

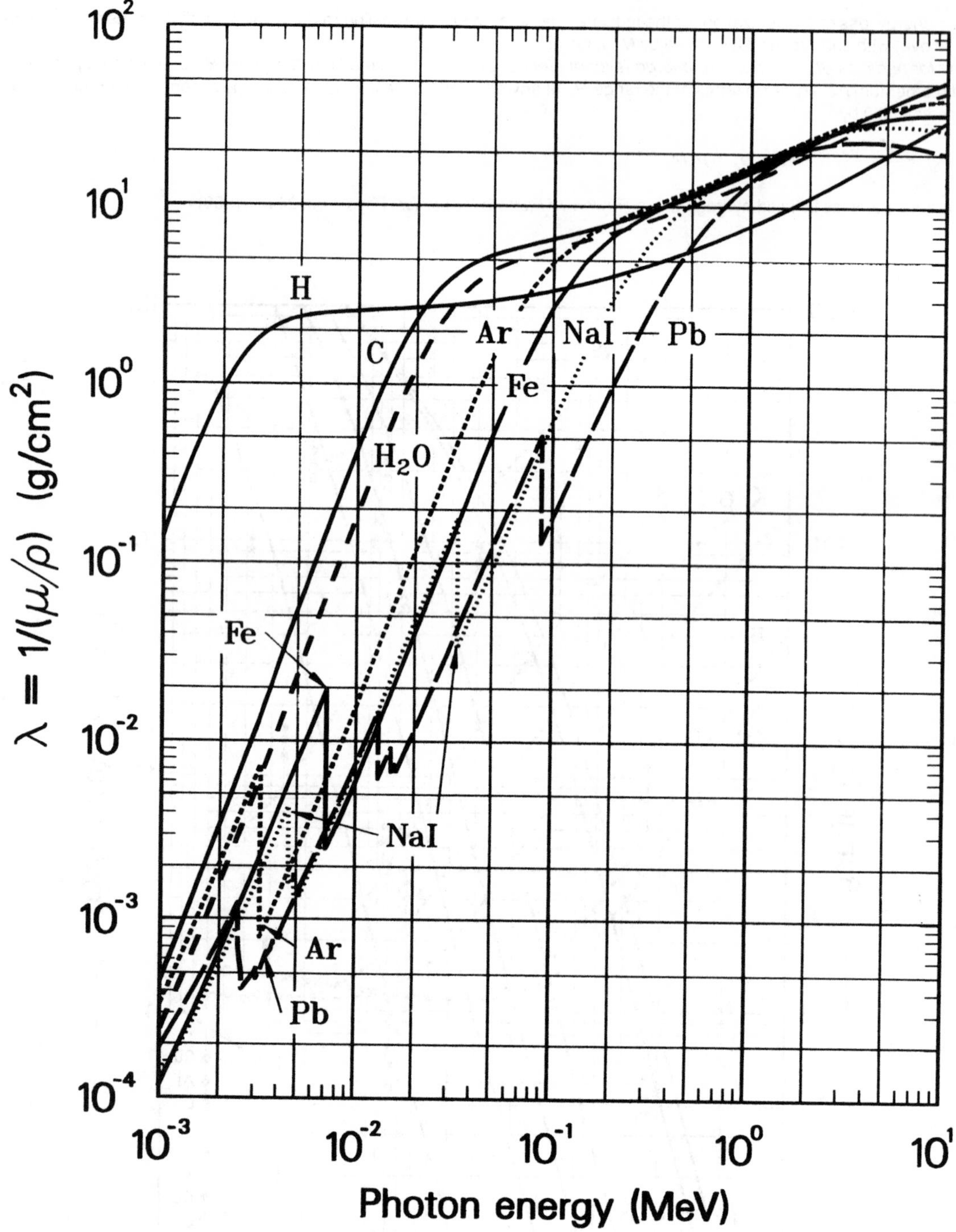

F. Plots: Photon attenuation length (high energy)

The photon mass attenuation length, high energy range (note that ordinate is linear scale). See caption on previous plot (9.05.E) for details. The attenuation length is constant beyond the range shown for at least two decades in energy.

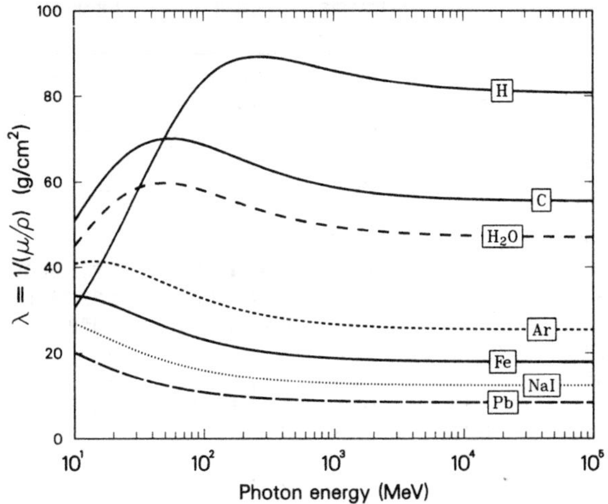

G. Plots: Fractional energy loss for e^+ and e^- in lead

Fractional energy loss per radiation length in lead as a function of electron or positron energy. Electron (positron) scattering is considered as ionization when the energy loss per collision is below 0.255 MeV, and as Moller (Bhabha) scattering when it is above.

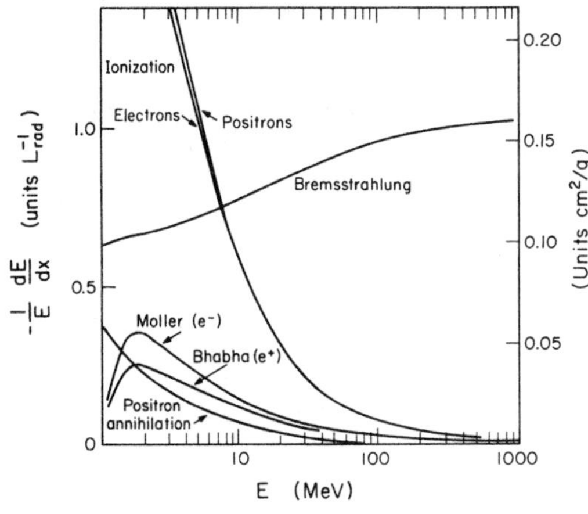

H. Plots: Contributions to photon cross section in lead

Photon total cross sections as a function of energy in lead, showing the contributions of different processes.

τ = Atomic photo-effect (electron ejection, photon absorption)
σ_{COH} = Coherent scattering (Rayleigh scattering—atom neither ionized nor excited)
σ_{INCOH} = Incoherent scattering (Compton scattering off an electron)
κ_n = Pair production, nuclear field
κ_e = Pair production, electron field
$\sigma_{PH.N.}$ = Photonuclear absorption (nuclear absorption, usually followed by emission of a neutron or other particle)

From Hubbell, Gimm, and Øverbø, J. Phys. Chem. Ref. Data **9**, 1023 (1980). The photon total cross section is assumed approximately flat for at least two decades beyond the energy range shown. Figure courtesy J. H. Hubbell.

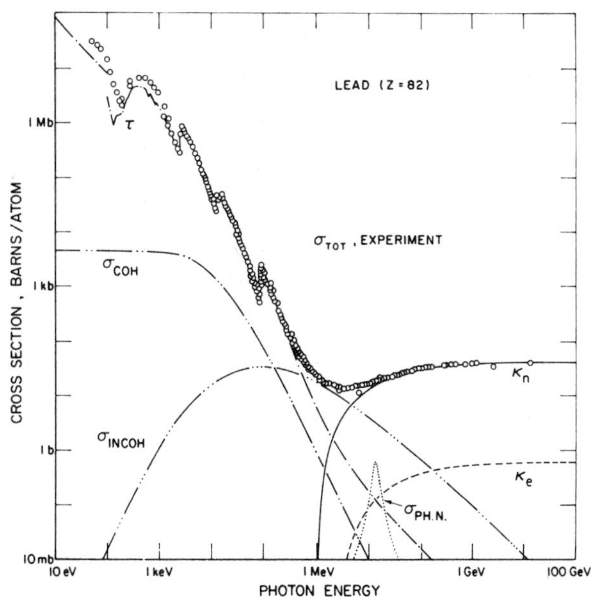

I. Table: Atomic and nuclear properties of materials*

Material	Z	A	Nuclear[a] total cross section σ_T [barn]	Nuclear[b] inelastic cross section σ_I [barn]	Nuclear[c] collision length λ_T [g/cm²]	Nuclear[c] interaction length λ_I [g/cm²]	dE/dx min[d] $\left[\frac{\text{MeV}}{\text{g/cm}^2}\right]$	Radiation length[e] L_{rad} [g/cm²]	Radiation length[e] L_{rad} [cm] ( ) is for gas	Density[f] [g/cm³] ( ) is for gas [g/l]	Refractive index n[f] ( ) is $(n-1)\times10^6$ for gas
H_2	1	1.01	0.0387	0.033	43.3	50.8	4.12	61.28	865	0.0708(0.090)	1.112(140)
D_2	1	2.01	0.073	0.061	45.7	54.7	2.07	122.6	757	0.162(0.177)	1.128
He	2	4.00	0.133	0.102	49.9	65.1	1.94	94.32	755	0.125(0.178)	1.024(35)
Li	3	6.94	0.211	0.157	54.6	73.4	1.58	82.76	155	0.534	—
Be	4	9.01	0.268	0.199	55.8	75.2	1.61	65.19	35.3	1.848	—
C	6	12.01	0.331	0.231	60.2	86.3	1.78	42.70	18.8	2.265[g]	—
N_2	7	14.01	0.379	0.265	61.4	87.8	1.82	37.99	47.0	0.808(1.25)	1.205(300)
O_2	8	16.00	0.420	0.292	63.2	91.0	1.82	34.24	30.0	1.14(1.43)	1.22(266)
Ne	10	20.18	0.507	0.347	66.1	96.6	1.73	28.94	24.0	1.207(0.90)	1.092(67)
Al	13	26.98	0.634	0.421	70.6	106.4	1.62	24.01	8.9	2.70	—
Si	14	28.09	0.660	0.440	70.6	106.0	1.66	21.82	9.36	2.33	—
Ar	18	39.95	0.868	0.566	76.4	117.2	1.51	19.55	14.0	1.40(1.78)	1.233(283)
Ti	22	47.88	0.995	0.637	79.9	124.9	1.51	16.17	3.56	4.54	—
Fe	26	55.85	1.120	0.703	82.8	131.9	1.48	13.84	1.76	7.87	—
Cu	29	63.55	1.232	0.782	85.6	134.9	1.44	12.86	1.43	8.96	—
Ge	32	72.59	1.365	0.858	88.3	140.5	1.40	12.25	2.30	5.323	—
Sn	50	118.69	1.967	1.21	100.2	163	1.26	8.82	1.21	7.31	—
Xe	54	131.29	2.120	1.29	102.8	169	1.24	8.48	2.77	3.057(5.89)	(705)
W	74	183.85	2.767	1.65	110.3	185	1.16	6.76	0.35	19.3	—
Pt	78	195.08	2.861	1.708	113.3	189.7	1.15	6.54	0.305	21.45	—
Pb	82	207.19	2.960	1.77	116.2	194	1.13	6.37	0.56	11.35	—
U	92	238.03	3.378	1.98	117.0	199	1.09	6.00	≈0.32	≈18.95	—
Air, 20°C, 1 atm. (STP in paren.)					62.0	90.0	1.82	36.66	(30420)	0.001205(1.29)	1.000273(293)
H_2O					60.1	84.9	2.03	36.08	36.1	1.00	1.33
Shielding concrete[h]					67.4	99.9	1.70	26.7	10.7	2.5	—
SiO_2 (quartz)					67.0	99.2	1.72	27.05	12.3	2.64	1.458
H_2 (bubble chamber 26 °K)					43.3	50.8	4.12	61.28	≈1000	≈0.063[i]	1.100
D_2 (bubble chamber 31 °K)					45.7	54.7	2.07	122.6	≈900	≈0.140[i]	1.110
H-Ne mixture (50 mole percent)[j]					65.0	94.5	1.84	29.70	73.0	0.407	1.092
Ilford emulsion G5					82.0	134	1.44	11.0	2.89	3.815	—
NaI					94.8	152	1.32	9.49	2.59	3.67	1.775
BaF_2					92.1	146	1.35	9.91	2.05	4.89	1.56
BGO ($Bi_4Ge_3O_{12}$)					97.4	156	1.27	7.98	1.12	7.1	2.15
Polystyrene, scintillator (CH)[k]					58.4	82.0	1.95	43.8	42.4	1.032	1.581
Lucite, Plexiglas ($C_5H_8O_2$)					59.2	83.6	1.95	40.55	≈34.4	1.16–1.20	≈1.49
Polyethylene (CH_2)					56.9	78.8	2.09	44.8	≈47.9	0.92–0.95	—
Mylar ($C_5H_4O_2$)					60.2	85.7	1.86	39.95	28.7	1.39	—
Borosilicate glass (Pyrex)[l]					66.2	97.6	1.72	28.3	12.7	2.23	1.474
CO_2					62.4	90.5	1.82	36.2	(18310)	(1.977)	(410)
Ethane C_2H_6					55.73	75.71	2.25	45.66	(34035)	0.509(1.356)[m]	(1.038)[m]
Methane CH_4					54.7	74.0	2.41	46.5	(64850)	0.423(0.717)	(444)
Isobutane C_4H_{10}					56.3	77.4	2.22	45.2	(16930)	(2.67)	(1270)
NaF					66.78	97.57	1.69	29.87	11.68	2.558	1.336
LiF					62.00	88.24	1.66	39.25	14.91	2.632	1.392
Freon 12 (CCl_2F_2) gas, 26 °C, 1 atm.[n]					70.6	106	1.62	23.7	4810	(4.93)	1.001080
Silica Aerogel[o]					65.5	95.7	1.83	29.85	≈150	0.1–0.3	$1.0+0.25\rho$
NEMA G10 plate[p]					62.6	90.2	1.87	33.0	19.4	1.7	—

I. Table—*(Continued)*

Material	Dielectric constant ( ) is $(\epsilon - 1) \times 10^6$ for gas	Young's modulus [10^6 psi]	Coeff. of thermal expansion [10^{-6} cm/cm-°C]	Specific heat [cal/g-°C]	Electrical resistivity [$\mu\Omega$-cm(@ °C)]	Thermal conductivity [cal/cm-°C-sec]
H_2	(253.9)	—	—	—	—	—
D_2	—	—	—	—	—	—
He	(64)	—	—	—	—	—
Li	—	—	56	0.86	8.55(0°)	0.17
Be	—	37	12.4	0.436	5.885(0°)	0.38
C	—	0.7	0.6-4.3	0.165	1375(0°)	0.057
N_2	(548.5)	—	—	—	—	—
O_2	(495)	—	—	—	—	—
Ne	—	—	—	—	—	—
Al	—	10	23.9	0.215	2.65(20°)	0.53
Si	—	16	2.8-7.3	0.162	—	0.20
Ar	(517)	—	—	—	—	—
Ti	—	16.8	8.5	0.126	50(0°)	—
Fe	—	28.5	11.7	0.11	9.71(20°)	0.18
Cu	—	16	16.5	0.092	1.67(20°)	0.94
Ge	—	—	5.75	0.073	—	0.14
Sn	—	6	20	0.052	11.5(20°)	0.16
Xe	—	—	—	—	—	—
W	—	50	4.4	0.032	5.5(20°)	0.48
Pt	—	21	8.9	0.032	9.83(0°)	0.17
Pb	—	2.6	29.3	0.038	20.65(20°)	0.083
U	—	—	36.1	0.028	29(20°)	0.064

* Table revised April 1988 by R. W. Kenney, σ_T, σ_I, λ_T, and λ_I are energy dependent. Values quoted apply to high energy range given in footnote a or b, where energy dependence is weak.

[a] σ_{total} at 80-240 GeV for neutrons ($\approx\sigma$ for protons) from Murthy *et al.*, Nucl. Phys. **B92**, 269 (1975). This scales approximately as $A^{0.77}$.

[b] $\sigma_{inelastic} = \sigma_{total} - \sigma_{elastic} - \sigma_{quasielastic}$; for neutrons at 60-375 GeV from Roberts *et al.*, Nucl. Phys., **B159**, 56(1979). For protons and other particles, see Carroll *et al.*, Phys. Lett. **80B**, 319 (1979); note that $\sigma_I(p) \approx \sigma_I(n)$. σ_I scales approximately as $A^{0.71}$.

[c] Mean free path between collisions (λ_T) or inelastic interactions (λ_I), calculated from $\lambda = A/(N \times \sigma)$, where N is the Avogadro number.

[d] For minimum-ionizing protons and pions. dE/dx is energy loss per g/cm^2 from Barkas and Berger, *Tables of Energy Losses and Ranges of Heavy Charged Particles*, NASA-SP-3013 (1964). For electrons and positrons see: M. J. Berger and S. M. Seltzer, *Stopping Powers and Ranges of Electrons and Positrons* (2nd Ed.), U. S. National Bureau of Standards report NBSIR 82-2550-A (1982).

[e] From Y. S. Tsai, Rev. Mod. Phys. **46**, 815 (1974); L_{rad} data for all elements up to uranium may be found here. Corrections for molecular binding applied for H_2 and D_2. Parentheses refer to gaseous form at STP (0°C, 1 atm.).

[f] Values for solids, or the liquid phase at boiling point, except as noted. Values in parentheses for gaseous phase at STP (0°C, 1 atm.). Refractive index given for sodium D line.

[g] For pure graphite; industrial graphite density may vary 2.1-2.3 g/cm^3.

[h] Standard shielding blocks, typical composition O_2 52%, Si 32.5%, Ca 6%, Na 1.5%, Fe 2%, Al 4%, plus reinforcing iron bars. The attenuation length, $l = 115 \pm 5$ g/cm^2, is also valid for earth (typical $\rho = 2.15$), from CERN-LRL-RHEL Shielding exp., UCRL-17841 (1968).

[i] Density may vary about $\pm 3\%$, depending on operating conditions.

[j] Values for typical working conditions with H_2 target: 50 mole percent, 29 °K, 7 atm.

[k] Typical scintillator; e.g., PILOT B and NE 102A have an atomic ratio H/C = 1.10.

[l] Main components: 80% SiO_2 + 12% B_2O_3 + 5% Na_2O.

[m] Solid ethane density at − 60 °C; gaseous refractive index at 0 °C, 546 mm pressure.

[n] Used in Cerenkov counters. Values at 26°C and 1 atm. Indices of refraction from E. R. Hayes, R. A. Schluter and A. Tamosaitis, ANL-6916 (1964).

[o] $n(SiO_2) + 2n(H_2O)$ used in Cerenkov counters, ρ = density in g/cm^3. From M. Cantin *et al.*, Nucl. Instr. Meth. **118**, 177 (1974).

[p] G10-plate, typical 60% SiO_2 and 40% epoxy.

J. Plot: Cross section ratio R in e^+e^- collisions

Selected measurements of $R \equiv \sigma(e^+e^- \to \text{hadrons})/\sigma(e^+e^- \to \mu^+\mu^-)$, where the annihilation in the numerator proceeds via one photon or via the Z^0. The denominator is the calculated QED single-photon process. Two curves are overlaid for $E_{cm} > 11$ GeV showing the theoretical prediction for R, including higher order QCD [M. Dine and J. Sapirstein, Phys. Rev. Lett. **43**, 668 (1979)] and electroweak corrections. The Λ values are for 5 flavors in the $\overline{\text{MS}}$ scheme and are $\Lambda^{(5)}_{\overline{\text{MS}}} = 60$ MeV (lower curve) and $\Lambda^{(5)}_{\overline{\text{MS}}} = 250$ MeV (upper curve). From the collection in *Review of Particle Properties*, Phys Lett. **204B** (1988).

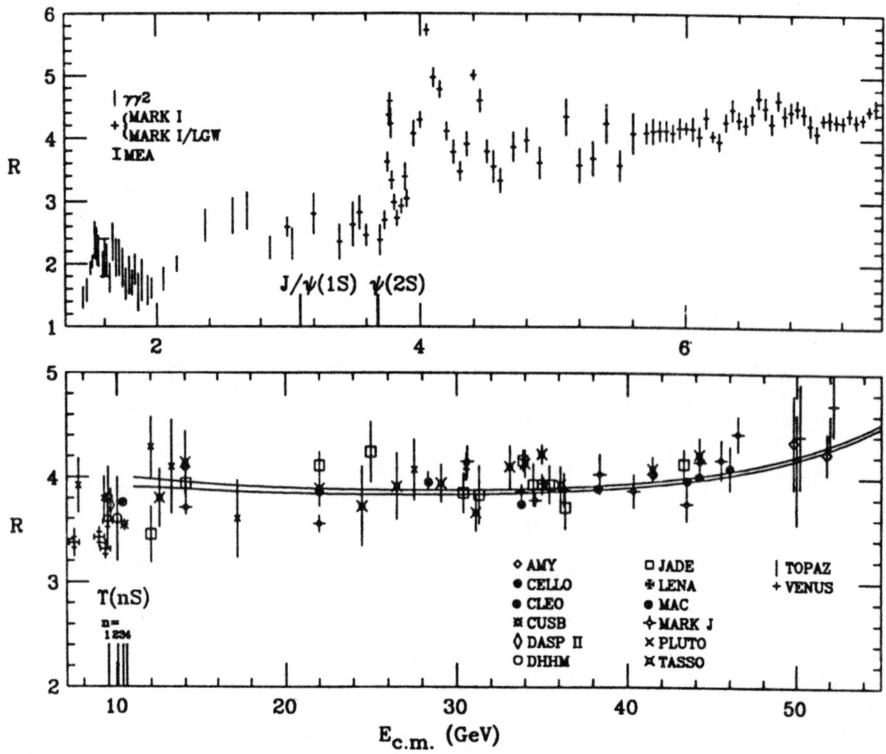

K. Plot: σ_T/E_ν for muon neutrinos and antineutrinos

σ_T/E_ν for the muon neutrino and antineutrino charged-current total cross section as a function of neutrino energy. The error bars include both statistical and systematic errors. The straight lines are averages for the CCFRR measurement. Note the change in the energy scale between 30 and 50 GeV. The data points on the right give averages for other high energy measurements. Courtesy M. H. Shaevitz, Columbia University (Nevis Laboratory). From the collection in *Review of Particle Properties*, Phys. Lett. **204B** (1988).

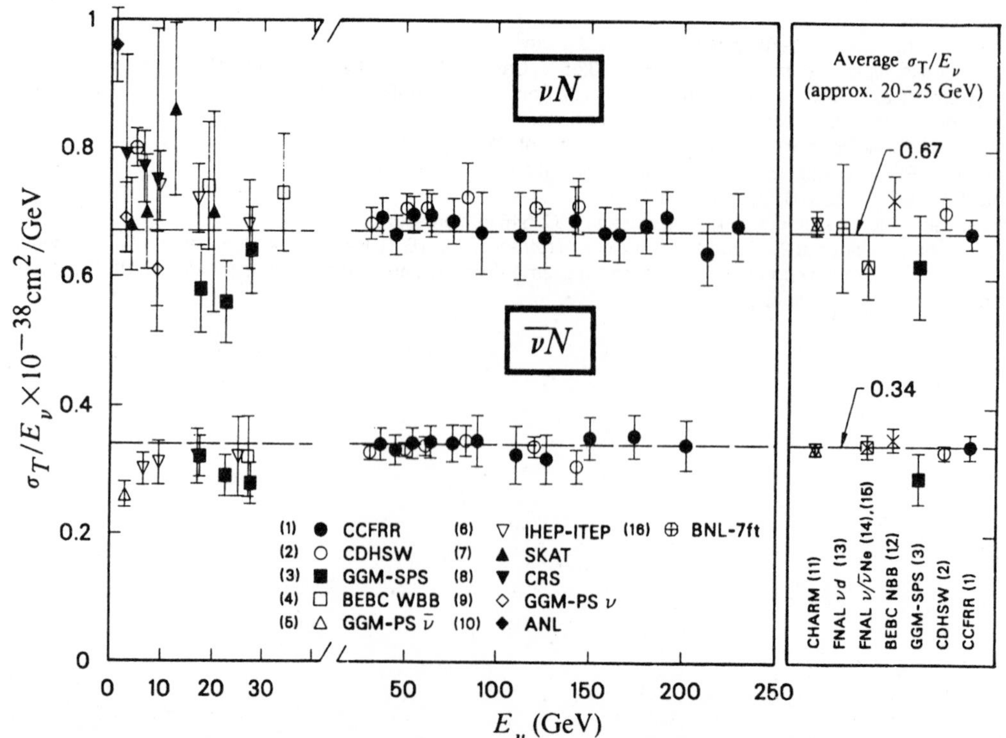

L. Table: Hadronic cross sections.

Approximate values for photonic and hadronic cross sections, in millibarns, at various momenta. Derived from the fits on p. 121 of the Review of Particle Properties, Phys. Lett. **204B** (1988). See *Total Cross-Sections for Reactions of High Energy Particles*, Landolt-Bornstein, New Series, Vols. 12a and 12b, H. Schopper, Ed. (1987).

Reaction	Momentum [GeV/c] 2	5	10	100	200
γp (total)	0.13	0.13	0.12	0.12	0.12
γd (total)	0.27	0.24	0.23	—	—
Λp (total)	—	—	52	—	—
Λp (elastic)	16	9	6	4	4
$\pi^+ p$ (total)	—	27	25	23	24
$\pi^+ p$ (elastic)	10	6	5	3	3
$\pi^+ d$ (total)	—	55	49	46	46
$\pi^- p$ (total)	36	29	27	24	24
$\pi^- p$ (elastic)	9	6	5	3	3
$\pi^- d$ (total)	67	54	49	46	47
$K^+ p$ (total)	18	17	17	19	20
$K^+ p$ (elastic)	7	4	3	2	3
$K^+ d$ (total)	36	34	34	37	39
$K^+ n$ (total)	19	18	18	19	20
$K^- p$ (total)	30	25	23	21	21
$K^- p$ (elastic)	8	4	3	2	3
$K^- d$ (total)	55	45	41	39	40
$K^- n$ (total)	17	15	13	11	12
pp (total)	55	41	40	39	39
pp (elastic)	22	13	10	7	7
pd (total)	86	80	76	73	74
pd (elastic)	13	11	10	7	6
np (total)	41	42	41	39	39
$\bar{p}p$ (total)	88	65	55	42	42
$\bar{p}p$ (elastic)	33	17	12	7	7
$\bar{p}d$ (total)	163	117	99	79	78
$\bar{p}n$ (total)	90	61	51	40	38
$\bar{p}n$ (elastic)	31	5	—	—	—

9.06. STATUS OF ELEMENTARY PARTICLES*

Elementary particle interactions have traditionally been classified in terms of four fundamental forces: strong, electromagnetic, weak, and gravitational. As gravitational effects are too weak to be observable in the laboratory, the field of high-energy physics concerns only the first three. At typical accelerator energies, these forces are characterized by marked differences in strength, range, and symmetry properties. Nevertheless, striking progress has been made in synthesizing these diverse phenomena in a way which suggests that they may simply be different manifestations of a single unified interaction.

According to a quantum-field-theoretic description, forces between "matter" fields arise from exchange of quanta of "radiation" fields. The first common feature of the strong, electromagnetic, and weak interactions is that they are characterized by spin-$\frac{1}{2}$ fermions as "matter" and by spin-1 vector bosons as "radiation." Since their structure is so similar, we may ask why the forces manifest themselves so differently in the laboratory.

In the static limit, a force between matter fields can be interpreted as the gradient of a potential of the form (we take throughout $\hbar = c = 1$)

$$|V(r)| = g^2 e^{-\mu r}/r,$$

where g is the effective charge, or "coupling constant," and μ is the mass of the exchanged quantum. In quantum electrodynamics (QED), the photon is apparently massless, resulting in a potential of infinite range. In contrast, the quanta (W,Z) exchanged in weak interactions have masses of nearly 100 GeV, which accounts for their limited range. In "high-energy" physics, one deals with scattering processes. The fermion–fermion scattering amplitude is given by the Fourier transform of the potential:

$$\mathscr{A}\,[f_1(p_1) + f_2(p_2) \to f_1(p_1') + f_2(p_2')] = \frac{g^2}{Q^2 + \mu^2},$$

where $-Q^2 = (p_1 - p_1')^2 = (p_2 - p_2')^2$ is the squared four-momentum transferred between fermions via the exchange of a quantum of mass μ. At low energies the momentum transfer is much smaller than the masses of the weak radiation quanta, so that weak interactions are suppressed relative to electromagnetic effects by a factor Q^2/μ^2. However, for momenta $Q^2 \gtrsim (100\text{ GeV})^2$, the W and Z masses become unimportant, and amplitudes for weak and electromagnetic interactions become comparable.

In quantum chromodynamics (QCD), the quantum field theory of strong interactions, forces are believed to be mediated by massless gluons. The finite range observed for the nuclear forces is attributed to the phenomenon of confinement, which is accepted as an experimental fact although it has not yet been demonstrated to be a property of the theory, in spite of encouraging indications. According to the doctrine of confinement, hadrons, the observed strongly interacting particles, are bound states of quarks and gluons. For example, a proton is a system of three valence quarks, immersed—owing to quantum fluctuations—in a sea of gluons and quark-antiquark pairs which are pemanently confined within some radius r_{conf}. For low-energy collisions, involving interactions over a distance $r \sim 1/E > r_{\text{conf}}$, a single gluon cannot escape its confining walls to generate a force between quarks in different protons. At best, an interaction can occur through the exchange of a pion, the lightest bound state, resulting in an effective range $r \sim 1/m_\pi \simeq 10^{-13}$ cm. At high energies, $E \gg 1/r_{\text{conf}}$, colliding protons can be sufficiently close that their spatial wave functions overlap, allowing the exchange of a single gluon between quarks. For a sufficiently "hard" process, with momentum transfer $Q \gg 1/r_{\text{conf}}$, the interacting quarks are ejected at a large angle with respect to the initial beam direction. As confinement forbids them to emerge as isolated particles, they radiate gluons and quark pairs, and drag along with them bits of the initial proton cloud. These quanta recombine to form two jets of hadrons flowing in the direction of the ejected quarks, while the excited spectator systems result in hadronic jets along the directions of the initial beams. Thus the kinematics of the elementary interacting quanta of QCD are reflected in the observed kinematics of hadronic jets.

* Prepared by Mary K. Gaillard.

This phenomenon has been dramatically illustrated by high-energy electron-positron annihilation into hadronic final states, which parallels the purely electromagnetic process $e^+e^- \to \mu^+\mu^-$. The dominant processes are those which occur at lowest order in the coupling constants via the exchange of a single photon:

$$e^+e^- \to \text{virtual photon} \to \mu^+ + \mu^- \text{ or } q + \bar{q} .$$

In the collision center of mass frame, the final states appear, respectively, as a back-to-back pair of muons and a back-to-back pair of hadronic jets. At the next to lowest order, one of the elementary fermions can radiate a hard photon, yielding an acollinear muon or jet pair, or, in the case of $q\bar{q}$ production, one of the quarks can radiate a hard gluon, yielding a three-jet final state. Events of the latter type, which apparently occur at the expected frequency relative to two-jet events, have been reported by the experimental groups at PETRA and PEP. Equally spectacular highly collimated jet events, reflecting hard scattering of quarks and gluons, have more recently been observed in $p\bar{p}$ collider experiments at CERN and Fermilab.

We are thus led to a picture where, viewed at sufficiently high energies, strong, weak, and electromagnetic interactions have a very similar structure, but are governed by different coupling constants: appropriately normalized, the strong interaction coupling is indeed the strongest, while the "weak" coupling constant is comparable to, but slightly stronger than, the electromagnetic charge. In fact, one never measures a coupling "constant," but rather the amplitude for some scattering process which, for energies $E \gg m_{W,Z}$ or $E \gg r_{\text{conf}}^{-1}$, can be parametrized by

$$\mathscr{A}(f_1f_2 \to f_1f_2) \equiv g_{\text{eff}}^2(Q^2)/Q^2 ,$$

where the Q dependence of the effective coupling constant reflects the effect of quantum fluctuations: the exchanged radiation quantum is accompanied by a cloud of virtual radiation and fermion-antifermion pairs which can screen or enhance the effective charge. This phenomenon, referred to as vacuum polarization, is well known in QED. A static charge will polarize the virtual electron-positron pairs which populate the vacuum state in such a way as to screen the net charge as measured at some distance r. With decreasing distance from the static charge, one probes less of the polarized vacuum, so the effective charge measured increases. When the momentum transferred in a scattering process is increased, shorter distances are probed, and the effective electric charge is an increasing function of Q^2. QED differs from the quantum field theories of strong and weak interactions in that the radiation fields of the latter couple directly to one another as well as to fermions. This provides an additional source of vacuum polarization which turns out to be dominant and to have the opposite effect: the effective charge diminishes with increasing momentum transfer. The effect is stronger for QCD, with its eight self-coupled gluons, than for the weak interactions with only three self-coupled vector bosons. As a result, the discrepancy in strength of the three basic interactions decreases with increasing momentum transfer.

This observation suggests naturally a hypothesis of unification: there is a single coupling governing strong, weak, and electromagnetic interactions. In this context the differences among observed phenomena are attributed to the effects of symmetry breaking. An illustration is provided by the partially unified theory of electromagnetic and weak interactions: the low-energy quantum-field-theory QED is embedded in the Glashow-Weinberg-Salam (GWS) electroweak theory which is manifest at energies $E \gg 100$ GeV. The GWS theory involves three self-coupled vectors ($W^\pm, W^0$) and an additional neutral vector (V^0) coupled only to fermions. Through symmetry breaking effects the charged W's acquire masses as does the Z^0, a linear combination of W^0 and V^0, while the orthogonal combination, the photon of QED, remains massless. The effects of masses are unimportant at high energy, but for $E \lesssim m_{W,Z}$ the effects of heavy boson exchange become suppressed.

The more general unification idea implies the existence of yet heavier vector bosons, X, whose contributions to the Q^2 dependence of the suitably normalized strong, electromagnetic, and weak coupling are arranged (via an underlying symmetry) in such a way that their evolution curves are identical for $Q \gg m_x$. However, at lower energies, $Q < m_x$, the X-pair contribution to the vacuum polarization becomes damped, and the different couplings evolve separately, accounting for the observed different values. This higher symmetry may entail couplings among quarks and leptons that can induce proton decay into leptons and pions or other mesons. Together with CP (charge conjugation times parity) violation, proton instability could account for the existence of matter in today's universe. Evidence for proton decay has been sought unsuccessfully in deep mine experiments.

The electroweak symmetry breaking is attributed to the existence of a field—known as the Higgs field—which distinguishes particles of different electric charge. Just as a photon passing through matter moves at a velocity less than the speed of light due to its interactions with atomic electrons, the W and Z interact with the Higgs field that permeates all space, and they cannot propagate with the speed of light; they acquire an "index of refraction," or equivalently, a rest mass. The photon does not interact with the Higgs field and hence remains massless.

The mass m_W acquired by the heavy gauge bosons is proportional to the strength g_w of their coupling to the Higgs field and to the strength v of that field. The field strength $v = m_w/g_w$ is known from the measurement of the free neutron lifetime. The coupling g_w has been independently determined by a series of measurements of weak transitions, including neutrino-induced interactions, and it was thus possible to predict accurately the masses of the W and Z particles before they were discovered at CERN in 1983. Electrically charged quarks and leptons also couple to the Higgs field, and acquire masses in a similar way. However, their coupling strengths cannot be independently determined.

While the standard model successfully accounts for a very large body of experimental data, the origin of electroweak symmetry breaking and the nature of the Higgs field are not understood. There is no understanding of the pattern of fermion masses, nor of flavor mixing parameters that determine how fast heavy quarks and leptons decay to lighter ones. All of these parameters are determined by the couplings of fermions to the Higgs field. In particular, these seem to be the only couplings in nature that are not invariant under *CP*.

The simplest possibility for understanding electroweak symmetry breaking is the existence of an elementary field that has a potential energy with a local maximum for vanishing field strength, and a global minimum at a constant value v of field strength. If such a field exists, it has spinless quantum excitations, that is, particles, called "Higgs bosons," expected to be less massive than a Tev. If they are more massive than a TeV, W and Z collisions will grow rapidly in rate when their total collisions energy exceeds a TeV. The existence of an elementary Higgs particle of a TeV or less in mass is problematic because the mass of a spinless particle gets large contributions from quantum fluctuations. An alternative possibility is that the Higgs field is a composite field induced by fermion-antifermion pairs. These new "technifermions," like quarks, would exist only in bound states that have masses of about a TeV. Alternatively, quantum corrections to the Higgs mass would be damped if the theory possesses a larger symmetry, called supersymmetry, which relates fermions (particles with half integral spin) to bosons (particles with integral spin). According to this conjecture, every known particle has a "superpartner" with identical properties (mass, electric charge,...) but differing in spin by one half unit. This symmetry is also broken in nature by mass splittings, but a Higgs mass of a TeV or less could be understood if superpartners also have masses on that order. Experiments at the proposed Superconducting Supercollider will probe the TeV energy region and therefore shed light on the mechanism for electroweak symmetry breaking.

Yet another possibility is that quarks and leptons are themselves composite and that some substructure will be revealed in higher energy experiments. A more fashionable idea at present is that elementary "particles" are not particles at all, but rather the lowest vibrational modes of tiny strings with an extension of the order of the Planck length, about 10^{-33} cm. When supersymmetry is included, this "superstring" theory provides the only known possibility for a consistent quantum theory of gravity. It suggests that space-time is actually ten dimensional, but with six dimensions curled up with a radius comparable to the Planck length. In the context of superstring theory, many new exotic particles (in addition to the superpartners of ordinary particles) are predicted.

9.07. REFERENCES

[1]*Methods of Experimental Physics*, edited by L. C. L. Yuan and C. -S. Wu (Academic Press, 1961), Vol. 5A, p. 127.

[2]R. K. Swank, Ann. Rev. Nucl. Sci. **4**, 137 (1954); and G. T. Wright, Proc. Phys. Soc. **B68**, 929 (1955).

[3]M. Laval *et al.*, Nucl. Instr. and Meth. **206**, 169 (1983).

[4]R. Allemand *et al.*, Communication LETI/MCTE/82-245, Grenoble, France (1982).

[5]M. Moszynski *et al.*, Nucl. Instr. and Meth. **A226**, 534 (1984).

[6]See also the Table of Atomic and Nuclear Properties (9.05.I).

[7]*Methods of Experimental Physics*, edited by L. C. L. Yuan and C. -S. Wu (Academic Press, 1961), Vol. 5A, p. 163.

[8]E. R. Hayes, R. A. Schluter, and A. Tamosaitis, "Index and Dispersion of Some Cerenkov Counter Gases," ANL-6916 (1964).

[9]Nuclear Enterprises Catalogue.

[10]D. Hitlin *et al.*, Nucl. Instr. and Meth. **137**, 225 (1976). See also W. J. Willis and V. Radeka, Nucl. Instr. and Meth. **120**, 221 (1974), for a more detailed discussion.

[11]E. B. Hughes *et al.*, IEEE Transactions on Nuclear Science **NS-19**, No. 3, 126 (1972).

[12]M. Holder *et al.*, Phys. Letters **40B**, 141 (1972); and J. S. Beale *et al.*, "A Lead-Glass Cerenkov Detector for Electrons and Photons," CERN Writeup, Intl. Conf. on Instrumentation in H.E.P., Frascati (1973).

[13]W. Hofmann *et al.*, DESY 81/045 (July 1981). See also S. L. Stone *et al.*, Nucl. Instr. and Meth. **151**, 387 (1978).

[14]R. L. Anderson *et al.*, "Tests of Proportional Wire Shower Counter and Hadron Calorimeter Modules," SLAC-PUB-2039 (1977).

[15]R. Wigmans, Nucl. Instr. and Meth. **A258** (1987).

[16]H. Abramowicz *et al.*, Nucl. Instr. and Meth. **A180**, 429 (1981); M. de Vincenzi *et al.*, Nucl. Instr. and Meth. **A243**, 348 (1986); and M. G. Catanesi *et al.*, Nucl. Instr. and Meth. **A260**, 43 (1987).

[17]T. Akesson *et al.*, Nucl. Instr. and Meth. **A262**, 243 (1987).

[18]E. Bernardi *et al.*, Nucl. Instr. and Meth. **A262**, 229 (1987).

[19]W. W. M. Allison and J. H. Cobb, "Relativistic Charged Particle Identification by Energy-Loss," Ann. Rev. Nucl. Part. Sci. **30**, 253 (1980), see p. 287.

[20]E. Shibamura *et al.*, Nucl. Instr. and Meth. **131**, 249 (1975).

[21]T. G. Ryan and G. R. Freeman, J. Chem. Phys. **68**, 5144 (1978).

[22]W. F. Schmidt, "Electron Migration in Liquids and Gases," HMI B156 (1974).

[23]A. O. Allen, "Drift Mobilities and Conduction Band Energies of Excess Electrons in Dielectric Liquids," NSRDS-NBS-58 (1976).

[24]R. L. Gluckstern, Nucl. Instr. and Meth. **24**, 381 (1963).

[25]T. Trippe, CERN NP Internal Report 69-18 (1969).

[26]S. Parker and R. Jones, LBL-797 (1972); and P. Morse and H. Feshbach, *Methods of Theoretical Physics* (McGraw-Hill, New York, 1953), p. 1236.

[27]B. Rossi, Rev. Mod. Phys. **20**, 537 (1948). See also C. Grupen, "News from Cosmic Rays at High Energies," Siegen University preprint SI-84-01, and Allkofer and Grieder, *Cosmic Rays on Earth*, Fachinformationszentrum, Karlsruhe (1984); flux ratio for protons at sea level from G. Brook and A. W. Wolfendale, Proc. of the Phys. Soc. of London, Vol. 83 (1964), p. 843.

[28]J. G Learned, F. Reines, and A. Soni, Phys. Rev. Lett. **43**, 907 (1979).

[29]M. F. Crouch *et al.*, Phys. Rev. **D18**, 2239 (1978).

[30]U. Fano, Ann. Rev. Nucl. Sci. **13**, 1 (1963).

[31]M. J. Berger and S. M. Seltzer, "Mean Excitation Energies for Use in Bethe's Stopping-Power Formula," pp. 57–74, Proceedings of Hawaii Conference on Charge States and Dynamic Screening of Swift Ions (1982).

[32]H. A. Bethe and J. Ashkin, *Experimental Nuclear Physics*, Vol. 1, edited by E. Segré (Wiley, New York, 1959).

[33]A. Crispin and G. N. Fowler, Rev. Mod. Phys. **42**, 290 (1970); R. M. Sternheimer, S. M. Seltzer, and M. J. Berger, "The Density Effect for the Ionization Loss of Charged Particles in Various Substances," Atomic Data & Nucl. Data Tables **30**, 261 (1984).

[34]For Z^3 calculations with $Z=1$, see J. D. Jackson and R. L. McCarthy, Phys. Rev. **B6**, 4131 (1972).

[35]For an approximate treatment of high-Z projectiles, see P. B. Eby and S. H. Morgan, Phys. Rev. **A5**, 2536 (1972).
[36]See, for instance, G. Sidenius, Det. Kong. Danske Viden. Selskab Mat.-Fysk. Med. **39**, No. 4 (1974).
[37]See, for instance, S. Datz, "Atomic Collisions in Solids" in "Structure and Collisions of Ions and Atoms" (Springer Verlag, Berlin, 1978), p. 309.
[38]See, for instance, K. A. Ispirian, A. T. Margarian, and A. M. Zverev, Nucl. Instr. and Meth. **117**, 125 (1974).
[39]L. V. Spencer "Energy Dissipation by Fast Electrons," Natl. Bureau of Standards Monograph No. 1 (1959).
[40]"Average Energy Required to Produce an Ion Pair," ICRU Report No. 31 (1979).
[41]N. Hadley *et al.*, "List of Poisoning Times for Materials," Lawrence Berkeley Lab Report TPC-LBL-79-8 (1981).
[42]B. Rossi, *High Energy Particles* (Prentice-Hall, Englewood Cliffs, NJ, 1952).
[43]For unit-charge projectiles, see E. A. Uehling, Ann. Rev. Nucl. Sci. **4**, 315 (1954). For highly charged projectiles, see J. A. Doggett and L. V. Spencer, Phys. Rev. **103**, 1597 (1956). A Lorentz transformation is needed to convert these center-of-mass data to knock-on energy spectra.
[44]N. F. Mott and H. S. W. Massey, *The Theory of Atomic Collisions* (Oxford Press, London, 1965).
[45]For a thorough discussion of simple formulae for single scatters and methods of compounding these into multiple-scattering formulae, see W. T. Scott, Rev. Mod. Phys. **35**, 231 (1963). For detailed summaries of formulae for computing single scatters, see J. W. Motz, H. Olsen, and H. W. Koch, Rev. Mod. Phys. **36**, 881 (1964).
[46]E. V Hungerford and B. W. Mayes, Atomic Data and Nuclear Data Tables **15**, 477 (1975).
[47]V. L. Highland, Nucl. Instr. and Meth. **129**, 497 (1975) and important modification Nucl. Instr. and Meth. **161**, 171 (1979).
[48]W. Lohmann, R. Kopp, and R. Voss, "Energy Loss of Muons in the Energy Range 1-10000 GeV," CERN Report 85-03 (1985).
[49]A. A. Petrukhin and V. V. Shestakov, Can. J. Phys. **46**, S377 (1968).
[50]A. G. Wright, J. Phys. A: Math. and Gen. **6**, 79 (1973).
[51]S. R. Kel'ner and Yu. D. Kotov, Sov. J. Nucl. Phys. **7**, 237 (1968).
[52]L. B. Bezrukov and E. V. Bugaev, Sov. J. Nucl. Phys. **33**, 635 (1981).
[53]A. Van Ginneken, Nucl. Instr. and Meth. **A251**, 21 (1986).
[54]U. Becker *et al.*, Nucl. Instr. and Meth. **A253**, 15 (1986).
[55]J. J. Eastman and S. C. Loken, "Muon Energy Loss at High Energy and Implications for Detector Design," LBL-24039 (1987).
[56]R. Ford and W. Nelson, SLAC-210 (1978).
[57]A similar form has been used by E. Longo and I. Sestili, Nucl. Instr. and Meth. **128**, 283 (1975), and J. Sass and M. Spiro, CERN $\bar{p}p$ Tech. Note 78-32 (1978).
[58]Y. S. Tsai, Rev. Mod. Phys. **46**, 815 (1974).
[59]K. -H. Weber, Nucl. Instr. and Meth. **25**, 261 (1964).
[60]M. J. Berger and S. M. Seltzer, NASA SP-3012 (1964) and SP-3036 (1966).
[61]P. Trower, UCRL-2426, Vol. III, Rev. (1966).
[62]S. M. Seltzer, "Transmission of Electrons through Foils," NBSIR **74**, 457 (1974).
[63]M. J. Berger and S. M. Seltzer, "Stopping Powers and Ranges of Electrons and Positrons" (2nd Ed.), U. S. National Bureau of Standards Report NBSIR 82-2550-A (1982).

10.00 Energy demand

ARTHUR H. ROSENFELD, ALAN K. MEIER, AND ROBERT J. MOWRIS

Lawrence Berkeley Laboratory

CONTENTS

10.01 INTRODUCTION

The following tables and figures present an overview of energy demand in the United States. We have tried to present data of interest to physicists rather than an exhaustive collection. The sources cited are generally good starting places for anyone interested in obtaining more detailed information. While we have made every attempt to present the most current data available, improved estimates and changing economic conditions will make some of the entries quickly obsolete.

For a good overview of energy conservation, see C. Flavin, A. Durning, *Building on Success: The Age of Energy Efficiency,* Worldwatch Paper 82, March 1988. For a physics perspective, see D. Hafemeister, *et al.*, *Energy Sources: Conservation and Renewables*, AIP Conference Proceedings 135, 1985. For an excellent overview of the energy conservation potential in the automobile sector, see D. L. Blevis, *The New Oil Crisis and Fuel Technologies: Preparing the Light Transporation Industry for the 1990s,* Quorum Books, 1988. Current energy consumption data for the United States are published by the Energy Information Agency in the *Monthly Energy Review* and the annual *State Energy Data Report.* The United Nations, OECD, and the World Bank all publish international energy statistics.

Energy data are typically approached from either a supply or demand perspective. This distinction becomes blurred as one examines the energy systems in greater detail, especially in the case of electricity, industrial energy use, and renewable energy sources. We begin with general supply tables, followed by more specific descriptions of the end uses of energy.

The precision of the data varies enormously and uncertainties are rarely listed. Energy that is taxed or regulated (such as oil or electricity) is very accurately known; those forms of energy or data not directly appearing in the marketplace are not. Renewable energy consumption and the precise end-use distribution of energy are not reliably known.

10.02 TRENDS IN ENERGY USE AND GNP

A. Figure: U.S. primary energy use: Actual vs. Predicted by GNP (1949–1987)

The figure shows energy consumption by source for the U.S. from 1949–1987. The upper dashed line is GNP in 1982 dollars scaled to go through 78.4 EJ in 1973, and illustrates how GNP and energy use were in lock-step before 1973. Energy conservation and improved efficiency have made a significant contribution to the U.S. economy since 1973 and currently provide annual savings of more than 28.4 EJ worth $150 billion. Adding the contribution from conservation to the other non-fossil sources gives 37.1 EJ—equal to one-third of the energy we would have used if we were still operating at 1973 efficiency levels. Sources: EIA.

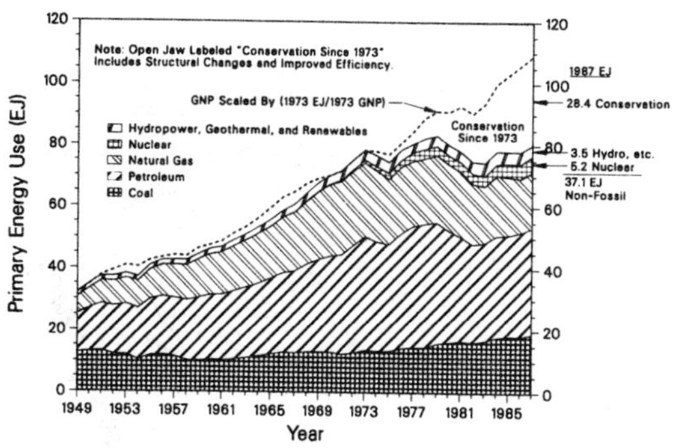

B. Figure: Energy use and GDP (1970–1985)

Each country is represented by a connected sequence of points, beginning in 1970 and ending in 1985. The conversion from local GDP to dollars is based on July 1, 1987 exchange rates; earlier points are plotted using individual national deflators. The lines labeled "U.S. energy use at 5%, 10%, 15%, and 20% of GDP" are based on an average price of primary energy in U.S. 1987 dollars, or $6.14/GJ. For oil, we use 1 ton-equivalent = 42.6 GJ; hydro and nuclear electricity are converted to primary energy using the EIA average generation efficiency of 38.5%, except for Japan where the typical value is 35.1%. Sources: DOE/EIA, OECD International Energy Agency, UN Demographic Yearbook.

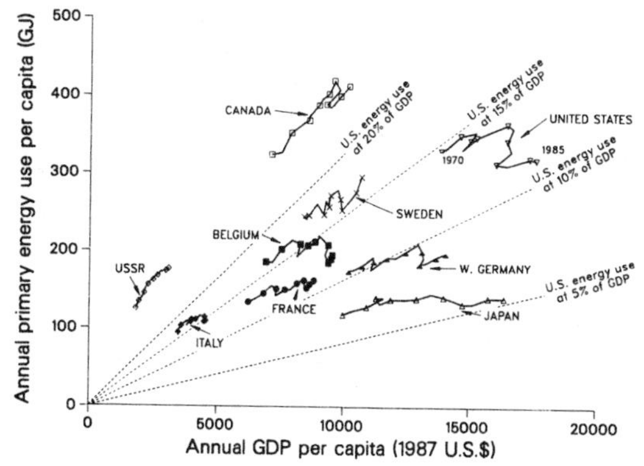

10.03 FIGURE: ENERGY FLOW IN THE U.S. ECONOMY

Although the figure gives a useful "snapshot" of current energy flows, we must emphasize that the distinctions between useful heat or work and waste are based on very rough estimates of average first law efficiencies in each sector which are continuously changing depending upon political, economic, environmental, and technical factors. Thus, the figure shows that 27 EJ of energy services are now coming from 80 EJ. If the present rate of improvement in energy efficiency continues over the next 50 years, the U.S. might require only 10 EJ (from a total of about 25–30 EJ) to obtain equivalent energy services (see Fig. 10.02. A, which shows a 35% improvement in just 14 years). The waste estimate is based on the current stock of buildings, cars, and factories, which use about twice as much energy as the economically efficient optimum at today's energy prices. For example, cars average 18.5 mpg, although the optimum at $1.00/gallon is 30–40 mpg, and many foreign prototypes achieve 80 mpg (see Figure 10.14). Note: Food energy is excluded, and all values are in 10^{18} J (EJ) Hydropower is computed as 100% efficient. Changes in inventories (relatively small) are lumped with net imports or exports. Industrial consumption includes non-fuel uses (e.g., asphalt, petrochemicals). Source: EIA, *Annual Energy Review 1987*.

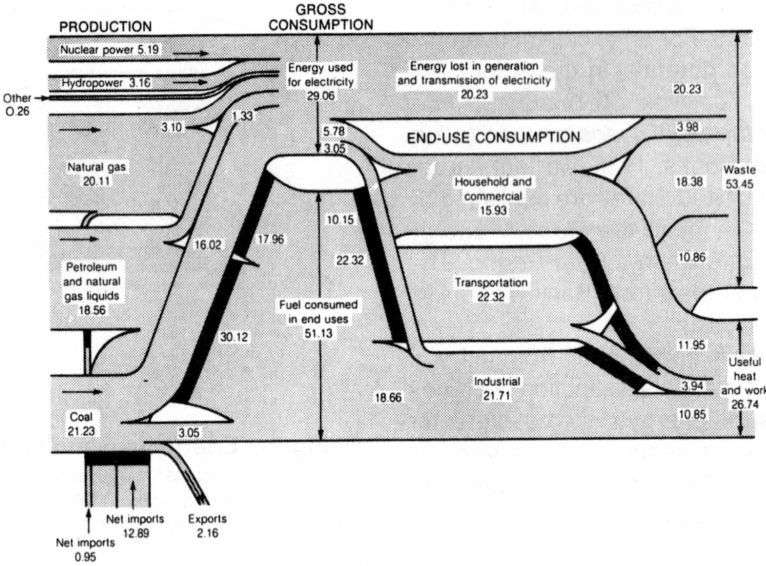

10.04 TABLE: PRIMARY ENERGY CONSUMPTION BY END USE (1985)

Total residential and commerical consumption are from the DOE Energy Information Agency (EIA), *State Energy Data Report: 1960–1985.* Distributions between end uses are based on the LBL Residential End-Use Model and the ORNL/PNL Commerical End-Use Model. Industrial and Transportation end-use distributions are from the DOE Office of Conservation *Energy Conservation Multi-Year Plan: FY 1989–93,* September 1987.

	Electricity (EJ)	Gas (EJ)	Oil (EJ)	Coal[a] (EJ)	Other[b] (EJ)	Total (EJ)	Percent (%)
Residental sector	**9.56**	**4.81**	**1.18**		**0.51**	**16.06**	**20.1**
Space Heating	1.87	3.12	1.07		0.42	6.48	40.3
Water Heating	1.66	0.89	0.11		0.06	2.72	16.9
Refrigerators	1.49					1.49	9.3
Lighting	1.08					1.08	6.7
Air Conditioners	1.12					1.12	7.0
Ranges/Ovens	0.66	0.22			0.03	0.91	5.7
Freezers	0.46					0.46	2.9
Other	1.22	0.58				1.8	11.2
Commercial sector	**8.39**	**2.65**	**0.91**		**0.30**	**12.25**	**15.4**
Space Heating	0.95	2.09	0.82		0.19	4.05	33.2
Lighting	3.11					3.11	25.5
Air Conditioning	1.06	0.12				1.18	9.6
Ventilation	1.47					1.47	12.0
Water Heaters	0.34	0.19	0.08			0.61	5.0
Other	1.46	0.25	0.01		0.11	1.83	14.8
Industrial	**9.96**	**7.47**	**8.13**	**2.92**	**1.78**	**30.26**	**37.9**
Chemicals	1.81	2.08	1.79	0.37	0.22	6.27	20.7
Petroleum Refining	0.43	2.55	2.67	0.01	0.11	5.77	19.1
Primary Metals	1.74	0.73	0.16	1.56	0.15	4.34	14.3
Pulp and Paper	0.97	0.41	0.28	0.26	0.49	2.41	7.9
Stone, Glass, and Clay	0.37	0.45	0.04	0.34	0.18	1.38	4.6
Food	0.52	0.45	0.09	0.12	0.15	1.33	4.4
Textiles Mills	0.29	0.08	0.03	0.03	0.06	0.49	1.6
Fabricated Metals	0.28	0.15	0.02	0.01	0.09	0.55	1.8
Machinery	0.40	0.13	0.02	0.03	0.08	0.66	2.2
Transportation Equipment	0.39	0.13	0.03	0.05	0.06	0.66	2.2
Other Manufacturing	1.19	0.26	0.09	0.07	0.14	1.75	5.8
Non-Manufacturing	1.57	0.05	2.91	0.07	0.05	4.65	15.4
Transportation	**0.05**	**0.55**	**20.63**			**21.23**	**26.6**
Automobiles			10.51			10.51	49.5
Light Trucks		0.55	2.40			2.95	13.9
Heavy Trucks			3.89			3.89	18.3
Air			1.78			1.78	8.4
Marine			1.46			1.46	6.9
Rail	0.05		0.59			0.64	3.0
TOTAL	**27.96**	**15.48**	**30.85**	**2.92**	**2.59**	**79.8**	**100.0**

[a] Coal is included with Other in residential and commercial.
[b] For residential: coal and LPG. For commerical: coal, LPG, and motor gasoline. Excludes estimated 0.84 EJ of energy from wood fuel in residential sector.

10.05.TABLE: PERFORMANCE OF SELECTED ENERGY CONVERSION DEVICES

Device	Typical range of efficiency or COP[a]	Comments
Fossil fuel power plant	30–40%	baseload at busbar: subtract 10% for transmission and distribution losses
Nuclear power plant	30–40%	baseload at busbar
Residential heat pump	1.0–3.0	typically 5 kW electrical input
Residential central air conditioner	1.5–3.3	typically 10 kW electrical input
Electric motor		
1/6 hp (125 W_{output})	18–50%	as might be found in a refrigerator
5 hp (3.7 kW_{output})	75–80%	as might be found in an industrial pump
Heat exchanger		
Air-to-air	60–90%	residential (100 m^3/h)
Liquid-to-air	65–95%	like in a car radiator
Residential natural gas furnace	60–92%	20 kW thermal output, 93% limit without condensation of combustion gas
Residential oil furnace	65–80%	20 kW thermal output
Lighting (efficacies in lm/W)[b]		
Incandescent	12–18	
Fluorescent	65–84	add about 10 W/unit for ballast
Metal halide	80	add about 25 W/unit for ballast
Low-pressure sodium	180	add about 30 W/unit for ballast
Sunlight	92	

[a]The efficiencies are generally taken at steady state rather than averaged over all operating conditions. Efficiency is defined as (useful energy output)/(energy input), COP, or "coefficient of performance," as (useful energy output)/(work input).

[b]Illumination efficacies (lumens/watt) are generally measured with respect to their electrical input. For sunlight, however, the efficacy is measured with respect to its thermal input.

10.06. TABLE: POTENTIAL FOR COGENERATION: SAMPLE PEFORMANCE OF ELECTRIC CONVERSION DEVICES SUPPLYING ELECTRICITY AND USABLE HEAT

Adapted from K. W. Ford *et al.*, AIP Conf. Proc. **25** (1974), Table 3.7.

Device	Electricity fraction[a]	Usable heat fraction	Temperature of heat (°C)	Units of fuel needed to produce same electricity and heat separately
Steam turbine	0.13	0.78	200	1.31
Gas turbine	0.20	0.70	200	1.40
Diesel	0.30	0.33	120	1.70
Fuel cell	0.25	0.65	150	1.72

[a]May be increased by sacrificing some of the usable heat.

10.07. TABLE: IMPROVEMENTS IN BUILDING ENERGY EFFICIENCY

A. Table: Space heat requirements for single-family dwellings

Data given for Northern Energy Homes are calculated. All other values are based on measured data. The incremental cost for the Northern Energy Homes compared to 1980 new construction is about $18/m² in 1987 dollars. Besides saving energy, other benefits of the Northern Energy Homes include uniform indoor temperatures and high quality control in construction with pre-fab panels. Source: Building Energy Compilation and Analysis (BECA), Lawrence Berkeley Laboratory, Berkeley, CA 94720.

	(KJ/m² HDD[a])	(Btu/ft² HDD[a])	Added Cost	Simple Payback[b]
Average, U. S. Housing Stock	160	7.8	-	-
New Construction, 1980	100	4.9	-	-
Measured Performance in Minnesota's Energy-efficient Housing Demonstration Program	51	2.5	$15/m²	13 years
Northern Energy Homes, Northeastern U. S.	15	0.8	$18/m²	9 years

[a] HDD = Heating Degree Days with a base of 18.3 °C or 65 °F.
[b] Assuming 4400 HDD (°C) and $5.31/GJ of natural gas.

B. Table: Energy efficiency improvements and potential for residential appliances and equipment, 1985.

Source: Howard Geller, American Council for an Energy-Efficient Economy.

Product	Current Stock Average	New Model Average	Best Commercial Model[a]	Estimated Cost-Effective Potential[b]	Potential Savings[c]
	(kWh/yr)	(kWh/yr)	(kWh/yr)	(kWh/yr)	(percent)
Refrigerator	1,500	1,100	750	200–400	87
Central Air Conditioner	3,600	2,900	1,800	900–1,200	75
Electric Water Heater	4,000	3,500	1,600	1,000–1,500	75
Electric Range	800	750	700	400–500	50
Electric Clothes Dryer	1,000	900	800	250–500	75
	(GJ/yr)[c]	(GJ/yr)	(GJ/yr)	(GJ/yr)	(percent)
Gas Furnace	77	65	50	30–50	59
Gas Water Heater	29	26	20	10–15	63
Gas Range	7	5	4	2–3	64
Gas Clothes Dryer	5	4	3.5	3–3.5	40

[a] Unit energy consumption for best available model sold in 1985.
[b] Unit energy consumption possible in new models by the mid-1900s if further cost-effective advances in energy efficiency is made.
[c] Annual energy savings: price of electricity = $0.075/kWh, price of gas = $5.3/GJ ($0.56/10^5 Btu). Source: *Monthly Energy Review, April 1988*, DOE/EIA-0035(88/04).

10.08. TABLE: ENGLISH–SI CONVERSION FACTORS FOR HEAT TRANSFER AND STORAGE

These conversions are only approximate. For greater accuracy, consult the ASHRAE *Handbook of Fundamentals* (American Society of Heating, Refrigerating and Air-Conditioning Engineers, 1977).

Common or English units	SI units	SI inverse	Process
1 Btu	1.05 kJ	0.952	energy
1 cal	4.18 J	0.239	energy
1 kW h	3.6 MJ	0.278	energy
1 Btu/h	0.243 W	3.41	power
1 hp	0.746 kW	1.34	power
1 Btu/h ft °F	1.73 W/m K	0.578	thermal conductivity
1 Btu in./h ft² °F	0.144 W/m K	6.94	thermal conductivity
1 Btu/h ft² °F	5.67 W/m² K	0.176	thermal conductance (U value)
1 Btu/h °F	0.527 W/K	1.90	thermal conductance
1 ft² °F h/Btu	0.176 m² K/W	5.67	thermal resistance (R value)
1 Btu/hr ft²	3.15 W/m²	0.317	heat flux
1 Btu/lb °F	4.18 kJ/kg K	0.239	specific heat
1 Btu/°F	1.90 kJ/K	0.527	thermal mass
1 Btu/°F	0.527 W h/K	1.90	thermal mass

10.09. TABLE: THERMAL PROPERTIES OF BUILDING MATERIALS

A. Table: Thermal properties of common building materials

Thermal resistances, or R values, for common building materials are given in metric units. To distinguish these values from the more common English values, we call the metric resistance "RSI values". Thermal resistances in series add directly like electrical resistances (see the example following the table). However, the thermal resistances of complete window assemblies are commonly expressed in terms of their conductivities, or U values (in the table we use the metric equivalent "USI value"). These conductivities include the thermal resistances of the associated air films. Note that conductivities in series do not add directly; they must be converted to resistances. To convert from the metric RSI value in m^2·°C/W to the English equivalent ft^2·F·hr/Btu, multiply by 5.67. Thus, RSI-3.3 corresponds to the more familiar R-19. To calculate the heat loss in (in watts) through a building surface, use the formula $dQ/dt = (USI_{total})(area)(\Delta T)$, where area is in m^2 and ΔT is in °C. For further information, consult the *ASHRAE Handbook 1985 Fundamentals,* revised every 4 years.

Material	RSI
Wood bevel siding, 1.27-cm × 20.3-cm	RSI-0.14
Wood siding shingles,	
40-cm, 20-cm exposure	RSI-0.15
Stucco, per cm	RSI-0.014
Building paper	RSI-0.011
1.27-cm nail-base insulation board	RSI-0.20
1.27-cm insulation board sheathing	RSI-0.23
2-cm insulation board sheathing	RSI-0.36
0.63-cm plywood	RSI-0.055
0.95-cm plywood	RSI-0.083
1.27-cm plywood	RSI-0.11
1.6-cm plywood	RSI-0.14
0.63-cm hardboard	RSI-0.032
Softwood, per cm	RSI-0.09
Softwood board, 1.9-cm thick	RSI-0.17
Concrete blocks, three oval cores	
Cinder aggregate, 10-cm thick	RSI-0.20
Cinder aggregare, 20-cm thick	RSI-0.30
Cinder aggregate, 30-cm thick	RSI-0.33
Sand-and-gravel aggregate, 20-cm thk	RSI-0.20
Lightweight aggregate, 20-cm thick	RSI-0.35
Concrete blocks, two rectangular cores	
Sand-and-gravel aggregate, 20-cm thk	RSI-0.18
Lightweight aggregate, 20-cm thick	RSI-0.38
Common brick, per cm	RSI-0.014
Face brick, per cm	RSI-0.007
Sand-and-gravel aggregate, per cm	RSI-0.006
Sand-and-gravel aggregate, 20-cm	RSI-0.11
1.27-cm gypsumboard	RSI-0.08
1.6-cm gypsumboard	RSI-0.1
1.27-cm lightweight-gypsum plaster	RSI-0.06
2-cm hardwood finish flooring	RSI-0.12
Asphalt, linoleum, or vinyl floor tile	RSI-0.01
Carpet and fibrous pad	RSI-0.22
Asphalt roof shingles	RSI-0.08
Wood roof shingles	RSI-0.17
Built-up roofing, 0.95-cm thick	RSI-0.06
Insulation	
5 to 6-cm thick	RSI-1.2
7 to 10-cm thick	RSI-1.9
12 to 18-cm thick	RSI-3.3
20 to 26-cm thick	RSI-5.3
Air spaces (2-cm)	
Heat flow up	
Nonreflective	RSI-0.15
Reflective, one surface[1]	RSI-0.39
Heat flow down	
Nonreflective	RSI-0.18
Reflective, one surface[1]	RSI-0.63
Heat flow horizontal	
Nonreflective (same for 5-cm thick)	RSI-0.18
Reflective, one surface[1]	RSI-0.61
Surface films	
Inside (still air)	
Heat flow up	
Nonreflective	RSI-0.11
Reflective	RSI-0.23
Heat flow down	
Nonreflective	RSI-0.16
Reflective	RSI-0.80
Heat flow horizontal	
Nonreflective	RSI-0.12
Outside	
Heat flow any direction or position	
14-m/s wind, winter	RSI-0.03
7-m/s wind, summer	RSI-0.04

[1]The addition of a second reflective surface facing the first increases thermal resistance of an air space 4 to 7%.

Example calculations (to determine the USI value of a typical residential exterior wall):

	1 (no insulation)		2 (insulation)	
Wall Construction	Between Stud (m^2·°C/W)	At Stud (m^2·°C/W)	Between Stud (m^2·°C/W)	At Stud (m^2·°C/W)
Outside surface film, 7 m/s wind	0.03	0.03	0.03	0.03
Wood bevel siding, lapped	0.14	0.14	0.14	0.14
Sheating board, 1.27-cm (0.5-in.)	0.23	0.23	0.23	0.23
Air space, 8.9-cm (3.5-in)	0.18	—	1.94	—
Wood stud, Nominal 5-cm × 10-cm (2-in × 4-in)	—	0.77	—	0.77
Gypsum wallboards, 1.27-cm (0.5-in)	0.08	0.08	0.08	0.08
Inside surface film, still air	0.12	0.12	0.12	0.12
Total thermal resistance (RSI)......	$RSI_i = 0.78$	$RSI_s = 1.37$	$RSI_i = 2.54$	$RSI_s = 1.37$

Construction 1: $USI_i = 1/0.78 = 1.28$; $USI_s = 1/1.37 = 0.73$. With 10% framing (2×4 studs on 16-in. centers),
$USI_1 = 0.9(1.28) + 0.1(0.73) = 1.23$.W/m^2·°C

Construction 2: $USI_i = 1/2.54 = 0.39$; $USI_s = 1/1.37 = 0.73$. With 10% framing (2×4 studs on 16-in. centers),
$USI_2 = 0.9(0.39) + 0.1(0.73) = 0.42$.W/m^2·°C

B. Table: Thermal properties of windows

Thermal conductivities, or USI values, are given for the center of the window and for the following complete window units: a 91-cm × 122-cm (3-ft × 4-ft) residential window with a center mullion and a 122-cm × 183-cm (4-ft × 6-ft) commercial window. The values given include the thermal resistances of associated air films. To convert from the metric USI value in W/m²·°C to the English equivalent Btu/ft²·F·hr, multiply by 0.176. Source: *ASHRAE Handbook 1989 Fundamentals.*

Window type	Center of Glass (W/m²·°C)	Residential Window (W/m²·°C)	Commercial Window (W/m²·°C)
Single pane			
Wood or viny frame	USI-6.3	USI-5.1	USI-5.6
Aluminum frame with thermal break	USI-6.3	USI-6.1	USI-6.2
Aluminum frame	USI-6.3	USI-7.4	USI-7.0
Double pane, 1.27-cm (0.5-in.) air space			
Wood or vinyl frame	USI-2.8	USI-2.8	USI-2.8
Aluminum frame with thermal break	USI-2.8	USI-3.6	USI-3.3
Aluminum frame	USI-2.8	USI-4.9	USI-4.1
Double pane, low-*E* (*E* = 0.15)			
Wood or vinyl frame	USI-1.9	USI-2.3	USI-2.1
Aluminum frame with thermal break	USI-1.9	USI-3.1	USI-2.6
Aluminum frame	USI-1.9	USI-4.3	USI-3.4
Double pane, low-*E* with argon gas (*E* = 0.15)			
Wood or vinyl frame	USI-1.6	USI-2.1	USI-1.9
Aluminum frame with thermal break	USI-1.6	USI-2.9	USI-2.4
Aluminum frame	USI-1.6	USI-4.2	USI-3.1
Triple glass, low-*E* with argon gas (*E* = 0.15), one low-*E* coating with argon gas per space			
Wood or vinyl frame	USI-0.85	USI-1.7	USI-1.4

10.10. TABLE: SOLAR GAIN AND THERMAL PROPERTIES OF WINDOWS

Information is given in the following tables for solar gain and thermal properties of windows. The interaction of solar gain and conduction loss through windows is very dynamic. Solar gain is presented both in terms of power (maximum solar gain for a clear day) and energy (daily solar gain for a clear day). Clear day values are reduced by about 50% in the winter due to cloudy skies over most of the U.S.. For comparison, daily conduction losses are tabulated for single-, double-, low-*E* double-, and low-*E* with argon gas double-glazed windows over a range of temperature differences. For further information, consult the American Society of Heating, Refrigerating, and Air-Conditioning Engineers, *ASHRAE Handbook 1985 Fundamentals.*

A. Table: Maximum solar gain through a clean single-pane window (40° N latitude)

Reduce values by 15% for standard double-pane, by 30% for low-*E*[a] double-pane, and by 30% for standard triple-pane windows. Adapted from *ASHRAE Handbook 1985 Fundamentals.*

Month	North (W/m²)	East/West (W/m²)	South (W/m²)	Horizontal (W/m²)
January	62	485	800	419
February	76	587	760	568
March	91	686	650	702
April	106	706	486	794
May	117	694	355	836
June	121	680	301	844
July	120	680	345	826
August	112	681	470	780
September	95	646	631	679
October	78	567	737	558
November	63	476	789	418
December	56	427	798	356

[a] Sputter-coated low-*E* glass with a shading coefficient of 0.7.

B. Table: Daily solar gain through a clean single-pane window (40° N latitude).

Reduce values by 15% for standard double-pane, by 30% for low-*E* double-pane, and by 30% for standard triple-pane windows. Values are rounded to nearest MJ. Adapted from *ASHRAE Handbook 1985 Fundamentals.* To convert to average watts it is convenient to note that 1 kWh = 3.6 MJ, and 1 kW × 24 hours = 86.4 MJ.

Month	North (W/m²)	East/West (W/m²)	South (W/m²)	Horizontal (W/m²)
January 21	1.4	5.8	18.5	8.0
February 21	1.9	8.3	18.7	12.5
March 21	2.6	10.7	15.8	17.4
April 21	3.5	12.6	11.1	21.7
May 21	4.9	13.6	8.1	24.6
June 21	5.8	13.9	7.2	25.6
July 21	5.1	13.5	8.0	24.6
August 21	3.7	12.3	10.8	21.5
September 21	2.7	10.3	15.3	16.8
October 21	2.0	8.1	18.0	12.3
November 21	1.4	5.8	18.1	8.0
December 21	1.2	4.8	17.6	8.4

C. Table: Daily conduction losses for residential windows

Based on the heat loss formula $dQ/dt = (\text{USI}_{\text{window}})(T_{in} - T_{out})$ (86 400). Based on a 1.27-cm(0.5 in.) air space for double-pane windows. Values given are for a wood or vinyl frame. For comparison the extra cost of a nominal 90-cm × 120-cm (36-in. × 48-in.) window with low-*E* sputter-coated glass from the major U.S. window manufacturers is currently about 10% more than standard double-pane glass, and the extra cost of argon gas-filled windows is an additional 2%.

Temperature Difference (°C)	Single (USI = 6.2) (MJ/m²)	Double (USI = 2.8) (MJ/m²)	Low-*E* Double (USI = 2.3) (MJ/m²)	Low-*E* with Argon Double (USI = 2.1) (MJ/m²)
5	2.7	1.2	1.0	0.9
10	5.3	2.4	2.0	1.8
20	10.7	4.8	3.9	3.6
40	21.4	9.7	7.9	7.3

10.11. TABLE: APPLIANCE ENERGY USE

Peak power and monthly electricity use for typical household appliances. The average cost of residential electricity is about 7.5¢/kW h, although users in the Northeast generally pay considerably more and those in the Northwest pay less.

Appliance	Peak power (W)	Monthly use (kW h)	Appliance	Peak power (W)	Monthly use (kW h)
Blender	350	1	Refrigerator-freezer (manual defrost, 12–14 ft³)	330	60
Broiler	1 430	8	Refrigerator-freezer (frostless, 15–16 ft³)	600	140
Coffeemaker	900	10	Toaster	1 200	5
Clothes dryer (five loads a week)	5 000	83	TV (black and white)		
Dishwasher (inc. hot water)	1 200	120	Tube type	240	30
Electric blanket	180	12	Solid state	55	10
Food freezer (15 ft³)			TV (color)		
Manual defrost	340	90	Tube type	350	44
Food freezer (15 ft³)			Solid state	200	36
Frostless	440	150	Vacuum cleaner	630	4
Food waste disposer	450	3	Washer (hot water not inc.)	500	8
Frypan	1 200	15	Water bed (heater)	370	150
Iron (hand)	1 000	12	Water heater	4 500	400
Microwave oven	1 400	16			
Radio-phono (stereo)	110	9			
Range	12 000	70			
Self-cleaning oven feature	4 000	4			

10.12. TABLE: TRANSPORTATION ENERGY INTENSITIES (1985)

Source: S. C. Davis *et al., Automated Transportation Energy Data Book*, Transportation Research, Oak Ridge National Laboratory, Draft, May 1988.

Passenger Transport	
Type	Intensity (MJ/passenger-km)[a]
Automobiles	2.77
Light Trucks	2.69
Motorcycles	1.60
Buses	0.40
Air	3.49
Rail	1.62
All Modes	2.35

Freight Transport	
Type	Intensity (MJ/tonne-km)[b]
Rail	0.35
Truck	1.36
Marine	0.32
All Modes	0.66

[a] To convert to Btu/passenger-mile, multiply by 1530.
[b] To convert to Btu/tonne-mile, multiply by 1390.

10.13. TABLE: ESTIMATED ENERGY FLOW FOR A 2500-lb (1140-kg) AUTOMOBILE OVER THE FEDERAL URBAN DRIVING CYCLE

Adated from K. W. Ford *et al.*, AIP Conf. Proc. **25** (1974).

Fuel, available work	100%
Fuel energy at commonly used lower heat of combustion (×0.96)	96%
Ideal gas air-cycle Otto engine (×0.56)	54%
Fuel-air Otto cycle (×0.75)	40%
Burning and cylinder wall losses (×0.8)	32%
Frictional losses (×0.8)	26%
Partial load factor (×0.75)	19%
Estimated loss due to accessories	11%
Automatic transmission (×0.75)	8%
Net output to rear wheels	8%

10.14 FIGURE: U.S. AUTOMOBILE FUEL COMSUMPTION

The figure shows that the new car fleet average fuel efficiency improved by 50% since 1973 going from 16.8 liter/100 km to 8.2 liter/100 km in 1988 (this translates to a 100% improvement in U.S. units, i.e., 13.8 mpg in 1973 to 28.5 mpg in 1988). The potential for further improvement is shown by the cluster of prototypes with demonstrated fuel efficiencies of 2 to 3 liter/100 km (75 to 100 mpg). One of these prototypes, the Volvo LCP-200, has been designed to withstand 56.3 km/hr (35 mph) frontal and side impacts and 48 km/hr (30 mph) rear impacts. For more information on the prototypes as well as further information on new automobile technologies, see D. L. Blevis, *The New Oil Crisis and Fuel Technologies: Preparing the Light Transportation Industry for the 1990s,* Quorum Books, 1988.

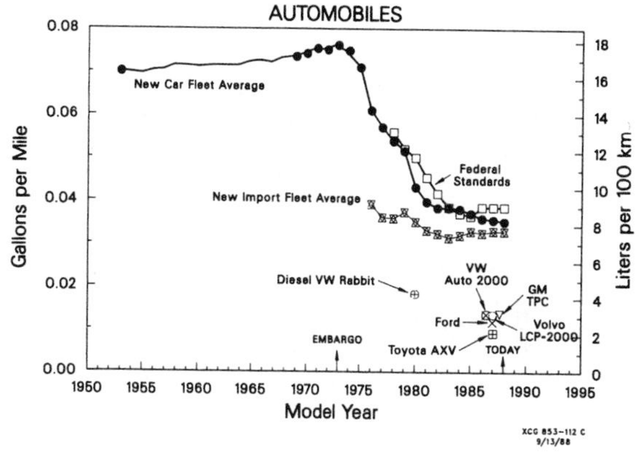

10.15. FIGURES: EFFECT OF FUEL-AIR RATIO ON THERMAL EFFICIENCY, PEAK CYLINDER PRESSURE, AND PEAK COMBUSTION TEMPERATURE FOR AN IDEAL FUEL-AIR CYCLE OTTO ENGINE

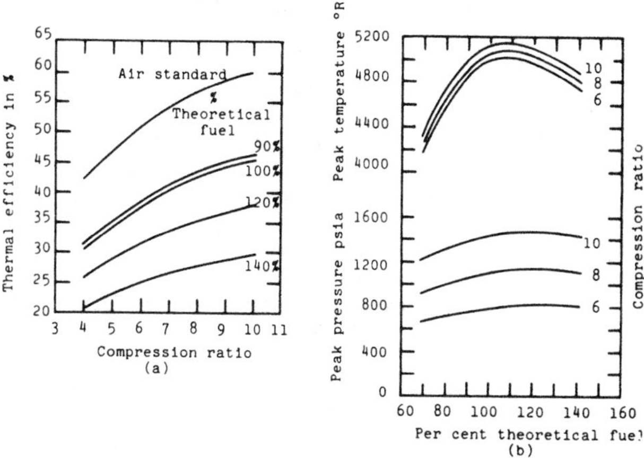

From K. W. Ford *et al.*, AIP Conf. Proc. **25** (1974).

11.00. Energy supply

HANS A. BETHE, *Cornell University*
DAVID BODANSKY, *University of Washington*

CONTENTS

11.01. CONVERSION OF UNITS[a]

General

1 short ton (ton) = 2000 lb = 0.907185 tonne
1 metric ton (tonne) = 1000 kg
1 barrel = 42 U.S. gallons = 159.0 litres

1 BTU (British thermal unit) = 1055 J (Joules)
1 kWh (kilowatt hour) = 3.6 MJ = 3412 BTU
1 kWh of electricity requires on average 10,261 BTU to produce, corresponding to a mean thermal efficiency of 33% (1987 U.S. fossil fuel average)

Large units

1 quadrillion BTU = 10^9 MBTU = 10^{15} BTU
1 exajoule (EJ) = 10^3 PJ = 10^{12} MJ = 10^{18} J
1 terawatt-yr (TWyr) = 10^9 kWyr = 8.76×10^{12} kWh

	Quad	EJ
1 Quad	1.000	1.055
1 EJ	0.948	1.000
1 TWyr (100% conversion)	29.89	31.54
1 TWyr (33% efficiency)	90.6	95.6
10^9 tonne coal equiv (Gtce)	27.76	29.29
10^9 barrel oil equiv (bboe)	5.80	6.12
10^9 tonne oil equiv (Gtoe)	42.43	44.76
10^9 tonne oil equiv (Gtoe)[b]	39.69	41.87

Fuel values

1 barrel of crude oil = 0.137 metric ton
1 million barrels per day of crude oil
= 2.12 quad/yr = 2.23 EJ/yr

	MBTU	GJ
Nominal or standard equivalents:		
1 barrel of crude oil (boe)	5.8	6.12
1000 cu. ft. of natural gas	1.000	1.055
1 short ton of coal	25.18	26.57
Average heat content (U.S. 1987):		
1 barrel of petroleum products	5.403	5.700
1000 cu. ft. of natural gas	1.030	1.087
1 short ton of coal	21.53	22.72
1 cord of dry wood (1.25 ton)	21.5	22.7
1 barrel of natural gas liquids	3.804	4.013
1 barrel of aviation gasoline	5.048	5.326
1 barrel of motor gasoline	5.253	5.542
1 barrel of distillate fuel oil	5.825	6.145
1 barrel of residual fuel oil	6.287	6.633

[a] Based on *Annual Energy Review 1987* (ref. 1), *Monthly Energy Review, March 1988* (ref. 2), and IIASA report, 1981 (ref. 3).
[b] Alternate equivalent, used by OECD (ref. 4).

11.02. UNITED STATES ENERGY CONSUMPTION

A. Table: Historical energy consumption, 1850-1987: total energy (*E* in exajoules), population (*N* in millions), and GNP (in billion 1982 $)

Series I includes wood, work animal feed, and direct wind and water power. Series II omits these, but includes geothermal power.

Year	*E* (EJ)	*N* (10^6)	GNP (10^9 $)	*E/N* (GJ/Cap)	*E*/GNP (MJ/$)
		Series I: 1850–1970[a]			
1850	3.51	23.3		151	
1860	4.80	31.5		152	
1870	5.65	39.9	53	142	107
1880	7.15	50.3	108	142	66
1890	9.82	63.1	160	156	61
1900	12.75	76.1	247	168	52
1910	20.36	92.4	357	220	57
1920	25.53	106.5	464	240	55.1
1930	26.99	123.1	601	219	44.9
1940	28.11	132.5	765	212	36.7
1950	37.91	151.9	1183	250	32.0
1960	48.41	180.0		269	
1970	72.24	203.8		354	
		Series II: 1950–1987[b]			
1950	34.90	151.3	1204	231	29.0
1955	40.96	165.1	1495	248	27.4
1960	46.21	179.3	1665	258	27.7
1965	55.58	193.5	2088	287	26.6
1970	70.08	203.2	2416	345	29.0
1975	74.43	215.5	2695	345	27.6
1980	80.14	226.5	3187	354	25.1
1985	78.01	238.7	3619	327	21.6
1986	78.32	241.1	3722	325	21.0
1987	80.97	243.4	3847	333	21.0

[a] From Fisher (ref. 5) and Bureau of the Census (ref. 6).
[b] From DOE/EIA (refs. 1 and 2).

B. Table: More recent consumption, by source, 1950–1987 (in exajoules per year)[a]

Year	Coal	Gas	Oil	Hydro	Nucl.	Other[b]
1950	13.03	6.30	14.05	1.52	0.00	
1955	11.78	9.50	18.20	1.49	0.00	
1960	10.38	13.07	21.02	1.75	0.01	
1965	12.22	16.64	24.53	2.17	0.04	
1970	12.93	22.99	31.14	2.80	0.25	
1975	13.36	21.05	34.53	3.40	2.00	0.09
1980	16.27	21.51	36.08	3.29	2.89	0.08
1985	18.44	18.81	32.62	3.54	4.38	0.21
1986	18.21	17.63	33.97	3.57	4.72	0.23
1987	19.01	18.64	34.67	3.19	5.19	0.27

[a] From DOE/EIA (refs. 1 and 2); see Section 10.02 for a graph of this information.
[b] Primarily geothermal energy for electricity; same exclusions as listed in Section 11.03.

11.03. SUPPLY AND DEMAND IN THE UNITED STATES

Data in Section 11.03 are obtained from DOE/EIA compilations (refs. 1 and 2), which exclude wood, waste, geothermal, wind, photovoltaic and solar thermal energy, except for small amounts used by electric utilities to generate electricity. The omitted energy sources are outside of the standard commercial markets. The largest omission is of approximately 3 EJ of energy from wood. (See Section 11.08 for information on these sources.)

A. Table: Energy by source and use, 1973 and 1987 (in exajoules)

Use	Coal	Gas	Oil	Source Elec.	Hydro[a]	Nucl.	Other[b]	Total
				1973				
Residential/Commericial	0.27	8.05	4.63	12.52				25.47
Industrial	4.27	10.96	9.61	8.39	0.04			33.27
Transportation	0.00	0.78	18.80	0.03				19.62
Electric Utilities	9.13	3.95	3.71	0.00	3.14	0.96	0.05	20.94*
Total	13.68	23.75	36.76	20.94*	3.18	0.96	0.04	78.37
				1987				
Residential/Commercial	0.17	7.32	2.75	19.00				29.24
Industrial	2.83	7.67	8.75	10.00	0.03			29.28
Transportation	0.00	0.56	21.86	0.05				22.46
Electric Utilities	16.02	3.10	1.33	0.00	3.16	5.19	0.26	29.05*
Total	19.01	18.64	34.67	29.05*	3.19	5.19	0.27	80.97

*Not included in totals.

[a] The electric energy is divided by the average efficiency of electric generation by fossil fuel thermal sources: 0.328 in 1973 and 0.333 in 1987 (ref. 2).

[b] Primarily geothermal energy for electricity; omissions are indicated in heading above.

B. Table: Fraction of total energy used for various sectors, 1950–1987 (in percent)

Year	Total Energy (EJ)	Fraction (%) Res. and Comm.	Industry	Trans-portation
1950	34.90	26.8	47.5	25.7
1955	40.96	26.8	48.6	24.6
1960	46.21	29.8	46.0	24.2
1965	55.58	30.4	46.0	23.6
1970	70.08	32.7	43.1	24.2
1975	74.43	33.9	40.3	25.9
1980	80.14	33.8	40.3	25.9
1985	78.01	36.2	36.6	27.2
1986	78.32	36.4	35.6	28.0
1987	80.97	36.1	36.2	27.7

C. Table: Fraction of primary energy used for electricity generation, by sector, 1950–1987 (in percent)[a]

Year	Res. and Comm.	Industry	Trans-portation	Total
1950	25.0	15.5	1.3	14.2
1955	29.0	18.2	0.7	16.7
1960	32.9	19.3	0.4	18.7
1965	37.6	20.6	0.2	20.9
1970	44.1	23.4	0.2	24.5
1975	51.5	28.3	0.2	28.8
1980	58.2	31.3	0.2	32.3
1985	63.3	35.1	0.2	35.8
1986	64.6	34.6	0.2	35.9
1987	65.0	34.2	0.2	35.9

[a] For each sector, fraction $= A/(A+B)$, where A = energy used to generate the electricity supplied to sector and B = nonelectric energy supplied to sector.

D. Table: Electricity generation by electric utilities, by energy source (in net terawatt-hours)

	TWhr		Percent	
	1973	1987	1973	1987
Coal	848	1464	45.6	56.9
Oil	314	118	16.9	4.6
Natural Gas	341	273	18.3	10.6
Hydro	272	250	14.6	9.7
Nuclear	83	455	4.5	17.7
Geothermal	2	11	0.1	0.4
Other[a]	0.3	1.5	0.0	0.1
Total	1861	2572	100.0	100.0

[a] Primarily wood and waste.

11.04. PREDICTIONS OF U.S. ENERGY NEEDS FOR THE FUTURE

A. Table: Projections of future energy consumption, 2000 and 2010 (in exajoules per year)

1987 consumption: 80.97 EJ

	Year	
Projection (with year made)[a]	2000	2010
Department of Interior (1972)	202	
Ford Foundation (1974)	106–197	
ERDA (1975)	128–173	
Inst. for Energy Analysis (1976)	107–133	125–168
Council for Env. Quality (1979)	91	
CONAES (1979)[b]		
I very aggressive		68–90
II aggressive		88–121
III moderate		108–148
IV unchanged		148–198
Nat. Energy Policy Plan (1981)	95–116	
GRI Baseline Projection (1987)	94	104
DOE/EIA (1988)	90–98	

[a] From refs. 7–15; where alternative scenarios are presented, the range of estimates is given.
[b] Different listed projections correspond to different assumed conservation policies; results are given for average annual GNP growth rates of 2% and 3%.

B. Table: CONAES projections of energy consumption in 2010 (in exajoules per year), with various assumptions on conservation[a]

			Projections 2010[c]		
Sector	1975[a]	1987[b]	A	B	C
Residential and Commercial	27.4	29.2	22.8	30.1	46.6
Industrial	27.5	29.3	38.0	44.6	61.0
Transportation	19.9	22.5	16.9	24.4	34.6
Total	74.7	81.0	77.6	99.1	142.2

[a] From CONAES report (ref. 12).
[b] From DOE/EIA (ref. 2).
[c] Scenario A: aggressive conservation, aimed at maximum efficiency plus minor life-style changes; Scenario B: moderate conservation policies; Scenario C: continuation of policies of late 1970s. These scenarios are similar to, but not identical to, Scenarios II, III, and IV of Table 11.04 A, for 2% annual growth in GNP.

C. Table: CONAES projections for energy supply in 2010 (in exajoules per year), for range of energy demand scenarios[a]

			Projections 2010[c]		
	1975[a]	1987[b]	I_2	III_2	IV_3
Oil	35	35	24	24	47
Oil shale	0	0	0	1	3
Coal[d]	14	19	16	40	77
Natural gas	21	19	9	17	24
Nuclear	2	5	6	14	31
Solar	0	0	6	4	2
Hydro	3	3	4	5	5
Geothermal	0	0	0	1	4
Other	0	0	2	3	5
Total	75	81	68	108	198
Imports (%)[e]	20	18	19	9	18
Electricity (%)	28	36	27	36	38

[a] From CONAES report (ref. 12).
[b] From DOE/EIA (ref. 2).
[c] Scenario: I_2: very aggressive conservation and 2% annual GNP growth; Scenario III_2: moderate conservation and 2% annual GNP growth; Scenario IV_3: continuation of conservation policies of late 1970s and 3% annual GNP growth.
[d] Includes synthetic oil and gas from coal.
[e] Oil and gas only.

11.05. ESTIMATES OF FOSSIL FUEL RESOURCES IN THE UNITED STATES AND WORLD

A. Table: Recoverable conventional crude oil, U.S. and world (in billion barrels)

The estimates presented below differ in their assumptions as to price and in the degree of inclusion of contributions from heavy oil, enhanced recovery, and natural gas liquids. These contributions are relatively small, under the self-imposed guidelines of the estimates.

Extensive additional resources may be obtained by increased recovery of the initial oil-in-place (at higher cost). Estimates of this potential vary widely for the U.S., extending to more than a doubling of the remaining resource.[a]

	U.S.	World
Estimates of original resource[b]		
IIASA, 1981		2500
USGS, 1981 (mean)	258	
Nehring (Rand), 1982		1800
Edmonds (IEA), 1985	241	2309
Riva (CRS), 1987	226	1765
Masters (USGS), 1987		
Low (95% level)[d]	206	1581
Average (mode)	220	1744
High (5% level)[d]	254	2245
Nominal average (rounded)	230	2000
Cumulative production, 12/31/87[c]	145	584
Remaining resource, 12/31/87	85	1400
Annual production, 1987	3.0	20.4
Remaining resource (EJ)[e]	520	8600

[a] See, e.g., Fisher (ref. 16).
[b] From refs. 3 and 17–21.
[c] Based on refs. 2 and 21.
[d] Corresponding to probabilities of 95% and 5% that resources exceed this level.
[e] Converted at 6.12 EJ per billion barrels.

B. Table: Geographical distribution of recoverable conventional crude oil (in billion barrels)[a]

Tabulated for each region: ultimate resources, cumulative production (to 12/31/86), remaining resources (1/1/87), and fraction of remaining world resources (in percent).

Region	Ultimate (bbo)	Prod. (bbo)	Remain. (bbo)	Fraction (%)
U.S. + Canada	273	154	119	10
United States	226	142	84	7
Canada	47	13	34	3
Latin America	204	68	136	11
Mexico	69	14	55	4
Venezuela + Trinidad	97	42	55	4
Other	38	12	26	2
Europe	77	19	59	5
Norway	30	2	28	2
United Kingdom	27	7	21	2
Other	20	9	10	1
Soviet Union	253	95	158	13
Africa	142	42	100	8
Libya	48	16	32	3
Nigeria	38	11	26	2
Other	57	15	42	3
Middle East	703	154	549	45
Saudi Arabia	263	55	209	17
Kuwait	122	24	97	8
Iran	114	35	79	6
Iraq	104	18	86	7
Un. Arab Emirates	61	10	51	4
Other	40	12	28	2
Asia/Oceania	140	33	107	9
China	71	13	59	5
Other	69	21	48	4
World Total	1792	565	1227	100

[a] Based on analyses of Riva (ref. 20) and Masters (ref. 21).

C. Table: Remaining U.S. recoverable natural gas, 12/31/86 (in trillion cubic feet, Tcf)[a]

Prices in 1987 $ per 1000 cubic feet
Cumulative production to 12/31/86: 667 Tcf[b]
Wellhead price, 1987 average: $1.71 per 1000 cubic feet[b]
Annual production, 1987: 16.35 Tcf[b]

	Technically Recoverable	Recoverable at given price < $3	$3–$5
Lower 48 states			
Reserves	267	267	
Reserve growth	180	86	29
Undiscovered resources	353	142	87
Unconventional	259	88	58
Alaska	129	12	2
Total (Tcf)	1188	595	176
Total (EJ)[c]	1253	628	186

[a] From OPPA (ref. 22).
[b] From Masters (ref. 21) and DOE/EIA (ref. 1).
[c] Converted at 1.055 EJ per Tcf.

D. Table: Geographical distribution of recoverable conventional natural gas (in Tcf)[a]

Tabulated for each region: ultimate resources, cumulative production (to 12/31/84), remaining resources (1/1/85), and fraction of remaining world resources (in percent).

Region	Ultimate[b] (Tcf)	Prod. (Tcf)	Remain. (Tcf)	Fraction (%)
U.S. + Canada	1717	692	1025	13
Latin America	506	42	464	6
Europe	517	131	386	5
Soviet Union	2819	234	2585	32
Africa	582	12	570	7
Middle East	2149	23	2126	27
Asia/Oceania	890	38	852	11
World (Tcf)[c,d]	9280	1173	8107	
World (EJ)	9790	1238	8553	

[a] From Masters (ref. 21).
[b] Corresponding to mode of probability distribution.
[c] Range of estimates, 95% to 5% levels (see Table 11.05.A): 7732–13672 Tcf.
[d] Total differs from regional sum due to analytic method.

E. Table: Coal resources, reserves, and production rates (in gigatonne coal equivalent and megatonne coal equivalent per year)

The geological resources and proved recoverable reserves listed below may be underestimates, due to incomplete reporting.

Country	Geol. resources[a] (Gtce)	Proved reserves[b] (Gtce)	Production 1986[c] (Mtce/yr)
Soviet Union	4860	186	503
United States	2570	215	702
China	1438	99	639
Australia	600	48	140
Canada	323	5	51
W. Germany	247	42	134
United Kingdom	190	4	89
Poland	140	35	196
India	81	34	134
South Africa	72	56	151
Sum	10521	724	2739
Estimated World	10750	779	3178
Energy Equiv. rounded:[d]	EJ	EJ	EJ/yr
United States	75000	6300	21
World	315000	23000	93

[a] From WOCOL (ref. 23).
[b] From WEC (refs. 24–26), converted at WOCOL rates: 1 tce = 1.05 tonne bituminous coal = 1.4 tonne subbituminous coal = 1.8 tonne lignite.
[c] From DOE/EIA (ref. 27), converted at 27.76 quad per Gtce.
[d] Converted at 29.29 EJ per Gtce.

F. Table: Geographical distribution of ultimate recoverable unconventional oil resources (in billion barrels)[a]

For the most part, the cost is greater for unconventional than for conventional oil. The distinction between bitumen (including tar sands) and heavy oil is on the basis of viscosity. With the exception of some heavy oil in the U.S., the production of unconventional oil to date is too small to have significantly depleted the ultimate resource.

Oil shale is not included in this tabulation. The ultimate resource is very large (over 10,000 bbo worldwide and over 5,000 bbo in North America), but the achievable annual extraction rate may be relatively low.

	Heavy Oil		Bitumen	
Region	(bbo)	(%)	(bbo)	(%)
U.S. + Canada	34	6	315	72
U.S.	31	5	7	2
Canada	3	0.5	308	71
Latin America	334	55		
Venezuela	312	52		
Europe	9	1		
Soviet Union	135	22	117	27
Africa	6	1	2	0.5
Middle East	72	12		
Asia/Oceania	14	2	2	0.5
World (bbo)	605	100	436	100

[a] From Masters (ref. 21).

11.06. CARBON DIOXIDE PRODUCTION

A. Table: Carbon dioxide in atmosphere[a]

1 gigatonne C (in CO_2) in atmosphere = 0.47 ppm CO_2 (by volume)
AF = (CO_2 retained in atmosphere)/(CO_2 produced) ~ 0.4–0.7.
Atmospheric CO_2 concentration (1988): 350 ppm
Annual increase (extrapolated to 1988)
From fossil fuel combustion: ~ 1.3 ppm
From terrestrial biosphere: 0–0.6 ppm
Time to reach 600 ppm (approximate)[b,c]
constant fossil fuel burning: 190 years
2% annual growth rate: 80 years

[a] Based on ref. 28.
[b] For AF = 0.5.
[c] Assuming zero net deforestation.

B. Table: Pools and fluxes of carbon[a]

Category	Amount
Pools (gigatonnes)	
Atmosphere (1988)	730
Oceans (97% deep inorganic)	39,000
Terrestrial biosphere (70% dead)	2,100
Fossil fuels (85% coal)	4,100
Gross flux (gigatonnes per year)	
Atmosphere↔ocean	105
Atmosphere↔terrestrial biosphere	120
Net flux to atmosphere (Gt/yr)	
Fossil fuel combustion (1985)[b]	5.5
Terrestrial biosphere (1970–1980)	0–2.6

[a] From refs. 28 and 29.
[b] See Table 11.06.D.

C. Table: Carbon dioxide emission rates for different fuels (in kg C [in CO_2] per GJ energy)[a]

Fuel	Kg C per GJ
Natural gas	13.5– 14.2
Liquid fuels from crude oil	18.2– 20.6
Bituminous coal	23.7– 23.9
Shale oil	28.4–104
Liquids from coal	30.5– 51.3
High BTU gas from coal	32.7– 40.7

[a] From G. Marland, in ref. 29.

D. Table: Geographical distribution of carbon dioxide emission from fossil fuel use, 1985 (in gigatonnes of carbon released to atmosphere)

	Carbon released	
Region	Gigatonnes[a]	Percent
U.S.	1.34	24
USSR	0.97	18
Western Europe	0.93	17
China	0.51	9
Eastern Europe	0.40	7
Japan	0.27	5
India	0.13	2
Canada	0.12	2
Other	0.86	16
World	5.53	100

[a] Based on energy data from DOE/EIA (ref. 27) and on IEA/ORAU carbon emission rates (ref. 19): 13.7 kg/GJ for gas, 19.2 kg/GJ for oil, and 23.8 kg/GJ for coal.

11.07. NUCLEAR ENERGY

A. Table: Geographical distribution of uranium resources (in million tonnes of U)

Cost relations: \$100 per pound U_3O_8 = \$260 per kg U equivalent to ~7 mills/kWh, for LWR without fuel recycle.

Uranium resources from lower grade deposits, are not included in tabulation. The amounts are large, subject to cost constraints.

Region	Category I[a]	Category II[b]
USSR and E. Europe[c]		4.2
Australia[d]	0.9	3.5-4.8
United States[e]	1.3	3.3
South Africa[d]	1.0	2.1
China[c]		2.1
Canada[d]	0.7	1.7
World total[c] (rounded)	30	

[a] Category I includes reasonably assured and estimated additional resources at costs of \$130/kg U or less (1987 \$).
[b] Category II includes reasonably assured, estimated additional, and speculative resources at costs of \$260/kg U or less (1987 \$).
[c] From IIASA (ref. 3); cost category redefined (doubled) for approximate translation to 1987 \$; world total reflects increased estimates for North America and Australia.
[d] From OECD (ref. 30); category II here includes only uranium at \$130/kg or less, except for speculative resources at \$260/kg or less.
[e] From DOE/EIA (ref. 1).

B. Table: Uranium requirements

Metric tons of U required by a million-kilowatt nuclear reactor for a 30-yr lifetime, operating at 70% of capacity.

Reactor type	Fuel recycle?	U required
Light water	No	4620
Light water	Yes	3000
High temperature, gas cooled (HTGR)	No	3320
HTGR, or heavy water, at 90% conversion efficiency	Yes	1360
Breeder	Yes	35

Note: This is a reproduction of Table 11.10.B from 1981 Vade Mecum.

C. Table: World nuclear power use, 1987

Gross capacity (gigawatts), gross generation (terawatt-hours), and nuclear fraction of electricity.

Country	Number[a] of Units	Capac[a] (GW)	Gen[a] (TWh)	Nuclear (Percent[b])
United States	106	97.2	479	18
France	53	52.1	265	70
USSR[c]	52	36.3	206	11
Japan	36	28.0	186	31
W. Germany	19	19.8	131	31
Canada	18	12.8	81	15
Sweden	12	10.1	67	45
United Kingdom	38	12.9	56	18
Belgium	7	5.7	42	66
Spain	8	5.8	41	31
South Korea	7	5.8	38	53
Taiwan	6	5.1	33	48
Others (14)[c]	50	24.5	128	
World Total[c]	412	316.2	1753	

[a] From *Nucleonics Week* (ref. 31).
[b] From USCEA (ref. 32).
[c] Data entries for USSR, Bulgaria, Czechoslovakia, and East Germany extrapolated from 1986 data (ref. 27) and USCEA (ref. 32).

11.08 RENEWABLE ENERGY

A. Table: Intensity of Solar Radiation

Solar constant (average normal intensity above atmosphere):[a] 1367 W/m^2

Peak normal intensity in U. S. southwest: ~1000 W/m^2

Average annual intensity for continental U.S.: ~200 W/m^2

Average intensity for selected U.S. cities (in W/m^2):[b]

City	Latitude deg N	Horizontal Surface[c] Dec	June	Year	Direct Normal Dec	June	Year
Brownsville, TX	25.9	113	278	203	127	228	183
El Paso, TX	31.8	135	353	249	236	367	303
Phoenix, AZ	33.4	123	360	246	203	366	288
Fresno, CA	36.7	75	359	225	96	381	255
Omaha, NB	41.3	67	278	174	110	251	186
Boston, MA	42.4	53	239	145	78	191	136
Madison, WI	43.1	51	256	157	75	208	155
Bismark, ND	46.8	49	271	164	89	252	186
Seattle, WA	47.5	28	236	136	25	181	118

[a] Frohlich (ref. 33).
[b] From Rapp (ref. 34).
[c] Includes direct and diffuse radiation.

B. Solar energy sources

In terms of present energy production, the major renewable energy sources are hydroelectric power, wood and other biomass, geothermal energy, and unmeasured amounts of solar energy for passive space heating (see Tables 11.08.C–F, below). In addition to possible further utilization of these sources, a number of solar energy sources may provide substantial amounts of energy in the future. Projections as to their potential vary widely. Possibilities include:

Solar thermal collectors. Solar thermal collectors can be used for space or water heating. For a mean flux of 200 W/m^2 and a 50% conversion to useful heat energy, the energy collected is approximately 300 MJ/ft^2 per year. Approximately 5 million ft^2 of solar collectors were produced in the United States in 1986.[a]. The primary use was for residential swimming pool heating. The most economical longer term prospects may involve summer collection of heat for a number of housing units (approximately 100), storage in a large pool, and utilization of the heat in the winter.

Wind power.[b] Wind turbine capacity in California at the end of 1987 totalled 1436 MW (rated at wind speeds of about 15 m/sec), representing almost the entire U.S. wind power capacity and most of the world capacity. Total generation was 1700 GWh, corresponding to a capacity factor of 0.14. Individual turbine capacity averaged approximately 100 kW. Wind turbine arrays use approximately 0.03 km^2 to 0.3 km^2 per MW capacity; most of this land is available for farming or grazing.

Biomass. Use of biomass can be increased above present levels with the utilization of farms for fast-growing plants which are harvested to be fermented into fuel. For example, sugar can be fermented into ethyl alcohol, which can be used as fuel for automobiles, either by itself or mixed with gasoline. Wood can be fermented into methyl alcohol. Typical yields (annual average): 0.1–1 W/m^2.

Solar photovoltaic power.[c] World shipment of photovoltaic modules in 1986 corresponded to a total capacity of about 22 MW; the U. S. share was roughly 30%. Most of these units are for power at remote installations or in small consumer products. Pilot plants for electricity generation, with capacities greater than 1 MW, have been constructed but costs are not competitive with other utility power sources.

Cost targets for competitive photovoltaic power are \$0.12/kWh for specialized situations and \$0.06/kWh for general use (in 1986 \$). The latter goal could be achieved with flat-plate systems at module efficiencies of 15% to 20% (roughly a doubling of 1987 efficiencies) and module costs of \$45/m^2 to \$80/m^2 (over a factor of five reduction in cost). With concentrator systems, the same electricity cost is achieved with higher efficiencies and higher module costs. The quoted module costs for \$0.06/kWh are equivalent to \$0.40 per peak watt (W_p) or less; for \$0.12/kWh, the module cost is about \$1/W_p. Present costs are roughly \$5/W_p.

[a] From DOE/EIA (ref. 1).
[b] From refs. 35 and 36.
[c] From refs. 1, 37, and 38.

C. Table: U. S. use of biomass energy, 1984 (in exajoules)[a]

This energy is not included in earlier tables (except for small electric utility usage).

Source	Energy (EJ)
Wood	
Industrial	1.77[b]
Residential	0.97
Commercial	0.02
Electric Utility	0.01
Total	2.78
Waste	0.22
Alcohol fuels	0.05
Total Biomass	3.04

[a] From DOE/EIA (ref. 1).
[b] 97% of this energy is used in the paper product, wood product, and allied industries.

D. Table: Biomass energy use, for selected countries (in exajoules per year)[a]

"Other" biomass sources include sugar cane, agricultural and other wastes, and dung.

Country	Year	Wood (EJ)	Other (EJ)
Brazil	1984	1.18	0.86
Canada	1982	1.04	
India	1981	2.62	1.38
Mexico	1984	0.32	
United States	1984	2.78	
USSR	1984	0.66	
Venezuela	1984	0.68	0.41

[a] From WEC (ref. 26).

E. Table: World hydroelectric generation and capability (in petawatt-hours per year)

Theoretical capability is based on the potential from all water flow at 100% efficiency; exploitable capability is based on limits of current technology and economic conditions.

Hydroelectric generation is converted to exajoules (EJ) using the nominal fossil fuel equivalence[a]:
1 kWh = 10,400 BTU = 10.97 MJ; 1PWh = 10^{12} kWh = 10.97 EJ

	Gen.[a] 1986 (PWh)	Estimated Capability[b] Theor (PWh)	Exploitable (PWh)	Fraction Utilized[c] (%)
Region[d]				
U. S.	0.294	1.06	0.64	46
OECD West[e]	0.759	3.10	1.30	59
OECD East[f]	0.112	0.71	0.28	40
USSR/E. Eur.	0.237	4.23	1.38	17
Mideast	0.010	0.18	0.17	6
Asia: CP[g]	0.121	2.50	1.60	8
Asia: other	0.104	3.05	1.16	9
Africa	0.049	4.28	2.03	2
Latin America	0.310	7.39	1.80	17
Total	1.995	26.51	10.35	19
Major countries[h]				
USSR	0.21		3.83	6
China	0.09	5.92	1.92	5
Brazil	0.18	3.02	0.93	19
Indonesia	0.01	3.39	0.71	1
Canada	0.31	1.09	0.41	75
World (PWh)[i]	1.995	26.51	10.35	19
World (EJ)[i]	21.9	291	114	19

[a] From DOE/EIA (ref. 27).
[b] Estimated capabilities by region and by country are not consistent, due to differences in dates of estimate (e.g., USSR).
[c] Calculated as ratio of 1986 generation to estimated exploitable capability.
[d] Capability from Edmonds (ref. 19), based on 1978 WEC study.
[e] W. Europe and Canada.
[f] Australia, Japan, and New Zealand.
[g] Centrally planned Asian economies, e.g., China and N. Korea.
[h] Countries reporting large exploitable capability for 1986 WEC study (ref. 26).
[i] Based on 1978 study (above); there have been subsequent changes in individual national estimates, but data are fragmentary.

F. Table: Geothermal energy use, for selected countries (in petajoules)[a]

Geothermal electric generation is converted to petajoules (1 PJ = 10^{15} J) using nominal fossil fuel equivalence: 1 kWh = 10,400 BTU = 10.97 MJ.

Country	Elec. 1984 (GWh)	Elec. 1984 (PJ)	Baths 1985 (PJ)	Other 1985 (PJ)
U. S.	7741	84.9	0.1	4.0
Italy	2900	31.8	4.3	2.3
Mexico	1424	15.6		
Japan	1390	15.2	50.8	3.5
New Zealand	1170	12.8		
Iceland	181	2.0	2.4	13.1
China	80	0.9	0.2	4.5
Hungary	0	0.0	6.6	11.5
USSR			4.1	10.1
France				7.4
Sum	14886	163.3	68	56
World			73	67

[a] From WEC (ref. 26).

11.09 WORLD ENERGY CONSUMPTION

A. Figure: World energy consumption, historical

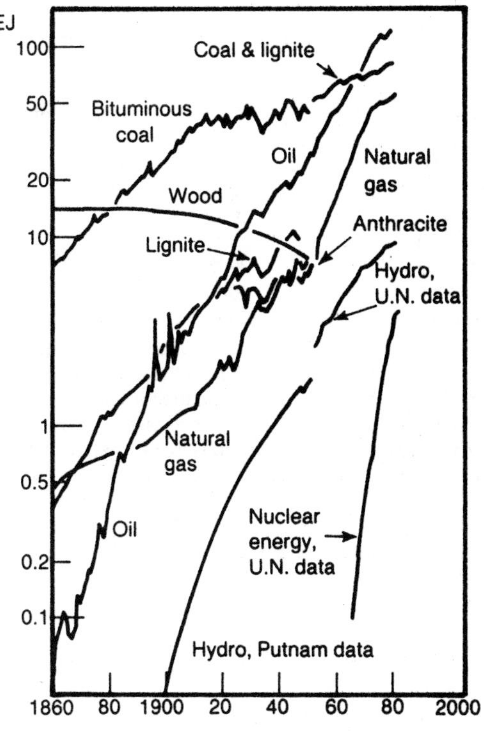

Note: The figure above is a reproduction of Figure 11.06 from the 1981 Vade Mecum.

B. Table: World energy production: 1976 and 1986 (in exajoules per year)[a]

	1976	1986	Ratio
Oil[b]	134.1	133.1	0.99
Coal	71.0	93.1	1.31
Natural Gas	48.2	66.1	1.37
Hydro	15.9	21.9	1.37
Nuclear	4.8	17.3	3.63
Total	274.0	331.5	1.21

[a] From DOE/EIA (ref. 27).
[b] Crude oil and natural gas liquids.

C. Table: Energy consumption patterns for major OECD countries, USSR and China, 1986

In the OECD source categories, (small) contributions from solar and geothermal energy are grouped with "hydro;" the "other" category includes peat, wood, wastes, and net electricity imports.

Country	Primary Energy (EJ)	Energy per cap. (US = 100)	Fractional Contribution from Sources (%) Coal	Oil	Gas	Nuclear	Hydro, etc.	Other	Percent for Electricity
OECD[a]									
United States	75.4	100	23	42	22	5	4	4	34
Japan	15.7	41	18	55	10	11	6	0	40
Germany	11.3	59	29	44	15	10	2	1	35
Canada	9.8	122	10	30	20	7	30	3	46
United Kingdom	8.6	49	32	37	23	6	1	0	34
France	8.4	49	10	43	12	28	7	−1	41
Italy	6.0	34	10	59	20	1	7	2	29
Australia	3.3	66	39	36	16	0	4	5	39
Spain	3.1	26	25	53	3	11	8	0	38
Netherlands	2.7	59	10	36	51	1	0	0	23
Sweden	2.3	89	5	32	0	28	25	9	57
Turkey	1.9	12	31	43	1	0	6	19	23
Belgium	1.9	60	19	45	15	20	1	0	30
Others (11)	8.1	46	15	45	8	5	21	7	40
OECD Total	158.5	62	21	43	18	8	7	3	36
USSR[b]	59	68	24	33	36	3	4		26
China[b]	24	7	76	17	3	0	4		16
World[c]	331	21	28	40	20	5	7		

[a] From OECD (ref. 4).
[b] Estimated from DOE/EIA (ref. 27) and WEC reports (ref. 39).
[c] From DOE/EIA (ref. 27).

Note: Compilations of data by the Organization for Economic Cooperation and Development (OECD), the U. S. Department of Energy (DOE), and the World Energy Conference (WEC) employ somewhat different definitions and conversion factors. Thus, numerical results from these sources are not precisely comparable.

11.10 PROJECTED WORLD ENERGY DEMAND AND RESOURCES UNTIL 2030

A. Table: IIASA estimates of energy consumption, by region (in exajoules per year)[a]

Two projected scenarios for primary energy consumption, based on the work of the Energy Systems Program of the International Institute for Applied Systems Analysis.

The average annual projected rates of increase from 1975 to 2000 for world energy consumption are 2.0% and 2.9% in the "low" and "high" scenarios, respectively. The actual rate from 1975 to 1986 (to 331 EJ in 1986) was 2.3%.

		High scenario		Low scenario	
Region[b]	1975	2000	2030	2000	2030
I (NA)	84	123	190	104	138
II (SU/EE)	58	116	231	104	158
III (WE/JANZ)	71	135	225	107	143
IV (LA)	11	42	116	31	73
V (Af/SEA)	10	45	147	34	84
VI (ME/NAf)	4	24	75	18	39
VII (C/CPA)	15	45	140	31	72
Total	259[c]	531	1124	429	706

[a] From ref. 3.
[b] The IIASA regions are:
- I United States and Canada
- II Soviet Union and E. Europe
- III W. Europe, Japan, Australia, N. Zealand, S. Africa, and Israel
- IV Latin America
- V Africa (except Northern Africa and S. Africa) and South and SE Asia
- VI Middle East and Northern Africa
- VII China and centrally planned Asian economies

[c] Includes 7 EJ of fuel used in international shipments of fuel.

B. Table: World primary energy consumption in alternative scenarios, for years near 2025 (in exajoules per year)[a]

		Projection		
Reference	Year considered	Low[b]	Base	High
Lovins (1981)[c]	2030	165		
Goldemberg (1985)[c]	2020	353		
Colombo (1978)[d]	2030	505		
Mintzer (1988)[c]	2025	250	520	710
Edmonds (1985)[e]	2025	533	921	1071
WEC (1983)[c]	2020	585		776
IIASA (1981)[f]	2030	706		1124

[a] The studies differ in their calculation of the energy equivalent of electricity from renewable sources and in their treatment of noncommercial energy (particularly biomass); these differences are responsible for only a small part of the differences in the final results.
[b] Includes single values when studies were oriented to low energy consumption.
[c] From refs. 40–43.
[d] U. Colombo, as reported in ref. 3.
[e] From ref. 19.
[f] From ref. 3.

C. Table: Alternative scenarios for world energy consumption patterns and carbon emission, for years near 2025

	IIASA[a] Base 1975	Actual[b] 1986	IIASA[a] High 2030	Edmonds[c] 2025	IIASA[a] Low 2030	Goldemberg[d] 2020
Energy, by source (EJ/yr)						
Oil	121	133	215	183	158	102
Gas	48	66	188	113	109	102
Coal	71	93	378	350	203	62
Nuclear	4	17	255	158	163	24
Hydro + solar	16	22	62	118	56	17[e]
Other[f]			26		16	47[g]
Total	259	331	1124	921	706	353
Energy by region (EJ/yr)						
Industrialized[h]	218	257	646	530	439	123
Developing	41	74	478	391	268	230
Total	259	331	1124	921	706	353
Carbon emission (Gt/yr)[i]	4.7	5.7	17	12.3	10	4.8

[a] From ref. 3.
[b] From ref. 27; 1986 regional consumption extrapolated from 1985 data.
[c] Base case from ref. 19; includes biomass with coal (as a solid fuel).
[d] Estimate by Goldemberg *et al.* (ref. 41), which emphasizes end-use efficiency and includes noncommercial biomass (wood, wastes).
[e] Energy of electricity produced (not of fuel replaced).
[f] Includes biomass and geothermal energy, except for ref. 19 which includes biomass with coal; noncommercial biomass (wood, wastes) only in ref. 41.
[g] Based on inclusion of noncommercial biomass energy (at same level as in 1980).
[h] Corresponds approximately to Europe incl. USSR (all) and non-European OECD countries.
[i] For columns 2, 3, and 7: calculated with conversion factors given in Table 11.06.D. For columns 4, 5, and 6: as given by authors.

11.11. REFERENCES

[1]*Annual Energy Review 1987*, Energy Information Administration Report DOE/EIA-0384(87) (U. S. DOE, Washington, DC, 1988).

[2]*Monthly Energy Review, July 1988*, Energy Information Adminstration Report DOE/EIA-0035(88/07) (U. S. DOE, Washington, DC, 1988).

[3]*Energy in a Finite World, A Global Systems Analysis*, W. Häfele, Program Leader, Report by the Energy Systems Program Group of the International Institute for Applied Systems Analysis (IIASA) (Ballinger, Cambridge, 1981).

[4]*Energy Balances of OECD Countries 1985/86*, International Energy Agency, Organization for Economic Co-operation and Development (OECD, Paris, 1988).

[5]J. C. Fisher, *Energy Crisis in Perspective* (Wiley, New York, 1974).

[6]*The Statistical History of the United States from Colonial Times to the Present*, U. S. Bureau of the Census (Fairfield, Stamford, 1965).

[7]W. G. Dupree, Jr. and J. A. West, *United States Energy Through the Year 2000* (U. S. Dept. of Interior, Washington, DC, 1972).

[8]*A Time to Choose, America's Energy Future*, Energy Policy Project of the Ford Foundation (Ballinger, Cambridge, 1974).

[9]*National Plan for Energy Research, Development, and Demonstration: Creating Energy Choices for the Future*, ERDA-48 (U. S. Energy Research and Development Adminstration, Washington, DC, 1975).

[10]*Economic and Environmental Impacts of a U. S. Nuclear Moratorium, 1985-2010*, Institute for Energy Analysis, ORAU (M.I.T. Press, Cambridge, MA, 1979).

[11]*The Good News About Energy* (Council on Environmental Quality, Washington, DC, 1979).

[12]*Energy in Transition, 1985-2010,* Final Report of the Committee on Nuclear and Alternative Energy Systems (CONAES), NRC/NAS (Freeman, San Francisco, 1980); H. Brooks and J. M. Hollander, Ann. Rev. Energy **4**, 1 (1979).

[13]*Energy Projections to the Year 2000*, DOE/PE-0029 (U. S. DOE, Washington, DC, 1981).

[14]P. D. Holtberg, T. J. Woods, and A. B. Ashby, *1987 GRI Baseline Projection of U. S. Energy Supply and Demand to 2010* (Gas Research Institute, Chicago, 1987).

[15]*Annual Energy Outlook 1987*, Energy Information Administration Report DOE/EIA-0383(87) (U.S. DOE, Washington, DC, 1988).

[16]W. L. Fisher, Science **236**, 1631 (1987).

[17]G. L. Dolton *et al.*, *Estimates of Undiscovered Recoverable Conventional Resources of Oil and Gas in the United States*, Geological Survey Circular 860 (USGS, Alexandria, 1981).

[18]R. Nehring, Ann. Rev. Energy **7**, 175 (1982).

[19]J. Edmonds and J. M. Reilly, *Global Energy: Assessing the Future* (Oxford University Press, New York, 1985).

[20]J. P. Riva, Jr., Report 87-414 SPR, Congressional Research Service Report for Congress (Library of Congress, Washington, DC, 1987).

[21]C. D. Masters, E. D. Attanasi, W. D. Dietzman, R. F. Meyer, R. W. Mitchell, and D. H. Root, Proceedings of the 12th World Petroleum Congress, 1987, Vol. 5, pp. 3–27 (Houston, 1987).

[22]*An Assessment of the Natural Gas Resource Base of the United States*, Office of Policy Planning and Analysis Report DOE/W/31109-H1 (U. S. DOE, Washington, DC, 1988).

[23]*Coal–Bridge to the Future*, C. L. Wilson, Project Director, Report of the World Coal Study (WOCOL) (Ballinger, Cambridge, 1980).

[24]*World Energy Resources 1985–2020*, World Energy Conference (IPC Science and Technology Press, Guildford, 1978).

[25]*1983 Survey of Energy Resources*, 12th Congress of the World Energy Conference (WEC, London, 1983).

[26]*1986 Survey of Energy Resources*, World Energy Conference (Holywell, Oxford, 1986).

[27]*International Energy Annual 1986*, Energy Information Administration Report DOE/EIA-0219(86) (U. S. DOE, Washington, DC, 1987).

[28]*Atmospheric Carbon Dioxide and the Global Carbon Cycle*, edited by J. R. Trabalka, Office of Energy Research Report DOE/ER-0239 (U. S. DOE, Washington, DC, 1985).

[29]*Carbon Dioxide Review: 1982*, edited by W. C. Clark, Institute for Energy Analysis, ORAU (Oxford University Press, New York, 1982).

[30]*Uranium: Resources, Production and Demand*, Joint Report of the OECD Nuclear Energy Agency and the International Atomic Energy Agency (OECD, Paris, 1988).

[31]*Nucleonics Week*, February 4, 1988, p. 9.

[32]*USCEA International Reactor Survey* (U. S. Council for Energy Awareness, Washington, DC, 1988).

[33]C. Frohlich and R. W. Brusa, Solar Phys. **74**, 209 (1981).

[34]D. Rapp, *Solar Energy* (Prentice-Hall, Englewood Cliffs, NJ, 1981).

[35]P. Gipe, "Wind Energy: No Longer an 'Alternative' Source of Energy" (to be published); Wind Energy Weekly **7**, No. 300 (April 24, 1988).

[36]D. R. Smith, Ann. Rev. Energy **12**, 145 (1987).

[37]*National Photovoltaics Program: 1987 Program Review*, Report DOE/CH10093-21 (Solar Energy Research Institute, Golden, CO, 1988).

[38]*Five Year Research Plan, 1987–1991, National Photovoltaics Program*, Report DOE/CH-10093-7 (U. S. DOE, Washington, DC, 1987).

[39]National Energy Data Profiles, USSR and China, 13th Congress of the World Energy Conference (1986).

[40]A. B. Lovins, L. H. Lovins, F. Krause, and W. Bach, *Least Cost Energy: Solving the CO_2 Problem* (Brick House, Andover, MA, 1981).

[41]J. Goldemberg, T. B. Johansson, A. K. N. Reddy, and R. H. Williams, Ann. Rev. Energy **10**, 613 (1985).

[42]I. M. Mintzer, *A Matter of Degrees: The Potential for Controlling the Greenhouse Effect* (World Resources Institute, Washington, DC, 1987).

[43]*Energy 2000–2020: World Prospects and Regional Stresses*, edited by J. R. Frisch, World Energy Conference Conservation Commission (Graham and Trotman, London, 1983).

12.00. Fluid dynamics

RUSSELL J. DONNELLY

University of Oregon

CONTENTS

12.01. INTRODUCTION

For most of the twentieth century the subject of fluid dynamics has been of research interest primarily to a relatively small group of practicing physicists. Although the APS Division of Fluid Dynamics has been in existence for more than sixty percent of the life of the AIP, it has always been relatively small, and is presently only a little over one-third of the size of the Division of Nuclear Physics, for example. The subject has not generally been part of the education of undergraduate physicists in the U.S., although it has been in some European countries. The only encounter most physicists might have had with the subject would be the occasional design problem which required searching a handbook for data. The situation has changed markedly in the last few years with the realization by mainstream physicists that fluids, in common with other nonlinear physical systems, exhibit various kinds of bifurcation and chaotic behavior; thus physicists and indeed chemists, mathematicians, and biologists are now giving more attention to what has hitherto been the purlieu of the engineering community and the oceanographic and atmospheric experts. This interdisciplinary interest is likely to grow as the AIP enters its second fifty years, and the data presented here have been selected as a starting point for investigators seeking information in the field without the extensive collections of books accumulated by the specialist.

Fluid dynamics is unusual in the great breadth of activity it encompasses from aerospace engineering to quantum fluid dynamics; some of these areas are represented by whole or part chapters of this volume: rheology, acoustics, high polymer physics, plasma physics, and cryogenics.

12.02. DIMENSIONLESS NUMBERS AND CRITICAL VALUES IN FLUID DYNAMICS

A. Definitions

Let L be a typical length, U a typical velocity, ν the kinematic viscosity, $\eta = \nu\rho$ the viscosity, ρ the density, c the velocity of sound, g the acceleration due to gravity, σ the surface tension, p the pressure, f the frequency, Ω the angular velocity of rotation, α the coefficient of expansion, k the thermal conductivity, T the temperature, C_p the specific heat, $\kappa = k/\rho C_p$ the thermal diffusivity (thermometric conductivity), H the transfer of heat per unit area, and F the total force on a body.

Coefficient of resistance for pipe flow. Defined from

$$\nabla p = (\lambda/d)\tfrac{1}{2}\rho U^2,$$

where ∇p is the pressure gradient, d the diameter of the pipe, and U the mean velocity of flow. For laminar flow $\lambda = 64/Re$.

Dean number. Measures the influence of curvature in laminar flow through pipes:

$$D = \tfrac{1}{2}Re(R/r)^{1/2},$$

where R is the radius of the cross section and r the radius of curvature of the pipe.

Drag coefficient. The ratio

$$C_D = F_D/\tfrac{1}{2}\rho U^2L^2,$$

where F_D is the drag force on a body of characteristic dimension L. The notation C_f is used when F_D arises from skin friction.

Ekman number. The ratio of viscous to Coriolis forces:

$$\nu/\Omega L^2.$$

Froude number. For long waves on a surface of a layer of depth L, the Froude number is the ratio of flow speed to wave speed:

$$Fr = U/(Lg)^{1/2}.$$

The "internal Froude number," in cases where free surface effects are absent, is the square root of the reciprocal of the Richardson number.

Grashof number. Indicates the relative importance of inertia and viscous forces in convection, but is not a simple ratio of two dynamical forces:

$$Gr = g\alpha\Delta TL^3/\nu^2.$$

Lift coefficient. The ratio

$$C_L = F_L/\tfrac{1}{2}\rho U^2L^2,$$

where F_L is the lift force on a body of characteristic dimension L.

Mach number. The ratio of the velocity to the velocity of sound:

$$Ma = U/c.$$

Marangoni number. Measures importance of the temperature dependence of surface tension in convection in thin layers of fluid:

$$B = \frac{\partial\sigma}{\partial T}\frac{\Delta TL}{\rho\nu\kappa}.$$

Nusselt number. Dimensionless rate of heat transfer H:

$$Nu = HL/k\Delta T.$$

Péclet number. The ratio of advection of heat to conduction of heat:

$$Pe = UL/\kappa.$$

Prandtl number. The ratio of the diffusivity of vorticity to the diffusivity of heat:

$$Pr = \nu/\kappa.$$

Pressure coefficient. The ratio

$$p/\tfrac{1}{2}\rho U^2.$$

The *Euler number* is one-half the pressure coefficient.

Rayleigh number. Dimensionless temperature gradient in horizontal convection, such as Bénard convection:

$$Ra = GrPr = g\alpha\Delta TL^3/\nu\kappa.$$

Reynolds number. The ratio of inertia forces to viscous forces in a flow:

$$Re = \frac{\rho U^2/L}{\eta U/L^2} = \frac{\rho UL}{\eta} = \frac{UL}{\nu}.$$

Richardson number. The stability of flows stratified

in the Z direction depends on the Richardson number,

$$Ri = -\frac{g}{\rho}\frac{d\rho}{dZ}\left(\frac{dU}{dZ}\right)^{-2} = \frac{1}{Fr^2},$$

which describes the stratification in terms of the velocity gradient at the wall. $Ri = 0$ corresponds to a homogeneous fluid; $Ri > 0$ and $Ri < 0$ refer to stable and unstable stratification. Fr is the internal Froude number.

Rossby number. The ratio of inertia to Coriolis forces:

$$U/\Omega L.$$

Roughness. The relative roughness is the ratio of the height of protrusions into a pipe k to the radius or hydraulic radius of the cross section. Equivalent sand roughness ks/R is valid only for roughness obtained with sand, and is discussed in Chap. 20 of Ref. 10.

Schmidt number:

$$\nu/\kappa_c,$$

analogous to the Prandtl number but for cases where a substance carried by the fluid has a concentration c and diffusion law

$$\frac{Dc}{Dt} = \kappa_c \nabla^2 c,$$

where κ_c is a diffusion coefficient depending on both the fluid and the diffusing substance.

Strouhal number. A nondimensional frequency of vortex shedding:

$$S = fL/U.$$

Taylor number. Dimensionless rotation rate:

$$Ta = \Omega^2 L^4/\nu^2,$$

for example. For flow between rotating cylinders with the inner cylinder rotating and radii R_1, R_2, the Reynolds number is

$$Re = \Omega_1 R_1 (R_2 - R_1)/\nu;$$
$$Ta = 2Re^2(1-\eta)/(1+\eta),$$

where $\eta = R_1/R_2$. There are many specialized definitions of Ta.

B. Typical observed and critical values of dimensionless numbers

1. Critical Reynolds numbers for pipes and channels

The situations in smooth pipes or channels depend very much on the entry conditions. With disturbed entry conditions, the disturbances die out after a certain length if Re is small enough. As Re is increased, a value Re_{crit} is reached such that for $Re > Re_{crit}$ disturbances no longer die out and the flow in the pipe is turbulent. There is a minimum value of Re_{crit} below which all disturbances are damped out downstream. This is of order 2000. With great care, laminar flow has been observed to $Re = 100\,000$. (See Refs. 2 and 10.) For various shaped pipes and channels see Ref. 10 and the data in Sec. 12.05.

2. Strouhal number

For $Re > 10^3$, the Strouhal number can be expressed as a function of drag coefficient for essentially two-dimensional bodies:

$$S = 0.21/C_D^{3/4}, \quad 0 < C_D < 2.$$

(See Secs. 3–6 of Ref. 4.)

The Strouhal number for the Kármán vortex sheet in the flow past a circular cyclinder in terms of the Reynolds number is shown in Figure 12.02.B.2.a.

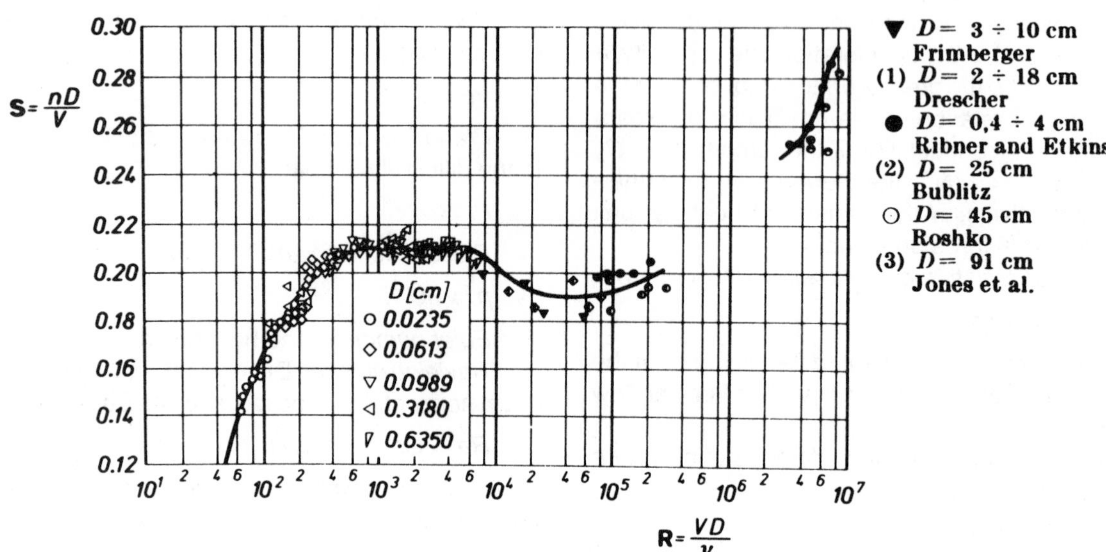

FIGURE 12.02.B.2.a. Strouhal number S for the Kármán vortex sheet in the flow past a circular cylinder in terms of Reynolds number R. From Ref. 10, Chap. 2.

3. Couette flow

The stability of flow between concentric cylinders of radii R_1 and R_2 has been investigated theoretically and experimentally. The theoretical critical angular velocities when the inner cylinder is rotating are expressed in terms of a Reynolds or Taylor number. Swinney and DiPrima[14] (Chap. 6) have compiled a table of critical Reynolds numbers $Rc = \Omega_c R_1(R_1 - R_2)/\nu$ as a function of radius ratio $\eta = R_1/R_2$. See Table 12.02.B.3.a.

TABLE 12.02.B.3.a. Critical Reynolds numbers

η	Rc	η	Rc
0.975	261.0	0.60	71.7
0.9625	213.2	0.50	68.19
0.950	185.0	0.40	68.29
0.925	151.5	0.36	69.54
0.900	131.6	0.35	70.0
0.875	118.2	0.30	77.83
0.85	108.3	0.28	75.10
0.80	94.7	0.25	78.75
0.75	85.8	0.20	88.15
0.70	79.5	0.15	105.06
0.65	75.0	0.10	141.05

4. Rayleigh number

The critical adverse temperature gradient across a horizontal layer signaling the onset of convection is measured by the critical Rayleigh number. Its exact value depends on the assumptions made about the surfaces; for two free surfaces

$$Ra_{\text{crit}} = 657.5,$$

for one rigid and one free surface

$$Ra_{\text{crit}} = 1100.6,$$

and for two rigid surfaces

$$Ra_{\text{crit}} = 1707.7.$$

12.03. PROPERTIES OF THE ATMOSPHERE

TABLE 12.03.A. Composition of dry air. Many other trace gases are present in the atmosphere, some of which (e.g., Xe, SO_2, NH_3, Rn, etc.) have been extensively studied, but none influences the radiation fluxes to a significant extent. In addition to these gases the atmosphere contains solid matter in suspension, whose concentration and composition are very variable. The figures below apply only to dry air. From Ref. 6, p.10.

Molecule	Fraction by volume in troposphere	Comments
N_2	7.8084×10^{-1}	Photochemical dissociation high in ionosphere. Mixed, at lower levels.
O_2	2.0946×10^{-1}	Photochemical dissociation above 95 km. Mixed at lower levels.
A	9.34×10^{-3}	Mixed up to 110 km; diffusive separation above.
CO_2	3.3×10^{-4}	Slightly variable. Mixed up to 100 km; dissociated above.
Ne	1.818×10^{-5}	Mixed up to 110 km; diffusive separation above.
He	5.24×10^{-6}	Mixed up to 110 km; diffusive separation above.
CH_4	1.6×10^{-6}	Mixed in troposphere; oxidized in stratosphere; dissociation in mesophere.
Kr	1.14×10^{-6}	Mixed up to 110 km; diffusive separation above.
H_2	5×10^{-7}	Mixed in troposphere and stratosphere, dissociated above.
N_2O	3.5×10^{-7}	Slightly variable at surface. Continuous dissociation in stratosphere and mesosphere.
CO	7×10^{-8}	Variable combustion product.
O_3	$\sim10^{-8}$	Highly variable; photochemical origin.
NO_2, NO	$0\text{–}2\times10^{-8}$	Industrial origin in troposphere. Photochemical origin in mesosphere and ionosphere.

TABLE 12.03.B. Isotopic abundances in nature. Almost all terrestrial hydrogen is combined in the form of water. Since HHO and HDO have different vapor pressures, the relative concentration D^2:H^1 can vary from phase to phase by as much as 10%. Small differences in the concentration of oxygen isotopes also occur. From Ref. 6, p. 11.

Isotope	Percentage relative abundance	Isotope	Percentage relative abundance
H^1	99.9851	O^{16}	99.758
D^2	0.0149	O^{17}	0.0373
		O^{18}	0.2039
C^{12}	98.892		
C^{13}	1.108	N^{14}	99.631
		N^{15}	0.369

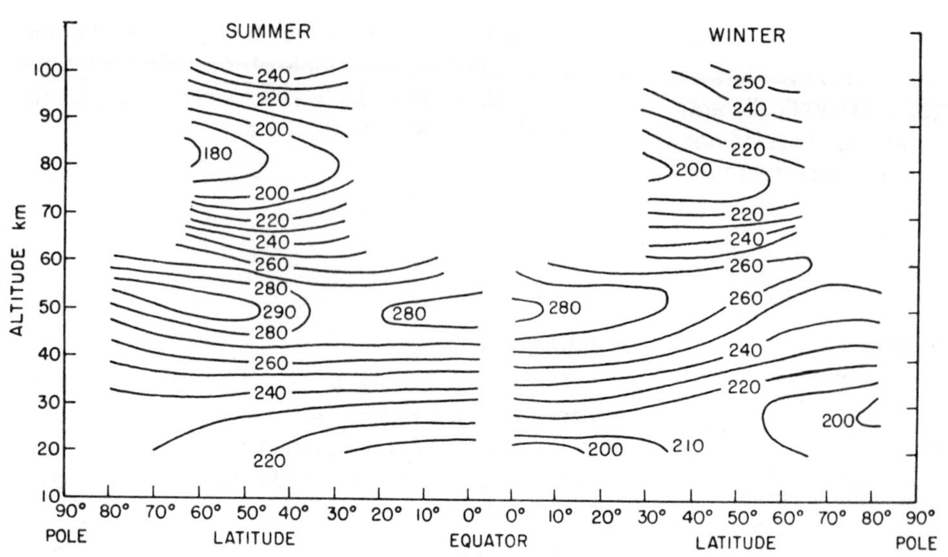

FIGURE 12.03.C. Mean temperature (K) between 20 and 100 km. From Ref. 6, p. 8; after Murgatroyd (1957).

TABLE 12.03.D. Standard atmosphere. From Ref. 1, p. 540 (with change of units).

Altitude (m)	Temp. (°C)	Pressure (mm Hg)	Density ρ/ρ_0 [a]	Speed of sound (m/s)	Altitude (m)	Temp. (°C)	Pressure (mm Hg)	Density ρ/ρ_0 [a]	Speed of sound (m/s)
304.80	13.0	732.85	0.9711	339.24	6 096.00	−24.6	349.18	0.5328	316.08
914.40	9.0	681.22	0.9151	336.80	6 705.60	−28.5	320.92	0.4976	313.64
1524.00	5.0	632.49	0.8617	334.67	7 315.20	−32.5	294.55	0.4642	311.20
2133.60	1.1	586.58	0.8106	332.23	7 924.80	−36.5	269.99	0.4325	308.46
2743.20	−2.8	543.36	0.7620	329.79	8 534.40	−40.4	247.11	0.4025	306.02
3352.80	−6.8	502.74	0.7156	327.36	9 144.00	−44.4	225.81	0.3741	303.28
3962.40	−10.7	464.60	0.6713	324.92	9 753.60	−48.4	205.98	0.3473	300.53
4572.00	−14.7	428.85	0.6292	322.48	10 363.20	−52.4	187.52	0.3220	298.09
5181.60	−18.6	395.37	0.5892	320.04	10 972.80	−55.3	170.32	0.2971	295.96
5791.20	−22.6	364.06	0.5511	317.30	11 277.60	−55.3	162.16	0.2844	295.96

[a] $\rho_0 = 0.001\,225$ g/cm^3.

12.04. PROPERTIES OF COMMON FLUIDS

Obviously there are an enormous number of fluids of interest in fluid dynamics. Space precludes even a representative listing of properties such as viscosity and kinematic viscosity, thermal conductivity of liquids and gases, specific heats of liquids and gases, and surface tension. All these properties depend on temperature, and many have significant variation with pressure. An important and convenient modern reference is *The Properties of Gases and Liquids,* by Reid, Prausnitz, and Sherwood.[7] Tables specific to liquids useful in fluid dynamics have been assembled by Dürst and Kline,[28] including comments on such matters as odor, toxicity, surface tension, vapor pressure, etc. Detailed information on "Freon" fluorocarbons may be obtained from the "Freon" Products Division, Du Pont Co., Wilmington, DE 19898. Similarly, information on Dow Corning silicone fluids may be obtained from the Dow Corning Corp., Midland, MI 48640. Viscosity and density tables for salt and fresh water in English and metric units are contained in Technical and Research Bulletin 1-25 (Feb. 1978), available from The Society of Naval Architects and Marine Engineers, One World Trade Center, Suite 1369, New York, NY 10048.

TABLE 12.04.A. Viscosity and kinematic viscosity of water. All values are at 1 atm. The percentage change in the viscosity of water between 1 and 100 atm seems to vary from about −1.2 at 0°C to about +0.7 at 75°C. From Ref. 8, Vol. I, pp. 5 and 6.

Temp. (°C)	100η (g/cm s)	100ν (cm^2/s)	Temp. (°C)	100η (g/cm s)	100ν (cm^2/s)
0	1.792	1.792	40	0.656	0.661
5	1.519	1.519	45	0.599	0.605
10	1.308	1.308	50	0.549	0.556
15	1.140	1.141	60	0.469	0.477
20	1.005	1.007	70	0.406	0.415
25	0.894	0.897	80	0.357	0.367
30	0.801	0.804	90	0.317	0.328
35	0.723	0.727	100	0.284	0.296

The properties of air for experiments in compressible flow are admirably summarized in Report 1135 of NASA Ames Research Center.[9]

Fluid properties are often subject to errors and anyone basing a critical experiment should seek independent verification of magnitudes.

TABLE 12.04.B. Viscosity (cP) of aqueous glycerol solutions. These mixtures have had much use in fluid mechanics to reach a variety of useful viscosities and also as viscosity standards. They are, however, hygroscopic. From *Handbook of Chemistry & Physics,* 43rd ed. (Chemical Rubber, Cleveland, 1976).

Glycerol (wt. %)	Temperature (°C) 0	10	20	30	40	50	60	70	80	90	100
0	1.792	1.308	1.005	0.8007	0.6560	0.5494	0.4688	0.4061	0.3565	0.3165	0.2838
10	2.44	1.74	1.31	1.03	0.826	0.680	0.575	0.500	...	...	...
20	3.44	2.41	1.76	1.35	1.07	0.879	0.731	0.635	...	...	...
30	5.14	3.49	2.50	1.87	1.46	1.16	0.956	0.816	0.690	...	...
40	8.25	5.37	3.72	2.72	2.07	1.62	1.30	1.09	0.918	0.763	0.668
50	14.6	9.01	6.00	4.21	3.10	2.37	1.86	1.53	1.25	1.05	0.910
60	29.9	17.4	10.8	7.19	5.08	3.76	2.85	2.29	1.84	1.52	1.28
65	45.7	25.3	15.2	9.85	6.80	4.89	3.66	2.91	2.28	1.86	1.55
67	55.5	29.9	17.7	11.3	7.73	5.50	4.09	3.23	2.50	2.03	1.68
70	76.0	38.8	22.5	14.1	9.40	6.61	4.86	3.78	2.90	2.34	1.93
75	132	65.2	35.5	21.2	13.6	9.25	6.61	5.01	3.80	3.00	2.43
80	255	116	60.1	33.9	20.8	13.6	9.42	6.94	5.13	4.03	3.18
85	540	223	109	58.0	33.5	21.2	14.2	10.0	7.28	5.52	4.24
90	1310	498	219	109	60.0	35.5	22.5	15.5	11.0	7.93	6.00
91	1590	592	259	126	68.1	39.8	25.1	17.1	11.9	8.62	6.40
92	1950	729	310	147	78.3	44.8	28.0	19.0	13.1	9.46	6.82
93	2400	860	367	172	89.9	51.5	31.6	21.2	14.4	10.3	7.54
94	2930	1040	437	202	105	58.4	35.4	23.6	15.8	11.2	8.19
95	3690	1270	523	237	121	67.0	39.9	26.4	17.5	12.4	9.08
96	4600	1585	624	281	142	77.8	45.4	29.7	19.6	13.6	10.1
97	5770	1950	765	340	166	88.9	51.9	33.6	21.9	15.1	10.9
98	7370	2460	939	409	196	104	59.8	38.5	24.8	17.0	12.2
99	9420	3090	1150	500	235	122	69.1	43.6	27.8	19.0	13.2
100	12070	3900	1412	612	284	142	81.3	50.6	31.9	21.3	14.8

TABLE 12.04.C. Viscosity and kinematic viscosity of air. From Ref. 8, Vol. I, p. 7.

Temp. (°C)	$10^4\eta$ (g/cm s)	ν (cm^2/s)
0	1.709	0.132
20	1.808	0.150
40	1.904	0.169
60	1.997	0.188
80	2.088	0.209
100	2.175	0.330
120	2.260	0.252
140	2.344	0.274
160	2.425	0.298
180	2.505	0.322
200	2.582	0.346
220	2.658	0.371
240	2.733	0.397
260	2.806	0.424
280	2.877	0.451
300	2.946	0.481
320	3.014	0.507
340	3.080	0.535
360	3.146	0.565
380	3.212	0.595
400	3.277	0.625
420	3.340	0.656
440	3.402	0.688
460	3.463	0.720
480	3.523	0.752
500	3.583	0.785

TABLE 12.04.D. Refractive index of dry air, $(n-1)\times10^8$. From Ref. 6, p. 389.

Wave-length (μ)	−30°C	0°C	+30°C
0.2	38 406	34 187	30 802
0.3	34 552	30 756	27 711
0.4	33 509	29 828	26 875
0.5	33 060	29 428	26 514
0.6	32 824	29 218	26 325
0.7	32 684	29 093	26 213
0.8	32 594	29 013	26 140
0.9	32 533	28 959	26 091
1.0	32 489	28 920	26 056
2.0	32 351	28 797	25 946
3.0	32 326	28 775	25 925
4.0	32 317	28 767	25 918
5.0	32 314	28 763	25 915
6.0	32 311	28 761	25 913
7.0	32 309	28 760	25 912
8.0	32 309	28 759	25 912
9.0	32 308	28 759	25 911
10.0	32 308	28 758	25 911
12.0	32 307	28 758	25 910
14.0	32 307	28 757	25 910
16.0	32 306	28 757	25 910
18.0	32 306	28 757	25 910
20.0	32 306	28 757	25 910
∞	32 305.7	28 756.5	25 909.2

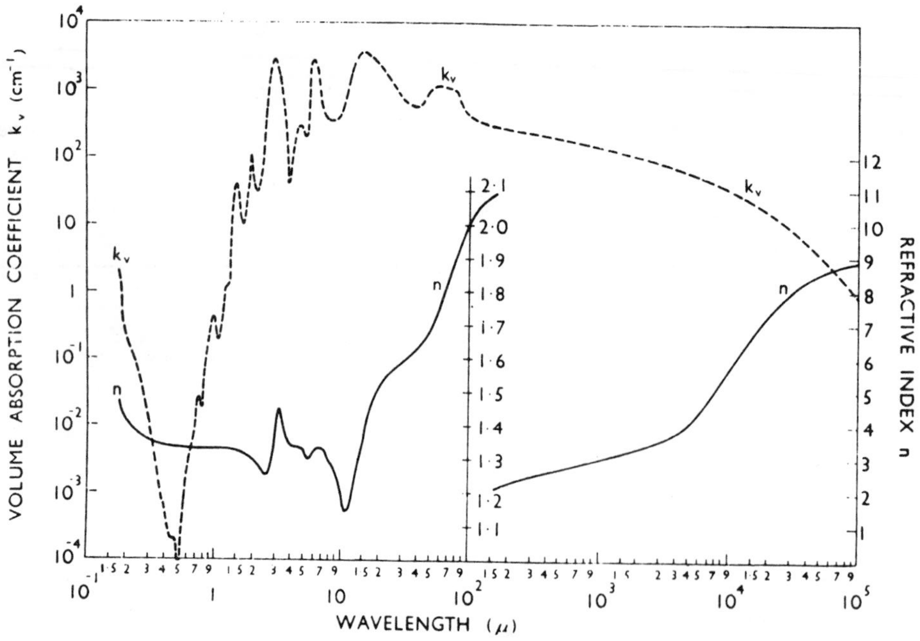

FIGURE 12.04.E. Optical properties of pure liquid water, at about 20°C and 1 atm. From Ref. 6, p. 415.

TABLE 12.04.F. Physical properties of mercury, water, and Dow Corning silicone oil relevant to Bénard convection experiments. The quantities listed in cgs units are density, expansion coefficient, viscosity (in poise), kinematic viscosity, thermal conductivity, specific heat, thermal diffusivity, and Prandtl number. From H. T. Rossby, J. Fluid Mech. **36**, 309 (1969). See article for discussion of accuracy.

				Mercury				
T	ρ	$\alpha\times10^3$	$\eta\times10^2$	$\nu\times10^3$	$k\times10$	$C\times10$	$\kappa\times10$	$P\times10$
20	13.55	0.181	1.550	1.144	0.208	0.3321	0.463	0.247
21	13.54	0.181	1.544	1.140	0.209	0.3321	0.464	0.246
22	13.54	0.181	1.538	1.136	0.209	0.3320	0.466	0.244
23	13.54	0.181	1.532	1.132	0.210	0.3319	0.467	0.242
24	13.54	0.181	1.526	1.127	0.211	0.3319	0.469	0.241
25	13.53	0.181	1.520	1.123	0.211	0.3318	0.470	0.239
				Water				
T	ρ	$\alpha\times10^3$	$\eta\times10^2$	$\nu\times10^2$	$k\times10^3$	C	$\kappa\times10^3$	P
20	0.9982	0.207	1.006	1.008	1.402	0.9991	1.406	7.17
21	0.9980	0.217	0.983	0.985	1.406	0.9989	1.410	6.99
22	0.9978	0.227	0.961	0.963	1.410	0.9988	1.415	6.81
23	0.9976	0.237	0.938	0.941	1.414	0.9987	1.419	6.63
24	0.9973	0.247	0.916	0.918	1.418	0.9986	1.424	6.45
25	0.9971	0.257	0.894	0.896	1.422	0.9985	1.428	6.28
				20 cSt silicone oil				
T	ρ	$\alpha\times10^3$	η	ν	$k\times10^3$	C	$\kappa\times10^3$	P
20	0.9603	1.07	0.2053	0.2137	0.337	0.346	1.014	211
21	0.9593	1.07	0.2006	0.2091	0.337	0.346	1.015	206
22	0.9582	1.07	0.1959	0.2044	0.337	0.346	1.016	201
23	0.9571	1.07	0.1912	0.1997	0.337	0.346	1.018	196
24	0.9561	1.07	0.1865	0.1951	0.337	0.346	1.019	192
25	0.9550	1.07	0.1818	0.1904	0.337	0.346	1.020	187

TABLE 12.04.G. (a) Viscosity and (b) kinematic-viscosity conversion factors. After Ref. 10, Chap. 1. kp and lbf are units of force.

(a) Absolute viscosity μ

	kp s/m²	kp h/m²	Pa s
kp s/m²	1	2.7778×10^{-4}	9.8067
kp h/m²	3600	1	3.5304×10^{4}
Pa s	1.0197×10^{-1}	2.8325×10^{-5}	1
kg/m h	2.8325×10^{-5}	7.8682×10^{-9}	2.778×10^{-4}
lbf s/ft²	4.8824	1.3562×10^{-3}	4.7880×10^{1}
lbf h/ft²	1.7577×10^{4}	4.8824	1.7237×10^{5}
lb/ft s	1.5175×10^{-1}	4.2153×10^{-5}	1.4882

	kg/m h	lbf s/ft²	lbf h/ft²	lb/ft s
kp s/m²	3.5316×10^{4}	2.0482×10^{-1}	5.6893×10^{-5}	6.5898
kp h/m²	127.1×10^{6}	7.3734×10^{2}	2.0482×10^{-1}	2.3723×10^{4}
Pa s	1	2.0885×10^{-2}	5.8015×10^{-6}	6.7197×10^{-1}
kg/m h	0.1724×10^{6}	5.8015×10^{-6}	1.6115×10^{-9}	1.8666×10^{-4}
lbf s/ft²	$620.8\cdot10^{6}$	1	2.7778×10^{-4}	3.2174×10^{1}
lbf h/ft²	5.358×10^{3}	3600	1	1.1583×10^{5}
lb/ft s		3.1081×10^{-2}	8.6336×10^{-6}	1

(b) Kinematic viscosity ν

	s/m²	m²/h	cm²/s	ft²/s	ft²/h
m²/s	1	3600	1×10^{4}	1.0764×10^{1}	3.8750×10^{4}
m²/h	2.7778×10^{-4}	1	2.778	2.9900×10^{-3}	1.0764×10^{1}
cm²/s (stokes)	1×10^{-4}	0.36	1	1.0764×10^{-3}	3.8750
ft²s	9.2903×10^{-2}	3.3445×10^{2}	9.2903×10^{2}	1	3600
ft²/h	2.5806×10^{-5}	9.2903×10^{-2}	2.5806×10	2.7778×10^{-4}	1

12.05. FRICTION AND DRAG

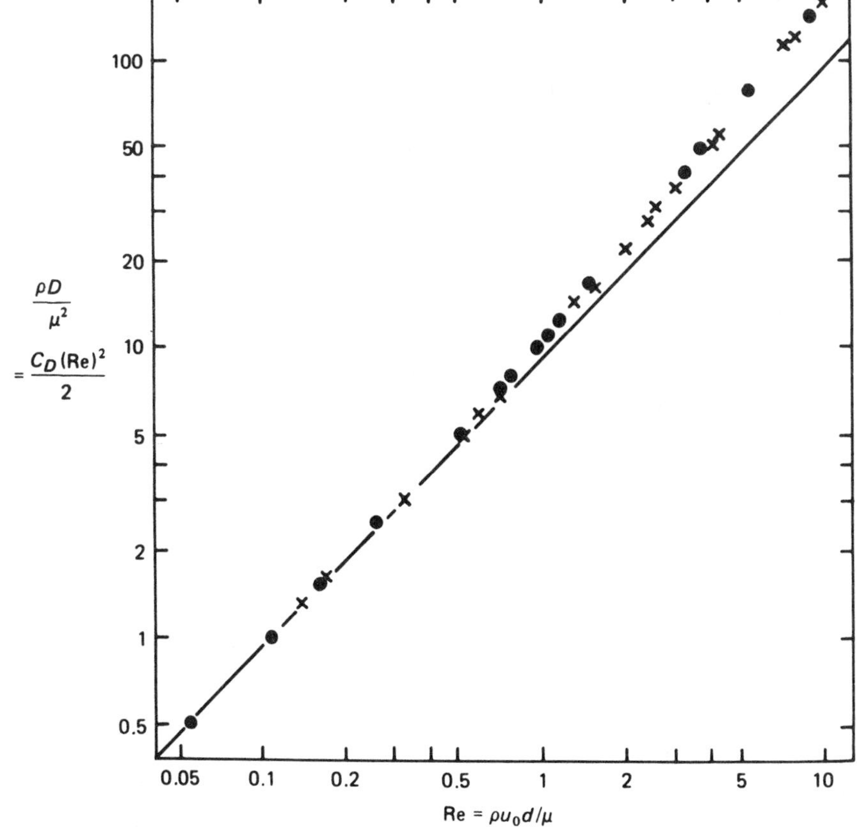

FIGURE 12.05.A. Low Reynolds number drag on a sphere. The solid line represents Stokes' formula $F=6\pi\eta aU$, where a is the radius of the sphere, or $C_D=24/Re$. After Ref. 2, p. 92.

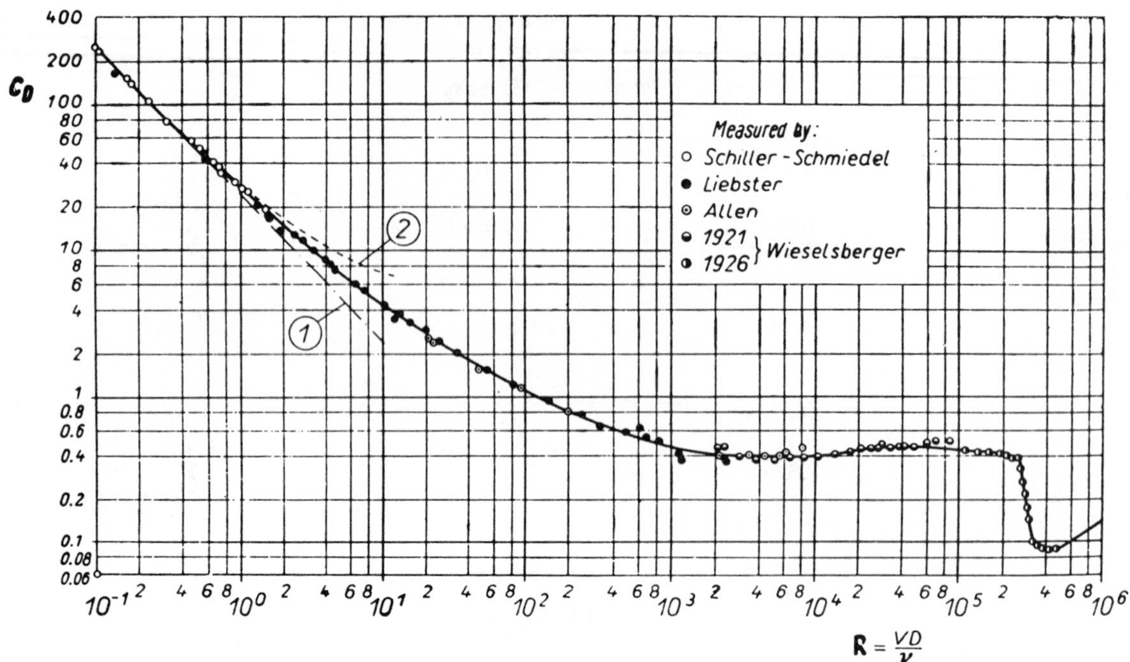

FIGURE 12.05.B. Drag coefficient for spheres as a function of Reynolds number. Curve (1) represents Stokes' theory as in Figure 12.05.A. Curve (2) represents Oseen's improvement $C_D = (24/Re)(1 + 3Re/16)$. After Ref. 10, p. 17.

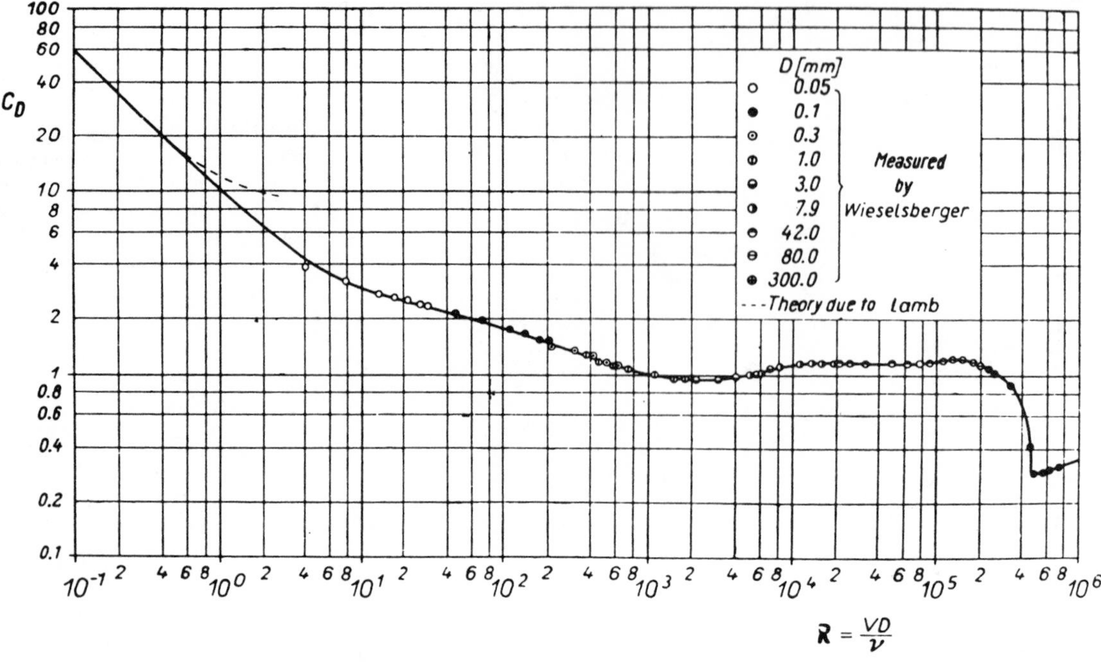

FIGURE 12.05.C. Drag coefficient for circular cylinders as a function of Reynolds number. After Ref. 10, p. 17. For more recent references see p. 29 of Ref. 2.

TABLE 12.05.D. Sinking speeds of particles in still air at 0°C at 1 atm for particles of the density of water. From Natl. Air Pollut. Control Adm. (U.S.) Publ. AP Ser. **49**.

Diameter (μm)	Settling velocity (cm/s)	Diameter (μm)	Settling velocity (cm/s)
0.1	0.000 08	100	25
1	0.004	1000	390
10	0.3		

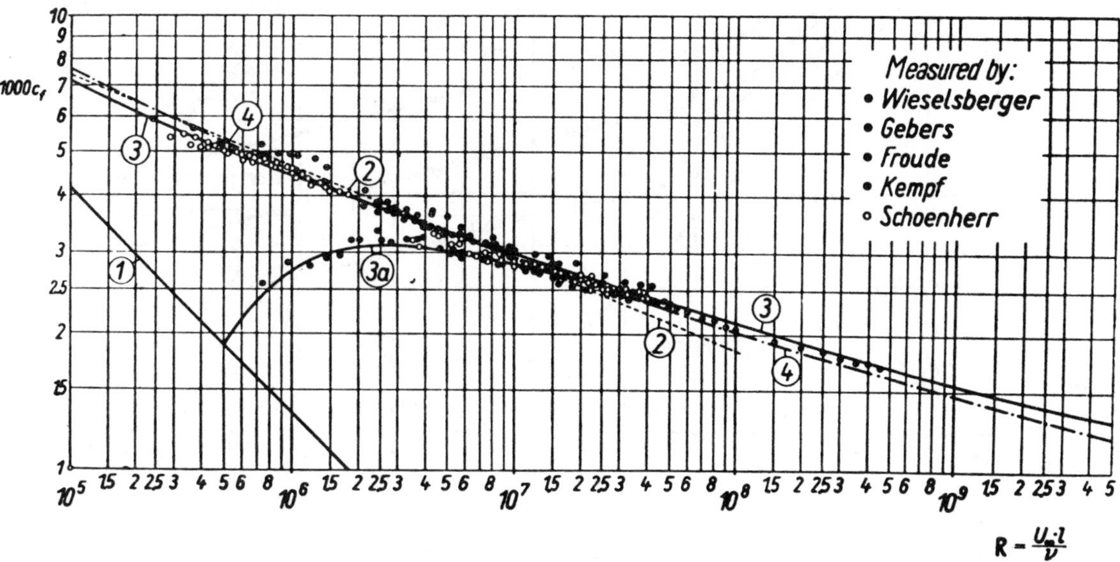

FIGURE 12.05.E. Dimensionless coefficient of skin friction for flow over a flat plate at zero incidence without separation. After Schlichting,[10] p. 639. The numbered curves are various theoretical approximations discussed by Schlichting.

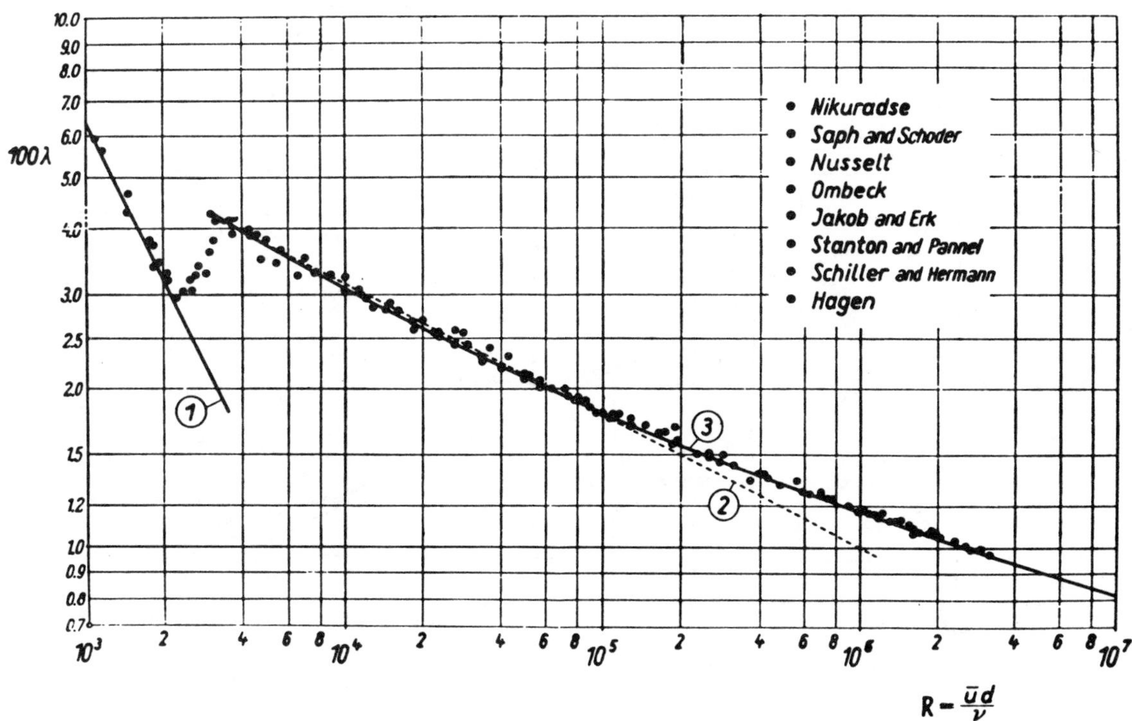

FIGURE 12.05.F. Dimensionless resistance coefficient for laminar and turbulent flow through smooth pipes. After Schlichting,[10] p. 598. The numbered curves are various theoretical approximations discussed by Schlichting. The quantity λ is defined for a pipe of length L, diameter d as $(p_1 - p_2)/L = (\lambda/d)(\rho/2)\bar{u}^2$, where ρ is the density and $\bar{u}$ is the average velocity.

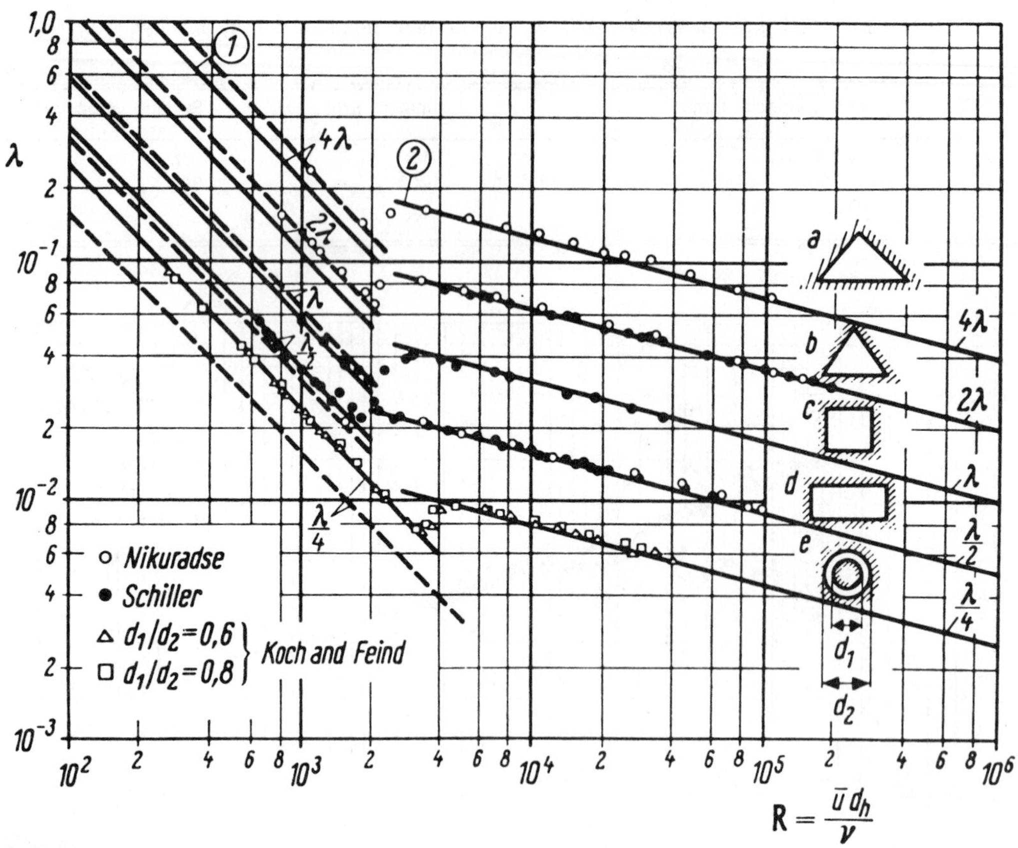

FIGURE 12.05.G. Resistance for flow through pipes of noncircular cross section. The characteristic length in the Reynolds number is the hydraulic diameter of the pipe $d_h = 4A/C$, where A is the area of the pipe and C the wetted perimeter. After Ref. 10, Chap. 20. See caption for Figure 12.05.F.

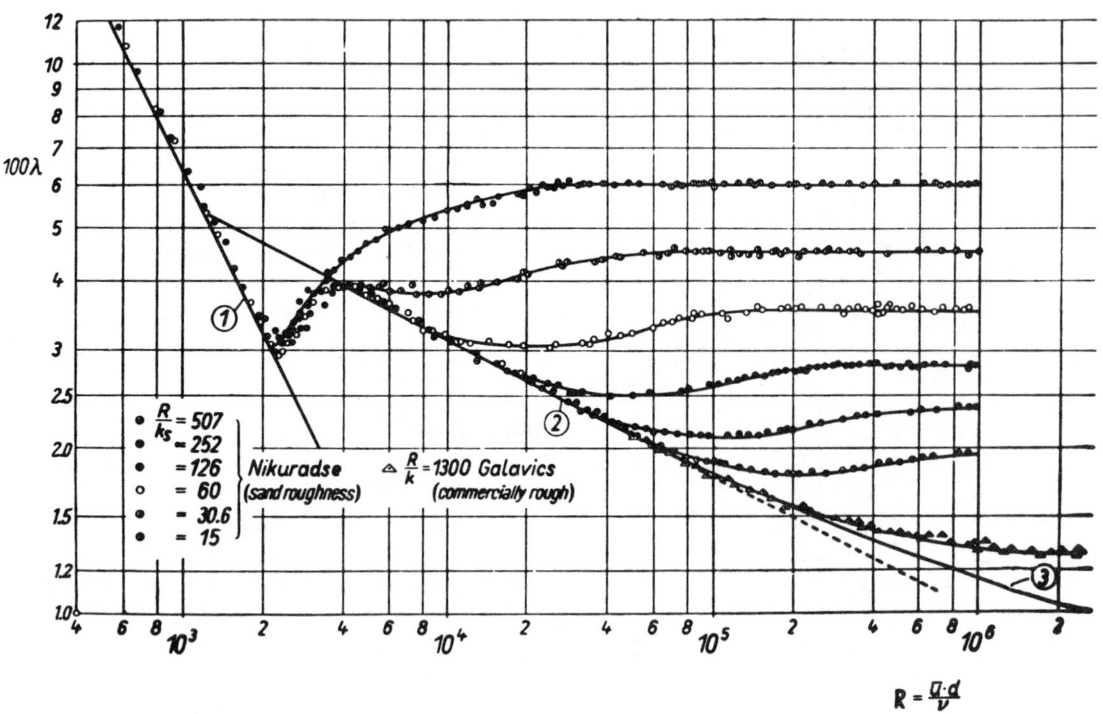

FIGURE 12.05.H. Resistance for flow through rough pipes. After Schlichting,[10] Chap. 20. The theoretical curves are discussed by Schlichting. See caption for Figure 12.05.F.

12.06. EXPERIMENTAL METHODS IN FLUID DYNAMICS

A brief general reference for experimental methods in fluid dynamics is contained in Tritton,[2] Chap. 23. The material here draws, in part, on that reference. Commercial products mentioned are representative and not endorsements by the AIP or the editors.

A. Pressure and temperature measurement

These techniques are common to many branches of physics. Highly sensitive methods have become particularly well developed in low-temperature physics, where capacitive pressure gauges have been developed which can measure very small displacements of a flexible diaphragm (down to angstroms!) and temperature sensors which can resolve a small fraction of a microdegree. This is why some fluid-dynamical experiments are being carried out at cryogenic temperatures.

At room temperature a very convenient high-resolution temperature standard is the Hewlett-Packard 2804 quartz thermometer, which can resolve 10^{-4} K and can be calibrated against a triple-point cell for absolute accuracy. Liquid in glass thermometers of good resolution can be purchased but need to be recalibrated from time to time. Manometer design is discussed by Bradshaw[16]; commercial instruments include the Texas Instruments Bourdon gage and the MKS Baratron, both of which can be calibrated accurately, resolve very small differences in pressure and have useful readout capabilities. Platinum resistance thermometers of secondary standard accuracy can be purchased from several sources, such as the Rosemount Co. and Leeds and Northrup Co. Specialized devices such as thermocouples and thermistors are available from a wide variety of sources.

B. Viscosity measurements

Viscosities of liquids represent a particularly demanding problem if precision is required. Many substances change their viscosity by as much as 2% per °C, and some, like glycerol-water solutions, are hygroscopic and hence change their concentration with time. A complete study may require an accurate thermometer, and calibrated viscosity and density devices. For viscosities in the centipoise range, calibrated capillary flow viscometers such as the Cannon type are available from scientific supply houses. Standard oils for calibration of such devices are available from Brookfield Engineering Laboratories and the Cannon Instrument Co.

C. Flow visualization

The photography of fluid-dynamic flows has become a highly refined art, and the presentation of good photographs and films is often a convincing proof of discovery of a new phenomenon. Since most fluids are homogeneous, various techniques have been developed to mark the flow. Smoke can be introduced into wind tunnels[16]; dyes such as potassium permanganate, gentian violet, and methyl blue have been used in liquids. Density changes in the fluid are to be avoided.

Markers can be produced electrically. The hydrogen bubble technique uses a wire stretched across the flow as a cathode, the anode being, for example, a wall of the channel. Typically 10 V are needed with tap water, larger voltage with deionized water. With fine (10–100 μm) wire the hydrogen bubbles produced at it are small and numerous enough to move with the flow and appear as a white dye. Sections of wire can be insulated and/or voltage pulsed to produce patches of dye, giving very effective pictures. A cathode made of, or coated with, metallic tellurium releases a dense brown dye, which is effective for flow visualization.[17] Care is needed to avoid toxic effects. Similarly, a white colloidal cloud is produced at an anode made of certain materials.[18] Those containing tin (in water containing salt and sodium carbonate) are particularly effective, for example, solder wire. This "electrolytic precipitation method" requires voltages of order 10 V. Finally, a method described by Baker[19] uses the fact that there is a change in pH in an electrolyte in the vicinity of an electrode. Hence if the working fluid contains an indicator such as thymol blue titrated close to the end point of that indicator, the application of 1–10 V between a fine wire and some other point in the flow produces a local color change at the wire. This method has two advantages: the production of the dye produces no density change or displacement of the flow, and the dye gradually reverts to its previous state, so the system can be run continuously without becoming filled with dye. Careful illumination is needed for all these flow visualization techniques.

An excellent general reference is the recent monograph on flow visualization by Merzkirch.[33]

Other techniques make use of suspended particles such as polystyrene beads and aluminum paint powder. The aluminum particles are small flakes which align in a shear flow. Two materials widely used in couette flow experiments are natural pearl essence (i.e., fish scales, titanium dioxide/mica, etc.), available from the Mearl Corp., Technical Service Dept., 41 E. 42nd St., New York, NY 10017; another is Kalliroscope platelets, available from Kalliroscope Corp., 145 Main St., Cambridge, MA 02142.

The last group of methods employ optical techniques making use of refractive index variations in the fluid which are associated with temperature or concentration variations.[20] In the shadowgraph method, parallel light enters the fluid and is deflected when there are refractive index variations. If the second spatial derivative of the refractive index is nonzero the amount of deflection varies, giving a pattern of light and dark regions related to the flow structure. In the schlieren method, parallel light is used again, but brought to a focus after passing through the fluid. A knife-edge or other stop at this focus blocks off some of the light so that light deflected in the fluid changes according to the first spatial derivative of the refractive index. In an interferometer, light that has passed through the fluid interferes with light from the same source which has not passed through the fluid.[34]

D. Velocity measurements

Pitot tubes have been widely used in air for speeds of 1 m/s upwards, and in water down to about 3 cm/s. They are slow in response and intrude into the flow.[21,22]

The hot-wire anemometer is an electrically heated wire cooled by the flow, the rate of cooling depending on the velocity. Hot wires have been widely and successfully used in gas flows down to about 30 cm/s. Hot-film anemometers have been used for liquids. These devices are small and rapidly responding and are the principal instruments for studying fluctuating flows such as transition and turbulence. Their output is electrical and can be processed electronically to give intensities, correlations, and spectra. Calibration and geometric combinations to measure different components of velocity fluctuations are a demanding scientific technique.[23–25]

The laser-Doppler anemometer measures velocity by measuring the Doppler shift of light scattered within the moving fluid. Scattering centers on tiny particles of dust present in any liquid or gas, or in the cases of liquids, micron-sized spheres. The techniques involved are rapidly responding, local, absolute, and may be extended to more than one velocity component. Several manufacturers produce lines of superb equipment. The principles are discussed in Refs. 26 and 27.

For optical work, one often wants index of refraction matching liquids for glass or perspex walls of an apparatus. Tables 12.06.D.1 and 12.06.D.2 list suitable liquids. Toxic and corrosive effects need to be watched carefully.

TABLE 12.06.D.1. Liquids to match index of refraction of glass. n_D is the index of refraction for the sodium D line, ρ the density in g/cm^3, and η the viscosity in centipoise. From Ref. 28.

Liquid	Formula	n_D (2π)	ρ	η (cP)
trans-decalin	$C_{10}H_{18}$	1.4696	0.87	2.4
Glycerol	$C_3H_5(OH)_3$	1.470	1.2563	3800
Geraniol	$(CH_3)_2CCH$	1.473 pure	0.87	7.95
	$(CH_2)_2C(CH_3)$	1.494 tech.		
Trichloroethylene	C_2HCl_3	1.4735	1.4556	0.55
Isophorone	$C_9H_{14}O$	1.474	0.923	2.4
Dimethylsulphoxide	$(CH_3)_2SO$	1.4783	1.1	2.3
cis-decalin	$C_{10}H_{18}$	1.481	0.886	2.41
Furfuryl alcohol	$C_4H_3OCH_2OH$	1.485	0.93	
Terpineol	$C_{10}H_{14}OH$	1.485	0.93	
Terpene solvent (Depanol®)		1.477	0.87/89	
Turpentine oil		1.465–1.478	0.86/89	1.487
Castor oil		1.477	0.961	986
Olive oil		1.468	0.918	84
Soybean oil		1.4729	0.927	70
Diesel oil		1.463	0.83	5.5
+ 36 vol % geraniol		1.472		5.35
+ 8 vol % perchloroethylene		1.473		3.75
Turpentine +				
60 vol % trichloroethylene		1.474		0.68
19 vol % xylene		1.474		1.25
22.5 vol % toluene		1.474		1.21
14 vol % benzene		1.474		1.33

TABLE 12.06.D.2. Liquids to match index of refraction of Perspex (Plexiglas/Lucite)$^{(P)}$ indicates a poor solvent action on Perspex especially when mixed. (See caption to Table 12.06.D.1.) From Ref. 28.

Liquid	Formula	n_D (2π)	ρ	η (cP)
Isophorone$^{(P)}$	$C_9H_{14}O$	1.4781	0.923	2.4
Dipentene	$C_{10}H_{16}$	1.480	0.842	1.7
cis-decaline	$C_{10}H_{18}$	1.481	0.886	2.41
Furfuryl alcohol	$C_4H_3OCH_2OH$	1.485	0.93	
Terpineol	$C_{10}H_{14}OH$	1.485	0.93	
Isobutylbenzoate	$C_{11}H_{14}O_2$	1.491	1.002	
Dibutylphtalate		1.4915		20.3
p-xylene$^{(P)}$	C_8H_{10}	1.495	0.861	0.65
m-xylene		1.4972	0.864	0.61
Toluene$^{(P)}$	$C_6H_5CH_3$	1.4969	0.867	0.58
Solvent naphta		1.50	0.86	
o-xylene$^{(P)}$	C_8H_{10}	1.5054	0.88	0.80
Diethylephtalate		1.514	1.189	10
Ethylbenzoate$^{(P)}$	$C_6H_5CO_2C_2H_5$	1.505	1.048	2–2.3
Dimethylphtalate	$C_{10}H_{10}O_4$	1.514	1.189	
Methylbenzoate$^{(P)}$	$C_6H_5COOCH_3$	1.514	1.094	2.07
Tetralin	$C_{10}H_{10}$	1.540/47	0.981	2.3
Monochlorobenzene	C_6H_5Cl	1.5248	1.106	0.80

E. Wind tunnel characteristics*

For turbulence research and transition studies, low initial (or "background") turbulence is required. Turbulence level or intensity is the ratio of either longitudinal rms velocity u' or transverse rms v' to mean velocity U. Generally $v'/U > u'/U$ in tunnels in which the turbulence has been damped in a low-velocity plenum chamber and accelerated by means of a contraction.[29,30] Values for v'/U of 0.05% are acceptable for grid turbulence (see below) but lower values (0.01% and less) are required for transition studies. Methods of achieving these levels by means of screened diffusers and plenum diameters and contractions are described in Ref. 29.

Grid turbulence. A rectangular array of grid bars (generally square or circular sectioned rods in a biplanar array[31]) provide the closest laboratory realization of isotropic turbulence ($u' = v' = w'$) although generally u' is 5% greater than v' and w', unless precautions are taken (see below). The grid is described in terms of its mesh length M (the spacing between the rods) and solidity σ (projected solid area per unit total area). Homogeneity in the transverse turbulence field is generally achieved at $x/M \sim 40$, where x is the downstream distance. However, if $\sigma > 0.4$, the initial flow is a series of jets and homogeneity is difficult to attain at all. Therefore $\sigma > 0.4$ is advisable. The turbulence intensity decays as a power law downstream from the grid:

$$\overline{u'^2}/U^2 = A(x/M - x_0/M)^{-n},$$

where A and n are constants and x_0 is the virtual origin. Detailed studies[31] suggest $n \sim 1.3$ ($\pm 10\%$), its dependence on Reynolds number Re being very weak and the trend not established over the moderate Re investigated (up to $Re \equiv UM/\nu \sim 10^4$). x_0/M and A depend on the grid geometry and mean speed. x_0/M varies from 0 to 5 in various studies; $\overline{u'^2}/U^2$ is typically about 10^{-4} at $x/M = 100$. However, the level depends on the nature of the grid, square rods giving slightly higher turbulence levels than round rods.[31] Turbulence levels may be further increased by means of active grids (e.g., by jet injection at the grid bars[32]) or using an array of circular disks.[31] Near perfect isotropy can be achieved by placing a gentle contraction dowstream from the grid.[31]

Similar comments could be made for the $\overline{v'^2}/U^2$ decay; its form is similar to that of $\overline{u'^2}$, but, as noted, the intensity is generally 5% lower.

The scale of the largest eddies is of order M and grows with x/M, the rate of growth being a function of n (if $n = 1$ the eddy size increases as $x^{1/2}$ and the turbulence Re remains constant, which is inconsistent with experiment). There is no precise theoretical determination of n (see, for example, Ref. 24).

Slightly divergent walls are required in the test section in order to keep the mean-stream velocity constant because of boundary layer growth. A grid of at least ten mesh lengths is required in order to provide a homogeneous core laterally in the flow.

*Prepared by Zellman Warhaft, Cornell University.

12.07. REFERENCES

[1]I. H. Shames, *Mechanics of Fluids* (McGraw-Hill, New York, 1962).

[2]D. H. Tritton, *Physical Fluid Dynamics* (Van Nostrand Reinhold, New York, 1977).

[3]H. Rouse and J. W. Howe, *Basic Mechanics of Fluids* (Wiley, New York, 1953).

[4]S. F. Hoerner, *Fluid Dynamic Drag* (Hoerner Fluid Dynamics, P.O. Box 342, Bricktown, NJ 08723, 1965).

[5]S. F. Hoerner and H. V. Borst, *Fluid Dynamic Lift* (Hoerner Fluid Dynamics, P.O. Box 342, Bricktown, NJ 08723, 1975).

[6]R. M. Goody, *Atmospheric Radiation* (Oxford University, New York, 1964).

[7]R. C. Reid, J. M. Prausnitz, and T. K. Sherwood, *The Properties of Gases and Liquids* (McGraw-Hill, New York, 1977).

[8]*Modern Developments in Fluid Dynamics,* edited by S. Goldstein (Oxford University, New York, 1957), Vols. I and II.

[9]*Equations, Tables and Charts for Compressible Flow,* U.S. Ames Aeronautical Laboratory, Moffett Field, CA (now NASA Ames Research Center), Report 1135.

[10]H. Schlichting, *Boundary-Layer Theory,* 7th ed., translated by J.Kestin (McGraw-Hill, New York, 1979).

[11]S. Chandrasekhar, *Hydrodynamic and Hydromagnetic Stability* (Clarendon, Oxford, 1961).

[12]L. D. Landau and E. M. Lifshitz, *Fluid Mechanics,* translated by J. B. Sykes and W. H. Reid (Pergamon, London, 1959).

[13]H. Tennekes and J. L. Lumley, *A First Course in Turbulence* (MIT, Cambridge, MA, 1974).

[14]H. L. Swinney and J. P. Gollub, *Hydrodynamics, Instabilities, and the Transition to Turbulence* (Springer, Berlin, 1981).

[15]*Handbook of Fluid Dynamics,* edited by V. L. Streeter (McGraw-Hill, New York, 1961).

[16]P. Bradshaw, *Experimental Fluid Mechanics* (Pergamon, New York, 1964).

[17]F. X. Wortmann, Z. Angew. Phys. **5**, 201 (1953).

[18]S. Taneda, H. Honji, and M. Tatsuno, J. Phys. Soc. Jpn. **37**, 784 (1974); H. Honji, S. Taneda, and M. Tatsuno, Rep. Inst. Appl. Mech. Kyushu Univ. **28**, No. 19 (1980).

[19]D. J. Baker, J. Fluid Mech. **26**, 573 (1966).

[20]D. W. Holder and R. J. North, Natl. Phys. Lab. Notes Appl. Sci. **31** (1963).

[21]C. Salter, J. H. Warsap, and D. G. Goodman, Aeronaut. Res. Council Rep. Memo. **3365** (1965).

[22]J. H. Preston, J. Phys. E **5**, 277 (1972).

[23]P. Bradshaw, *Introduction to Turbulence and Its Measurement* (Pergamon, New York, 1971).

[24]J. O. Hinze, *Turbulence,* 2nd ed. (McGraw-Hill, New York, 1975).

[25]A. E. Perry and G. L. Morrison, J. Fluid Mech. **47**, 577 (1971).

[26]F. Dürst, A. Melling, and J. H. Whitelaw, *Principles and Practice of Laser-Doppler Anemometry* (Academic, London, 1976).

[27]T. S. Durrani and C. A. Greated, *Laser Systems in Flow Measurement* (Plenum, New York, 1977).

[28]F. Dürst and R. Kline, Institut für Hydromechanik, University of Karlsruhe, Kaiserstrasse 12, B75 Karlsruhe 1, Germany.

[29]H. L. Dryden and G. B. Schubauer, J. Aeronaut. Sci. **14**, 221 (1947).

[30]M. S. Uberoi, J. Aeronaut. Sci. **23**, 754 (1956).

[31]G. Comte-Bellot and S. Corrsin, J. Fluid Mech. **25**, 657 (1966).

[32]M. Gad-El-Hak and S. Corrsin, J. Fluid Mech. **62**, 115 (1974).

[33]W. Merzkirch, *Flow Visualization* (Academic, London, 1974).

[34]L. H. Tanner, J. Sci. Instrum. **43**, 878 (1966).

13.00. High polymer physics

RONALD K. EBY

The Johns Hopkins University

CONTENTS

This chapter has been prepared by a number of authors. The primary responsibilities for the sections are as follows: 13.01 and 13.02, R. K. Eby (The Johns Hopkins University); 13.03–13.05, I. C. Sanchez (The University of Texas at Austin); 13.06, C. C. Han and I. C. Sanchez; 13.07, D. L. VanderHart and B. M. Fanconi; 13.08, G. T. Davis; 13.09 and 13.11, F. Khoury; 13.10, B. M. Fanconi; 13.12, S. S. Chang; 13.13, E. A. DiMarzio and J. E. McKinney; 13.14–13.16, E. Passaglia; 13.17, H. Markovitz (Carnegie-Mellon University) and L. J. Zapas; 13.18, M. G. Broadhurst and F. I. Mopsik; 13.19, S. S. Chang, J. H. Flynn, and L. E. Smith; 13.20, P. N. Prasad (University at Buffalo, SUNY); and 13.21, B. L. Farmer (University of Virginia). Except as noted, the authors are from the National Institute of Standards and Technology.

13.01. INTRODUCTION

Polymer science as a *coherent* subject is barely fifty years old. Therefore, many of the concepts and data are not in a final state. In this chapter, we present equations and data which we believe to have wide acceptance. Space limitation precluded completeness and required us to omit a large amount of material which we wished to include. We believe, however, that the included material and references will prove useful to those working in polymer physics.

13.02. POLYMER MOLECULES

In the simplest cases, polymers are made by the chemical combination of small chemical units which are known as monomers. The name is usually derived from that of the monomer. Thus polyethylene has the structural formula

$$[\text{—}CH_2\text{—}CH_2\text{—}]_N,$$

where N represents the degree of polymerization or the number of times the unit is repeated in the molecule. Degrees of polymerization for polymers are often in the range 10^3–10^5. The range can even be exhibited within one sample. Therefore, polymers are often characterized by an average degree of polymerization or molecular weight. Other important structural variations, including geometrical, stereochemical, and composition, are discussed by Bovey and Winslow,[1] who also give structural formulas for many polymers.

13.03. MOLECULAR-WEIGHT AVERAGES

kth moment of a molecular-weight distribution $P(M)$:

$$\langle M^k\rangle \equiv \int_0^\infty M^k P(M)\,dM.$$

Number average molecular weight:

$$M_n \equiv \langle M\rangle.$$

Weight average molecular weight:

$$M_w \equiv \langle M^2\rangle/\langle M\rangle.$$

Z-average molecular weight:

$$M_z \equiv \langle M^3\rangle/\langle M^2\rangle.$$

Viscosity average molecular weight:

$$M_v \equiv (\langle M^{1+\alpha}\rangle/\langle M\rangle)^{1/\alpha}.$$

α is solvent dependent and usually lies between 0.5 and 0.8.

Polydispersity ratio:

$$R \equiv M_w/M_n = \langle M^2\rangle/\langle M\rangle^2.$$

Values of R are a measure of the width of the molecular-weight distribution.

13.04. CHAIN DIMENSIONS

A schematic representation of a homopolymer chain of $N+1$ mass points is shown in Figure 13.04.A.

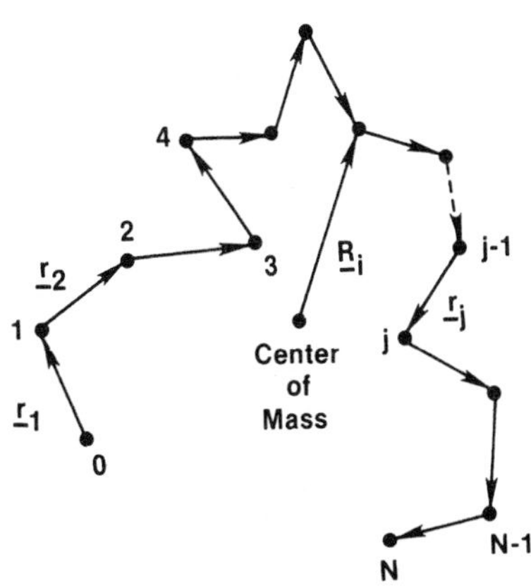

FIGURE 13.04.A. Schematic representation of homopolymer chain of $N+1$ mass points

End-to-end vector:

$$\mathbf{R} \equiv \sum_{i=1}^{N} \mathbf{r}_i.$$

Mean-square end-to-end distance:

$$\langle R^2\rangle \equiv \sum_{i=1}^{N}\sum_{j=1}^{N} \langle \mathbf{r}_i\cdot\mathbf{r}_j\rangle.$$

Mean-square radius of gyration (homopolymer chain):

$$\langle S^2\rangle \equiv (N+1)^{-1}\sum_{i=0}^{N}\langle \mathbf{R}_i^2\rangle$$

$$\equiv (N+1)^{-2}\sum_{i<j}^{N}\sum^{N}\langle \mathbf{R}_{ij}^2\rangle,\quad \mathbf{R}_{ij} \equiv \mathbf{R}_i - \mathbf{R}_j$$

$$\equiv (N+1)^{-1}\sum_{i=1}^{N}\sum_{j=1}^{N} g_{ij}\langle \mathbf{r}_i\cdot\mathbf{r}_j\rangle,$$

$$g_{ij} = \begin{cases} i - ij/(N+1), & i \leqslant j \\ j - ij/(N+1), & i > j. \end{cases}$$

Special values of $\langle S^2\rangle$ valid for large N:

(1) Linear, freely jointed chain:

$$\langle S^2\rangle = \langle R^2\rangle/6 = Nb^2/6,$$

where b is the bond length.

(2) Cyclic, freely jointed chain:

$$\langle S^2\rangle = Nb^2/12.$$

(3) Freely jointed chain whose end-to-end distance is fixed at R:

$$\langle S^2\rangle = (Nb^2 + R^2)/12.$$

(4) Linear, freely rotating chain with fixed bond angles of π-θ:

$$\langle \mathbf{r}_i\cdot\mathbf{r}_{i+k}\rangle = b^2\cos^k\theta,\quad \langle S^2\rangle = \frac{Nb^2}{6}\,\frac{1+\cos\theta}{1-\cos\theta}.$$

(5) Rigid rod:

$$\langle S^2\rangle = (Nb)^2/12.$$

Characteristic ratio[2]:

$$C_\infty \equiv \lim_{N\to\infty} \langle R^2\rangle/Nb^2.$$

TABLE 13.05.A. Selected Θ solvents and temperatures. From Ref. 3.

Polymer	Solvent	Θ temp. (°C)
Polyisobutene	benzene	24
	ethylbenzene	−24
	toluene	−13
Polyethylene	diphenylmethane	142
	n-decanol	153
	diphenylether	162
Polypropylene		
Atactic	*i*-amyl acetate	34
	n-butyl acetate	58.5
	cyclohexanone	92
Isotactic	*i*-amyl acetate	70
Polystyrene (atactic)	cyclohexane	35
	methylcyclohexane	60

For a freely jointed chain $C_\infty = 1$, and for a freely rotating chain

$$C_\infty = (1+\cos\theta)/(1-\cos\theta).$$

The characteristic ratio is a measure of chain backbone rigidity or stiffness.

13.05. Θ SOLVENTS AND TEMPERATURES

A thèta (Θ) solvent is the generic name given to those polymer solvents in which upper critical solution temperatures are observed (typically near room temperature). With increasing polymer molecular weight, the upper critical solution temperature (UCST) approaches a limiting value called the Θ temperature.

A solvent is termed a *good solvent* if the temperature is well above the UCST. Θ solvents are sometimes referred to as *poor solvents.* For large N, chain dimensions in solution vary as N^γ. In a Θ solvent $\gamma = 1$, whereas in a good solvent $\gamma \simeq 1.2$.

Selected Θ solvents and temperatures are given in Table 13.05.A.

13.06. MOLECULAR-WEIGHT CHARACTERIZATION

A. Solution viscosity

Specific viscosity:

$$\eta_{sp} \equiv (\eta - \eta_0)/\eta_0,$$

where η is the solution viscosity and η_0 the solvent viscosity.

Limiting viscosity number or intrinsic viscosity:

$$[\eta] \equiv \lim_{c\to 0} \eta_{sp}/c,$$

where c is the polymer concentration.

Dependence on concentration:

$$\eta_{sp}/c = [\eta] + k[\eta]^2 c + \cdots,$$

where k is the Huggins constant; $0.3 < k < 0.4$ for most good solvents. Values of $[\eta]$ and k for polystyrene dissolved in toluene at 30°C are given in Table 13.06.A.1.

Mark-Houwink-Sakurada equation (semiempirical):

$$[\eta] = KM^\alpha.$$

K and α are constants for a given polmer/solvent system and are molecular-weight independent over a broad range of molecular weight ($\alpha = 0.5$ for a Θ solvent). Tabulated values of K and α are available for many polymer/solvent pairs.[3]

TABLE 13.06.A.1. Values of $[\eta]$ and k for polystyrene dissolved in toluene at 30°C. From F. Danusso and G. Moraglio, J. Polym. Sci. **24**, 161 (1957).

M_n (g/mol)	$[\eta]$ (cm^3/g)	k
76 000	38.2	0.31
135 000	59.2	0.33
163 000	69.6	0.33
336 000	105.4	0.35
440 000	129.2	0.34
556 000	165.0	0.31
850 000	221.0	0.31

B. Osmotic pressure π

Concentration, c, dependence:

$$\pi/cRT = 1/M_n + A_2 c + \cdots.$$

A_2 is the second virial coefficient.

van't Hoff's law:

$$\lim_{c\to 0} \pi/cRT = 1/M_n.$$

C. Ultracentrifugation

An ultracentrifuge can be used to measure M_w and, under appropriate conditions, M_z.[4] The method is based on the fact that heavier particles sediment more rapidly than lighter particles.

D. Static light scattering [2,5,6]

Consider a polymer solution of concentration c scattering light of incident intensity I_0. If the incident light is unpolarized, the intensity $I(\theta)$, at a distance r from a unit volume of solution, of light scattered in a direction that makes an angle θ with the incident beam is

$$\frac{I(\theta)}{I_0} = \frac{K(1+\cos^2\theta)c(1 - q^2\langle S^2\rangle/3 + \cdots)}{r^2(1/M_w + 2A_2 c + \cdots)},$$

where $K \equiv 2\pi^2 n^2 (dn/dc)^2/N_A\lambda^4$, n is the refractive index of the solution, λ the wavelength of incident light in vacuum, N_A Avogadro's number, A_2 the second virial coefficient, $\langle S^2\rangle$ the mean-square radius of gyration, $q \equiv (4\pi/\lambda)\sin(\theta/2)$, and M_w is the weight average molecular weight.

Rayleigh's ratio is defined as

$$R_\theta \equiv r^2 I(\theta)/I_0(1+\cos^2\theta).$$

To determine molecular parameters Rayleigh's ratio can be rewritten as

$$\frac{Kc}{R_\theta} = \left(\frac{1}{M_w} + 2A_2 c + \cdots\right)\left(1 + \frac{q^2\langle S^2\rangle}{3} - \cdots\right).$$

M_w, A_2, and $\langle S^2\rangle$ can be determined from experimental data through various graphical and regression procedures such as a Zimm plot.[6]

In practice an absolute value of R_θ is not measured. Instead a standard such as benzene is used to determine R_θ:

$$R_\theta = [I_\theta/I_B(90)](1+\cos^2\theta)R_B,$$

where $I_B(90)$ is the scattering intensity and R_B the Rayleigh ratio of benzene at a scattering angle of 90°.

E. Dynamic (quasielastic) light scattering [7–9]

This method can be used, under appropriate conditions, to determine the diffusion coefficient, D, of the center of mass of polymer molecules in dilute solution. As the concentration, c, approaches zero, the diffusion coefficient and molecular weight of a flexible monodisperse linear polymer are related by

$$\lim_{c\to 0} D(c) = AM^{-\nu}.$$

A is a constant and $\nu = 1/2$ for the Θ condition. In general, ν varies between 0.5 and 0.6.

13.07. CHARACTERIZATION BY SPECTROSCOPIC TECHNIQUES

A. NMR

Data on solutions can be acquired such that relative intensities are proportional to the relative number of nuclei having a resonance frequency. Absolute concentrations are obtained by comparison against known standards. Resonances are influenced by the magnetic environment of the nuclei (chemical shifts) providing information on chemical functionality and physical structure. Chemical shift ranges are available.[10,11] Proton-decoupled ^{13}C spectra usually provide the broadest information about geometrical, stereochemical, and compositional variations.[1,12]

B. Vibrational spectroscopy

Vibrational spectroscopy (I. R., Raman) provides information on chemical composition, physical structure, and parameters of inter- and intramolecular potential energy functions.[13] Of particular importance is the use of vibrational frequencies and intensities to determine tacticity, degree of branching, crystallinity, and chainstem lengths in lamellar crystals.[1]

13.08. MELTING AND CRYSTALLIZATION

Variation of melting point of thin crystals with thickness:

$$T_m = T_m^0(1 - 2\sigma_e/\Delta h_f l).$$

Spherulitic growth rate controlled by secondary nucleation:

$$G = G_0 \exp[-u^*/R(T-T_\infty)]\exp(-nb\sigma\sigma_e/\Delta f kT),$$

where T is the temperature, T_m the observed melting point, T_m^0 the equilibrium melting point of infinitely thick crystal, T_∞ the hypothetical temperature where viscous flow ceases ($\cong T_g - 30$), T_g the glass transition temperature, l the thickness of crystal (assumed to be small relative to the lateral dimensions), Δh_f the heat of fusion, σ the lateral surface energy, empirically approximated as $0.1\Delta h_f(ab)^{1/2}$, σ_e the surface energy of large surface of crystal, Δf the free energy of fusion $[\cong \Delta h_f(T_m^0 - T)/T_m^0]$, G the growth rate of spherulite, G_0 the preexponential factor involving terms not strongly temperature dependent, u^* the activation energy for transport of polymer in melt, R the molar gas constant, n a constant, equal to 4 when rapid growth follows nucleation, to 2 when nucleation is comparable to or greater than subsequent growth, b the thickness of crystallizing layer, a the width of crystalline molecular segment, and k the Boltzmann constant.

Specific values for the above parameters are subject to data selection and analysis. Typical values are given in Table 13.08.A. A general reference is available.[14]

Avrami equation to describe kinetics of phase changes:

$$\ln[1/(1-x)] = kt^n,$$

where x is the fraction transformed from liquid to crystal phase, t the time, n the exponent dependent upon growth processes (see Table 13.08.B), and k a constant dependent upon nucleation rate, geometry of growing center, and growth rate. For spheres

$$k = k_s = (\pi\rho_c\dot{N}G^3)/(3\rho_l).$$

For disks

$$k = k_d = (\pi l_c\rho_c\dot{N}G^2)/(3\rho_l),$$

where ρ_c is the density of crystal phase, ρ_l the density of liquid phase, $\dot{N}$ the steady-state rate of nucleation, G the linear growth rate of growing center, and l_c the thickness of disk. A general reference is available.[15]

TABLE 13.08.A. Typical values of crystallization parameters for crystalline polymers

Polymer	T_m^0 (K)	T_g (K)	u^* (J/mol)	Δh_f (J/m^3)	σ_e (J/m^2)	b (nm)
Polyethylene	419.2	231	2.93×10^4	2.80×10^8	0.101	0.415
Polystyrene	515.2	363.5	6.53×10^3	9.11×10^7	0.035	0.55
Polyoxymethylene	459.2	213	6.28×10^3	1.86×10^8	0.061	0.386
Poly(ethylene oxide)	348.4	206	6.28×10^3	2.45×10^8	0.037	0.465
trans 1-4 polyisoprene	360.2	211	6.28×10^3	1.97×10^8	0.109	0.395

TABLE 13.08.B. Values of n in Avrami equation for various types of nucleation and growth. Steady-state nucleation rate in a completely crystallizable system is assumed. Impingement of growing centers has been accounted for.

	Homogeneous nucleation	
Growth habit	Linear growth	Diffusion-controlled growth
Three-dimensional	4	5/2
Two-dimensional	3	2
One-dimensional	2	3/2

TABLE 13.09. CRYSTAL STRUCTURE

The crystal structures of a number of representative polymers are listed. Other extensive listings are available.[3,16,17] Note that polymers exhibit polymorphism.

Polymer	Crystal system space group	Unit cell axes (nm) and angles (deg)	Monomer units in unit cell	Chain conformation[a]	Density ρ_c (g/cm^3)
Polyethylene[b] $[CH_2—CH_2—]_n$	orthorhombic *Pnam*	$a = 0.740$ $b = 0.493$ $c = 0.2534$	2	2* 1/1	1.000
Polypropylene[c] $[CH_2—CHCH_3—]_n$	monoclinic *C2/c* or *Cc*	$a = 0.665$ $b = 2.096$ $c = 0.650$ $\beta = 99°20'$	12	2* 3/1	0.936
Polystyrene[d] $[CH_2—CHC_6H_5—]_n$	trigonal $R\bar{3}c$ or $R3c$	$a = 2.19$ $c = 0.665$	18	2* 3/1	1.126
Poly(vinylidene fluoride) $[CH_2—CF_2—]_n$					
α phase, form II[e]	orthorhombic *P2cm*	$a = 0.496$ $b = 0.964$ $c = 0.462$	4	4(*TGTG'*)[f]	1.924
β phase, form I[g]	orthorhombic *Cm2m*	$a = 0.847$ $b = 0.490$ $c = 0.256$	2	2* 1/1	2.001
γ phase, form III[h]	orthorhombic *C2cm*	$a = 0.497$ $b = 0.966$ $c = 0.918$	8	8(*TTTGTTTG'*)	1.929
δ phase, form IV[i]	orthorhombic $P2_1cn$	$a = 0.496$ $b = 0.964$ $c = 0.462$	4	4(*TGTG'*)	1.924
Polytetrafluoroethylene (above 19° C)[j] $[—CF_2—]_n$	trigonal	$a = 0.566$ $c = 1.950$	15	1* 15/7	2.302
trans 1,4-Polybutadiene[k] $[CH_2—CH{=}CH—CH_2—]_n$	monoclinic $P2_1/a$	$a = 0.863$ $b = 0.911$ $c = 0.483$ $\beta = 114°$	4	4* 1/1	1.036
cis 1,4-Polybutadiene[l] $[CH_2—CH{=}CH—CH_2—]_n$	monoclinic *C2/c*	$a = 0.460$ $b = 0.950$ $c = 0.860$ $\beta = 109°$	4	8* 1/1	1.01
Poly[1,2-bis(*p*-tolylsulphonyloxymethylene)-1-butene-3-ynylene][m] $[CR—C{\equiv}C—CR{=}]_n$ $(R = —CH_2—O—SO_2—C_6H_4—CH_3)$	monoclinic $P2_1/c$	$a = 1.4493$ $b = 0.4910$ $c = 1.4936$ $\beta = 118.14°$	2	4* 1/1	1.483
Polyoxymethylene[n] $[CH_2—O—]_n$	trigonal $P3_1$ or $P3_2$	$a = 0.447$ $c = 17.39$	9	2* 9/5	1.49
Poly(ethylene oxide)[o] $[CH_2—CH_2—O—]_n$	monoclinic $P2_1/a$	$a = 0.805$ $b = 1.304$ $c = 1.948$ $\beta = 125.4°$	28	3* (7/2)	1.228
Poly(ethylene terephthalate)[p] $[(CH_2—)_2O—CO—C_6H_4—CO—O—]_n$	triclinic $P\bar{1}$	$a = 0.456$ $b = 0.594$ $c = 1.075$ $\alpha = 98.5°$ $\beta = 118°$ $\gamma = 112°$	1	12* 1/1	1.455
Poly(hexamethylene adipamide) (α form)[q] $[(CH_2—)_6NH—CO—(CH_2—)_4CO—NH—]_n$	triclinic $P\bar{1}$	$a = 0.49$ $b = 0.54$ $c = 1.72$ $\alpha = 48.5°$ $\beta = 77°$ $\gamma = 63.5°$	1	14* 1/1	1.24

[a] The notation n^* p/q specifies the number (n) of skeletal atoms in the asymmetric unit of the chain and the number of such units (p) in q turns of the helix in the crystallographic repeat.
[b] C. W. Bunn, Trans. Faraday Soc. **35**, 482 (1939).
[c] G. Natta and P. Corradini, Nuovo Cimento Suppl. **15**, 40 (1960).
[d] G. Natta and P. Corradini, Nuovo Cimento Suppl. **15**, 68 (1960).
[e] M. Bachmann and J. B. Lando, Macromolecules **14**, 40 (1981).
[f] *T* = *trans*, *G* = *gauche*.
[g] J. B. Lando, H. G. Olf, and A. Peterlin, J. Polym. Sci. Part A1 **4**, 941 (1966).
[h] S. Weinhold, M. H. Litt, and J. B. Lando, Macromolecules **13**, 1178 (1980).
[i] M. Bachmann, W. L. Gordon, S. Weinhold, and J. B. Lando, J. Appl. Phys. **51**, 5095 (1980).
[j] E. S. Clark and L. T. Muus, Z. Kryst. **117**, 119 (1962).
[k] S. Iwayanagi, I. Sakurai, T. Sakurai, and T. Seto, J. Macromol. Sci. Phys. B **2**, 163 (1968).
[l] G. Natta and P. Corradini, Nuovo Cimento Suppl. **15**, 111 (1960).
[m] D. Kobelt and E. F. Paulus, Acta Crystallogr. Sect. B **30**, 232 (1974).
[n] T. Uchida and H. Tadokoro, J. Polym. Sci. Part A 2 **5**, 63 (1967).
[o] Y. Takahashi and H. Tadokoro, Macromolecules **6**, 672 (1973).
[p] R. de P. Daubeny, C. W. Bunn, and C. J. Brown, Proc. R. Soc. London Ser. A **226**, 531 (1954).
[q] C. W. Bunn and A. V. Garner, Proc. R. Soc. London Ser. A **189**, 39 (1947).

TABLE 13.10. BOND LENGTHS AND ANGLES FOR REPRESENTATIVE POLYMERS

Polymer	Bond or angle	Bond length or angle (nm or deg)
Polyethylene[a]	C C	0.1532
	C H	0.1058
	C C C	112.01
	H C H	109.29
	C C H	108.88
Polytetrafluoroethylene[b]	C C	0.1553
	C F	0.136
	C C C	113.85
	F C F	108
	C C F	108.7
Poly amides[c] (C′ denotes amide carbon)	C N	0.147
	N C′	0.132
	C′ O	0.124
	N H	0.100
	N C C′	109.7
	N C′ C	115.4
	C′ N C	120.9
	O C′ C	121.0
	O C′ N	123.6
	C′ N H	123.0
	C N H	116.1
Polyoxymethylene[d]	C O	0.142
	C H	0.109
	C O C	112.4
	O C O	110.8
	H C H	109.5
	H C O	108.5
		109.8
Poly[1,2-bis(*p*-tolylsulphonyloxy-methylene)-1-butene-3-ynylene][e] (C′ on —C≡C—C=C—)	C—C	0.1428
	C=C	0.1356
	C≡C	0.1191
	C—C=C	121.9
	C′—C=C	120.3
	C—C≡C	177.6
Polystyrene[f] $—C_1—C_2—C_1'$, C_2–C_3, C_3–C_4	C_1 C_2	0.154
	C_2 C_3	0.154
	C_3 C_4	0.140
	C_1 C_2 C_1'	116
	C_3 C_2 C_1	108
	C_3 C_2 C_1'	111
Polypropylene[g] $C_1—C_2—C_1'$, C_2–C_3	C_1 C_2	0.154
	C_2 C_3	0.154
	C_1 C_2 C_1'	114
	C_1 C_2 C_3	110
Poly(ethylene oxide)[h]	C C	0.154
	C O	0.143
	C H	0.109
	C C O	110
	C O C	112
	H C H	109.5
Poly(ethylene terephthalate)[i] $C_{5b}—C_{5a}—O_7—C_2—C_1$, C_2–O_6, C_1=C_4, C_1–C_3	C_1 C_2	0.149
	C_1 C_3	0.134
	C_1 C_4	0.136
	C_2 O_6	0.127
	C_2 O_7	0.134
	C_{5a} O_7	0.144
	C_{5a} C_{5b}	0.149
	C_2 C_1 C_3	125
	C_2 C_1 C_4	118
	C_1 C_2 O_6	127
	C_1 C_2 O_7	110
	O_6 C_2 O_7	122
	C_2 O_7 C_{5a}	114
	O_7 C_{5a} C_{5b}	104
	C_4 C_1 C_3	117
Poly(vinylidene fluoride)[j]	C C	0.1541
	C F	0.1344
	C H	0.109
	F C F	109.5
	C C C	112.3
Polybutadiene[k] $C_1—C_2—C_3═C_4$	C_1 C_2	0.153
	C_2 C_3	0.154
	C_3 C_4	0.115
	C_1 C_2 C_3	121
	C_2 C_3 C_4	142

[a]J. D. Barnes and B. M. Fanconi, J. Phys. Chem. Ref. Data **7**, 1309 (1978).
[b]M. J. Hannon, F. H. Boerio, and J. L. Koenig, J. Chem. Phys. **50**, 2829 (1969).
[c]S. Arnott, S. D. Dover, and A. Elliott, J. Mol. Biol. **30**, 201 (1967).
[d]Reference n of Table 13.09.A.
[e]Reference m of Table 13.09.A.
[f]Reference d of Table 13.09.A.
[g]Reference c of Table 13.09.A.
[h]Reference o of Table 13.09.A.
[i]Reference p of Table 13.09.A.
[j]References g and h of Table 13.09.A.
[k]Reference l of Table 13.09.A.

13.11. OPTICAL PROPERTIES

Orientation birefringence Δn in amorphous polymers[18]:

$$\Delta n = \frac{2}{45}\pi \frac{(\bar{n}^2+2)^2}{\bar{n}} N(\lambda^2 - 1/\lambda)(b_1 - b_2),$$

where N is the number of network chains per unit volume, $\bar{n}$ the average refractive index of the system, λ the extension ratio, and b_1 and b_2 the polarizabilities parallel (b_1) and perpendicular (b_2) to axis of cylindrical statistical chain segments.

Stress optical coefficient C:

$$C = \frac{\Delta n}{\sigma_a} = \frac{2}{45}\frac{\pi}{kT}\frac{(\bar{n}^2+2)^2}{\bar{n}}(b_1 - b_2),$$

where σ_a is the true stress, k the Boltzmann constant, and T the absolute temperature.

Form birefringence Δn_f in two-phase systems.[19] For a two-phase system of optically *isotropic* rods (phase 1, volume fraction ϕ_1, refractive index n_1) parallel to one another and separated by an *isotropic* matrix material (phase 2, volume fraction ϕ_2, refractive index n_2), with the inter-rod distance between surfaces being small relative to the wavelength of light,

$$n_\parallel^2 = \phi_1 n_1^2 + \phi_2 n_2^2, \quad n_\perp^2 = \frac{n_2^2(\phi_1+1)n_1^2 + \phi_2 n_2^4}{(\phi_1+1)n_2^2 + \phi_2 n_1^2},$$

$$\Delta n_f = n_\parallel - n_\perp,$$

where $n_\parallel$ and $n_\perp$ are the apparent refractive indices parallel and perpendicular to the rods.

Birefringence Δn of oriented crystalline polymers[18]:

$$\Delta n = \Delta n_f + \phi_{cr} f_{cr} \Delta \mathring{n}_{cr}^0 + \phi_{am} f_{am} \Delta \mathring{n}_{am}^0,$$

where Δn_f is the form birefringence, ϕ_{cr} and ϕ_{am} are the volume fractions of crystalline and amorphous components [$\phi_{cr} = 1 - \phi_{am}$ and $\phi_{cr} = (\rho - \rho_a)/(\rho_c - \rho_a)$], ρ is the sample density, ρ_a the density of polymer in the amorphous state, ρ_c the calculated density based on unit cell structure of polymer, $\Delta \mathring{n}_{cr}^0$, and $\Delta \mathring{n}_{am}^0$ are the intrinsic birefringences of crystalline and amorphous components, and f_{cr} and f_{am} the orientation functions of crystalline and amorphous components.

A general reference is available.[18]

Birefringence of spherulites Δn_{sph}:

$$\Delta n_{sph} = n_\parallel - n_\perp,$$

where $n_\parallel$ and $n_\perp$ are the refractive indices parallel and perpendicular to the spherulite radius.

A general reference is available.[20] Also a compendium of refractive indices of polymers is available.[3]

TABLE 13.12.A. Estimated heat capacities of crystalline and amorphous polyethylene. From Ref. 21.

Temp. (K)	Crystalline C_p ($J\,K^{-1}\,g^{-1}$)	Amorphous C_p ($J\,K^{-1}\,g^{-1}$)
1	0.000 008	0.000 028
5	0.000 98	0.003 7
10	0.007 6	0.024 3
25	0.087 8	0.138
50	0.319	0.436
100	0.678	0.694
150	0.900	1.043
200	1.103	1.407
		Glass transformation
250	1.335	(2.01)
300	1.623	(2.20)
350	1.965	(2.36)
400	2.44	(2.51)
	Fusion	
450		2.67
500		2.82
550		2.97
600		3.13

TABLE 13.12.B. Linear temperature dependence or proportionality of heat capacity

Polymer	C_p/T ($J\,K^{-2}\,g^{-1}$)	Range (K)	% deviation from linearity
Nylon 6	0.005 2	150–450	10
Phenolic resin (cured)	0.004 1	150–350	2
Polycarbonate	0.004 6	200–350	5
Polyethylene (crystalline)	0.005 6	160–375	5
Poly(methyl methacrylate)	0.004 6	200–350	5
Polystyrene	0.004 1	150–350	2
Poly(vinyl chloride)	$0.002\,66 + 0.16/T$	80–340	1

13.12. HEAT CAPACITY

The heat capacity of polymers with a mass fraction of crystalline component χ may be obtained from the crystalline heat capactiy $C_{p,c}$ and amorphous heat capacity $C_{p,a}$:

$$C_{p\chi} = \chi C_{p,c} + (1-\chi) C_{p,a},$$

$$\chi = \frac{\rho_c}{\rho}\frac{\rho - \rho_a}{\rho_c - \rho_a},$$

where ρ is the sample density, ρ_c the calculated density of the unit cell, and ρ_a the density of amorphous state.

The estimated heat capacities of crystalline and amorphous polyethylene are given in Table 13.12.A.

Some general features applicable to glassy, $C_{p,g}$, and crystalline heat capacities:

$T < 1$ K, $C_{p,g} \propto T$,

$T \lesssim 10$ K, $C_{p,g}/T^3$ reaches maximum, $C_{p,g} \gg C_{p,c}$,

$T \sim 25$ K, $C_{p,g} - C_{p,c}$ reaches maximum,

$T \sim 70$ K $< T_g$, $C_{p,g} \sim C_{p,c}$,

$T \sim T_g$, $C_{p,g} > C_{p,c}$.

Thermal relaxations in glassy polymers begin to be observable at $\sim T_g - 50$ K.

In many cases C_p/T of solid polymer is in the vicinity of 4.2 mJ $K^{-2}\,g^{-1}$ or $C_p = T/1000$ cal $K^{-1}\,g^{-1}$ for a wide range of temperature.

The linear temperature dependence or proportionality of the heat capacity is given in Table 13.12.B.

Glassy polymers are expected to have residual entropies:

$$S_{0,g}=\int_0^{T_m}\frac{C_{p,c}}{T}dT+\Delta S_m-\int_{T_g}^{T_m}\frac{C_{p,l}}{T}dT-\int_0^{T_g}\frac{C_{p,g}}{T}dT,$$

where the subscripts l and m denote liquid, and melting, respectively. Similarly, $H_{0,g}-H_{0,c}>0$. When calculating Gibbs free energy as $G_T-H_{0,c}$ for glassy and semicrystalline polymers, contributions from $RS_{0,g}$ should be included. Typical values for $S_{0,g}$ are between $(R\ \ln 2)/2$ and $R\ \ln 2$ per chain atom. Some useful references are available.[21,22]

13.13. GLASS TRANSITION

Amorphous polymer materials display phenomenologically a glass transition which appears to be a second-order transition in the Ehrenfest sense.[23] The apparent transition temperature, T_g, depends upon the experimental technique and the time scale by which it is determined. High viscosities and long relaxation times characterize glasses (below T_g). The viscosity above T_g is given approximately by the universal WLF equation (13.17.B.5). Table 13.13.A lists the glass temperature for pure linear homopolymers of high molecular weight along with the three thermodynamic susceptibilities. The glass temperature shows a strong variation as a function of molecular weight, amount and kind of diluent (plasticizer), pressure, composition of copolymers, composition of polymer blends, number of crosslinks in a rubber, and stretch ratio in a rubber.[24,25] The state of a glass depends upon the vitrification history. A glass formed isobarically at an elevated pressure will have a greater density than one formed at atmospheric pressure and the same cooling rate. Data are available.[26–30] Proceedings of a recent conference gives a good indication of the state of knowledge of polymer glasses.[31]

13.14. STRESS σ_{ij} AND DISPLACEMENT u_i AT CRACK TIPS

Three modes of crack tip deformation for which a linear elastic stress field may be established are given in Ref. 32.

A coordinate system (r,α) is selected with the origin at the crack tip and with $r\ll a$; a is the crack length.

Mode I:

$$\begin{Bmatrix}\sigma_{xx}\\ \sigma_{xy}\\ \sigma_{yy}\end{Bmatrix}=\frac{K_{\rm I}}{(2\pi r)^{1/2}}\cos\left(\frac{\alpha}{2}\right)\begin{Bmatrix}1-\sin\frac{\alpha}{2}\sin\frac{3\alpha}{2}\\ \sin\frac{\alpha}{2}\cos\frac{3\alpha}{2}\\ 1+\sin\frac{\alpha}{2}\sin\frac{3\alpha}{2}\end{Bmatrix},$$

$$\begin{Bmatrix}u_x\\ u_y\end{Bmatrix}=\frac{K_{\rm I}}{2G}\left(\frac{r}{2\pi}\right)^{1/2}\begin{Bmatrix}\cos\frac{\alpha}{2}\left(\kappa-1+2\sin^2\frac{\alpha}{2}\right)\\ \sin\frac{\alpha}{2}\left(\kappa+1-2\cos^2\frac{\alpha}{2}\right)\end{Bmatrix}.$$

Mode II:

$$\begin{Bmatrix}\sigma_{xx}\\ \sigma_{xy}\\ \sigma_{yy}\end{Bmatrix}=\frac{K_{\rm II}}{(2\pi r)^{1/2}}\begin{Bmatrix}-\sin\frac{\alpha}{2}\left(2+\cos\frac{\alpha}{2}\cos\frac{3\alpha}{2}\right)\\ \cos\frac{\alpha}{2}\left(1-\sin\frac{\alpha}{2}\sin\frac{3\alpha}{2}\right)\\ \sin\frac{\alpha}{2}\cos\frac{\alpha}{2}\cos\frac{3\alpha}{2}\end{Bmatrix},$$

$$\begin{Bmatrix}u_x\\ u_y\end{Bmatrix}=\frac{K_{\rm II}}{2G}\left(\frac{r}{2\pi}\right)^{1/2}\begin{Bmatrix}\sin\frac{\alpha}{2}\left(\kappa+1+2\cos^2\frac{\alpha}{2}\right)\\ -\cos\frac{\alpha}{2}\left(\kappa-1-2\sin^2\frac{\alpha}{2}\right)\end{Bmatrix}.$$

Mode III:

$$\begin{Bmatrix}\sigma_{xz}\\ \sigma_{yz}\end{Bmatrix}=\frac{2K_{\rm III}}{(2\pi r)^{1/2}}\begin{Bmatrix}-\sin\frac{\alpha}{2}\\ \cos\frac{\alpha}{2}\end{Bmatrix},$$

$$u_z=\frac{K_{\rm III}}{G}\left(\frac{r}{2\pi}\right)^{1/2}\sin\frac{\alpha}{2}.$$

TABLE 13.13.A. Thermodynamic properties at the glass temperature[a]

Material/chemical formula	Expansion coefficient (10^{-4}/K) Liquid, α_e	Discontinuity, $\Delta\alpha$	Compressibility (10^{-5}/MPa) Liquid, β_e	Discontinuity, $\Delta\beta$	Specific heat discontinuity ΔC_p (J/g K)	Glass temperature (K)
Polyethylene	5.1	3.2	50	20	0.60	140[b]
Polypropylene	6.8	4.4	38	9	0.48	244
Polyisobutylene	6.2	4.7	40	10	0.40	198
Poly(vinyl chloride)	5.7	3.7	44	20	0.30	350
Poly(vinyl acetate)	7.1	4.3	50	21	0.41	304
Poly(methylmethacrylate)	5.8	3.1	58	28	0.30	378
Polystyrene	5.1	2.9	61	29	0.34	362
Poly(α-methylstyrene)	5.5	3.1	64	32	0.32	440
Polyisoprene	5.8	3.9	51	26	0.47	201
Polydimethylsiloxane	9.9	6.9	60	30	0.42	150
Poly(phenylene oxide)	5.3	3.2	50	20	0.24	480
Poly(ethylene terephthalate)	5.8	3.5	55	25	0.33	337
Polycarbonate[c]	6.0	3.44	51	18	0.23	424

[a]J. M. O'Reilley, J. Appl. Phys. **48**, 4047 (1977).
[b]Note that there is a diversity of opinion on the value of T_g for polyethylene.
[c]P. Zoller, J. Polym. Sci. Polym. Phys. Ed. **16**, 1261 (1978).

TABLE 13.15.A. Data on internal friction peaks for important polymers

	α peak			β peak			γ peak		
Polymer	T (K)	tanδ	ν (Hz)	T (K)	tanδ	ν (Hz)	T (K)	tanδ	ν (Hz)
High-density polyethylene [a]	343	0.2	1.5	271	0.03	3.1	149	0.05	4.8
Low-density polyethylene [a]	339	0.25	1.8	262	0.03	6.8	147	0.05	13.7
Polypropylene [b]	323–378	0.11–0.15	0.25–0.38	277–285	0.06–0.11	0.6–1.0	213–223	0.016	0.8–1.3
6-6 nylon [c]	355	0.10	590	250	0.04	1050	165	0.041	1300
Poly(ethylene terephthalate) [d]	360	0.06 [e]	1.0	213	0.03–0.05	1.0		(not observed)	
Poly(chlorotrifluoroethylene) [f]	406	0.09	0.59	362–365	0.16–0.29	0.4–0.8	241	0.08	1.0–1.8

[a]A. A. Flocke, Kolloid Z. **180**, 118 (1962).
[b]E. Passaglia and G. M. Martin, J. Res. Natl. Bur. Stand. Sec. A **68**, 519 (1964).
[c]A. E. Woodward, J. A. Sauer, C. W. Deeley, and D. E. Kline, J. Colloid Sci. **12**, 363 (1957).
[d]K. H. Illers and H. Breuer, J. Colloid Sci. **18**, 1 (1963).
[e]For high-crystallinity specimen; varies greatly with crystallinity.
[f]J. M. Crissman and E. Passaglia, J. Polym. Sci. **14**, 237 (1966).

K is the stress intensity factor $[=\sigma_0(\pi a)^{1/2}f(a/b)$, with $f(a/b)$ determined by geometry of specimen and mode, σ_0 the applied stress, and b the width of the specimen], G the shear modulus, and

$$\kappa=(3-\nu)/(1+\nu) \quad \text{for plane stress}$$

$$=3-4\nu \quad \text{for plane strain,}$$

where ν is Poisson's ratio.

13.15. INTERNAL FRICTION PEAKS IN SEMICRYSTALLINE POLYMERS

A frequently used measure of internal friction is the tangent of the phase angle δ between stress and strain under sinusoidal deformation. When measurements are made at constant frequency as a function of temperature, three peaks associated with various relaxation processes generally appear. The highest temperature peak is usually termed α, the next lower β, etc., although this nomenclature is not always followed. Other peaks are sometimes observed, either at lower temperature or as shoulders on these. The temperature of the peaks vary with the measure of mechanical loss used [tanδ, G'', or J'' (see 19.00)]. The temperature and magnitude of the peaks are influenced by the frequency of measurement as well as by thermal treatment, moisture content, and other factors that influence the molecular mobility.

The activation energy for the γ process is generally of the order of 15 kcal/mol. The β process is associated with the glass transition, and the frequency-temperature change of the maximum approximately follows the Williams-Landel-Ferry equation.[28] The activation energy of the α process varies from 40 to 80 kcal/mol depending on which portion of the process is investigated. Values as high as 150 kcal/mol have been reported.[33] Data for some important polymers are given in Table 13.15.A. A general reference is available.[34]

13.16. REPRESENTATIVE MECHANICAL PROPERTIES OF SOME COMMON STRUCTURAL POLYMERS

Table 13.16.A gives mechanical properties often used for design purposes for some common structural polymers. The values are for room temperature and are obtained by various testing methods of the American Society for Testing and Materials. The values are strongly

TABLE 13.16.A. Mechanical properties of common structural polymers [a]

Polymer	Density (g/cm^3)	Yield strength (MN/m^2)	Young's modulus (GN/m^2)	Hardness (Rockwell)	Impact strength (kJ/m^2)
Polystyrene	1.04	35–70	3.3–3.5	M72	0.83 [b]
Poly(methyl methacrylate)	1.17–1.29	84–120	2.5–3.5	M80–M102	1.33 [b]
Polycarbonate	1.20	58	2.4	M70	4.8 [b]
Polyethylene (low density)	0.91–0.925 [c]	4.2–16 [c]	0.1–0.25 [c]	R10 [c]	34.5 [b]
Polypropylene	0.900–0.910	34–36	1.1–1.5	R80–R110	
6-6 nylon	1.13–1.15	60–90	2.7–3.3	R108–R118	5.1 [b]
Polyethylene (high density)	0.941–0.965 [c]	22–38 [c]	0.4–1.3 [c]	D60–D70 (shore)	3.3 [b]

[a]Data from *Handbook of Materials Science,* edited by Charles T. Lynch (Chemical Rubber, Cleveland, 1975).
[b]J. G. Williams, *Advances in Polymers* (Springer, Berlin, 1978).
[c]*Handbook of Plastics and Elastomers,* edited by Charles A. Harper (McGraw-Hill, New York, 1975).

TABLE 13.17.B.1.a. Parameters for zero-shear viscosity relations[a]

Polymer	T_0 (K)	$10^4\alpha$ (K^{-1})	$\ln\zeta_0$	v_2 (cm^3/g)	$10^{18}\langle S^2\rangle_0/M$ (cm^2 mol/g)	$10^{17}X_c$ (cm^5 mol/g^2)	Z_c
Polybutadiene[b]	128	7.12	−10.90	1.11	12.6	360	330
Polydimethylsiloxane	30	7.12	−10.20	1.04	7.2	460	660
Polyethylene	0[c]	2.75[c]	−11.80[c]	1.307	17.0	350	270
Polyisobutylene	122	3.20	−11.74	1.123	8.7	420	540
Poly(methyl methacrylate)	308	3.88	−11.40	0.880	6.2	390	550
Polystyrene	313	5.57	−11.05	1.038	7.6	435	600
Poly(vinyl acetate)	248	4.58	−11.95	0.880	5.7	370	570

[a]From G. C. Berry and T. G. Fox, Adv. Polym. Sci. **5**, 261 (1968).
[b]50% *cis*.
[c]These values are uncertain, but represent the data over the limited temperature range for which they are available.

affected by temperature, time, the thermal history of the material, and other characteristics (such as molecular weight). For careful experiment work, the original literature should be consulted.

13.17. RHEOLOGY

A. Introduction

Chapter 19.00 presents definitions and additional discussion of basic rheological terms, experiments, and material properties.

B. Linear viscoelasticity[28]

1. (Zero-shear) viscosity[35]

The limiting value $\eta(0)$ of the viscosity function $\eta(\dot\gamma)$, is known as the zero-shear viscosity η. For linear random coil polymers and their concentrated solutions, the empirical dependence on the absolute temperature T, the weight-average number of atoms (or groups) in the chain backbone, Z_w, and the volume fraction of polymer ϕ_2 is

$$\eta = F(X)\zeta,$$

where F is the *structure factor* and ζ the *friction factor per chain atom,*

$$F(X) = (N_A/6)X, \quad X < X_c$$
$$= (N_A/6X_c^{2.4})X^{3.4}, \quad X > X_c$$
$$X = Z_w\phi_2(\langle S^2\rangle_0/M)v_2,$$
$$X_c = 400\times10^{-17} \pm 10\%,$$
$$\ln\zeta = \ln\zeta_0 + 1/\alpha(T - T_0), \quad T_g < T < T_g + 100,$$

where N_A is Avogadro's number, $\langle S^2\rangle_0$ the unperturbed radius of gyration, M the molecular weight, v_2 the specific volume of the polymer, and ζ_0, α, and T_0 are constants whose values for some polymers are listed in Table 13.17.B.1.a together with other pertinent properties: v_2, $\langle S^2\rangle_0/M$, and Z_c, the value of Z_w corresponding to X_c. The values of α and T_0 listed are for an undiluted polymer of high molecular weight ($Z > 80$).

2. Steady-state compliance J_e^0

For a linear viscoelastic fluid,

$$J_R(\infty) = J'(0) = \lim_{\omega\to0}\frac{G'(\omega)/\omega^2}{\eta^2} = J_e^0.$$

For a nonlinear viscoelastic fluid in steady flow, $N_1(\dot\gamma)$ at low values of $\dot\gamma$ is given by $N_1(\dot\gamma) = 2\eta^2J_e^0\dot\gamma^2 + O(\dot\gamma^4)$.

For monodisperse linear random coil polymers and their concentrated solutions,[36]

$$J_e^0 = \frac{0.4M}{cRT}\left[1 + \left(\frac{cM}{\rho M_c'}\right)^2\right]^{-1/2},$$

where M is the molecular weight, R the gas constant, T the absolute temperature, c the concentration of polymer, ρ the density of the polymer, and M_c' a characteristic

TABLE 13.17.B.2.a. Characteristic viscoelastic values[a]

	Rubber plateau				From steady-state compliance		
Polymer	°C	$\log J_N^0$[a] (cm^2/dyn)	M_e (g/mol)	Z_e	°C	$\log J_e^0$[b] (cm^2/dyn)	M_c' (g/mol)
Hevea rubber	25	−6.76	6 100	360	−30	−5.85	60 000
Polybutadiene ≈50% *cis.*	25	−7.06	1 900	140	25	−6.60	13 800
Polydimethylsiloxane	25	−6.30	12 000	330	20	−6.00	61 000
Polyisobutylene	25	−6.46	7 600	270			
Poly(methyl methacrylate)	170	−6.94	4 700	94			
Polystyrene	140	−6.31	17 300	333	200	−5.76	130 000
					160	−5.85	
Poly(vinyl acetate)	60	−6.55	9 100	210	30	−5.90	86 000

[a] J. D. Ferry, *Viscoelastic Properties of Polymers*, 3rd. ed. (Wiley, New York, 1980), and W. W. Graessley, Adv. Polym. Sci. **16**, 1 (1974).
[b] J_e^0 for monodisperse polymer with $M > M_c'$.

molecular weight, values of which are included in Table 13.17.B.2.a.

For $M \gg M_c'$, J_e^0 is independent of molecular weight. Values for various undiluted monodisperse polymers are listed in Table 13.17.B.2.a. J_e^0 is highly sensitive to molecular-weight distribution, a dependence on $M_z M_{z+1}/M_w^2$ often being useful.

3. Rubber or entanglement plateau

For high-molecular-weight linear amorphous polymers, the compliance functions have a plateau J_N^0. The modulus functions have a corresponding plateau G_N^0 $(=1/J_N^0)$. The value of J_N^0 is independent of molecular weight M for $M > \approx 2 \times 10^4$ and is associated with an entanglement network characterized by M_e $(=\rho RTJ_N^0)$, the average molecular weight between entanglement coupling point, or by Z_e, the corresponding number of chain atoms. The plateau occurs only if the molecular weight is considerably greater than M_e. Values of J_N^0, M_e, and Z_e, for several polymers are listed in Table 13.17.B.2.a.

4. Glass-rubber transition (or dispersion) zone

The loss functions have maxima in this zone, while other functions [$J(t)$, $G(t)$, $J'(\omega)$, and $G'(\omega)$] change by several orders of magnitude for amorphous polymers. There is little dependence on molecular weight. Table 12.I. of Ref. 28 lists values for some typical points in these functions.

5. Time-temperature shift factor a_{0T}

For many materials, the logarithmic plot of a viscoelastic function at the temperature T may be obtained from that at the temperature T_0 by shifting the curve along the $\log t$ or $\log \omega$ axis by the amount $\log a_{0T}$. Various expressions for a_{0T} are:

Arrhenius: $\ln a_{0T} = \ln A + B/RT$,

Vogel or Fulcher: $\ln a_{0T} = \ln A + 1/\alpha(T - T_\infty)$,

WLF (Williams-Landel-Ferry):

$$\log a_{0T} = -c_1^0(T - T_0)/(c_2^0 + T - T_0),$$

where $T_\infty = T_0 - c_2^0$. The WLF equation and its equivalent, the Vogel equation, are valid for temperatures between T_g and $T_g + 100$ K. They have been interpreted in terms of the Doolittle free volume expression

$$\ln a_{0T} = B(f^{-1} - f_0^{-1}),$$

where B is an empirical constant usually assumed to be approximately unity, f and f_0 are the fractional free volumes at T and T_0, respectively, and f is assumed to be a linear function of T:

$$f = f_0 + \alpha_f(T - T_0),$$

where α_f is the thermal expansion coefficient for the free volume:

$$f_0 = B/2.303c_1^0, \quad \alpha_f = B/2.303c_1^0 c_2^0.$$

See Table 13.17.B.5.a, which also lists f_g/B, the fractional free volume at T_g, which is 0.025 ± 0.005 for most systems. In the absence of other information, for a rough estimate of a_{0T}, T_0 may be chosen as T_g, $c_1^0 \approx 17$ and $c_2^0 \approx 52$ K, which are called "universal WLF parameters."

6. Rouse-Zimm theories for viscoelastic properties of dilute solutions of random coil macromolecules in Θ solvents

The molecule (degree of polymerization P, molecular weight M) is represented as N submolecules, where the end-to-end distance σ of the submolecule represents one statistical length.

$$\eta'(\omega) = \eta_s + \frac{cRT}{M}\sum_{p=1}^{N}\frac{\tau_p}{1+\omega^2\tau_p^2},$$

$$G'(\omega) = \frac{cRT}{M}\sum_{p=1}^{N}\frac{\omega^2\tau_p^2}{1+\omega^2\tau_p^2},$$

where c is the concentration of polymer in the solvent in g/ml, and η_s the viscosity of the solvent. In the Rouse (free draining) model,

$$\tau_p \cong \sigma^2 N^2 f_0/6\pi^2 p^2 kT,$$

where f_0 is the friction coefficient of the submolecule. In the Zimm theory, the perturbation of the velocity of the solvent is taken into account approximately by introduc-

TABLE 13.17.B.5.a. Parameters characterizing temperature dependence of a_{0T}[a]

Polymer	T_0 (K)	c_1^0	c_2^0 (K)	T_g (K)	f_g/B	α_f/B (10^4 K^{-1})	T_∞ (K)
Butyl rubber[b]	298	9.03	201.6	205	0.026	2.4	96
Hevea rubber	248	8.86	101.6	200	0.026	4.8	146
	298	5.94	151.6				146
Polybutadiene[c]	298	3.64	186.5	172	0.039	6.4	112
Polydimethylsiloxane	303	1.90	222	150	0.071	10.3	81
Polyisobutylene	298	8.61	200.4	205	0.026	2.5	101
Poly(methyl methacrylate)	388	32.2	80	388	0.013	1.7	308
Polystyrene	373	13.7	50.0	373	0.032	6.3	323
Poly(vinyl acetate)	349	8.86	101.6	305	0.028	5.9	258
Styrene-butadiene copolymer[d]	298	4.57	113.6	210	0.021	8.2	184

[a]From J. D. Ferry, *Viscoelastic Properties of Polymers*, 3rd ed. (Wiley, New York, 1980).
[b]Lightly vulcanized with sulfur.
[c]*cis:trans:*vinyl = 43:50:7.
[d]Styrene:butadiene = 23.5:76.5 random (by weight).

TABLE 13.18.A.1. Dipole moments of polymers in solution, expressed as $(\bar{\mu}^2/N)^{1/2}$, the dipole moment per monomer unit. $\phi = \bar{\mu}^2/N\mu_0^2$, where μ_0 is the dipole moment of the isolated monomer unit and N the number of monomers per molecule. 1 D $= 10^{-18}$ esu cm $= 3.335\,64 \times 10^{-30}$ Cm. From *Polymer Handbook*, 2nd ed., edited by J. Brandrup and E. H. Immergut (Wiley, New York, 1975).

Polymer	Solvent	Temp. (°C)	$(\bar{\mu}^2/N)^{1/2}$ (D)	ϕ
Polystyrene	carbon tetrachloride	25	0.26	0.56
Poly(*p*-chlorostyrene)	benzene	30, 25	1.45	0.56
Poly(vinyl acetate)	benzene	20	1.70	0.84
Poly(vinyl chloride)	dioxane	20, 40	1.62	0.59
Poly(methyl acrylate)	benzene	25	1.41–1.44	0.64–0.67
Poly(methyl methacrylate)				
Isotactic	benzene	25–65	1.40–1.44	0.77–0.81
Atactic	benzene	25–65	1.33–1.41	0.69–0.78
Syndiotactic	benzene	25–65	1.34–1.41	0.70–0.78
cis 1,4-polyisoprene	benzene	25	0.28	0.70
trans 1,4-polyisoprene	benzene	25	0.31	0.82

ing a parameter h^* (a measure of the hydrodynamic interaction). Numerical values of τ_p must be obtained by computer calculation. Chapter 9 of Ref. 28 contains graphs of G' and G'' calculated from the Rouse and Zimm theories.

13.18. ELECTRICAL PROPERTIES

A. Dipole moments

Dipole moments of polymers in solution are given in Table 13.18.A.1.

B. Typical electrical properties

Table 13.18.B.1 gives some typical electrical property data. The values are to be considered estimates only. Dielectric properties are particularly sensitive to ionic impurities and may reflect effects of moisture content, thermal history, chemical variations, and details of the test method. For careful experimental work, the original literature should be consulted.

13.19. DIFFUSION AND PERMEATION

A. Diffusion into plane sheet[37]

1. General solution

For Fickian-type diffusion, the amount of diffusant transferred, M_t, at time t between a sheet of thickness $2l$ and a stirred liquid of finite volume V_s is given by

$$\frac{M_t}{M_\infty} = 1 - \sum_{n=1}^{\infty} \frac{2\alpha(1+\alpha)}{1+\alpha+\alpha^2 q_n^2} e^{-q_n^2 \tau}.$$

M_∞ is the amount of diffusant transferred at equilibrium, $\tau = Dt/l^2$, D is the diffusion coefficient, K is the partition

TABLE 13.18.B.1. Typical electrical properties of selected polymers at room temperature. From *Handbook of Plastics and Elastomers,* edited by C. A. Harper (McGraw-Hill, New York, 1975).

Polymer	Volume resistivity (Ω cm)	Dielectric constant at 60 Hz	Dielectric strength, 3 mm thickness (kV/cm)	Dissipation factor at 60 Hz
Cellulose acetate	10^{10}–10^{12}	3.2–7.5	120–240	0.01–0.1
Cellulose acetate butyrate	10^{10}–10^{12}	3.2–6.4	100–160	0.01–0.04
Cellulose proprionate	10^{12}–10^{16}	3.3–4.2	120–180	0.01–0.05
Polytetrafluoroethylene	$>10^{18}$	2.1	160	<0.0001
Nylon 6	10^{14}–10^{15}	6.1	140	0.5
Nylon 6/6	10^{14}–10^{15}	3.6–4.0	140	0.014
Nylon 6/10	10^{14}–10^{15}	4.0–7.6	140	0.05
Polycarbonate	6×10^{15}	3.0	160	0.0001–0.0005
Polyethylene	10^{15}–10^{18}	2.3	180–400	0.0001–0.006
Polyimide	10^{16}–10^{17}	3.5	160	0.003
Polypropylene	10^{15}–10^{17}	2.1–2.7	180–260	0.0007–0.005
Polystyrene	10^{17}–10^{21}	2.5–2.65	200–280	0.0001–0.0005
Polystyrene, high impact	10^{10}–10^{17}	2.5–3.5	200	0.003–0.005
Poly(vinyl chloride)				
Flexible	10^{11}–10^{15}	5–9	120–400	0.08–0.15
Rigid	10^{12}–10^{16}	3.4	170–400	0.01–0.02
Poly(vinylidene chloride), rigid	10^{15}	3.1	480–610	0.02
Poly(4-methylpentene-1)	$>10^{16}$	2.12	280	0.001
Poly(aryl sulfone)	3×10^{16}	3.9	150	0.003
Poly(phenylene sulfide)	10^{16}	3.1	240	0.0004
Poly(phenylene oxide)	10^{17}	2.6	160–200	0.0004
Polysulfone	5×10^{16}	2.8	160	0.007
Poly(ethersulfone)	10^{17}–10^{18}	3.5	160	0.001

TABLE 13.19.B.1. Typical values of diffusion coefficients and temperature dependence. $D = A\exp(-E/RT)$, where R is the molar gas constant and T the temperature. Diffusion coefficients are often strongly dependent on morphology and concentration. From J. H. Flynn, Polymer **13**, 1325 (1982).

Diffusing molecule	Molecular weight ($g\ mol^{-1}$)	Low-density polyethylene D (30°C) ($10^{-8}\ cm^2s^{-1}$)	$-\log_{10}A$	E ($kJ\ mol^{-1}$)	High-density polyethylene D (30°C) (10 ^{8} cm^2s^{-1})	$-\log_{10}A$	E ($kJ\ mol^{-1}$)	Polyisobutylene D (30°C) ($10^{-8}\ cm^2s^{-1}$)	$-\log_{10}A$	E ($kJ\ mol^{-1}$)
Methane	16.0	29	1.42	47	7.4	0.38	44			
Ethane	30.1	9.2	2.92	58	2.6	2.11	57			
Propane	44.1	4.7	2.26	56	0.72	1.66	57	0.30		
n-pentane	72.2	1.6	5.02	73	0.55	4.93	77	0.19	3.32	70
Benzene	78.1	2.0	3.28	66						
p-dioxane	88.1	0.84	4.90	75						
Bromoethane	95.0	8.3	1.84	52						
n-heptane	100.2	1.6	−1.09	45				0.20		
o-xylene	106.2	20	1.60	48	5.9	0.73	46			
p-xylene	106.2	56	−0.41	34	18	−0.25	44			
n-octane	114.2	0.66	2.42	62				0.19	4.37	76
n-decane	142.3	0.58	3.34	67						
Carbon tetrachloride	153.8	0.68	7.44	83						

TABLE 13.19.B.2. Self-diffusion of polyolefins. $D = A\exp(-E/RT)$, where R is the molar gas constant and T the temperature. Asterisks denote extrapolated values. From J. H. Flynn, Polymer **13**, 1325 (1982).

Diffusant	Molecular weight ($g\ mol^{-1}$)	D (30°C) ($10^{-8}\ cm^2\ s^{-1}$)	D (140°C) ($10^{-8}\ cm^2\ s^{-1}$)	$-\log_{10}A$	E ($kJ\ mol^{-1}$)
n-octadecane	254.5	310*	1800	2.67	175
n-dotriacontane	450.9	52*	630	2.22	24
Polyisobutylene	(~700)	5.9*	64	3.34	23
High-density polyetheylene	(~4100)	1.3*	14	4.04	22
Low-density polyethylene	(~5800)	1.8*	10	4.91	16

coefficient (ratio of diffusant concentration in the polymer to its concentration in the solvent), $\alpha = V_s/KV_p$, V_p is the volume of polymer, and q_n the nonzero positive roots of $\tan q_n = -\alpha q_n$.

2. Approximation for $\tau < 0.1$

$$M_t/M_\infty = (1+\alpha)[1 - e^{\tau/\alpha^2}\mathrm{erf}(\tau^{1/2}/\alpha)]$$

3. Infinite bath for $\tau \leqslant 0.1$

$$M_t/M_\infty = 2(\tau/\pi)^{1/2}.$$

B. Diffusion data

See Tables 13.19.B.1 and 13.19.B.2.

C. Gas transmission

The product of the diffusion constant and the solubility is the permeability coefficient P, useful for calculating the flux of gas through a unit area of polymer:

$$\dot{Q} = P(P_1 - P_0)/l,$$

where $\dot{Q}$ is the rate of gas transmission, P_1, P_0 the pressure on opposite sides of the membrane, and l the membrane thickness. The diffusion coefficient is defined by Fick's law and the solubility coefficient by Henry's law. Equations for calculating transport under non-steady-state conditions and for a variety of geometries are available.[37] Tabulations of permeability coefficients can be found.[3,38]

13.20. NONLINEAR OPTICAL PROPERTIES

Polymers are an important class of materials for nonlinear optical processes.[39] Materials exhibit nonlinear optical effects when subjected to an intense oscillating electric field. Polarization generated in the material due to electric dipole interaction with the electric field can be expressed by a Taylor series expansion[40]

$$\vec{P} = \overset{\rightrightarrows}{\chi}{}^{(1)}\cdot\vec{E} + \overset{\rightrightarrows}{\chi}{}^{(2)}:\vec{E}\vec{E} + \overset{\rightrightarrows}{\chi}{}^{(3)}:\vec{E}\vec{E}\vec{E}\cdot$$

$$= \overset{\rightrightarrows}{\chi}_{\text{eff}}\cdot\vec{E}\ . \tag{1}$$

In the equation, $\overset{\rightrightarrows}{\chi}{}^{(1)}$ is the linear susceptibility which is generally adequate to describe the optical response to a weak field. It describes linear absorption, refraction and scattering. The terms $\overset{\rightrightarrows}{\chi}{}^{(2)}$ and $\overset{\rightrightarrows}{\chi}{}^{(3)}$ are the second and third order nonlinear optical susceptibilities which de-

TABLE 13.20.A. $\chi^{(3)}$ values for some polymers.

Common Name	Structure	Measurement Technique	λ(nm) or $h\nu$(eV)	$\chi^{(3)}$ (esu)	Reference
p-toluene sulfonate (PTS) polydiacetylene	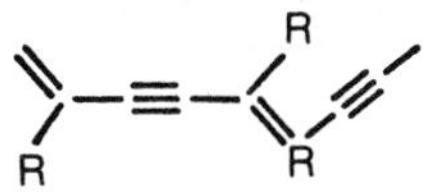 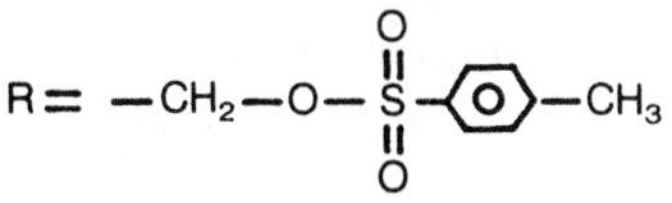 $R = -CH_2-O-SO_2-C_6H_4-CH_3$	THG[a]	0.66 eV	8.5×10^{-10} (parallel to chains)	43
poly-4-BCMU polydiacetylene	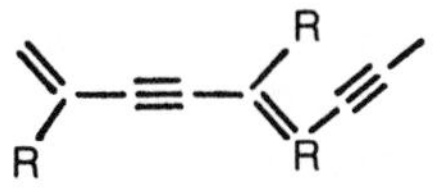	DFWM[b]	2.07 eV	4×10^{-10} (Red form)	44
polyacetylene		THG	0.65 eV 1.5 eV	13×10^{-10} 0.5×10^{-10}	45
poly (*p*-phenylene benzobisthiazole) (PBZT)	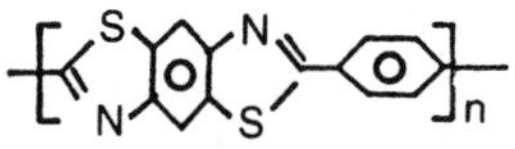	DFWM	602 nm	$\sim10^{-11}$	46
		THG	0.67 eV	7.8×10^{-12}	47
poly (*p*-phenylene vinylene) (PPV)	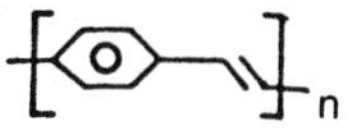	DFWM	602 nm	4×10^{-10} (parallel to draw direction)	48
polythiophene	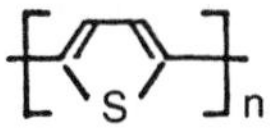	DFWM	602 nm	$\sim5\times10^{-10}$	49

[a] THG = Third harmonic generation

[b] DFWM = Degenerate four wave mixing

scribe the nonlinear response. At optical frequencies, the refractive index, η, and dielectric constant, $\epsilon(\omega)$ are related by:

$$\eta^2(\omega) = \epsilon(\omega) = 1 + 4\pi\chi(\omega) \quad (2)$$

For a plane wave, the wave vector $k = \eta\omega/c$ and the phase velocity $v = c/\eta$. In a nonlinear medium, $\overleftrightarrow{\chi}(\omega) = \overleftrightarrow{\chi}_{\text{eff}}$ of equation (1) is dependent on E; therefore, η, k and v are all dependent on E. Two consequences of the second order nonlinearity described by $\chi^{(2)}$ are second harmonic generation and Pockel's electrooptic effect. In second harmonic generation, an intense beam of frequency ω passing through a nonlinear medium of $\chi^{(2)} \neq 0$ generates an optical wave at frequency 2ω. The electrooptic effect describes the phase shift introduced by an applied low frequency electric field. In other words, the refractive index of a material with nonvanishing $\chi^{(2)}$ can be modulated by the application of a DC or low frequency AC field. Two consequences of the third order optical nonlinearities represented by $\chi^{(3)}$ are third harmonic generation and intensity dependent refractive index. Third harmonic generation describes the process in which an incident photon field of frequency ω generates through nonlinear polarization in the medium, a coherent optical field at 3ω. Through $\chi^{(3)}$ interaction, the refractive index of a nonlinear medium is given as $\eta = \eta_0 + \eta_2 I$ where η_2 describes the intensity dependence of refractive index; I is the intensity of the laser pulse.

Since χ^2 is a third-rank tensor, it vanishes in a centrosymmetric or isotropic medium. Therefore, second order nonlinear optical processes can be observed only in noncentrosymmetric media such as piezoelectric and ferroelectric systems. In contrast, there is no symmetric restriction on the third-order processes which can occur in all media.

At the microscopic level, the nonlinearity of the organic structures is described by the electric dipole interaction of the radiation field with the molecules. The resulting induced dipole moment is given as

$$\vec{\mu}_{\text{ind}} = \overleftrightarrow{\alpha} \cdot \vec{E} + \overleftrightarrow{\beta} : \vec{E}\vec{E} + \overleftrightarrow{\gamma} \vdots \vec{E}\vec{E}\vec{E} \cdot \quad (3)$$

In the equation $\overleftrightarrow{\alpha}$ is the linear polarizability. The terms $\overleftrightarrow{\beta}$ and $\overleftrightarrow{\gamma}$, called first and second hyperpolarizabilities, describe the nonlinear optical interactions and are microscopic analogues of $\overleftrightarrow{\chi}^{(2)}$ and $\overleftrightarrow{\chi}^{(3)}$.

In the weak coupling limit, as in the case for most organic systems, each molecule can be treated as an independent source of nonlinear effects. Then the macroscopic susceptibilities $\chi^{(3)}$ are derived from the microscopic nonlinearities β and γ by simple orientationally averaged site sums using appropriate local field correction factors which relate the applied field to the local field at the molecular site.

For second-order nonlinearity in organic systems, a system which consists of a π-electron structure containing an electron donor group on one side and an electron acceptor group on another side has been identified[41] as giving a large value of β. A typical example is

A—⟨O⟩—D

in which A and D are, respectively, the electron acceptor and donor groups.

Second-order nonlinear processes have been observed in many types of organic structures and bulk systems. One does not need a polymeric structure. Small molecules can be used either in the crystalline form, or in the form of Langmuir–Blodgett films. Although piezoelectric polymers such as polyvinylidene fluoride give rise to second-order effects, a polymeric medium has been used mainly as a passive host in which the nonlinearily active organic molecules are dispersed as a guest or attached to the polymeric structure by a flexible spacer (such as in a side-chain liquid crystalline polymers).[42] The second-order active nonlinear group is in the side-chain. The noncentrosymmetric arrangement of dipoles in the polymeric structure is created by electric field poling. The polymer is heated to its glass temperature, an electric ield is applied to orient the dipoles and then the polymer is cooled to freeze the oriented dipoles.

All microscopic theoretical models predict the largest nonresonant third-order optical nonlinearity from π-electrons. Therefore, conjugated polymeric structures which provide effective π-electron delocalization have become an important class of $\chi^{(3)}$ materials. In the nonresonant regime, the response of $\chi^{(3)}$ derived from the delocalized π-electrons is extremely fast (femtoseconds).

Some conjugated polymers, with their reported χ^3 values are listed in Table 13.20.A along with the method of measurement and the wavelength used.

13.21. THERMAL EXPANSION

The linear thermal expansion coefficient, α:

$$\alpha = \frac{\Delta L}{L\,\Delta T}$$

where ΔL is the change in length for a temperature change ΔT, and L is the original length. The expansion coefficient depends on degree of crystallinity, thermal history, and mechanical history.[50] The value changes at the glass transition temperature. Values for polymer are in the range 0.5×10^{-5} to 30×10^{-5}/°C, typically an order of magnitude larger than for other materials (fused silica: 0.54×10^{-6}/°C; aluminum: 23×10^{-6}/°C). Mismatched thermal expansion properties often contribute to the premature failure of objects manufactured from dissimilar materials.[50] Table 13.21.A lists values for several common polymers. More extensive tabulations are available, both for pure polymers[50,51,52] and for engineering formulations.[52]

TABLE 13.21.A. Linear thermal expansion of polymers.[50]

Polymer	α °C^{-1}; $\times 10^5$
Polyethylene, high density	11–13
low density	18–20
Polypropylene	5.8–10.2
Polytetrafluoroethylene	10
Poly(vinyl chloride), rigid	5–18
flexible	7–25
Polystyrene	6–8
Poly(methyl methacrylate)	4.5
Nylon-6	5.9
Nylon-66	8
Nylon-6,10	9
Phenol-Formaldehyde	6.8
Epoxy Resin (Shell 828-Z)	5.6

13.22. REFERENCES

[1]*Macromolecules: An Introduction to Polymer Science,* edited by F. A. Bovey and F. H. Winslow. See also "List of Standard Abbreviations for Synthetic Polymers and Polymer Materials," Pure Appl. Chem. **40**, 473 (1974).

[2]P. J. Flory, *Statistical Mechanics of Chain Molecules* (Interscience, New York, 1969).

[3]*Polymer Handbook,* edited by J. Brandrup and E. H. Immergut (Wiley, New York, 1975).

[4]C. H. Chervenka, *A Manual of Methods for the Analytical Ultracentrifuge* (Beckman Instruments, Palo Alto, 1973).

[5]*Light Scattering From Polymer Solutions,* edited by M. G. Huglin (Academic, New York, 1972).

[6]*Light Scattering From Dilute Polymer Solutions,* edited by D. McIntyre and F. Gornick (Gordon and Breach, New York, 1964).

[7]B. Chu, *Laser Scattering* (Academic, New York, 1974).

[8]B. Berne and R. Pecora, *Dynamic Light Scattering* (Wiley, New York, 1976).

[9]A. Z. Akcasu, M. Benmouna, and C. C. Han, Polymer **21**, 866 (1980).

[10]E. D. Becker, *High Resolution NMR* (Academic, New York, 1969).

[11]R. K. Jensen and L. Petrakis, J. Magn. Reson. **7**, 105 (1972).

[12]J. C. Randall, in *Carbon-13 NMR in Polymer Science,* edited by W. M. Pasika, ACS Symp. Ser. **103** (1979), Chap. 14.

[13]H. W. Siesler and K. Holland-Moritz, *Practical Spectroscopy, Vol. 4, Infrared and Raman Spectroscopy of Polymers* (Dekker, New York, 1980).

[14]J. D. Hoffman, G. T. Davis, and J. I. Lauritzen, Jr., *Treatise on Solid State Chemistry,* edited by N. B. Hannay (Plenum, New York, 1976), Vol. 3.

[15]L. Mandelkern, *Crystallization of Polymers* (McGraw-Hill, New York, 1964).

[16]H. Tadokoro, *Structure of Crystalline Polymers* (Wiley, New York, 1979).

[17]B. Wunderlich, *Macromolecular Physics* (Academic, New York, 1973), Vol. I.

[18]R. S. Stein and G. L. Wilkes, *Structure and Properties of Oriented Polymers,* edited by I. M. Ward (Wiley, New York, 1975); and L. R. G. Treloar, *The Physics of Rubber Elasticity,* 2nd ed. (Oxford University, Oxford, 1958).

[19]M. V. Folkes and A. Keller, Polymer (London) **12**, 222 (1971).

[20]F. Khoury and E. Passaglia, *Treatise on Solid State Chemistry,* edited by N. B. Hannay (Plenum, New York, 1976), Vol. 3.

[21]B. Wunderlich and H. Bauer, Adv. Polymer Sci. **7**, 151 (1970); V. Gaur and B. Wunderlich, J. Phys. Chem. Ref. Data **10**, 19 (1981).

[22]S. S. Chang, ACS Org. Coatings Plast. Preprints **35**, 364 (1975).

[23]P. Ehrenfest, Leiden Comm. Suppl. **756** (1933).

[24]E. A. DiMarzio, Annals N. Y. Acad. Sci. **371**, 1 (1981).

[25]P. R. Couchman, Macromolecules **11**, 1156 (1978).

[26]J. E. McKinney and M. Goldstein, J. Res. Natl. Bur. Stand. **78A**, 331 (1974).

[27]J. E. McKinney and H. V. Belcher, J. Res. Natl. Bur. Stand. **67A**, 43 (1962).

[28]J. D. Ferry, *Viscoelastic Properties of Polymers,* 3rd ed. (Wiley, New York, 1980).

[29]M. Goldstein, J. Phys. Chem. **77**, 667 (1973).

[30]J. E. McKinney and R. Simha, J. Res. Natl. Bur. Stand. **81A**, 283 (1977).

[31]*Conference on Structure and Mobility in Molecular and Atomic Glasses Proceedings,* Annals N. Y. Acad. Sci. **371**, 7 (1981).

[32]James R. Rice, *Fracture, an Advanced Treatise,* edited by H. Liebowitz (Academic, New York, 1968), Vol. II.

[33]H. Kramer and K. E. Helf, Kolloid Z. **180**, 115 (1962).

[34]A general reference is N. G. McCrum, B. E. Read, and G. Williams, *Anelastic and Dielectric Effects in Polymeric Solids* (Wiley, New York, 1967).

[35]G. C. Berry and T. G. Fox, Adv. Polymer Sci. **5**, 261 (1968).

[36]W. W. Graessley, Adv. Polymer Sci. **16**, 1 (1974).

[37]J. Crank, *The Mathematics of Diffusion* (Clarendon, Oxford, 1976).

[38]S. T. Hwang *et al.*, Sep. Sci. **9**, 461 (1974).

[39]*Nonlinear Optical and Electroactive Polymers*, edited by P. N. Prasad and D. R. Ulrich (Plenum Press, New York, 1988).

[40]*The Principles of Nonlinear Optics,* Y. R. Shen (Wiley, New York, 1984).

[41]D. J. Williams, Ang. Chem. **23**, 690 (1984).

[42]A. C. Griffin, A. M. Bhatti, and R. S. L. Hung in *Nonlinear Optical and Electroactive Polymers*, edited by P. N. Prasad and D. R. Ulrich (Plenum Press, New York, 1988), p. 375.

[43]C. Sauteret, J. P. Hermann, R. Frey, F. Pradire, J. Ducuing, R.Baughman, and R. R. Chance, Phys. Rev. Lett. **36**, 956 (1976).

[44]D. N. Rao, P. Chopra, S. K. Gheshal, J. Swiatkiewicz, and P. N. Prasad, J. Chem. Phys. **84**, 7049 (1986).

[45]F. Kajzar, S. Etemad, G. L. Baker, and J. Messier, Solid State Commun. **63**, 1113 (1987).

[46]D. N. Rao, J. Swiatkiewicz, P. Chopra, S. K. Cheshal, and P. N. Prasad, Appl. Phys. Lett. **48**, 1187 (1986).

[47]T. Kaino, K. I. Kobedera, S. Tomaru, T. Kurihara, S. Saito, T. Tsutsui, and S. Tokito, Electron. Lett. **23**, 1095 (1987).

[48]B. P. Singh, P. N. Prasad, and F. E. Karasz, Polymer **29**, 1940 (1988).

[49]P. N. Prasad, M. K. Casstereus, J. Pfleger, and P. Logsdon, Symposium on Multifunctional Materials. SPIE Proceedings Vol. 878 (1988), p. 106.

[50]*Encyclopedia of Polymer Science and Technology*, edited by H. F. Mark, N. G. Gaylord, and N. M. Bikales (Wiley, New York, 1970), Vol. 13, p. 780.

[51]*Polymer Handbook*, edited by J. Brandrup and E. H. Immergut (Wiley, New York, 1967).

[52]*Handbook of Plastics and Elastomers*, C. E. Harper (McGraw-Hill, New York, 1975).

14.00. Medical physics

THOMAS N. PADIKAL

National Institutes of Health

CONTENTS

14.01. SOME DEFINITIONS AND USEFUL RELATIONSHIPS

Absorbed dose D. Absorbed dose is defined as the quotient of dE by dm, where dE is the mean energy imparted by ionizing radiation to the matter in a volume element and dm is the mass of the matter in that volume element. The unit of absorbed dose is the gray (Gy). One gray equals one joule per kilogram. The special unit of absorbed dose is the rad. One rad equals 10^{-2} J/kg (1 Gy = 100 rads).

It is conventional to measure the absorbed dose at a point with an ion chamber having a calibration factor (traceable to the National Bureau of Standards) to measure *exposure* in roentgens. An exposure-to-dose conversion factor is then used to arrive at the absorbed dose at that point.[1,2]

Exposure X. Exposure is a measure of x- or γ-ray radiaton based upon the ionization produced in air by x rays or γ rays. It is defined as the quotient dQ/dm, where dQ is the net electrical charge of one sign produced in air when all the electrons (liberated by photons) in a volume element of air whose mass is dm is completely stopped in the air. This definition applies only to photons. The special unit of exposure is the roentgen (R). One roentgen is equal to 2.58×10^{-4} C/kg; it is also equivalent to 1 esu/cm^3 of air at STP.

Figure: *f* factor as a function of photon energy for water, muscle, and bone[1]

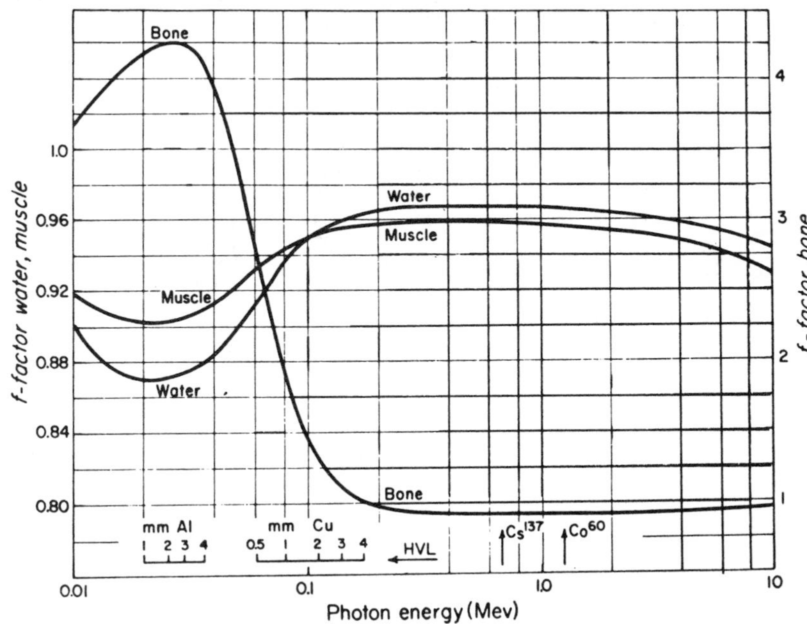

f factor. From the first two definitions in Sec. 14.01. it follows that if the mean energy required to produce an ion pair in air is W, the energy deposited in the air is

$$dE = Wn = W(dQ/e) = W(X\,dm)/e,$$

where n is the total number of electrons (each of charge e) released. Hence, the absorbed dose to air is given by $D_{\rm air} = X(W/e)$. The dose to an alternate medium subjected to the same energy fluence will therefore be

$$D_m = X(W/e)\frac{(\mu_{\rm en}/\rho)_{\rm med}}{(\mu_{\rm en}/\rho)_{\rm air}},$$

where $\mu_{\rm en}/\rho$ is the mass energy absorbtion coefficient (Sec. 14.03). If one uses the special unit of rad for absorbed dose and roentgen for exposure and uses 33.7 eV[1] for W, then

$$D_m = \left(0.869\frac{(\mu_{\rm en}/\rho)_{\rm med}}{(\mu_{\rm en}/\rho)_{\rm air}}\right)X \text{ rads.}$$

The quantity in the large parentheses is referred to as the *f* factor. (See also Sec. 14.03).

14.02. TABLE: MASS DENSITY AND ELECTRON DENSITY OF SELECTED MATERIALS[5,6]

N_A is the Avogadro number (6.022×10^{23}/mol); $N_g = N_A\langle Z/A\rangle_{\rm av}$, the number of electrons per gram; and ρN_g is the electron density in electrons/cm^3.

Material	ρ (g/cm^3)	ρ (N_g/N_A)
Water (H_2O)	1.00	0.556
Polyethylene (C_2H_4)	0.92	0.526
Polystyrene (C_8H_8)	1.05	0.565
Nylon ($C_6H_{11}NO$)	1.15	0.631
Lexan ($C_{16}H_{14}O_3$)	1.20	0.633
Plexiglas ($C_5H_8O_2$)	1.19	0.643
Bakelite ($C_{43}H_{38}O_7$)	1.34	0.708
Teflon (C_2F_4)	2.20	1.056
Brain	1.03	0.567[a]
Muscle	1.04	0.573
Kidney	1.05	0.567
Liver	1.05	0.583
Heart	1.03	0.568
Pancreas	1.05	0.581
Adipose	0.92	0.513
Blood	1.06	0.584

[a]Values for biological materials based on weight fractions given in Table 14.04.B.

14.03. TABLE: MASS ENERGY ABSORPTION COEFFICIENTS AND *f* FACTORS OF SELECTED MATERIALS[7,8]

Mass energy absorption coefficient is defined as the fraction of photon energy absorbed per unit length (g/cm^2) of absorber. The signed two-digit number following each value is its power-of-ten multiplier.

	Mass energy absorption coefficient μ_{en}/ρ (m^2/kg) (multiply by 10 for cm^2/g)								$f = 0.869(\mu_{en}/\rho)_{med}/(\mu_{en}/\rho)_{air}$		
Photon energy	Air	Water (H_2O)	Polystyrene (C_8H_8)	Lucite ($C_5H_8O_2$)	Polyeth (CH_2)	Bakelite ($C_{43}H_{38}O_7$)	Compact bone	Muscle	Water	Compact bone	Muscle
10 keV	4.648 − 01	4.839 − 01	1.849 − 01	2.943 − 01	1.717 − 01	2.467 − 01	1.900 + 00	0.496 + 00	0.912	3.54	0.925
15	1.304 − 01	1.340 − 01	5.014 − 02	8.081 − 02	4.662 − 02	6.741 − 02	0.589 + 00	0.136 + 00	0.889	3.97	0.916
20	5.266 − 02	5.364 − 02	2.002 − 02	3.231 − 02	1.868 − 02	2.692 − 02	0.251 + 00	0.544 − 01	0.881	4.23	0.916
30	1.504 − 02	1.519 − 02	6.056 − 03	9.385 − 03	5.754 − 03	7.904 − 03	0.743 − 01	0.154 − 01	0.869	4.39	0.910
40	6.706 − 03	6.800 − 03	3.190 − 03	4.498 − 03	3.128 − 03	3.898 − 03	0.305 − 01	0.677 − 02	0.878	4.14	0.919
50	4.038 − 03	4.153 − 03	2.387 − 03	3.019 − 03	2.410 − 03	2.711 − 03	0.158 − 01	0.409 − 02	0.892	3.58	0.926
60	3.008 − 03	3.151 − 03	2.153 − 03	2.505 − 03	2.218 − 03	2.316 − 03	0.979 − 02	0.312 − 02	0.905	2.91	0.929
80	2.394 − 03	2.582 − 03	2.152 − 03	2.292 − 03	2.258 − 03	2.191 − 03	0.520 − 02	0.255 − 02	0.932	1.91	0.930
100	2.319 − 03	2.539 − 03	2.292 − 03	2.363 − 03	2.419 − 03	2.288 − 03	0.386 − 02	0.252 − 02	0.948	1.45	0.918
150	2.494 − 03	2.762 − 03	2.631 − 03	2.656 − 03	2.788 − 03	2.593 − 03	0.304 − 02	0.276 − 02	0.962	1.05	0.956
200	2.672 − 03	2.967 − 03	2.856 − 03	2.872 − 03	3.029 − 03	2.808 − 03	0.302 − 02	0.297 − 02	0.973	0.979	0.983
300	2.872 − 03	3.192 − 03	3.088 − 03	3.099 − 03	3.275 − 03	3.032 − 03	0.311 − 02	0.317 − 02	0.966	0.938	0.957
400	2.949 − 03	3.279 − 03	3.174 − 03	3.185 − 03	3.367 − 03	3.117 − 03	0.316 − 02	0.325 − 02	0.966	0.928	0.954
500	2.966 − 03	3.298 − 03	3.195 − 03	3.205 − 03	3.389 − 03	3.137 − 03	0.316 − 02	0.327 − 02	0.966	0.925	0.957
600	2.952 − 03	3.284 − 03	3.181 − 03	3.191 − 03	3.375 − 03	3.123 − 03	0.315 − 02	0.326 − 05	0.966	0.925	0.957
800	2.882 − 03	3.205 − 03	3.106 − 03	3.115 − 03	3.295 − 03	3.049 − 03	0.306 − 02	0.318 − 02	0.965	0.920	0.956
1.0 MeV	2.787 − 03	3.100 − 03	3.005 − 03	3.014 − 03	3.188 − 03	2.950 − 03	0.297 − 02	0.308 − 02	0.965	0.922	0.956
1.5	2.545 − 03	2.831 − 03	2.744 − 03	2.752 − 03	2.911 − 03	2.693 − 03	0.270 − 02	0.281 − 02	0.964	0.920	0.958
2.0	2.342 − 03	2.604 − 03	2.522 − 03	2.530 − 03	2.675 − 03	2.476 − 03	0.248 − 02	0.257 − 02	0.968	0.921	0.954
3.0	2.055 − 03	2.279 − 03	2.196 − 03	2.208 − 03	2.325 − 03	2.160 − 03	0.219 − 02	0.225 − 02	0.962	0.928	0.954
4.0	1.868 − 03	2.064 − 03	1.978 − 03	1.993 − 03	2.089 − 03	1.950 − 03	0.199 − 02	0.203 − 02	0.958	0.930	0.948
5.0	1.739 − 03	1.914 − 03	1.822 − 03	1.842 − 03	1.919 − 03	1.801 − 03	0.186 − 02	0.188 − 02	0.954	0.934	0.944
6.0	1.646 − 03	1.805 − 03	1.707 − 03	1.730 − 03	1.793 − 03	1.691 − 03	0.178 − 02	0.178 − 02	0.960	0.940	0.949
8.0	1.522 − 03	1.658 − 03	1.548 − 03	1.578 − 03	1.618 − 03	1.541 − 03	0.165 − 02	0.163 − 02	0.958	0.950	0.944
10	1.445 − 03	1.565 − 03	1.445 − 03	1.480 − 03	1.503 − 03	1.445 − 03	0.159 − 02	0.154 − 02	0.935	0.960	0.929
15	1.347 − 03	1.440 − 03	1.302 − 03	1.346 − 03	1.341 − 03	1.313 − 03					
20	1.306 − 03	1.384 − 03	1.233 − 03	1.284 − 03	1.206 − 03	1.251 − 03					

14.04. HUMAN BODY COMPOSITION

A. Table: Chemical composition, adult human body[9]

Element	Amount (g)	Percent of total body weight	Element	Amount (g)	Percent of total body weight
Oxygen	43 000	61	Lead	0.12	0.000 17
Carbon	16 000	23	Copper	0.072	0.000 10
Hydrogen	7 000	10	Aluminum	0.061	0.000 09
Nitrogen	1 800	2.6	Cadmium	0.050	0.000 07
Calcium	1 000	1.4	Boron	<0.048	0.000 07
Phosphorus	780	1.1	Barium	0.022	0.000 03
Sulfur	140	0.20	Tin	<0.017	0.000 02
Potassium	140	0.20	Manganese	0.012	0.000 02
Sodium	100	0.14	Iodine	0.013	0.000 02
Chlorine	95	0.12	Nickel	0.010	0.000 01
Magnesium	19	0.027	Gold	<0.010	0.000 01
Silicon	18	0.026	Molybdenum	<0.009 3	0.000 01
Iron	4.2	0.006	Chromium	<0.001 8	0.000 003
Fluorine	2.6	0.003 7	Cesium	0.001 5	0.000 002
Zinc	2.3	0.003 3	Cobalt	0.001 5	0.000 002
Rubidium	0.32	0.000 46	Uranium	0.000 09	0.000 000 1
Strontium	0.32	0.000 46	Beryllium	0.000 036	
Bromine	0.20	0.000 29	Radium	3.1×10^{-11}	

B. Table: Elemental composition, atomic number, and atomic mass of significant components of selected human soft tissue organs [5,28]

Water is included for reference purposes. The signed two-digit number following each value indicates the power-of-ten multiplier.

Element			Organ mass (g)								
Name	Atomic No.	Atomic mass	Adipose 15 055	Blood 5394	Brain 1400	Heart 330	Kidney 310	Liver 1800	Muscle 28 000	Pancreas 100	Water 1
Ca	20	40.1	3.4 − 01	3.1 − 01	1.2 − 01	1.2 − 02	2.9 − 02	9.0 − 02	8.7 − 01	9.1 − 03	
C	6	12.0	9.6 + 03	5.4 + 02	1.7 + 02	5.4 + 01	4.0 + 01	2.6 + 02	3.0 + 03	1.3 + 01	
Cl	17	35.5	1.8 + 01	1.5 + 01	3.2 + 00	5.4 − 01	7.4 − 01	3.6 + 00	2.2 + 01	1.6 − 01	
H	1	1.0	1.8 + 03	5.5 + 02	1.5 + 02	3.4 + 01	3.2 + 01	1.8 + 02	2.8 + 03	9.7 + 00	1.1 − 01
Fe	26	55.8	3.6 − 01	2.5 + 00	7.4 − 02	1.5 − 02	2.3 − 02	3.2 − 01	1.1 + 00	3.9 − 03	
Mg	12	24.3	3.0 − 01	2.1 − 01	2.1 − 01	5.4 − 02	4.0 − 02	3.1 − 01	5.3 + 00	1.6 − 02	
N	7	14.0	1.2 + 02	1.6 + 02	1.8 + 01	8.8 + 00	8.5 + 00	5.1 + 01	7.7 + 02	2.1 + 00	
O	8	16.0	3.5 + 03	4.1 + 03	1.0 + 03	2.3 + 02	2.3 + 02	1.2 + 03	2.1 + 04	6.7 + 01	8.9 − 01
P	15	31.0	2.2 + 00	1.9 + 00	4.8 + 00	4.8 − 01	5.0 − 01	4.7 + 00	5.0 + 01	2.3 − 01	
K	19	39.1	4.8 + 00	8.8 + 00	4.2 + 00	7.2 − 01	5.9 − 01	4.5 + 00	8.4 + 01	2.3 − 01	
Na	11	23.0	7.6 + 00	1.0 + 01	2.5 + 00	4.0 + 00	6.2 − 01	1.8 + 00	2.1 + 01	1.4 − 01	
S	16	32.1	1.1 + 00	5.5 + 00	2.4 + 00	5.4 − 01	0.0 + 00	5.2 + 00	6.7 + 01		
Zn	30	65.4	2.7 − 02	3.4 − 02	1.7 − 02	8.4 − 03	1.5 − 02	8.5 − 02	1.5 + 00	2.5 − 03	

14.05. CHARACTERISTICS OF SELECTED RADIATION DETECTORS

A. Table: Basic properties of solid scintillators [13]

Material	Wavelength of maximum emission (nm)	Decay constant (μs) [a]	Scintillation cutoff wavelength (nm)	Index of refraction [b]	Density (g/cm^3)	Hygroscopic	γ scintillation conversion efficiency (%) [c]
NaI(Tl)	410	0.23	320	1.85	3.67	Yes	100
CaF_2(Eu)	435	0.94	405	1.44	3.18	No	50
CsI(Na)	420	0.63	300	1.84	4.51	Yes	85
CsI(Tl)	565	1.0	330	1.80	4.51	No	45
^{6}LiI(Eu)	470–485 [d]	1.4	450	1.96	4.08	Yes	35
TlCl(Be,I)	465	0.2	390	2.4	7.00	No	2.5
CsF	390	0.005	220	1.48	4.11	Yes	5
BaF_2	325	0.63	134	1.49	4.88	No	10
$Bi_4Ge_3O_{12}$	480	0.30	350	2.15	7.13	No	8
KI(Tl)	426	0.24/2.5 [e]	325	1.71	3.13	Yes	24
$CaWO_4$	430	0.9–20 [e]	300	1.92	6.12	No	14–18
$CdWO_4$	530	0.9–20 [f]	450	2.2	7.90	No	17–20

[a] Room temperature, best single exponential decay constant, $I_0 e^{-\lambda t}$.
[b] At wavelength of maximum emssion.
[c] Referred to NaI(Tl) with S-11 photocathode response.
[d] Primarily used for neutron detection.
[e] KI(Tl) has two scintillation decay components for γ excitation.
[f] Several decay components have been reported for the tungstates.

B. Table: Comparison of some properties of semiconductor, gas-filled, and NaI(Tl) detectors[10–12]

Parameter	Detector type		
	Semiconductor	Gas-filled	NaI(Tl)
Energy required to form electron hole or ion pair (eV)	3.5(Si), 2.9(Ge)	~25–40	~50; ~1000 to obtain photoelectron at photocathode of PM tube
Examples of FWHM (% energy resolution) at 140 keV	~0.6 keV (0.4%)	~5 keV (3.6%)	~18 keV (13%)
Response time (s)	~10^{-9}	~10^{-3}	~10^{-7}–10^{-6}

C. Figure: Total linear attenuation coefficients of Ge, Si, CdTe, NaI(Tl), and CsI detectors[14,15]

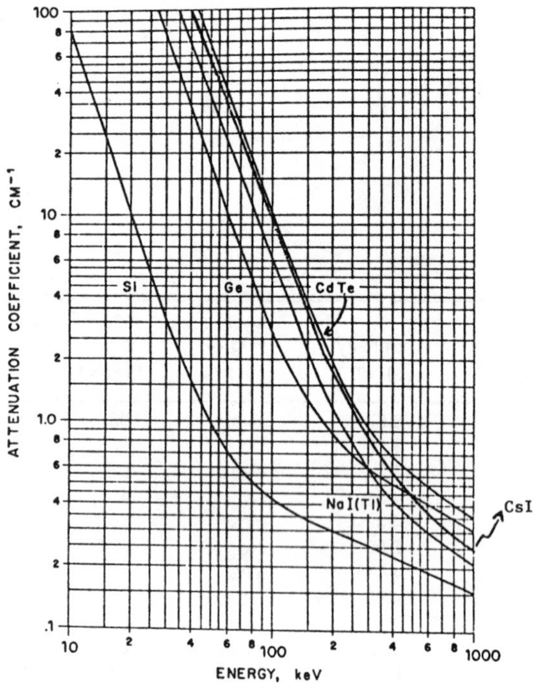

14.06. PHYSICAL CHARACTERISTICS OF SCREENS AND FILMS USED IN DIAGNOSTIC RADIOLOGY[16]

The numbers in the body of the following tables give the relative speeds of the various film-screen combinations, normalized to a value of 0.8 for par-speed screens used with XRP film, for a peak potential of 80 kV (three-phase), 3.5 mm aluminum filtration, "standard processing," and narrow-beam geometry.

The average gradient represents the slope of the portion of the *D*-ln*E* (characteristic) curve between net densities of 0.25 and 2.0 above base fog density. f_{50} and f_{10} represent those values of the spatial frequency (cycles/mm) at which the modulation transfer function is 0.5 and 0.1, respectively.

A. Table: Kodak X-omatic and DuPont screens and blue-sensitive film

The manufacturers' names and average gradient (in parentheses) for each film type are: A—Kodak XG (3.0) and DuPont Cronex 7 (3.0); B—Kodak XRP (2.8), DuPont Cronex 4 (3.0), DuPont Cronex 6 (2.2), DuPont Cronex 6+ (2.6), and 3M type R (2.4); C—Kodak XS (2.6); and D—Kodak XR (2.4).

Screen	f_{50}, f_{10}	Film speed			
		A	B	C	D
X-omatic fine ($BaPbSO_4$) Yellow dye	3.1, 9.7	0.15	0.3	0.45	0.6
X-omatic regular ($BaSrSO_4$:Eu)	1.5, 4.0	0.8	1.6	2.4	3.2
Detail ($CaWO_4$) Yellow dye	2.5, 10.5	0.1	0.2	0.3	0.4
Fast Detail ($CaWO_4$)	1.8, 6.5	0.2	0.4	0.6	0.8
Par ($CaWO_4$)	1.4, 5.0	0.4	0.8	1.2	1.6
Hi-Speed ($BaPbSO_4$)	1.0, 3.2	0.6	1.2	1.8	2.4
Hi-Plus ($CaWO_4$)	1.1, 3.3	0.8	1.6	2.4	3.2
Lightning-Plus ($CaWO_4$)	0.9, 3.2	1.2	2.4	3.6	4.8
Quanta II (BaFCl)	1.1, 3.3	1.6	3.2	4.8	6.4
Quanta III (La_2O_2Br)	0.9, 3.0	3.2	6.4	9.6	12.8

B. Table: Kodak and 3M rare earth screens and green-sensitive film

The manufacturers' names and average gradient (in parentheses) for each film type are: A—Kodak Ortho G (SO-225) (2.4); B—3M XD (2.9); and C—3M XM (2.2).

Screen	f_{50}, f_{10}	Film speed		
		A	B	C
Kodak Lanex Fine (La_2O_2S:Tb, Gd_2O_2S:Tb)	2.1, 7.8	1.0	1.3	3.3
3M Alpha 4 (La_2O_2S:Tb, Gd_2O_2S:Tb) Pink dye	1.7, 5.8	1.5	2.0	5.0
3M Alpha 8 (La_2O_2S:Tb, Gd_2O_2S:Tb)	1.2, 3.3	3.0	4.0	10.0
Kodak Lanex Regular (SO-359) (La_2O_2S:Tb, Gd_2O_2S:Tb)	1.1, 3.3	3.5	4.7	11.7

C. Table: U.S. Radium and G.E. rare earth screens and blue-sensitive film

The manufacturers' names and average gradient (in parentheses) for each film type are: A—Kodak XG (3.0) and DuPont Cronex 7 (3.0); B—Kodak XRP (2.8), DuPont Cronex 4 (3.0), DuPont Cronex 6 (2.2), DuPont Cronex 6+ (2.6), and 3M type R (2.4); C—Kodak XS (2.6); and D—Kodak XR (2.4).

Screens	f_{50}, f_{10}	Film speed			
		A	B	C	D
U.S.R. Rarex BG Detail (75% Gd_2O_2S:Tb, 25% Y_2O_2S:Tb)	1.4, 3.9	0.5	1.0	1.5	2.0
U.S.R. Rarex BG Mid-Speed (75% Gd_2O_2S:Tb, 25% Y_2O_2S:Tb)	1.1, 3.5	1.0	2.0	3.0	4.0
U.S.R. Rarex BG-Hi-Speed (75% Gd_2O_2S:Tb, 25% Y_2O_2S:Tb)	0.8, 2.3	2.0	4.0	6.0	8.0
G.E. Blue Max 1 (La_2O_2Br)	1.5, 4.3	1.0	2.0	3.0	4.0
G.E. Blue Max 2 (La_2O_2Br)	1.2, 3.2	2.0	4.0	6.0	8.0

14.07. AVERAGE EXPOSURE RATES AS A FUNCTION OF PEAK POTENTIAL AND DISTANCE[17]

A. Figure: Exposure rate in air (mR/mA s) at 40 in. as a function of total filtration for various values of peak tube potential (full-wave-rectified)

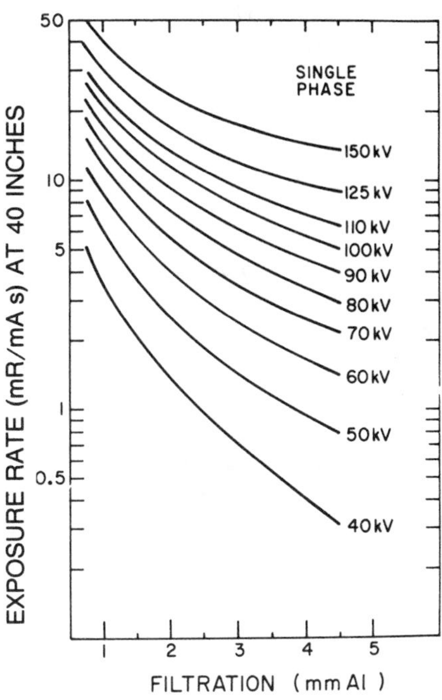

B. Figure: Exposure rate in air (mR/mA s) at 40 in. as a function of tube potential for various amounts of total aluminum filtration

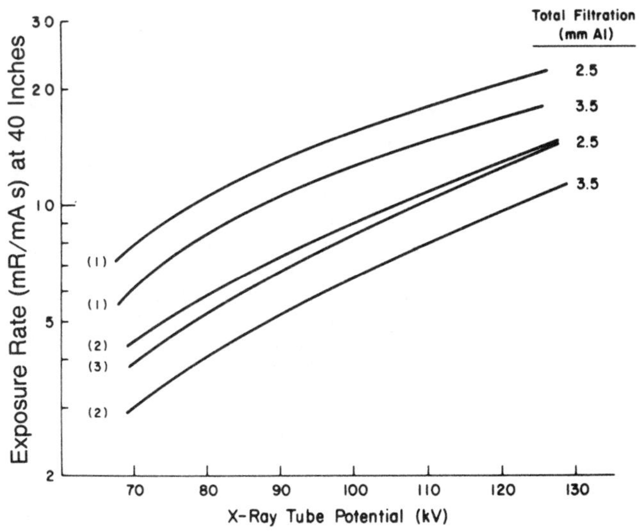

(1) Three-phase, ±15% ripple; (2) single-phase, full-wave rectification (calculated); (3) single-phase, full-wave rectification.

14.08. TABLE: HALF-VALUE LAYER AS A FUNCTION OF TUBE POTENTIAL AND FILTRATION[17]

Half-value layers (mm of Al) are given as a function of (a) filtration and tube potential for diagnostic units (full-wave-rectified single-phase potential) and (b) tube potential for three-phase generators. The half-value layer is defined as the thickness of an absorber required to attenuate half of the incident radiation.

(a) As a function of filtration and tube potential for diagnostic units

Total filtration (mm Al)	Peak potential (kV) 30	40	50	60	70	80	90	100	110	120
0.5	0.36	0.47	0.58	0.67	0.76	0.84	0.92	1.00	1.08	1.16
1.0	0.55	0.78	0.95	1.08	1.21	1.33	1.46	1.58	1.70	1.82
1.5	0.78	1.04	1.25	1.42	1.59	1.75	1.90	2.08	2.25	2.42
2.0	0.92	1.22	1.49	1.70	1.90	2.10	2.28	2.48	2.70	2.90
2.5	1.02	1.38	1.69	1.95	2.16	2.37	2.58	2.82	3.06	3.30
3.0	...	1.49	1.87	2.16	2.40	2.62	2.86	3.12	3.38	3.65
3.5	...	1.58	2.00	2.34	2.60	2.86	3.12	3.40	3.68	3.95

(b) As a function of tube potential for three-phase generators

Total filtration (mm Al)	Peak potential (kV) 60	70	80	90	100	110	120	130	140
2.5	2.2	2.4	2.7	3.1	3.3	3.6	4.0	...	...
3.0	2.3	2.6	3.0	3.3	3.6	4.0	4.3	4.6	5.0
3.5	2.6	2.9	3.2	3.6	3.9	4.3	4.6	...	...

14.09. SCATTER FRACTION IN GENERAL RADIOGRAPHY

A. Definitions

Scatter fraction $= s/(s+p)$, where s and p are the intensities of scattered and primary radiation, respectively.

Thickness (g/cm^2) = thickness (cm) × density (g/cm^3).

Air gap = distance between phantom and radiation receptor.

B. Table: Dependence of scatter fraction on field size, phantom thickness, and air gap[18]

Scatter fractions for 78 kV(peak) (single-phase), 2.1 mm Al half-value layer, and fixed 400 cm source-to-phantom (rear surface) distance.

Field size (cm×cm)	Phantom thickness (g/cm^2)	Air gap (cm) 0	5	10	20	30	40	50
5×5	9	0.42	0.22	0.17	0.11	0.08	0.08	0.07
	18.5	0.52	0.34	0.24	0.13	0.12	0.12	0.12
10×10	9	0.54	0.41	0.32	0.20	0.16	0.13	0.12
	18.5	0.68	0.56	0.46	0.31	0.25	0.19	0.18
20×20	13	0.71	0.65	0.58	0.45	0.37	0.31	0.27
	22	0.82	0.77	0.69	0.60	0.51	0.45	0.38
30×30	18.5	0.80	0.79	0.75	0.65	0.56	0.50	0.44
	27.5	0.88	0.86	0.84	0.78	0.71	0.63	0.59

C. Dependence on beam quality

Typically a 4–10% increase in the scatter fraction was observed as the kV(peak) was changed from 60 to 105 for polychromatic beams or as the photon energy was changed from 32 to 69 keV for monoenergetic x rays.[19]

14.10. IMAGE INTENSIFIERS

A. Definitions

Conversion factor Gx = ratio of the luminance of the output screen to the input exposure rate of applied x radiation. The luminance is expressed in units of candela per square meter (cd/m^2) and the input exposure rate in milliroentgens per second (mR/s). *Gx* is expressed in units of Cd s/mR m^2. *Gx* is measured with 70 $\pm$ 1 cm SID, an x-ray beam of a half-value layer of 7 $\pm$ 0.2 mm Al, and an input screen exposure rate of 1 $\pm$ 0.1 mR/s.

Contrast ratio = ratio between the luminance in the center of the output screen when the total entrance field is irradiated and the luminance of the output screen when a concentric circular area of 10% of the entrance field is completely shielded by a lead mask. The measurement is made at a peak potential of 50 kV and an input screen exposure rate of 1 mR/s.

Resolution = the greatest number of line pairs per millimeter visually resolved under optimal measuring conditions.

B. Table: Typical image intensifier performance data

Based on information supplied by Philips, Picker, and Siemens Companies.

Input screen diameter	*Gx* (Cd s/mR m^2)	Contrast ratio	Resolution (line pairs/mm)
23 cm (9 in.)	70–200	10:1–20:1	2.5–5.4
15 cm (6 in.)	40–200	10:1–20:1	3.0–6.0

14.11. COMPARISON OF REPRESENTATIVE CENTRAL AXIS DEPTH DOSES USED IN RADIATION THERAPY

A. Figure: Photons[1]

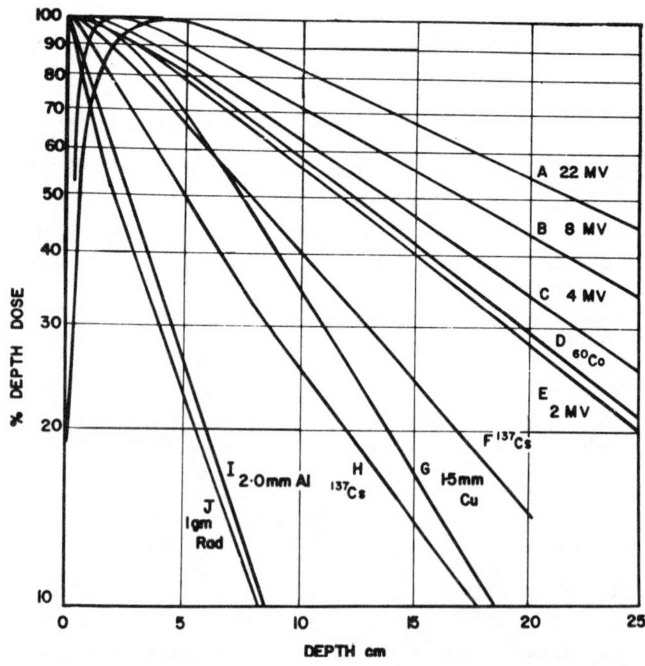

Radiation dose on central axis is measured with a dosimeter at various depths in water, while maintaining a constant source-surface distance (SSD). (A) 22 MV radiation with copper compensating filter, 10×10 cm field, SSD 70 cm. (B) 8 MV radiation from linear accelerator, 10×10 cm field, SSD 100 cm. (C) 4 MV radiation from linear accelerator, 10×10 cm field, SSD 100 cm. (D) Cobalt-60, 10×10 cm field, SSD 80 cm. (E) 2 MV Van de Graaff, 10×10 cm field, SSD 100 cm. (F) Cesium-137, 10×10 cm field, SSD 35 cm. (G) 200 kV(peak), 10×10 cm field, half-value layer 1.5 mm Cu, SSD 50 cm. (H) Cesium-137, 10 cm circle, SSD 15 cm. (I) 120 kV(peak), half-value layer 2.0 mm Al, area 100 cm^2, SSD 15 cm. (J) 1 g radium unit, 5 cm circle, SSD 5 cm.

B. Figure: Electrons[3,4]

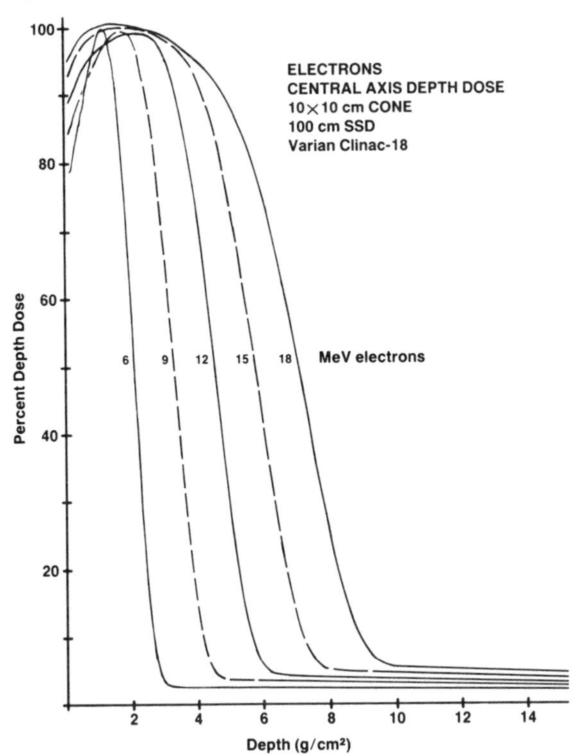

C. Figure: Neutrons, pions, and other charged particles[20]

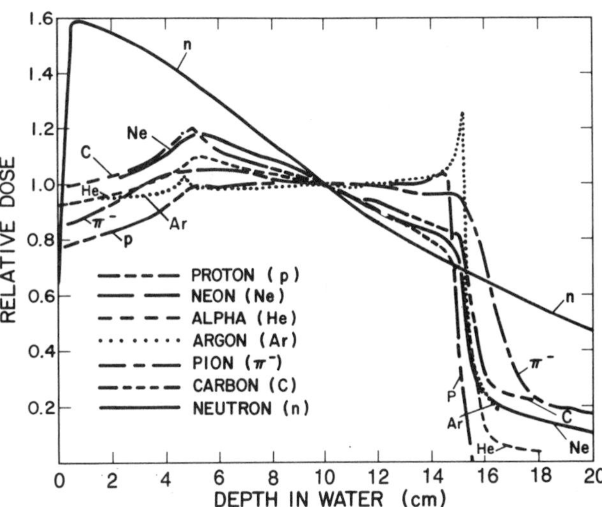

This graph is arbitrarily normalized to a depth of 10 cm in water.

14.12. FIGURE: ATTENUATION PROCESSES OF X RAYS IN A 10-cm LAYER OF WATER[21]

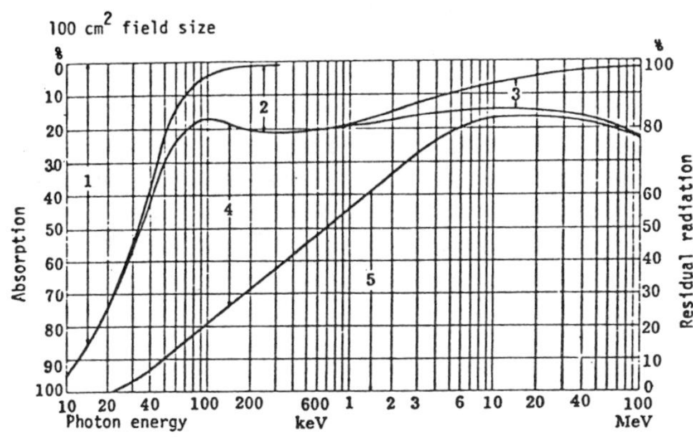

(1) Photoelectric absorption, (2) Compton absorption, (3) pair production, (4) scattering, (5) transmitted primary radiation. For example, if a 100-keV photon beam is allowed to go through a 10-cm slab of water, (a) 20% of the photons will be transmitted, (b) 63% will undergo scattering, (c) 12% will suffer Compton interaction, and (d) 5% will suffer photoelectric absorption.

14.13. TABLE: LINEAR SOURCE TABLES FOR RADIUM[22]

Rads per milligram hour in tissue delivered at various distances by linear radium sources. Dose rate is given as a function of r, the radial distance, and z, the axial distance from the center of the source. Active length = 1.5 cm. Filtration = 0.5 mm Pt. (Dose rates are omitted where γ rays traverse more than 7 mm Pt).

Perpendicular distance from source (cm)	Distance along source axis (cm from center) 0	0.5	1.0	1.5	2.0	2.5	3.0	3.5	4.0	4.5	5.0
0.25	50.67	43.75	11.94	3.34	1.48	0.81	0.50	...	...	...	...
0.5	20.26	16.95	8.18	3.38	1.70	1.00	0.64	0.44	0.31	0.23	0.18
0.75	10.84	9.29	5.67	2.99	1.67	1.03	0.68	0.48	0.35	0.27	0.21
1.0	6.67	5.89	4.10	2.52	1.55	1.01	0.69	0.50	0.37	0.28	0.22
1.5	3.20	2.96	2.38	1.74	1.24	0.89	0.65	0.48	0.37	0.29	0.23
2.0	1.85	1.76	1.52	1.23	0.96	0.74	0.57	0.45	0.35	0.28	0.23
2.5	1.20	1.15	1.04	0.89	0.74	0.60	0.49	0.40	0.32	0.26	0.22
3.0	0.83	0.81	0.75	0.67	0.58	0.49	0.41	0.34	0.29	0.24	0.21
3.5	0.61	0.60	0.57	0.52	0.46	0.40	0.35	0.30	0.26	0.22	0.19
4.0	0.47	0.46	0.44	0.41	0.37	0.33	0.29	0.26	0.23	0.20	0.17
4.5	0.37	0.36	0.35	0.33	0.30	0.28	0.25	0.22	0.20	0.18	0.16
5.0	0.30	0.29	0.28	0.27	0.25	0.23	0.21	0.19	0.17	0.16	0.14

14.14. FIGURE: DOSE/EXPOSURE CURVES IN WATER FOR VARIOUS POINT SOURCES[23]

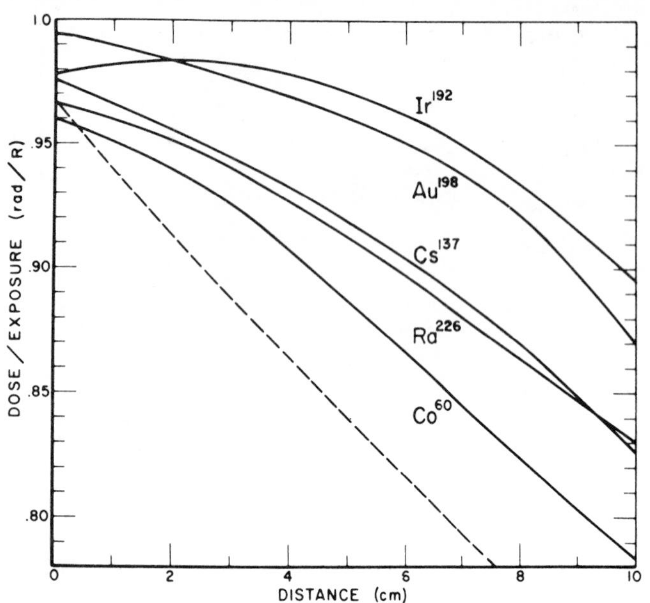

Dose per unit exposure vs distance for a point source of γ radiation in water. The ordinates give the ratio of the absorbed dose to water at a given point, to the exposure in air at the same point in the absence of the water. The broken curve is calculated for exponential absorption with buildup, using the narrow-beam attenuation coefficient 0.028 cm^{-1}.

14.15. TABLE: PROPERTIES OF SELECTED RADIONUCLIDES USED IN BRACHYTHERAPY[29]

Property	^{222}Rn	^{226}Ra (both in equilibrium with decay products)	^{60}Co	^{125}I	^{137}Cs	^{182}Ta	^{192}Ir	^{198}Au
Half-life	3.823 days	1604 years	5.26 years	60.25 days	30.0 years	115.0 days	74.2 days	2.698 days
γ-ray energies (MeV)		0.047–2.44 Principal: 0.61, 0.77, 0.94, 1.12, 1.24, 1.42, 1.77, 2.09	1.173, 1.332	0.0355	0.662	0.043–1.453 Principal: 0.100, 0.152, 0.156, 0.179, 0.222, 0.264, 1.12, 1.19, 1.22	0.136–1.062 Principal: 0.30, 0.31, 0.32, 0.47, 0.61	0.412–1.088 Principal: 0.412
Average γ-ray energy (MeV)		0.83	1.25	0.0284 (incl. x rays)	0.662	0.67	0.38	0.416
β-ray spectra E_{max} (MeV)		0.017–3.26	0.313	None	0.514, 1.17	0.18–0.514	0.24–0.67	0.96
Other primary radiation	α	α		x rays, 0.0272, 0.0275, 0.0310, 0.0318 MeV				
Specific γ ray constant (R cm^2 h^{-1} mCi^{-1})	9.178[a] 8.35[c]	9.068[b] 8.25[c]	13.07	0.0423 0.052	3.226	7.692 6.8	4.89 4.57	2.327
Exposure rate constant (R cm^2 h^{-1} mCi^{-1})	10.27[a]	10.15[b]	13.07	1.089[d]	3.275	7.815	4.64 4.72	2.376
Rads/roentgen (muscle)		0.957	0.957	0.905	0.957	0.96	0.96	0.957
Rads/roentgen (compact bone)		0.921	0.923	4.2	0.924	0.92	0.93	0.929
Half-value layer in water (narrow beam) (cm)		10.6	10.8	2.0	8.2	10	6.3	7.0
Tenth-value layer in lead (broad beam) (cm)		4.2	4.6	~0.01	2.2	3.9	1.2	1.0

[a]No filtration. This value is obtained from the corresponding value for ^{226}Ra, since 0.988 mCi of ^{222}Rn is in equilibrium with 1 mg of ^{226}Ra.
[b]No filtration.
[c]0.5 mm Pt filtration.
[d]From Schulz, Chandra, and Nath.[31]

14.16. FIGURE: TENTH-VALUE LAYERS FOR BROAD-BEAM X RAYS IN CONCRETE[25]

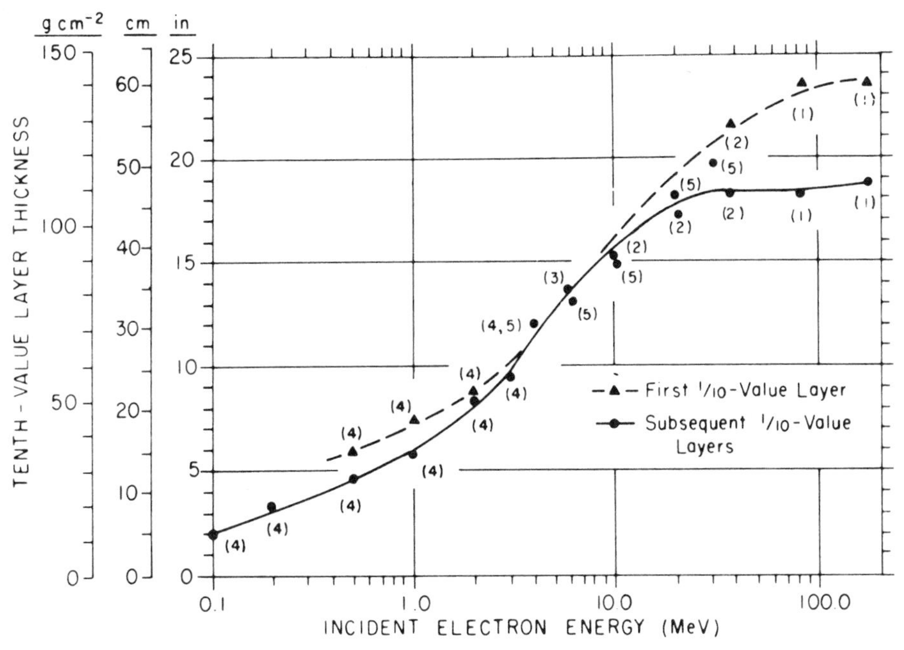

Dose-equivalent index tenth-value layers in ordinary concrete (density 2.35 g/cm^3) for thick-target x rays under broad-beam conditions, as a function of the energy of electrons incident on the target. The dotted curve refers to the first tenth-value layer; the solid curve refers to subsequent or "equilibrium" tenth-value layers. Both curves are empirically drawn through data points derived from several sources.

14.17. TABLE: DOSE-LIMITING RECOMMENDATIONS[26]

The indicated values are for the limited scope of this article. NCRP[a] Rep. **39** (1971b) *should* be consulted for more complete information.

Maximum permissible dose equivalent for occupational exposure	
Combined whole-body occupational exposure	
Prospective annual limit	5 rems in any one year
Retrospective annual limit	10–15 rems in any one year
Long-term accumulation to age N years	$(N-18)\times 5$ rems
Skin	15 rems in any one year
Hands	75 rems in any one year
Forearms	30 rems in any one year
Lenses of eyes	5 rems in any one year
Gonads	5 rems in any one year
Red bone marrow	5 rems in any one year
Other systems, tissues, and organ systems	15 rems in any one year
Fertile women (with respect to fetus)	0.5 rem in gestation period
Dose limits for the public, or occasionally exposed individuals	
Individual or occasional	0.5 rem in any one year
Students	0.1 rem in any one year
Population dose limits (averaged over the population)	
Genetic	0.17 rem in a year
Somatic	0.17 rem in a year

[a]National Council on Radiation Protection and Measurements.

14.18. TABLE: EMBRYO (UTERINE) DOSES FOR SELECTED X-RAY PROJECTIONS PER EXAMINATION[17]

Average dose (mrad) to the uterus per unit skin exposure (R) without backscatter.

Projection	View[a]	SID[b] (in.)	Image receptor size (in.)[c]	Beam quality (HVL, mm Al) 1.5	2.0	2.5	3.0	3.5	4.0
Pelvis, lumbopelvic	AP	40	17×14	142	212	283	353	421	486
	LAT	40	14×17	13	25	39	56	75	97
Abdominal[d]	AP	40	14×17	133	199	265	330	392	451
	PA	40	14×17	56	90	130	174	222	273
	LAT	40	14×17	13	23	37	53	71	91
Lumbar spine	AP	40	14×17	128	189	250	309	366	419
	LAT	40	14×17	9	17	27	39	53	68
Hip	AP (one)	40	10×12	105	153	200	244	285	324
	AP (both)	40	17×14	136	203	269	333	395	454
Full spine (chiropractic)	AP	40	14×36	154	231	308	384	457	527
Urethrogram cystography	AP	40	10×12	135	200	265	327	386	441
Upper G.I.	AP	40	14×17	9.5	16	25	34	45	56
Femur (one side)	AP	40	7×17	1.6	3.0	4.8	6.9	9.4	12
Cholecystography	PA	40	10×12	0.7	1.5	2.6	4.1	6.0	8.3
Chest	AP	72	14×17	0.3	0.7	1.3	2.0	3.1	4.3
	PA	72	14×17	0.3	0.6	1.2	2.0	3.0	4.5
	LAT	72	14×17	0.1	0.3	0.5	0.8	1.2	1.8
Ribs, barium swallow	AP	40	14×17	0.1	0.3	0.5	0.9	1.4	2.0
	PA	40	14×17	0.1	0.3	0.5	0.9	1.5	2.2
	LAT	40	14×17	0.03	0.08	0.2	0.3	0.4	0.6
Thoracic spine	AP	40	14×17	0.2	0.4	0.8	1.4	4.1	3.0
	LAT	40	14×17	0.04	0.1	0.2	0.4	0.5	0.8
Skull, cervical spine, scapula, shoulder, humerus	...	40	...	<0.01	<0.01	<0.1	<0.1	<0.1	<0.1

[a]AP, PA, and LAT refer to anterio-posterior, posterio-anterior, and lateral views, respectively.
[b]Source image distance.
[c]Field size is collimated to the image receptor size.
[d]Includes retrograde pyelogram, KUB, barium enema, lumbosacral spine, intra-venous pyelogram, and renal arteriogram.

14.19. TABLE: CHARACTERISTICS OF TYPICAL ENVIRONMENTAL RADIATION FIELD AT SEA LEVEL (1 m ABOVE GROUND)[17]

Radiation	Energy (MeV)	Source	Absorbed dose rate (μrad/h) Free air	Gonads
α	1–9	radon (atm)	2.7	~0
β	0.1–2	radon (atm)	0.2	~0
	0.1–2	K, U, Th, Sr (soil)	2.5	~0
	2–200	cosmic rays	0.7	0.3
γ	<2.4	radon (atm)	0.2	0.1
	<1.5	K (soil)	2.0	1.3
	<2.4	U (soil)	1.0	0.7
	<2.6	Th (soil)	2.4	1.8
	<0.8	Cs + other fallout (soil)	0.3	0.2
n	0.1–100	cosmic rays	0.1	0.2
p	10–2000	cosmic rays	0.1	0.1
μ	100–30 000	cosmic rays	2.3	2.3
Total:			14.5	7.0

14.20. TABLE: ACOUSTICAL PROPERTIES OF SELECTED BIOLOGIC MATERIALS

Acoustic velocity and attenuation for human tissues.[27] All values are for 37°C, fresh tissues.

Tissue	Velocity (m/s)	Attenuation coefficient α at 1 MHz (dB/cm)	Approximate frequency dependence of α (dB/cm MHz)
Amniotic fluid	1510 ± 3	5.1×10^{-3}	1.6
Blood[a]	1581 at 40% Hct	0.13	1.33
Brain, fetus	1520–1540	0.63	1.27
Breast			
Postmenopause	1465 ± 5	...	...
Premenopause	1529 ± 5		
Eye			
Lens	1638.4 ± 3	0.8	1.0
Vitreous	1531.7 ± 0.9	...	...
Fat	1479	0.6	1.0
Liver	1540	0.9	1.0
Muscle	1500–1610	1.3 for gastronemius muscle perpendicular to fibers	1.0

[a]In general, $V = 1541.8 + 0.98$ (% Hct), where Hct is the hematocrit.

14.21. REFERENCES

[1]H. E. Johns and J. Cunningham, *The Physics of Radiology* (Thomas, Springfield, IL, 1983).

[2]*Determination of Absorbed Dose in a Patient Irradiated by Beams of X- or Gamma Rays in Radiotherapy Procedures,* ICRU Rep. **24** (1976).

[3]T. N. Padikal, unpublished data from a Varian Clinac-18 accelerator (1981).

[4]F. M. Khan, *The Physics of Radiation Therapy* (Williams & Wilkins, Baltimore, MD, 1984)

[5]E. McCollugh, Med. Phys. **2**, 307 (1975), and references therein.

[6]M. E. Phelps, M. H. Gado, and E. J. Hoffman, Radiology **117**, 585 (1975).

[7]*Physical Aspects of Irradiation,* ICRU Rep. **10b** (1962).

[8]J. H. Hubbel, Radiat. Res. **70**, 58 (1977).

[9]W. S. Snyder *et al., Report of the Task Group on Reference Man,* ICRP Rep. **23** (1975).

[10]J. G. Hine, *Instrumen. Nucl. Med.* **1** (1976).

[11]M. M. Ter-Pogossian and M. E. Philips, Semin. Nucl. Med. **3**, 345 (1973).

[12]F. D. Rollo, *Detection and Measurement of Nuclear Radiation* (Mosby, St. Louis, 1977).

[13]*Harshaw Scintillation Phosphors,* 3rd ed. (Harshaw Chemical, Solon, OH, 1975).

[14]*Semiconductor Detectors in the Future of Nuclear Medicine,* edited by P. B. Hoffer *et al.* (The Society of Nuclear Medicine, New York, 1971).

[15]*Scintillation Phosphors* (Harshaw Chemical, Cleveland, 1962).

[16]G. U. V. Rao *et al.*, Invest. Radiol. **13**, 460 (1978).

[17]NCRP Rep. **54** (1977) (NCRP, P.O. Box 30175, Washington, DC, 20014).

[18]C. E. Dick, C. G. Soares, and J. W. Motz, Phys. Med. Biol. **23**, 243 (1978).

[19]K. H. Reiss and B. Steinle, *Tabellen zur Roentgen Diagnostic II* (Siemens AG, Erlangen, 1973).

[20]M. R. Raju, *Heavy Particle Radiotherapy* (Academic, New York, 1980).

[21]F. Wachsmann and G. Drexler, *Graphs and Tables for Use in Radiotherapy* (Springer, Berlin, 1975).

[22]R. Shalek and M. Stovall, Am. J. Roentgenol. **102**, 662 (1968).

[23]R. Loevinger, in *Proceedings of Afterloading in Radiotherapy* (US HEW 72-8024, 1971), p. 199.

[24]NCRP Rep. **49** (1976), Table 28.

[25]NCRP Rep. **51** (1976).

[26]NCRP Rep. **39** (1971).

[27]S. A. Goss, R. L. Johnston, and F. Dunn, J. Acoust. Soc. Am. **64**, 423 (1978).

[28]Y. S. Kim, Radiat. Res. **57**, 38 (1974); **60**, 361 (1974).

[29]M. Cohen, in *Radiation Dosimetry* (AAPM, Burlington, VT, 1976).

[30]T. N. Padikal, editor, *Medical Physics Data Book,* (AAPM, Chicago, 1980).

[31]R. J. Schulz, P. Chandra, and R. Nath, Med. Phys. **7**, 4 (1980).

15.00. Molecular spectroscopy and structure

MARLIN D. HARMONY

University of Kansas

CONTENTS

15.01. ROTATIONAL ENERGIES AND SPECTRA

A. Diatomic molecules

To a good approximation, the rotational energy in the vth vibrational state for $^1\Sigma$ molecules is

$$E_r/h = B_v J(J+1) - D_v J^2(J+1)^2, \quad (1)$$

where the rotational and centrifugal distortion constants are given by

$$B_v = B_e - \alpha_e(v+\tfrac{1}{2}), \quad (2)$$

$$D_v = D_e + \beta_e(v+\tfrac{1}{2}). \quad (3)$$

β_e is small compared to D_e (which itself is a small correction) and may therefore be neglected in many cases.

$J = 0, 1, 2, 3\ldots,$

$v = 0, 1, 2, 3\ldots.$

The equilibrium rotational constant B_e is related to the molecular moment of inertia I_e by

$$B_e = \frac{h}{8\pi^2 I_e} \quad \mathrm{s}^{-1}. \quad (4)$$

For a molecule with atomic masses m_1 and m_2 and internuclear distance r_e, I_e is given by

$$I_e = \mu r_e^2 = \frac{m_1 m_2}{m_1 + m_2} r_e^2. \quad (5)$$

With electric dipole selection rules $\Delta J = \pm 1$, the $J \to J+1$ transition of a *polar* diatomic molecule is given by

$$\nu = 2B_v(J+1) - 4D_v(J+1)^3. \quad (6)$$

For more precise work, the Dunham expansion for the energy levels,

$$\frac{E_r}{h} = \sum_{l,m} Y_{l,m}(v+\tfrac{1}{2})^l J^m (J+1)^m, \quad (7)$$

leads to the transition frequency ($J \to J+1$)

$$\nu = 2Y_{01}(J+1) + 2Y_{11}(v+\tfrac{1}{2})(J+1) + 2Y_{21}(v+\tfrac{1}{2})^2(J+1) + 4Y_{02}(J+1)^3 + 4Y_{12}(v+\tfrac{1}{2})(J+1)^3 + \cdots. \quad (8)$$

The Dunham coefficients are related to the parameters of Eqs. (2) and (3) by[1,2]

$$Y_{01} \approx B_e, \quad Y_{02} \approx -D_e,$$

$$Y_{11} \approx -\alpha_e, \quad Y_{12} \approx -\beta_e,$$

and

$$B_v \approx \sum_l Y_{l,1}(v+\tfrac{1}{2})^l.$$

For numerical purposes, if B is taken in MHz and I in amu Å^2,

$B = 505\,379/I.$

Also, 1 cm^{-1} = 29 979.25 MHz, so B is easily obtained in cm^{-1}. The rotational constants (B_e, α_e, D_e) and the internuclear distance (r_e) for selected diatomic molecules are given in Table 15.01.A.1.

TABLE 15.01.A.1. Rotational, vibrational, and structural parameters of diatomic molecules in the ground electronic state. All species are $^1\Sigma$ except for OH, NO, CN, and O_2, which are $^2\Pi$, $^2\Pi$, $^2\Sigma^+$, and $^3\Sigma_g^-$, respectively. More extensive tabulations are given by Huber and Herzberg[8] and by Lovas and Tiemann.[9] Data refer to the most abundant isotopic species. See Eqs. (1)–(5), (23), and (28).

	B_e (MHz)	D_e (10^{-2} MHz)	α_e (MHz)	r_e (Å)	ω_e (cm^{-1})	$\omega_e x_e$ (cm^{-1})
HCl	317 582.7	1594	9209	1.275	2990.9	52.82
KCl	3 856.38	0.326	23.68	2.667	281	1.3
BaO	9 371.93	0.817	41.73	1.940	669.8	2.03
SnO	10 664.19	0.798	64.24	1.832	814.6	3.73
CO	57 898.35	18.35	524.6	1.128	2169.8	13.29
OH	566 932	5810	2.171×10^4	0.970	3737.8	84.88
NO	50 123.8	1.62	513	1.151	1904.2	14.08
CN	56 953	19.19	520.7	1.172	2068.6	13.09
H_2	1.8243×10^6	1.41×10^5	9.180×10^4	0.741	4401.2	121.34
N_2	59 905.8	17.3	519.2	1.098	2358.6	14.32
O_2	43 337.9	14.51	474.8	1.208	1580.2	11.98
C_2	54 557	20.7	529	1.243	1854.7	13.34
Cl_2	7 314.6	0.56	44.7	1.988	559.7	2.68

B. Linear polyatomic molecules

For $^1\Sigma$ molecules in nondegenerate vibrational states, Eq. (1) applies, but now

$$B_v = B_e - \sum_i \alpha_i(v_i + \tfrac{1}{2}d_i), \quad (9)$$

$$D_v = D_e + \sum_i \beta_i(v_i + \tfrac{1}{2}d_i), \quad (10)$$

where the sums are over all unique vibrational modes, and d_i is the degeneracy of the ith mode. The $J \to J+1$ transition frequency for polar molecules is then given by (6), and B_e by (4), but I_e is given by

$$I_e = \sum_j m_j z_j^2, \quad (11)$$

with z_j the c.m. coordinate of atom j.

In degenerate bending modes, vibrational angular momentum and Coriolis coupling lead to a splitting of each J, v state. For the linear XYZ molecule with $v_2 = 1$, the two states are given by

$$E_r^{\pm}/h = B_v[J(J+1) - 1] - D_v[J(J+1) - 1]^2 \pm \tfrac{1}{2}q_2 J(J+1), \qquad (12)$$

where q_2 is the Coriolis coupling constant, and $J = 1,2,3\cdots$ includes the vibrational angular momentum $|l| = 1$. The two electric dipole transitions for each $J \rightarrow J+1$ are

$$\nu^{\pm} = 2B_v(J+1) - 4D_v(J+2)(J+1)J \pm q_2(J+1). \qquad (13)$$

C. Symmetric rotor molecules

For nonlinear polyatomic molecules we adopt the convention that

$$A \geqslant B \geqslant C,$$

with

$$A = \frac{h}{8\pi^2 I_a},\quad B = \frac{h}{8\pi^2 I_b},\quad C = \frac{h}{8\pi^2 I_c}. \qquad (14)$$

In the principal (a, b, c) inertial axis system, the moments of inertia are defined by

$$I_a = \sum_j m_j(b_j^2 + c_j^2),\quad I_b = \sum_j m_j(a_j^2 + c_j^2),\quad I_c = \sum_j m_j(a_j^2 + b_j^2), \qquad (15)$$

and the products of inertia vanish:

$$I_{ab} = \sum_j m_j a_j b_j = 0, \text{ etc.} \qquad (16)$$

For the *prolate* symmetric rotor ($A > B = C$), the rotational energies may be written

$$E_r/h = BJ(J+1) + (A-B)K^2 - D_J J^2(J+1)^2 - D_{JK}J(J+1)K^2 - D_K K^4. \qquad (17)$$

The rotational and centrifugal distortion constants have vibrational dependence of the same type as given in (8) and (9), e.g.,

$$A_v = A_e - \sum_i \alpha_i^a(v_i + \tfrac{1}{2}d_i),\quad B_v = B_e - \sum_i \alpha_i^b(v_i + \tfrac{1}{2}d_i),\quad C_v = C_e - \sum_i \alpha_i^c(v_i + \tfrac{1}{2}d_i). \qquad (18)$$

However, we shall henceforth suppress the subscript v labels. Also (17) must be modified for excited states of degenerate vibrational modes, which suffer from vibration-rotation interactions similar to the l doubling of linear molecules.[1,2]

For the *oblate* symmetric rotor ($A = B > C$), the rotational energies are

$$E_r/h = BJ(J+1) + (C-B)K^2 - D_J J^2(J+1)^2 - D_{JK}J(J+1)K^2 - D_K K^4. \qquad (19)$$

In (17) and (19), K takes values $J, J-1\cdots 0\cdots -J+1, -J$.

Using electric dipole selection rules $\Delta J = 0, \pm 1$ and $\Delta K = 0$, the allowed transition frequencies ($J \rightarrow J+1$) for the symmetric rotor (prolate or oblate) are given by

$$\nu = 2B(J+1) - 2D_{JK}(J+1)K^2 - 4D_J(J+1)^3. \qquad (20)$$

Note that ν is independent of A or C.

D. Spherical rotor molecules

In this case, $A = B = C$, and the energy states in the simplest approximation are given by Eqs. (1), (9), and (10). No pure rotational spectrum normally exists, since such molecules are nonpolar. Vibration-rotation interactions lead to additional contributions to the rotational energy in higher approximations.[3]

E. Asymmetric rotor molecules

In this case, all rotational constants [as in (14)] are unique. Molecular asymmetry is conveniently classified by

$$\kappa = \frac{2B - A - C}{A - C}, \qquad (21)$$

which reduces to -1 and $+1$ for the limiting prolate and oblate symmetric rotor cases, respectively. Closed-form expressions for the rotational energy states are not generally available. One convenient formulation is (neglecting centrifugal distortion)

$$E_r(J_{K_{-1}K_{+1}})/h = \tfrac{1}{2}(A+C)J(J+1) + \tfrac{1}{2}(A-C)E^J_{K_{-1}K_{+1}}(\kappa), \qquad (22)$$

where tabulations of $E(\kappa)$, the eigenvalues of a reduced energy matrix, are available,[1,4] and the states are labeled by J and the limiting prolate (K_{-1}) and oblate (K_{+1}) quantum numbers. $E(\kappa)$ can be expressed analytically for a few low-J states as illustrated in the following table.

$J_{K_{-1}K_{+1}}$	$E(\kappa)$
0_{00}	0
1_{10}	$\kappa+1$
1_{11}	0
1_{01}	$\kappa-1$
2_{20}	$2[\kappa+(\kappa^2+3)^{1/2}]$
2_{21}	$\kappa+3$
2_{11}	4κ
2_{12}	$\kappa-3$
2_{02}	$2[\kappa-(\kappa^2+3)^{1/2}]$

Rotational spectra under electric dipole selection rules include P, Q, and R branch transitions corresponding to $\Delta J = -1$, 0, and $+1$, respectively. In addition, group-theoretic classification of the states permits the identification of transitions according to the active dipole moment component as follows.

Dipole component	Transitions
a	$ee \leftrightarrow eo$
	$oo \leftrightarrow oe$
b	$ee \leftrightarrow oo$
	$eo \leftrightarrow oe$
c	$ee \leftrightarrow oe$
	$oo \leftrightarrow eo$

Here e and o refer to the evenness and oddness of the K_{-1} and K_{+1} labels.

The rotational constants for selected triatomic, symmetric rotor, and asymmetric rotor molecules are listed in Tables 15.01.E.1, 2, and 3, respectively.

TABLE 15.01.E.1. Rotational, vibrational, and structural parameters of triatomic molecules. Asterisks indicate units of cm^{-1}. For more extensive tabulation see Lovas,[10] Herzberg,[4,5] Shimanouchi,[11] Harmony *et al.*,[12] and Landolt-Bornstein.[13(a)] Data refer to the most abundant isotopic species. See Eqs. (9), (14), (18), and (32).

	Mode	α^a (MHz)	α^b (MHz)	α^c (MHz)	ν_0† (cm^{-1})	Rotational constant (MHz)	Equilibrium structure
SO_2	1	33.60	50.42	42.75	1101	$A_e = 60\,502.05$	$r_e = 1.431$ Å
(C_{2v})	2	−1127.41	−2.39	15.98	497	$B_e = 10\,359.24$	$\theta_e = 119.3°$
	3	612.44	34.24	32.03	1318	$C_e = 8\,845.57$	
O_3	1	−91.06	76.16	69.47	1103	$A_e = 106\,654.36$	$r_e = 1.272$ Å
(C_{2v})	2	−1603.04	37.64	69.12	701	$B_e = 13\,466.90$	$\theta_e = 116.8°$
	3	1590.74	119.25	108.24	1042	$C_e = 11\,958.48$	
F_2O	1	−430.95	72.08	39.32	928	$A_e = 58\,744.90$	$r_e = 1.405$ Å
(C_{2v})	2	−699.02	42.36	53.28	461	$B_e = 10\,985.28$	$\theta_e = 103.1°$
	3	585.01	69.56	115.25	831	$C_e = 9\,255.27$	
H_2O	1	0.750*	0.238*	0.2018*	3650	$A_0 = 835\,833$	$r_e = 0.957$ Å
(C_{2v})	2	−2.941*	−0.160*	0.1392*	1588	$B_0 = 435\,094$	$\theta_e = 104.5°$
	3	1.253*	0.078*	0.1445*	3742	$C_0 = 278\,372$	
OCS	1		18.13		2062		$(CO)_e = 1.157$ Å
$(C_{\infty v})$	2		−10.59		520	$B_e = 6\,099.22$	$(CS)_e = 1.561$ Å
	3		36.43		859		
CO_2	1		0.001 26*		1388		
$(D_{\infty h})$	2		0.000 76*		667	$B_e = 0.391\,625$*	$r_e = 1.160$ Å
	3		0.003 09*		2349		

†Fundamental frequencies.

TABLE 15.01.E.2. Rotational constants of symmetric rotor molecules ($v = 0$). More extensive tabulations are given by Cord *et al.*[14] and Landolt-Bornstein.[13(a)] Data refer to the most abundant isotopic species. See Eqs. (14) and (17)–(20).

	B_0 (MHz)	D_J (kHz)	D_{JK} (kHz)		B_0 (MHz)	D_J (kHz)	D_{JK} (kHz)
CH_3Cl	13 292.86	18.1	198	$VOCl_3$	1 741.72	0.53	−0.9
$CHCl_3$	3 302.08	1.52	−2.5	CH_3CN	9 198.90	3.81	176.9
CF_3CCH	2 877.95	0.24	6.3	SF_5Cl	1 824.59	0.11	0.19
CH_3CCH	8 545.88	2.96	162.9	$XeOF_4$	2 786.34	0.58	1.47
NF_3	10 681.02	14.53	−22.69				

TABLE 15.01.E.3. Rotational constants and barriers to internal rotation for asymmetric rotor molecules ($v = 0$). More complete tabulations are given by Wacker *et al.*,[15] Cord *et al.*,[14] and Landolt-Bornstein.[13(a)] Data refer to the most abundant isotopic species. See Eqs. (14) and (22).

	A (MHz)	B (MHz)	C (MHz)	V_3† (cal/mol)
Acetaldehyde	56 920.5	10 165.1	9 100.0	1162
Acetone	10 165.20	8 515.27	4 910.15	778
Fluoroethane	36 070.30	9 364.54	8 199.74	3306
Propane	29 207.36	8 446.07	7 458.98	3400
Nitromethane	13 277.5	10 542.7	5 876.7	6.0‡
Ethanol	33 326.64	9 112.96	8 019.39	3329
Acetic acid	11 335.5	9 478.64	5 325.01	481
Methanol	127 602.57	24 629.2	23 761.0	1070
Methylamine	22 626.1	21 703.1	10 314.8	1976
Methyl nitrate	11 795.06	4 707.52	3 438.29	2321
Methyl phosphine	71 869.5	11 792.6	11 677.7	1960
Benzonitrile	5 656.7	1 546.84	1 214.41	...
Difluoromethane	49 138.4	10 603.89	9 249.20	...
Fluorobenzene	5 663.54	2 570.64	1 767.94	...
Carbonyl chloride	7 918.14	3 474.72	2 412.07	...
Vinyl cyanide	49 847.1	4 971.12	4 513.88	...
Cyclopropene	30 063.7	21 825.6	13 795.7	...
Ethylene oxide	25 483.7	22 120.9	14 098.0	...
Ethyleneimine	22 736.1	21 192.3	13 383.3	...
Formamide	72 716.12	11 373.75	9 833.72	...
Pyrrole	9 130.53	9 001.30	4 532.09	...
Bicyclobutane	17 311.98	9 313.51	8 393.52	...

†See Refs. 1 and 2 for internal rotation theory.
‡V_6 barrier.

15.02. VIBRATIONAL ENERGIES AND SPECTRA

A. Diatomic molecules

The energy states can be expanded generally in a power series in $v+\frac{1}{2}$, where the vibrational quantum number takes values $v = 0, 1, 2, 3, \ldots$:

$$E_v/hc = \omega_e(v+\tfrac{1}{2}) - \omega_e x_e (v+\tfrac{1}{2})^2 + \omega_e y_e (v+\tfrac{1}{2})^3 + \cdots, \tag{23}$$

which is conventionally expressed in cm^{-1}, and where the first anharmonicity constant ($\omega_e x_e$) is always positive. The classical force constant is given as

$$k_e = 4\pi^2\omega_e^2 c^2 \mu = 4\pi^2 \nu_e^2 \mu = 5.891\times 10^{-2}\omega_e^2\mu \text{ dyn/cm}, \tag{24}$$

where the numerical result uses μ [defined in Eq. (5)] in amu.

For the Morse oscillator, with $\xi = r - r_e$,

$$V(\xi) = D_e[1-\exp(-\beta\xi)]^2, \tag{25}$$

where D_e is the dissociation energy [not to be confused with the centrifugal distortion constant in (3)]. It can be shown that

$$\beta = (2\pi^2\omega_e^2 c^2 \mu/D_e)^{1/2} \tag{26}$$

and

$$k_e = 2\beta^2 D_e. \tag{27}$$

If D_e is in cm^{-1} and μ in amu, (26) becomes

$$\beta = 0.1218\omega_e(\mu/D_e)^{1/2}\ \text{Å}^{-1}.$$

With electric dipole selection rules in the harmonic oscillator limit, $\Delta v = \pm 1$ gives the vibrational spectrum for a *polar* molecule. Owing to anharmonicity, overtones ($\Delta v = \pm 2, \pm 3\cdots$) appear at lower intensity. For the *fundamental* transition ($v=0 \to v=1$) in cm^{-1},

$$\nu_0 = \omega_e - 2\omega_e x_e + \cdots, \tag{28}$$

or, for the $v \to v+1$ transition,

$$\nu_v = \omega_e - 2\omega_e x_e (v+1) + \cdots. \tag{29}$$

In the gas phase a *rotation-vibration* spectrum will result, with the selection rules $\Delta v = \pm 1$ and $\Delta J = \pm 1$. If we write

$$E(v,J) = E_r + E_v,$$

with E_r as in (1) but neglecting the D_v term, the $v \to v+1$ transition yields $P(J \to J-1)$ and $R(J \to J+1)$ branches

$$\nu_P = P(J) = \nu_v + B_{v+1}J(J-1) - B_v J(J+1),$$
$$\nu_R = R(J) = \nu_v + B_{v+1}(J+1)(J+2) - B_v J(J+1), \tag{30}$$

with B in cm^{-1}.

The vibrational constants (ω_e and $\omega_e x_e$) for selected diatomic molecules are given in Table 15.01.A.1.

B. Polyatomic molecules

Including degeneracies, we have $3N-6$ vibrational modes for a nonlinear case or $3N-5$ for linear systems. The modes may be classified generally according to the symmetry point group of the equilibrium configuration. According to this classification, modes may have degeneracies of 1, 2, or 3. For nondegenerate cases, the vibrational energies are

$$\frac{E_v}{hc} = \sum_i \omega_i(v_i + \tfrac{1}{2}) + \sum_i \sum_{j \geq i} x_{ij}(v_i + \tfrac{1}{2})(v_j + \tfrac{1}{2}) + \cdots. \quad (31)$$

Herzberg[5] has discussed degenerate cases for which the second term above is modified to account for degeneracies [as in (9) and (10)], and additional terms are added which partially or completely remove the degeneracy of the vibrational states.

In the harmonic approximation, the dipole selection rules are $\Delta v_i = \pm 1$ and $\Delta v_{k \neq i} = 0$; that is, the vibrational quantum number may change by unity for *only* one mode. In addition, $(\partial\mu/\partial Q_i)_0$, where μ is the electric dipole moment operator and Q_i the normal coordinate, must be nonvanishing. In lowest order, the infrared spectrum will consist of the fundamental frequencies ($v = 0 \rightarrow v = 1$)

$$\nu_i = \omega_i + \cdots \quad (32)$$

for all *active* modes $[(\partial\mu/\partial Q_i)_0 \neq 0]$. In higher order, overtones ($\Delta v_i = 2, 3 \cdots$), combinations (more than one quantum number increasing), and difference bands (a mixture of increasing and decreasing quantum numbers) all may occur.

Group theory selection rules are available based on the classification of the vibrational states in terms of the symmetry point group of the equilibrium nuclear configuration.[6] This classification is based on the symmetry of the $3N - 6$ (or 5) normal coordinates Q_i which form bases for the irreducible representations of the group. For the fundamentals, the frequency ν_i will appear (i.e., be infrared *active*) only if

$$\Gamma(Q_i) = \Gamma(x, y, \text{or } z). \quad (33)$$

Fundamental frequencies for selected molecules are given in Tables 15.02.B.1–15.02.B.3.

Normal-coordinate analysis. In the limit of infinitesimal vibrations (normal modes) the fundamental frequencies $\omega_i = \nu_i/c$ can be related to molecular force constants.

$\underset{R}{Y-}X\underset{R}{-Y}$ case. Let S_1 and S_2 be the left and right stretching coordinates, S_3 and S_4 the bending coordinates, $\mu_Y = m_Y^{-1}$, $\mu_X = m_X^{-1}$,

$\rho = 1/R$, and

$$V = k_R(S_1^2 + S_2^2) + 2k_{12}S_1S_2 + k_\theta(S_3^2 + S_4^2).$$

The fundamental frequencies are given by

$$\boxed{\begin{aligned} &\lambda_1 = \mu_Y(k_R + k_{12}) \\ &\lambda_2 = 2k_\theta\,\rho^2(\mu_Y + 2\mu_X) \ \text{(twice)} \\ &\lambda_3 = (k_R - k_{12})(2\mu_X + \mu_Y) \\ &\lambda_i = 4\pi^2\nu_i^2 = 4\pi^2c^2\omega_i^2. \end{aligned}} \quad (34)$$

$Y \overset{R}{\diagup} \overset{X}{\theta} \overset{R}{\diagdown} Y$ case. S_1 and S_2 are stretches, S_3 is the bend. If

$$2V = k_R(S_1^2 + S_2^2) + 2k_{12}S_1S_2 + k_\theta S_3^2,$$

then

$$\boxed{\begin{aligned} \lambda_1 + \lambda_2 &= (\mu_Y + 2\mu_X\cos^2\tfrac{1}{2}\theta)(k_1 + k_{12}) \\ &\quad + 2(\mu_Y + 2\mu_X\sin^2\tfrac{1}{2}\theta)k_\theta\,\rho^2 \\ \lambda_1\lambda_2 &= 2\mu_Y(\mu_Y + 2\mu_X)(k_1 + k_{12})k_\theta\,\rho^2 \\ \lambda_3 &= (\mu_Y + 2\mu_X\sin^2\tfrac{1}{2}\theta)(k_1 - k_{12}) \end{aligned}} \quad (35)$$

$X\overset{R_1}{-}Y\overset{R_2}{-}Z$ case. Using the same symbolism, with

$$2V = k_1S_1^2 + k_2S_2^2 + k_\theta(S_3^2 + S_4^2),$$

$$\boxed{\begin{aligned} &\lambda_1 + \lambda_3 = k_1(\mu_X + \mu_Y) + k_2(\mu_Y + \mu_Z) \\ &\lambda_1\lambda_3 = (\mu_Y\mu_Z + \mu_X\mu_Z + \mu_X\mu_Y)k_1k_2 \\ &\lambda_2 = k_\theta[\rho_1^2\mu_X + (\rho_1 + \rho_2)^2\mu_Y + \rho_2^2\mu_Z] \ \text{(twice)} \end{aligned}} \quad (36)$$

TABLE 15.02.B.1. Fundamental vibrational frequencies (cm^{-1}) of polyatomic molecules (tetratomic). More extensive tabulations are given by Herzberg[4,5] and Shimanouchi.[11] Data refer to the most abundant isotopic species. See Eqs. (32), (33), and (38). Representative symmetry species are given in parentheses.

Symmetry	Molecule	ν_1	ν_2	ν_3	ν_4	ν_5	ν_6
D_{3h}	BF_3	888 (a_1')	719 (a_2'')	1503 (e')	482 (e')		
	SO_3	1065	498	1391	530		
C_{3v}	NH_3	3337 (a_1)	950 (a_1)	3444 (e)	1627 (e)		
	NF_3	1032	647	905	493		
	PH_3	2323	992	2328	1118		
	PF_3	892	487	860	344		
C_{2v}	CH_2O	2766 (a_1)	1746 (a_1)	1501 (a_1)	2843 (b_1)	1247 (b_1)	1164 (b_2)
	CF_2O	1928	965	584	1249	626	774
C_s	SF_2O	1333 (a')	808 (a')	530 (a')	378 (a')	747 (a'')	393 (a'')
$D_{\infty h}$	C_2H_2	3374 (σ_g^+)	1974 (σ_g^+)	3289 (σ_u^+)	612 (π_g)	730 (π_u)	
	C_2N_2	2330	846	2158	503	234	

TABLE 15.02.B.2. Fundamental vibrational frequencies (cm^{-1}) of polyatomic molecules (pentatomic). See caption to Table 15.02.B.1.

Symmetry	Molecule	ν_1	ν_2	ν_3	ν_4	ν_5	ν_6	ν_7	ν_8	ν_9
T_d	CH_4	2917 (a_1)	1534 (e)	3019 (f_2)	1306 (f_2)					
	CCl_4	458	218	776	314					
	$TiCl_4$	389	114	498	136					
C_{3v}	CH_3Cl	2937 (a_1)	1355 (a_1)	732 (a_1)	3042 (e)	1452 (e)	1017 (e)			
	CH_3F	2930	1464	1049	3006	1467	1182			
	SiH_3F	2206	990	872	2196	956	728			
	$VOCl_3$	1042	408	163	502	246	125			
C_{2v}	CH_2Cl_2	2999 (a_1)	1467 (a_1)	717 (a_1)	282 (a_1)	1153 (a_2)	3040 (b_1)	898 (b_1)	1268 (b_2)	758 (b_2)
	SiH_2Cl_2	2224	954	527	188	710	2237	602	876	590
$D_{\infty h}$	C_3O_2	2196 (σ_g^+)	786 (σ_g^+)	2258 (σ_u^+)	1573 (σ_u^+)	573 (π_g)	550 (π_u)	61 (π_u)		
$C_{\infty v}$	ClCCCN	2297 (σ^+)	2194 (σ^+)	1093 (σ^+)	527 (σ^+)	483 (π)	333 (π)	145 (π)		

TABLE 15.02.B.3. Fundamental vibrational frequencies (cm^{-1}) of polyatomic molecules (six or seven atoms). See caption to Table 15.02.B.1.

Symmetry	Molecule	ν_1	ν_2	ν_3	ν_4	ν_5	ν_6	ν_7	ν_8	ν_9	ν_{10}
D_{3h}	PF_5	816 (a_1')	648 (a_1')	947 (a_2'')	575 (a_2'')	1024 (e')	533 (e')	174 (e')	520 (e'')		
	PCl_5	395	370	465	299	592	273	100	261		
	VF_5	718	608	784	331	810	282	110	336		
C_{3v}	CH_3CN	2965 (a_1)	2267 (a_1)	1400 (a_1)	920 (a_1)	3009 (e)	1454 (e)	1041 (e)	361 (e)		
	CF_3CN	2275	1227	802	522	1214	618	463	196		
O_h	SF_6	772 (a_{2g})	642 (e_g)	932 (f_{1u})	613 (f_{1u})	522 (f_{2g})	344 (f_{2u})				
	TeF_6	697	670	751	327	314	197				
C_{3v}	CH_3CCH	3334 (a_1)	2918 (a_1)	2142 (a_1)	1382 (a_1)	931 (a_1)	3008 (e)	1452 (e)	1053 (e)	633 (e)	328 (e)
	CF_3CCH	3327	2165	1253	812	536	1179	686	612	453	171

15.03. RAMAN SPECTRA

A. Vibrational Raman spectra

In the ordinary Raman effect with incident photons of frequency ν', scattered photons of frequency ν appear according to

$$\nu = \nu' \pm \Delta\nu. \tag{37}$$

The frequency shifts $\Delta\nu$ correspond to changes in vibrational energy and satisfy the same selection rules on v as in the infrared vibrational absorption spectrum. Thus, in the harmonic approximation, the frequency shifts $\Delta\nu$ for fundamental transitions are given for diatomic and polyatomic molecules by Eqs. (29) and (32). However, a transition is Raman *active* only if the polarizability α_{ij} changes during the normal vibration, i.e., $\partial\alpha_{ij}/\partial Q_k \neq 0$. Thus both homo- and heteronuclear diatomic molecules exhibit a Raman spectrum. For polyatomic molecules, the activity of the normal modes is again best determined by recourse to group theory.[6] The result is that the ith normal (fundamental) frequency is Raman *active* only if

$$\Gamma(Q_i) = \Gamma(\alpha). \tag{38}$$

That is, Q_i must transform like components of the polarizability tensor. One general result of this is that molecules having inversion symmetry (such as CO_2 or SF_6) have mutually exclusive infrared and Raman fundamental spectra.

B. Rotational Raman spectra

For $^1\Sigma$ *linear* molecules in nondegenerate vibrational states, the pure rotational spectral shifts $\Delta\nu$ satisfy the selection rules $\Delta J = 0, \pm 2$, so for $J \to J + 2$ we get from Eq. (1):

$$\Delta\nu = (4B_v - 6D_v)(J + \tfrac{3}{2}) - 8D_v(J + \tfrac{3}{2})^3. \tag{39}$$

Both polar and nonpolar linear molecules exhibit a Raman rotational spectrum.

The rotational Raman spectrum of the *symmetric rotor* obeys $\Delta J = 0, \pm 1, \pm 2$, except $\Delta J \neq \pm 1$ for $K = 0$. Thus, the R-branch ($J \to J + 1$) transitions are given by Eq. (20), while the S-branch ($J \to J + 2$) is obtained from Eq. (17) as

$$\Delta\nu = (4B_v - 6D_J)(J + \tfrac{3}{2}) - 4D_{JK}K^2(J + \tfrac{3}{2}) - 8D_J(J + \tfrac{3}{2})^3. \tag{40}$$

The *spherical* rotor exhibits no pure rotational Raman spectrum, while the *asymmetric* rotor spectrum must satisfy $\Delta J = 0, \pm 1, \pm 2$. Spectra of asymmetric rotors with axes of C_2 symmetry must also satisfy the requirements $ee \longleftrightarrow ee$, $eo \longleftrightarrow eo$, $oe \longleftrightarrow oe$, and $oo \longleftrightarrow oo$.[5]

15.04. ELECTRONIC STATES AND SPECTRA OF DIATOMIC MOLECULES

Molecular electronic states may be classified generally according to the symmetry point group of the equilibrium nuclear configuration. For homonuclear and heteronuclear diatomic molecules the appropriate point groups are therefore $D_{\infty h}$ and $C_{\infty v}$, respectively. Electric dipole selection rules for electronic transitions of diatomic molecules are as follows:

$C_{\infty v}$	$D_{\infty h}$
$\Sigma^+ \longleftrightarrow \Sigma^+$	$\Sigma_g^+ \longleftrightarrow \Sigma_u^+$
$\Sigma^- \longleftrightarrow \Sigma^-$	$\Sigma_g^- \longleftrightarrow \Sigma_u^-$
$\Pi \longleftrightarrow \Sigma^\pm$	$\Pi_g \longleftrightarrow \Sigma_u^\pm$, $\Pi_u \longleftrightarrow \Sigma_g^\pm$
$\Pi \longleftrightarrow \Pi$	$\Pi_g \longleftrightarrow \Pi_u$
$\Pi \longleftrightarrow \Delta$	$\Pi_g \longleftrightarrow \Delta_u$, $\Pi_u \longleftrightarrow \Delta_g$
$\Delta \longleftrightarrow \Delta$	$\Delta_g \longleftrightarrow \Delta_u$
.	.
.	.
.	.

For the rotationless molecule, electronic orbital angular momentum is quantized along the internuclear axis with values $\pm\Lambda\hbar$. $\Lambda = 0, 1, 2, 3, \ldots$ correspond to the group theory symbols $\Sigma, \Pi, \Delta, \Phi, \ldots$. If the molecule has spin angular momentum **S**, and if $\Lambda \neq 0$, it will have components along the axis of $\Sigma\hbar$, where $\Sigma = S, S-1, \ldots, -S$, and total electronic angular momentum of $\Omega\hbar$, with $\Omega = |\Lambda + \Sigma|$. The electronic state symbol is then written as the group theory symbol with a left superscript of $2S+1$, e.g., ${}^1\Sigma_g^+$, ${}^3\Sigma_u^-$, ${}^2\Pi$, ${}^3\Pi$, etc. The value of $\Lambda + \Sigma$ is often written as a right subscript (except when $\Lambda = 0$) to designate the substate. In this case the g and u symbols are dropped (but not forgotten). Thus a ${}^3\Pi_u$ state has substates ${}^3\Pi_2$, ${}^3\Pi_1$, and ${}^3\Pi_0$. Finally, in the absence of spin-orbit coupling, the electric dipole selection rule $\Delta S = 0$ can be added to those given in the previous table.

At moderate resolution, electronic spectra of diatomic molecules will consist of a series of bands corresponding to changes in vibrational quantum number v. There are no selection rules on v, but the intensity of a transition $v' \rightarrow v''$ is given by

$$I_{v',v''} \propto \left(\int \psi_{v'} \psi_{v''} \, d\tau \right)^2. \tag{41}$$

At high resolution (in the gas phase) the vibrational bands are seen to consist of individual rotational lines which satisfy the additional selection rule $\Delta J = 0, \pm 1$, where J is the total angular momentum exclusive of that arising from nuclear spin. ($\Delta J = 0$ is forbidden when $J = 0$.) For ${}^1\Sigma \longleftrightarrow {}^1\Sigma$ transitions, the rotational energies are those of Sec. 15.01.A, and the vibrational bands will consist of P, Q, and R branches. When Λ and Σ are nonzero, the rotational energies of Sec. 15.01.A are in general modified by the effects of spin-spin, spin-rotation, and spin-orbit interactions. The standard references should be consulted for the details of the rotational and electronic spectra of non-${}^1\Sigma$ molecules.[4,7]

15.05. ROTATIONAL STARK EFFECT

A. Linear molecules

The first-order contribution to the energy vanishes, and the second-order term is

$$E^{(2)} = \frac{\mu^2 \mathscr{E}^2}{2hB} \frac{J(J+1) - 3M^2}{J(J+1)(2J-1)(2J+3)}, \tag{42}$$

with $M = J, J-1 \cdots 0 \cdots -J$, μ the electric dipole moment, and $\mathscr{E}$ the external applied electric field. With $\Delta M = 0$ selection rules ($\Delta M = \pm 1$ allowed also) for the $J \rightarrow J+1$ transition, the frequency shifts are

$$\Delta\nu^{(2)}(M) = \frac{\mu^2 \mathscr{E}^2}{h^2 B} \times \frac{3M^2(8J^2 + 16J + 5) - 4J(J+1)^2(J+2)}{J(J+1)(J+2)(2J-1)(2J+1)(2J+3)(2J+5)}. \tag{43}$$

B. Symmetric rotor molecules

The leading term is

$$E^{(1)} = \frac{-\mu \mathscr{E} KM}{J(J+1)}. \tag{44}$$

When $K = 0$, $E^{(2)}$ in Eq. (42) becomes the leading term. For $\Delta M = 0$, $J \rightarrow J+1$, and $K \neq 0$,

$$\Delta\nu^{(1)}(M) = \frac{2\mu\mathscr{E}}{h} \frac{KM}{J(J+1)(J+2)}. \tag{45}$$

For numerical purposes, if $\mu\mathscr{E}/h$ is replaced with $0.5034\mu\mathscr{E}$ in Eqs. (43) and (45), then μ is in D, $\mathscr{E}$ in V/cm, and B and ν in MHz. Note: 1 D = 1 debye = 1×10^{-18} esu cm.

C. Asymmetric rotor molecules

Analytical expressions are not generally available. In the absence of near-degeneracies (which are not rare), the Stark shifts are second order and may be written

$$\Delta\nu^{(2)}(M) = \mathscr{E}^2 \sum_g \mu_g^2 (\mathscr{A}^{(g)} + \mathscr{B}^{(g)} M^2), \tag{46}$$

where $g = a, b, c$, and $\mathscr{A}$ and $\mathscr{B}$ depend upon $J_{K_{-1}K_{+1}}$ for the upper and lower states and upon A, B, and C. Methods are available for evaluating $\mathscr{A}$ and $\mathscr{B}$ numerically.[1,2] Note also that the total dipole moment is given by

$$\mu^2 = \sum_g \mu_g^2. \tag{47}$$

Table 15.05.C.1 presents selected dipole moment values.

15.06. NUCLEAR QUADRUPOLE INTERACTIONS

For nuclei having $I \geqslant 1$, the nuclear electric quadrupole moment Q is generally nonvanishing. The electrostatic interaction of a nucleus with the extranuclear charge distribution leads to molecular energy contributions proportional to eQq_{ij}, which is known as the *quadrupole coupling constant*. q_{ij} is an element of the field-gradient tensor:

$$q_{ij} = \frac{\partial^2 V}{\partial i \partial j}, \quad i,j = x,y,z. \tag{48}$$

The tensor is traceless,

$$\frac{\partial^2 V}{\partial x^2} + \frac{\partial^2 V}{\partial y^2} + \frac{\partial^2 V}{\partial z^2} = 0, \tag{49}$$

and in the principal axis system of the field gradient

TABLE 15.05.C.1. Molecular electric dipole moments in the ground electronic state. See Nelson *et al.*[16] and McClellan[17] for more extensive tabulations. See Eqs. (43) and (45)–(47).

Molecule	μ (D)	Molecule	μ (D)	Molecule	μ (D)	Molecule	μ (D)
AgCl	5.73	HI	0.44	N_2O	0.167	H_2O	1.85
ClF	0.88	LiH	5.88	NH_3	1.47	H_2S	0.97
HCl	1.08	OH	1.66	O_3	0.53	OCS	0.712
NaCl	9.00	NO	0.153	SO_2	1.63	CH_3Cl	1.87
SO	1.55	CO	0.112	CH_3F	1.85	CH_3Br	1.81
AsF_3	2.59	NF_3	0.235	FCN	2.17	CH_2O	2.33
F_2O	0.297	NO_2	0.316				

Molecule	Name	μ (D)
CH_2O_2	formic acid	1.41
CH_4O	methanol	1.70
CH_5N	methylamine	1.31
C_2H_3Cl	chloroethylene	1.45
C_2H_4O	acetaldehyde	2.69
$C_2H_4O_2$	acetic acid	1.74
C_2H_4O	ethylene oxide	1.89
C_2H_6O	ethanol	1.69

$$\frac{\partial^2 V}{\partial x \partial y} = \frac{\partial^2 V}{\partial x \partial z} = \frac{\partial^2 V}{\partial y \partial z} = 0. \tag{50}$$

Coupling constants are often symbolized by $\chi_{ij} = eQq_{ij}$ and commonly expressed in units of MHz or kHz. Because of (49), only two diagonal elements are independent, and these are often chosen to be

$$q \equiv \frac{\partial^2 V}{\partial z^2}$$

and (51)

$$\eta = \frac{\partial^2 V/\partial x^2 - \partial^2 V/\partial y^2}{\partial^2 V/\partial^2 z}.$$

A. Pure quadrupole resonance in solids

1. Axial field-gradient case ($\eta = 0$). The energy for a single nucleus of spin I is (in frequency units)

$$E_Q = eQq \frac{3M^2 - I(I+1)}{4I(2I-1)}, \tag{52}$$

with $M = I, I-1 \cdots -I+1, -I$. The magnetic dipole selection rule $\Delta M = \pm 1$ leads for $M \to M+1$ to frequencies

$$\nu = \frac{3eQq(2M+1)}{4I(2I-1)}. \tag{53}$$

2. Nonaxial field-gradient case. For $I = 1$, the energies are

$$E_Q^0 = -\tfrac{1}{2}eQq,$$
$$E_Q^{\pm} = \tfrac{1}{4}eQq(1 \pm \eta), \tag{54}$$

where the latter two states arise by mixing of $M = \pm 1$. Three transitions are possible, given by

$$\nu^{\pm} = \tfrac{1}{4}eQq(3 \pm \eta),$$
$$\nu = \tfrac{1}{2}eQq\eta. \tag{55}$$

For $I = \tfrac{3}{2}$, the energies are

$$E_Q(M = \pm \tfrac{3}{2}) = \tfrac{1}{4}eQq(1 + \tfrac{1}{3}\eta^2)^{1/2},$$
$$E_Q(M = \pm \tfrac{1}{2}) = -\tfrac{1}{4}eQq(1 + \tfrac{1}{3}\eta^2)^{1/2}, \tag{56}$$

and the transition ($\Delta M = \pm 1$) is

$$\nu = \tfrac{1}{2}eQq(1 + \tfrac{1}{3}\eta^2)^{1/2}. \tag{57}$$

B. Hyperfine splitting in rotational spectra

The rotational states of a symmetric rotor with a quadrupolar nucleus on the symmetry axis have the additional contribution

$$E_Q = eQq\left(\frac{3K^2}{J(J+1)} - 1\right) f(I, J, F), \tag{58}$$

where $F = J+I, J+I-1 \cdots |J-I|$ and

$$f(I, J, F) = \frac{\tfrac{3}{4}C(C+1) - I(I+1)J(J+1)}{2I(2I-1)(2J-1)(2J+3)}, \tag{59}$$

with

$$C = F(F+1) - I(I+1) - J(J+1). \tag{60}$$

For the linear molecule, (58) applies with $K = 0$. The spectral lines are determined by the selection rules $\Delta K = 0, \Delta I = 0$, and $\Delta F = 0, \pm 1$ for the rotational transition $J \to J+1$.

For asymmetric rotors, the contribution to the rotational energy may be written

$$E_Q = \frac{f(I, J, F)}{J(J+1)} \left[\chi_{aa}\left(J(J+1) + E(\kappa) - (\kappa+1)\frac{\partial E(\kappa)}{\partial \kappa}\right) + 2\chi_{bb}\frac{\partial E(\kappa)}{\partial \kappa}\right.$$

TABLE 15.06.B.1. Quadrupole coupling constants (MHz) in molecules. All field gradients are axial. More extensive tabulations are given by Lovas and Tiemann,[9] Lovas,[10] and Landolt-Bornstein.[13(a)] See Eqs. (48) and (51).

^{14}N		^{127}I		^{35}Cl		^{39}K		^{79}Br		^{2}H	
NO	−1.85	KI	−86.8	AgCl	−36.5	KCl	−5.67	AgBr	297.1	HBr	0.147
HCN	−4.71	IF	−3438	ClF	−145.8	KI	−4.12	BrF	1089.0	HCl	0.187
FCN	−2.67	CH_3I	−1934	HCl	−67.6			HBr	532.3	HCN	0.202
ClCN	−3.37	CF_3I	−2150	KCl	0.06			CH_3Br	557.2	HC_2F	0.212
NH_3	−4.08			NaCl	−5.65			SiH_3Br	336		
NF	−7.07			CH_3Cl	−74.8			$(CH_3)_3CBr$	511.6		
				SiH_3Cl	−40.0						
				CF_3Cl	−78.0						

$$+\chi_{cc}\left(J(J+1)-E(\kappa)+(\kappa-1)\frac{\partial E(\kappa)}{\partial\kappa}\right)\Bigg], \quad (61)$$

where κ and $E(\kappa)$ are as defined earlier. The observed hyperfine structure is given by $\Delta I=0$ and $\Delta F=0, \pm 1$.

15.07. MOLECULAR STRUCTURE

A. Equilibrium (r_e) structures

Structures based upon the equilibrium rotational constants [see, e.g., Eqs. (18)] and corresponding equilibrium moments of inertia are the most desirable, since they locate the atomic configurations at the minima of the Born-Oppenheimer potential surfaces. Such structures are invariant to isotopic substitution to a high approximation, and are free of zero-point vibration effects. Moreover, the equilibrium moments satisfy various rigid body relations, e.g., for a planar molecule,

$$I_a^e + I_b^e = I_c^e. \quad (62)$$

For diatomic molecules the equilibrium structure is given by (5). For small polyatomic molecules, the structure is obtained by solution of the moment of inertia equations [e.g., Eq. (15)] along with the first moment equation

$$\sum_i m_i \mathbf{r}_i = 0. \quad (63)$$

Since measurements are needed for the ground vibrational state plus at least one excited state for each mode, such structures are seldom available for molecules with more than three atoms. An additional problem is that data are generally needed for more than one isotopic form of the molecule of interest. When single isotopic substitution is used, the a coordinate of the substituted atom measured in the principal axis system of the normal (unsubstituted) species is given by

$$|a| = \left[\frac{\Delta P_a^e}{\mu}\left(1+\frac{\Delta P_b^e}{I_a^e - I_b^e}\right)\left(1+\frac{\Delta P_c^e}{I_a^e - I_c^e}\right)\right]^{1/2}, \quad (64)$$

and $|b|$ and $|c|$ are given by cyclic permutation of subscripts. The ΔP_g^e are the changes in equilibrium *planar second moments* upon isotopic substitution, e.g.,

$$\Delta P_a^e = \tfrac{1}{2}(\Delta I_b^e + \Delta I_c^e - \Delta I_a^e), \text{ etc.}, \quad (65)$$

and

$$\mu = M\Delta m/(M+\Delta m), \quad (66)$$

where M is the mass of the normal molecule and Δm the change in mass of the substituted atom. In the case of the linear molecule, or atoms on the symmetry axis of the symmetric rotor, (64) becomes

$$|z| = (\Delta I_b^e/\mu)^{1/2}. \quad (67)$$

Note that in (64), (65), and (67) all moments are *equilibrium* moments.

B. Effective (r_0) structures

If the moment of inertia and first moment equations are solved with a minimal amount of *ground state* ($v=0$) data, zero-point vibration effects remain and the structure is not generally isotopically invariant. Moreover, relations such as (62) apply only approximately. Such structures are termed "r_0" or "effective" structures.

C. Substitution (r_s) structures

In the special case where *ground state* data are used, but every atom is located by isotopic substitution using Eq. (64), the resulting structure is known as a *substitution* structure. Such structures are relatively free of zero-point vibration effects and are considered generally to be the most reliable approximations to *equilibrium* structures. Because *substitution* coordinates satisfy Eq. (63) to a good approximation, this equation may be used to locate one of the atoms in a molecule if all remaining atoms have already been determined by substitution.

TABLE 15.07.C.1. Structures of polyatomic molecules (distances in Å, angles in degrees). Asterisks represent equilibrium values; all other values are substitution parameters. More extensive tabulations are given by Harmony *et al.*[12] and Landolt-Bornstein.[13(b)] See Sec. 15.07.

Molecule	Parameter	Value
BF_3 (D_{3h})	BF	1.307*
$BrSiH_3$ (C_{3v})	SiBr	2.210
	SiH	1.481
	HSiBr	107.9
$ClSiH_3$ (C_{3v})	SiCl	2.049
	SiH	1.485
	HSiCl	107.9
NH_3 (C_{3v})	NH	1.012*
	HNH	106.7*
HCN ($C_{\infty v}$)	CN	1.153*
	CH	1.065*
HNO_3 (C_s)	HO_1	0.964
	NO_1	1.406
	NO_2	1.211
	NO_3	1.199
	HO_1N	102.2
	O_1NO_2	115.9
	O_1NO_3	113.8
	O_2NO_3	130.3
H_2CO (C_{2v})	CO	1.208
	CH	1.116
	HCH	116.6
CH_4 (T_d)	CH	1.092
CH_3F (C_{3v})	CF	1.383
	CH	1.100
	HCH	110.6
CH_3OH (C_s)	CH	1.094
	OH	0.963
	CO	1.421
	HCH	108.5
	COH	108.0
	ϕ^a	3.2
C_2H_2 ($D_{\infty h}$)	CH	1.061*
	CC	1.203*
CH_3CN (C_{3v})	CN	1.157
	CC	1.458
	CH	1.104
	HCC	109.5
C_2H_5Cl (C_s)	CC	1.520
	CCl	1.788
	CH (methyl)	1.091
	CH (methylene)	1.089
	CCCl	111.0
	HCH (methylene)	109.2
	HCH (methyl)	108.5
	CCH (methylene)	111.6
CH_3Cl (C_{3v})	CCl	1.778*
	CH	1.086*
	HCH	110.7*
C_2H_4O (C_{2v})	CC	1.466
	CO	1.431
	CH	1.085
	HCH	116.6
	θ^b	22.0
C_2H_6O (C_s)	CC	1.512
	CO	1.431
	OH	0.971
	C_2H	1.098
	C_1H_a	1.091
	C_1H_s	1.088
	CCO	107.8
	COH	105.4
	C_1C_2H	110.7
	HC_2H	108.0
	$C_2C_1H_a$	110.1
	$C_2C_1H_s$	110.5
	$H_aC_1H_a$	108.4
CH_9B_5 (C_s)	B_3B_4	1.759
	B_4B_5	1.830
	B_1B_4	1.781
	B_1B_3	1.782
	$B_3B_1B_4$	59.11
	$B_4B_1B_5$	65.57
	$B_3B_4B_5$	103.9
$(CH_3)_2CO$ (C_{2v})	CC	1.507
	CO	1.222
	CH	1.085
	CCC	117.2
	HCH	108.7
	$2\theta^c$	119.9
C_4H_6 (C_{2v})	C_1C_3	1.497
	C_2C_3	1.498
	C_2H_9	1.093
	C_2H_7	1.093
	C_1H_5	1.071
	$C_1C_3C_2$	60.0
	$C_1C_2C_3$	60.0
	$C_2C_3C_4$	98.3
	HCH	115.6
	$H_9C_2C_3$	116.9
	$H_7C_2C_3$	118.1
	$C_2C_3H_6$	129.9
	$C_1C_3H_6$	128.4
C_6H_5Cl (C_{2v})	C_1C_2	1.399
	C_2C_3	1.386
	C_3C_4	1.398
	C_1Cl	1.725
	C_2H	1.080
	C_3H	1.081
	C_4H	1.081
	$C_1C_2C_3$	119.7
	$C_2C_3C_4$	120.2
	$C_3C_4C_5$	120.0
	C_1C_2H	119.5
	C_2C_3H	119.6
C_3H_4 (C_{2v})	C_1C_2	1.509
	$C_2C_{2'}$	1.296
	C_2H	1.072
	C_1H	1.088
	$C_1C_2C_{2'}$	50.8
	$C_2C_{2'}H$	149.9
	HCH	114.6

[a]Tilt angle (toward oxygen lone pair) of CH_3 group.
[b]Angle between CC bond and bisector of angle HCH.
[c]Angle between symmetry axes of methyl groups.

15.08. REFERENCES

[1]C. H. Townes and A. L. Schawlow, *Microwave Spectroscopy* (McGraw-Hill, New York, 1955).
[2]W. Gordy and R. L. Cook, *Microwave Molecular Spectra* (Wiley-Interscience, New York, 1970).
[3]H. C. Allen and P. C. Cross, *Molecular Vib-Rotors* (Wiley, New York, 1963).
[4]G. Herzberg, *Electronic Spectra of Polyatomic Molecules* (Van Nostrand, New York, 1966).
[5]G. Herzberg, *Infrared and Raman Spectra of Polyatomic Molecules* (Van Nostrand, New York, 1945).
[6]E. B. Wilson, J. C. Decius, and P. C. Cross, *Molecular Vibrations* (McGraw-Hill, New York, 1955).
[7]G. Herzberg, *Spectra of Diatomic Molecules* (Van Nostrand, New York, 1950).
[8]K. P. Huber and G. Herzberg, *Constants of Diatomic Molecules* (Van Nostrand Reinhold, New York, 1979).
[9]F. J. Lovas and E. Tiemann, J. Phys. Chem. Ref. Data **3**, 609 (1974).
[10]F. J. Lovas, J. Phys. Chem. Ref. Data **7**, 1445 (1978).
[11]T. Shimanouchi, Natl. Bur. Stand. Ref. Data Ser. **39** (1972); J. Phys. Chem. Ref. Data **6**, 993 (1977).
[12]M. D. Harmony, V. W. Laurie, R. L. Kuczkowski, R. H. Schwendeman, D. A. Ramsay, F. J. Lovas, W. J. Lafferty, and A. G. Maki, J. Phys. Chem. Ref. Data **8**, 619 (1979).
[13]*Landolt-Bornstein Numerical Data and Functional Relationships in Science and Technology,* New Series, Group II (Springer, Berlin), (a) Vol. 6 (1972); (b) Vol. 7 (1976).
[14]M. S. Cord, J. D. Peterson, M. S. Lojko, and R. H. Haas, Natl. Bur. Stand. (U.S.) Monogr. **70** (1968), Vol. IV.
[15]P. F. Wacker, M. S. Cord, D. G. Burkhard, J. D. Peterson, and R. F. Kukol, Natl. Bur. Stand. (U.S.) **70** (1969), Vol. III.
[16]R. D. Nelson, D. R. Lide, and A. A. Maryott, Natl. Bur. Stand. Ref. Data Ser. **10** (1967).
[17]A. L. McClellan, *Tables of Experimental Dipole Moments* (Rahara Enterprises, El Cerrito, CA, 1974).

16.00. Nuclear physics

JAGDISH K. TULI

National Nuclear Data Center
Brookhaven National Laboratory

CONTENTS

16.01. NUCLEAR SPECTROSCOPY STANDARDS

(1) *γ-ray calibration sources.* Table 16.01.A gives some of the commonly used calibration sources. Only the strong γ rays emitted by the sources are shown. See Ref. 1 for some of the relatively weaker lines also useful for γ-ray energy calibrations. For relative and absolute photon intensities see also Refs. 2–4.

(2) *Recommended α energy and intensity values.* See Ref. 5.

(3) *Electron energy standards.* See Ref. 6.

(4) *Electron intensity standards.* See Ref. 2, Appendix 6.

(5) *Internal conversion coefficient standards.* See Ref. 2, Appendix 6.

(6) *Thermal-neutron cross-section standards.* See Table 16.01.B. The values given refer to a neutron energy of 0.0253 eV (neutron velocity 2200 m/s). Representative values are given from the extensive tabulation of Mughabghab *et al.*[9] The capture cross section σ_γ refers to the (n,γ) process and is given in barns ($=10^{-24}$ cm^2). The epicadmium resonance integral is given by

$$I=\int_{E_c}^{\infty}\sigma_R(E)\,\frac{dE}{E},$$

where $E_c \simeq 0.5$ eV is determined by the cadmium cutoff energy and σ_R is the reaction cross section. If the resonances are not located close to E_c, the cross section may be described by a sum of single-level Breit-Wigner contributions.

The coherent scattering cross section is related to the coherent scattering amplitude by

$$\sigma_{\text{coh}}=4\pi a_{\text{coh}}^2=4\pi(g_+a_+ + g_-a_-)^2,$$

where a_+ and a_- refer to the different amplitudes for spin states $I+\frac{1}{2}$ and $I-\frac{1}{2}$, I being the spin of the target nucleus, and g_+ and g_- are the corresponding states' spin statistical factors.

TABLE 16.01.A. γ-ray calibration sources. Uncertainties given are in the last significant figures.

Sources $(T_{1/2})^3$	Energy[1] (keV)	Intensity[7] (%)	Sources $(T_{1/2})^3$	Energy[1] (keV)	Intensity[7] (%)
^{7}Be(53.29 d)	477.605 3	10.43 5	^{124}Sb(60.20 d)	602.730 3	97.89 5
^{22}Na(2.602 y)	1274.542 7	99.937 15	^{137}Cs(30.17 y)	661.660 3	85.21 7
^{24}Na(15.020 h)	1368.633 6	99.994 2	^{152}Eu(13.542 y)	121.7824 4	28.40 15
	2754.030 14	99.876 8		244.6989 10	7.54 5
^{46}Sc(83.81 d)	889.277 3	99.984 1		344.2811 19	26.52 18
	1120.545 4	99.987 1		964.055 4	14.60 8
^{51}Cr(27.702 d)	320.0842 9	9.85 9		1112.087 6	13.56 6
^{54}Mn(312.12 d)	834.843 6	99.975 2		1408.011[b] 14	20.80 12
^{59}Fe(44.496 d)	1099.251 4	56.3 10	^{153}Gd(241.6 d)	97.4316 3	27.6 15
	1291.596 7	43.4 8		103.1807 3	19.6 12
^{56}Co(77.12 d)	846.764 6	99.926 6	^{182}Ta(114.5 d)	67.75001 20	41.2 22
	1238.287 6	66.8 7		100.10653 30	14.0 5
	1771.350 15	15.48 4		152.4308 5	7.18 19
	2598.46 1	16.95 4		222.1099 6	7.60 24
	3253.417 14	7.60 15		264.0755 8	3.62 11
^{57}Co(270.8 d)	122.06135 30	85.68 13		1121.301 5	34.7 6
^{60}Co(5.2714 y)	1173.238 4	99.90 2		1189.050 5	16.5 4
	1332.502 5	99.9824 5		1221.408 5	27.3 7
^{65}Zn(243.9 d)	1115.546 4	44.15 15		1231.016 5	11.6 3
^{88}Y(106.64 d)	898.042 4	94.1 5	^{192}Ir(73.831 d)	295.9582 8	28.7 1
	1836.063 13	99.36 5		308.4569 8	29.8 1
^{95}Zr(64.02 d)	724.199 5	44.15 15		316.5080 8	83.0 3
	756.729[a] 12	54.50 25		468.0715 12	47.8 1
^{94}Nb(20.3×10^3 y)	702.645 6	99.814 6	^{198}Au(2.696 d)	411.8044 11	95.56 7
	871.119 4	99.892 3	^{203}Hg(46.60 d)	279.1967 12	81.56 8
^{95}Nb(34.97 d)	765.807 6	99.80 2	^{207}Bi(32.2 y)	569.702 2	97.8 5
^{108m}Ag(127 y)	433.936 4	90.5 6		1063.662 4	74.9 15
	614.281 4	89.8 19	^{228}Th(1.9131 y)	238.632 2	43.5 4
	722.929 4	90.8 19	+ daughters	583.191 2	30.6 2
^{110m}Ag(249.76 d)	657.7622 20	94.37 10		860.564 5	4.50 4
	763.944 3	22.45 7		2614.533 13	35.88 6
	884.685 3	72.7 3	^{241}Am(432.2 y)	26.345 1	2.4 1
	937.493 4	34.26 12		59.537 1	35.9 4
	1384.300 4	24.2 1			
	1505.040 5	13.05 6			

[a] From Ref. 8.
[b] From Ref. 3.

Table 16.01.B. Thermal-neutral cross-section standards. From Ref. 9. (See also Table 16.09.C.1.)

Target nuclide	Capture cross section (10^{-24} cm)	Capture resonance integrals (10^{-24} cm^2)
^{1}H	0.333	
^{55}Mn	13.3	14.0
^{59}Co	37.2	75.5
^{197}Au	98.7	1550
^{232}Th	7.37	85
^{238}U	2.68	277
	"free" neutron scattering cross section (10^{-24} cm^2)	
C	4.75	
V	4.93	
Ni	17.3	
	Bound coherent scattering amplitude (10^{-13} cm)	
N	9.38	
O	5.804	
F	5.64	
Si	4.1491	
Ni	10.3	
	Fission cross section (10^{-24} cm^2)	Fission integral (10^{-24} cm^2)
^{235}U	582.6	275
^{239}Pu	748.1	301
^{241}Pu	1011	570
	Neutrons/fission $\bar{\nu}$	
^{252}Cf	3.74	

16.02. PHOTON LIFETIMES

The estimates for photon transition probability $T_{SP} = (\ln 2)/t_{1/2}$, the so-called Weisskopf and Moszkowski estimates, are given in Table 16.02.A and Figure 16.02.B. Here E_γ is the transition energy in MeV, A the mass number and Z the proton number. The statistical factor is omitted.

TABLE 16.02.A. Single-particle estimates. From Refs. 10–12.

Multi-polarity	T_{SP} (s^{-1}) (Weisskopf)	T_{SP} (s^{-1}) (Moszkowski)
$E1$	$1.0\times10^{14}A^{2/3}E_\gamma^3$	
$M1$	$3.1\times10^{13}E_\gamma^3$	$2.9\times10^{13}E_\gamma^3$
$E2$	$7.4\times10^{7}A^{4/3}E_\gamma^5$	
$M2$	$2.2\times10^{7}A^{2/3}E_\gamma^5$	$8.4\times10^{7}A^{2/3}E_\gamma^5$
$E3$	$3.5\times10^{1}A^{2}E_\gamma^7$	
$M3$	$1.1\times10^{1}A^{4/3}E_\gamma^7$	$8.7\times10^{1}A^{4/3}E_\gamma^7$
$E4$	$1.1\times10^{-5}A^{8/3}E_\gamma^9$	
$M4$	$3.3\times10^{-6}A^{2}E_\gamma^8$	$4.8\times10^{-5}A^{2}E_\gamma^9$
$E5$	$2.4\times10^{-12}A^{10/3}E_\gamma^{11}$	
$M5$	$7.4\times10^{-13}A^{8/3}E_\gamma^{11}$	$1.7\times10^{-11}A^{8/3}E_\gamma^{11}$

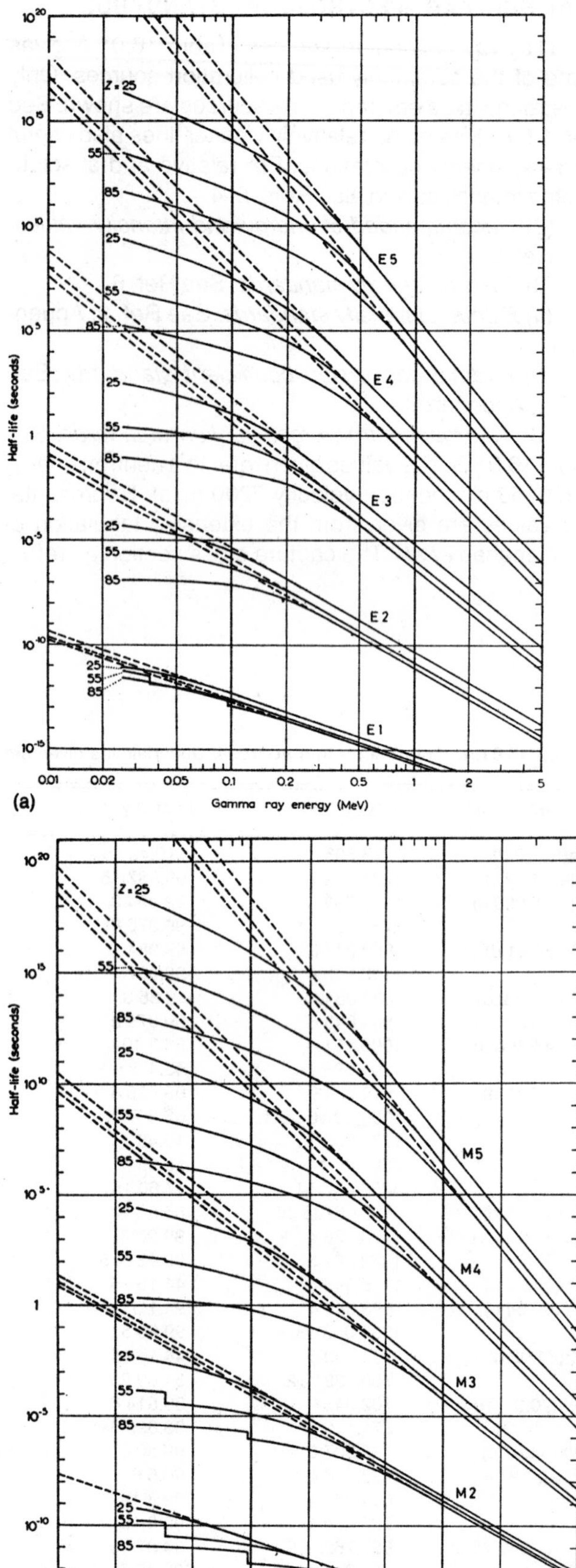

FIGURES 16.02.B. Half-lives of (a) electric and (b) magnetic transitions (Weisskopf estimate). Solid curve, with correction for internal conversion; dashed curve, without. From Ref. 10.

16.03. INTERNAL CONVERSION COEFFICIENTS

Internal conversion is the process in which an excited nucleus deexcites through the ejection of an atomic electron. This process competes with the nuclear decay through γ emission. The ratio of the rates of the two processes is the conversion coefficient. The internal conversion coefficient for an atomic subshell i is given by $\alpha_i = \lambda_{e_i}/\lambda_\gamma$, where λ_{e_i} and λ_γ are the respective decay probabilities.

The values of α_i depend on nuclear charge, Z, nuclear transition energy, atomic subshell, and multipolarity of the nuclear transition.

The total internal conversion coefficient for a transition is given by

$$\alpha = \sum_i \alpha_i = \alpha_K + \alpha_L + \alpha_M + \cdots.$$

Figures 16.03.A–16.03.C give α_K, the conversion coefficients for electrons in K shell, for $E1$, $E2$, and $M1$ transitions as calculated by Hager and Seltzer.[13]

Some of the tables of conversion coefficients available in literature:

Z	Energy (MeV)	Shells	Multi-pole	Ref.
30–103	0.001–1.5	K–$M5$	1–4	13
30–104	0.002–5	All	1–4	14
3–30	0.015–6	K–$L3$	1–5	15
60–100	0.001–0.5	$N1$–$N5$	1–4	16
37–100	0.05–0.5	$N+O+P+$	1–4	17
30–104	0.002–6	K–$M5$	5	18

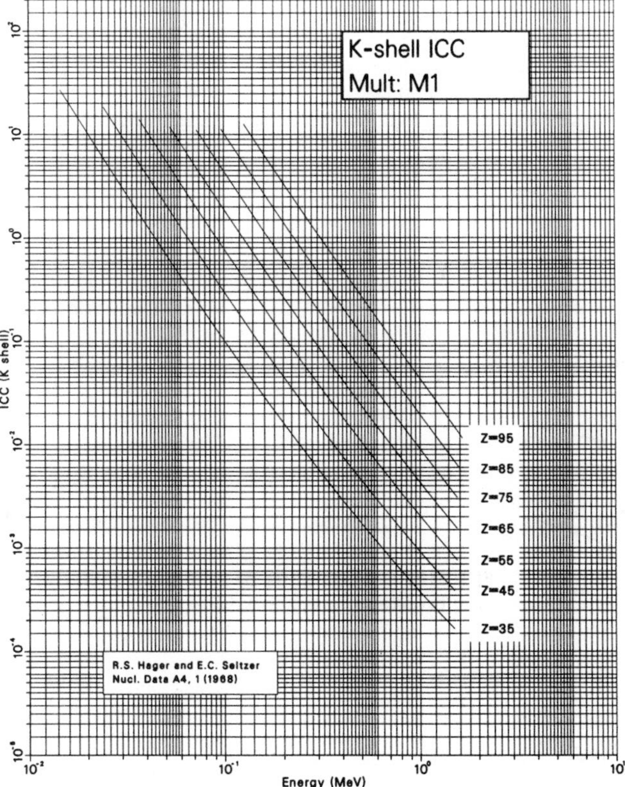

FIGURE 16.03.A. K-shell ICC-$M1$

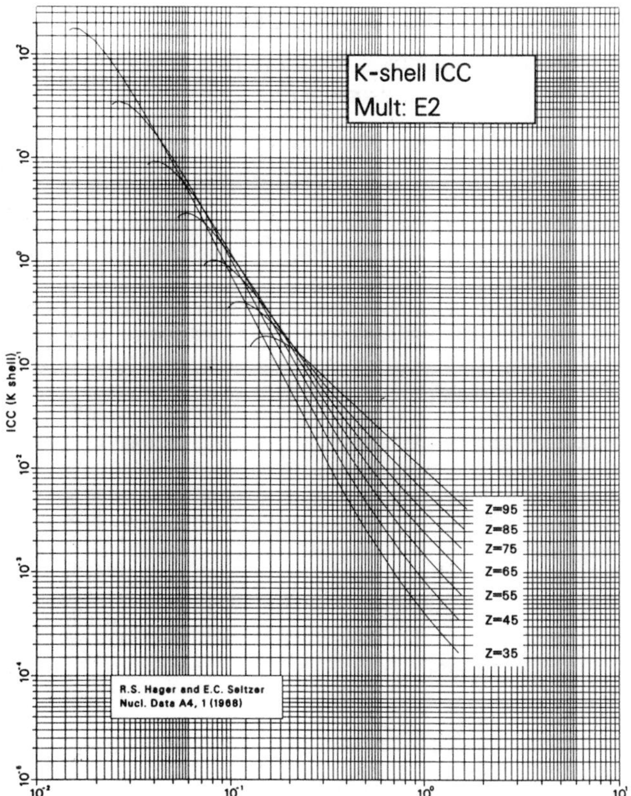

FIGURE 16.03.B. K-shell ICC-$E2$

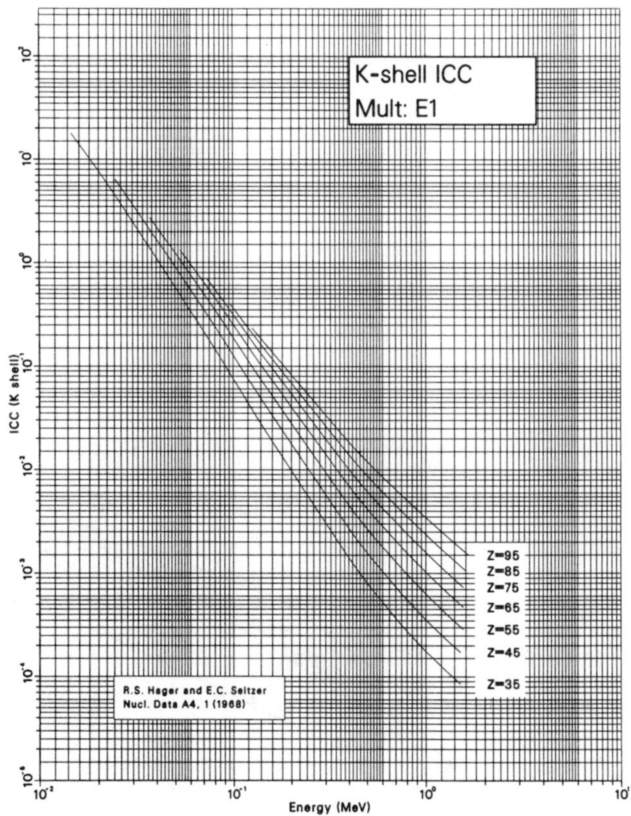

FIGURE 16.03.C. K-shell ICC-$E1$

16.04. PROPERTIES OF NUCLIDES

Selected examples for some properties of nuclides are given in Table 16.04.A. Figure 16.04.B shows lines of stability against various breakup modes.

TABLE 16.04.A. Selected examples for some properties of nuclides. For properties of other nuclei: see Refs. 2 and 3 for half-life, spin, and parity, Ref. 20 for abundances, Ref. 19 for mass excess, and Ref. 2 (Appendix) for magnetic dipole and electric quadrupole moments.

Nucleus	Half-life[2] or abundance[20](%)	Mass excess[19] (MeV)	Spin-parity[2]	Magnetic dipole moment[2,3](nm)	Electric quadrupole moment[2,3](b)
^{1}n	10.3 m	8.071	1/2 +	− 1.9130431	
^{1}H	99.985%	7.289	1/2 +	+ 2.7928467	
^{2}H	0.015%	13.136	1 +	+ 0.8574376	+ 0.002875
^{3}H	12.33 y	14.950	1/2 +	+ 2.978962	
^{3}He	0.000138%	14.931	1/2 +	− 2.127624	
^{6}Li	7.5%	14.086	1 +	+ 0.8220467	− 0.000644
^{7}Li	92.5%	14.907	3/2 −	+ 3.256424	− 0.034
^{12}C	98.90%	0	0 +		
^{9}Be	100%	11.348	3/2 −	− 1.1778	+ 0.053
^{10}B	20%	12.051	3 +	+ 1.80065	+ 0.0847
^{14}N	99.63%	2.863	1 +	+ 0.4037607	0.016
^{19}F	100%	− 1.487	1/2 +	+ 2.628866	
^{23}Na	100%	− 9.531	3/2 +	+ 2.217520	+ 0.101
^{27}Al	100%	− 17.197	5/2 +	+ 3.641504	+ 0.140
^{36}Cl	3.01×10^{5} y	− 29.522	2 +	+ 1.28547	− 0.0180
^{80}Br	17.68 m	− 75.891	1 +	0.5140	0.199
^{99}Tc	2.111×10^{5} y	− 87.324	9/2 +	+ 5.6847	0.34
^{138}La	1.05×10^{11} y, 0.09%	− 86.531	5 +	+ 3.713633	+ 0.51
^{176}Lu	3.6×10^{10} y, 2.60%	− 53.395	7 −	+ 3.19	+ 8.0
^{181}Ta	99.988%	− 48.445	7/2 +	2.371	+ 3.9
^{197}Hg	64.1 h	− 30.565	1/2 −	+ 0.527341	
^{197m}Hg	23.8 h	− 30.266	13/2 +	− 1.027684	+ 1.47
^{235}U	7.038×10^{8} y, 0.72%	40.916	7/2 −	− 0.43	+ 4.9
^{239}Pu	2.412×10^{4} y	48.585	1/2 +	+ 0.203	
^{241}Am	432.2 y	52.931	5.2 −	+ 1.59	+ 4.9

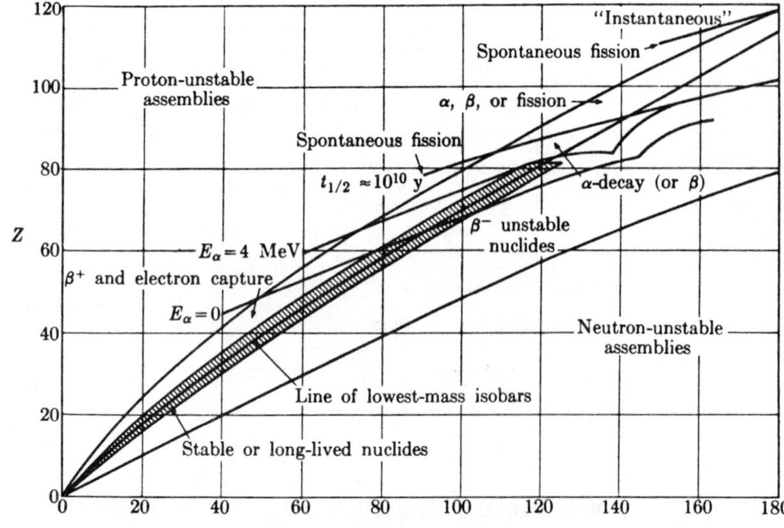

FIGURE 16.04.B. Lines of stability against various breakup modes. From Ref. 21.

16.05. NON RELATIVISTIC KINEMATICS OF NUCLEAR REACTIONS AND SCATTERING[22]

(See Sec. 9 for relativistic mechanics.)

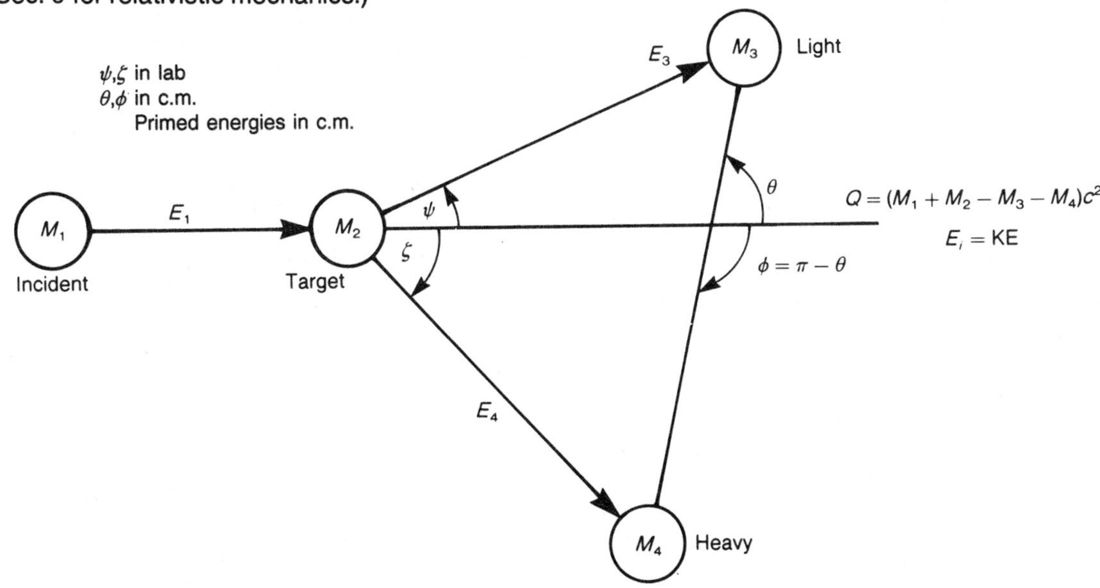

$$E_3^{1/2}=\frac{(M_1M_3E_1)^{1/2}}{M_3+M_4}\cos\psi\left(1\pm\left\{1+\frac{1+M_4/M_3}{\cos^2\psi}\left[\frac{M_4}{M_1}\left(1+\frac{Q}{E_1}\right)-1\right]\right\}^{1/2}\right),$$

$$E_3=\frac{M_1M_3E_1}{(M_3+M_4)^2}\left\{2\cos^2\psi+\frac{M_4(M_3+M_4)}{M_1M_3}\left(\frac{Q}{E_1}-\frac{M_1}{M_4}+1\right)\pm 2\cos\psi\left[\cos^2\psi+\frac{M_4(M_3+M_4)}{M_1M_3}\left(\frac{Q}{E_1}-\frac{M_1}{M_4}+1\right)\right]^{1/2}\right\},$$

$$Q=\frac{M_3+M_4}{M_4}E_3-\frac{M_4-M_1}{M_4}E_1-\frac{2(M_1M_3E_1E_3)^{1/2}}{M_4}\cos\psi.$$

If a particle x with mass M_1 produces a (x,n) reaction on a stationary target nucleus of atomic mass M_2, the nonrelativistic expression connecting the threshold energy E_{th} and Q value Q_0 is

$$|Q_0|=\frac{M_2}{M_1+M_2}E_{th},$$

where $Q_0<0$.

Define

$$E_T=E_1+Q=E_3+E_4,$$

$$A=\frac{M_1M_4(E_1/E_T)}{(M_1+M_2)(M_3+M_4)},$$

$$B=\frac{M_1M_3(E_1/E_T)}{(M_1+M_2)(M_3+M_4)},$$

$$C=\frac{M_2M_3}{(M_1+M_2)(M_3+M_4)}\left(1+\frac{M_1Q}{M_2E_T}\right)=\frac{E_4'}{E_T},$$

$$D=\frac{M_2M_4}{(M_1+M_2)(M_3+M_4)}\left(1+\frac{M_1Q}{M_2E_T}\right)=\frac{E_3'}{E_T}.$$

Note that $A+B+C+D=1$ and $AC=BD$.

Lab energy of light product:

$$E_3/E_T=B+D+2(AC)^{1/2}\cos\theta,$$
$$=B[\cos\psi\pm(D/B-\sin^2\psi)^{1/2}]^2.$$

Use only plus sign unless $B>D$, in which case $\zeta_{max}=\sin^{-1}(D/B)^{1/2}$.

Lab energy of heavy product:

$$E_4/E_T=A+C+2(AC)^{1/2}\cos\phi$$
$$=A[\cos\zeta\pm(C/A-\sin^2\zeta)^{1/2}]^2.$$

Use only plus sign unless $A>C$, in which case $\zeta_{max}=\sin^{-1}(C/A)^{1/2}$.

Lab angle of heavy product:

$$\sin\zeta=(M_3E_3/M_4E_4)^{1/2}\sin\psi.$$

Center-of-mass angle of light product:

$$\sin\theta=\frac{E_3/E_T}{D}\sin\psi.$$

Intensity or solid-angle ratio for light product:

$$\frac{\sigma(\theta)}{\sigma(\psi)}=\frac{I(\theta)}{I(\psi)}=\frac{\sin\psi\,d\psi}{\sin\theta\,d\theta}=\frac{\sin^2\psi}{\sin^2\theta}\cos(\theta-\psi)$$
$$=\frac{(AC)^{1/2}(D/B-\sin^2\psi)^{1/2}}{E_3/E_T}.$$

Intensity or solid-angle ratio for heavy product:

$$\frac{\sigma(\phi)}{\sigma(\zeta)}=\frac{I(\phi)}{I(\zeta)}=\frac{\sin\zeta\,d\zeta}{\sin\phi\,d\phi}=\frac{\sin^2\zeta}{\sin^2\phi}\cos(\phi-\zeta)$$
$$=\frac{(AC)^{1/2}(C/A-\sin^2\zeta)^{1/2}}{E_4/E_T}.$$

Intensity or solid-angle ratio for associated particles in the lab system:

$$\frac{\sigma(\zeta)}{\sigma(\psi)}=\frac{I(\zeta)}{I(\psi)}=\frac{\sin\psi\,d\psi}{\sin\zeta\,d\zeta}=\frac{\sin^2\psi\cos(\theta-\psi)}{\sin^2\zeta\cos(\phi-\zeta)}.$$

Elastic scattering:

$M_4 = M_2$ (recoiling nucleus),
$M_3 = M_1$ (scattering particle),
E_1 = incident particle energy,
E_3 = scattered particle energy,
E_4 = recoil nucleus energy,
$E_T = E_1$,
$Q = 0$.

16.06. SINGLE-PARTICLE STATES

Figure 16.06.A gives the calculated energies of neutron orbits using the following potential[24]:

$$U(r) = Vf(r) + V_{ls}(\mathbf{l}\cdot\mathbf{s})r_0^2\frac{1}{r}\frac{d}{dr}f(r),$$

where $V = -51 + 33(N-Z)/A$ MeV, with $N-Z$ for each A chosen to correspond to the minimum in the valley of β stability, $V_{ls} = -0.44V$,

$$f(r) = \left[1 + \exp\left(\frac{r-R}{a}\right)\right]^{-1},$$

$$R = r_0 A^{1/3},\ r_0 = 1.27 \text{ fm},\ a = 0.67 \text{ fm}.$$

Figure 16.06.B gives the neutron energy levels as calculated by Nilsson,[25] as a function of deformation δ and defined as

$$\delta \equiv \frac{a-b}{\bar{R}} = 2\frac{a-b}{a+b},$$

where a and b are semimajor and semiminor axes of the ellipsoidal nucleus and $\bar{R} \equiv (a+b)/2$ is the mean radius.

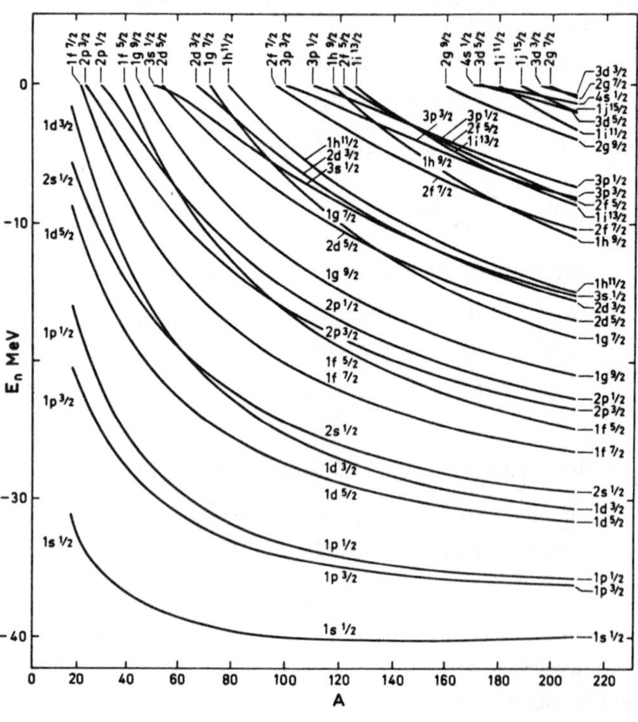

FIGURE 16.06.A. Single-particle neutron states—Shell model. From Ref. 24.

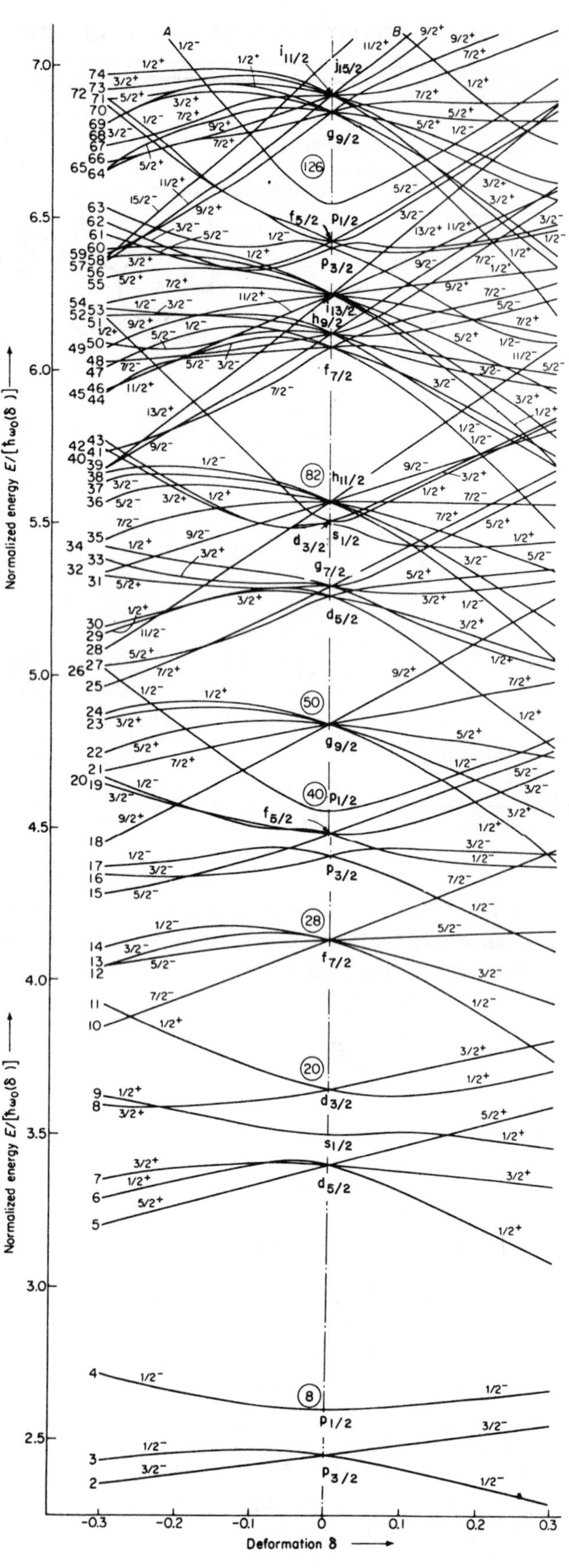

FIGURE 16.06.B. Single-particle neutron energy levels as a function of deformation δ predicted by the Nilsson model. The harmonic oscillator quantum $\hbar\omega_0 \sim A^{-1/3}$. From Ref. 25.

16.07. SOME USEFUL FORMULAS AND DEFINITIONS[26]

A. Glossary

R	nuclear radius
A	nucleon number/mass number
Z	proton number/atomic number
BE	nuclear binding energy
E	energy
M_{H}	mass of H atom
m_p	proton mass
m_n	neutron mass
m_e	electron mass
M_A	atomic mass of nucleus with mass number A
u	atomic mass unit (unified scale)
e	electron charge
ze	charge of a particle
α	fine-structure constant
ρ	density
σ	cross section
$d\sigma/d\Omega$	differential cross section

B. Nuclear properties

1. Nuclear electric charge density distribution

$$\rho(r) \approx \frac{\rho_1}{1+\exp[(r-R_e)/Z_1]},$$

and

$$R_e = 1.07A^{1/3}\ \text{fm},\quad Z_1 \approx 0.545\ \text{fm}.$$

Nuclear surface thickness t over which the density drops from 0.9 of its value at $r=0$ to 0.1 of that value is given by

$$t = 4Z_1 \ln 3 \approx 2.4\ \text{fm}.$$

ρ_1 is determined from normalization

$$Ze = \int_0^\infty \rho(r) 4\pi r^2\, dr.$$

2. Nuclear binding energy

$$BE = [ZM_{\mathrm{H}} + (A-Z)m_n - M_A]c^2,$$

neglecting small correction due to the electronic binding energy.

$$BE/A \approx 8.5\ \text{MeV/nucleon for } A \geqslant 12.$$

3. Semiempirical mass formula (Weizsäcker's)

$$M_A = ZM_{\mathrm{H}} + (A-Z)m_n - a_vA + a_sA^{2/3} + a_cZ^2/A^{1/3} + a_a(A-2Z)^2/A + \Delta,$$

where

$$\Delta = \begin{cases} a_pA^{-3/4} & \text{for } o\text{-}o \text{ nuclei} \\ 0 & \text{for } e\text{-}o \text{ or } o\text{-}e \text{ nuclei} \\ -a_pA^{-3/4} & \text{for } e\text{-}e \text{ nuclei.} \end{cases}$$

The coefficients, a's, are determined empirically. A representative set[26]:

$$a_v = 1.51\times10^{-2}\ \text{u} \simeq 14.1\ \text{MeV},$$
$$a_s = 1.40\times10^{-2}\ \text{u} \simeq 13\ \text{MeV},$$
$$a_c = 6.39\times10^{-4}\ \text{u} \simeq 0.595\ \text{MeV},$$
$$a_a = 2.04\times10^{-2}\ \text{u} \simeq 19\ \text{MeV},$$
$$a_p = 3.60\times10^{-2}\ \text{u} \simeq 33.5\ \text{MeV},$$
$$M_{\mathrm{H}} = 1.007\,825\ \text{u} \simeq 938.767\ \text{MeV}.$$

4. Magnetic dipole moment

In the extreme single-particle model the magnetic dipole moment of an odd-A nucleus is given by the unpaired particle. This is given by the Schmidt formulas

$$\mu^{\pm} = j\left(g^{(l)} \pm \frac{g^{(s)} - g^{(l)}}{2l+1}\right)\mu_N,\quad j = l \pm \tfrac{1}{2}$$

$$g^{(l)} = 0 \quad \text{for neutron}$$
$$= 1 \quad \text{for proton,}$$
$$g^{(s)} = -3.8260842^{18} \quad \text{for neutron}^2$$
$$= 5.5856912^{22} \quad \text{for proton}^2.$$

5. Electric quadrupole moment

Quadrupole moment of a nucleus, assumed to be a homogeneously charged ellipsoid of revolution whose major and minor semiaxes are a and b and whose symmetry axis is in the direction of its spin vector $\mathbf{J}$, is given by

$$Q_0 = \tfrac{2}{5}Ze(a^2-b^2) = \tfrac{4}{5}\bar{R}^2 Ze\delta_D,$$

where

$$\bar{R} \equiv (a+b)/2,\quad \delta_D \equiv (a-b)/\bar{R}.$$

In terms of β_D, defined by $R = \bar{R}[1+\beta_D Y_{20}(\theta)]$,

$$Q_0 = [3/(5\pi)^{1/2}]\bar{R}^2 Ze\beta_D[1+\beta_D(5/64\pi)^{1/2}].$$

Observed quadrupole moment, which corresponds to nonalignment of spin and symmetry axes, is given by

$$Q = \frac{3K^2 - J(J+1)}{(J+1)(2J+3)}Q_0,\quad J = K, K+1, \ldots.$$

The ground-state quadrupole moment is given by

$$Q_{\text{g.s.}} = \frac{J(2J-1)}{(J+1)(2J+3)}Q_0.$$

C. Some scattering formulas

1. Rutherford scattering (Coulomb scattering)

The differential cross section for classical elastic Coulomb scattering of a particle of charge z_1e and velocity v from a stationary target of charge z_2e, in center-of-mass system, is

$$\frac{d\sigma'}{d\Omega'} = \left(\frac{z_1z_2e^2}{\frac{1}{2}\mu v^2}\frac{1}{4\sin^2(\theta'/2)}\right)^2,$$

where $\mu \equiv m_1 m_2/(m_1 + m_2)$ is the reduced mass. This holds rigorously if $2z_1z_2/[137(v/c)] \gg 1$. The Rutherford formula also holds quantum mechanically for Coulomb scattering except in the case of indistinguishable like particles.

2. Mott scattering formula

For Coulomb scattering of two equal pointlike particles with an initial relative velocity v and reduced mass $\mu = \frac{1}{2}m$, in center-of-mass system,

$$\left(\frac{d\sigma'}{d\Omega'}\right)_{\text{Mott}} = \left(\frac{z^2e^2}{\frac{1}{2}\mu v^2}\right)^2 \left(\frac{1}{16\sin^4(\theta'/2)} + \frac{1}{16\cos^4(\theta'/2)} + \frac{\Phi}{\sin^2(\theta'/2)\cos(\theta'/2)}\right),$$

where

$$\Phi \equiv K\cos\left(\frac{z^2e^2}{\hbar v}\ln\tan^2(\theta'/2)\right),$$

$K = \frac{1}{8}$ for collision of spin-0 particles

$= -\frac{1}{16}$ for collision of spin-$\frac{1}{2}$ particles.

In the high-energy nonrelativistic limit for electrons

$$\left(\frac{d\sigma'}{d\Omega'}\right)_{\text{Mott}} \approx \left(\frac{e^2}{\frac{1}{2}\mu v^2}\right)^2 \frac{4 - 3\sin^2\theta'}{4\sin^4\theta'}.$$

3. Born collision formula

Elastic scattering cross section, brought about by interaction of strength H', under Born approximation is

$$\frac{d\sigma}{d\Omega} = \left|\frac{2m}{\hbar^2}\int_0^\infty H'(r)r^2\frac{\sin kr}{kr}dr\right|^2,$$

where $\mathbf{k} = \mathbf{k}_i - \mathbf{k}_f$, with $\mathbf{k}_i$ and $\mathbf{k}_f$ the initial and final wave-number vectors.

D. Passage of electromagnetic radiation through matter

The processes through which the electromagnetic radiation interacts with matter are summarized in Table 16.07.D.1.

The intensity of the electromagnetic beam after traversing a thickness x of a homogeneous material is given by

$$I(x) = I_0\exp(-\mu x),$$

where I_0 is the initial beam intensity, i.e., at $x = 0$, and μ is the total linear attentuation coefficient of the material for the energy of the traversing beam.

Mass attenuation coefficient:

$$\mu_m = \mu/\rho,$$

where ρ is the density of the attenuator.

Half thickness is defined as

$$x_{1/2} = 0.693/\mu,$$

where $\mu = n\sigma$, with n the number of atoms per unit volume of the attenuator; σ, the total atomic cross section per atom, consists of three major components—Compton, photoelectric, and pair production cross sections:

$$\sigma = \sigma_C + \sigma_{\text{PE}} + \sigma_{\text{PP}}.$$

At high incident energies such that the electron binding energy is negligible,

$$\sigma_C = Z\sigma_C^e,$$

where σ_C^e is the Compton cross section per electron.

The photoelectric cross section due to both K-shell electrons, for nonrelativistic energies away from K-shell binding energy, is given by

$$\sigma_{\text{PE}}^K = \sqrt{\frac{32}{(E_\gamma/m_ec^2)^7}}\,\alpha^4 Z^5 \sigma_{\text{Th}}\ \text{cm}^2/\text{atom},$$

TABLE 16.07.D.1 Processes whereby electromagnetic radiation interacts with matter

Process	Interaction	Significant region (MeV)	Approximate Z dependence
Scattering from electrons			
Coherent scattering			
Rayleigh scattering	with bound atomic electrons	<1 (mainly at forward angles)	Z^2 (small angles) Z^3 (large angles)
Thomson scattering	with "free electrons"	independent of E	Z^2 (small angles) Z^3 (large angles)
Incoherent scattering	with bound atomic electrons	<1 (mainly at large angles)	Z
Compton scattering	with "free electrons"	≈1 (↓ as E↑)	Z
Photoelectric effect	with bound atomic electrons, causing ejection of electrons	0–0.5 (↓ as E↑)	Z^5
Nuclear photoeffect	with nucleus as a whole, causing emission of photons, particles, and (above threshold) mesons	≳10	
Interaction with Coulomb field			
Pair production	with nuclear Coulomb field (elastic pair production)	>1 especially at 5–10 MeV	Z^2
	with electron Coulomb field (inelastic pair and triplet production)	>2 (↑ as E↓)	Z
Delbrück scattering	in nuclear Coulomb field		Z^4
Nuclear scattering			
Coherent scattering	Mössbauer, resonance, Thomson		
Incoherent scattering	Compton, with individual nucleons	≳100	

where σ_{Th} is the Thomson scattering cross section per electron. For $E_\gamma \gg m_e c^2$,

$$\sigma_{\mathrm{PE}}^{K} \approx \frac{1.5}{E_\gamma/m_e c^2}\alpha^4 Z^5 \sigma_{\mathrm{Th}} \ \mathrm{cm^2/atom},$$

$$\sigma_{\mathrm{Th}} = \tfrac{8}{3}\pi(e^2/m_e c^2)^2.$$

The total photoelectric cross section due to all shells is

$$\sigma_{\mathrm{PE}} \sim \tfrac{5}{4}\sigma_{\mathrm{PE}}^{K}.$$

For $1 \ll E_\gamma/m_e c^2 \ll 1/\alpha Z^{1/3}$,

$$\sigma_{PP} = \alpha Z^2\left(\frac{e^2}{m_e c^2}\right)^2\left(\frac{28}{9}\ln\frac{2E_\gamma}{m_e c^2} - \frac{218}{27}\right) \ \mathrm{cm^2/atom}.$$

For $E_\gamma/m_e c^2 \gg 1/\alpha Z^{1/3}$,

$$\sigma_{\mathrm{PP}} = \alpha Z^2\left(\frac{e^2}{m_e c^2}\right)^2\left(\frac{28}{9}\ln\frac{183}{Z^{1/3}} - \frac{2}{27}\right) \ \mathrm{cm^2/atom}.$$

Scattering of electromagnetic radiation from free electrons—Compton effect. Compton shift:

$$\lambda' - \lambda = \lambda_C(1 - \cos\theta_\gamma).$$

Kinetic energy of the Compton recoil electron:

$$E_{\mathrm{kin}} = E_\gamma \frac{(E_\gamma/m_e c^2)(1-\cos\theta_\gamma)}{1 + (E_\gamma/m_e c^2)(1-\cos\theta_\gamma)}$$

$$= E_\gamma \frac{2(E_\gamma/m_e c^2)\cos^2\theta_e}{(1 + E_\gamma/m_e c^2)^2 - (E_\gamma/m_e c^2)^2\cos^2\theta_e},$$

where E_γ is the incident photon energy and θ_e, the electron scattering angle, is related to photon scattering angle θ_γ by

$$\cot\theta_e = (1 + E_\gamma/m_e c^2)\tan(\theta_\gamma/2).$$

Compton scattering cross section—Klein-Nishina formula. The differential Compton cross section per electron for a linearly polarized incident plane electromagnetic wave is

$$\frac{d\sigma_C^e}{dr}\Big|_{\mathrm{pol}} = \tfrac{1}{4}\left(\frac{e^2}{m_e c^2}\right)^2\left(\frac{E'_\gamma}{E_\gamma}\right)^2 \times\left(\frac{E_\gamma}{E'_\gamma} + \frac{E'_\gamma}{E_\gamma} + 4\cos^2\Theta - 2\right) \ \mathrm{cm^2\,sr^{-1}/electron},$$

where Θ is the angle between the directions of polarization of the incident and scattered waves. For unpolarized incident waves,

$$\frac{d\sigma_C^e}{d\Omega}\Big|_{\mathrm{unpol}} = \tfrac{1}{2}\left(\frac{e^2}{m_e c^2}\right)^2\left(\frac{E'_\gamma}{E_\gamma}\right)^2 \times\left(\frac{E_\gamma}{E'_\gamma} + \frac{E'_\gamma}{E_\gamma} - \sin^2\theta_\gamma\right) \ \mathrm{cm^2\,sr^{-1}/electron}.$$

Thomson scattering. In the case of low energies the expression for unpolarized incident wave reduces to the Thomson differential cross section per electron

$$\frac{d\sigma_{\mathrm{Th}}}{d\Omega} = \tfrac{1}{2}\left(\frac{e^2}{m_e c^2}\right)^2(1 + \cos^2\theta_\gamma).$$

E. Interaction of charged particles with matter

1. Stopping power—Bethe equation

Stopping power, the rate of energy loss per unit path length, for heavy particle of charge ze is given by the Bethe equation

$$-\frac{dE}{dx} = \frac{4\pi z^2 e^4}{m_e v^2} nZ \times\left(\ln\frac{2m_e v^2}{I} - \ln(1-\beta^2) - \beta^2 - \frac{C_K}{Z}\right) \ \mathrm{ergs\,cm^{-1}},$$

where $\beta \equiv v/c$, with v the particle velocity, n is the number of atoms per unit volume of absorber, and the mean ionization potential

$$I \approx 9.1Z(1 + 1.9Z^{-2/3}) \ \mathrm{eV},$$

and $C_K \equiv C_K(E,Z)$ is a correction term to be included only at comparatively low energies.

Mass stopping power:

$$-\frac{1}{\rho}\frac{dE}{dx} \underset{\sim}{\propto} \frac{Z}{A},$$

where ρ is the density of the absorber.

2. Range

Mean range of protons with energy between a few MeV and 200 MeV in dry air is approximately given by

$$\bar{R}_{\mathrm{air}} \approx [E(\mathrm{MeV})/9.3]^{1.8} \times 100 \ \mathrm{cm}.$$

Mean range in a material of density ρ and atomic weight A with respect to its mean range $\bar{R}_s$ in a standard material of density ρ_s and atomic weight A_s is given by the Bragg-Kleeman rule

$$\frac{\bar{R}}{\bar{R}_s} \approx \frac{\rho_s}{\rho}\sqrt{\frac{A}{A_s}}.$$

3. Straggling

Energy straggling about the mean value $\bar{E}$ is given by

$$\frac{N(E)dE}{N} = \frac{1}{\alpha_s\sqrt{\pi}}\exp\left(-\frac{(E-\bar{E})^2}{\alpha_s^2}\right),$$

where straggling parameter α_s is given by

$$\alpha_s^2 = 4\pi z^2 e^4 nZx_0\left(1 + \frac{kI}{m_e v^2}\ln\frac{2m_e v^2}{I}\right),$$

$N(E)$ is the number of particles having energies between E and $E + dE$, out of total number of N particles whose mean energy is $\bar{E}$, after traversing a thickness x_0 of the absorber. k is a constant dependent upon the electron shells of the absorber.

Range straggling is

$$\frac{N(R)}{N}dR = \frac{1}{\alpha_s\sqrt{\pi}}\exp\left(-\frac{(R-\bar{R})^2}{\alpha_s^2}\right).$$

Angle straggling (broadening of originally collimated beam) is given by mean angle of divergence $\bar{\theta}$:

$$\overline{\theta^2} = \frac{2\pi z^2 e^4}{\bar{E}^2} nZ^2 x_0 \ln\left(\frac{\bar{E}}{zZ^{4/3}e^2}\frac{\hbar^2}{m_e e^2}\right).$$

F. Energy loss of electrons in matter

$$-\frac{dE}{dx}\approx\left(-\frac{dE}{dx}\right)_i+\left(-\frac{dE}{dx}\right)_r,$$

where i stands for energy loss due to ionization and r for energy loss due to radiation.

Energy loss due to scattering from atomic electrons is given by Møller scattering:

$$-\left(\frac{dE}{dx}\right)_i=\frac{4\pi e^4}{m_e v^2}nZ$$
$$\times\left[\ln\frac{2m_e c^2}{I}+\ln(\gamma-1)+\tfrac{1}{2}\ln(\gamma+1)\right.$$
$$\left.-\left(3+\frac{2}{\gamma}-\frac{1}{\gamma^2}\right)\ln\sqrt{2}+\frac{1}{16}-\frac{1}{8\gamma}+\frac{9}{16\gamma^2}\right],$$

where $\gamma=1/[1-(v/c)^2]^{1/2}$. At low energies ($\gamma\approx1$)

$$-\left(\frac{dE}{dx}\right)_i=\frac{4\pi e^4}{m_e v^2}$$
$$\times nZ\left(\ln\frac{2m_e v^2}{I}-1.2329\right)\ \text{ergs cm}^{-1}.$$

At very high energies ($\gamma\gg1$),

$$-\left(\frac{dE}{dx}\right)_i\approx\frac{4\pi e^4}{m_e v^2}nZ\left(\frac{1}{2}\ln\frac{E^3}{2m_e c^2 I^2}+\frac{1}{16}\right).$$

Bremsstrahlung. At relativistic velocities the energy loss due to radiation becomes significant. It is given by

$$-\left(\frac{dE}{dx}\right)_r=nE\sigma_{\text{rad}},$$

where

$$\sigma_{\text{rad}}=\tfrac{16}{3}\alpha Z^2\left(\frac{e^2}{m_e c^2}\right)^2\ \text{cm}^2/\text{nucleus if } E_{\text{kin}}\ll m_e c^2$$
$$=8\alpha Z^2\left(\frac{e^2}{m_e c^2}\right)^2\left(\ln\frac{E}{m_e c^2}-\frac{1}{6}\right)\ \text{cm}^2/\text{nucleus}$$
$$\text{if } m_e c^2\ll E_{\text{kin}}\ll m_e c^2 Z^{-1/3}/\alpha$$
$$=4\alpha Z^2\left(\frac{e^2}{m_e c^2}\right)^2[\ln(183Z^{-1/3})+\tfrac{1}{18}]\ \text{cm}^2/\text{nucleus}$$
$$\text{if } E_{\text{kin}}\gg m_e c^2 Z^{-1/3}/\alpha$$

= complicated power-series formula

for $E_{\text{kin}}\approx m_e c^2$.

Radiation length. The absorber thickness needed to reduce the electron energy by radiation loss to $1/e$ of its original value.

Critical energy. The energy at which the radiation loss equals the collision loss for that substance.

Range. The concept of range for electrons has only limited validity. One of the empirical expressions for mean range in aluminum is

$$\bar{R}\approx0.412E^n\ \text{g cm}^{-2},$$

with $n=1.265-0.0954\ln E$ for $0.01<E<3$ MeV,

$$\bar{R}\approx0.530E-0.106\ \text{g cm}^{-2}\quad(2.5<E<20\ \text{MeV}).$$

G. Radioactive decay

$$N=N_0e^{-\lambda t},$$

where N and N_0 are the number of radioactive nuclei at time t and $t=0$, and λ is the decay constant $\lambda\equiv1/\tau$, where τ is the mean life.

Energy level width of the level with mean life τ is given by

$$\Gamma=\hbar/\tau.$$

For successive decay, e.g., $A\xrightarrow{\lambda_A}B\xrightarrow{\lambda_B}C$ (stable),

$$N_A=N_{A_0}e^{-\lambda_A t},$$
$$N_B=\frac{\lambda_A}{\lambda_B-\lambda_A}N_{A_0}(e^{-\lambda_A t}-e^{-\lambda_B t})+N_{B_0}e^{-\lambda_B t},$$
$$N_C=N_{C_0}+N_{B_0}(1-e^{-\lambda_B t})+N_{A_0}$$
$$\times\left(1+\frac{\lambda_A}{\lambda_B-\lambda_A}e^{-\lambda_B t}-\frac{\lambda_B}{\lambda_B-\lambda_A}e^{-\lambda_A t}\right).$$

Activity:

$$-\frac{dN}{dt}=\lambda N.$$

Specific activity: activity per unit mass of radioactive source.

1. α decay[23]

Decay constant:

$$\lambda=\lambda_0e^{-G}.$$

Here $\lambda_0=v_0/R$, where v_0 is the α-particle velocity inside the nucleus and G is called the *Gamow factor.* For α energy $E\ll E_c$ ($\equiv Zze^2/R$, ze being the α-particle charge),

$$G\approx\frac{(2mE_c)^{1/2}}{\hbar}R\left[\pi\left(\frac{E_c}{E}\right)^{1/2}-4\right].$$

2. β decay

Q value:

$$Q_{\beta-}=[M(Z,A)-M(Z+1,A)]c^2,$$
$$Q_{\text{EC}}=[M(Z,A)-M(Z-1,A)]c^2,$$
$$Q_{\beta+}=[M(Z,A)-M(Z-1,A)-2m_e]c^2.$$

The β-momentum spectrum is given by

$$N(p)dp=\frac{g^2}{2\pi^3c^3\hbar^7}|M_{fi}|^2F(E,Z)(E_0-E)^2p^2dp,$$

where E_0 is the β end-point energy, M_{fi} the transition matrix element, $F(E,Z)$ the *Fermi function,* and g the weak-interaction coupling constant;

$$\sqrt{\frac{N(p)}{p^2F(E,Z)}}=C(E_0-E).$$

For allowed transition C is energy independent, and hence the plot of the expression on the left-hand side versus the kinetic energy of β particles, called the *Kurie plot,* is a straight line.

The *ft* value, also called *reduced half-life,* is given by

$$ft=\frac{2\pi^3}{g^2}\frac{\hbar^7}{m_e^5c^4}\frac{\ln2}{|M_{fi}|^2},$$

where t is the half-life and f a calculable function

$$f \equiv f(W_0, Z) = \int_1^{W_0} F(W,Z)(W^2-1)^{1/2}(W_0-W)^2 W\,dW,$$

where $W \equiv (E+m_ec^2)/m_ec^2$, $W_0 \equiv (E_0+m_ec^2)/m_ec^2$.

3. γ decay

Selection rules for 2^L-pole γ transition between levels of spin-parity $J_i\pi_i$ and $J_f\pi_f$ are

$$|J_f - J_i| \leqslant L \leqslant J_f + J_i,$$

$$\pi_i\pi_f = (-1)^L \quad \text{for } EL \text{ transition}$$

$$= (-1)^{L+1} \quad \text{for } ML \text{ transition.}$$

Transition probability for a nuclear transition from a state i to state f through γ emission of multipolarity 2^L is given by

$$T(XL, J_i \to J_f) \equiv \frac{1}{\tau} = \frac{4\pi}{\hbar}\left(\frac{E_\gamma}{\hbar c}\right)^{2L+1} SB(XL, J_i \to J_f),$$

where τ is the partial γ-ray lifetime, S is a statistical factor, $B(XL)$ is the reduced transition probability, and XL is EL or ML depending upon the transition being electric or magnetic multipole 2^L.

The single-particle estimates for photon lifetimes are given in Sec. 16.02.

Mixing ratio δ gives the admixture of electric multipolarity $L' = L+1$ in a mixed transition $ML + EL'$. δ^2 gives the intensity ratio $I(L')/I(L)$ of the mixed multipolarities.

H. Nuclear reaction cross section

For reaction $A(a,b)B$, where the target is at rest and the residual nucleus is almost at rest, the reaction cross section, in the absence of resonances and polarization, is given by

$$\sigma = \frac{1}{\pi\hbar^4 n_a}\frac{p_b^2}{v_a v_b} V|H'_{fi}|^2(2j_b+1)(2J_B+1),$$

where v is the velocity, p the momentum, H'_{fi} the interaction matrix element, V the volume of normalization, and n_a the density of incident particles. This expression refers to center-of-mass system, but if A and B are sufficiently massive the formula will also hold for laboratory velocities and momenta.

Principle of detailed balance:

$$\frac{\sigma[A(a,b)B]}{\sigma[B(b,a)A]} = \frac{p_b^2(2j_b+1)(2J_B+1)}{p_a^2(2j_a+1)(2J_A+1)}.$$

In this expression the cross sections and momenta are in center-of-mass system.

Some general trends for $\sigma[A(a,b)B]$ in the region away from resonances[23,26]:

(1) $\sigma(n,n) \approx$ const for $E_n \lesssim 5$ eV.

(2) $\sigma(n,b)$ [exothermic reaction, e.g., (n,p), (n,α), (n,γ), (n,f), (fission), etc., with E_n = low energy ($\approx$few eV)]: $\sigma(n,b) \propto 1/v_n$, where v is the velocity.

(3) $\sigma(n,n') \underset{\sim}{\propto} (E_n - E^*)^{1/2}$ for neutron energy near threshold E^*.

(4) $\sigma(n,b)$ (endothermic reaction; b is a charged particle) $\sim e^{-G_b}(E_n - E^*)^{1/2}$. The factor e^{-G_b} causes the cross-sectional curve to become convex to the energy axis.

(5) Charged incoming particle, uncharged outgoing particle, e.g., in reactions (p,n), (p,γ), (α,n), (α,γ), etc.: for incident energy $\ll Q$, $\sigma(a,b) \propto (1/v_a)e^{-G_a}$.

1. Resonance reactions—Breit-Wigner formula

Cross section for formation of a compound nucleus near a single resonance of energy E_0 is given by

$$\sigma^l_{CN} = \pi\lambda\!\!\!^-{}^2(2l+1)\frac{\Gamma_a\Gamma}{(E-E_0)^2+(\Gamma/2)^2},$$

where $\lambda\!\!\!^-$ is the rationalized de Broglie wavelength for incident particle and l is orbital angular momentum. Γ is the total width (FWHM) of the compound state, Γ_a the partial width for formation of the Compound nucleus through the channel corresponding to particle a.

Reaction cross section for reaction (a,b) near resonance:

$$\sigma^l_{(a,b)} = \sigma^l_{CN}\frac{\Gamma_b}{\Gamma} = \pi\lambda\!\!\!^-{}^2(2l+1)\frac{\Gamma_a\Gamma_b}{(E-E_0)^2+(\Gamma/2)^2}.$$

This expression does not include contributions due to the nonresonant part of the reaction.

$$\sigma_{tot} = \sigma_r + \sigma_{el},$$

where σ_r, called the reaction cross section, is the sum of cross sections over all resonance exit channels, for a particular entrance channel, excepting the elastic channel

$$\sigma_{el}(\theta) = \sigma_{compound\ el}(\theta) + \sigma_{shape\ el}(\theta).$$

Compound elastic, also called *resonance scatttering,* proceeds via an excited compound state. *Shape elastic,* also called *potential scattering,* is the process in which no intermediate complex state is formed.

Absorption cross section is defined as

$$\sigma_{absorption} \equiv \sigma_r + \sigma_{compound\ el}.$$

16.08. ELASTIC INTERACTION OF PIONS WITH NUCLEI*

Elastic scattering of pions from nuclei is evaluated equivalently within optical model contexts and within the isobar-hole model. The most extensive and careful recent analyses have been done within the former framework at low energies (pion kinetic energy T_π less than 50 MeV) and within the latter framerwork in the resonance region ($50 \lesssim T_\pi \lesssim 250$ MeV).

A. Lower energy

Pionic atom and low-energy scattering has been analyzed[27] within the context of the optical model

$$(-\nabla^2 + 2\bar{\omega}U_{opt} + \text{Coulomb})\psi(r) = k_0^2\psi(r), \quad (1)$$

*Prepared by Mikkel Johnson.

where k_0 is the incident pion momentum, and "Coulomb" refers to the Coulomb potential appropriate to the Klein-Gordon equation.[28] The optical potential is

$$2\bar{\omega}U_{\rm opt} = -4\pi[b(r)+B(r)] + 4\pi\nabla\cdot\{L(r)[c(r)+C(r)]\}\nabla - 4\pi\left(\frac{p_1-1}{2}\nabla^2 c(r) + \frac{p_2-1}{2}\nabla^2 C(r)\right), \quad (2)$$

where

$$b(r) = p_1[\bar{b}_0\rho(r) - \epsilon_\pi b_1\delta\rho(r)],$$
$$\bar{b}_0 = b_0 - (3/2\pi)1.4(b_0^2 + 2b_1^2),$$
$$c(r) = p_1^{-1}[c_0\rho(r) - \epsilon_\pi c_1\delta\rho(r)], \quad (3)$$
$$B(r) = p_2 B_0\rho^2(r),$$
$$C(r) = p_2^{-1}C_0\rho^2(r),$$
$$L(r) = \{1 + (4\pi/3)\lambda[c(r)+C(r)]\}^{-1},$$

and

$$\delta\rho(r) = \rho_n(r) - \rho_p(r). \quad (4)$$

Here p_1 and p_2 are kinematic factors,

$$p_1 = 1 + \omega/M, \quad p_2 = 1 + \omega/2M, \quad (5)$$

TABLE 16.08.A.1. Parameters which reproduce pionic atom and elastic scattering data at 30 MeV for Eqs. (1)–(5) in text

b_0 (fm)	-0.046^a
b_1 (fm)	-0.134
ReB_0 (fm^4)	0.007
ImB_0 (fm^4)	0.19^a
c_0 (fm^3)	0.66
c_1 (fm^3)	0.428
ReC_0 (fm^6)	0.287
ImC_0 (fm^6)	0.93^a
λ	1.4^a

^{a}Fitted to pionic atom data.

TABLE 16.08.A.2. Coefficients β_n and γ_n in Eqs. (7) and (8).

$\beta_0 = (-0.222,\ 0.206)$fm^4
$\beta_1 = (\ 0.724, -0.140)$fm^4 MeV^{-1}
$\beta_2 = (\ 0.146, -1.016)$fm^4 MeV^{-2}
$\beta_3 = (-0.233,\ 0.054)$fm^4 MeV^{-3}

$\gamma_0 = (\ 0.638,\ 0.422)$fm^6
$\gamma_1 = (\ 1.378,\ 1.099)$fm^6 MeV^{-1}
$\gamma_2 = (-4.183, -0.155)$fm^6 MeV^{-2}
$\gamma_3 = (\ 2.611,\ 0.937)$fm^6 MeV^{-3}

ω is the pion total energy, ϵ_π the pion charge, and M the nucleon mass. The densities ρ_n, ρ_p, and ρ are normalized to N, Z, and A, respectively. Fits to the widths and positions of atomic levels give the results in Table 16.08.A.1.

The parameters $\bar{b}_0$, b_1, c_0, and c_1 are the values taken from the free pion-nucleon scattering amplitude evaluated in the form

$$f_{\pi N} = b_0 + b_1\mathbf{t}\cdot\boldsymbol{\tau} + (c_0 + c_1\mathbf{t}\cdot\boldsymbol{\tau})\mathbf{k}\cdot\mathbf{k}', \quad (6)$$

extrapolated to zero energy and corrected for the Pauli effect.[27] The same parameter set reproduces fairly well π^+ and π^- scattering at 30 MeV throughout the periodic table. The energy dependence of the parameters of the optical potential becomes important by 30 MeV. This means that the b_0, b_1, c_0, and c_1 values should be taken from the pion-nucleon phase shifts according to Eq. (6), and the B_0 and C_0 parameters must be fit to the pionic atom and elastic scattering data as a function of energy. We give in Table 16.08.A.2 the coefficients β_n and γ_n of a recent[29] polynomial fit of the energy dependence of B_0 and C_0 having the following form:

$$B_0(T_\pi) = \sum_{n=0}^{3} \beta_n(\epsilon)^n, \quad (7)$$

$$C_0(T_\pi) = \sum_{n=0}^{3} \gamma_n(\epsilon)^n, \quad (8)$$

where $\epsilon = 0.01\ T\pi$, with $T\pi$ the pion laboratory kinetic energy in MeV. The corresponding value of λ is 1.6. This fit reproduces scattering data for $T\pi$ up to 115 MeV. The pion optical potential has also been generalized to include the charge-exchange reaction, see Ref. 30.

B. Resonance energy

The most recent and careful phenomenological analysis in the resonance region has been carried out by Horikawa, Thies, and Lenz[31] in the context of the isobar-hole model. Unfortunately, this method is difficult to use for routine data analysis. The phenomenological input is the isobar-nucleus shell potential, taken to have the form

$$W_{\rm sp} = W_0(E)\rho(R) + 2\mathbf{L}_\Delta\cdot\Sigma_\Delta V_{\rm LS}^0\ \mu r^2 e^{-\mu r^2}, \quad (9)$$

where L_Δ and Σ_Δ are the isobar orbital angular momentum and spin operators, respectively. Values for the spin-orbit interaction are given in Table 16.08.B.1, and $W_0(E)$ in Figure 16.08.B.2. Only light nuclei can be analyzed in this method owing to limitations of computer sizes.

TABLE 16.08.B.1. Strength $V_{\rm LS}^0$ and range parameters μ for spin-orbit potential

	μ (fm^{-2})	$V_{\rm LS}^0$ (MeV)
π-^{4}He	0.25	-4.6–$i1.8$
π-^{12}C	0.35	$-10\ -i4$
π-^{16}O	0.3	$-10\ -i4$

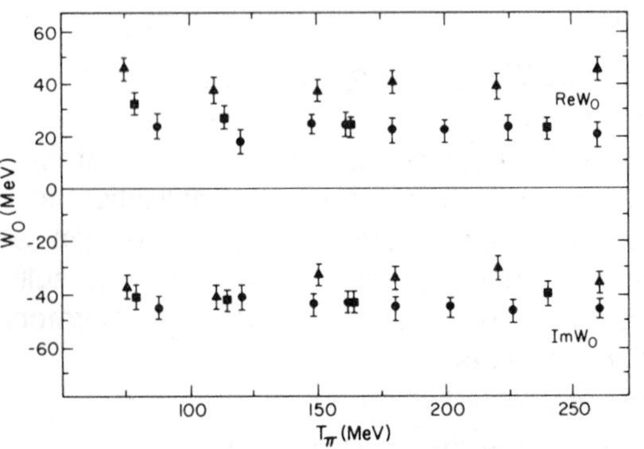

FIGURE 16.08.B.2. Strength of central part of spreading potential π-^{4}He (triangles), π-^{12}C (circles), and π-^{16}O (squares)

16.09. SLOW NEUTRONS[23]*

A. Slowing down of neutrons

An important parameter in slowing down theory is the Fermi age,

$$\tau = \frac{\lambda^2}{3\xi(1-2/3A)}(u-u_0) = \tfrac{1}{6}\langle R^2\rangle,$$

where λ is the mean free path for scattering,

$$u = \ln(E_0/E),$$

*Prepared by Herbert L. Anderson.

TABLE 16.09.A.1. Fermi age of 1.4-eV and thermal neutrons for various moderators

	τ (1.4 eV)	τ (0.0253 eV)
H_2O	31 cm^2	33 cm^2
D_2O	109	120
Be	80	98
Graphite	311	350

with $E_0 = 10$ MeV as the lethargy, ξ as the average value of the logarithmic energy loss,

$$\xi = \langle \ln(E_0/E_1) \rangle = 1 + \frac{(A-1)^2}{2A} \ln \frac{A-1}{A+1},$$

A as the atomic weight of the slowing-down medium, and $\langle R^2 \rangle$ as the mean-square distance between the neutron and its source in passing from lethargy u_0 to lethargy u.

Table 16.09.A.1 gives the Fermi age of 1.4-eV and thermal neutrons for various moderators.

B. Diffusion of thermal neutrons

In the useful moderators, once the neutrons reach thermal energies they diffuse according to the equation

$$\nabla^2 n - (3/\lambda\Lambda)n + 3q/\lambda v = 0,$$

TABLE 16.09.B.1. Density, diffusion length, and mean free path for scattering and absorption of common moderators

	ρ (g/cm^3)	L (cm)	λ (cm)	Λ (cm)
H_2O	1.00	2.70	0.43	51.8
D_2O	1.10	1.02	2.4	13 400
Be	1.84	22.2	2.1	705
Graphite	1.62	47.3	2.7	2 480

where n is the neutron density, number per cm^3; λ is the scattering mean free path, Λ the mean free path for absorption, v the neutron velocity, and q the number per cm^3 per second of slowing neutrons that reach thermal energies. The quantity $L = (\lambda\Lambda/3)^{1/2}$ is called the diffusion length.

C. Chain reactors

The values of nuclear constants important for chain reactors are given in Table 16.09.C.1. σ_a, σ_f, and σ_γ are absorption, fission, and capture cross sections, respectively; $\alpha = \sigma_\gamma/\sigma_f$, ν is the average number of neutrons produced per fission, η the average number produced per neutron absorbed.

Table 16.09.C.1. Nuclear constants important for chain reactors[9]: cross sections in barns (10^{-24} cm^2) Neutron energy=0.025 eV, Velocity=2200 m/s

	^{233}U	^{235}U	^{239}Pu	^{241}Pu
σ_a	574.7 ± 1.0	680.9 ± 1.1	1017 ± 3	1369 ± 8
σ_f	529.1 ± 1.2	582.6 ± 1.1	748.1 ± 2.0	1011 ± 6
σ_γ	45.5 ± 0.7	98.3 ± 0.8	269 ± 3	358 ± 5
α	0.086 ± 0.002	0.169 ± 0.002	0.360 ± 0.003	0.354 ± 0.006
η	2.296 ± 0.004	2.075 ± 0.003	2.115 ± 0.005	2.169 ± 0.008
$\bar{\nu}$	2.493 ± 0.004	2.425 ± 0.003	2.877 ± 0.006	2.937 ± 0.007

16.10. Compilations in Nuclear Physics (See Refs. 32, 33, and 34)

16.11. REFERENCES

[1]R. G. Helmer, P. H. M. Van Assche, and C. Van Der Leun, At. Data Nucl. Data Tables **24**, 39 (1979).

[2]E. Browne *et al.* and A. A. Shihab-Eldin *et al.*, in *Table of Isotopes,* 7th ed., edited by C. M. Lederer and V. S. Shirley (Wiley, New York, 1978).

[3]*Evaluated Nuclear Structure Data File*, edited and maintained by the National Nuclear Data Center, Brookhaven National Laboratory, on behalf of the International Network for Nuclear Structure Data Evaluation. For individual *A*-chain evaluations see Nucl. Data Sheets.

[4]R. A. Meyer, "Multigamma-ray Calibration Sources," Lawrence Livermore Laboratory Report M-100, 1978.

[5]A. Rytz, At. Data Nucl. Data Tables **23**, 507 (1979).

[6]K. Siegbahn, in *Alpha-, Beta- and Gamma-Ray Spectroscopy,* edited by K. Siegbahn (North-Holland, Amsterdam, 1965).

[7]Evaluation by R. G. Helmer (private communication).

[8]R. G. Helmer, R. C. Greenwood, and R. J. Gehrke, Nucl. Instrum. Methods **155**, 189 (1978).

[9]S. F. Mughabghab *et al.*, *Neutron Resonance Parameters* (Academic Press, Orlando, 1984), Vols. I and II.

[10]A. H. Wapstra, G. J. Nijgh, and R. van Lieshout, *Nuclear Spectroscopy Tables* (North-Holland, Amsterdam, 1959).

[11]J. M. Blatt and V. F. Weisskopf, *Theoretical Nuclear Physics* (Wiley, New York, 1952).

[12]S. A. Moszkowski, in *Alpha-, Beta- and Gamma-Ray Spectroscopy,* edited by K. Siegbahn (North-Holland, Amsterdam, 1965).

[13]R. S. Hager and E. C. Seltzer, Nucl. Data Sect. A **4**, 1 (1968).

[14]F. Rosel, H. M. Fries, K. Alder, and H. C. Pauli, At. Data Nucl. Data Tables **21**, 92 (1978).

[15]I. M. Band, N. B. Trzhaskovskaya, and M. A. Listengarten, At. Data Nucl. Data Tables **19**, 433 (1977).

[16]O. Dragoun, H. C. Pauli, and F. Schmutzler, Nucl. Data Sect. A **6**, 235 (1969).

[17]O. Dargoun, Z. Plajner, and F. Schmutzler, Nucl. Data Sect. A **9**, 119 (1971).

[18]I. M. Band, M. B. Trzhaskovskaya, and M. A. Listengarten, At. Data. Nucl. Data Tables **21**, 1 (1978).

[19]A. H. Wapstra and G. Audi, Nuclear Physics A**432**, 1 (1985).

[20]N. E. Holden, "Isotopic Composition of the Elements and their Variation in Nature: A Preliminary Report," National Nuclear Data Center, Brookhaven National Laboratory, Report BNL-NCS-50605, 1977 (unpublished); private communication from the author.

[21]H. A. Enge, *Introduction to Nuclear Physics* (Addison-Wesley, Reading, MA, 1966).

[22]J. B. Marion and F. C. Young, *Nuclear Reaction Analysis Graphs and Tables* (North-Holland, Amsterdam, 1968).

[23]E. Segrè, *Nuclei and Particles,* 2nd ed. (Benjamin, New York, 1977).

[24]A. Bohr and B. R. Mottelson, *Nuclear Structure* (Benjamin, New York, 1969), Vol. I.

[25]S. G. Nilsson, K. Dan. Vidensk. Selsk. Mat.-Fys. Medd. **29**, No. 16 (1955).

[26]P. Marmier and E. Sheldon, *Physics of Nuclei and Particles* (Academic, New York, 1969), Vols. I and II.

[27]K. Stricker, J. A. Carr, and H. McManus, Phys. Rev. C **22**, 2043 (1980); K. Stricker, H. McManus, and J. A. Carr, *ibid.* **19**, 929 (1979).

[28]M. D. Cooper, R. H. Jeppesen, and M. B. Johnson, Phys. Rev. C **20**, 696 (1979).

[29]P. W. F. Alons, M. J. Leitch, and E. R. Siciliano (unpublished).

[30]E. R. Siciliano, M. D. Cooper, M. B. Johnson, and M. J. Leitch, Phys. Rev. C **34**, 267 (1986).
[31]Y. Horikawa, M. Thies, and F. Lenz, Nucl. Phys. A **345**, 386 (1980).
[32]F. Ajzenberg-Selove, *Nuclear Spectroscopy and Reactions*, Part C (Academic Press, 1974), p. 551.
[33]T. W. Burrows and N. E. Holden, Rept. BNL-NCS-50702 (1978) (unpublished).
[34]H. Behrens *et al.*, Physik Daten, 3–5 (1985), FIZ, Karlsruhe.

17.00. Optics

John N. Howard

Air Force Geophysics Laboratory

CONTENTS

17.01. SPECTRUM, TERMS, AND UNITS

A. Figure: Electromagnetic spectrum[1]

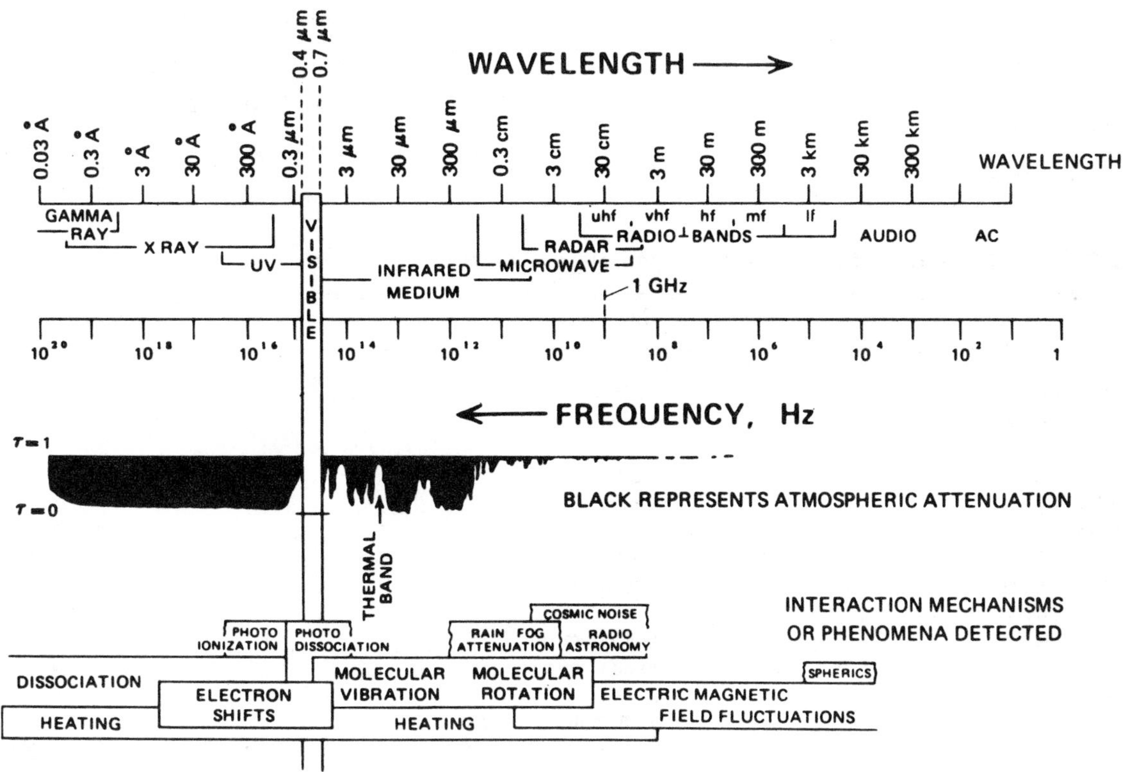

B. Terms and constants

Optics. The science of light (originally visible light, but now including also ultraviolet and infrared radiation).

Visible light. Electromagnetic radiation of wavelength 380 to 750 nm.

Ultraviolet radiation. Light of shorter wavelength than the eye can see ($\lambda < 380$ nm). λ below 190 nm called vacuum ultraviolet, VUV; below 100 nm called extreme ultraviolet, XUV.

Infrared radiation. Light of longer wavelength than the eye can see. 760 nm $< \lambda < 1\ \mu$m called near IR; $\lambda > 50$ μm called far IR or submillimeter radiation.

Photon. A quantum of radiant energy possessing energy $h\nu$.

Speed of light c (in vacuum):

$$c = 2.997\,924\,58 \times 10^8 \text{ m/s} \quad \text{(error 0.004 ppm)}.$$

Optical radiators and sources

Blackbody. An ideal body of uniform temperature that perfectly absorbs all incident radiation. A blackbody at temperature T also emits thermal radiation to its surroundings according to the Planck radiation formula.

Emissivity. A measure of how closely the flux from a given temperature radiator approaches that from a blackbody at the same temperature.

Planck's radiation formula for blackbody emission:

$$M = c_1 \lambda^{-5} \Delta\lambda / (\epsilon^{c_2/\lambda T} - 1),$$

where M is the spectral radiant exitance at λ and in range $\Delta\lambda$,

$$c_1 = 2\pi hc^2$$
$$= 3.741\,832 \times 10^{-16} \text{ W m}^2$$
$$= 3.741\,832 \times 10^{10}\ \mu\text{W}\,\mu\text{m}^4\text{ cm}^{-2},$$
$$c_2 = hc/k$$
$$= 1.438\,786 \times 10^{-2} \text{ m K}.$$

Wien's displacement law. Numerically, the peak wavelength, in nm, is

$$\lambda_m = 2.8978 \times 10^6 / T,$$

where T is the absolute temperature in kelvin.

Kirchhoff's law of radiation:

$$W/A = W_b,$$

where W_b is the radiant exitance of a blackbody at same temperature at which W and A are measured, W the radiant emittance, and A absorptance.

Stefan-Boltzmann law. Total radiation emitted by a blackbody at temperature T:

$$M_b = \int_0^\infty M_\lambda \, d\lambda = \sigma T^4,$$

where

$$\sigma = \frac{\pi^2}{60} \frac{k^4}{h^3 c^2} = 5.670\,32 \times 10^{-8} \text{ W m}^{-2}\text{ K}^{-4}.$$

Candela. The unit of luminous intensity. 1/60 of 1 cm^2 of projected area of a blackbody radiator operating at the temperature of solidification of platinum. The candela emits one lumen per steradian: $1 cd = 1 lm\ sr^{-1}$.

Other radiometric definitions

Absorptance:

$\alpha = Q$ absorbed/Q incident.

Reflectance:

$\rho = Q$ reflected/Q incident.

Transmittance:

$\tau = Q$ transmitted/Q incident.

Emissance:

$\epsilon = Q$ emitted/Q blackbody.

Here Q is the appropriate radiant quantity in Table 17.01.C.

C. Table: Radiometric and photometric terms, symbols, and units[2]

	Symbol	Unit	Abbreviation
Energy	Q		
Radiant energy	Q_e	joule	J
Luminous energy	Q_v	lumen second	lm s
Power (flux)	ϕ		
Radiant power	ϕ_e	watt	W
Luminous power	ϕ_v	lumen	lm
Areance (exitance)	M		
Radiant areance	M_e	watt/meter^2	W m^{-2}
Luminous areance	M_v	lumen/meter^2	lm m^{-2}
Pointance (intensity)	I		
Radiant pointance	I_e	watt/steradian	W sr^{-1}
Luminous pointance	I_v	lumen/steradian (candela)	lm sr^{-1} (cd)
Sterance	L		
Radiant sterance (radiance)	L_e	watt/meter^2 steradian	W m^{-2} sr^{-1}
Luminous sterance (luminance)	L_v	lumen/meter^2 steradian (candela/meter^2) (nit)	lm m^{-2} sr^{-1} (cd m^{-2})
Areance (irradiance/illuminance)	E		
Radiant areance	E_e	watt/meter^2	W m^{-2}
Luminous areance	E_v	lumen/meter^2 (lux)	lm m^{-2} (lx)

17.02. ABSORPTION LAWS

Power transmitted through an absorbing sample is given as

$$\phi_2 = \phi_1 e^{-\alpha x c},$$

which is the *exponential law of absorption;* α is the absorption coefficient, x the length of the absorption path, and c the concentration the of absorber.

$$A = \log_{10}(1/T),$$

$$A = 0.4343\alpha x c,$$

$$T = 10^{-A} = 1/10^A,$$

$$\alpha = 2.3026 A/xc.$$

Index of refraction. The ratio of the velocity of light in a medium to the velocity of light *in vacuo, c.*

17.03. REFLECTION

When electromagnetic radiation interacts with an interface consisting of a change in refractive index, reflection takes place. Specular reflection occurs from a flat abrupt surface, giving rise to reflected radiation at equal angles. Some portion of the incident light is transmitted at angle θ_2. The laws of specular radiation are derived from Maxwell's equations, hold for all wavelengths, and are determined by the refractive indices of the two media.

When both media are transparent the angle of the transmitted ray is given by Snell's law,

$$n_1 \sin\theta_1 = n_2 \sin\theta_2.$$

The reflectivity and transmission of the interface is given by the ratio of the electric field of the reflected ray E_1 and transmitted ray E_2 to the incident ray E_0. The reflectivity is different for light polarized in the plane of incidence (the plane of the drawing in Figure 17.03.A) and light polarized normal to the plane of incidence. Incident light polarized normal to the plane of incidence has an electric field component transverse to the plane of incidence and is called TE (transverse electric). Light polarized in the plane of the drawing has a magnetic field component transverse to the plane of incidence and is called TM (transverse magnetic). The electric field ratios are given by Fresnel's reflection formulas:

$$\text{TE:} \quad \frac{E_1}{E_0} = \frac{\cos\theta - (n^2 - \sin^2\theta)^{1/2}}{\cos\theta + (n^2 - \sin^2\theta)^{1/2}},$$

$$\frac{E_2}{E_0} = \frac{2\cos\theta}{\cos\theta + (n^2 - \sin^2\theta)^{1/2}};$$

$$\text{TM:} \quad \frac{E_1}{E_0} = \frac{n^2\cos\theta - (n^2 - \sin^2\theta)^{1/2}}{n^2\cos\theta + (n^2 - \sin^2\theta)^{1/2}},$$

$$\frac{E_2}{E_0} = \frac{2n\cos\theta}{n^2\cos\theta + (n^2 - \sin^2\theta)^{1/2}},$$

where $n = n_2/n_1$ and $\theta = \theta_1$.

The reflectivity is defined as the ratio of the reflected light intensity to the incident light intensity:

$$R = |E_1^2/E_2^2|.$$

The transmission is given by the ratio of the transmitted light intensity to the incident light intensity:

$$T = |E_1^2/E_0^2|.$$

If there are no losses, $T + R = 1$. At normal incidence there is no distinction between TE and TM polarizations, and

$$R = \left(\frac{n_1 - n_2}{n_1 + n_2}\right)^2.$$

For the TM polarization there is one angle, called Brewster's angle, at which there is no reflection. This is given by

$$\tan\theta_B = n_2/n_1.$$

At this angle the reflection is completely TE polarized and there is total transmission of TM polarization.

When medium 2 is not transparent, its refractive index can be considered complex,

$$n = n' + ik,$$

and the previous Fresnel's formulas still hold, but with a complex refractive index. The limiting case of an absorbing medium is metallic reflection, in which the reflectivity is large at all angles.

A. Figure: Reflection and refraction[3]

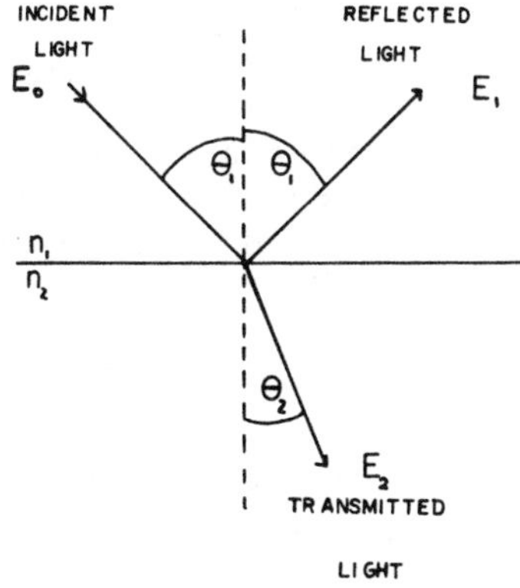

B. Geometrical optics

Although light is a form of electromagnetic radiation and its propagation can be described as a wave motion satisfying Maxwell's equations, for many purposes a simpler description called geometrical optics can be used. Here the normal to the wave front is defined as a light ray. Light travels through a vacuum in straight lines at a constant velocity c irrespective of color.

C. Sign convention

In general, there exist several sign conventions, and in principle we are free to choose whichever convention we please. But once we have made a choice, we must be consistent and adhere to the convention.

The two major groups of sign conventions in optics are the *rational system,* based on Cartesian, *x-y*, coordinates (which we are using here), and the *empirical system,* where the distance of a real object is taken as positive. Neither convention is completely satisfactory but the rational system has the advantage that it connects easily with the concept of vergence. It also is used in all modern methods of lens design. In the rational system[4]:

(1) All figures are drawn with the light traveling from left to right.

(2) Most distances are measured *from* a reference surface, such as a wave front (or a lens). All distances measured to the *left* of this surface, opposite to the direction of the light, are considered negative. All distances measured to the *right,* in the direction of the light, are *positive.*

(3) The refractive *power* of a surface that makes light more convergent, or less divergent, is positive. Such a surface is converging; its (second) *focal length* is positive also. The power and the (second) focal length of a diverging surface are negative.

(4) Since all distances are measured in Cartesian coordinates, the *distance of a real object is negative.*

(5) For the same reason, the *distance of a real image is positive.* The distance of a virtual image is negative.

(6) *Heights* above the optic axis are positive, distances below the axis negative.

(7) *Angles* measured clockwise, beginning at the optic axis or at the normal erected on a surface at a point of incidence, are positive. Angles measured counterclockwise are negative.

1. Figure: Sign convention[5]

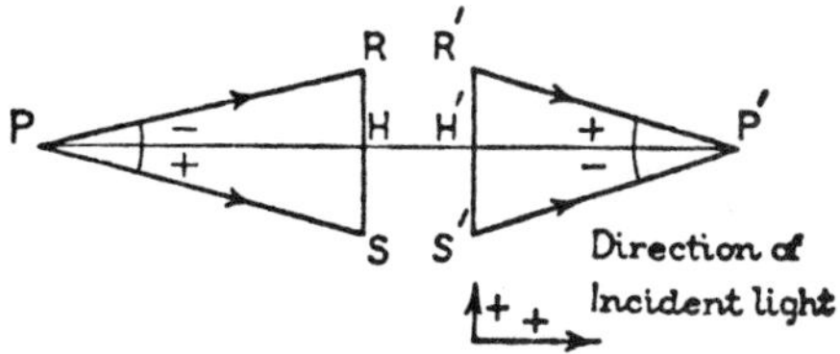

17.04. PHYSICAL OPTICS

$$P_{\text{lens}} = (n_{\text{lens}} - n_0)(1/R_1 - 1/R_2),$$

which is the *lens-makers formula.*

A. Gauss's thin-lens equation

$$n_1/o + n_2/f_2 = n_2/i,$$

where n_2 refers to the index of the medium to the right of the lens, and *not* to the index of the material of the lens. In air,

$$1/o + 1/f_2 = 1/i;$$

$$N_0 N_i = f_1 f_2,$$

which is *Newton's lens equation.* Here N_0 and N_i are the distances of the object and image from the focal points f_1 and f_2, respectively.

B. Aberrations

A simple lens with spherical surfaces suffers from a number of well-recognized aberrations. The most familiar of these are:

(1) *Spherical:* different zones of the lens form images at different distances along the lens axis.

(2) *Coma:* the different zones of the lens form images of different sizes.

(3) *Chromatic:* the images in different colors are formed at different distances from the lens.

(4) *Lateral color:* the images in different colors are of different sizes.

(5) *Astigmatism:* radial lines and tangential lines in the object are imaged at different distances from the lens.

C. Rayleigh criterion for resolution

The Rayleigh criterion for resolution is that two adjacent, equal-intensity, point sources can be considered resolved if the first dark ring of the diffraction pattern of one point image coincides with the center of the other pattern. The intensity of crossing point of the two diffraction patterns is then 0.8106 of either pattern relative to the maximum. While it is possible under favorable conditions to resolve sources closer than this separation, this definition, proposed by Lord Rayleigh, is a convenient measure of resolution.

In 1879 Rayleigh also expressed the conclusion that an optical system would give an image only slightly inferior to that produced by an absolutely perfect system if all the light arrived at the focus with differences of phase not exceeding one-quarter of a wavelength, "for then the resultant cannot differ much from the maximum." This practical criterion is called the Rayleigh limit.

D. Figure: Diffraction grating[6]

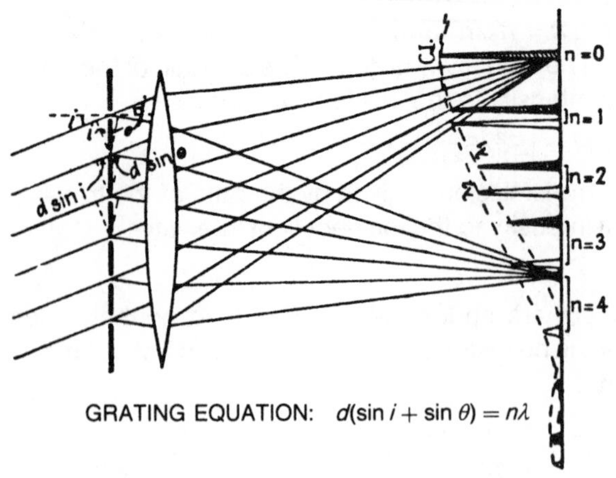

Positions and intensities of the principal maxima from a grating, where light containing two wavelengths is incident at an angle i and diffracted at various angles θ.

E. Figure: Michelson interferometer[7]

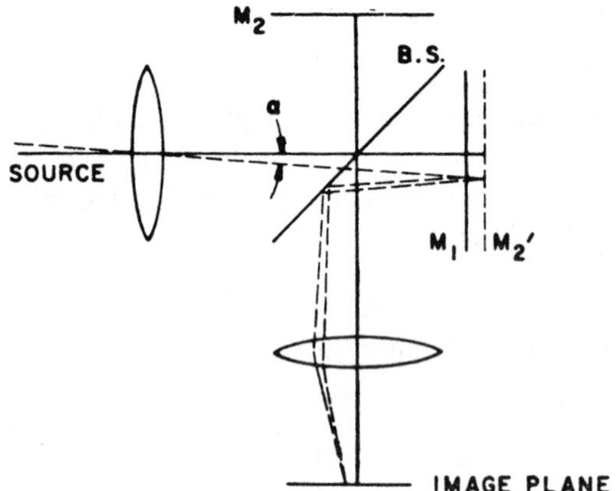

M_1 and M_2 are the end mirrors, M_2' the image of M_2 as seen through the beam splitter B.S. When adjusted for circular fringes, maxima are $2t\cos\phi = m\lambda$. Fringe shift due to a displacement $t' - t$ of movable mirror, $m' - m = 2(t' - t)/\lambda$.

17.05. FIGURES AND TABLES

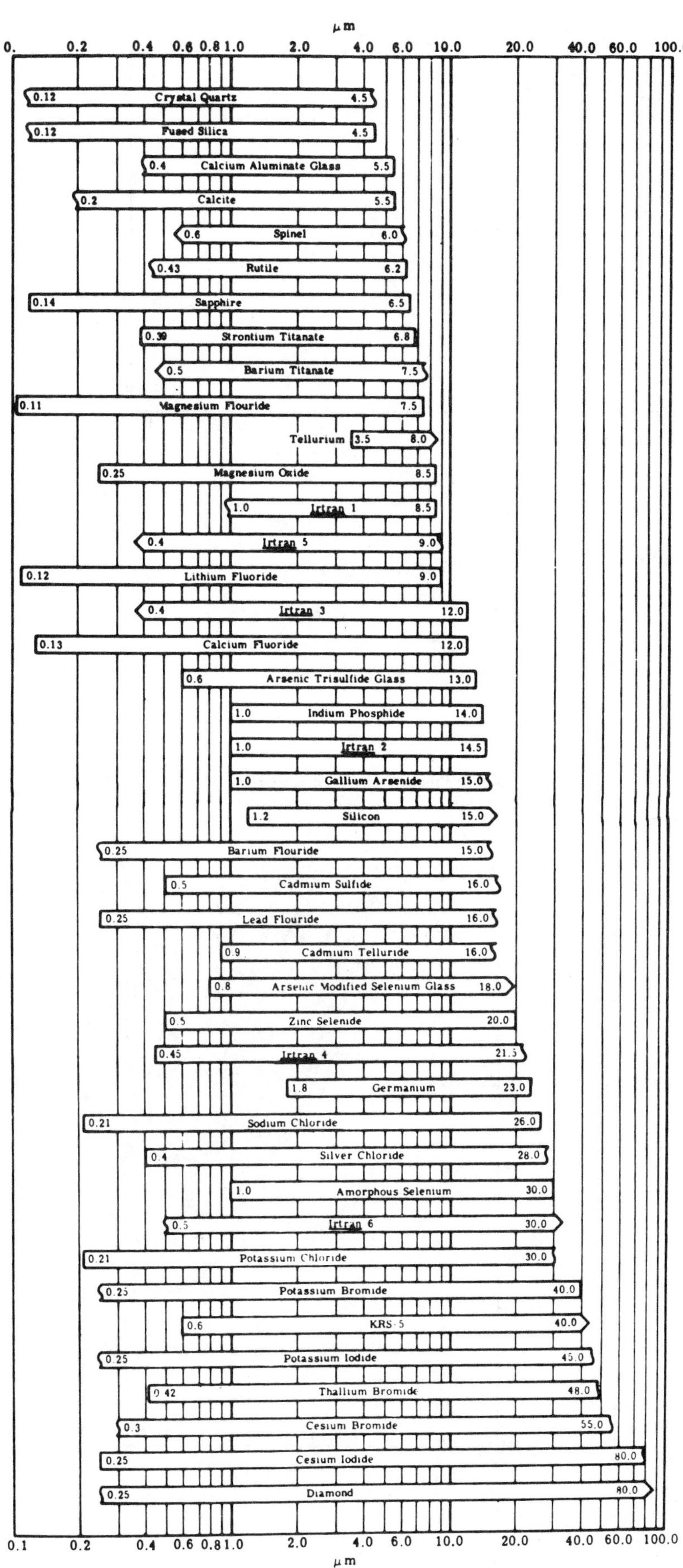

A. Figure: Transmission regions of various optical materials[8]

Limiting wavelengths, for both long and short cutoff, are shown as those wavelengths at which a sample 2 mm thick has 10% transmission.

B. Figure: Spectral reflectance of certain metals[9]

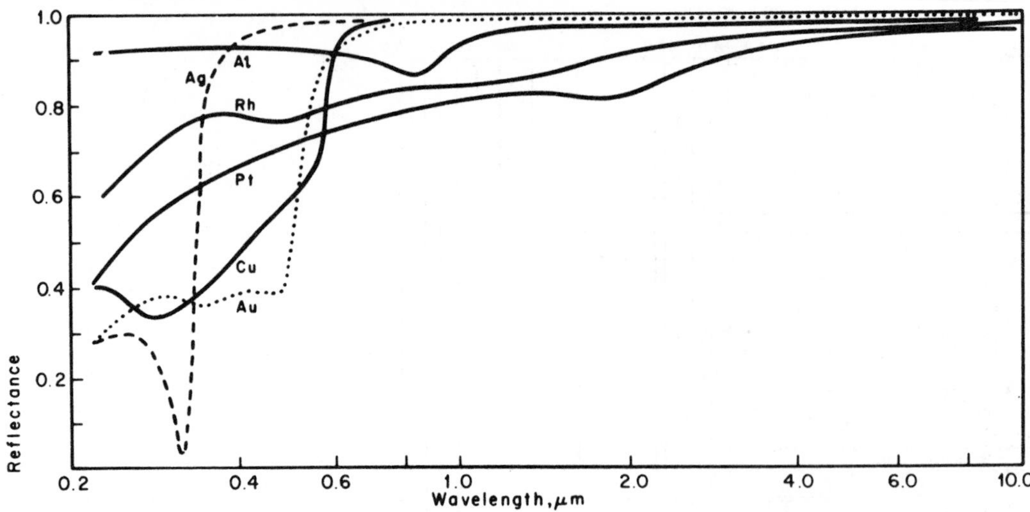

C. Figure: Refractive index vs wavelength for several optical materials[8]

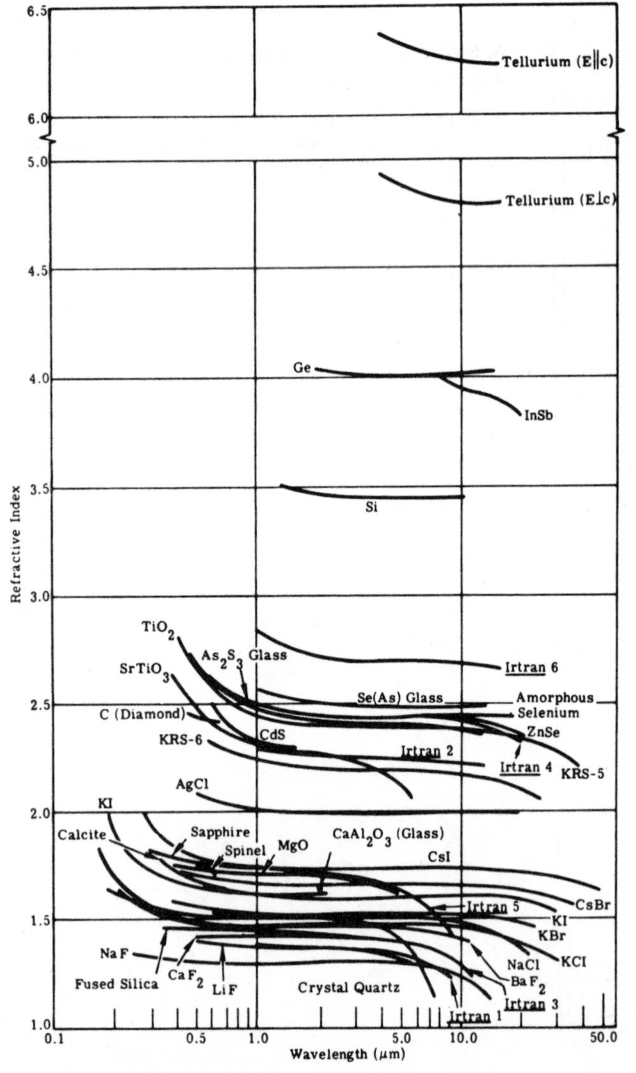

D. Figure: Dispersion vs wavelength for several optical materials[8]

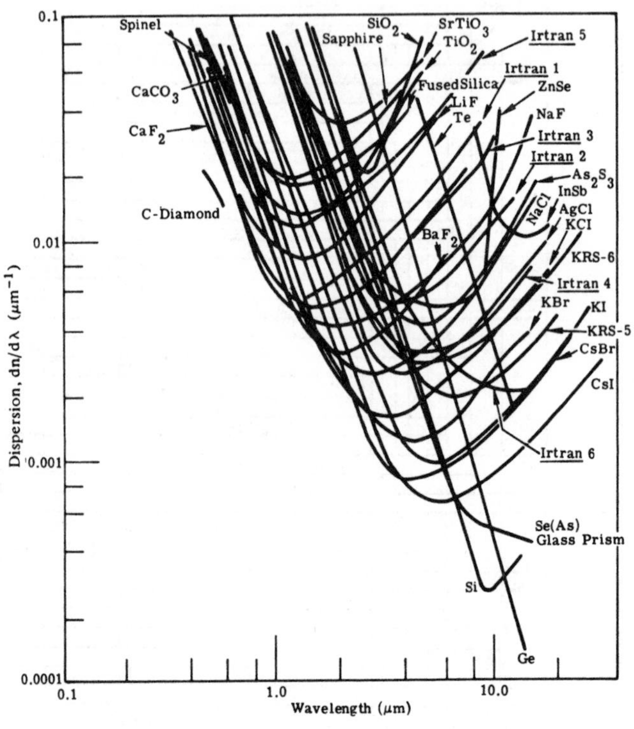

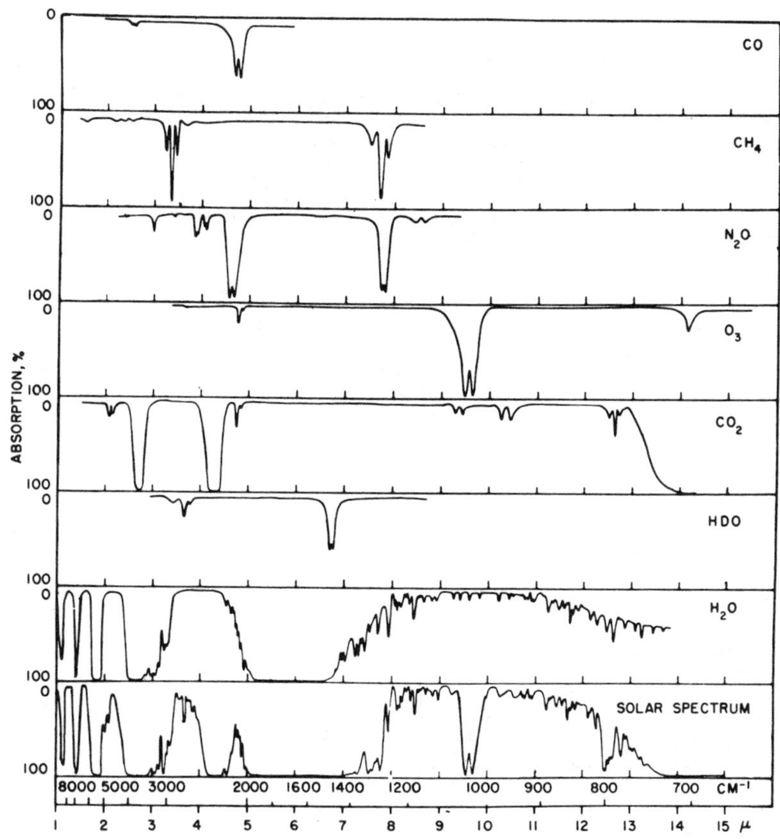

E. Figure: Low-resolution solar spectrum[10]

Low-resolution solar spectrum (bottom panel) from 1 to 15 μm. The other panels show spectra at the same resolution of various trace gases found in the atmosphere.

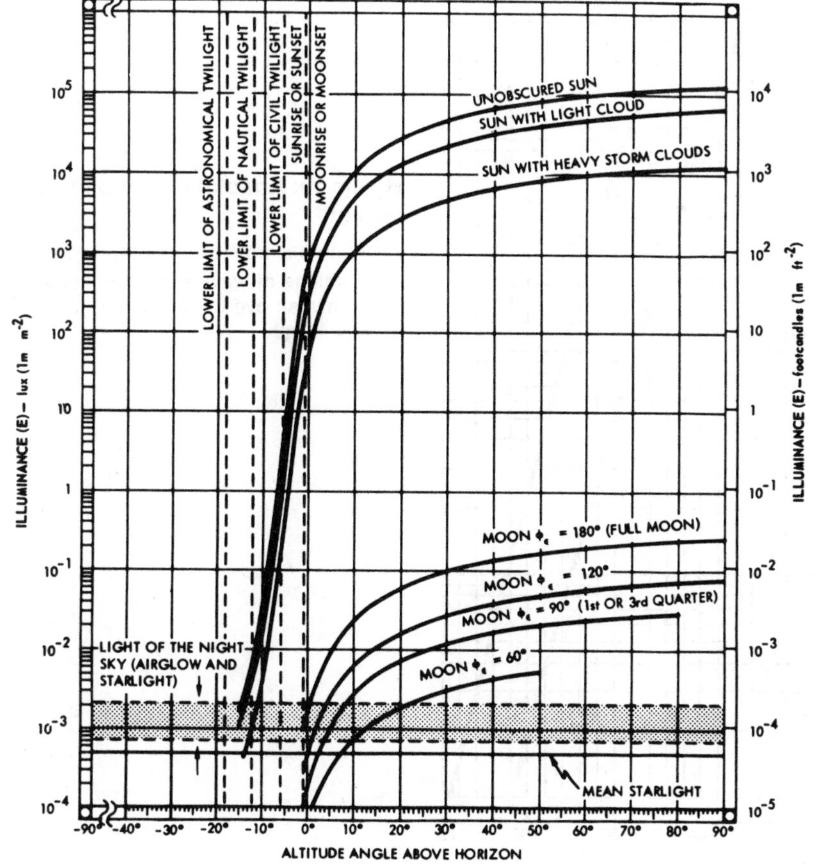

F. Figure: Illuminance levels on the surface of the Earth due to the Sun, Moon, and sky[11]

G. Figure: Range of natural illuminance levels[11]

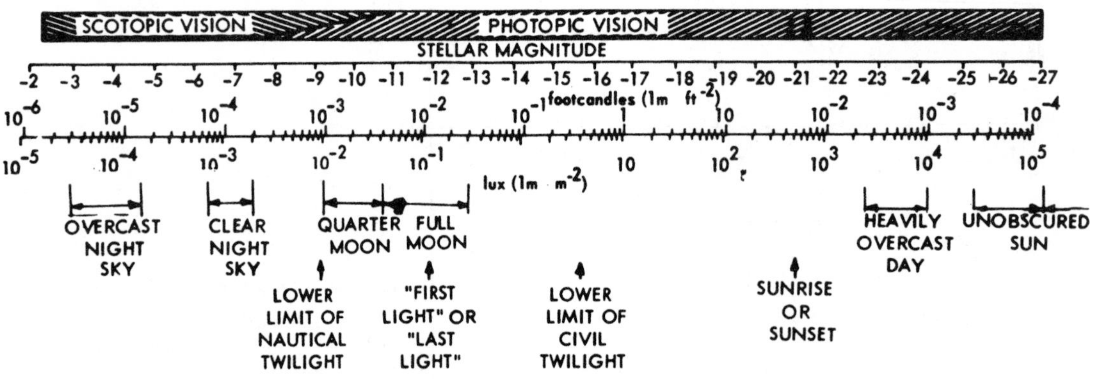

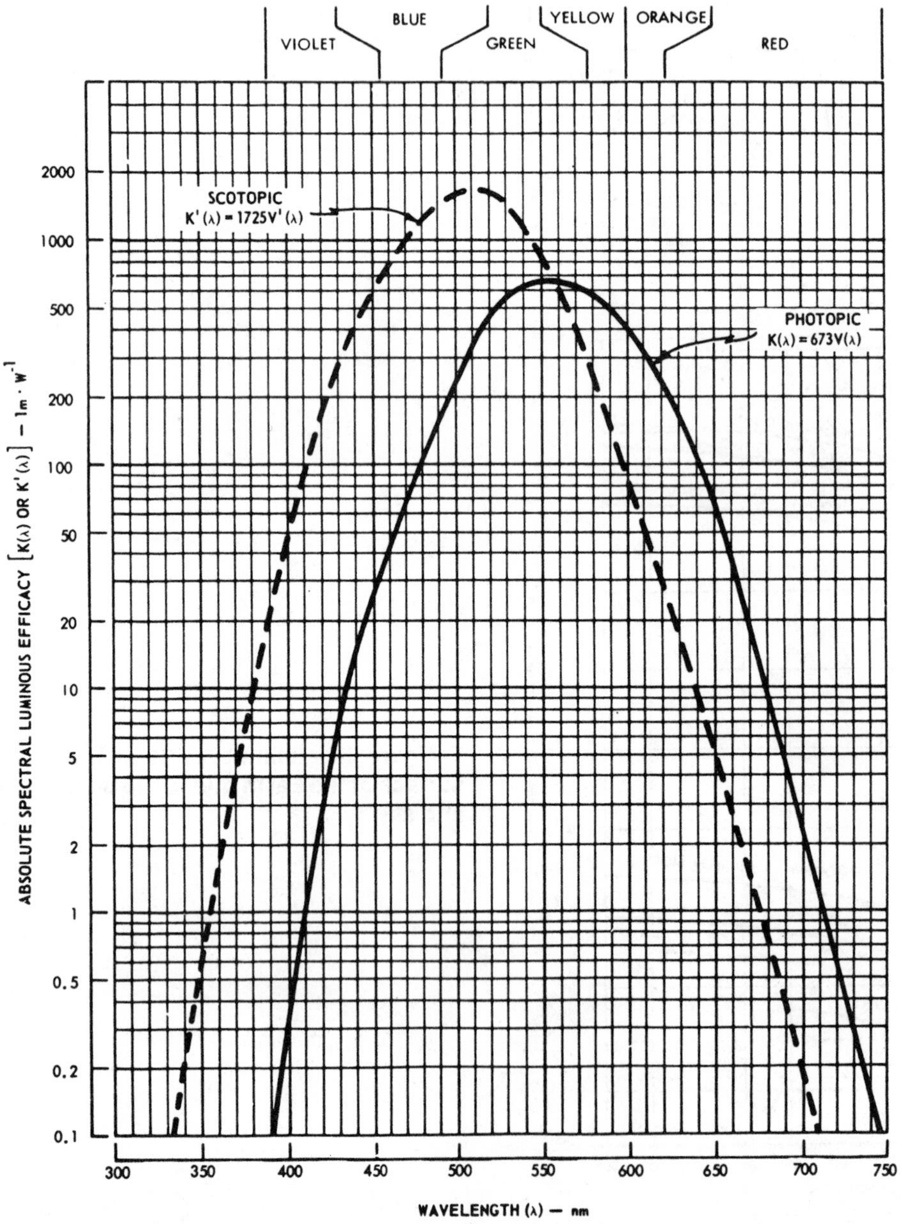

VIOLET BLUE GREEN YELLOW ORANGE RED

SCOTOPIC $K'(\lambda) = 1725V'(\lambda)$

PHOTOPIC $K(\lambda) = 673V(\lambda)$

ABSOLUTE SPECTRAL LUMINOUS EFFICACY $[K(\lambda)$ OR $K'(\lambda)]$ — lm · W^{-1}

2000 1000 500 200 100 50 20 10 5 2 1 0.5 0.2 0.1

300 350 400 450 500 550 600 650 700 750

WAVELENGTH (λ) — nm

H. Figure: Absolute luminosity curves[11]

Absolute luminosity curves K_λ and K'_λ as functions of wavelength (response of the human eye to radiation of a given wavelength).

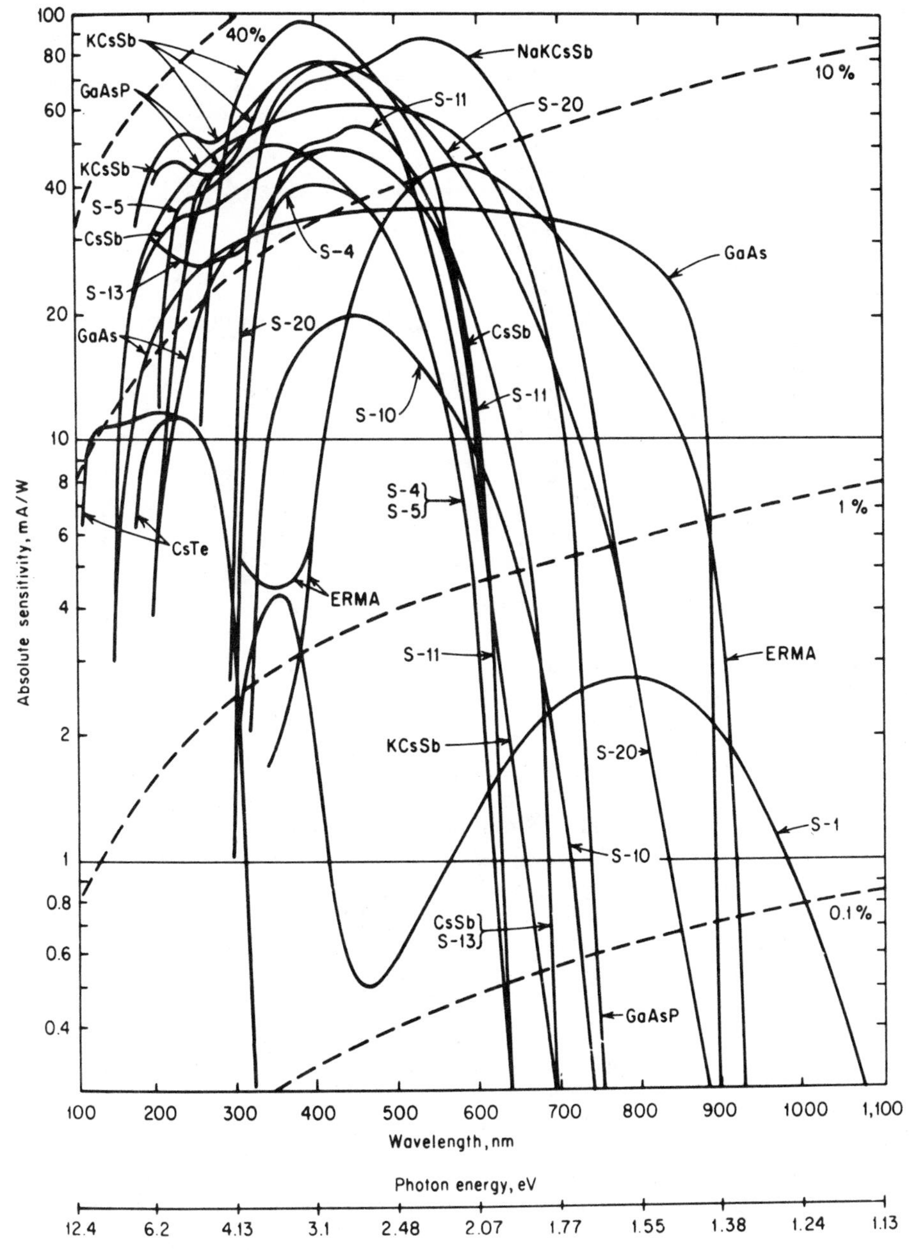

I. Figure: Spectral sensitivity of various photoemitters in the ultraviolet, visible, and near-infrared regions[9]

The dotted lines indicate quantum efficiency. S-1 to S-20 are designations of commercially available photodetectors, mostly cathodes of Cs combined with Ag, Sb, and Bi. (From RCA Electronic Component chart PIT-701B.)

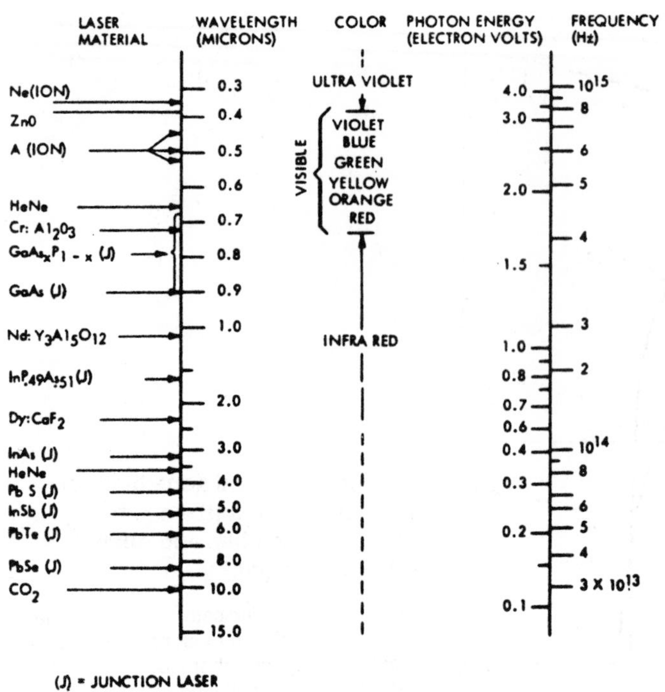

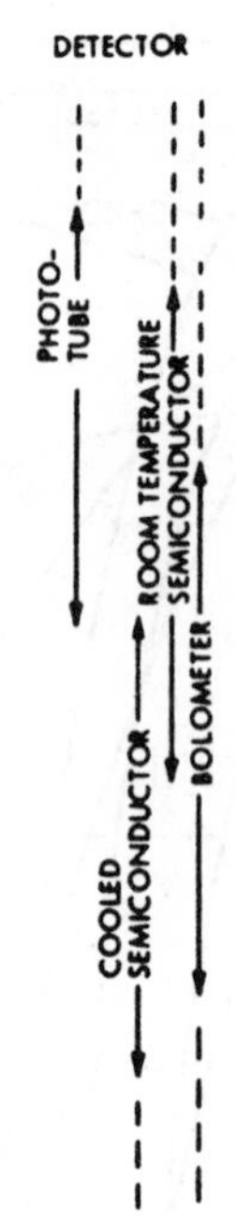

J. Figure: User's guide to lasers

K. Table: Common laser wavelengths

Wavelength (nm)	Type	Wavelength (μm)	Type
337.1	Pulsed N_2	1.06	Nd^+
488.0	Ar^+	2.10	YAG-Ho
514.5	Ar^+	3.39	He-Ne
540.1	Pulsed Ne	3.507	He-Xe
568.2	Kr^+	9.0	He-Xe
632.8	Ne	10.6	CO_2
647.0	Kr		
694.3	Pulsed ruby		
844.6	GaAs		

L. Table: Laser ions in glass

Ion	Host	Wavelength (μm)	Inversion for 1% gain/cm (cm^3)
Nd^{3+}	K-Ba-Si	1.06	0.7×10^{18}
	La-Ba-Th-B	1.37	
	Na-Ca-Si	0.92	3.5×10^{18}
Nd^{3+}	YAG	1.065	1.1×10^{16}
Yb^{3+}	Li-Mg-Al-Si	1.015	2.8×10^{18}
	K-Ba-Si	1.06	11.0×10^{19}
Ho^{3+}	Li-Mg-Al-Si	2.1	
Er^{3+}	Yb-Na-K-Ba-Si	1.543	1.8×10^{18}
	Li-Mg-Al-Si	1.55	
	Yb-Al-Zn-P_2O_3	1.536	9×10^{17}
	Yb-fluorophosphate	1.54	
Tm^{3+}	Li-Mg-Al-Si	1.85	
	Yb-Li-Mg-Al-Si	2.015	

M. Table: Semiconductor lasers

Material	Photon energy (eV)	Method of excitation
ZnS	3.82	Electron beam, optical
ZnO	3.30	
CdS	2.50	
GaSe	2.09	Electron beam
CdS_xSe_{1-x}	1.80–2.50	
CdSe	1.82	
CdTe	1.58	
$Ga(As_xP_{1-x})$	1.47–1.95	*p-n* junction, electron beam, optical, avalanche
GaAs	1.47	
InP	1.37	*p-n* junction, electron beam
$In_xGa_{1-x}As$	1.5	
GaSb	0.82	
InP_xAs_{1-x}	1.40	*p-n* junction, electron beam, optical
InAs	0.40	
InSb	0.23	*p-n* junction, electron beam, optical
Te	0.34	Electron beam
PbS	0.29	*p-n* junction, electron beam
PbTe	0.19	*p-n* junction, electron beam, optical
PbSe	0.145	*p-n* junction, electron beam
$Hg_xCd_{1-x}Te$	0.30–0.33	Optical
$Pb_xSn_{1-x}Te$	0.075–0.19	

17.06. DETECTORS

Optical radiation is detected when it impinges on a photosensitive element, which may produce a change in temperature, or a voltage or current change that can be measured.

Thermal detectors absorb incident radiation and produce a temperature change. Examples are thermocouples (a junction of two dissimilar metals which develops a thermoelectric emf) and bolometers (a blackened slab whose impedance is temperature dependent).

Quantum detectors respond to incident photon rates. The sensitivity of a detector is the degree to which small amounts of radiation can be sensed. Detectivity is also limited by noise, either thermal noise due to random motion of charge carriers, or shot noise resulting from random arrival of charge carriers. Many detectors are specified in terms of D^*, a figure of merit in which the detectivity is normalized to a standard area, electrical bandwidth, chopping frequency, and field of view.

17.07. COLOR

Color is that aspect of visual perception by which an observer can distinguish between two fields of view of the same size, shape, and structure, such as may be caused by differences in the spectral composition of the radiation concerned in the observation.

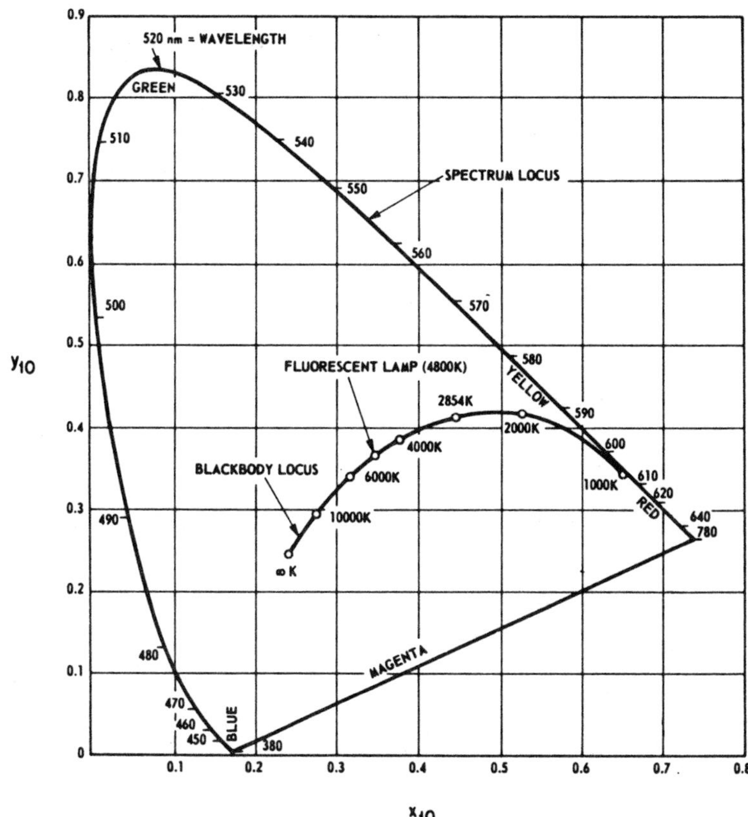

A. Figure: CIE 1964 (x_{10},y_{10}) chromaticity diagram[9]

Spectrum locus (chromaticity points of monochromatic stimuli of $\lambda = 380$ to 780 nm) and "purple" line connecting the ends of the spectrum locus are shown. Each point on the diagram specifies chromaticity (hue and saturation) independent of luminance. The standard CIE (International Commission on Illumination) primaries are represented by the points $x=0$, $y=1$; $x=0$, $y=0$; and $x=1$, $y=0$. The subscript 10 refers to the standard 1964 colorimetric curves.

17.08. REFERENCES

[1]*Manual of Remote Sensing* (American Society of Photogrammetry, Falls Church, VA, 1975).

[2]The symbols used in this table are those in *The Infrared Handbook,* edited by W. L. Wolfe and Geo. J. Zissis (Environmental Research Institute of Michigan, Ann Arbor, 1978). This book also contains a chapter on how to use hand-held electronic calculators to compute blackbody radiation functions.

[3]*Encyclopedia of Physics,* edited by R. G. Lerner and G. L. Trigg (Addison-Wesley, Reading, MA, 1981).

[4]For a detailed discussion of these sign conventions and application to lens design see J. R. Meyer-Arendt, *Introduction to Classical and Modern Optics* (Prentice-Hall, Englewood Cliffs, NJ, 1972). For application of pocket calculators to lens design see, for example, T. D. Settimi, *Optical Design and Ray-Tracing by Pocket Calculator* (Innovative Optical Material, Riverside, CA, 1977), or the columns by H. W. Straat in Opt. Spectra (1980 and 1981).

[5]Ditchburn, *Light.*

[6]F. Jenkins and H. White, Physical Optics.

[7]G. Vanasse, in Proceedings of the Aspen International Conference on Fourier Spectroscopy, 1979, Air Force Cambridge Research Labs, Bedford, MA.

[8]*American Institute of Physics Handbook,* 3rd ed., edited by Dwight E. Gray (McGraw-Hill, New York, 1972).

[9]*OSA Handbook of Optics* (McGraw-Hill, New York, 1978).

[10]*Infrared Handbook,* edited by W. L. Wolfe and G. L. Zissis (Environmental Research Institute, Ann Arbor, 1978).

[11]*Electro-Optics Handbook* (RCA Solid State Division, Lancaster, PA, 1974).

18.00. Plasma physics

DAVID L. BOOK

United States Naval Research Laboratory

CONTENTS

The material in this chapter is taken from David L. Book, *NRL Plasma Formulary* (Naval Research Laboratory, Washington, DC, 1987).

18.01. FUNDAMENTAL PLASMA PARAMETERS

All quantities are in Gaussian units except temperature (T_e, T_i, T) expressed in eV and ion mass (m_i) expressed in units of proton mass, $\mu = m_i/m_p$; Z is charge state; k is Boltzmann's constant; K is wavelength; γ is the adiabatic index; $\ln \Lambda$ is the Coulomb logarithm.

A. Frequencies

Electron gyrofrequency:

$$f_{ce} = \omega_{ce}/2\pi = 2.80\times 10^6 B \text{ Hz},$$

$$\omega_{ce} = eB/m_e c = 1.76\times 10^7 B \text{ rad/s}.$$

Ion gyrofrequency:

$$f_{ci} = \omega_{ci}/2\pi = 1.52\times 10^3 Z\mu^{-1} B \text{ Hz},$$

$$\omega_{ci} = eB/m_i c = 9.58\times 10^3 Z\mu^{-1} B \text{ rad/s}.$$

Electron plasma frequency:

$$f_{pe} = \omega_{pe}/2\pi = 8.98\times 10^3 n_e^{1/2} \text{ Hz},$$

$$\omega_{pe} = (4\pi n_e e^2/m_e)^{1/2}$$

$$= 5.64\times 10^4 n_e^{1/2} \text{ rad/s}.$$

Ion plasma frequency:

$$f_{pi} = \omega_{pi}/2\pi = 2.10\times 10^2 Z\mu^{-1/2} n_i^{1/2} \text{ Hz},$$

$$\omega_{pi} = (4\pi n_i Z^2 e^2/m_i)^{1/2}$$

$$= 1.32\times 10^3 Z\mu^{-1/2} n_i^{1/2} \text{ rad/s}.$$

Electron trapping rate:

$$\nu_{Te} = (eKE/m_e)^{1/2} = 7.26\times 10^8 K^{1/2} E^{1/2} \text{ s}^{-1}.$$

Ion trapping rate:

$$\nu_{Ti} = (eKE/m_i)^{1/2} = 1.69\times 10^7 K^{1/2} E^{1/2} \mu^{-1/2} \text{ s}^{-1}.$$

Electron collision rate:

$$\nu_e = 2.91\times 10^{-6} n_e \ln\Lambda \, T_e^{-3/2} \text{ s}^{-1}.$$

Ion collision rate:

$$\nu_i = 4.78\times 10^{-8} n_i Z^2 \ln\Lambda \, T_i^{-3/2} \text{ s}^{-1}.$$

B. Lengths

Electron de Broglie length:

$$\lambda = \hbar (m_e k T_e)^{1/2} = 2.76\times 10^{-8} T_e^{-1/2} \text{ cm}.$$

Classical distance of minimum approach:

$$e^2/kT = 1.44\times 10^{-7} T^{-1} \text{ cm}.$$

Electron gyroradius:

$$r_e = v_{Te}/\omega_{ce} = 2.38 T_e^{1/2} B^{-1} \text{ cm}.$$

Ion gyroradius:

$$r_i = v_{Ti}/\omega_{ci} = 1.02\times 10^2 \mu^{1/2} Z^{-1} T_i^{1/2} B^{-1} \text{ cm}.$$

Plasma skin depth:

$$c/\omega_{pe} = 5.31\times 10^5 n^{-1/2} \text{ cm}.$$

Debye length:

$$\lambda_D = (kT/4\pi n e^2)^{1/2} = 7.43\times 10^2 T^{1/2} n^{-1/2} \text{ cm}.$$

C. Velocities

Electron thermal velocity:

$$v_{Te} = (kT_e/m_e)^{1/2} = 4.19\times 10^7 T_e^{1/2} \text{ cm/s}.$$

Ion thermal velocity:

$$v_{Ti} = (kT_i/m_i)^{1/2} = 9.79\times 10^5 \mu^{-1/2} T_i^{1/2} \text{ cm/s}.$$

Ion sound velocity:

$$c_s = (\gamma Z k T_e/m_i)^{1/2} = 9.79\times 10^5 (\gamma Z T_e/\mu)^{1/2} \text{ cm/s}.$$

Alfvén velocity:

$$v_A = B/(4\pi n_i m_i)^{1/2}$$

$$= 2.18\times 10^{11} \mu^{-1/2} n_i^{-1/2} B \text{ cm/s}.$$

D. Dimensionless

(Electron/proton mass ratio)$^{1/2}$:

$$(m_e/m_p)^{1/2} = 2.33\times 10^{-2} = 1/42.9.$$

Number of particles in Debye sphere:

$$\tfrac{4}{3}\pi n \lambda_D^3 = 1.72\times 10^9 T^{3/2} n^{-1/2}.$$

Alfvén velocity/speed of light:

$$v_A/c = 7.28\mu^{-1/2} n_i^{-1/2} B.$$

Magnetic/ion rest energy ratio:

$$B^2/8\pi n_i m_i c^2 = 26.5\mu^{-1} n_i^{-1/2} B^2.$$

Electron plasma/gyrofrequency ratio:

$$\omega_{pe}/\omega_{ce} = 3.21\times 10^{-3} n_e^{1/2} B^{-1}.$$

Ion plasma/gyrofrequency ratio:

$$\omega_{pi}/\omega_{ci} = 0.137\mu^{1/2} n_i^{1/2} B^{-1}.$$

Thermal/magnetic energy ratio:

$$\beta = 8\pi n k T/B^2 = 4.03\times 10^{-11} n T B^{-2}.$$

E. Miscellaneous

Bohm diffusion coefficient:

$$D_B = \frac{ckT}{16eB} = 6.25\times 10^6 T B^{-1} \text{ cm}^2\text{/s}.$$

Transverse Spitzer resistivity:

$$\eta_\perp = 1.15\times 10^{-14} Z \ln\Lambda \, T^{-3/2} \text{ s}$$

$$= 1.03\times 10^{-2} Z \ln\Lambda \, T^{-3/2} \text{ ohm cm}.$$

Anomalous collision rate due to low-frequency ion sound turbulence:

$$\nu^* \approx \omega_{pe} W/kT = 5.64\times 10^4 n^{-1/2} W/kT \text{ s}^{-1},$$

where W is the total energy of waves with $\omega/K < v_{Ti}$.

Magnetic pressure:

$$P = B^2/8\pi = 3.98\times 10^6 B^2 \text{ dyn/cm}^2$$

$$= 3.93(B/B_0)^2 \text{ atm},$$

where $B_0 = 10$ kG $= 1$ T.

Energy of detonation of 1 kiloton of high explosive:

$$W_{kT} = 10^{12} \text{ cal} = 4.2\times 10^{19} \text{ ergs}.$$

18.02. PLASMA DISPERSION FUNCTION

Definition (first form valid only for $\mathrm{Im}\zeta > 0$):

$$Z(\zeta) = \pi^{-1/2}\int_{-\infty}^{\infty}\frac{dt\, e^{-t^2}}{t-\zeta} = 2ie^{-\zeta^2}\int_{-\infty}^{i\zeta} dt\, e^{-t^2}.$$

Physically $\zeta = x + iy$ is the ratio of phase to thermal velocity.[1]

Differential equation:

$$\frac{dZ}{d\zeta} = -2(1+\zeta Z), \quad Z(0) = i\pi^{1/2}$$

$$\frac{d^2Z}{d\zeta^2} + 2\zeta\frac{dZ}{d\zeta} + 2Z = 0.$$

Real argument ($y = 0$):

$$Z(x) = e^{-x^2}\left(i\pi^{1/2} - 2\int_0^x dt\, e^{t^2}\right).$$

Imaginary argument ($x = 0$):

$$Z(iy) = i\pi^{1/2}\exp(y^2)\,[1 - \mathrm{erf}(y)].$$

Power series (small argument):

$$Z(\zeta) = i\pi^{1/2}\exp(-\zeta^2)$$
$$-2\zeta(1 - 2\zeta^2/3 + 4\zeta^4/15 - 8\zeta^6/105 + \cdots).$$

Asymptotic series (large argument, $x > 0$)[2]:

$$Z(\zeta) = i\pi^{1/2}\sigma\exp(-\zeta^2)$$
$$-\zeta^{-1}(1 + 1/2\zeta^2 + 3/4\zeta^4 + 15/8\zeta^6 + \cdots),$$

$$\sigma = \begin{cases} 0 & y > |x| \\ 1 & |y| < |x| \\ 2 & -y > 1\ |x|. \end{cases}$$

Symmetry properties ($\zeta^* = x - iy$):

$$Z(\zeta^*) = -[Z(-\zeta)]^*,$$

$$Z(x-iy)$$
$$= [Z(x+iy)]^* + 2i\pi^{1/2}\exp[-(x-iy)^2] \quad (y>0).$$

Two-pole approximations for ζ in upper half plane (good except when $y < \pi^{1/2}x^2 e^{-x^2}, x \gg 1$)[3]:

$$Z(\zeta) \approx \frac{0.50 + 0.81i}{a+\zeta} - \frac{0.50 - 0.81i}{a^* + \zeta}, \quad a = 0.51 - 0.81i$$

$$Z'(\zeta) \approx \frac{0.50 + 0.96i}{(b+\zeta)^2} + \frac{0.50 - 0.96i}{(b^*+\zeta)^2}, \quad b = 0.48 - 0.91i.$$

18.03. COLLISIONS AND TRANSPORT

Temperatures are in eV; the corresponding value of Boltzmann's constant is $k = 1.60\times10^{-12}$ ergs/eV; masses μ, μ' are in units of the proton mass; $e_\alpha = Z_\alpha e$ is the charge of species α. All other units are cgs except where noted.

A. Relaxation rates

Rates are associated with four relaxation processes arising from the interaction of test particles (labeled α) streaming through a background of field particles (labeled β):

slowing down: $\quad \dfrac{d\mathbf{v}_\alpha}{dt} = -\nu_s^{\alpha/\beta}\mathbf{v}_\alpha;$

transverse diffusion: $\quad \dfrac{d}{dt}(\mathbf{v}_\alpha - \bar{\mathbf{v}}_\alpha)_\perp^2 = \nu_\perp^{\alpha/\beta}v_\alpha^2;$

parallel diffusion: $\quad \dfrac{d}{dt}(\mathbf{v}_\alpha - \bar{\mathbf{v}}_\alpha)_\parallel^2 = \nu_\parallel^{\alpha/\beta}v_\alpha^2;$

energy loss: $\quad \dfrac{d}{dt}v_\alpha^2 = -\nu_\epsilon^{\alpha/\beta}v_\alpha^2,$

where the averages are performed over an ensemble of test particles and a Maxwellian field particle distribution. The exact formulas may be written[4]

$$\nu_s^{\alpha/\beta} = (1 + m_\alpha/m_\beta)\psi(x^{\alpha/\beta})\nu_0^{\alpha/\beta}\ \mathrm{s}^{-1},$$
$$\nu_\perp^{\alpha/\beta} = 2[\psi(x^{\alpha/\beta})(1 - 1/2x^{\alpha/\beta}) + \psi'(x^{\alpha/\beta})]\nu_0^{\alpha/\beta}\ \mathrm{s}^{-1},$$
$$\nu_\parallel^{\alpha/\beta} = [\psi(x^{\alpha/\beta})/x^{\alpha/\beta}]\nu_0^{\alpha/\beta}\ \mathrm{s}^{-1},$$
$$\nu_\epsilon^{\alpha/\beta} = 2[(m_a/m_\beta)\psi(x^{\alpha/\beta}) - \psi'(x^{\alpha/\beta})]\nu_0^{\alpha/\beta}\ \mathrm{s}^{-1},$$

where

$$\nu_0^{\alpha/\beta} = 4\pi e_\alpha^2 e_\beta^2\lambda_{\alpha\beta}n_\beta/m_\alpha^2 v_\alpha^3\ \mathrm{s}^{-1}, \quad x^{\alpha/\beta} = m_\beta v_\alpha^2/2kT_\beta,$$

$$\psi(x) = \frac{2}{\pi^{1/2}}\int_0^x dt\, t^{1/2}e^{-t}, \quad \psi'(x) = \frac{d\psi}{dx},$$

and $\lambda_{\alpha\beta} = \ln\Lambda_{\alpha\beta}$ is the Coulomb logarithm (see below). Limiting forms of ν_s, $\nu_\perp$, and $\nu_\parallel$ are given in the following table. All the expressions shown have units cm^3/s. Test particle energy ϵ and field particle temperature T are both in eV; $\mu = m_i/m_p$, where m_p is the proton mass; Z is ion charge state; for electron-electron and ion-ion encounters, field particle quantities are distinguished by a prime. The two expressions given for each rate hold for very slow ($x^{\alpha/\beta} \ll 1$) and very fast ($x^{\alpha/\beta} \gg 1$) test particles, respectively.

	Slow	Fast
Electron-electron		
$\nu_s^{e/e'}/n_{e'}\lambda_{ee'}$	$\approx 5.8\times 10^{-6}T^{-3/2}$	$\rightarrow 7.7\times 10^{-6}\epsilon^{-3/2}$
$\nu_\perp^{e/e'}/n_{e'}\lambda_{ee'}$	$\approx 5.8\times 10^{-6}T^{-1/2}\epsilon^{-1}$	$\rightarrow 7.7\times 10^{-6}\epsilon^{-3/2}$
$\nu_\parallel^{e/e'}/n_{e'}\lambda_{ee'}$	$\approx 2.9\times 10^{-6}T^{-1/2}\epsilon^{-1}$	$\rightarrow 3.9\times 10^{-6}T\epsilon^{-5/2}$
Electron-ion		
$\nu_s^{e/i}/n_iZ^2\lambda_{ei}$	$\approx 0.23\mu^{3/2}T^{-3/2}$	$\rightarrow 3.9\times 10^{-6}\epsilon^{-3/2}$
$\nu_\perp^{e/i}/n_iZ^2\lambda_{ei}$	$\approx 2.5\times 10^{-4}\mu^{1/2}T^{-1/2}\epsilon^{-1}$	$\rightarrow 7.7\times 10^{-6}\epsilon^{-3/2}$
$\nu_\parallel^{e/i}/n_iZ^2\lambda_{ei}$	$\approx 1.2\times 10^{-4}\mu^{1/2}T^{-1/2}\epsilon^{-1}$	$\rightarrow 2.1\times 10^{-9}\mu^{-1}T\epsilon^{-5/2}$
Ion-electron		
$\nu_s^{i/e}/n_eZ^2\lambda_{ie}$	$\approx 1.6\times 10^{-9}\mu^{-1}T^{-3/2}$	$\rightarrow 1.7\times 10^{-4}\mu^{1/2}\epsilon^{-3/2}$
$\nu_\perp^{i/e}/n_eZ^2\lambda_{ie}$	$\approx 3.2\times 10^{-9}\mu^{-1}T^{-1/2}\epsilon^{-1}$	$\rightarrow 1.8\times 10^{-7}\mu^{1/2}\epsilon^{-3/2}$
$\nu_\parallel^{i/e}/n_eZ^2\lambda_{ie}$	$\approx 1.6\times 10^{-9}\mu^{-1}T^{-1/2}\epsilon^{-1}$	$\rightarrow 1.7\times 10^{-4}\mu^{1/2}T\epsilon^{-5/2}$
Ion-ion		
$\nu_s^{i/i'}/n_iZ^2Z'^2\lambda_{ii'}$	$\approx 6.8\times 10^{-8}(\mu'^{1/2}/\mu)(1+\mu'/\mu)T^{-3/2}$	$\rightarrow 9.0\times 10^{-8}\mu^{-1/2}(1+\mu/\mu')\epsilon^{-3/2}$
$\nu_\perp^{i/i'}/n_iZ^2Z'^2\lambda_{ii'}$	$\approx 1.4\times 10^{-7}\mu'^{1/2}\mu^{-1}T^{-1/2}\epsilon^{-1}$	$\rightarrow 1.8\times 10^{-7}\mu^{-1/2}\epsilon^{-3/2}$
$\nu_\parallel^{i/i'}/n_iZ^2Z'^2\lambda_{ii'}$	$\approx 6.8\times 10^{-8}\mu'^{1/2}\mu^{-1}T^{-1/2}\epsilon^{-1}$	$\rightarrow 9.0\times 10^{-8}\mu^{1/2}\mu'^{-1}T\epsilon^{-5/2}$

In the same limits, the energy transfer rate follows from the identity

$$\nu_\epsilon = 2\nu_s - \nu_\perp - \nu_\parallel,$$

except for the case of fast electrons or fast ions scattered by ions, where the leading terms cancel. Here the appropriate forms are

$$\nu_\epsilon^{e/i} \rightarrow 4.2\times 10^{-9} n_i Z^2 \lambda_{ei} [\epsilon^{-3/2}\mu^{-1} - 8.9\times 10^4 (\mu/T)^{1/2}\epsilon^{-1}\exp(-1836\mu\epsilon/T)] \text{ s}^{-1}$$

and

$$\nu_\epsilon^{i/i'} \rightarrow 1.8\times 10^{-7} n_i Z^2 Z'^2 \lambda_{ii'} [\epsilon^{-3/2}\mu^{1/2}/\mu' - 1.1(\mu'/T)^{1/2}\epsilon^{-1}\exp(-\mu'\epsilon/T)] \text{ s}^{-1}.$$

In general, the energy transfer rate $\nu_\epsilon^{\alpha/\beta}$ is positive for $\epsilon > \epsilon_\alpha^*$ and negative for $\epsilon < \epsilon_\alpha^*$, where $x^* = (m_\beta/m_\alpha)\epsilon_\alpha^*/T_\beta$ is the solution of $m_\beta/m_\alpha = \psi(x^*)/\psi'(x^*)$.

The ratio $\epsilon_\alpha^*/T_\beta$ is given for a number of specific α, β in the following table:

$\alpha/\beta =$	i/e	e/e	i/i	e/p	e/D	e/T, e/He^3	e/He^4
$\epsilon_\alpha^*/kT_\beta =$	1.5	0.98	0.98	4.8×10^{-3}	2.6×10^{-3}	1.8×10^{-3}	1.4×10^{-3}

When both species are near Maxwellian with $T_i \lesssim T_e$, there are just two characteristic collision rates. For $Z = 1$,

$$\nu_e = 2.9\times 10^{-6} n\lambda T_e^{-3/2} \text{ s}^{-1},$$

$$\nu_i = 4.8\times 10^{-8} n\lambda T_i^{-3/2}\mu^{-1/2} \text{ s}^{-1}.$$

B. Thermal equilibration

If the components of a plasma have different temperatures, but no relative drift, equilibration is described by

$$dT_\alpha/dt = \sum_\beta \bar{\nu}_\epsilon^{\alpha/\beta}(T_\beta - T_\alpha),$$

where

$$\bar{\nu}_\epsilon^{\alpha/\beta} = 1.8\times 10^{-19} \frac{(m_\alpha m_\beta)^{1/2} Z_\alpha^2 Z_\beta^2 n_\beta \lambda_{\alpha\beta}}{(m_\alpha T_\beta + m_\beta T_\alpha)^{3/2}} \text{ s}^{-1}.$$

For electrons and ions with $T_e \sim T_i = T$, this implies

$$\bar{\nu}_\epsilon^{e/i}/n_i = \bar{\nu}_\epsilon^{i/e}/n_e = 3.2\times 10^{-9} Z^2\lambda/\mu T^{3/2} \text{ cm}^3/\text{s}.$$

C. Temperature anisotropy

Isotropization is described by

$$dT_\perp/dt = -\tfrac{1}{2} dT_\parallel/dt = -\nu_T^\alpha (T_\perp - T_\parallel),$$

where, if $A = T_\perp/T_\parallel - 1 > 0$,

$$\nu_T^\alpha = \frac{2\pi^{1/2} e_\alpha^2 e_\beta^2 n_\alpha \lambda}{m_\alpha^{1/2}(kT_\parallel)^{3/2}} A^{-2} \times \left(-3 + (A+3)\frac{\tan^{-1}A^{1/2}}{A^{1/2}}\right) \text{ s}^{-1}.$$

If $A < 0$, $(\tan^{-1}A^{1/2})/A^{1/2}$ is replaced by

$$[\tanh^{-1}(-A)^{1/2}]/(-A)^{1/2}.$$

For $T_\perp \approx T_\parallel = T$,

$$\nu_T^e = 8.2\times 10^{-7} n\lambda T^{-3/2} \text{ s}^{-1},$$

$$\nu_T^i = 1.9\times 10^{-8} n\lambda Z^2/\mu T^{3/2} \text{ s}^{-1}.$$

D. Coulomb logarithm

For test particles of mass m_α, charge $e_\alpha = Z_\alpha e$, scattering off field particles of mass m_β, charge $e_\beta = Z_\beta e$, the Coulomb logarithm is defined as $\lambda = \ln\Lambda = \ln(r_{max}/r_{min})$. Here r_{min} is the larger of $e_\alpha e_\beta / m_{\alpha\beta}\bar{u}^2$ and $\hbar/2m_{\alpha\beta}\bar{u}$, averaged over both particle velocity distributions, where $m_{\alpha\beta} = m_\alpha m_\beta/(m_\alpha + m_\beta)$ and $\mathbf{u} = \mathbf{v}_\alpha - \mathbf{v}_\beta$; $r_{max} = (4\pi\Sigma n_\gamma e_\gamma^2/kT_\gamma)^{-1/2}$, where the summation extends over all species γ for which $\bar{u}^2 < v_{T\gamma}^2$, with $V_{T\gamma} = (kT_\gamma)/m_\gamma)^{1/2}$. If this inequality cannot be satisfied or if either $\bar{u}\,\omega_{c\alpha}^{-1} < r_{max}$ or $\bar{u}\,\omega_{c\beta}^{-1} < r_{max}$, the theory breaks down. Typically $\lambda \approx 10$–20. Corrections to the transport coefficients are $O(\lambda^{-1})$; hence the theory is good only to $\sim 10\%$ and fails when $\lambda \sim 1$.

The following cases are of particular interest.

(a) Thermal electron-electron collisions:

$$\lambda_{ee} = 23 - \ln(n_e^{1/2} T_e^{-3/2}), \quad T_e \leqslant 10 \text{ eV}$$
$$= 24 - \ln(n_e^{1/2} T_e^{-1}), \quad T_e \gtrsim 10 \text{ eV}.$$

(b) Electron-ion collisions:

$$\lambda_{ei} = \lambda_{ie}$$
$$= 23 - \ln(n_e^{1/2} Z T_e^{-3/2}), \quad 10Z^2 \text{ eV} > T_e > T_i m_e/m_i$$
$$= 24 - \ln(n_e^{1/2} T_e^{-1}), \quad T_e > 10Z^2 \text{ eV} > T_i m_e/m_i$$
$$= 30 - \ln(n_i^{1/2} T_i^{-3/2} Z^2 \mu^{-1}), \quad T_i > T_e m_i/m_e Z.$$

(c) Mixed ion-ion collisions:

$$\lambda_{ii'} = \lambda_{i'i} = 23 - \ln\left[\frac{ZZ'(\mu+\mu')}{\mu T_i' + \mu' T_i}\left(\frac{n_i Z^2}{T_i} + \frac{n_i' Z'^2}{T_i'}\right)^{1/2}\right].$$

(d) Counterstreaming ions (relative velocity $v_D = \beta_D c$) in the presence of warm electrons, $kT_e/m_e > v_D^2 > kT_i/m_i$, $kT_{i'}/m_{i'}$:

$$\lambda_{ii'} = \lambda_{i'i} = 35 - \ln\left[\frac{ZZ'(\mu+\mu')}{\mu\mu'\beta_D^2}\left(\frac{n_e}{T_e}\right)^{1/2}\right].$$

E. Fokker-Planck equation

$$\frac{Df^\alpha}{Dt} = \frac{\partial f^\alpha}{\partial t} + \mathbf{v}\cdot\nabla f^\alpha + \mathbf{F}\cdot\nabla_\mathbf{v} f^\alpha = \left(\frac{\partial f^\alpha}{\partial t}\right)_{coll},$$

where F is an external force field. The general form of the collision integral is $(\partial f^\alpha/\partial t)_{coll} = -\Sigma_\beta \nabla_\mathbf{v}\cdot\mathbf{J}^{\alpha/\beta}$, with

$$\mathbf{J}^{\alpha/\beta} = 2\pi\lambda_{\alpha\beta}\frac{e_\alpha^2 e_\beta^2}{m_\alpha}\int d^3v'(u^2\mathbf{I} - \mathbf{uu})u^{-3}$$
$$\times\left(\frac{1}{m_\beta} f^\alpha(\mathbf{v})\nabla_{\mathbf{v}'} f^\beta(\mathbf{v}') - \frac{1}{m_\alpha} f^\beta(\mathbf{v}')\nabla_\mathbf{v} f^\alpha(\mathbf{v})\right)$$

(Landau form), where $\mathbf{u} = \mathbf{v}' - \mathbf{v}$ and $\mathbf{I}$ is the unit dyad, or alternatively

$$\mathbf{J}^{\alpha/\beta} = 4\pi\lambda_{\alpha\beta}\frac{e_\alpha^2 e_\beta^2}{m_\alpha^2}$$
$$\times\{f^\alpha(\mathbf{v})\nabla_\mathbf{v} H(\mathbf{v}) - \tfrac{1}{2}\nabla_\mathbf{v}\cdot[f^\alpha(\mathbf{v})\nabla_\mathbf{v}\nabla_\mathbf{v} G(\mathbf{v})]\},$$

where the Rosenbluth potentials are

$$G(\mathbf{v}) = \int f^\beta(\mathbf{v}')u\, d^3v',$$
$$H(\mathbf{v}) = \left(1 + \frac{m_\alpha}{m_\beta}\right)\int f^\beta(\mathbf{V}')u^{-1}\, d^3v'.$$

If species α is a weak beam (number and energy density small compared with background) streaming through a Maxwellian plasma, then

$$\mathbf{J}^{\alpha/\beta} = -\nu_s^{\alpha/\beta}\mathbf{v} f^\alpha - \tfrac{1}{2}\nu_\perp^{\alpha/\beta} v^2 \nabla_\mathbf{v} f^\alpha$$
$$+ \tfrac{1}{2}(\nu_\perp^{\alpha/\beta} - \nu_\parallel^{\alpha/\beta})\mathbf{vv}\cdot\nabla_\mathbf{v} f^\alpha.$$

F. BGK collision operator

For distribution functions with no large gradients in velocity space, the Fokker-Planck collision terms can be approximated according to

$$\frac{Df_e}{Dt} = \nu_{ee}(F_e - f_e) + \nu_{ei}(\bar{F}_e - f_e),$$
$$\frac{Df_i}{Dt} = \nu_{ie}(\bar{F}_i - f_i) + \nu_{ii}(F_i - f_i).$$

The respective slowing-down rates $\nu_s^{\alpha/\beta}$ given in Sec. 18.03.A can be used for $\nu_{\alpha\beta}$, assuming slow ions and fast electrons, with ϵ replaced by T_α. (For ν_{ee} and ν_{ii}, $\nu_\perp$ can equally well be used, and the result is insensitive to whether the slow- or fast-test-particle limit is employed.) The Maxwellians F_j and $\bar{F}_j$ are given by

$$F_j = n_j\left(\frac{m_j}{2\pi kT_j}\right)^{3/2}\exp\left(-\frac{m_j(\mathbf{v}-\mathbf{u}_j)^2}{2kT_j}\right),$$
$$\bar{F}_j = n_j\left(\frac{m_j}{2\pi k\bar{T}_j}\right)^{3/2}\exp\left(-\frac{m_j(\mathbf{v}-\bar{\mathbf{u}}_j)^2}{2k\bar{T}_j}\right),$$

where n_j, $\mathbf{u}_j$, and T_j are the number density, mean drift velocity, and effective temperature obtained by taking moments of f_j. Some latitude in the definition of $\bar{T}_j$ and $\bar{u}_j$ is possible[5]; one choice is $\bar{T}_e = T_i$, $\bar{T}_i = T_e$, $\bar{\mathbf{u}}_e = \mathbf{u}_i$, $\bar{\mathbf{u}}_i = \mathbf{u}_e$.

G. Transport coefficients

Transport equations for a multispecies plasma:

$$\frac{D^\alpha n_\alpha}{Dt} + n_\alpha\nabla\cdot\mathbf{v}_\alpha = 0,$$

$$m_\alpha n_\alpha\frac{D^\alpha\mathbf{v}_\alpha}{Dt} + \nabla p_\alpha + \nabla\cdot\mathbf{P}_\alpha$$
$$= Z_\alpha e n_\alpha\left(\mathbf{E} + \frac{1}{c}\mathbf{v}_\alpha\times\mathbf{B}\right) + \mathbf{R}_\alpha,$$

$$\tfrac{3}{2}n_\alpha k\frac{D^\alpha T_\alpha}{Dt} + p_\alpha\nabla\cdot\mathbf{v}_\alpha + \nabla\cdot\mathbf{q}_\alpha + \mathbf{P}_\alpha\cdot\nabla\mathbf{v}_\alpha = Q_\alpha.$$

Here $D^\alpha/Dt = \partial/\partial t + \mathbf{v}_\alpha\cdot\nabla$; $p_\alpha = n_\alpha kT_\alpha$, where k is Boltzmann's constant; $\mathbf{R}_\alpha = \Sigma_\beta\mathbf{R}_{\alpha\beta}$ and $Q_\alpha = \Sigma_\beta Q_{\alpha\beta}$, where $\mathbf{R}_{\alpha\beta}$ and $Q_{\alpha\beta}$ are, respectively, the momentum and energy gained by the αth species through collisions with the βth; $\mathbf{P}_\alpha$ is the stress tensor, and $\mathbf{q}_\alpha$ is the heat flow.

The transport coefficients in a simple two-component (electrons and singly charged ions) plasma are tabulated below. Here $\|$ and $\perp$ refer to the direction of the magnetic field $\mathbf{B}=\mathbf{b}B$; $\mathbf{u}=\mathbf{v}_e-\mathbf{v}_i$ is the relative streaming velocity; $n_e=n_i=n$; $\mathbf{j}=-ne\mathbf{u}$ is the current; $\omega_{ce}=1.76\times10^7 B$ and $\omega_{ci}=(m_e/m_i)\,\omega_{ce}$ are the electron and ion gyrofrequencies, respectively; and the basic collisional times are taken to be

$$\tau_e=\frac{3m_e^{1/2}(kT_e)^{3/2}}{4(2\pi)^{1/2}n\lambda e^4}=3.44\times10^5\,\frac{T_e^{3/2}}{n\lambda},$$

where λ is the Coulomb logarithm, and

$$\tau_i=\frac{3m_i^{1/2}(kT_i)^{3/2}}{4\pi^{1/2}n\lambda e^4}=2.09\times10^7\,\frac{T_i^{3/2}}{n\lambda}\mu^{1/2}.$$

In the limit of large fields ($\omega_{ci}\tau_i\gg1$) the transport processes may be summarized as follows[6]:

Momentum transfer:

$$\mathbf{R}_{ei}=-\mathbf{R}_{ie}=\mathbf{R}=\mathbf{R}_{\mathbf{u}}+\mathbf{R}_T.$$

Frictional force:

$$\mathbf{R}_{\mathbf{u}}=ne(\mathbf{j}_\perp/\sigma_\perp+\mathbf{j}_\|/\sigma_\|).$$

Conductivities:

$$\sigma_\|=2.0\sigma_\perp=2.0\,\frac{ne^2\tau_e}{m_e}.$$

Thermal force:

$$\mathbf{R}_T=-0.71nk\nabla_\| T_e-\frac{3}{2}\frac{nk}{\omega_{ce}\tau_e}\mathbf{b}\times\nabla_\perp T_e.$$

Ion heating:

$$Q_i=3\frac{m_e}{m_i}\frac{nk}{\tau_e}(T_e-T_i).$$

Electron heating:

$$Q_e=-Q_i-\mathbf{R}\cdot\mathbf{u}.$$

Ion heat flux:

$$\mathbf{q}_i=-\kappa_\|^i\nabla_\| kT_i-\kappa_\perp^i\nabla_\perp kT_i+\kappa_\wedge^i\,\mathbf{b}\times\nabla_\perp kT_i.$$

Ion thermal conductivities:

$$\kappa_\|^i=3.9\,\frac{nkT_i\tau_i}{m_i},\quad \kappa_\perp^i=2\,\frac{nkT_i}{m_i\omega_{ci}^2\tau_i},\quad \kappa_\wedge^i=\frac{5}{2}\frac{nkT_i}{m_i\omega_{ci}}.$$

Electron heat flux:

$$\mathbf{q}^e=\mathbf{q}_{\mathbf{u}}^e+\mathbf{q}_T^e.$$

Frictional heat flux:

$$\mathbf{q}_{\mathbf{u}}^e=0.71nkT_e\mathbf{u}_\|+\frac{3}{2}\frac{nkT_e}{\omega_{ce}\tau_e}\mathbf{b}\times\mathbf{u}_\perp.$$

Thermal gradient heat flux:

$$\mathbf{q}_T^e=-\kappa_\|^e\nabla_\| kT_e-\kappa_\|^e\nabla_\perp kT_e-\kappa_\wedge^e\,\mathbf{b}\times\nabla_\perp kT_e.$$

Electron thermal conductivities:

$$\kappa_\|^e=3.2\,\frac{nkT_e\tau_e}{m_e},\quad \kappa_\perp^e=4.7\,\frac{nkT_e}{m_e\omega_{ce}^2\tau_e},$$

$$\kappa_\wedge^e=\frac{5}{2}\frac{nkT_e}{m_e\omega_{ce}}.$$

Stress tensor (both species):

$$P_{xx}=-\tfrac{1}{2}\eta_0(W_{xx}+W_{yy})-\tfrac{1}{2}\eta_1(W_{xx}-W_{yy})-\eta_3W_{xy},$$
$$P_{yy}=-\tfrac{1}{2}\eta_0(W_{xx}+W_{yy})+\tfrac{1}{2}\eta_1(W_{xx}-W_{yy})+\eta_3W_{xy},$$
$$P_{xy}=P_{yx}=-\eta_1W_{xy}+\tfrac{1}{2}\eta_3(W_{xx}-W_{yy}),$$
$$P_{xz}=P_{zx}=-\eta_2W_{xz}-\eta_4W_{yz},$$
$$P_{yz}=P_{zy}=-\eta_2W_{yz}+\eta_4W_{xz},$$
$$P_{zz}=-\eta_0W_{zz}$$

(here the z axis is defined parallel to $\mathbf{B}$).

Ion viscosity:

$$\eta_0^i=0.96nkT_i\tau_i,\quad \eta_1^i=\frac{3}{10}\frac{nkT_i}{\omega_{ci}^2\tau_i},$$

$$\eta_2^i=\frac{6}{5}\frac{nkT_i}{\omega_{ci}^2\tau_i},\quad \eta_3^i=\frac{1}{2}\frac{nkT_i}{\omega_{ci}},\quad \eta_4^i=\frac{nkT_i}{\omega_{ci}}.$$

Electron viscosity:

$$\eta_0^e=0.73nkT_e\tau_e,\quad \eta_1^e=0.51\,\frac{nkT_e}{\omega_{ce}^2\tau_e},$$

$$\eta_2^e=2.0\,\frac{nkT_e}{\omega_{ce}^2\tau_e},\quad \eta_3^e=-\frac{1}{2}\frac{nkT_e}{\omega_{ce}},\quad \eta_4^e=-\frac{nkT_e}{\omega_{ce}}.$$

For both species the rate-of-strain tensor is defined as

$$W_{jk}=\frac{\partial v_j}{\partial x_k}+\frac{\partial v_k}{\partial x_j}-\tfrac{2}{3}\delta_{jk}\nabla\cdot\mathbf{v}.$$

When $\mathbf{B}=0$ the following simplifications occur:

$$\mathbf{R}_{\mathbf{u}}=ne\mathbf{j}/\sigma_\|,\quad \mathbf{R}_T=-0.71n\nabla kT_e,\quad \mathbf{q}_i=-\kappa_\|^i\nabla kT_i,$$
$$\mathbf{q}_{\mathbf{u}}^e=0.71nkT_e\mathbf{u},\quad \mathbf{q}_T^e=-\kappa_\|^e\nabla kT_e,\quad P_{jk}=-\eta_0W_{jk}.$$

When $\omega_{ce}\tau_e\gg1\gg\omega_{ci}\tau_i$, the high-field expressions are obeyed by the electrons and the zero-field expressions by the ions.

Collisional transport theory is applicable when (1) macroscopic time rates of change satisfy $d/dt\ll1/\tau$, where τ is the longest collisional time scale, and (in the absence of a magnetic field) (2) macroscopic length scales L satisfy $L\gg l$, where $l=\bar{v}\tau$ is the mean free path. In a strong field, $\omega_{ce}\tau\gg1$, condition (2) is replaced by $L_\|\gg l$ and $L_\perp\gg(lr_e)^{1/2}$ ($L_\|\gg r_e$ in a uniform field), where $L_\|$ is a macroscopic scale parallel to the field $\mathbf{B}$ and $L_\perp$ is the smaller of $(\nabla_\perp B/B)^{-1}$ and the transverse plasma dimension. In addition, the standard transport coefficients are valid only when (3) the Coulomb logarithm satisfies $\lambda\gg1$; (4) the electron gyroradius satisfies $r_e\gg\lambda_D$ or $B^2\ll8\pi n_em_ec^2$; (5) relative drifts $\mathbf{u}=\mathbf{v}_\alpha-\mathbf{v}_\beta$ between two species are small compared with the thermal velocities, $u\ll(kT_\alpha/m_\alpha)^{1/2}$, $(kT_\alpha/m_\beta)^{1/2}$; and (6) anomalous transport processes owing to microinstabilities are negligible.

H. Weakly ionized plasmas

Collision frequency for scattering of charged particles by neutrals is

$$\nu_\alpha=n_0\sigma_{0\alpha}(kT_\alpha/m_\alpha)^{1/2},$$

where n_0 is the neutral density and $\sigma_{0\alpha}$ is the cross section, typically $\sim 5\times10^{-15}$ cm^2 and weakly dependent on temperature.

When the system is small, $L \ll \lambda_p$, the charged particle diffusion coefficients are

$$D_\alpha = kT_\alpha / m_\alpha \nu_\alpha .$$

In the opposite limit, both species diffuse at the ambipolar rate

$$D_A = \frac{\mu_i D_e - \mu_e D_i}{\mu_i - \mu_e} = \frac{(T_i + T_e) D_i D_e}{T_i D_e + T_e D_i},$$

where $\mu_\alpha = e_\alpha / m_\alpha \nu_\alpha$ is the mobility. The conductivity σ_α satisfies $\sigma_\alpha = n_\alpha e_\alpha \mu_\alpha$.

In the presence of a magnetic field **B**, μ and σ become tensors

$$\mathbf{J}^\alpha = \boldsymbol{\sigma}_\alpha \cdot \mathbf{E} = \sigma_\parallel^\alpha \mathbf{E}_\parallel + \sigma_\perp^\alpha \mathbf{E}_\perp + \sigma_\wedge^\alpha \mathbf{E}\times\mathbf{b},$$

where $\mathbf{b} = \mathbf{B}/B$ and

$$\sigma_\parallel^\alpha = \frac{n_\alpha e_\alpha^2}{m_\alpha \nu_\alpha}, \quad \sigma_\perp^\alpha = \frac{\sigma_\parallel^\alpha \nu_\alpha^2}{\nu_\alpha^2 + \omega_{c\alpha}^2}, \quad \sigma_\wedge^\alpha = \frac{\sigma_\parallel^\alpha \nu_\alpha \omega_{c\alpha}}{\nu_\alpha^2 + \omega_{c\alpha}^2}.$$

Here $\sigma_\perp$ and $\sigma_\wedge$ are the Pedersen and Hall conductivities, respectively.

18.04. TABLE: APPROXIMATE MAGNITUDES IN SOME TYPICAL PLASMAS

Plasma type	n (cm^{-3})	T (eV)	ω_{pe} (s^{-1})	λ_D (cm)	$n\lambda_D^3$	ν_{ei} (s^{-1})
Interstellar gas	1	1	6×10^4	7×10^2	4×10^8	7×10^{-5}
Gaseous nebula	10^3	1	2×10^6	20	10^7	6×10^{-2}
Solar corona	10^6	10^2	6×10^7	7	4×10^8	6×10^{-2}
Diffuse hot plasma	10^{12}	10^2	6×10^{10}	7×10^{-3}	4×10^5	40
Solar atmosphere, gas discharge	10^{14}	1	6×10^{11}	7×10^{-5}	40	2×10^9
Warm plasma	10^{14}	10	6×10^{11}	2×10^{-4}	10^3	10^7
Hot plasma	10^{14}	10^2	6×10^{11}	7×10^{-4}	4×10^4	4×10^6
Thermonuclear plasma	10^{15}	10^4	2×10^{12}	2×10^{-3}	10^7	5×10^4
Theta pinch	10^{16}	10^2	6×10^{12}	7×10^{-5}	4×10^3	3×10^8
Dense hot plasma	10^{18}	10^2	6×10^{13}	7×10^{-6}	4×10^2	2×10^{10}
Laser plasma	10^{20}	10^2	6×10^{14}	7×10^{-7}	40	2×10^{12}

The accompanying diagram gives comparable information in graphical form.

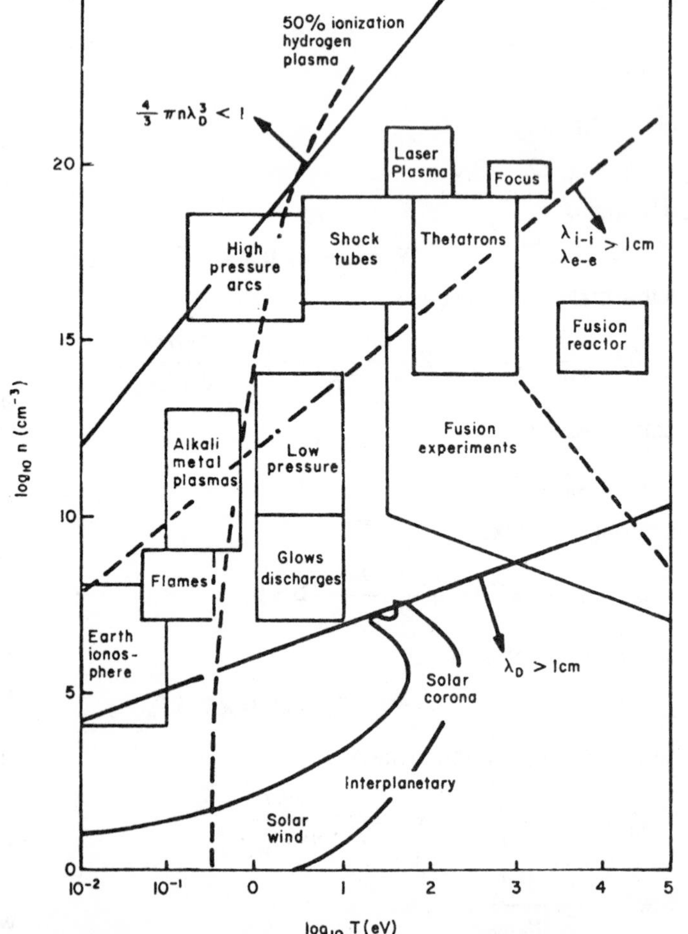

18.05. TABLE: IONOSPHERIC PARAMETERS[8]

Average nighttime values. Where two numbers are entered, the first refers to the lower and the second to the upper portion of the layer.

Quantity	E region	F region
Altitude (km)	90–160	160–500
Number density (m^{-3})	1.5×10^{10}–3.0×10^{10}	5×10^{10}–2×10^{11}
Height-integrated number density (m^{-2})	9×10^{14}	4.5×10^{15}
Ion-neutral collision frequency (s^{-1})	2×10^{3}–10^{2}	0.5–0.05
Ion gyro-/collison frequency ratio κ_i	0.09–2.0	4.6×10^{2}–5.0×10^{3}
Ion Pederson factor $\kappa_i/(1+\kappa_i^2)$	0.09–0.5	2.2×10^{-3}–2×10^{-4}
Ion Hall factor $\kappa_i^2/(1+\kappa_i^2)$	8×10^{-4}–0.8	1.0
Electron-neutral collision frequency	1.5×10^{4}–9.0×10^{2}	80–10
Electron gyro-/collision frequency ratio κ_e	4.1×10^{2}–6.9×10^{3}	7.8×10^{4}–6.2×10^{5}
Electron Pedersen factor $\kappa_e/(1+\kappa_e^2)$	2.7×10^{-3}–1.5×10^{-4}	10^{-5}–1.5×10^{-6}
Electron Hall factor $\kappa_e^2/(1+\kappa_e^2)$	1.0	1.0
Mean molecular weight	28–26	22–16
Ion gyrofrequency (s^{-1})	180–190	230–300
Neutral diffusion coefficient (m^2/s)	30–5×10^{3}	10^{5}

The terrestrial magnetic field in the lower ionosphere at equatorial latitudes is approximately $B_0=0.35\times10^{-4}$ T. The Earth's radius is $R_E=6371$ km.

18.06. SOLAR PHYSICS PARAMETERS

A. Table: General solar parameters[9]

Parameter	Symbol	Value	Units
Total mass	$M_\odot$	1.99×10^{33}	g
Radius	$R_\odot$	6.96×10^{10}	cm
Surface gravity	$g_\odot$	2.74×10^{4}	cm s^{-2}
Escape speed	v_∞	6.18×10^{7}	cm s^{-1}
Upward mass flux in spicules	...	1.6×10^{-9}	g cm^{-2} s^{-1}
Vertically integrated atmospheric density	...	4.28	g cm^{-2}
Sunspot magnetic field strength	B_{max}	2500–3500	G
Surface temperature	T_0	6420	K
Radiant power	$\mathscr{L}_\odot$	3.90×10^{33}	erg s^{-1}
Radiant flux density	$\mathscr{F}$	6.41×10^{10}	erg cm^{-2} s^{-1}
Optical depth at 500 nm measured from photosphere	τ_{500}	0.99	...
Astronomical unit (radius of Earth's orbit)	AU	1.50×10^{13}	cm
Solar constant (radiant flux density at 1 AU)	f	1.39×10^{6}	erg cm^{-2} s^{-1}

B. Table: Chromosphere and corona[10]

Parameter	Quiet Sun	Coronal hole	Active region
Chromospheric radiation losses (erg cm^{-2} s^{-1})			
Low chromosphere	2×10^{6}	2×10^{6}	$\gtrsim10^{7}$
Middle chromosphere	2×10^{6}	2×10^{6}	10^{7}
Upper chromosphere	3×10^{5}	3×10^{5}	2×10^{6}
Total	4×10^{6}	4×10^{6}	$\gtrsim2\times10^{7}$
Transition layer pressure (dyn cm^{-2})	0.2	0.07	2
Coronal temperature (K, at $1.1R_\odot$)	1.1–1.6×10^{6}	10^{6}	2.5×10^{6}
Coronal energy losses (erg cm^{-2} s^{-1})			
Conduction	2×10^{5}	6×10^{4}	10^{5}–10^{7}
Radiation	10^{5}	10^{4}	5×10^{6}
Solar wind	$\lesssim5\times10^{4}$	7×10^{5}	$<10^{5}$
Total	3×10^{5}	8×10^{5}	10^{7}
Solar wind mass loss (g cm^{-2} s^{-1})	$\lesssim2\times10^{-11}$	2×10^{-10}	$<4\times10^{-11}$

18.07. CTR[11]

Natural abundance of isotopes:

hydrogen $n_D/n_H = 1.5\times10^{-4}$

helium $n_{He^3}/n_{He^4} = 1.3\times10^{-6}$

lithium $n_{Li^6}/n_{Li^7} = 0.08$

Mass ratios:

$m_e/m_D = 2.72\times10^{-4} = 1/3670,$

$(m_e/m_D)^{1/2} = 1.65\times10^{-2} = 1/60.6,$

$m_e/m_T = 1.82\times10^{-4} = 1/5496,$

$(m_e/m_T)^{1/2} = 1.35\times10^{-1/2} = 1/74.1.$

Fusion reactions (branching ratios are correct for energies near the cross-section peaks; a negative yield means the reaction is endothermic)[12]:

$$D + D \underset{50\%}{\rightarrow} T(1.01\ \text{MeV}) + p(3.02\ \text{MeV}) \quad (1a)$$

$$\underset{50\%}{\rightarrow} He^3(0.82\ \text{MeV}) + n(2.45\ \text{MeV}), \quad (1b)$$

$$D + T \rightarrow He^4(3.5\ \text{MeV}) + n(14.1\ \text{MeV}), \quad (2)$$

$$D + He^3 \rightarrow He^4(3.6\ \text{MeV}) + p(14.7\ \text{MeV}), \quad (3)$$

$$T + T \rightarrow He^4 + 2n + 11.3\ \text{MeV}, \quad (4)$$

$$He^3 + T \underset{51\%}{\rightarrow} He^4 + p + n + 12.1\ \text{MeV} \quad (5a)$$

$$\underset{43\%}{\rightarrow} He^4(4.8\ \text{MeV}) + D(9.5\ \text{MeV}) \quad (5b)$$

$$\underset{6\%}{\rightarrow} He^5(2.4\ \text{MeV}) + p(11.9\ \text{MeV}), \quad (5c)$$

$$p + Li^6 \rightarrow He^4(1.7\ \text{MeV}) + He^3(2.3\ \text{MeV}), \quad (6)$$

$$p + Li^7 \underset{\sim 20\%}{\rightarrow} 2He + 17.3\ \text{MeV} \quad (7a)$$

$$\underset{\sim 80\%}{\rightarrow} Be^7 + n - 1.6\ \text{MeV}, \quad (7b)$$

$$D + Li^6 \rightarrow 2He^4 + 22.4\ \text{MeV}, \quad (8)$$

$$p + B^{11} \rightarrow 3He^4 + 8.7\ \text{MeV}, \quad (9)$$

$$n + Li^6 \rightarrow He^4(2.1\ \text{MeV}) + T(2.7\ \text{MeV}). \quad (10)$$

The total cross section in barns as a function of E, the energy in keV of the incident particle [the first ion on the left side of Eqs. (1)–(5)], assuming the target ion at rest, can be fitted by[13]

$$\sigma_T(E) = \frac{A_5 + [(A_4 - A_3E)^2 + 1]^{-1}A_2}{E[\exp(A_1/E^{1/2}) - 1]},$$

where the Duane coefficients A_j for the principal fusion reactions are as follows:

	D-D (1a)	D-D (1b)	D-T (2)	D-He^3 (3)	T-T (4)	T-He^3 (5)
A_1	46.097	47.88	45.95	89.27	38.39	123.1
A_2	372	482	5.02×10^4	2.59×10^4	448	1.125×10^4
A_3	4.36×10^{-4}	3.08×10^{-4}	1.368×10^{-2}	3.98×10^{-3}	1.02×10^{-3}	0
A_4	1.220	1.177	1.076	1.297	2.09	0
A_5	0	0	409	647	0	0

Reaction rates σv (cm^3/s), averaged over Maxwellian distributions:

Temperature (keV)	D-D (1a) + (1b)	D-T (2)	D-He^3 (3)	T-T (4)	T-He^3 (5a) + (5b) + (5c)
1.0	1.5×10^{-22}	5.5×10^{-21}	3×10^{-26}	3.3×10^{-22}	10^{-28}
2.0	5.4×10^{-21}	2.6×10^{-19}	1.4×10^{-23}	7.1×10^{-21}	10^{-25}
5.0	1.8×10^{-19}	1.3×10^{-17}	6.7×10^{-21}	1.4×10^{-19}	2.1×10^{-22}
10.0	1.2×10^{-18}	1.1×10^{-16}	2.3×10^{-19}	7.2×10^{-19}	1.2×10^{-20}
20.0	5.2×10^{-18}	4.2×10^{-16}	3.8×10^{-18}	2.5×10^{-18}	2.6×10^{-19}
50.0	2.1×10^{-17}	8.7×10^{-16}	5.4×10^{-17}	8.7×10^{-18}	5.3×10^{-18}
100.0	4.5×10^{-17}	8.5×10^{-16}	1.6×10^{-16}	1.9×10^{-17}	2.7×10^{-17}
200.0	8.8×10^{-17}	6.3×10^{-16}	2.4×10^{-16}	4.2×10^{-17}	9.2×10^{-17}
500.0	1.8×10^{-16}	3.7×10^{-16}	2.3×10^{-16}	8.4×10^{-17}	2.9×10^{-16}
1000.0	2.2×10^{-16}	2.7×10^{-16}	1.8×10^{-16}	8.0×10^{-17}	5.2×10^{-16}

For low energies ($T \leqslant 25$ keV) the data may be represented by

$$(\overline{\sigma v})_{DD} = 2.33\times10^{-14}T^{-2/3}\exp(-18.76T^{-1/3})\ \text{cm}^3/\text{s},$$

$$(\overline{\sigma v})_{DT} = 3.68\times10^{-12}T^{-2/3}\exp(-19.94T^{-1/3})\ \text{cm}^3/\text{s},$$

where T is measured in keV.

The power density released in the form of charged particles is

$$P_{DD} = 3.3\times10^{-13}n_D^2(\overline{\sigma v})_{DD}\ \text{W/cm}^3$$

(including subsequent D-T reaction),

$$P_{DT} = 5.6\times10^{-13}n_D n_T(\overline{\sigma v})_{DT}\ \text{W/cm}^3,$$

$$P_{DHe^3} = 2.9\times10^{-12}n_D n_{He^3}(\overline{\sigma v})_{DHe^3}\ \text{W/cm}^3.$$

The curie (abbreviated Ci) is a measure of radioactivity: 1 Ci $= 3.7\times10^{10}$ disintegrations/s. Absorbed radiation dose is measured in rads: 1 rad $= 10^2$ ergs/g.

18.08. TABLE: LASERS

Efficiencies and power levels are approximately state-of-the art (1987).[31]

Type	Wavelength (μm)	Efficiency	Power levels available (W) Pulsed	CW
CO_2	10.6	0.01–0.02 (pulsed)	$>2\times10^{13}$	$>10^5$
CO	5	0.4	$>10^9$	>100
Holmium	2.06	0.03	$>10^7$	30
Iodine	1.315	0.003	$>10^{12}$	-
Nd-glass, YAG	1.06	0.001–0.06	$\sim10^{14}$ (10-beam system)	$1–10^3$
*Color center	1–4	10^{-3}	$>10^6$	1
*OPO	0.7–0.9	10^{-3}	10^6	1
Ruby	0.6943	$<10^{-3}$	10^{10}	1
He–Ne	0.6328	10^{-4}	–	$1–50\times10^{-3}$
*Argon ion	0.45–0.60	10^{-3}	5×10^4	1–10
N_2	0.3371	0.001–0.05	$10^5–10^6$	-
*Dye	0.3–1.1	10^{-3}	$>10^6$	140
Kr–F	0.26	0.08	$>10^9$	-
Xenon	0.175	0.02	$>10^8$	-

*Tunable sources

18.09. LASER FORMULAS

An electromagnetic wave with $\mathbf{k}\|\mathbf{B}$ has an index of refraction given by

$$n_\pm = [1 - \omega_{pe}^2/\omega(\omega \mp \omega_{ce})]^{1/2},$$

where $\pm$ refers to the helicity. The rate of change of polarization angle θ as a function of displacement s (Faraday rotation) is given by

$$d\theta/ds = (k/2)(n_- - n_+) = 2.36\times10^4 NBf^{-2} \text{ cm}^{-1},$$

where N is the electron number density, B is the field strength, and f is the wave frequency, all in cgs.

The quiver velocity of an electron in an electromagnetic field of angular frequency ω is

$$v_0 = eE_{\max}/m\omega = 25.6 I^{1/2}\lambda_0 \text{ cm/s}$$

in terms of the laser flux $I = cE_{\max}^2/8\pi$, with I in W/cm^2, laser wavelength λ_0 in μm.

The ratio of quiver energy to thermal energy is

$$W_{qu}/W_{th} = m_e v_0^2/2kT = 1.81\times10^{-13}\lambda_0^2 I/T,$$

T in eV. For example, if $I = 10^{15}$ W/cm^2, $\lambda_0 = 1\ \mu$m, $T = 2$ keV, then $W_{qu}/W_{th} \approx 0.1$.

The ponderomotive force is

$$\mathbf{f} = N\ \nabla\langle E^2\rangle/8\pi N_c,$$

where

$$N_c = 1.1\times10^{21}\lambda_0{}^{-2} \text{ cm}^{-3}.$$

For uniform illumination of a lens with f number F, the diameter d at focus (85% of the energy) and the depth of focus l (distance to first zero in intensity) are given by

$$d \approx 2.44 F\lambda\theta/\theta_{DL} \quad \text{and} \quad l \approx \pm 2F^2\lambda\theta/\theta_{DL}.$$

Here θ is the beam divergence containing 85% of energy and θ_{DL} is the diffraction-limited divergence

$$\theta_{DL} = 2.44\lambda/b,$$

where b is the aperture. These formulas are modified for nonuniform (such as Gaussian) illumination of the lens or for pathological laser profiles.

18.10. RELATIVISTIC ELECTRON BEAMS

Here $\gamma = (1-\beta^2)^{-1/2}$ is the relativistic scaling factor; in analytic formulas units are mks or cgs, as indicated; in numerical formulas, I is in amps, B in gauss, electron density N in cm^{-1}, and temperature, voltage, and energy in MeV; $\beta_z = v_z/c$; k is Boltzmann's constant.

Relativistic electron gyroradius:

$$r_e = (mc^2/eB)(\gamma^2-1)^{1/2} \text{ (cgs)}$$

$$= 1.70\times10^3(\gamma^2-1)^{1/2}B \text{ cm}.$$

Relativistic electron energy:

$$W = mc^2\gamma \text{ (cgs)} = 0.511\gamma \text{ MeV}.$$

Bennett pinch condition:

$$I^2 = 2Nk(T_e + T_i)c^2 \text{ (cgs)} = 3.20\times10^{-4}N(T_e+T_i) \text{ A}^2.$$

Alfvén-Lawson limit:

$$I_A = (mc^3/e)\beta_z\gamma \text{ (cgs)}$$

$$= (4\pi mc/\mu_0 e)\beta_z\gamma \text{ (mks)} = 1.70\times10^4\beta_z\gamma \text{ A}.$$

The ratio of net current to I_A is

$$I/I_A = \nu/\gamma,$$

where $\nu = Nr_e$, with $r_e = e^2/mc^2 = 2.82\times10^{-13}$ cm. Beam electron number density is

$$n_b = 2.08\times10^8 J/\beta \text{ cm}^{-3},$$

where J is current density in A/cm^2. For a uniform beam of radius a (in cm),

$$n_b = 6.63\times10^7 I/\beta a^2 \text{ cm}^{-3},$$

and

$$2r_e/a = \nu/\gamma.$$

Child's law: (nonrelativistic) space-charge-limited current density between parallel plates with voltage drop V and separation d in cm,

$$J = 2.34\times10^3 V^{3/2}d^{-2} \text{ A cm}^{-2}.$$

The saturated parapotential current (magnetically self-limited flow along equipotentials in pinched diodes and transmission lines) is[14]

$$I_p = 8.5\times10^3 G\gamma \ln[\gamma + (\gamma^2-1)^{1/2}] \text{ A},$$

where G is a geometrical factor depending on the diode structure:

$$G = w/2\pi d$$

for parallel plane cathode and anode of width w, separation d;

$$G=[\ln(R_2/R_1)]^{-1}$$

for cylinders of radii R_1 (inner) and R_2 (outer);

$$G=R_c/d_0$$

for conical cathode of radius R_c, maximum separation d_0 (at $r=R_c$) from plane anode. For $\beta \to 0$ ($\gamma \to 1$), both I_A and I_p vanish.

The condition for suppression of filamentation in a beam of current density J A cm^{-1} by a longitudinal magnetic field B_z is

$$B_z > 47\beta_z(\gamma J)^{1/2} \text{ G}.$$

Voltage registered by Rogowski coil of minor cross-sectional area A, n turns, major radius a, inductance L, external resistance R, and capacitance C (all in mks):

externally integrated: $V=(1/RC)(nA\mu_0 I/2\pi a)$;

self-integrating: $V=(R/L)(nA\mu_0 I/2\pi a)=RI/n$.

X-ray production for target with average atomic number Z ($V \lesssim 5$ MeV):

$$\eta = \text{x-ray power/beam power} = 7\times 10^{-4} ZV.$$

X-ray dose at 1 m generated by an electron beam depositing total charge Q coulombs while $V \geqslant 0.84 V_{max}$ in material with charge state Z:

$$D=150 V_{max}^{2.8} Q Z^{1/2} \text{ rads}.$$

18.11. TABLE: BEAM INSTABILITIES[15]

Subscripts e, i, d, and b stand for "electron," "ion," "drift," and "beam," respectively. Thermal velocities are denoted by a bar. In addition:

m	electron mass	λ_D	Debye length
M	ion mass	r_e, r_i	gyroradius
V	velocity	β	plasma/magnetic energy density ratio
T	temperature	V_A	Alfvén speed
n_e, n_i	number density	Ω_e, Ω_i	gyrofrequency
n	harmonic number	Ω_H	hybrid gyrofrequency, $\Omega_H^2=\Omega_e\Omega_i$
$C_s=(T_e/M)^{1/2}$	ion sound speed	U	relative drift velocity of two ion species
ω_e, ω_i	plasma frequency		

	Parameters of most unstable mode					
Name	Conditions	Growth rate	Frequency	Wave number	Group velocity	Saturation mechanism
Electron-electron	$V_d > V_{ej}$, $j=1,2$	$\frac{1}{2}\omega_e$	0	$(3^{1/2}/2)\omega_e/V_d$	0	Electron trapping until $\overline{V}_e \sim V_d$
Beam-plasma	$(n_b/n_p)^{1/3} > \overline{V}_b/V_b$	$(3^{1/2}/2^{4/3}) \times (n_b/n_p)^{1/3}\omega_e$	$\omega_e[1-(1/2^{4/3}) \times (n_b/n_p)^{1/3}]$	ω_e/V_d	$\frac{2}{3}V_b$	Trapping of beam electrons
Buneman	$V_d > (M/m)^{1/3}\overline{V}_i$, $V_d > \overline{V}_e$	$(3^{1/2}/2^{4/3}) \times (m/M)^{1/3}\omega_e$	$(1/2^{4/3})\omega_e \times (m/M)^{1/3}$	ω_e/V_d	$\frac{2}{3}V_d$	Electron trapping until $\overline{V}_e \sim V_d$
Weak beam-plasma	$\overline{V}_b/V_b > (n_b/n_p)^{1/3}$	$\frac{1}{2}(n_b/n_p) \times (V_b/\overline{V}_b)^2\omega_e$	ω_e	ω_e/V_b	$3\overline{V}_e^2/V_b$	Quasilinear or nonlinear (mode coupling)
Beam-plasma (hot electron)	$\overline{V}_e > V_b > \overline{V}_b$, $(n_b/n_p)^{1/2}\overline{V}_e > \overline{V}_b$	$(n_b/n_p)^{1/2} \times (\overline{V}_e/V_b)\omega_e$	$(V_b/V_e)\omega_e$	λ_D^{-1}	V_b	Quasilinear or nonlinear
Ion acoustic	$T_e \gg T_i$, $V_d > C_s$	$(m/M)^{1/2}\omega_i$	$(C_s/V_e)\omega_e$	λ_D^{-1}	C_s	Quasilinear; ion tail formation; nonlinear scattering; or resonance broadening
Anisotropic temperature (hydro)	$T_\parallel^e/T_\perp^e < \frac{1}{2}$	Ω_e	$\omega_e\cos\theta \sim \Omega_e$	r_e^{-1}	$\overline{V}_{e\perp}$	Isotropization
Ion cyclotron	$V_d > 20\overline{V}_i$ for $T_e \approx T_i$	$0.1\Omega_i$	$1.2\Omega_i$	r_i^{-1}	$\frac{1}{3}\overline{V}_i$	Ion heating
Beam cyclotron (hydro)	$V_d > C_s$	$(1/2^{1/2})\Omega_e$	$n\Omega_e$	$(1/2^{1/2})\lambda_D^{-1}$	$V_d \lesssim V_g \lesssim C_s$	Resonance broadening
Modified two-stream (hydro)	$(1+\beta)^{1/2}V_A > V_d > C_s$	$\frac{1}{2}\Omega_H$	$(3^{1/2}/2)\Omega_H$	$3^{1/2}\Omega_H/V_d$	$\frac{1}{2}V_d$	Trapping
Ion-ion (equal beam)	$U < 2V_A \times (1+\beta)^{1/2}$	$(1/2 \cdot 2^{1/2})\Omega_H$	0	$(\frac{3}{2})^{1/2}\Omega_H/U$	0	Ion trapping
Ion-ion (equal beam)	$U < 2C_s$	$(1/2 \cdot 2^{1/2})\omega_i$	0	$(\frac{3}{2})^{1/2}\omega_i/U$	0	Ion trapping

18.12 ATOMIC PHYSICS AND RADIATION

Energies and temperatures are in eV; all other units are cgs except where noted. Z is the charge state ($Z=0$ refers to a neutral atom); the subscript e labels electrons. N refers to number density, n to principal quantum number. Asterisk superscripts on level population densities denote local thermodynamic equilibrium (LTE) values. Thus N_n^* is the LTE number density of atoms (or ions) in level n.

Characteristic atomic collision cross section:

$$\pi a_0^2 = 8.80\times10^{-17}\ \text{cm}^2. \tag{1}$$

Binding energy for outer electron in level labeled by quantum numbers n, l:

$$E_\infty^Z(n,l) = -Z^2E_\infty^{\text{H}}/(n-\Delta_l)^2, \tag{2}$$

where $E_\infty^{\text{H}} = 13.6$ eV is the hydrogen ionization energy and $\Delta_l = 0.75l^{-5}$, $l \gtrsim 5$, is the quantum defect.

A. Excitation and decay

Cross section (Bethe approximation) for electron excitation by dipole-allowed transition $m \to n$[17,18]:

$$\sigma_{mn} = 2.36\times10^{-13}\frac{f_{nm}g(n,m)}{\epsilon\Delta E_{nm}}\ \text{cm}^2, \tag{3}$$

where f_{nm} is the oscillator strength, $g(n,m)$ is the Gaunt factor, ϵ is the incident electron energy, and $\Delta E_{nm} = E_n - E_m$.

Electron excitation rate averaged over Maxwellian velocity distribution[19,20]:

$$X_{mn} = N_e\langle\sigma_{mn}v\rangle = \frac{1.6\times10^{-5}f_{mn}\langle g(n,m)\rangle N_e}{\Delta E_{nm}T^{1/2}} \times\exp\left(-\frac{\Delta E_{nm}}{T_e}\right)\ \text{s}^{-1}, \tag{4}$$

where $\langle g(n,m)\rangle$ denotes the thermal-averaged Gaunt factor (generally ~1 for atoms, ~0.2 for ions).

Rate for electron collisional deexcitation:

$$Y_{nm} = (N_m^*/N_n^*)X_{mn}\ \text{s}^{-1}. \tag{5}$$

Here $N_m^*/N_n^* = (g_m/g_n)\exp(\Delta E_{nm}/T_e)$ is the Boltzmann relation for level population densities, where g_n is the statistical weight of level n.

Rate for spontaneous decay $n \to m$ (Einstein A coefficient)[19]:

$$A_{nm} = 4.3\times10^7(g_n/g_m)f_{nm}(\Delta E_{nm})^2\ \text{s}^{-1}, \tag{6}$$

Intensity emitted per unit volume from the transition $n \to m$ in an optically thin plasma:

$$I_{nm} = 1.6\times10^{-19}A_{nm}N_n\Delta E_{nm}\ \text{W/cm}^3. \tag{7}$$

Condition for steady state in a corona model:

$$N_0N_e\langle\sigma_{0n}v\rangle = N_nA_{n0}, \tag{8}$$

where the ground state is labeled by a subscript zero. Hence for a transition $n \to m$ in ions, where $\langle g(n,0)\rangle \approx 0.2$,

$$I_{nm} = 5.1\times10^{-25}f_{nm}(g_0/g_m)(\Delta E_{nm}/\Delta E_{n0})^3 \times N_eN_0T_e^{-1/2}\exp(-\Delta E_{n0}/T_e)\ \text{W/cm}^3. \tag{9}$$

B. Ionization and recombination

In a general time-dependent situation the number density of the charge state Z satisfies

$$\frac{dN(Z)}{dt} = N_e[-S(Z)N(Z) - \alpha(Z)N(Z) + S(Z-1) \times N(Z-1) + \alpha(Z+1)N(Z+1)]. \tag{10}$$

Here $S(Z)$ is the ionization rate. The recombination rate $\alpha(Z)$ has the form $\alpha(Z) = \alpha_r(Z) + N_e\alpha_3(Z)$, where α_r and α_3 are the radiative and three-body recombination rates, respectively.

Classical ionization cross section[21] for any atomic shell k:

$$\sigma_i = 6\times10^{-14}b_kg_k(x)/U_k^2\ \text{cm}^2. \tag{11}$$

Here b_k is the number of shell electrons, U_k is the binding energy of the ejected electron, $x = \epsilon/U_k$, where ϵ is the incident electron energy, and g is a universal function with a maximum value ≈0.2 at $x \approx 4$.

Ionization from ion ground state, averaged over Maxwellian electron distribution[20]:

$$S(Z) = \frac{10^{-5}(T_e/E_\infty^Z)^{1/2}}{(E_\infty^Z)^{3/2}(6 + T_e/E_\infty^Z)}\exp(-E_\infty^Z/T_e)\ \text{cm}^3/\text{s}, \tag{12}$$

where E_∞^Z is the ionization energy.

Electron-ion radiative recombination rate $[e + N(Z) \to N(Z-1) + h\nu]$[22]:

$$\alpha_r(Z) = 5.2\times10^{-14}Z(E_\infty^Z/T_e)^{1/2}[0.43 + \tfrac{1}{2}\ln(E_\infty^Z/T_e) + 0.469(E_\infty^Z/T_e)^{-1/3}]\ \text{cm}^3/\text{s}, \tag{13}$$

For $10 < T_e/Z^2 < 150$ keV, this becomes approximately[20]

$$\alpha_r(Z) = 2.7\times10^{-13}Z^2T_e^{-1/2}\ \text{cm}^3/\text{s}. \tag{14}$$

Collisional (three-body) recombination rate for singly ionized plasma[23]:

$$\alpha_3 = 8.75\times10^{-27}T_e^{-4.5}\ \text{cm}^6/\text{s}. \tag{15}$$

Photoionization cross section for ions in level n, l (short-wavelength limit):

$$\sigma_{\text{ph}}(n,l) = 1.64\times10^{-16}Z^5/n^3K^{7+2l}, \tag{16}$$

where K is wave number in Rydberg units (1 Rydberg $= 1.0974\times10^5\ \text{cm}^{-1}$).

C. Ionization equilibrium model

Saha equilibrium[24]:

$$\frac{N_eN_1^*(Z)}{N_n^*(Z-1)} = 6.0\times10^{21}\frac{g_1^ZT_e^{3/2}}{g_n^{Z-1}}\exp\left(-\frac{E_\infty^Z(n,l)}{T_e}\right)\ \text{cm}^{-3}, \tag{17}$$

where g_n^Z is the statistical weight for level n for charge state Z and $E_\infty^Z(n,l)$ is the ionization energy of the neutral atom initially in level (n,l). In a steady state at high electron density

$$N_e N^*(Z)/N^*(Z-1) = S(Z-1)/\alpha_3 \text{ cm}^{-3}, \quad (18)$$

a function only of T.

Conditions for LTE[24]: (a) Collisional and radiative excitation rates for a level n must satisfy

$$Y_{nm}/A_{nm} > 10. \quad (19)$$

(b) Electron density must satisfy

$$N_e \geqslant 7\times 10^{18} Z^7 n^{-17/2} (T/E_\infty^Z)^{1/2} \text{ cm}^{-3}. \quad (20)$$

Steady state condition in corona model:

$$N(Z-1)/N(Z) = \alpha_r/S(Z-1). \quad (21)$$

Corona model is applicable if[25]

$$10^{12} t_i^{-1} < N_e < 10^{16} T_e^{7/2} \text{ cm}^{-3}, \quad (22)$$

where t_i^{-1} is the inverse ionization time.

D. Radiation

Energies and temperatures are in eV; all other units are cgs except where noted. Z is the charge state ($Z=0$ refers to a neutral atom); the subscript e labels electrons. N refers to number density.

Average radiative decay rate of state with principal quantum number n:

$$A_n = \sum_{m<n} A_{nm} = 1.6\times 10^{10} Z^4 n^{-9/2} \text{ s}. \quad (23)$$

Natural linewidth (ΔE in eV):

$$\Delta E \Delta t = h = 4.14\times 10^{-15} \text{ eV s}, \quad (24)$$

where Δt is the lifetime of the line.

Doppler width:

$$\Delta\lambda/\lambda = 7.7\times 10^{-5} (T/\mu)^{1/2}, \quad (25)$$

where μ is the emitting atom or ion mass in units of the proton mass.

Optical depth for a Doppler-broadened line[24]:

$$\tau = 1.76\times 10^{-13} \lambda (Mc^2/kT)^{1/2} Nl$$
$$= 5.4\times 10^{-9} \lambda (\mu/T)^{1/2} Nl, \quad (26)$$

where λ is wavelength and l the physical depth of the plasma; M, N, and T are mass, number density, and temperature of the absorber; μ is M divided by the proton mass. Optically thin means $\tau < 1$.

Resonance absorption cross section at center of line:

$$\sigma_{\lambda=\lambda_c} = 5.6\times 10^{-13} \lambda^2/\Delta\lambda \text{ cm}^2. \quad (27)$$

Wien displacement law: wavelength of maximum blackbody emission is given by

$$\lambda_{\max} = 2.50\times 10^{-5} T^{-1} \text{ cm}. \quad (28)$$

Radiation from surface of black body of temperature T:

$$W = 1.03\times 10^5 T^4 \text{ W/cm}^2. \quad (29)$$

Bremsstrahlung from hydrogenlike plasma[11]:

$$P_{\text{br}} = 1.69\times 10^{-32} N_e T_e^{1/2} \sum [Z^2 M(Z)] \text{ W/cm}^3, \quad (30)$$

where the sum is over all ionization states Z.

Bremsstrahlung optical depth[26]:

$$\tau = 5.0\times 10^{-38} N_e N_i Z^2 \bar{g} l T^{-7/2}, \quad (31)$$

where $\bar{g} \approx 1.2$ is an average Gaunt factor and l is the physical path length.

Inverse bremsstrahlung absorption coefficient[27] for radiation of angular frequency ω:

$$K = 3.1\times 10^{-7} Z n_e^2 (\ln\Lambda)/\omega^2 T^{3/2} (1 - \omega_p^2/\omega^2)^{1/2} \text{ cm}^{-1},$$

where $\Lambda = v_{Te}/V$, with V equal to the maximum of ω and ω_p, multiplied by the maximum of Ze^2/kT and $\hbar/(mkT)^{1/2}$.

Recombination (free-bound) radiation:

$$P_r = 1.69\times 10^{-32} N_e T_e^{1/2} \sum \left(Z^2 N(Z) \frac{E_\infty^{Z-1}}{T_e} \right) \text{ W/cm}^3. \quad (32)$$

Cyclotron radiation[11] in magnetic field **B**:

$$P_c = 6.21\times 10^{-28} B^2 N_e T_e \text{ W/cm}^3. \quad (33)$$

For $N_e kT_e = N_i kT_i = B^2/16\pi$ ($\beta = 1$, isothermal plasma),[11]

$$P_c = 5.00\times 10^{-38} N_e^2 T_e^2 \text{ W/cm}^3. \quad (34)$$

Cyclotron radiation energy loss e-folding time for a single electron[26]:

$$t_c \approx \frac{9.0\times 10^8 B^{-2}}{2.5+\gamma} \text{ s}, \quad (35)$$

where γ is the energy divided by the rest energy mc^2.

Number of cyclotron harmonics[26] trapped in a medium of finite depth l:

$$m^* = (57\beta Bl)^{1/6}, \quad (36)$$

where $\beta = NkT/8\pi B^2$.

Line radiation is given by summing Eq. (9) over all species in the plasma.

18.13. REFERENCES

[1]The Z function is tabulated in B. D. Fried and S. D. Conte, *The Plasma Dispersion Function* (Academic, New York, 1961).

[2]R. W. Landau and S. Cuperman, J. Plasma Phys. **6**, 495 (1971).

[3]D. B. Fried, C. L. Hedrick, and J. McCune, Phys. Fluids **11**, 249 (1968).

[4]B. A. Trubnikov, in *Reviews of Plasma Physics*, edited by M. A. Leontovich, Series 1 (Plenum, New York, 1965), p. 105.

[5]J. M. Greene, Phys. Fluids **16**, 2022 (1973).

[6]S. I. Braginskii, in *Reviews of Plasma Physics,* edited by M. A. Leontovich, Series 1 (Plenum, New York, 1965), p. 205.

[7]John Sheffield, *Plasma Scattering of Electromagnetic Radiation* (Academic, New York, 1975), p. 6 (after J. W. Paul).

[8]K. H. Lloyd and G. Härendel, J. Geophys. Res. **78**, 7389 (1973).

[9]C. W. Allen, *Astrophysical Quantities,* 2nd ed. (University of London, Athone Press, London, 1963), p. 161.

[10]G. L. Withbroe and R. W. Noyes, Ann. Rev. Astrophys. **15**, 363 (1977).

[11]S. Glasstone and R. H. Lovberg, *Controlled Thermonuclear Reactions* (Van Nostrand, New York, 1960), Chap. 2.

[12]References to experimental measurements of branching ratios and cross sections are listed in F. K. McGowan *et al.*, Nucl. Data Tables **A6**, 353 (1969); **A8**, 199 (1970). The yields listed in the table are calculated directly from the mass defect.

[13]G. H. Miley, H. Towner, and N. Ivich, *Fusion Cross Section and Reactivities,* University of Illinois Report COO-2218-17, 1974.
[14]J. M. Creedon, J. Appl. Phys. **46**, 2946 (1975).
[15]See, for example, A. B. Mikhailovskii, *Theory of Plasma Instabilities* (Consultants Bureau, New York, 1974), Vol. I. Table 18.11 was compiled by K. Papadopoulos.
[16]Table prepared from data compiled by J. M. McMahon (personal communication, 1987).
[17]M. J. Seaton, in *Atomic and Molecular Processes,* edited by D. R. Bates (Academic, New York, 1962), Chap. 11.
[18]H. Van Regemorter, Astrophys. J. **136**, 906 (1962).
[19]A. C. Kolb and R. W. P. McWhirter, Phys. Fluids **7**, 519 (1974).
[20]R. W. P. McWhirter, in *Plasma Diagnostic Techniques,* edited by R. H. Huddlestone and S. L. Leonard (Academic, New York, 1965).
[21]M. Gryzinski, Phys. Rev. **138**, 336A (1965).
[22]M. J. Seaton, Mon. Not. R. Astron. Soc. **119**, 81 (1959).
[23]Ya. B. Zel'dovich and Yu. P. Raizer, *Physics of Shock Waves and High-Temperature Hydrodynamic Phenomena* (Academic, New York, 1966), Vol. I, p. 407.
[24]H. R. Griem, *Plasma Spectroscopy* (Academic, New York, 1966).
[25]T. F. Stratton, in *Plasma Diagnostic Techniques,* edited by R. H. Huddlestone and S. L. Leonard (Academic, New York, 1965).
[26]G. Bekefi, *Radiation Processes in Plasmas* (Wiley, New York, 1966).
[27]T. W. Johnston and J. M. Dawson, Phys. Fluids **16**, 722 (1973).

Additional material can also be found in D. L. Book, NRL Memorandum Report 3332 (1977).

19.00. Rheology

HERSHEL MARKOVITZ

Carnegie-Mellon University

CONTENTS

The author expresses his appreciation to various colleagues fo[r] their suggestions, and regrets that space restrictions make it im[possible] to use so many of them. He especially wishes to than[k] Professor Donald J. Plazek for permission to present graphs of hi[s] data and calculations on polystyrene, some of which have no[t] been published. This work was partially supported by NSF (Polyme[r] Program) Grant DMR 79-19853 and AFOSR Grant 77-3404A.

19.01. INTRODUCTION

Rheology, in its broadest sense, is that part of mechanics which deals with the deformation of material and thus is essentially synonymous with the term *continuum mechanics.* However, the word *rheology* as generally used refers to mechanical behavior which does not follow the classical models of the perfect or inviscid fluid, linear viscous or Newtonian fluid, linear elastic or Hookean solid, and plasticity. The classical models are not adequate to describe phenomena such as creep, stress relaxation, ultrasonic attenuation, anelasticity, and nonlinear stress-deformation relations. Nonlinear elasticity serves as a useful model for some large equilibrium deformations of materials such as rubber; linear viscoelasticity for certain time-dependent behavior (e.g., creep, stress relaxation, sinusoidal deformation) if the stress is not too large; nonlinear viscoelasticity for time-dependent behavior and steady-state flow at large stresses.

Some notes on notation. Cartesian coordinates (indicated by symbols such as x, y, z, or alternatively x_1, x_2, x_3) are used throughout except where another type of orthogonal coordinate system is indicated. All indices may take on the values 1, 2, or 3 unless otherwise indicated. The Einstein summation convention is employed; i.e., a repeated index in any term is taken to mean summation over that index, e.g.,

$$e_{kk} \equiv e_{11} + e_{22} + e_{33},$$

$$\frac{\partial x_k}{\partial x_i}\frac{\partial x_k}{\partial x_j} \equiv \frac{\partial x_1}{\partial x_i}\frac{\partial x_1}{\partial x_j} + \frac{\partial x_2}{\partial x_i}\frac{\partial x_2}{\partial x_j} + \frac{\partial x_3}{\partial x_i}\frac{\partial x_3}{\partial x_j},$$

$\bar{f}(p)$ is the Laplace transform of $f(t)$:

$$\bar{f}(p) = \int_0^\infty f(t)e^{-pt}dt.$$

19.02. DEFORMATION

A. Deformation tensors

Let (X_1, X_2, X_3), or more briefly X_i, be the coordinates in a specified Cartesian coordinate system of the position of a given particle in the reference configuration of a body. The motion of the particle is specified by the function $x_i(t)$, the coordinates of the same particle at time t in the same coordinate system:

$$x_i(t) = x_i(X_1, X_2, X_3, t) = x_i(X_k, t).$$

The velocity of the particle $\mathbf{v}$ is

$$v_i \equiv dx_i(X_k, t)/dt.$$

The symmetric tensors $\mathbf{C}$ and $\mathbf{B}$ whose components are

$$C_{ij} \equiv (\partial x_k/\partial X_i)(\partial x_k/\partial X_j), \quad B_{ij} \equiv (\partial x_i/\partial X_k)(\partial x_j/\partial X_k)$$

may be used to specify the deformation. $\mathbf{C}$ is the *right Cauchy-Green tensor* and $\mathbf{B}$ the *left Cauchy-Green tensor.*

1. Infinitesimal deformations

To within terms quadratic in the magnitude of the strain, the infinitesimal strain tensor $\mathbf{e}$ is given by

$$e_{ij} = \tfrac{1}{2}(\partial u_i/\partial X_j + \partial u_j/\partial X_i) \approx \tfrac{1}{2}(C_{ij} - \delta_{ij}),$$

where the displacement $\mathbf{u}$ is

$$u_i = x_i - X_i.$$

B. Relative deformation tensors

The configuration of a body at time t is taken to be the reference configuration. The particle which occupies the position x_i,

$$x_i(t) = x_i(X_k, t),$$

in this configuration has the location $\xi_i(\tau) = \xi_i(x_k, \tau)$ at time τ.

$$C_{(t)ij}(\tau) \equiv (\partial \xi_k/\partial x_i)(\partial \xi_k/\partial x_j)$$

is the *right relative Cauchy-Green tensor* or *Cauchy tensor.*

The *relative strain history* corresponding to $C_{(t)ij}(\tau)$ is

$$E_{(t)ij}(s) = C_{(t)ij}(t-s) - \delta_{ij}.$$

C. Rate of deformation tensor

$$d_{ij} \equiv \tfrac{1}{2}\left(\frac{\partial v_i}{\partial x_j} + \frac{\partial v_j}{\partial x_i}\right).$$

D. Special deformations

1. Simple shear

Let a Cartesian coordinate system be chosen so that

the planes normal to the x_2 axis move in the x_1 direction:

$$x_1(t) = X_1 + \gamma(t)X_2, \quad \xi_1(\tau) = x_1(t) + [\gamma(\tau) - \gamma(t)]\, x_2(t),$$

$$x_2(t) = X_3, \quad \xi_2(\tau) = x_2(t),$$

$$x_3(t) = X_3, \quad \xi_3(\tau) = x_3(t).$$

Infinitesimal deformation:

$$e_{12} = e_{21} = \tfrac{1}{2}\gamma(t),$$

$$e_{11} = e_{22} = e_{33} = e_{13} = e_{23} = e_{31} = e_{32} = 0.$$

2. Extension

$$x_1(t) = \lambda_1(t)X_1, \quad x_2(t) = \lambda_2(t)X_2, \quad x_3(t) = \lambda_2(t)X_3.$$

For an incompressible material: $\lambda_2(t) = [\lambda_1(t)]^{1/2}$.

Infinitesimal extension. If $\lambda_1 - 1 \equiv \epsilon(t)$, $\lambda_2 - 1 \equiv -\beta(t)$, and $|\epsilon(t)| \ll 1$,

$$e_{11} = \epsilon(t), \quad e_{22} = e_{33} = -\beta(t), \quad e_{ij} = 0, \; i \neq j.$$

3. Isotropic expansion or isotropic compression

$$x_1(t) = \mu(t)X_1, \quad x_2(t) = \mu(t)X_2, \quad x_3(t) = \mu(t)X_3.$$

Infinitesimal isotropic expansion or compression. The fractional change of volume $\Delta V/V$, $|\Delta V/V| \ll 1$:

$$\Delta V/V \approx 3[\mu(t) - 1].$$

4. Torsion of a cylinder (radius a and height l)

A cylindrical coordinate system is chosen with z axis coincident with the axis of the cylinder and origin at the base of the cylinder. During the deformation a material point originally at (r,θ,z) moves to $(r,\theta + \Theta(z,t),z)$. We call $\Theta(l,t) \equiv \phi(t)$ and $\psi(t) \equiv \phi(t)/l$.

E. Viscometric flows

This class of flows, in which lamellae of the fluid slide over one another, is often used to determine material properties. The following examples are the most widely used.

1. Simple shearing flow

In a Cartesian coordinate system, the components of the velocity are

$$v_1 = v(x_2,t), \quad v_2 = v_3 = 0; \quad \dot\gamma = dv/dx_2.$$

The flow is assumed to occur between two large parallel plates, e.g., at $x_2 = 0$ and $x_2 = h$, with the former stationary and the latter moving with velocity $V(t)$. If the flow is steady, and to a good approximation for an unsteady flow slow enough for inertial effects to be small,

$$\dot\gamma = V/h.$$

2. Couette flow

In a cylindrical coordinate system, the physical components of the velocity are

$$v_r = r\omega(r,t), \quad v_\theta = v_z = 0; \quad \dot\gamma = r\, d\omega/dr.$$

The flow is assumed to occur between two long coaxial cylinders, e.g., at $r = R_1$ and $r = R_2 > R_1$, one of which rotates at an angular speed Ω. If $(R_2 - R_1)/R_1 \ll 1$ and the flow is steady (and to a good approximation for an unsteady flow slow enough for inertial effects to be small),

$$\dot\gamma \approx \Omega R_2/(R_2 - R_1).$$

3. Poiseuille flow

In a cylindrical coordinate system, the physical components of the velocity are

$$v_z = v(r,t), \quad v_\theta = v_r = 0; \quad \dot\gamma = dv/dr.$$

The flow is assumed to take place in a long cylinder of radius R.

4. Cone-plate flow

In a spherical coordinate system, the physical components of the velocity are

$$v_\phi = r \sin\theta\, \omega(\theta,t), \quad v_r = v_\theta = 0; \quad \dot\gamma = (d\omega/d\theta)\sin\theta.$$

This flow generally takes place in the region $0 < r < R$ between a cone and disk. The disk lies in the plane $\theta = \pi/2$ and the cone is the surface $\theta = \pi/2 \pm \alpha$, where $\alpha \ll 1$. Either the cone or disk rotates with angular velocity $\Omega(t)$. If the flow is slow enough for inertial effects to be negligible,

$$\dot\gamma \approx \Omega/\alpha.$$

F. Uniaxial extensional or elongational flow

$$v_1 = \dot\epsilon x_1, \quad v_2 = -\tfrac{1}{2}\dot\epsilon x_2, \quad v_3 = -\tfrac{1}{2}\dot\epsilon x_3.$$

If the body is a circular cylinder of length $L(t)$ and radius $R(t)$, and $\dot\epsilon$ is constant,

$$L = L_0 \exp(\dot\epsilon t), \quad R = R_0 \exp(-\dot\epsilon t/2).$$

19.03. STRESS

A. Body forces and stress tensor

It is assumed that two types of forces act in the body. The external body forces (e.g., due to gravity) are characterized by $\mathbf{b}$, a force density per unit mass. The body force acting on an infinitesimal volume dV is $\mathbf{b}\,\rho dV$, where ρ is the density.

The other type of force is the contact force, the force exerted by one part of the body on a neighboring part across the surface separating them; it can be characterized by a vector field $\mathbf{t}(\mathbf{x},\mathbf{n})$ called the *stress vector,* which is the contact force per unit area. It depends on the position $\mathbf{x}$ and the orientation of the separating surface at that point, as indicated by $\mathbf{n}$, the external unit normal vector on the surface at $\mathbf{x}$. It can be shown that there exists a stress tensor $\mathbf{T}(\mathbf{x})$ such that

$$\mathbf{t}(\mathbf{x},\mathbf{n}) = \mathbf{T}(\mathbf{x})\mathbf{n}, \quad t_i(x_l,n_k) = T_{ij}(x_l)n_j,$$

and that this tensor is symmetric:

$$T_{ij}(x_l) = T_{ji}(x_l).$$

If a cylindrical coordinate system is used, the physical components of the stress are indicated by T_{rr}, $T_{r\theta}$, T_{rz}, etc.; if spherical coordinates by T_{rr}, $T_{r\theta}$, $T_{r\phi}$, etc.

B. Special stresses

1. Isotropic pressure

$$T_{11}(t) = T_{22}(t) = T_{33}(t) \equiv -p(t); \quad T_{ij} = 0, \quad i \neq j$$

where p is called the pressure.

2. Shear stress

There exists a Cartesian coordinate system such that the only nonzero components of **T** are

$$T_{12}(t) = T_{21}(t) \equiv S(t).$$

3. Tensile stress

There exists a Cartesian coordinate system such that the only nonzero component of the stress is

$$T_{11}(t) \equiv \sigma(t),$$

where $\sigma(t)$ is called the tensile stress.

19.04. NONLINEAR ELASTICITY

A. Introduction

Many vulcanized rubbers and similar materials can undergo large deformations on application of a force and then return to the original shape when the force is removed. They exhibit elastic behavior, but the classical theory of linear elasticity, which deals with very small deformations, is clearly not applicable. Based on the assumption that the stress is an essentially arbitrary function of the deformation, the theory of finite, or nonlinear, elasticity leads to a constitutive equation of relatively simple form relating the stress and deformation (e.g., **B**) tensors. For the case of the *incompressible* isotropic elastic material (to which we limit ourselves here), it is possible to obtain exact solutions for many types of deformation.

To specify the properties of the material requires two functions W_1 and W_2, each of which is a function of two variables. One consequence of nonlinearity is the existence of normal stress effects; to perform a torsion on a cylindrical rod requires only a torque in the case of linear elasticity but demands also compressive normal forces on the ends in nonlinear elasticity.

B. Incompressible isotropic elastic materials, constitutive equation

The general constitutive equation for an incompressible isotropic elastic material can be written

$$\mathbf{T} = -p\mathbf{1} + W_1\mathbf{B} - W_2\mathbf{B}^{-1},$$

where

$$W_i = W_i(I_B, II_B), \quad i = 1, 2$$

where I_B and II_B are the first and second principal invariants of **B**.

If the material has a *strain energy function* $W(I_B, II_B)$ then

$$W_1 = \partial W/\partial I_B, \quad W_2 = \partial W/\partial II_B.$$

The value of W is the work required per unit volume to perform the deformation. Figure 19.04.B.1 shows graphs of W_1 and W_2 for a vulcanized rubber.

C. Special deformations

1. Extension

$$\frac{f}{A_0} = \frac{T_{xx}}{\lambda} = 2\left(\lambda - \frac{1}{\lambda^2}\right)\left(W_1 + \frac{1}{\lambda}W_2\right),$$

where f is the applied force along the axis and A_0 the original cross-sectional area. Figure 19.04.C.1.a shows a plot of force-extension data for a series of vulcanized rubber, plotted as $\frac{1}{2} f/A_0(\lambda - \lambda^{-2})$ vs λ^{-1}.

2. Simple shear

The shear stress is a nonlinear function of the shear strain γ:

$$T_{xy} = 2\gamma(W_1 + W_2), \quad T_{yz} = T_{xz} = 0,$$

and, in addition, normal stresses are required, such that

$$T_{xx} - T_{zz} = 2\gamma^2 W_1, \quad T_{yy} - T_{zz} = -2\gamma^2 W_2,$$
$$T_{xx} - T_{yy} = \gamma T_{xy}.$$

3. Torsion of a solid cylinder

Forces are applied only on the plane end surfaces; not only a torque,

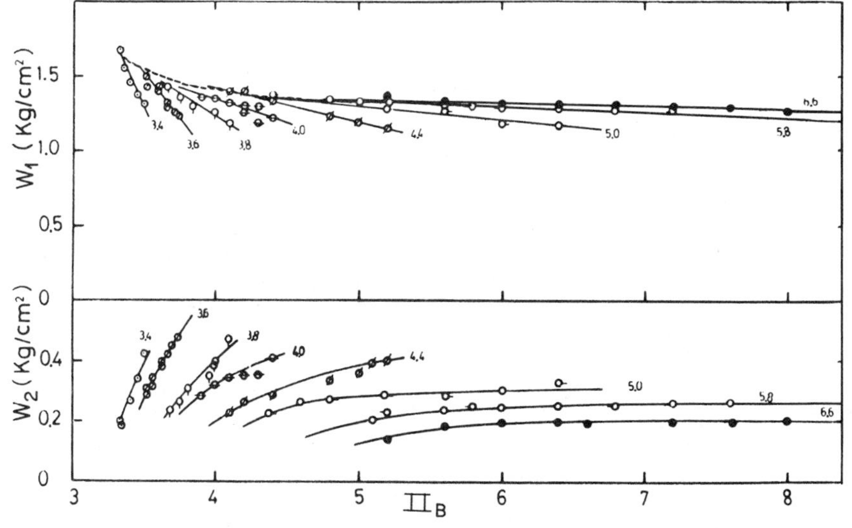

FIGURE 19.04.B.1. Plots of W_1 and W_2 for rubber as a function of II_B for various values of I_B (indicated on each curve). From Y. Obata, S. Kawabata, and H. Kawai, J. Polym. Sci. Part A2 **8**, 903 (1970).

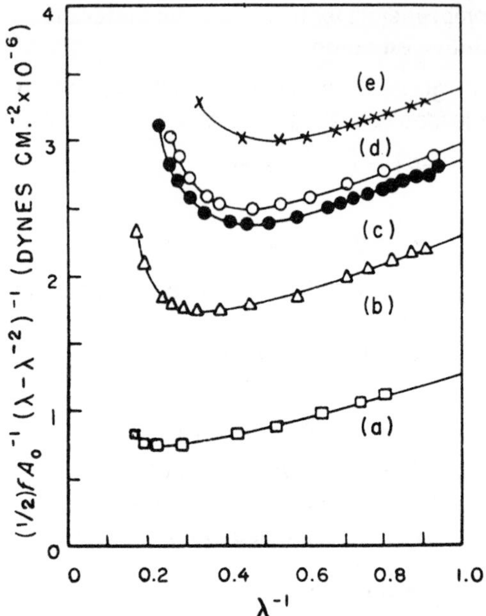

FIGURE 19.04.C.1.a. Experimental force-extension data for peroxide cross-linked natural rubber containing (a) 1, (b) 2, (c) 3, (d) 4, and (e) 5% peroxide. From L. Mullins, J. Appl. Polym. Sci. **2**, 257 (1959).

$$M = 4\pi\psi \int_0^a r^3 (W_1 + W_2) dr,$$

must be supplied, but also a normal force,

$$N = -\pi\psi^2 \int_0^a 2r^3 (W_1 + 2W_2) dr.$$

D. Mooney-Rivlin nonlinear solid

The first terms in an expansion of $W(I_B, II_B)$ give rise to the Mooney-Rivlin equation, expected to be valid for small deformations ($I_B \approx 3$, $II_B \approx 3$, $\lambda \approx 1$, etc.):

$$W = C_1(I_B - 3) + C_2(II_B - 3), \quad W_1 = C_1, \quad W_2 = C_2,$$

where C_1 and C_2 are constants:

$$\mathbf{T} = -p\mathbf{1} + C_1\mathbf{B} - C_2\mathbf{B}^{-1}.$$

Extension:

$$f/A_0 = 2(\lambda - 1/\lambda^2)(C_1 + C_2/\lambda).$$

This expression has led to the widely used plots, such as seen in Figure 19.04.C.1.a of $f/2\,A_0(\lambda - \lambda^{-2})$ vs λ^{-1}. Often a straight line is obtained for λ between ≈ 1.1 and 2. However, the slope and intercept of such a line cannot be identified with the C_2 and C_1 of the small deformation approximation.

19.05. LINEAR VISCOELASTICITY

A. Introduction

The basic concept of the theory of viscoelasticity is that the stress at a given time depends on the deformation over all previous time. Mathematically, this is expressed by setting the stress equal to a functional of the previous history of the deformation.

In linear viscoelasticity, which is a useful model if the stresses are small enough, this functional is linear, so that the stress can be written in terms of integrals which require a knowledge of the previous history of the strain. The properties of the material are specified by functions of time (or their equivalents).

The material functions in linear viscoelasticity are generally defined in terms of simple experiments in shear, isotropic compression, and extension. In addition, specific stress and strain histories are used to define the material functions. Here, we discuss the various commonly used basic histories in terms of simple shear (see Figures 19.05.A.1): the creep experiment in which a shear stress is suddenly imposed and then held constant, the stress relaxation experiment in which a shear strain is suddenly imposed and then held constant, and the sinusoidal shear strain. An alternative fourth method of specifying the material properties is the relaxation and retardation spectra.

Analogous material functions are based on corresponding experiments in isotropic compression and extension. For an isotropic material, the theory of linear viscoelasticity gives a relation between tensile, shear, and isotropic compression material functions.

Given an arbitrary strain history, it is possible to calculate the corresponding stress history (and vice versa) by use of Boltzmann's constitutive equation of linear viscoelasticity.

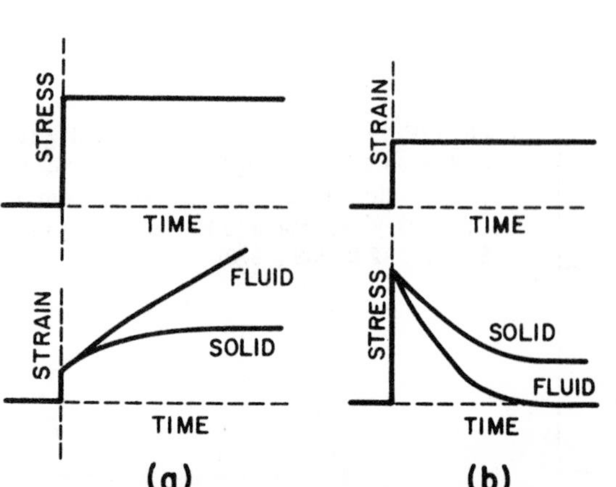

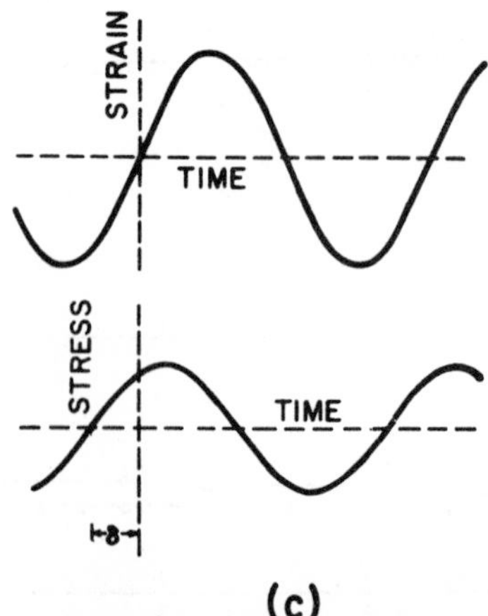

FIGURES 19.05.A.1. Schematic diagrams of basic experiments in viscoelasticity: (a) creep, (b) stress relaxation, and (c) steady-state sinusoidal stress and strain.

B. Basic experiments and viscoelastic functions for isotropic materials

1. Simple shear: Creep, recovery

If a shear stress is suddenly imposed at $t=\theta$ and then held fixed at S_0, *creep* is said to occur, i.e.,

$$S(t)=0,\quad t<\theta$$
$$=S_0,\quad t>\theta;$$

the shear strain is

$$\gamma(t)=0,\quad t<\theta$$
$$=S_0\,J(t-\theta),\quad t>\theta$$

where $J(t)$ is a non-negative, smooth, monotonically increasing function of t, defined only for positive values of its argument. It is called the *shear creep compliance.* The limiting value $J(0^+)=J_0$ is called the *instantaneous shear creep compliance.*

If the limit $J(\infty)$ exists, it is called J_e, the *equilibrium shear creep compliance;* the material is a viscoelastic solid.

If $J(t)$ asymptotically approaches a linear dependence on t as t increases indefinitely, the material is a viscoelastic fluid. It is convenient to write $J(t)$ as

$$J(t)=J_R(t)+t/\eta,$$

where η is the *viscosity,* and $J_R(t)$, the *recoverable shear compliance,* is a non-negative, monotonically increasing function of its argument. The limiting value $J_R(\infty)$ is the *steady-state shear compliance* J_e^0.

Plots of $J(t)$ are seen in Figure 19.05.B.1.a for a high-molecular-weight (592 000) polystyrene of narrow molecular-weight distribution.

Time-temperature superposition. For many materials the logarithmic plot of $J(t)$ data for the temperature T can be superimposed on that for the temperature T_0 by a translation by log a_{0T} along the log$t$ axis and by a much smaller translation, logb_{0T}, along the log$J$ axis. In such a case, using the notation $J(t,T)$ to indicate the function $J(t)$ at the temperature T, we can then write

$$J(t,T)/b_{0T}=J(t/a_{0T},T_0).$$

The material is said to be *thermorheologically simple,* and *time-temperature superposition* (or time-temperature equivalence) is said to be applicable. If the curves for all temperatures are superposed on the log$J(t,T_0)$ graph, T_0 is called the *reference temperature,* t/a_{0T} is the *reduced time* t_r, and $J(t,T_0)/b_{0T}$ is the *reduced shear creep compliance* $J_r(t,T)$, so that we can write

$$J_r(t,T)\equiv J(t,T)/b_{0T}=J(t/a_{0T},T_0)\equiv J(t_r,T_0).$$

The graph of J_r vs t_r resulting from such a superposition is called a *reduced curve, composite plot,* or *master curve.*

The composite plot obtained from the $J(t)$ data for the polystyrene of Figure 19.05.B.1.a is shown in Figure 19.05.B.1.b with $T_0=100°C$, $b_{0T}=1$, and a_{0T} given by the WLF equation

$$\log a_{0T}=-13.412+456/(T-66.0°C).$$

The composite plot for $J_R(t)$ is shown in the same figure.

Recovery is said to occur after the stress is removed following a creep experiment of duration θ. The strain during the recovery is

$$\gamma(\theta+s)=S_0[J(\theta+s)-J(s)],\quad s>0$$

where s is the time since the recovery experiment was begun. The recovery curve $R(\theta,s)$ is defined as

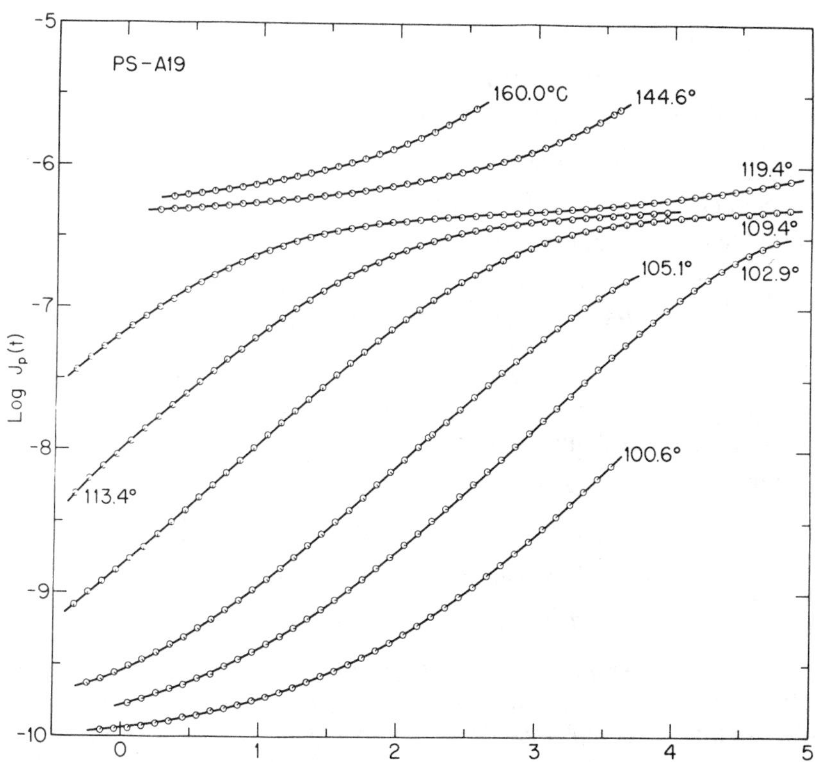

FIGURE 19.05.B.1.a. Logarithmic plots of $J(t)$ for polystyrene (molecular weight 592 000) of narrow molecular-weight distribution at the indicated temperature. Unpublished data of D. J. Plazek.

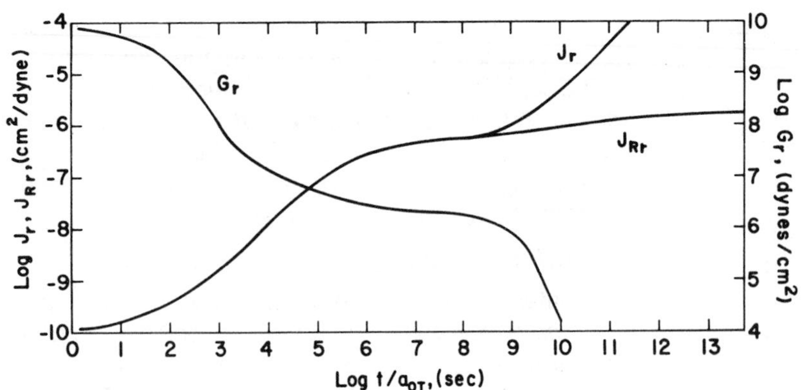

FIGURE 19.05.B.1.b. Logarithmic reduced plots of $J(t)$, $J_R(t)$, and $G(t)$ for polystyrene of Figure 19.05.B.1.a. $T_0 = 100°C$. Unpublished data of D. J. Plazek.

$$R(\theta,s) \equiv [\gamma(\theta^-) - \gamma(\theta+s)]/S_0,$$
$$R(\theta,s) = J(\theta) - J(\theta+s) + J(s),$$
$$R(\infty,s) = J(s) \text{ (solid)}$$
$$= J(s) - s/\eta = J_R(s) \text{ (fluid)},$$
$$R(\infty,\infty) = J_e^0 \text{ (fluid)}.$$

2. Simple shear: Stress relaxation

If a shear strain is suddenly imposed at $t=\theta$ and then held fixed at γ_0, i.e.,

$$\gamma(t) = 0, \quad t<\theta$$
$$= \gamma_0, \quad t>\theta$$

the shear stress is a decreasing function of time:

$$S(t) = 0, \quad t<\theta$$
$$= \gamma_0 G(t-\theta), \quad t>\theta$$

where $G(t)$ is a non-negative, smooth, monotonically decreasing function of time, defined only for positive values of its argument. It is called the *shear relaxation modulus.* The limiting value $G(0^+) = G_0$ is called the *instantaneous shear relaxation modulus.*

If $G(\infty)=0$, the material is called a *viscoelastic fluid;* if $G(\infty) = G_e$, where G_e is nonzero, the material is called a *viscoelastic solid* and G_e the *equilibrium shear relaxation modulus.* In this chapter equations containing G_e apply to the viscoelastic solid; the corresponding equation for the fluid may be obtained by setting $G_e = 0$.

Time-temperature superposition. If the material is thermorheologically simple, then

$$G_r(t,T) \equiv b_{0T} G(t,T) = G(t/a_{0T},T_0) \equiv G(t_r,T_0).$$

A reduced plot of the shear relaxation modulus is shown in Figure 19.05.B.1.b for the same polystyrene whose reduced creep compliance (using the same a_{0T}, b_{0T}, and T_0) is also shown in that figure.

3. Simple shear: Sinusoidal strain

If the sinusoidal strain

$$\gamma(t) = \gamma_0 \sin\omega t$$

is imposed, then the stress, in the steady state, is also sinusoidal but there is a phase difference δ between the stress and strain:

$$S(t) = S_0 \sin(\omega t + \delta).$$

There are several equivalent ways of expressing the relation between stress and strain and of defining viscoelastic properties:

$$S(t) = \gamma_0[G'(\omega)\sin\omega t + G''(\omega)\cos\omega t],$$

where the viscoelastic functions $G'(\omega)$ and $G''(\omega)$ are called the *shear storage modulus* and *shear loss modulus,* respectively. Using complex notation,

$$\gamma^*(t) = \gamma_0 e^{i\omega t}, \quad S^*(t) = S_0 e^{i(\omega t+\delta)}.$$

Complex shear dynamic modulus:

$$S^*(t)/\gamma^*(t) \equiv G^*(\omega) = G'(\omega) + iG''(\omega).$$

Absolute dynamic shear modulus:

$$S_0/\gamma_0 \equiv G_d(\omega) = \{[G'(\omega)]^2 + [G''(\omega)]^2\}^{1/2}.$$

Dynamic shear viscosity:

$$\eta'(\omega) = G''(\omega)/\omega.$$

Complex dynamic shear compliance:

$$\gamma^*(t)/S^*(t) \equiv J^*(\omega) = 1/G^*(\omega)$$
$$= J'(\omega) - iJ''(\omega).$$

Shear storage compliance:

$$J'(\omega) = G'(\omega)/[G_d(\omega)]^2.$$

Shear loss compliance:

$$J''(\omega) = G''(\omega)/[G_d(\omega)]^2.$$

Shear loss tangent:

$$\tan\delta = G''(\omega)/G'(\omega) = J''(\omega)/J'(\omega).$$

As a group, these material properties are called *dynamic mechanical properties in shear.*

Time-temperature superposition. Reduced dynamic mechanical properties may be defined in accordance with the relations

$$G_r^\dagger(\omega,T) \equiv b_{0T} G^\dagger(\omega,T) = G^\dagger(\omega a_{0T},T_0) \equiv G^\dagger(\omega_r,T_0),$$
$$J_r^\dagger(\omega,T) \equiv J^\dagger(\omega,T)/b_{0T} = J^\dagger(\omega a_{0T},T_0) \equiv J^\dagger(\omega_r,T_0),$$
$$\tan\delta_r(\omega,T) \equiv \tan\delta(\omega,T) = \tan\delta(\omega a_{0T},T_0),$$
$$\eta_r'(\omega,T) \equiv \eta'(\omega,T)/\eta'(0,T) = \eta'(\omega a_{0T},T_0)/\eta'(0,T_0),$$

where † may stand for any one of $'$, $''$, $*$, or $_d$.

For the same polystyrene whose reduced creep compliance is shown in Figure 19.05.B.1.b (using the same a_{0T}, b_{0T}, and T_0), reduced plots of G', G'', J', and J''

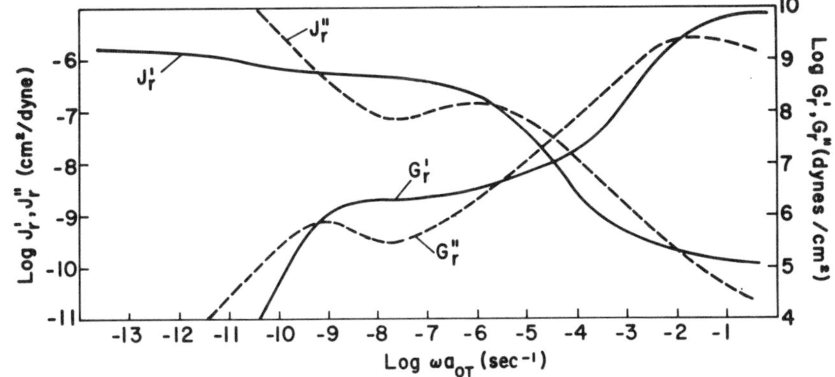

FIGURE 19.05.B.3.a. Logarithmic reduced plots of $G'(\omega)$, $G''(\omega)$, $J'(\omega)$, and $J''(\omega)$ for polystyrene of Figure 19.05.B.1.a. $T_0 = 100°C$. Unpublished data of D. J. Plazek.

are shown in Figure 19.05.B.3.a.

Work. Work per unit volume between t and $t+\theta$:

$$W(\theta,t) = \tfrac{1}{2}\omega\gamma_0^2 G''(\omega)\theta + \tfrac{1}{2}G_d(\omega)\gamma_0^2 \times \sin\omega\theta \sin(\omega\theta + 2\omega t + \delta).$$

Energy dissipated per cycle per unit volume:

$$\mathscr{E}_d = \pi\gamma_0^2 G''(\omega) = \pi S_0^2 J''(\omega).$$

4. Simple shear: Spectra

If $G(t)$ can be represented by an expression of the form

$$G(t) - G_e = \int_{-\infty}^{\infty} H(\tau)\exp(-t/\tau)\, d\ln\tau,$$

where $H(\tau)$ is smooth and non-negative, $H(\tau)$ is called the *(continuous) relaxation spectrum* and τ a *relaxation time.* If $G(t)$ can be represented by a sum of the form

$$G(t) - G_e = \sum_{i=1}^{N} G_i \exp(-t/\tau_i),$$

where G_i and τ_i are positive constants, it is said to possess a *(discrete) relaxation spectrum* where G_i is the strength of the ith relaxation time τ_i.

If $J(t)$ can be represented by an expression of the form

$$J(t) - J_0 - t/\eta = \int_{-\infty}^{\infty} L(\lambda)[1 - \exp(-t/\lambda)]\, d\ln\lambda,$$

where $L(\lambda)$ is smooth and non-negative, $L(\lambda)$ is called the *(continuous) retardation spectrum* and λ a *retardation time.* If $J(t)$ can be represented by an expression of the form

$$J(t) - J_0 - t/\eta = \sum_{i=1}^{n} J_i[1 - \exp(-t/\lambda_i)],$$

where the J_i and λ_i are positive constants, it is said to possess a *(discrete) retardation spectrum* where J_i is the strength of the retardation time λ_i.

Time-temperature superposition. If the material is thermorheologically simple, reduced spectra may be defined:

$$H_r(\tau,T) \equiv b_{0T}H(\tau,T) = H(\tau/a_{0T},T_0) \equiv H(\tau_r,T_0),$$
$$L_r(\lambda,T) \equiv L(\lambda,T)/b_{0T} = L(\lambda/a_{0T},T_0) \equiv L(\lambda_r,T_0).$$

5. Simple shear: Viscoelastic constants

The material constants characterize the limiting behavior of the viscoelastic functions.

$$G_0 J_0 = 1,$$
$$G'(0) = G_e = 1/J_e = 1/J'(0) \quad \text{(solid)},$$
$$J'(0) = J_e^0 \quad \text{(fluid)},$$
$$1/\eta = \lim_{t\to\infty} dJ(t)/dt = \lim_{\omega\to 0}\omega J''(\omega) \quad \text{(fluid)},$$
$$\eta = \eta'(0) = \int_0^{\infty} G(t)dt \quad \text{(fluid)},$$
$$\eta'(0) = \int_0^{\infty}[G(t) - G_e]dt = \int_{-\infty}^{\infty}\tau H(\tau)\, d\ln\tau$$
$$= \sum_i G_i\tau_i = \frac{2}{\pi}\int_0^{\infty}[G'(\omega) - G_e]\frac{d\omega}{\omega^2},$$
$$\eta^2 J_e^0 = \int_0^{\infty} tG(t)dt = \lim_{\omega\to 0}\frac{G'(\omega)}{\omega^2}$$
$$= \int_{-\infty}^{\infty}\tau^2 H(\tau)\, d\ln\tau = \sum_i G_i\tau_i^2 \quad \text{(fluid)},$$
$$G_0 = G'(\infty) = G_e + \sum_i G_i = G_e + \int_{-\infty}^{\infty} H(\tau)\, d\ln\tau.$$

6. Simple shear: Relations among viscoelastic functions

There is only one independent viscoelastic function for shear. Interrelations among the functions defined above can be obtained from the theory of linear viscoelasticity. [From the expressions for a fluid, corresponding expressions for a viscoelastic solid are obtained by substituting J_e *for* J_e^0 and $J(t)$ for $J_R(t)$ and by omitting terms containing η.]

$$\int_0^t J(t-\theta)G(\theta)d\theta = t,$$
$$p\bar{J}(p) = 1/p\bar{G}(p),$$
$$G'(\omega) = G_e + \omega\int_0^{\infty}[G(t) - G_e]\sin\omega t\, dt,$$
$$G''(\omega) = \omega\int_0^{\infty}[G(t) - G_e]\cos\omega t\, dt,$$
$$J'(\omega) = J_e^0 - \omega\int_0^{\infty}[J_e^0 - J_R(t)]\sin\omega t\, dt, \quad \text{(fluid)}$$
$$J''(\omega) = \frac{1}{\omega\eta} + \omega\int_0^{\infty}[J_e^0 - J_R(t)]\cos\omega t\, dt, \quad \text{(fluid)}$$

$$G(t) - G_e = \frac{2}{\pi}\int_0^{\infty} \frac{G'(\omega) - G_e}{\omega}\sin\omega t\, d\omega$$

$$= \frac{2}{\pi}\int_0^{\infty} \frac{G''(\omega)}{\omega}\cos\omega t\, d\omega,$$

$$J(t) = J_0 + \frac{2}{\pi}\int_0^{\infty} \frac{J'(\omega) - J_0}{\omega}$$

$$\times \sin\omega t\, d\omega + t/\eta, \quad \text{(fluid)}$$

$$J(t) = J_0 + \frac{2}{\pi}\int_0^{\infty} \frac{J''(\omega) - 1/\omega\eta}{\omega}$$

$$\times (1 - \cos\omega t) d\omega + t/\eta, \quad \text{(fluid)}.$$

$$G'(\omega) - G_e = \int_{-\infty}^{\infty} \frac{H(\tau)\omega^2\tau^2}{1 + \omega^2\tau^2}\, d\ln\tau,$$

$$G''(\omega) = \int_{-\infty}^{\infty} \frac{H(\tau)\omega\tau}{1 + \omega^2\tau^2}\, d\ln\tau,$$

$$J'(\omega) = J_0 + \int_{-\infty}^{\infty} \frac{L(\lambda)}{1 + \omega^2\lambda^2}\, d\ln\lambda,$$

$$J''(\omega) = \frac{1}{\omega\eta} + \int_{-\infty}^{\infty} \frac{\omega\lambda L(\lambda)}{1 + \omega^2\lambda^2}\, d\ln\lambda \quad \text{(fluid)}.$$

7. Isotropic pressure, isotropic compression

In the equations of Sec. 19.05.B replace S with $-p$, γ with $\Delta V/V$, G with K [e.g., $G(t)$ with $K(t)$, G_0 with K_0, $G''(\omega)$ with $K''(\omega)$], and J with B. $K(t)$ is called the *bulk relaxation modulus*, $B'(\omega)$ the *bulk storage compliance*, K_0 the *instantaneous bulk modulus*, etc.

8. Extension, tensile stress

In the equations of Sec. 19.05.B replace S with the tensile stress σ, G with E, J with D, H with H_e, L with L_e, η with η_e, and γ with ϵ. The function $E(t)$ is called the *tensile relaxation modulus*, $D(t)$ is the *tensile creep compliance*, etc. For the concurrent lateral contraction $\beta(t)$, another set of equations is required. For example, for the creep experiment, defined analogously to Sec. 19.05.B.1, under the stress σ_0,

$$\epsilon(t) = \sigma_0 D(t - \theta), \quad t > \theta$$

$$\beta(t) = \sigma_0[\tfrac{1}{2} J(t - \theta) - D(t - \theta)], \quad t > \theta.$$

For a tensile stress relaxation experiment, where a strain ϵ_0 is imposed at $t = \theta$, the concurrent lateral contraction is

$$\beta(t) = \epsilon_0 \nu(t - \theta), \quad t > \theta$$

and the tensile stress

$$\sigma(t) = \epsilon_0 E(t - \theta), \quad t > \theta$$

where $\nu(t)$ is the *viscoelastic Poisson's ratio.*

Corresponding to the sinusoidal tensile strain $\epsilon^*(t) = \epsilon_0^{i\omega t}$, the lateral contraction is $\beta^*(t) = \nu^*(\omega)\epsilon^*(t)$ and the tensile stress $\sigma^*(t) = E^*(\omega)\epsilon^*(t)$, where $\nu^*(\omega)$ is the *complex Poisson's ratio.*

9. Relations among shear, tensile, and bulk properties

$$\bar{E}(p) = 9\bar{K}(p)\bar{G}(p)/[3\bar{K}(p) + \bar{G}(p)],$$

$$D(t) = \tfrac{1}{3} J(t) + \tfrac{1}{9} B(t),$$

$$D^*(\omega) = \tfrac{1}{3} J^*(\omega) + \tfrac{1}{9} B^*(\omega) = J^*/2(1 + \nu^*).$$

If $B(t) \ll J(t)$, or equivalently $K(t) \gg G(t)$,

$$D(t) \approx \tfrac{1}{3} J(t), \quad E(t) \approx 3G(t),$$

$$E^*(\omega)/G^*(\omega) = J^*(\omega)/D^*(\omega) \approx 3.$$

C. Boltzmann constitutive equation of linear viscoelasticity

Arbitrary stress and strain histories:

$$T_{ij}(t) = 2\int_{-\infty}^{t} G(t - \theta)\frac{\partial e_{ij}(\theta)}{\partial \theta}\, d\theta$$

$$+ \delta_{ij}\int_{-\infty}^{t} [K(t - \theta) - \tfrac{2}{3}G(t - \theta)]\, \frac{\partial e_{kk}(\theta)}{\partial \theta}\, d\theta,$$

$$e_{ij}(t) = \tfrac{1}{2}\int_{-\infty}^{t} J(t - \theta)\frac{\partial T_{ij}(\theta)}{\partial \theta}\, d\theta$$

$$- \delta_{ij}\int_{-\infty}^{t} [\tfrac{1}{6}J(t - \theta) - \tfrac{1}{9}B(t - \theta)]\, \frac{\partial T_{kk}(\theta)}{\partial \theta}\, d\theta.$$

1. Simple shear, linear viscoelasticity

Continuous strain history $\gamma(t)$:

$$S(t) = \int_{-\infty}^{t} G(t - \theta)\frac{d\gamma(\theta)}{d\theta} d\theta,$$

$$\bar{S}(p) = p\bar{G}(p)\bar{\gamma}(p),$$

$$S(t) = G_0\gamma(t) + \int_0^{\infty} \gamma(t - s)\frac{dG(s)}{ds}\, ds.$$

Continuous stress history $S(t)$:

$$\gamma(t) = \int_{-\infty}^{t} J(t - \theta)\frac{dS(\theta)}{d\theta}\, d\theta.$$

D. Simple linear viscoelastic models

1. Maxwell fluid

If a viscoelastic fluid has a discrete spectrum with only one relaxation time τ of strength G, the material is called a *Maxwell fluid.*

$$dS/dt + S/\tau = G d\gamma/dt,$$

$$S(t) = \int_{-\infty}^{t} Ge^{-(t - \theta)/\tau}\frac{d\gamma(\theta)}{d\theta}\, d\theta,$$

$$G(t) = Ge^{-t/\tau}, \quad J(t) = 1/G + t/\tau G.$$

2. Voigt solid

A viscoelastic solid with a discrete retardation spectrum having only one retardation time λ of strength J and with $J_0 = 0$ is called a *Voigt solid.*

$$\lambda d\gamma/dt + \gamma = SJ,$$

$$\gamma(t) = \int_{-\infty}^{t} J(1 - e^{-(t - \theta)/\lambda})\frac{dS(\theta)}{d\theta}\, d\theta,$$

$$J(t) = J(1 - e^{-t/\lambda}), \quad G(t) = (\lambda/J)\delta(t) + 1/J,$$

where $\delta(t)$ is the Dirac δ function.

3. Standard linear solid or standard inelastic solid

$$a_1 dS/dt + a_0 S = b_1 d\gamma/dt + b_0\gamma,$$

$$G(t) = \frac{b_0}{a_0} + \left(\frac{b_1}{a_1} - \frac{b_0}{a_0}\right)\exp\left(-\frac{a_0}{a_1}t\right),$$

$$J(t) = \frac{a_0}{b_0} + \left(\frac{a_1}{b_1} - \frac{a_0}{b_0}\right)\exp\left(-\frac{b_0}{b_1}t\right).$$

E. Time-temperature superposition, shift factors

The shift factors can be expressed in terms of the viscoelastic constants:

$$a_{0T} = J_e^0(T)\eta(T)/J_e^0(T_0)\eta(T_0) \quad \text{(fluid)},$$

$$b_{0T} = J_e^0(T)/J_e^0(T_0) \quad \text{(fluid)},$$

$$a_{0T} = J_e(T)\eta'(0,T)/J_e(T_0)\eta'(0,T) \quad \text{(solid)},$$

$$b_{0T} = J_e(T)/J_e(T_0) \quad \text{(solid)},$$

where $\eta(T)$ is the value of η at T, etc. For some polymeric materials

$$b_{0T} = T_0\rho(T_0)/T\rho(T)$$

is widely used, and for polymeric fluids

$$a_{0T} = \eta(T)T_0\rho(T_0)/\eta(T_0)T\rho(T).$$

The empirical WLF equation

$$\log a_{0T} = -c_1^0(T-T_0)/(c_2^0 + T - T_0)$$

is often useful. See Sec. 13.17.B.5 for a table and discussion.

F. Linear viscoelastic properties of high polymers

The viscoelastic functions exhibited above for a polystyrene have features typical of other high-molecular-weight linear amorphous polymers. These characteristic features and the relevant viscoelastic constants are discussed here. Tables and more discussion are found in Sec. 13.17.

1. Flow or terminal region

At long times the logarithmic plot of $J(t)$ for this viscoelastic fluid has a unit slope corresponding to the dominance of the t/η term in $J(t)$ when t is large. This range of times is referred to as the *flow* or *terminal region.* In this time region, the $J_R(t)$ graph approaches its asymptotic limit J_e^0 and the $G(t)$ graph has a steep negative slope as $G(t)$ approaches its asymptotic zero limit. For the dynamic mechanical properties, there is a corresponding behavior at low frequencies as indicated in the relations of Sec. 19.05.B.5.

2. Rubber or entanglement plateau

For high-molecular-weight polymers there is a plateau in the logarithmic plots of $J(t)$, $J_R(t)$, and $G(t)$ which occurs at times just before the flow region; $G'(\omega)$ and $J'(\omega)$ have corresponding plateaus at frequencies somewhat higher than the flow region. The level of the plateau, called J_N^0 for the compliance functions and G_N^0 $(=1/J_N^0)$ for the moduli, is independent of the molecular weight M of the polymer for $M > \approx 2\times10^4$ and is associated with an entanglement network.

3. Glassy region

At very short times $J(t)$ approaches J_g, the *glasslike shear compliance,* and $G(t)$ approaches G_g $(=1/J_g)$, the *glasslike shear modulus,* generally of the order of 10^{10} dyn/cm^2. At high frequencies, $J'(\omega)$ and $G'(\omega)$ approach J_g and G_g, respectively.

4. Glass-rubber transition (or dispersion) zone

Between the glassy region and the rubber plateau, $J(t)$, $G(t)$, $J'(\omega)$, and $G'(\omega)$ change by several orders of magnitude; the loss functions [$G''(\omega)$, $\tan\delta$, $J''(\omega)$] have maxima at successively lower frequencies in this zone.

5. Secondary relaxations

If data are obtained on many polymers at frequencies higher than the glassy region, $\tan\delta$ exhibits one or more maxima lower than the maximum associated with the glass-rubber transition. They are said to arise from *secondary relaxations, dispersions,* or *transitions.* Almost always the data are obtained at a single frequency (often in the neighborhood of 1 Hz) over a range of temperatures. See Sec. 13.15.

19.06. NONLINEAR VISCOELASTICITY

A. Introduction

When deformations become large, the theory of linear viscoelasticity is no longer adequate. For example, in a shear stress relaxation experiment the ratio $S(t)/\gamma_0$ is no longer independent of γ_0 and there are normal stresses which are functions of time.

The theory of nonlinear viscoelasticity has received the greatest development for the case of the incompressible fluid, and we shall restrict our attention to this model, known as the *simple fluid.* It has been shown that, under very general assumptions, the stress can be expressed (to within an isotropic pressure) as a nonlinear functional of the relative strain tensor $\mathbf{E}_{(t)}(s)$ over all previous time.

To characterize the fluid in steady simple shearing and other viscometric flows, it can be shown that three functions of the shear rate are required. One of these functions characterizes the nonlinear shear-rate dependence of the shear stress. The other two relate to the normal stresses which, in contrast to the case of the Newtonian fluid, must be applied to maintain a viscometric flow.

B. Constitutive equation for incompressible fluid: Simple fluid

$$\mathbf{T}(t) = -p\mathbf{1} + \underset{s=0}{\overset{\infty}{\mathfrak{Q}}}\, \mathbf{E}_{(t)}(s),$$

where p is a constant and $\mathfrak{Q}$ an isotropic, smooth functional.

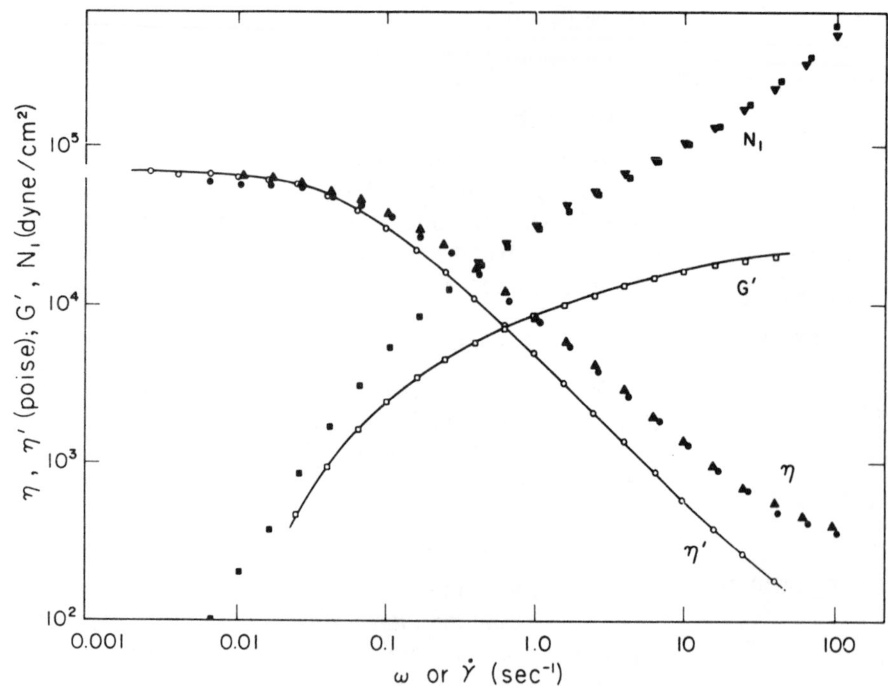

FIGURE 19.06.C.1.a. Viscometric functions for a 12% solution of polystyrene (MW $= 1.8\times10^6$) in tri-cresyl phosphate. Linear viscoelastic properties $G'(\omega)$ and $\eta'(\omega)$ are also shown. From W. W. Graessley, W. S. Park, and R. L. Crawley, Rheol. Acta **16**, 291 (1979).

C. Steady viscometric flows of incompressible simple fluids

1. Steady simple shearing flow, viscometric functions

To maintain steady simple shearing flow requires the application of a shear stress and a set of normal forces which are determined by three material functions called the *viscometric functions:* the *shear stress function* $S(\dot\gamma)$ or *viscosity function* $\eta(\dot\gamma)$, the *first* and *second normal stress functions* $N_1(\dot\gamma)$ and $N_2(\dot\gamma)$, or the *normal stress coefficients* $\Psi_1(\dot\gamma)$ and $\Psi_2(\dot\gamma)$:

$$T_{xy} = S(\dot\gamma) \equiv \dot\gamma\eta(\dot\gamma),$$
$$T_{xx} - T_{yy} = N_1(\dot\gamma) \equiv \dot\gamma^2\Psi_1(\dot\gamma),$$
$$T_{yy} - T_{zz} = N_2(\dot\gamma) \equiv \dot\gamma^2\Psi_2(\dot\gamma).$$

See Figure 19.06.C.1.a. The inverse function of $S(\dot\gamma)$ is indicated as $\dot\gamma(S)$. The normal stress functions can also be expressed as functions of S: $\hat N_1(S)$ and $\hat N_2(S)$.

Limiting values of viscometric functions. From the theory of the simple fluid, it can be shown that

$$\eta(0) = \eta, \quad \Psi_1(0) = 2\int_0^\infty sG(s)ds = 2\eta^2 J_e^0,$$

where η and J_e^0 are the linear viscoelastic constants discussed in Sec. 19.05.B.5 Because η is the limiting value of $\eta(\dot\gamma)$, it is often called the *zero shear viscosity* and given the symbol η_0 if this point is to be emphasized.

2. Steady Couette flow

$$T_{r\theta} = S(\dot\gamma), \quad T_{\theta\theta} - T_{rr} = N_1(\dot\gamma), \quad T_{rr} - T_{zz} = N_2(\dot\gamma).$$

The shear stress is related to M/l, the moment per unit height exerted on the fluid across the cylindrical surface, $r=$ const:

$$T_{r\theta} = M/2\pi r^2 l = S(r).$$

If the annulus between the cylinders is thin (i.e., $R_2 - R_1 \ll R_1$), $\dot\gamma(S_1) \approx \Omega R_1/(R_2 - R_1)$ is the shear rate corresponding to $S_1 \equiv S(R_1)$. $\hat N_1(S)$ can be determined from measurements of $T_{rr}(R_2) - T_{rr}(R_1)$, the difference between the normal stresses on the two cylindrical walls, since

$$T_{rr}(R_2) - T_{rr}(R_1) + \int_{R_1}^{R_2} \rho r[\omega(r)]^2 dr = \int_{R_1}^{R_2} \frac{\hat N_1(S)}{r} dr,$$

where ρ is the density and $\omega(r)$ the angular velocity of a fluid particle.

3. Steady Poiseuille flow

This type of flow is the basis for a large variety of capillary viscometers which are widely used for determining $\dot\gamma(S)$. The shear stress is given by

$$T_{rz} = rp/2l = S(r),$$

where p/l is the gradient of the total driving force (due to applied pressure and body forces). From data on Q, the volume rate of flow, as a function of p, it is possible to calculate $\dot\gamma(S)$ from a logarithmic plot of $D \equiv 4Q/\pi R^3$ vs $S_R \equiv S(R)$ by using the relation

$$\dot\gamma(S_R) = D\left(\frac{3}{4} + \frac{1}{4}\frac{d\log D}{d\log S_R}\right).$$

4. Steady cone-plate flow

$S(\dot\gamma)$ is easily determined from the torque M required to produce the angular velocity Ω of the moving element (cone or plate):

$$S(\dot\gamma) = 3M/2\pi R^3.$$

Two types of normal stress measurements are made. The normal stress $T_{\theta\theta}(r)$ on the plate, after applying the centrifugal force correction $\mathscr{I}_{CP} = 0.15\rho\Omega^2 r^2$, varies linearly with $\ln r$:

$$T_{\theta\theta}(r) + \mathscr{I}_{CP} = [N_1(\dot\gamma) + 2N_2(\dot\gamma)]\ln r + \text{const.}$$

More commonly, $N_1(\dot\gamma)$ is determined from the total normal force F acting on the plate in excess of that due to the ambient pressure:

$$N_1(\dot\gamma) = 2F/\pi R^2 + 0.15\rho R^2\Omega^2.$$

D. Steady uniaxial extensional flow of a circular cylinder

In the absence of body forces and of inertial effects

$$T_{11} - T_{22} = \dot\epsilon\eta_T(\dot\epsilon),$$

where $T_{11} - T_{22}$ is ordinarily the steady force per unit area imposed axially, and the *extensional* (or *Trouton*) viscosity $\eta_T(\dot\epsilon)$ is determined by the functional $\mathfrak{Q}$. In the limit of low $\dot\epsilon$,

$$\eta_T(0) = 3\eta = \eta_e.$$

See Figure 19.06.D.1.

E. Approximate constitutive equations

1. Second-order viscoelasticity

Among the approximations to the simple fluid are those known as *nth-order viscoelasticity.* Qualitatively these may be considered approximations useful when the deformation has been small in the recent past. *First-order viscoelasticity* is the Boltzmann equation of linear viscoelasticity (Sec. 19.05.C). The constitutive equation for a *second-order viscoelastic* incompressible fluid is

$$T_{ij}(t) = -p\delta_{ij} + \int_0^\infty \frac{dG(s)}{ds}E_{(t)ij}(s)ds$$
$$+ \int_0^\infty\int_0^\infty [a(s,r)E_{(t)ik}(s)E_{(t)kj}(r)$$
$$+ b(s,r)E_{(t)kk}(s)E_{(t)ij}(r)]dsdr,$$

where $a(s,r)$ and $b(s,r)$ are material functions determined by the functional $\mathfrak{Q}$. This expression includes all terms up to the square of the norm of the strain history $\mathbf{E}_{(t)}$.

If a simple shear history $\gamma(t)$ is imposed, the corresponding shear stress history $T_{12}(t)$ is given by the Boltzmann equation in terms of the material property $G(t)$. In addition, normal stresses must be imposed. The material property required for the normal stress difference $T_{11}(t) - T_{22}(t) \equiv N_1(t)$ is also $G(t)$. Other normal stress differences depend on the material function $a(t,s)$.

Stress relaxation in shear:

$$T_{12}(t) = \gamma_0 G(t),$$
$$N_1(t) \equiv T_{11}(t) - T_{22}(t) = \gamma_0 T_{12}(t) = \gamma_0^2 G(t).$$

2. Empirical constitutive equations for incompressible fluids

Lodge rubberlike liquid:

$$T_{ij}(t) = -p\delta_{ij} + \int_{-\infty}^{t} m_L(t-\tau)C_{(t)ij}^{-1}d\tau,$$

where $m_L(t-\tau)$ is a material function.

BKZ (Bernstein-Kearsley-Zapas) fluid:

$$T_{ij}(t) = -p\delta_{ij} + 2\int_{-\infty}^{t}\left(\frac{\partial U}{\partial I_B}B_{(t)ij}(\tau) - \frac{\partial U}{\partial II_B}B_{(t)ij}^{-1}(\tau)\right)d\tau,$$

where U is a function of I_B, II_B, and $t-\tau$. $I_B(t,\tau)$ and $II_B(t,\tau)$ are the first two principal invariants of $B_{(t)ij}$.

Oldroyd models:

$$T_{ij} + \lambda_1\frac{\mathscr{D}T_{ij}}{\mathscr{D}t} + \mu_0 T_{kk}d_{ij} - \mu_1(T_{ik}d_{kj} + d_{ik}T_{kj}) + \nu_1 T_{kl}d_{kl}\delta_{ij}$$
$$= 2\eta_0\left(d_{ij} + \lambda_2\frac{\mathscr{D}d_{ij}}{\mathscr{D}t} - 2\mu_2 d_{ik}d_{kj} + \nu_2 d_{kl}d_{kl}\delta_{ij}\right),$$

where the Jaumann derivative $\mathscr{D}/\mathscr{D}t$ of any tensor q_{ij} is defined as

$$\frac{\mathscr{D}q_{ij}}{\mathscr{D}t} \equiv \frac{\partial q_{ij}}{\partial t} + v^k\frac{\partial q_{ij}}{\partial x^k} - \omega_{ik}q_{kj} + \omega_{kj}q_{ik}.$$

Special cases of this eight-constant model are:

(1) Six-constant model: $\nu_1 = \nu_2 = 0$.

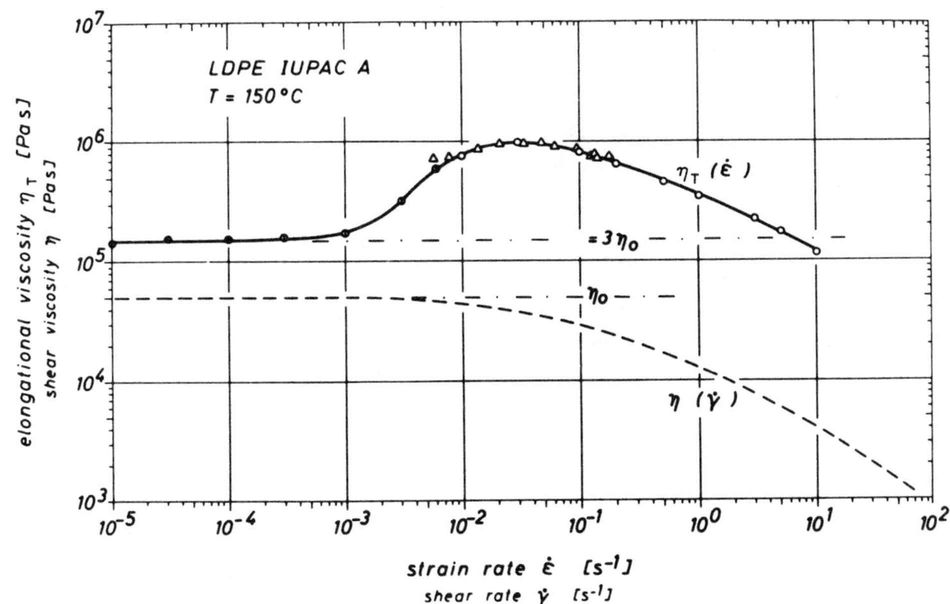

FIGURE 19.06.D.1. Shear and elongational viscosity functions $\eta(\dot\gamma)$ and $\eta_T(\dot\epsilon)$ for a low-density polyethylene at 150°C. From H. M. Laun and H. Münstedt, Rheol. Acta **17**, 415 (1978).

(2) Four-constant model: $\nu_1 = \nu_2 = 0$, $\mu_1 = \lambda_1$, and $\mu_2 = \lambda_2$.

(3) Fluid A: $\nu_1 = \nu_2 = \mu_0 = 0$, $\mu_1 = -\lambda_1$, and $\mu_2 = -\lambda_2$.

(4) Fluid B: $\nu_1 = \nu_2 = \mu_0 = 0$, $\mu_1 = \lambda_1$, and $\mu_2 = \lambda_2$.

19.07. HISTORICAL NOTES

Stress and deformation. Our concept of the stress vector comes from L. Euler (1757) and A.-L. Cauchy (1823). Expressions for finite deformation were formulated by Cauchy (1823), G. Piola (1836), and G. Green (1839).

Nonlinear elasticity. Constitutive equations for nonlinear elasticity were independently obtained by J. Finger (1894), M. Reiner (1948), and R. S. Rivlin (1948). In 1905, J. H. Poynting performed a calculation which indicated that unequal normal stresses in addition to shear stress where required for simple shear; in 1909 and 1912 he measured the change in length and volume produced on torsion. Rivlin, beginning in 1948, showed that assuming incompressibility made it possible to obtain exact solutions for many important types of deformation valid for an arbitrary strain-energy function. In cooperation with others, he performed a wide range of experiments on vulcanized rubbers to verify his theory.

Linear viscoelasticity. Wilhelm Weber in 1835 reported tensile creep and recovery experiments on silk threads and referred to the phenomenon as *Nachwirkung.* He deduced that stress relaxation should occur and that the same effect is responsible for the damping of vibrations. In 1863 Friedrich Kohlrausch established the linearity of the phenomenon. In 1874, Ludwig Boltzmann wrote down the integral representation of linear viscoelasticity in its full three-dimensional generality and discussed the now standard experiments. In 1867 James Clerk Maxwell, as an empirical explanation for the existence of shear stresses in flowing gases, introduced the first-order differential equation relating shear stress and deformation. In 1888 E. Wiechert and Joseph J. Thomson independently introduced a distribution of relaxation times. In 1905, F. T. Trouton and E. S. Andrews performed torsional stress relaxation, creep, and recovery measurements on fluids such as hot glass and pitch. Herbert Leaderman in 1943 first applied time-temperature superposition in linear viscoelasticity.

Nonlinear steady flow behavior. In 1889 Theodore Schwedoff of Odessa, on the basis of measurements with a Couette viscometer, concluded that the viscosity of his gelatin solutions depended on the rate of shear. But it wasn't until after Emil Hatschek independently came to the same conclusion regarding various colloidal solutions in 1913 that the existence of this phenomenon slowly gained acceptance. In 1929, B. Rabinowitsch published K. Weissenberg's derivation of the equation for $\dot{\gamma}(S)$ in Sec. 19.06.C.3, which made it possible to deduce $\dot{\gamma}(S)$ for a fluid from data obtained in capillary viscometers. The active experimental study of normal stress effects began as classified research during World War II in England by F. H. Garner, A. H. Nissan, and G. F. Wood and by K. Weissenberg, who published some of their results in 1950 and 1949, respectively.

Nonlinear viscoelasticity: General constitutive equations. A. E. Green and Rivlin in 1957 (and independently W. Noll in 1958) formulated properly invariant theories for finite deformations based on the assumption that the stress depends on the previous history of the deformation. Noll also gave a precise statement of the principle of material objectivity which such constitutive equations must obey.

19.08. BIBLIOGRAPHY

G. Astarita and G. Marucci, *Principles of Non-Newtonian Fluid Mechanics* (McGraw-Hill, London, 1974).

B. Bernstein, "Time-dependent behavior of an incompressible elastic fluid. Some homogeneous deformation histories," Acta Mech. **2**, 329 (1966).

G. C. Berry and T. G. Fox, "The viscosity of polymers and their concentrated solutions," Adv. Polym. Sci. **5**, 261 (1968).

R. B. Bird, R. C. Armstrong, and O. Hassager, *Fluid Mechanics,* Vol. 1 of *Dynamics of Polymeric Liquids* (Wiley, New York, 1977).

R. B. Bird, O. Hassager, R. C. Armstrong, and C. F. Curtiss, *Kinetic Theory,* Vol. 2 of *Dynamics of Polymeric Liquids* (Wiley, New York, 1977).

B. D. Coleman, H. Markovitz, and W. Noll, *Viscometric Flows of Non-Newtonian Fluids* (Springer, Berlin, 1966).

B. D. Coleman and H. Markovitz, "Asymptotic relations between shear stresses and normal stresses in general incompressible fluids," J. Polym. Sci. Polym. Phys. Ed. **12**, 2195 (1974).

John D. Ferry, *Viscoelastic Properties of Polymers,* 3rd ed. (Wiley, New York, 1980).

W. W. Graessley, "The entanglement concept in polymer rheology," Adv. Polym. Sci. **16**, 1 (1974).

Bernhard Gross, "*Mathematical Structure of the Theories of Viscoelasticity* (Hermann, Paris, 1953).

H. Markovitz, "Nonlinear steady-flow behavior," in *Rheology, Theory and Applications,* edited by F. R. Eirich (Academic, New York, 1967), Vol. 4, Chap. 6.

R. S. Rivlin, "Viscoelastic fluids," in *Research Frontiers in Fluid Dynamics,* edited by R. J. Seeger and G. Temple (Interscience, New York, 1965), Chap. 5.

A. J. Staverman and F. Schwarzl, "Linear deformation behavior of high polymers," in *Die Physik der Hochpolymeren,* edited by H. A. Stuart (Springer, Berlin, 1965), Vol. 4, Chap. 1.

L. R. G. Treloar, *The Physics of Rubber Elasticity,* 3rd ed. (Clarendon, Oxford, 1975).

C. Truesdell and W. Noll, *The Non-Linear Field Theories of Mechanics,* Vol. III/3 of *Handbuch der Physik,* edited by S. Flügge (Springer, Berlin, 1965).

20.00. Solid state physics

HANS P. R. FREDERIKSE

National Institute of Standards and Technology

CONTENTS

20.01. ELECTRONIC PROPERTIES OF SOLIDS; DEFINITIONS AND FORMULAS

A. Energy band structure

Electrons in a crystalline solid can be described by waves in a periodic lattice. The most important parameters are the electronic energy E and the electronic wave vector **k**.

The components of the *wave vector* **k** are given by

$$k_x = \pi n_x/L, \quad k_y = \pi n_y/L, \quad \text{and} \quad k_z = \pi n_z/L,$$

where n_x, n_y, and n_z are integers and L^3 is the volume of a cube of the crystal. k_x, k_y, and k_z are the coordinates in k *space* or *reciprocal space.*

In a periodic lattice the electronic energies are continuous within certain allowed bands. The wave vector **k** at the edge of such a band in k space is related to the lattice constant a (in real space) according to the expression $\mathbf{k}\cdot\mathbf{n} = p\pi/a$, where **n** is the unit vector normal to the lattice plane and p is an integer. (This relation is identical to the Bragg condition for the constructive interference of x rays reflected from a set of lattice planes separated by a distance a.) Hence, the planes of the periodic lattice in real space lead to a number of similar planes in k space. The smallest volume enclosed by this set of intersecting planes (in k space) is called the first *Brillouin zone*. The first Billouin zone for an fcc lattice is shown in Fig. 20.01.A.1.

For free electrons the energy is a quadratic function of the wave vector $E = \hbar^2k^2/2m$, where m is the mass of the electron. In a periodic array of atoms (crystalline solid) not all values of E are permitted. The range of energy covered when k varies across the zone is called an *allowed energy band.* Because the energies are usually plotted within a reduced zone, multiple energy values are possible for a given k value. Consequently, there may be several energy surfaces or bands in the first zone; these energy bands may or may not overlap. In the latter case there are certain ranges of energies that cannot be assumed by the electrons. Such an energy range is called a *forbidden energy gap,* E_g. The entire picture of the electron energy as a function of wave vector is known as the (electronic) *energy band structure*.

The occupation of the higher energy levels by the available conduction electrons (several times $10^{22}/\text{cm}^3$ for metals and orders of magnitude smaller for semiconductors) is statistically determined. The probability that a given state of energy E is occupied is given by

$$f = \{\exp[(E - E_F)/k_BT] + 1\}^{-1}.$$

This is called the *Fermi-Dirac distribution function.* E_F is

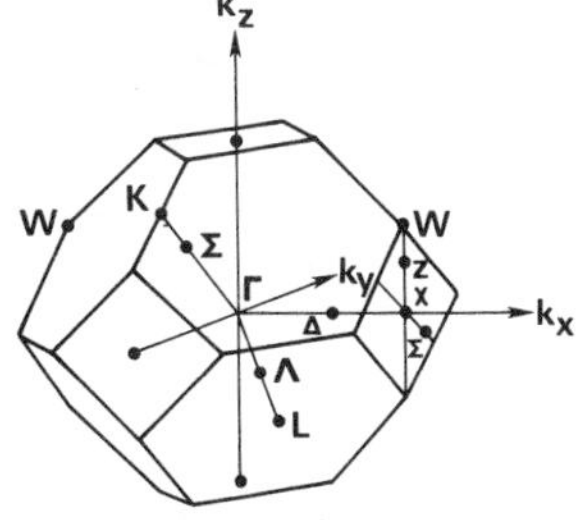

FIGURE 20.01.A.1. Brillouin zone for the fcc lattice. Points and lines of symmetry are indicated. After Joseph Callaway, *Energy Band Theory* (Academic, New York, 1964), p. 8.

the *Fermi energy*, and is equal to the chemical potential; k_B is the Boltzmann constant. E_F at absolute zero, E_F^0, has the significance of a cutoff energy. All states with energy less than E_F^0 are occupied, and all states with energy greater than E_F^0 are vacant. The distribution is called degenerate when $E_F \gg k_BT$ and nondegenerate when $E_F \ll k_BT$. In the latter case the distribution function becomes

$$f = e^{(E_F - E)/k_BT} = Ae^{-E/k_BT}.$$

This is known as the *Maxwell-Boltzmann* or *classical distribution function.* The *density of states* (or number of states with energies between E and $E + dE$) per unit volume is given by

$$g(E)dE = \frac{4\pi(2m)^{3/2}E^{1/2}}{h^3}dE$$

(for spherical energy surfaces). The Fermi energy or Fermi level is determined by the total number of electrons per unit volume (n_0). One calculates for Fermi-Dirac statistics

$$E_F \simeq E_F^0[1 - \tfrac{1}{12}\pi^2(k_BT/E_F^0)^2 + \cdots],$$

$$E_F^0 = (h^2/2m)(3n_0/8\pi)^{2/3},$$

and for Maxwell-Boltzmann statistics

$$E_F = k_BT \ln[n_0h^3/2(2\pi mkT)^{3/2}].$$

The Fermi energy can also be expressed in terms of the *Fermi wave vector* k_F, *Fermi velocity* v_F, or *Fermi temperature* T_F:

$$E_F = \hbar^2k_F^2/2m = \tfrac{1}{2}mv_F^2 = k_BT_F,$$

where m is the free-electron mass.

Near the top or the bottom of a band the energy is generally a quadratic function of the wave vector, so that by analogy with the expression $E = p^2/2m = \hbar^2k^2/2m$ for free electrons we can define an *effective mass* m^* such that $\partial^2E/\partial k^2 = \hbar^2/m^*$ (p is the momentum, k the wave vector, and $\hbar$ Planck's constant$\times 1/2\pi$). The effective mass of electrons is positive. Near the top of a band m^* is negative, so that the motion corresponds to that of a positive charge (*hole*). In a real crystal the above expressions for $g(E)dE$, E_F, and E_F^0 must be modified by substituting m^* for the free-electron mass m.

For each partially filled band there will be at $T = 0$ K a surface in k space separating the occupied from the unoccupied energy levels. The set of all these surfaces in k space is known as the *Fermi surface.*

The three kinds of crystalline solids can be distinguished on the basis of the energy band theory[1–7]:

Metal: A material in which the highest occupied energy band is only partly filled. The resistivity of metals increases with temperature; the temperature dependence is close to linear except at low temperatures.

Semiconductor: A material in which the highest occupied energy band (valence band) is completely filled at absolute zero. The energy gap between the valence band and the next higher band (conduction band) is between zero and 4 or 5 eV. In certain temperature ranges the resistivity decreases exponentially with increasing temperature.

Insulator: A material in which the highest occupied energy band is completely filled. The distinction between insulators and semiconductors is one of degree. Materials with energy gaps larger than 4 or 5 eV are usually called insulators. The resistivity of pure insulators at room temperature is extremely high. At elevated temperature ionic conduction often dominates electrical conduction.

Examples of the energy bands in three transition metals are shown in Figure 20.01.A.2. A similar structure for a semiconductor, silicon, is presented in Figure 20.01.A.3. The Fermi surface of copper is shown in Figure 20.01.A.4. The Fermi parameters for a number of representative metals are listed in Table 20.01.A.5. Energy band parameters for some 30 semiconductors are presented in Table 20.01.A.6.

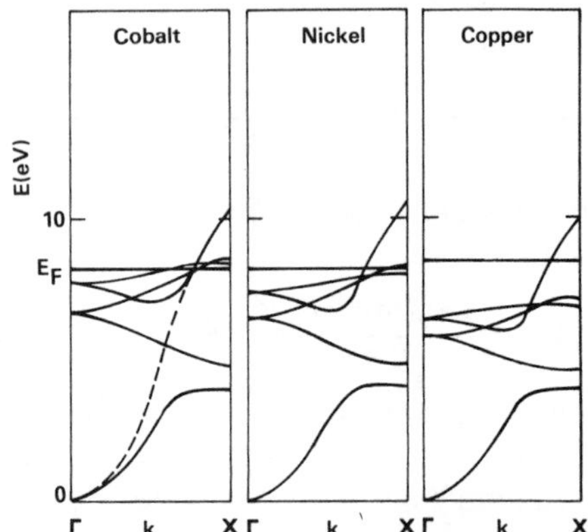

FIGURE 20.01.A.2. Energy bands in fcc Co, Ni, and Cu for k along (1,0,0). These may be regarded crudely as made up of fairly flat d bands with an s band crossing them. If the d bands are neglected the s band would be as shown (dashed) in the first figure. The Fermi energy cuts the d bands for Co and Ni, giving them a complicated Fermi surface. In Cu it cuts only the s band. After W. M. Lomer and W. A. Gardner, Prog. Mater. Sci. **14**, 117 (1969).

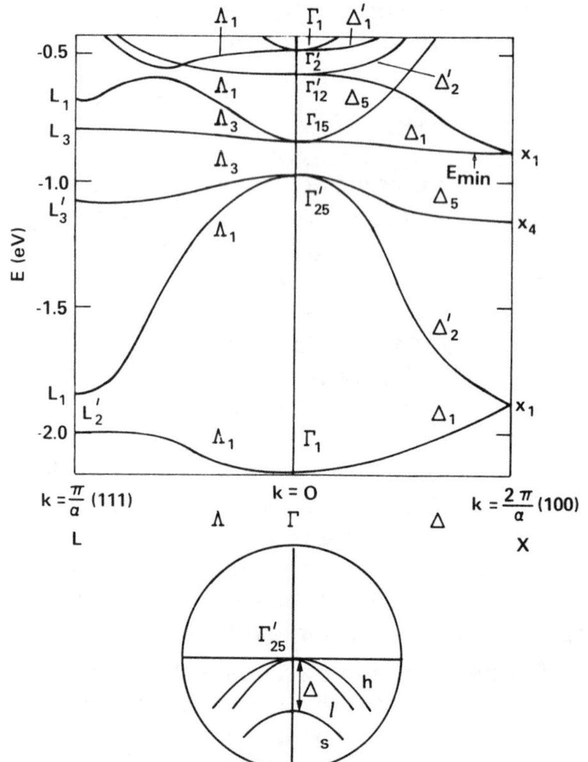

FIGURE 20.01.A.3. Energy band structure of Si. The four valence bands (Λ_3 and Δ_5 are doubly degenerate) are built from 3s and 3p bonding states. The conduction bands show d electron states as well as the 3s and 3p antibonding states. The insert shows the spin orbit splitting at the top of the valence bands on an expanded scale. Heavy-hole band h and light-hole band l are separated by spin orbit splitting $\Delta = 0.04$ eV from split-off band s. After W. Kleinmann and J. C. Phillips, Phys. Rev. **118**, 1164 (1960).

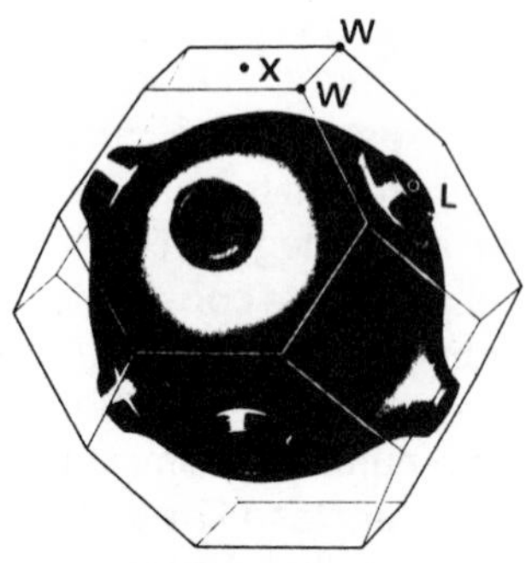

FIGURE 20.01.A.4. Fermi surface of copper showing "necks" where it reaches out to the zone boundary. Letters label special points of the zone, X at the midpoint of a square face, L at the midpoint of a hexagonal face, and W at the corners. After Ref. 3, p. 110.

TABLE 20.01.A.5. Number of electrons, Fermi energies, Fermi temperatures, Fermi wave vectors, and Fermi velocities for representative metals. From Ref. 2, pp. 5 and 38.

Element	n (10^{22}/cm^3)	E_F (eV)	T_F (10^4 K)	k_F (10^8 cm^{-1})	v_F (10^8 cm/s)
Li	4.70	4.74	5.51	1.12	1.29
Na	2.65	3.24	3.77	0.92	1.07
K	1.40	2.12	2.46	0.75	0.86
Rb	1.15	1.85	2.15	0.70	0.81
Cs	0.91	1.59	1.84	0.65	0.75
Cu	8.47	7.00	8.16	1.36	1.57
Ag	5.86	5.49	6.38	1.20	1.39
Au	5.90	5.53	6.42	1.21	1.40
Be	24.7	14.3	16.6	1.94	2.25
Mg	8.61	7.08	8.23	1.36	1.58
Ca	4.61	4.69	5.44	1.11	1.28
Sr	3.55	3.93	4.57	1.02	1.18
Ba	3.15	3.64	4.23	0.98	1.13
Nb	5.56	5.32	6.18	1.18	1.37
Fe	17.0	11.1	13.0	1.71	1.98
Mn	16.5	10.9	12.7	1.70	1.96
Zn	13.2	9.47	11.0	1.58	1.83
Cd	9.27	7.47	8.68	1.40	1.62
Hg	8.65	7.13	8.29	1.37	1.58
Al	18.1	11.7	13.6	1.75	2.03
Ga	15.4	10.4	12.1	1.66	1.92
In	11.5	8.63	10.0	1.51	1.74
Ti	10.5	8.15	9.46	1.46	1.69
Sn	14.8	10.2	11.8	1.64	1.90
Pb	13.2	9.47	11.0	1.58	1.83
Bi	14.1	9.90	11.5	1.61	1.87
Sb	16.5	10.9	12.7	1.70	1.96

TABLE 20.01.A.6. Electronic parameters in several elemental and compound semiconductors. Data from O. Madelung, in *Landolt-Boernstein Numerical Data and Functional Relationships in Science and Technology, New Series, Vols. 17a–d* (Springer, Berlin, 1982).

Substance	Structure	E_g (eV) at	T (K)	Dir./ind.	$m_{n\parallel}(m_0)$	$m_{n\perp}(m_0)$	$m_{p,l}(m_0)$	$m_{p,h}(m_0)$	μ_n (cm²/V s, 300 K)	μ_p (cm²/V s, 300 K)
C (diamond)	diamond	5.48	300	ind.			0.7	2.18	1 800	1 600
Si	diamond	1.170	0	ind.	0.916	0.191	0.153	0.537	1 500	500
Ge	diamond	0.744	1.5	ind.	1.588	0.815	0.043	0.352	3 900	1 900
Sn (grey)	diamond	0	...	zero gap	0.024 (light el.)		0.2–0.45		1.2×10^5 (l. el., 100 K)	3 000 (270 K)
					0.21 (heavy el.)				3×10^3 (h. el., 270 K)	
SiC	zinc blende	2.60	0	ind.	0.65	0.24			900	
	wurtzite	2.86	300	ind.	1.5	0.25	1.0		260	50
BN	zinc blende	6–8	300	ind.					below 4	
BP	zinc blende	2	300	ind.					120	500
AlN	wurtzite	6.3	5	dir.						14
AlP	zinc blende	2.5	2	ind.					60	
AlAs	zinc blende	2.23	2	ind.	5.8	0.19	0.26	0.5	294	
AlSb	zinc blende	1.696	4.2	ind.	1	0.26	0.11	0.5–0.9	200	400
GaN	wurtzite	3.503	1.6	dir.	0.20–0.27		0.8		440	400
GaP	zinc blende	2.350	0	ind.	7.25	0.21	0.17	0.67	190	150
GaAs	zinc blende	1.519	2	dir.	0.067		0.087	0.45	8 900	450
GaSb	zinc blende	0.811	2	dir.	0.042		0.047	0.36	3 800	1 000
InN	wurtzite	1.95	300	dir.	0.11				250	
InP	zinc blende	1.424	2	dir.	0.079		0.12	0.45–0.65	5 400	150
InAs	zinc blende	0.418	4.2	dir.	0.023		0.025	0.41	33 000	450
InSb	zinc blende	0.2352	1.7	dir.	0.0136		0.016	0.40	77 000	850
ZnO	wurtzite	3.438	1.6	dir.	0.28		0.59			
ZnTe	zinc blende	2.39	4	dir.	0.28		0.6		330	
CdTe	zinc blende	1.61	2	dir.	0.11		0.4–0.6		60 000 (10 K)	60
HgS	trigonal	2.275	4.2	dir.					45	
	zinc blende	−0.2		inverted gap	0.05				23	
HgSe	zinc blende	−0.20	80	inverted gap	0.03		0.78		15 000	
HgTe	zinc blende	−0.303	4.2	inverted gap	0.03		0.42		30 000	200
CuCl	zinc blende	3.202	4.2	dir.	0.42		2–20			
CuBr	zinc blende	2.96	8	dir.	0.23		2–20			
CuI	zinc blende	3.06	4.2	dir.	0.30		1.4–2.4			

B. Transport properties

1. Electrical conductivity

In a solid in which ohmic conduction takes place, the current density **J** is given by

$$\mathbf{J} = \sigma \mathbf{E},$$

where σ is the conductivity and **E** the applied electric field. In a homogeneous isothermal crystal σ is a tensor whose form is determined by the symmetry of the crystal.

2. Mobility

The drift mobility of charge carriers is defined as the drift velocity per unit applied electric field (v_D/E). The relation to the collision time τ_c is given by

$$\mu^{(D)} = e\tau_c/m^*.$$

3. Hall effect

When a magnetic field is applied to a conductor carrying a current density **J**, an electric field $\mathbf{E}_H$ (Hall field) is developed, given by

$$\mathbf{E}_H = R\mathbf{J}\times\mathbf{B}.$$

R is called the *Hall coefficient* and **B** is the magnetic induction. When the current is in the length direction of the sample (J_x) and the magnetic field in the z direction, the Hall coefficient (for electrons or holes) is

$$R = \mp r/ne = \mp \mu/\sigma,$$

where the units are as follows: n, carriers/cm³; e, 1.6×10^{-19} C; μ, cm²/V s; σ, $(\Omega\ \text{cm})^{-1}$; r, a scattering factor of the order of 1. The parameter $\mu = R\sigma$ is the *Hall mobility*.

If both electrons and holes are present (e.g., in an intrinsic semiconductor), the expression for the Hall coefficient becomes

$$R = \frac{-n\mu_n^2 + p\mu_p^2}{(n\mu_n + p\mu_p)^2}\,\frac{3\pi}{8ec},$$

where n is the number of electrons, p the number of holes, and μ_n and μ_p their Hall mobilities. Values of the Hall mobility of electrons and/or holes in a number of semiconductors are given in Table 20.01.A.6.

4. Einstein relation

The mobility μ is related to the *diffusion constant D* by

$$\mu = eD/k_B T.$$

5. Magnetoresistance

The resistance of a metal or semiconductor is altered by the presence of a magnetic field. The relative change in resistance is

$$\Delta\rho/\rho = aB^2/(1+\mu^2B^2).$$

The theory for a single isotropic energy band gives no change in resistance for metals. For semiconductors (with one type of carrier scattered by acoustical lattice vibrations) one finds at low fields that

$$a = 0.38\mu^2 10^{-16},$$

where the mobility μ is measured in cm²/V s and B in Oe.

6. Seebeck effect (thermoelectric power)

If two different conductors 1 and 2, are joined together at both ends and the two junctions kept at different temperatures, an electromotive force is set up which is proportional to the temperature difference (for small ΔT). The thermoelectromotive force per degree centigrade is called the thermoelectric power (Q).

For metals,

$$Q_{12} = \frac{\pi^2 k_B^2 T}{3e}\left(\frac{\partial \log\sigma(E)}{\partial E}\right)_{E=E_F},$$

where $\sigma(E)$ is the electrical conductivity due to charge carriers of energy E.

For semiconductors the thermoelectric power Q (measured in μV/deg) depends on the temperature, the number of carriers, the statistics, and the scattering mechanism. Some of the most common formulas (neglecting phonon-drag effects) are given below.

Extrinsic range (one type of isotropic carriers, classical statistics):

$$Q_{12} = \pm\frac{k_B}{e}\left(r + \frac{E_F}{k_B T}\right) = \pm\frac{k_B}{e}\left(r - \ln\frac{nh^3}{2(2\pi M^{(N)} k_B T)^{3/2}}\right),$$

where $+$ refers to p-type, $-$ to n-type semiconductors. $r = 2$ for acoustical lattice scattering, 4 for ionized impurity scattering (actually $r = 3.2$), and 3 for polar mode scattering. n is the concentration of carriers (cm^{-3}), $m^{(N)}$ the density-of-states effective mass, E_F the Fermi energy measured from the bottom of the conduction band or the top of the valence band (positive in both cases) (eV), and k_B the Boltzmann constant (eV/deg).

Intrinsic range (classical statistics):

$$Q_{12} = -\frac{k_B}{e}\frac{c-1}{c+1}\left(\frac{E_g}{2k_B T} + r + \frac{3}{4}\frac{c+1}{c-1}\ln\frac{m_n^{(N)}}{m_p^{(N)}}\right),$$

where $E_g = E_0 + aT$ is the energy gap at temperature T, c the mobility ratio (μ_n/μ_p), and a a constant.

7. Thomson effect

When an electric current **J** passes between two points of a homogeneous conductor, with a temperature difference ΔT existing between these points, an amount of heat $\sigma_T \mathbf{J}\Delta T$ is emitted or absorbed in addition to the Joule heat. The parameter σ_T is called the Thomson coefficient.

8. Peltier effect

If two conductors 1 and 2, are joined together and kept at a constant temperature while a current **J** passes through the junction, heat is generated or absorbed at the junction in addition to the Joule heat. The Peltier coefficient Π_{12} is defined so that the heat emitted or absorbed per second at the junction is $\Pi_{12} J$.

9. Kelvin relations

$$Q_{12} = \Pi_{12}/T, \quad T\frac{dQ_{12}}{dT} = \sigma_{T_1} - \sigma_{T_2}.$$

10. Nernst effect

If a temperature gradient is maintained in an electronic conductor ($\mathbf{J} = 0$) in the presence of a transverse magnetic field, a transverse electric field develops which is given by

$$\mathbf{E}_t = Q_N \nabla T \times \mathbf{B}.$$

Q_N is called the isothermal Nernst coefficient. For semiconductors (one type of carrier, classical statistics, and acoustical lattice scattering),

$$Q_N = -\tfrac{3}{16}\pi(k_B/e)\mu.$$

11. Ettinghausen effect

If a temperature difference is allowed across an electronic conductor perpendicular both to a current of density **J** and to a magnetic field **B**, a transverse temperature gradient will be established:

$$\nabla_t T = P\mathbf{J}\times\mathbf{B}.$$

P is called the Ettinghausen coefficient. The Ettinghausen coefficient P, Nernst coefficient Q_N, and thermal conductivity κ are related by

$$\kappa P = TQ_N.$$

12. Righi-Leduc effect

If a longitudinal temperature difference is maintained in an electronic conductor in which $\mathbf{J} = 0$, in the presence of a magnetic field a transverse temperature gradient is established:

$$\nabla_t T = S\mathbf{B}\times\nabla T.$$

S is called the Righi-Leduc coefficient.

C. Magnetic properties of electrons

1. Cyclotron resonance

Current carriers in a solid when accelerated by a microwave electric field perpendicular to an externally applied static magnetic field H will spiral about the magnetic field. For sufficiently large mean free path l or collision time τ—the condition is $\omega_c\tau > 1$—a resonance absorption is observed for a frequency

$$\omega_c = eH/m^* c,$$

where c is the velocity of light. This technique provides a direct measurement of the effective mass of electrons (or holes) m^*.

2. Magnetic susceptibility of charge carriers

Charge carriers contribute a diamagnetic effect through their translational motion and a paramagnetic effect due to their spin. For nondegenerate conductors (semiconductors), the magnetic susceptibility is

$$\chi_c = (n\mu_B^2/k_B T)(1 - m^2/3m^{*2}),$$

where n is the concentration of free carriers and μ_B the Bohr magneton. If m^* is small (Ge), the susceptibility is mainly diamagnetic. If m^* is large (TiO_2), the paramagnetic effect dominates.

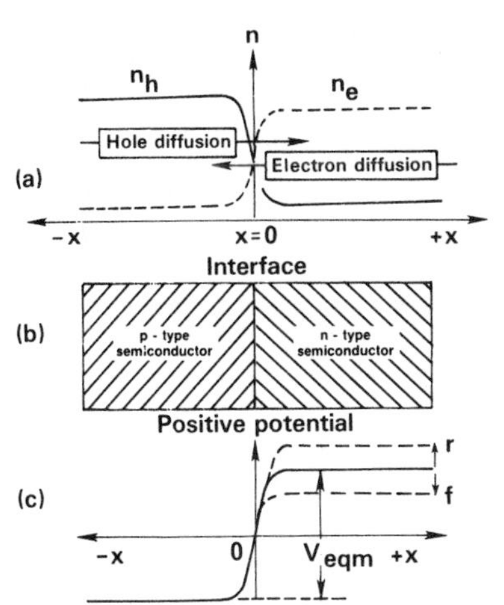

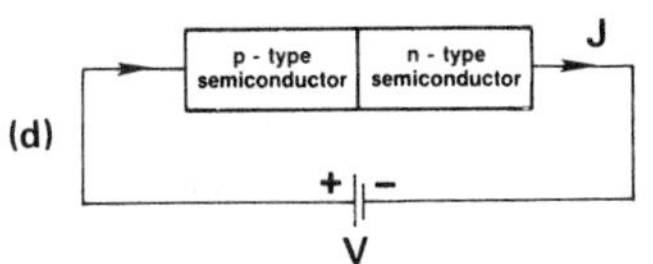

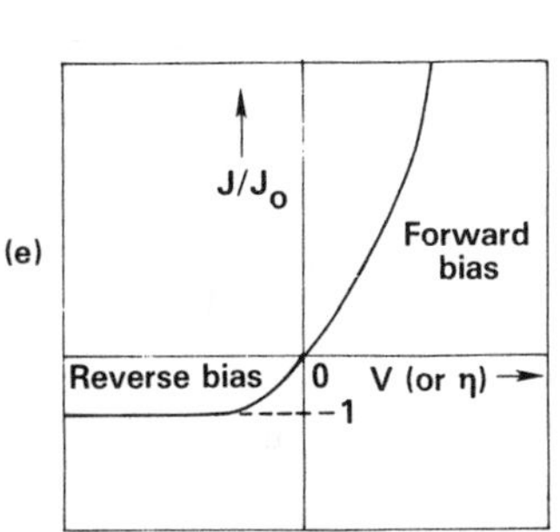

FIGURE 20.01.D.1. ***p-n* junction.** (a) Concentration gradient of holes and electrons across the interface leading to hole and electron diffusion; (b) formation of a charged interface; (c) potential vs position plotted across the junction (solid curve, equilibrium; dashed curves, under applied field; r = reverse bias, f = forward bias); (d) *p-n* junction with forward bias; (e) rectification characteristics. After Ref. 6, pp. 530–531.

For degenerate conductors (metals, semimetals, and impure semiconductors) at low temperature,

$$\chi_c = 3n\frac{\mu_B^2}{2E_F}(1 - m^2/3m^{*2})$$

$$= (4m^*\mu_B^2/h^2)(3\pi^2 n)^{1/3}(1 - m^2/3m^{*2}).$$

Transition metals have a large m^*, and consequently show a high magnetic susceptibility (*Pauli paramagnetism*); semimetals with small m^* (e.g., Bi) have a diamagnetic susceptibility.

3. Knight shift

Polarization of conduction electrons will produce a shift in frequency at which nuclear magnetic resonance absorption will occur for a given type of nucleus in a metal relative to a particular nonmetallic solid.

D. Semiconductor junction: Rectification

The current-voltage relation for a *p-n* junction is given by

$$J = J_0(e^{qV/k_BT} - 1),$$

where q is the electronic charge, V the applied field, J the total electric current density, J_0 the sum of the electron and hole generated currents.

See Figure 20.01.D.1.

20.02. LATTICE VIBRATIONS

A. Normal modes

A crystal lattice can be treated as a three-dimensional array of particles joined by harmonic springs. The classical motion of such a system consists of small oscillations, which can be described by a set of *normal modes* each with its own characteristic frequency. The energy of these modes (or waves) is quantized just as the energy of electromagnetic waves (photons). The quanta of the lattice vibrational field are known as *phonons.* The average number of phonons with frequency ω in thermal equilibrium at temperature T is given by the *Bose-Einstein distribution function*

$$n(\omega) = [\exp(h\omega/k_BT) - 1]^{-1}.$$

The energy of the lattice vibrations $h\omega$ will vary with the wave vector **q** (dispersion). The dispersion curves can be divided into several branches: the *acoustic branches* [because for small q the frequency ω ($= cq$) is characteristic of *sound* waves] and *optical branches* (because in ionic crystals these modes are largely responsible for the *optical* behavior of the solid in the infrared). In a crystal with p atoms per unit cell there will be three acoustic branches, two transverse (TA) and one longitudinal (LA), and $3p - 3$ optical branches, $2p - 2$ transverse (TO) and $p - 1$ longitudinal (LO).

The phonon dispersion can be measured by the technique of *inelastic neutron scattering.* An example of such a vibrational spectrum is shown in Figure 20.02.A.1.

The slopes of the acoustic phonon branches (for small q) are related to the *elastic constants* C_{kl}. Table 20.02.A.2 presents values of these constants for a number of cubic crystals.

The ratio of the longitudinal and transverse optical mode frequencies (for $q = 0$) is related to the high- and low-frequency dielectric constants as follows (in simple polar crystals):

$$\frac{\omega_L^2}{\omega_T^2} = \frac{\epsilon(0)}{\epsilon(\infty)}.$$

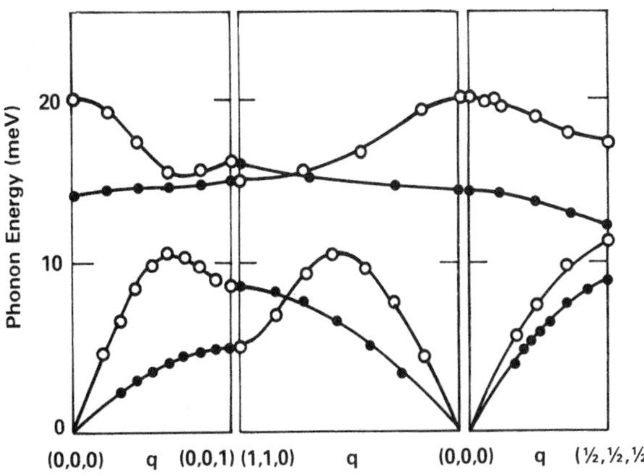

FIGURE. 20.02.A.1. Vibrational spectrum in KBr at 90 K. The open circles are longitudinal modes, the closed circles transverse modes (doubly degenerate). Note the large splitting between these for the optic branch at $q = 0$. The Brillouin zone is as in Figure 20.01.A.1. After B. N. Brockhouse *et al.*, Phys. Rev. **131**, 1025 (1963).

TABLE 20.02.A.2. Elastic constants for some cubic crystals (in 10^{11} N/m² at 300 K). From Ref. 2, p. 447.

Substance	C_{11}	C_{12}	C_{44}
Li (78 K)	0.148	0.125	0.108
Na	0.070	0.061	0.045
Cu	1.68	1.21	0.75
Ag	1.24	0.93	0.46
Au	1.86	1.57	0.42
Al	1.07	0.61	0.28
Pb	0.46	0.39	0.144
C (diamond)	10.76	1.25	5.76
Ge	1.29	0.48	0.67
Si	1.66	0.64	0.80
V	2.29	1.19	0.43
Ta	2.67	1.61	0.82
Nb	2.47	1.35	0.287
Fe	2.34	1.36	1.18
Ni	2.45	1.40	1.25
LiCl	0.494	0.228	0.246
NaCl	0.487	0.124	0.126
KF	0.656	0.146	0.125
RbCl	0.361	0.062	0.047
InSb	0.672	0.367	0.302
InAs	0.833	0.453	0.396
GaAs	1.188	0.538	0.594

This equation is known as the *Lyddane-Sachs-Teller relation.* Values for the frequencies of the optical lattice modes and for the low- and high-frequency dielectric constants are listed in Table 20.02.B.1.

B. Dielectric properties

The dielectric constants ϵ are dependent on the *polarizabilities* α of the constituent ions: for low frequencies (static dielectric constant),

$$\frac{\epsilon_{st}-1}{\epsilon_{st}+2}=\frac{4\pi}{3v}\left(\alpha^{+}+\alpha^{-}+\frac{e^2}{M\bar{\omega}^2}\right),$$

and for high frequencies (optical dielectric constant),

$$\frac{\epsilon_{op}-1}{\epsilon_{op}+2}=\frac{4\pi}{3v}(\alpha^{+}+\alpha^{-}),$$

TABLE 20.02.B.1. Lattice vibrational parameters, dielectric constants, and effective charges of selected inorganic crystals. From S. S. Mitra, in *Handbook on Semiconductors, Vol. 1,* edited by W. Paul (North-Holland, Amsterdam, 1982) and from *Handbook of Optical Constants of Solids,* edited by E. D. Palik (Academic, Orlando, FL, 1985).

Crystal	ω_T (cm^{-1})	ω_L (cm^{-1})	ϵ_{st}	ϵ_{op} ($=n^2$)	e^*/e
C (diamond)	1332	...	5.7	5.7	0
Si	520	...	12	12	0
Ge	301	...	16	16	0
InSb	185	197	17.88	15.68	0.42
InAs	219	243	15.15	12.25	0.56
InP	304	345	12.61	9.61	0.66
GaSb	231	240	15.69	14.44	0.33
GaAs	269	292	12.9	10.9	0.51
GaP	367	403	10.18	8.46	0.58
AlSb	319	340	17.88	15.68	0.48
ZnS(Z)	274	350	8.3	5.0	0.96
CdS(W)‖	235	306	$\epsilon_{11}=9.0$	5.32	0.91
⊥	242	304	$\epsilon_{33}=8.47$	5.32	0.87
PbS	65	223	202	17.2	...
PbSe	...	44	284	22.9	...
PbTe	31	114	420	32.8	0.55
LiF	306	659	8.8	1.92	0.87
NaF	244	418	5.1	1.7	0.93
NaCl	164	264	5.9	2.25	0.74
NaBr	134	209	6.4	2.62	0.69
KF	190	326	5.5		
KCl	142	214	4.85	2.13	0.80
KBr	113	165	4.9	2.33	0.76
KI	101	139	5.1	2.69	0.69
RbF	156	286	6.5	1.90	...
RbCl	116	173	4.9	2.2	0.84
TlCl	63	158	31.9	5.1	0.80
TlBr	43	101	29.8	5.4	0.82
AgCl	106	196	12.3	4.0	0.71
AgBr	79	138	13.1	4.6	0.70
CuCl	156	220	9.8	3.7	...
CuBr	...	...	8.0	...	...
MgO	401	718	9.64	3.01	0.88
Al_2O_3 ⊥	...	...	9.34	2.66	
‖	...	...	11.54	2.82	
Cu_2O	...	...	7.6	~6	
TiO_2(a)[a]	183	373	89	5.9	
	388	458			
	500	806			
(c)	167	811	173	7.2	

[a]D. M. Eagles, J. Phys. Chem. Solids **25**, 1243 (1964).

TABLE 20.02.B.2. Electronic polarizabilities of ions in 10^{-24} cm³. Values from L. Pauling, Proc. R. Soc. London Ser. A **114**, 181 (1927); J. Pirenne and E. Kartheuser, Physica **20**, 2005 (1964); and J. Tessman, A. Kahn, and W. Shockley, Phys. Rev. **92**, 890 (1953). The PK and TKS polarizabilities are at the frequency of the *D* lines of sodium. The values are in cgs; to convert to SI, multiply by $\frac{1}{9}\times10^{-15}$. From Ref. 1, p. 462.

			He	**Li^+**	**Be^{2+}**	**B^{3+}**	**C^{4+}**
Pauling			0.201	0.029	0.008	0.003	0.0013
PK				0.029			
	O^{2-}	**F^-**	**Ne**	**Na^+**	**Mg^{2+}**	**Al^{3+}**	**Si^{4+}**
Pauling	3.88	1.04	0.390	0.179	0.094	0.052	0.0165
PK (TKS)	(2.4)	0.87		0.312			
	S^{2-}	**Cl^-**	**Ar**	**K^+**	**Ca^{2+}**	**Sc^{3+}**	**Ti^{4+}**
Pauling	10.2	3.66	1.62	0.83	0.47	0.286	0.185
PK (TKS)	(5.5)	3.06		1.14	(1.1)		(0.19)
	Se^{2-}	**Br^-**	**Kr**	**Rb^+**	**Sr^{2+}**	**Y^{3+}**	**Zr^{4+}**
Pauling	10.5	4.77	2.46	1.40	0.86	0.55	0.37
PK (TKS)	(7)	4.28		1.76	(1.6)		
	Te^{2-}	**I^-**	**Xe**	**Cs^+**	**Ba^{2+}**	**La^{3+}**	**Ce^{4+}**
Pauling	14.0	7.10	3.99	2.42	1.55	1.04	0.73
PK (TKS)	(9)	6.52		3.02	(2.5)		

TABLE 20.02.B.3. Atomic and ionic radii. Units are 1 Å = 10^{-10} m. From Ref. 1, p. 129. (For other estimates of radii see Ref. 8.)

←Standard radii for ions in inert gas (filled shell) configuration→
←Radii of atoms when in tetrahedral covalent bonds→
←Radii of ions in valence state indicated in superscript→

H 2.08 0.28																	**He** 1.48
Li 0.68	**Be** 1.06 1.06											**B** 0.16 0.88	**C** 0.77	**N** 0.70	**O** 1.46 0.66	**F** 1.33 0.64	**Ne** 1.58
Na 0.98	**Mg** 0.65 1.40											**Al** 0.45 1.26	**Si** 0.38 1.17	**P** 1.10	**S** 1.90 1.04	**Cl** 1.81 0.99	**Ar** 1.88
K 1.33	**Ca** 0.94	**Sc** 0.68	**Ti** 0.60 $^{2+}$0.90	**V** $^{2+}$0.88	**Cr** $^{2+}$0.84	**Mn** $^{2+}$0.80	**Fe** $^{2+}$0.76	**Co** $^{2+}$0.74	**Ni** $^{2+}$0.72	**Cu** 1.35 $^{+}$0.96	**Zn** 1.31 $^{2+}$0.83	**Ga** 1.26 $^{3+}$0.62	**Ge** 1.22 $^{4+}$0.44	**As** 1.18 $^{3+}$0.69	**Se** 1.14	**Br** 1.95 1.11	**Kr** 2.00
Rb 1.48	**Sr** 1.10	**Y** 0.88	**Zr** 0.77	**Nb** 0.67	**Mo**	**Tc**	**Ru**	**Rh**	**Pd** $^{2+}$0.86	**Ag** 1.52 $^{+}$1.13	**Cd** 1.48 $^{2+}$1.03	**In** 1.44 $^{3+}$0.92	**Sn** 1.40 $^{4+}$0.74	**Sb** 1.36 $^{3+}$0.90	**Te** 1.32	**I** 2.16 1.28	**Xe** 2.17
Cs 1.67	**Ba** 1.29	**La** 1.04	**Hf**	**Ta**	**W**	**Re**	**Os**	**Ir**	**Pt**	**Au** $^{+}$1.37	**Hg** 1.48 $^{2+}$1.12	**Ti** $^{3+}$1.05	**Pb** $^{4+}$0.84	**Bi**	**Po**	**At**	**Rn**
Fr 1.75	**Ra** 1.37	**Ac** 1.11															
			Ce 0.92 $^{3+}$1.11	**Pr**	**Nd** $^{3+}$1.08	**Pm**	**Sm** $^{3+}$1.04	**Eu**	**Gd** $^{3+}$1.02	**Tb**	**Dy** $^{3+}$0.99	**Ho**	**Er** $^{3+}$0.96	**Tm**	**Yb** $^{3+}$0.94	**Lu**	
			Th 0.99	**Pa** 0.90	**U** 0.83	**Np** $^{4+}$1.05	**Pu**	**Am**	**Cm**	**Bk**	**Cf**	**Es**	**Fm**	**Md**	**No**	**Lw**	

where α^+ and α^- are the electronic polarizabilities, M is the reduced ionic mass ($1/M = 1/M_1 + 1/M_2$), v the volume of the unit cell, and $\overline{\omega}$ the infrared absorption frequency ($= \omega_T$). These expressions are two forms of the *Clausius-Mosotti relation.*

The electronic polarizabilities of a number of ions are given in Table 20.02.B.2.

In conjunction with the Clausius-Mosotti expression one can define an *effective charge* (or *Szigeti charge*) e^* based on the experimental values of ϵ_{st}, ϵ_{op}, and ω_T:

$$\frac{(e^*)^2}{Mv} = \frac{q(\epsilon_{st} - \epsilon_{op})}{4\pi(\epsilon_{op} + 2)^2}\,\omega_T^2.$$

Values for the effective charge of the ions in a number of compounds are listed in the last column of Table 20.02.B.1. For the above expression to hold true depends on the validity of the assumptions of nondeformable and nonoverlapping ions. Hence, the results are very sensitive to the values of the ionic radii. Atomic and ionic radii are presented in Table 20.02.B.3.

A somewhat more chemical manner of describing the ionic (or covalent) character is through the concept of *electronegativity.* Pauling has developed an empirical scale of relative electronegativities (Table 20.02.B.4). The larger the difference in electronegativities, the more ionic is the compound.

TABLE 20.02.B.4. Electronegativities of the elements. From Ref. 6, p. 55.

H 2.1																
Li 1.0	**Be** 1.5											**B** 2.0	**C** 2.5	**N** 3.0	**O** 3.5	**F** 4.0
Na 0.9	**Mg** 1.2											**Al** 1.5	**Si** 1.8	**P** 2.1	**S** 2.5	**Cl** 3.0
K 0.8	**Ca** 1.0	**Sc** 1.3	**Ti** 1.5	**V** 1.6	**Cr** 1.6	**Mn** 1.5	**Fe** 1.8	**Co** 1.8	**Ni** 1.8	**Cu** 1.9	**Zn** 1.5	**Ga** 1.6	**Ge** 1.8	**As** 2.0	**Se** 2.4	**Br** 2.8
Rb 0.8	**Sr** 1.0	**T** 1.2	**Zr** 1.4	**Nb** 1.6	**Mo** 1.8	**Tc** 1.9	**Ru** 2.2	**Rh** 2.2	**Pd** 2.2	**Ag** 1.9	**Cd** 1.7	**In** 1.7	**Sn** 1.8	**Sb** 1.9	**Te** 2.1	**I** 2.5
Cs 0.7	**Ba** 0.9	**La** 1.1	**Hf** 1.3	**Ta** 1.5	**W** 1.7	**Re** 1.9	**Os** 2.2	**Ir** 2.2	**Pt** 2.2	**Au** 2.4	**Hg** 1.9	**Ti** 1.8	**Pb** 1.8	**Bi** 1.9	**Po** 2.0	**At** 2.2

TABLE 20.02.C.1. Selected ferroelectric crystals. To obtain P_s in the SI unit of C m^{-2}, divide the value given in cgs by 3×10^5. To obtain P_s in μC/cm^2, divide the value in cgs by 3×10^3. From Ref 1, p. 476.

		T_C (K)	P_s (esu cm^{-2})	at T (K)
KDP	KH_2PO_4(KDP)	123	16 000	96
	KD_2PO_4	213	13 500	...
	RbH_2PO_4(RDP)	147	16 800	90
	RbH_2AsO_4(RDA)	111	...	...
	KH_2AsO_4(KDA)	96	15 000	80
	KD_2AsO_4(KDDA)	162	...	...
	CsH_2AsO_4(CDA)	143	...	...
	CsD_2AsO_4(CDDA)	212	...	...
TGS type	Triglycine sulfate	322	8 400	293
	Triglycine selenate	295	9 600	273
Perovskites	$BaTiO_3$	406	78 000	296
	$KNbO_3$	712	90 000	523
	$PbTiO_3$	765	> 150 000	300
	$LiTaO_3$	833–891	70 000	720
	$LiNbO_3$	1483	900 000	...
Rochelle salt	$NaKC_2H_4O_6.4H_2O$	255, 297	...	
	$NaKC_2H_2D_2O_6.4D_2O$	251, 308	...	

TABLE 20.02.C.2. Selected antiferroelectric crystals. From Ref. 1, p. 491.

Crystal	Transition temperature to antiferroelectric state (K)	Crystal	Transition temperature to antiferroelectric state (K)
WO_3	1010	$ND_4D_2PO_4$(DDP)	242
$NaNbO_3$	793, 911	$NH_4H_2AsO_4$(ADA)	216
$PbZrO_3$	506	$ND_4D_2AsO_4$(DDA)	304
$PbHfO_3$	488	$(NH_4)_2H_3IO_6$	254
$NH_4H_2PO_4$(ADP)	148		

C. Ferroelectric and antiferroelectric crystals

There are two classes of ionic crystals which show a transition to a spontaneously polarized state below a certain temperature (*Curie temperature* T_C). In one case, *ferroelectrics,* the crystal as a whole shows an electric dipole moment. In the other case, *antiferroelectrics,* periodic arrays of ions are polarized in one direction while the adjacent arrays are polarized in the antiparallel direction. The crystal as a whole does *not* show a dipole moment.

The (static) dielectric constant often reaches very large values in the neighborhood of the Curie temperature. The transition is due to a lattice instability. The frequency of a low-lying optic mode at $q=0$ will approach $\omega=0$ (*soft mode*); consequently the ions will be displaced and the crystal transforms to a new structure at the transition temperature.

Some selected ferroelectric and antiferroelectric compounds with their transition temperatures and spontaneous polarizations (P_s) are listed in Tables 20.02.C.1 and 20.02.C.2.

20.03. MODES OF EXCITATION AND MIXED MODES IN CRYSTALS

A. Definitions

1. Exciton

An excited electronic state of a crystal (insulator or semiconductor) which has an energy less than the energy gap is called an exciton. In semiconductor language: ionization of an ion or atom in a crystal will produce an electron (in the conduction band) and a hole (in the valence band) which are independent of each other. A slightly lower energy ($<E_{\text{gap}}$) will produce an *electron bound to a hole.* This entity, which is known as an exciton, does not carry current. The electron and hole attract each other through a Coulomb interaction screened by the dielectric medium. The radius of the exciton (similar to the Bohr radius a_0) is given by

$$a_{\text{ex}} = (\hbar^2\epsilon/m^*e^2) = (\epsilon m/m^*)a_0,$$

where ϵ is the dielectric constant, m^* is the reduced effective mass ($1/m^* = 1/m_e + 1/m_h$), and the exciton binding energy is

$$E_{\text{ex}} = \frac{m^*}{m\epsilon^2}\frac{e^2}{2a_0} = \frac{m^*}{m\epsilon^2}13.6\ \text{eV}.$$

[These expressions hold for $a_{\text{ex}} \gg a_0$, because the (static) dielectric constant is defined only for dimensions large compared with the ion. This nonlocalized exciton is known as the *Mott-Wannier exciton.* Strongly bound excitons—e.g., in molecular crystals—have received the name *Frenkel excitons.*]

2. Polaron

An electron plus its induced lattice polarization is called a polaron.

TABLE 20.03.A.2.a. Polaron coupling constants α, masses m^*_{pol}, and band masses m^* for electrons in the conduction band (in units of the mass of the free electron). From Ref. 1, p. 391.

Crystal	KCl	KBr	AgCl	AgBr	ZnO	PbS	InSb	GaAs
α	3.97	3.52	2.00	1.69	0.85	0.16	0.014	0.06
m^*_{pol}	1.25	0.93	0.51	0.33	...	...	0.014	...
m^*	0.50	0.43	0.35	0.24	...	...	0.014	...
m^*_{pol}/m^*	2.5	2.2	1.5	1.4	...	...	1.0	...

When an electron (or hole) is introduced into an ionic lattice the charge carrier will polarize (and hence displace) the ions in its immediate neighborhood. One says that the electron (or hole) is "dressed." This dressed electron (or hole) still moves freely through the crystals. The effective mass of this entity (the polaron) is larger than that of the "undressed" charge carrier:

$$m_{pol} = m^*\left(\frac{1 - 0.0008\alpha^2}{1 - \frac{1}{6}\alpha + 0.0034\alpha^2}\right),$$

where α is the *polaron coupling constant.* One distinguishes two regimes:

(1) $\alpha > 6$: small polaron. This is the regime in which the quasifree electron theory and band structure are not valid anymore. The transport mechanism for the charge carrier is that of *hopping*.

(2) $1 < \alpha < 6$: large polaron. The electronic band picture is still valid. However, the mobility of the dressed electron (or hole) is small.

Polaron parameters for a number of strongly and weakly ionic crystals are given in Table 20.03.A.2.a.

3. Polariton

A transverse electromagnetic wave interacting with the optical lattice vibrations of a diatomic ionic crystal will produce a mixed mode which has two branches, as shown in Figure 20.03.A.3.a. At low and at high k the two branches are clearly photonlike and optical-phonon-like, respectively. In the intermediate region both modes are of a mixed photon-phonon nature. The coupled modes are called polaritons. No radiation can propagate in an ionic crystal in the range $\omega_L > \omega > \omega_T$.

4. Plasmon

A collective excitation of the electron gas is called a plasmon. High-frequency radiation can propagate through a metal (or degenerate semiconductor) if the frequency ω is larger than the *plasma frequency* ω_p and $\omega\tau \ll 1$.

$$\omega_p = \sqrt{4\pi ne^2/m^*},$$

where n is the charge carrier concentration, m^* the effective mass, and τ the collision time. This mode describes a charge density wave, known as a plasma oscillation or plasmon.

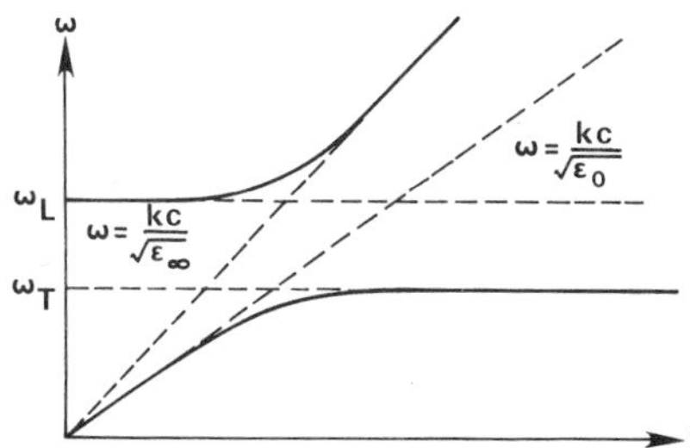

FIGURE 20.03.A.3.a. Mixed modes resulting from interaction of an e.m. wave and optical phonons

5. Helicon

If a metal is subjected to a uniform magnetic field **H** and a circularly polarized electric field $\mathbf{E}e_{i\omega t}$, there will be a low-frequency wave (for ω much smaller than the cyclotron frequency $\omega_c = eH/mc$) which is known as a helicon. For this mode the relation between ω and k is given by $\omega = \omega_c(k^2c^2/\omega_p^2)$, where c is the velocity of light and ω_p the plasma frequency.

6. Magnon

The collective excitation of the magnetic spins of the assembly of ions in a ferromagnetic or antiferromagnetic crystal is called a spin wave or magnon. This collective state $|\mathbf{k}\rangle$ has a wave vector **k** and an energy $E(\mathbf{k})$. Superposition of spin waves with wave vectors $\mathbf{k}_1, \mathbf{k}_2, \ldots, \mathbf{k}_N$ will yield the magnetization $M(T)$ of the crystal (at temperature T).

20.04. DEFECTS AND MASS TRANSPORT IN CRYSTALS

A. Color centers

In Figure 20.04.A.1., F is an electron trapped at a negative-ion vacancy, F' an F center plus an extra electron, F_A an F center with one adjacent impurity, M a

Na^+ Cl^- Na^+ Cl^- Na^+ Cl^- Na^+ Cl^- Na^+ Cl^-
F F_A
Cl^- Na^+ e Na^+ Cl^- Na^+ Cl^- Na^+ e K^+
F'
Na^+ Cl^- Na^+ Cl^- Na^+ 2e Na^+ Cl^- Na^+ Cl^-
Cl^- Na^+ e Na^+ Cl^- Na^+ Cl^- Na^+ Cl^- Na^+
M V_K h
Na^+ e Na^+ Cl^- Na^+ Cl^- Na^+ Cl^- Na^+ Cl^-
Cl^- Na^+ Cl^- Na^+ Cl^- Na^+ Cl^- Na^+ Cl^- Na^+
H
Na^+ Cl^- Na^+ Cl e Cl Na^+ Cl^- Na^+ Cl^- Na^+ Cl^-
Cl^- Na^+ Cl^- Na^+ Cl^- Na^+ Cl^- Na^+ Cl^- Na^+

FIGURE 20.04.A.1. Various types of color centers in an ionic binary crystal (e.g., NaCl)

TABLE 20.05.A.1. Transition points (T_C or T_N) of ferromagnetic and antiferromagnetic elements and compounds. From Ref. 7, Sec. 5, pp. 144, 151, and 168.

Substance	T_C (K)	T_N (K)	Substance	T_C (K)	T_N (K)
Cr		311	Fe_3O_4	858	
Mn		100	Co_3O_4		40
Fe	1043		$CoFe_3O_4$	793	
Co	1388		$NiFe_2O_4$	728	
Gd	293		CrO_2	386	
Tb	222	229	Cr_2O_3		318
Dy	85	179	α-Fe_2O_3		948
Ho	20	131	FeO		198
Er	20	84	MnO		120
Tm	25	56	CoO		291
			NiO		530
			CuO		230
			EuO		77
			MnF_2		67
			$MnCl_2$		2
			FeF_2		78
			CoF_2		38

bound pair of F centers in the (100) plane, V_K a hole trapped by a pair of negative ions in two sites (Cl-Cl^-), H a singly ionized negative-ion molecule squeezed in *one* site (Cl_2^-), and R a cluster of three F centers in the (111) plane (*not* shown in the figure).

Schottky defect: vacancy (anion or cation).

Frenkel defect: vacancy-interstitial pair.

B. Mass transport (diffusion)

Fick's law:

$$J = -D\,\mathrm{grad}N.$$

where J is the atom flux, N the concentration, and D the diffusion constant.

For a single diffusion mechanism,

$$D = D_0 \exp(-W/kT).$$

In the intrinsic case $W = E_m + E_F$ (E_m = activation energy for defect formation and E_F = activation energy for motion); in the extrinsic case $W = E_m$. Furthermore, $D_0 = \gamma \nu a$, where γ is a geometry and entropy factor, ν the attempt frequency ($\sim$Debye frequency), and a the jump distance ($\sim$lattice constant).

If the transport mechanism is that of vacancies moving through the solid, the conductivity will be

$$\sigma = ne\mu = (Ne^2 D_{\text{tracer}}/k_B T f),$$

where f is a *correlation factor* ($= D_{\text{tracer}}/D_{\text{cond}}$).

(Diffusion measurements employing radioactive tracers involve correlations in their random-walks motions which differ from the motions of the vacancies producing the conductivity.)

20.05. MAGNETISM

A. Definitions

Diamagnetism. Substances whose magnetic susceptibility $\chi = (\partial M/\partial B)_{B\to 0}$ is negative are called diamagnetic.

Paramagnetism. Substances with a positive susceptibility are known as paramagnetics. The magnetization of these materials is given by

$$M = N\mu L(\chi),$$

where N is the number of atoms per unit volume, μ the magnetic moment, L the Langevin function ($\mathrm{ctnh}\chi - 1/\chi$), $\chi = \mu B/k_B T$, and B is the magnetic field. For $\chi \ll 1$ (low field, high temperature)

$$\chi = M/B = C/T.$$

This is the *Curie law* and $C = N\mu^2/3k_B$ is the *Curie constant.*

The magnetic moment of an atom or ion in free space is

$$\mu = \gamma \hbar \mathbf{J} = g\mu_B \mathbf{J},$$

where $\hbar\mathbf{J}$ is the total angular moment (orbital plus spin), γ the gyromagnetic ratio (magnetic moment/angular moment), g the spectroscopic splitting factor, and the Bohr magneton

$$\mu_B = e\hbar/2mc = 0.927\,410 \times 10^{-20}\ \text{erg/Oe}.$$

Ferromagnetism. A magnetically ordered system that has a spontaneous magnetic moment is known as a ferromagnet below T_C. Above T_C the susceptibility follows the *Curie-Weiss law,* $\chi = C/(T-\theta)$.

The *Curie temperature* T_C is the temperature above which the spontaneous magnetization vanishes. The intercept θ is usually slightly larger than T_C.

An antiferromagnet is a material in which the magnetic spins are ordered antiparallel (with zero net moment) at temperatures below the ordering temperature T_N (the *Néel temperature*). In this case, $\chi = C/(T+\theta)$.

A material with two sublattices of ordered antiparallel spins which still has a net magnetization is called *ferrimagnetic.*

Transition temperatures (T_c or T_N) of ferromagnetic and antiferromagnetic elements and compounds are given in Table 20.05.A.1.

20.06. REFERENCES

[1] *Charles Kittel, Introduction to Solid State Physics,* 4th ed. (Wiley, New York, 1971).

[2] Neil W. Ashcroft and N. David Mermin, *Solid State Physics* (Holt, Rinehart and Winston, New York, 1976).

[3] R. J. Elliott and A. F. Gibson, *Solid State Physics and Its Applications* (Harper and Row and Barnes and Noble Import Division, New York, 1974).

[4] A. H. Wilson, *Theory of Metals,* 2nd ed. (Cambridge University, London, 1954).

[5] J. M. Ziman, *Principles of the Theory of Solids* (Cambridge University, London, 1964).

[6] K. M. Ralls, T. H. Courtney, and J. Wulff, *Introduction to Materials Science and Engineering* (Wiley, New York, 1976).

[7] *American Institute of Physics Handbook,* 3rd ed., edited by Dwight E. Gray (McGraw-Hill, New York, 1972), Sec. 9.

[8] R. D. Shannon, Acta Cryst. A **32**, 751 (1976).

21.00. Surface physics

HOMER D. HAGSTRUM

AT&T Bell Laboratories

CONTENTS

The surface of a crystalline solid in vacuum is here defined as comprising the few, approximately three, top atomic layers of the solid that differ appreciably in geometrical and electronic structure from the bulk, including any foreign atoms adsorbed to it or incorporated into it. A good introduction to the physics of solid surfaces is to be found in Refs. 1 through 4.

21.01. SURFACE GEOMETRICAL STRUCTURE

A. Definitions and notation[5–9]

1. Diperiodic nets

Analogous to the fourteen Bravais lattices of the triperiodic crystal there are five diperiodic nets. The area unit of a diperiodic net analogous to the triperiodic unit cell is called the unit mesh (Figure 21.01.A.1.a). In terms of the unit mesh vectors $\mathbf{a}_1,\mathbf{a}_2$ the area of the unit mesh is $A=|\mathbf{a}_1\times\mathbf{a}_2|$. The inter-row spacings d_{hk} between rows having two-dimensional Miller indices h,k in the diperiodic nets are given in Table 21.01.A.1.b.

2. Superposition of nets

The surface net $(\mathbf{c}_1,\mathbf{c}_2)$ is the superposition of the net $(\mathbf{b}_1,\mathbf{b}_2)$ of the outermost layer of a reconstructed surface or of an adsorbed monolayer on the reference or substrate net $(\mathbf{a}_1,\mathbf{a}_2)$, which is the net of a bulk plane parallel to the surface. In column vector and matrix notation the primitive translations of these superposed nets are related by

$$\begin{pmatrix}\mathbf{b}_1\\ \mathbf{b}_2\end{pmatrix}=\begin{pmatrix}m_{11} & m_{12}\\ m_{21} & m_{22}\end{pmatrix}\begin{pmatrix}\mathbf{a}_1\\ \mathbf{a}_2\end{pmatrix}=M\begin{pmatrix}\mathbf{a}_1\\ \mathbf{a}_2\end{pmatrix}.$$

In terms of det M, the area of $(\mathbf{b}_1,\mathbf{b}_2)$ divided by the area of $(\mathbf{a}_1,\mathbf{a}_2)$, three cases can be distinguished: simple superposition, coincidence site structure, and incoherent superposition for which det M is an integer, a rational fraction, or an irrational fraction, respectively. The net $(\mathbf{c}_1,\mathbf{c}_2)$, which may be written

$$\begin{pmatrix}\mathbf{c}_1\\ \mathbf{c}_2\end{pmatrix}=P\begin{pmatrix}\mathbf{a}_1\\ \mathbf{a}_2\end{pmatrix}=\begin{pmatrix}p_{11} & p_{12}\\ p_{21} & p_{22}\end{pmatrix}\begin{pmatrix}\mathbf{a}_1\\ \mathbf{a}_2\end{pmatrix}=Q\begin{pmatrix}\mathbf{b}_1\\ \mathbf{b}_2\end{pmatrix},$$

formed by superposition of $(\mathbf{a}_1,\mathbf{a}_2)$ and $(\mathbf{b}_1,\mathbf{b}_2)$, is the net with the smallest unit mesh for which det P and det Q are integers having no common factor.

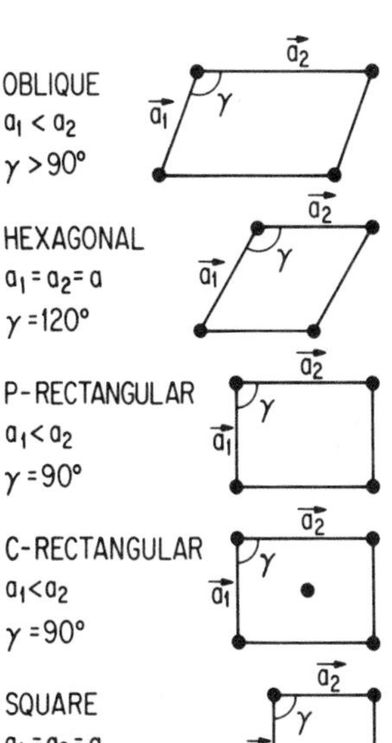

FIGURE 21.01.A.1.a. The five diperiodic unit meshes. From Ref. 5. By convention the unit mesh vectors are chosen with $a_1 < a_2$ and $\gamma \geqslant 90°$, with the difference $\gamma - 90°$ as small as possible.

TABLE 21.01.A.1.b. Inter-row spacing d_{hk} in the five diperiodic nets. From Ref. 5.

Oblique	$\frac{1}{d_{hk}^2}=\frac{h^2}{a_1^2\sin^2\gamma}+\frac{k^2}{a_2^2\sin^2\gamma}-\frac{2hk\cos\gamma}{a_1a_2\sin^2\gamma}$
Rectangular (p and c)	$1/d_{hk}^2=(h/a_1)^2+(k/a_2)^2$
Hexagonal	$1/d_{hk}^2=\frac{4}{3}[(h^2+hk+k^2)/a^2]$
Square	$1/d_{hk}^2=(h^2+k^2)/a^2$

3. Surface structure notation

In the *matrix notation* the matrix P above is used to denote the superposition net $(\mathbf{c}_1,\mathbf{c}_2)$ with respect to the reference structure $(\mathbf{a}_1,\mathbf{a}_2)$.

The more commonly used *Wood notation* can be used only if the included angles of the superposed unit meshes are equal. It expresses the relation of the mesh $(\mathbf{c}_1,\mathbf{c}_2)$ to the reference mesh $(\mathbf{a}_1,\mathbf{a}_2)$ as

$$\left(\frac{c_1}{a_1}\times\frac{c_2}{a_2}\right)R\alpha$$

in terms of the lengths of the unit mesh vectors and the angle α of relative rotation R of the two meshes. If $\alpha=0$, the angle designation is omitted.

A third notation is based on a nonprimitive unit mesh and is exemplified by a superposition on a rectangular substrate net for which

$$P=\begin{pmatrix}1 & -1\\ 1 & 1\end{pmatrix},\quad Q=\begin{pmatrix}1 & 0\\ 0 & 1\end{pmatrix}.$$

Except for $a_1=a_2$ no Wood notation is possible because the included angles of the superposed meshes differ. However, the unit mesh of the superposition differs from the unit mesh of the $(2\mathbf{a}_1,2\mathbf{a}_2)$ net only by the presence of a net point at the center. The net $(\mathbf{c}_1,\mathbf{c}_2)$ is designated $c(2\times2)$ for centered (2×2) and is equivalent for $a_1=a_2$ to the Wood notation $(\sqrt{2}\times\sqrt{2})R45°$. In the absence of a centered net point the Wood notation would be (2×2), often written $p(2\times2)$ to indicate that the net of the superposition is primitive. Table 21.01.A.3.a gives some equivalent notations.

TABLE 21.01.A.3.a. Some equivalent surface structural notations.

Wood	Matrix	Substrate
$p(n,m)$, n,m integers$\geqslant1$	$\begin{pmatrix}n & 0\\ 0 & m\end{pmatrix}$	all
$(\sqrt{2}\times\sqrt{2})R45°[=c(2\times2)]$	$\begin{pmatrix}1 & -1\\ 1 & 1\end{pmatrix}$	fcc, bcc(100), fcc(110)
$(\sqrt{3}\times\sqrt{3})R30°$	$\begin{pmatrix}2 & 1\\ -1 & 1\end{pmatrix}$	fcc(111), $\gamma=120°$

A *notation for stepped surfaces* having high Miller indices is based on their geometry of relatively close-packed terraces or treads of low Miller indices separated by risers also with low Miller indices. The notation $(S)[m(h,k,l)\times n(h',k',l')]$ indicates that the stepped surface (S) has a tread or terrace of (h,k,l) orientation m atoms wide and a riser of (h',k',l') orientation n atoms high.

4. Reciprocal net

The reciprocal net unit vectors $\mathbf{a}_1^*,\mathbf{a}_2^*$ are related to the direct net unit vectors $\mathbf{a}_1,\mathbf{a}_2$ by the equations

$$\mathbf{a}_1\cdot\mathbf{a}_2^* = \mathbf{a}_2\cdot\mathbf{a}_1^* = 0,\quad \mathbf{a}_1\cdot\mathbf{a}_1^* = \mathbf{a}_2\cdot\mathbf{a}_2^* = 1 \ \text{(or } 2\pi).$$

Here the 1 (or 2π) indicates that two conventions are in use. These are carried through in this form in the subsequent discussion. $\mathbf{a}^*(\mathbf{a}_2^*)$ is perpendicular to $\mathbf{a}_2(\mathbf{a}_1)$ and of magnitude $a_2A^{-1}(a_1A^{-1})$. A general reciprocal net vector $\mathbf{u}$ may be written

$$\mathbf{u} = \mathbf{u}_{hk} = nh\mathbf{a}_1^* + nk\mathbf{a}_2^*,$$

where n is the integer and h, k are the row indices having no common factor, the two-dimensional analogs of Miller indices.

B. Low-energy electron diffraction[2,3,8,9] (LEED)

1. Diffraction conditions

The primary electron beam is represented by a plane wave $\exp(i\mathbf{K}^0\cdot\mathbf{r})$, where the propagation vector $\mathbf{K}^0$ has magnitude K^0 and components $\mathbf{K}_\parallel^0$ and $\mathbf{K}_\perp^0$, parallel and perpendicular to the surface, respectively. $K^0 = 2\pi/\lambda$, in which the electron wavelength λ in Å is $(150/E)^{1/2}$ with E in eV. The electron energy is $E = (\hbar^2/2m)(K^0)^2$. The propagation vector of a back-diffracted beam is $\mathbf{K}'$ of magnitude K'.

For *elastic diffraction from a 2D grating*, $K' = K^0$ expresses conservation of energy and $\mathbf{K}_\parallel' = \mathbf{K}_\parallel^0 + 2\pi\mathbf{u}$ (or $\mathbf{K}_\parallel^0 + \mathbf{u}$) expresses conservation of parallel momentum in which $\mathbf{u}$ is a reciprocal net vector. The interrelations of the vectors in these equations are visualized by the Ewald sphere construction.[6] Conservation of parallel momentum yields the *plane grating formula* (the Laue equation for the diperiodic case)

$$n\lambda = d_{hk}(\sin\psi - \sin\psi_0),$$

with n an integer specifying the order of diffraction, d_{hk} the distance between rows in the 2D grating or net (Table 21.01.A.1.b), and ψ and ψ_0 the angles with respect to the normal of diffracted and incident beams, respectively.

2. Debye–Waller factor[9]

This factor, $\exp(-M)$, expresses the exponential decrease in diffracted beam intensity $I(T)$ with increasing temperature T in the range $T\gg\theta_D$:

$$I(T) = \exp(-M)I(T=0),$$

$$M = 3|\mathbf{K}^0 - \mathbf{K}'|^2T/2mk_B\theta_D^2,$$

in which $|\mathbf{K}^0 - \mathbf{K}'|$ is the momentum transfer on diffraction, T the temperature in kelvins, m the atomic mass in units of electron mass, k_B Boltzmann's constant (3.17×10^{-6} hartrees/K), and θ_D the Debye temperature (kelvins). Experimental θ_D tend toward the bulk θ_D for high-energy electrons (>300 eV for Pd) which penetrate into the crystal. For electrons of low energy (~50 eV) that penetrate less deeply into the crystal the measured θ_D are about 0.6 of the bulk θ_D, indicating enhanced vibrational amplitude perpendicular to the surface for surface atoms.

3. Nature of LEED results

Several types of information can be obtained directly from observation and measurement of the geometrical features of the LEED pattern of diffracted beam terminations on a fluorescent screen or by means of a movable electron collector. For an ideally flat uniform surface the *spot pattern* presents the two-dimensional reciprocal net. In many instances one can reconstruct from the experimental LEED pattern the direct net using the equations of Sec. 21.01.A.4, yielding its size and orienation and revealing superlattices induced by reconstruction or by adsorbed monolayers (Figure 21.01.B.3.a) and the possible presence of symmetrical domains. Surface defects such as foreign or missing atoms, steps or superstructure domain boundaries, and faceting affect spot shape and motion, and background intensity between spots.[10]

Of the order of 1000 surface structures have been identified by LEED on clean and adsorbate covered surfaces. These require specification by about 800 surface structure notations that range from (1×1) through $c(2\times4)$, (7×7), to the likes of $(\sqrt{\tfrac{7}{3}}\times\sqrt{\tfrac{7}{3}})R49.1°$. Most of these identified surfaces involve the adsorption of foreign atoms or molecules in ordered arrays.

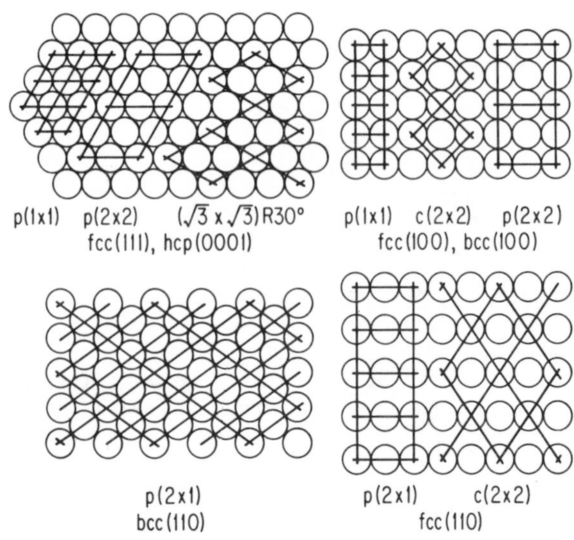

FIGURE 21.01.B.3.a. Surface nets. Circles represent top substrate atoms, lines represent some superlattices produced by ordered overlayers. LEED pattern alone cannot specify whether intersections of lines representing positions of adsorbed atoms should be placed as shown or shifted to bridge or hollow sites. From Ref. 11.

Tables of observed LEED patterns for monolayers on metals, and for nonmetallic adsorbates on metallic and nonmetallic substrates and on stepped surfaces are given in Ref. 7.

The second use of LEED data, called LEED crystallography,[9,11,12] constitutes the attempt to determine specific atomic positions within the surface unit mesh or in a three-dimensional unit cell including more than one layer at the surface. This is done by comparing the predictions of large-scale computer calculations applied to specific structural models with experimental intensity–voltage (*I–V*) curves for several diffraction beams. Structural and nonstructural parameters of the model are varied to produce the best comparisons as quantified by the use of reliability (*R*) factors.[12] Position coordinates perpendicular to the surface are generally reliable to ± 0.1 Å, those parallel to the surface to ± 0.2 Å unless the atom in question actually occupies a position of high, two-dimensional symmetry.[11] The resultant atomic positions determine registry of the surface net with respect to the substrate net, bond lengths, and directions, and can be used in conjunction with measured work function change to estimate charge transfer in adsorbate systems.

The instrumentation for LEED is discussed in Ref. 13. A recent apparatus improvement is position-sensitive detection which has the advantage of parallel rather than serial detection of diffraction beams and permits digital records of LEED patterns with consequent ease of storage, retrieval, and analysis.[14]

The diffraction of electrons photoemitted from core levels can, in a method known as photoelectron diffraction, be used to determine surface geometry. See Secs. 5 and 6 of Ref. 115. Reflection, high-energy, electron diffraction[155,156] (RHEED) is performed in an electron microscope. See Sec. 21.01.F.2.

C. LEED crystallography results

The surface structures that have been determined by the several variants of dynamical LEED are given in a "Handbook of Surface Structures" published by the Surface Crystallographic Information Service.[15] This estimable compilation also includes what are considered to be reliable structural results obtained by all other methods, thus enabling one to compare different methods of surface structural determination. Also available, in addition, to the Handbook are "Database and Graphing Programs" and a "User's Manual" which enable one to display structures on a personal computer screen.[16]

1. Clean metals[9,11]

The surfaces of clean metals may be classified as either reconstructed or unreconstructed. In unreconstructed surfaces the atomic arrangement is that of the bulk up to and including the top layer except for layer relaxations perpendicular to the surface near the surface. Table 21.01.C.1.a gives spacing changes of surface layers for clean, unreconstructed surfaces. Reconstructions and relaxations at metal surfaces are reviewed in Refs. 17 and 18.

2. Adsorption on metals[9,11]

Atomic adsorption on metals takes place preferentially but with some exceptions into surface sites (so-called hollow sites) that maximize the number of nearest substrate neighbors. Adatom–metal bond lengths tend to be near those in bulk compounds but with substantial scatter. Bond angles are primarily determined by bond lengths in hard-ball packing geometry with bond bending more easily accommodated than bond stretching. Adsorption tends to elongate underlying metal–metal bonds in cases where bonds are shortened on the clean surface.

Metal adatoms at low coverages on metals predominantly form ordered monolayers determined by substrate periodicity. At higher coverages substrate periodicity may dominate but incoherent or coincidence structures with new periodicities unrelated to substrate periodicity may form. The ordering of large metallic adatoms (especially K, Rb, and Cs) shows relatively little dependence on substrate lattice and tends to be hexagonal close packed on any metal substrate. At high coverages mole-

TABLE 21.01.C.1.a. Relaxation without registry change of normal layer spacings of clean, unreconstructed metal surfaces. Each entry consists of: chemical symbol (percent changes from the bulk spacing of surface spacings, proceeding into the bulk). All entries are LEED results unless specified otherwise. Other data by: MEES, medium-energy electron diffraction; SPLEED, spin-polarized, low-energy electron diffraction [Sec. 21.03.B.6]; RBS, Rutherford back scattering; HEIS, MEIS, LEIS, high, medium, and low energy ion scattering [Sec. 21.01.G.1]. From layer-spacing data of Ref. 15.

fcc(100)	Al(0.0; MEED 1.5), Au(0.0), Co(− 4.0), Cu(− 1.1, 1.7, 1.5), Ir(− 3.6), Ni(1.1; RBS − 8.9), Pd(0.3), Pt(0.0; RBS 0.2), Rh(0.5)
fcc(110)	Ag(− 6.6; HEIS − 7.6, 4.0), Al(− 8.4, 5.6, 2.3, 1.7), Cu(− 8.5, 2.3; HEIS − 5.3, 3.2), Ir(− 7.4), Ni(− 8.6, 3.1, − 0.4; HEIS − 4.8, 2.4), Pd(− 5.8, 0.7), Rh(− 0.7)
fcc(111)	Al(0.9), Co(0.0), Cu(− 0.7), Ir(− 2.6), Ni(− 1.2; HEIS 0.14), Pd(HEIS 0.0), Pt(1.1; SPLEED 0.49; MEIS 1.4), Rh(0.0)
fcc(311)	Al(− 13), Cu(− 5.0), Ni(− 16, 4.1, − 1.7)
bcc(100)	Fe(− 1.6), Mo(− 9.5, − 1.0), Ta(− 11, 1.2), V(− 6.6, 1.3), W(− 7.6)
bcc(110)	Fe(0.5), Mo(− 1.6), Na(0.0), V(− 0.5)
bcc(111)	Fe(− 17, − 9.3, 4.0, − 2.1)
bcc(210)	Fe(− 22, − 11, − 4.8)
bcc(211)	Fe(− 10, 5.1, − 1.7)
bcc(310)	Fe(− 16, 13, − 4.0)
hcp(0001)	Co(0.0), Na(0.0), Ru(− 1.9), Sc(− 1.9), Ti(− 2.1), Zn(− 2.0), Zr(− 1.2)

cules such as CO, for example, can also form close-packed structures that ignore the substrate periodicity.

The structures of atomic chemisorption on metals determined by LEED are listed in Ref. 15. Surface structure determinations by EXAFS methods for atomic adsorption on metals and semiconductors are discussed in Sec. 21.01.D.3. Reference 15 includes surface structure determinations and graphical representations for a number of molecules adsorbed on metal surfaces. The study by LEED of the physisorption of noble gases on metals and the interactions between adatoms has been reviewed in Ref. 19.

3. Nonmetals[20–23]

Although LEED has provided much information on the superstructures formed at clean and covered nonmetal surfaces, the determination of specific atom positions has been much more difficult for these surfaces than for metals. An important reason for this is pervasive surface reconstruction resulting from the energetically favorable rearrangement of broken covalent or ionic bonds at the surface. Atom displacements from positions of the truncated bulk solid can amount to 0.5 Å in the surface layer accompanied by corresponding movement of atoms in a "selvage" extending several atom layers into the crystal.

Of the 275 surface structures described in Ref. 15 only 36 involve the elemental, III–V, and II–VI semiconductors, clean or with atomic adsorption. The structures of these surfaces were determined by LEED in 24 cases, by SEXAFS in six, ion scattering in four, and x-ray standing waves in two. A review of experimental and theoretical structural techniques and an evaluation of them as applied to semiconductor surfaces is given in Ref. 23.

D. X-ray methods

1. X-ray diffraction

The advent of synchrotron radiation has transformed x-ray diffraction on individual monolayers of atoms "from the verge of feasibility to the routinely possible."[24] X-ray diffraction differs from low-energy electron diffraction in two important respects.[25] Electrons interact strongly with matter making LEED surface specific but at the cost of multiple scattering that complicates the interpretation. X rays interact weakly with matter, thereby reducing surface specificity but simplifying interpretation by permitting the use of kinematical scattering theory. With the development of intense synchrotron x-ray sources and the use of grazing angles of incidence, the classical methods of bulk x-ray crystallography can be used in many cases to solve surface atomic structures.[26] An example[27] is the use of 120 independent superlattice reflections of the clean reconstructed Si(111)7$\times$7 surface to refine a full set of in-plane structural coordinates and to determine within 0.2 Å the atomic positions in the model of Takayanagi *et al.* described in Sec. 21.01.F.1.

2. X-ray standing waves

A relatively recent technique for determining parameters of surface structure is that employing x-ray standing waves. For strong Bragg scattering there is a Bragg band gap region in which the incident and diffracted x rays are of nearly equal intensity and couple to create x-ray standing waves. These have the periodicity of the diffracting Miller planes and a phase that varies by π as the incidence angle is varied from one side of the total reflectivity region to the other. The demonstration of the existence of x-ray standing waves and of the fact that their wave field excites host atom or impurity fluorescence became the basis of the x-ray standing wave method of determining atom positions at surfaces.[28] The pattern of fluorescence yield versus x-ray incident angle varies systematically with position of the fluorescing atom plane relative to the extrapolated position of bulk atomic planes. The maximum (minimum) of fluorescence scattering is observed when the standing-wave antinodal (nodal) plane passes through the impurity position. Comparison of the phase of the observed fluorescence relative to that predicted by x-ray dynamical theory yields the position of the fluorescing atoms relative to extrapolated bulk crystal planes. Recent studies employing this method report position measurements of Ga and As atoms at monolayer coverages of heteroepitaxial GaAs on clean Si(111)[29] and of As on passivated Si(111).[30]

3. Surface extended x-ray absorption fine structure[31,32] *(SEXAFS)*

SEXAFS developed as a possible surface structural tool because of the prior development of EXAFS, the advent of high brightness synchrotron radiation, and the recognition that EXAFS becomes surface specific if detection of Auger electrons, the nonradiative decay component, is employed. SEXAFS is used primarily for investigating structure of surface complexes formed by the reaction of gases with surfaces and is not well suited to the study of clean surfaces. The distance between an adsorbed atom and its nearest substrate neighbors is determined from the oscillatory fine structure in the x-ray adsorption cross section above an absorption edge of the adsorbed atoms. These oscillations result from interference between the wave of a photoelectron ejected from a core level and electron waves backscattered from the adsorbate's nearest substrate neighbors. An empirical processing of the data yields a spectrum that is approximated by a sinusoidal function whose argument involves the interatom distance. The EXAFS method determines the difference in bond length between two atoms in the surface adsorbed site and these atoms in a bulk compound where the bond length is known. This yields bond lengths with an uncertainty of about $\pm$ 0.03 Å, at least a factor of 5 more accurate than LEED determinations. Specific characteristics of the spectra and the measurement of next-nearest-neighbor bond lengths are useful in determining the nature of the adsorption site (atop, bridge, hollow). Table 21.01.D.3.a lists position parameters obtained by various EXAFS methods for atoms adsorbed on metals and semiconductor surfaces.

Recently, a new use of SEXAFS has been demonstrated in the study of the anisotropy of surface vibrational amplitudes.[33]

TABLE 21.01.D.3.a. Nearest-neighbor bond lengths and normal layer spacings of atomic adsorption on metals and semiconductors determined by the various EXAFS methods. The specific methods, as recorded in Ref. 15, are here indicated by letters:S, SEXAFS; E, Conventional or Standard EXAFS; N, NEXAFS; X, XANES (NEXAFS). Bond lengths, BL, are from the original literature referenced in Ref. 15. Normal layer spacings, NLS, calculated from the bond lengths, are those listed in Ref. 15. For several of these surfaces there is more than one entry indicating how well data from different investigations agree.

Surface	Mtd.	BL	NLS	Surface	Mtd.	BL	NLS
Ag(110)(2×1)0	S	2.06	0.2	Ni(100)*p*(2×2)0	E	1.96	0.86
Al(100)0	E	1.92	0.98	Ni(100)*p*(2×2)	X	1.98	0.90
Al(111)(1×1)0	S	1.75	0.6	Ni(100)*c*(2×2)0	X	1.98	0.9
Al(111)[Al – O – Al – Al]	S	1.75	0.6	Ni(100)*c*(2×2)0	E	1.96	0.85
Cu(100)(2×2)Te	S	2.67	1.90	NI(100)*c*(2×2)0	E	1.96	0.86
Cu(100)(2×2)I	S	2.69	1.98	Ni(100)*c*(2×2)S	S	2.23	1.39
Cu(100)*c*(2×2)0	S	1.94	0.7	Ni(100)*c*(2×2)S	S	2.22	1.36
Cu(100)*c*(2×2)Cl	S	2.37	1.59	Si(111)(1×1)Cl	S	1.98	1.98
Cu(110)HCOOH[a] [C – O]	N	1.25	0.53	Si(111)(7×7)Cl	S	2.03	2.03
[O – Cu]		1.98	1.51	Si(111)(7×7)Te	S	1.506	1.506
Cu(111)($\sqrt{3}\times\sqrt{3}$)*R*30°	S	2.66	2.21	Si(111)(7×7)I	S	2.44	2.44
Ge(111)(1×1)Cl	S	2.07	2.07				

[a] Disordered.

4. Near-edge x-ray absorption fine structure (NEXAFS)

An x-ray spectroscopy related to EXAFS and SEXAFS is NEXAFS, also known a XANES, x-ray absorption near edge structure.[34] An example of its application to surface problems is its use to yield information on molecular orientation, bond-length change upon chemisorption, and molecular orbital rehybridization.[35]

E. Scanning microscopies

1. Scanning tunneling microscopy (STM)

An important advance in the instrumentation of surface science was made with the development of a scanning microscope having scanning resolution at atomic dimensions.[36] Its basic element is a sharp metal tip placed at a separation of about 6 Å from a sample surface so that wave functions of sample and tip overlap causing a small electron tunneling current to flow when a voltage in the range 0–2 eV is applied. The tunneling current, of the order of nanoamperes, varies exponentially with separation of tip and sample and is thus a very sensitive probe of tip to sample separation. A quantum-mechanical theory of electron tunneling at small voltage between a surface and a spherical tip concludes that the tunneling conductance is proportional to the local density at the tip position of electron states of the sample at its Fermi level.[37] This conclusion has been confirmed by an exact calculation of single-atom imaging involving as tip a single sulfur atom adsorbed on a jellium surface, facing a single Na atom adsorbed on a second jellium surface as sample.[38]

The position of the tip in the scanning tunneling microscope is controlled by three mutually perpendicular *x, y, z* piezoelectric transducers each of which can move the tip a maximum of one micrometer. The *x, y* transducers are used in scanning over the surface; the *z* transducer controls tip to sample separation. Coarser tip movement is also provided so that the *x, y* scanning area can be changed and the tip brought approximately to its proper separation from the sample. In the topographic or current-imaging mode the tunneling current is kept constant by feeding back to the *z* transducer a correction voltage that changes the tip to surface separation as the tip is scanned over the surface.[39,40] The feedback signal is thus a measure of the corrugations of the surface revealing its atomicity. One presentation of these data is a plot of feedback signal for raster scanning along *x* at different settings of *y*. In another presentation the *z*-correction voltage is used to produce a gray-scale image of the surface corrugations. See Fig. 21.01.E.1.a.

Approximating the surface local density of Fermi-level states at the tip position by a superposition of atomic charge densities[42] has made possible theoretical simulation of gray-scale STM images like that of Fig. 21.01.E.1.a. This simulation, applied to the calculation of images for the several competing atomic models of the

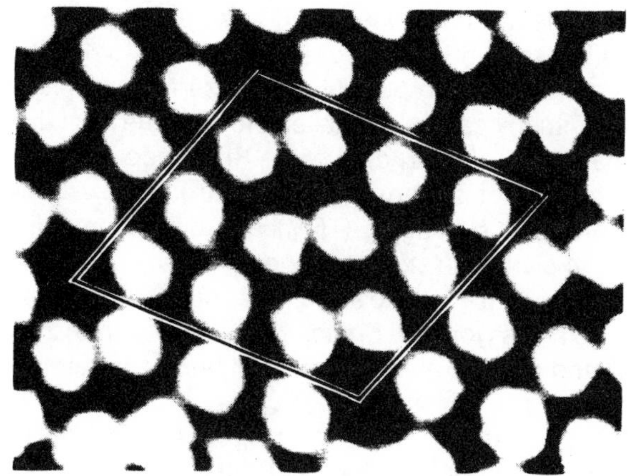

FIGURE 21.01.E.1.a. A gray-scale STM image of the Si(111)7×7 surface at positive bias. This was obtained at a tip bias of +2 eV when electron tunneling occurs from the conduction band of the metal tip into the empty states of the semiconductor that lie up to 2 eV above its Fermi level. The white areas indicate surface protrusions with the black-to-white range corresponding to a 2 Å *z*-displacement of the tip. The unit mesh outlined in the figure is seen to include twelve protrusions that are due to the twelve outermost Si atoms of the surface, the so-called "adatoms" of the DAS model. This figure is Fig. 19.1 of Ref. 41. Further discussion of this and related images is in Sec. 21.03.B.3.

Si(111)7$\times$7 surface,[43] shows that an adatom model is required and that among these models only that of Takayanagi *et al.* (Sec. 21.01.F.1) gives almost perfect agreement with the measured STM image throughout the unit mesh. Thus the STM image was shown to provide detailed surface-structural information. It has also been demonstrated that STM presents evidence that surface states characteristic of the Si(111)7$\times$7 superlattice are laterally localized and correlate with the differing substructures of the two halves of the unit mesh.[44] Features of steps on the Si(111)7$\times$7 surface are observed by STM.[45] References 39, 40, and 46–48 are to reviews of STM.

2. Atomic force microscopy (AFM)

The atomic force microscope is a novel instrument in which interatomic forces are used to record images of surfaces on an atomic scale. In one form of the apparatus[49] the tip opposite the sample surface is at a corner of the free end of a tiny cantilever beam of metal-coated SiO_2. As the tip is scanned over the sample surface the force between it and the surface atoms is counterbalanced by the restoring force of bending of the cantilever. In the most widely used version of the apparatus a signal proportional to the bending of the beam is obtained from an STM tip facing the side of the cantilever opposite the sample surface. The characteristics of a typical cantilever are: length 240 μ, width 20 μ, thickness 2.5 μ, spring constant 0.07 N/m.[50] Whereas STM is limited to the study of metal surfaces, ATM can be used to study nonmetallic as well as metallic surfaces. The hexagonal structure of the insulator boron oxide, which has a lattice constant of 2.504 Å, has been resolved.[51]

F. Electron microscopy methods

1. Transmission methods (TEM) (TED)

With the application of ultra-high vacuum techniques (UHV) to electron microscopes it has been possible to reduce background pressures to the range 3×10^{-8} to 1×10^{-9} Torr making possible clean surface structural studies in many cases. A detailed study of the growth of epitaxial layers of $NiSi_2$ on a clean Si surface illustrates the capability of a transmission electron microscope, used in both the imaging (TEM) and diffraction (TED) modes, to clarify the structural changes that occur in the epitaxial film as it grows to a thickness of about 10 monolayers.[52] In this work the Si substrate was less than 2000 Å thick and 100 keV electrons normally incident were employed.

When phase contrast imaging is employed with specimens that are less than 100 Å thick resolution approaches atomic dimensions. This is the realm of high resolution electron microscopy (HREM).[53] A plan view is obtained if the beam interacts along the surface normal, profile view if the beam passes the somewhat curved end of the sample parallel to the surface. In the latter case the imaging reveals the projected arrangement of atomic columns along the surface. A number of surfaces have been studied by this profile imaging at atomic resolution.[54]

TED in the selected-area diffraction mode of a 100-keV UHV electron microscope is particularly well suited to the solution of surface structural problems.[55] TED has the advantage over LEED in that reflections with scattering vectors parallel or nearly parallel to surfaces can be observed readily. These reflections are those most sensitive to atom positions parallel to the surface. Furthermore, single-scattering, kinematical theory applies whereas it does not with LEED. A particularly elegant application of TED to the investigation of surface structure is that of Takayanagi *et al.* in determining the structure of the long-studied Si(111)7$\times$7 reconstructed surface.[55] This work has brought to a successful conclusion the efforts of many investigators using essentially all of the surface structural tools of surface science in attempting to determine the atomic structure of this surface. Takayanagi *et al.* proposed on the basis of their work the so-called dimer, adatom, stacking-fault (DAS) model which consists of nine atom pairs (dimers) on the sides of the two triangular subcells of the 7$\times$7 unit surface unit cell, 12 adatoms, and a layer with different stackings in the two subcells.

2. Reflection electron microscopy[56] (REM)

In REM the microscope is configured as it is for reflection high energy electron diffraction (RHEED). The electron beam interacts with the surface at grazing incidence, one to two degrees off parallelism with the surface. REM images one of the diffracted beams. Although the image is foreshortened in the direction of the incident beam by a factor of 20 to 50, the amplification in the lateral direction is inthe 10^5 to 10^6 range revealing monolayer detail of steps, terraces, and defects.

3. Low-energy electron microscopy[57] (LEEM)

In this mode of electron spectroscopy, initial electron-beam focussing and final magnification and imaging of the surface or its diffraction pattern are done at high energy (20 keV) but electron diffraction at the surface occurs at low energy (<200 eV). This is achieved by placing a short (<5 mm), apertured cathode lens immediately in front of the specimen. Diffraction at low energy provides increased surface specificity and diffracted intensity but does not preclude Fourier transformation in the imaging lens to produce the magnified object function. Increased accessibility to the target face facilitates target cleaning procedures and permits operation in emission modes with electrons ejected from the specimen by incident uv photons or electrons. Resolution of 15 nm has been achieved; calculations by Bauer indicate that resolution of 2 nm should be feasible.

G. Other structural methods

1. Ion scattering[3,58,59]

Methods based on the interactions with surfaces of well-directed ion beams with a wide range of kinetic energies provide important tools determining atomic position and character at solid surfaces. Ion velocity equal to the Bohr velocity, the orbital velocity of the hydrogen electron, 2.2×10^8 cm s^{-1} corresponding to a 2.5 keV proton or a 100-keV He^+ ion, separates the phenomena into high and low energy regions.[58] The high energy region includes the high and medium energy ion scattering, HEIS and MEIS. The low energy region subsumes low energy ion scattering, LEIS, and impact collision ion scattering spectroscopy, ICISS. The probability of incident ion neutralization changes rapidly at the Bohr velocity, being negligible above it and large below it for noble gas ions. In the high energy regime, scattering is pure Coulomb in character in which outer electrons of the impacted atoms are involved.

Rutherford back scattering in the energy range 0.1 to 5 MeV can be used for surface studies because the energy of ions backscattered from surface atoms is greater than that of ions backscattered deeper in the crystal.[58–61] With the incident beam directed along perfectly aligned atom rows perpendicular to the surface of an ideal crystal the so-called surface peak consists primarily of scattering from the outermost atom and appears at an energy characteristic of the mass of the struck atom. The surface peak then consists of the backscattering from one atom row of the crystal. Lateral movement of the top atom in the surface plane increases the magnitude of the surface peak as more and more scattering is detected from the second atom. When the second atom is completely outside the shadow cone of the first atom the surface peak corresponds to two atoms per atomic row. Computer simulation enables one to determine the magnitude of lateral surface atom displacement between these limits. Adsorbed atoms of different mass can be identified by the energy placement of the surface peak even for perfect alignment of the rows. Displacement of surface atoms normal to the surface is measured by the magnitude of the surface peak for incident beam alignment along major crystal directions not parallel to the surface normal. He ion scattering from atoms deeper in the bulk than the surface atoms is typically two orders of magnitude smaller than the surface peak from the first monolayer and appears at lower energies. Thus high-energy Rutherford backscattering provides a nondestructive, quantitative measurement of surface atom position.

In the low-energy ICISS procedure 1 keV ions backscattered near 180° from their incident direction are detected.[62,63] It is observed that this backscattering drops to zero as the incidence angle relative to the surface is reduced to the value at which each surface atom enters the shadow cone of its nearest neighbor atom. Knowledge of atom spacings on the surface enables one to determine the shape of the shadow cone in the regions where cutoff occurs. Measured incidence angles for extinction of backscattering from second layer atoms and the known shape of the shadow cone determines the distance between the first and second layer atoms of the crystal.[62]

Scattering with noble gas ions in the low energy range is subject to the beam intensity uncertainty caused by ion neutralization when the projectile is near the solid surface. Alkali ions, however, are not neutralized in this way.[64] Use of time-of-flight m/e analysis and a detector sensitive to both ions and neutral atoms has also circumvented the neutralization problem.[65] This combined with computer simulation has made possible not only accurate positioning of atoms in the first three layers of a crystal but also quantitative chemical analysis of each layer in mixed atom crystals such as Cu_3Au.

In a variant of low-energy scattering known as direct recoil scattering information on the binding of hydrogen on a stepped surface has been obtained by measuring the yield of desorbed recoil hydrogen atoms under bombardment with 10-keV Ar^+ ions.[66]

In the medium energy range of MEIS, 50–500 keV, a method employing what is known as double alignment has proven to be a very sensitive probe of the geometrical structure of surfaces.[67,68] One alignment employs the phenomenon of channeling used in Rutherford MeV scattering. The second alignment, known as blocking, determines directions in which the backscattered ions are blocked from leaving the crystal, thereby giving better energy and depth resolution than MeV scattering for single crystal targets. Thus, finding the angular position of the blocking cone of first-layer atoms for particles scattered by second-layer atoms yields the orientation of the intermolecular axis between atoms in the first and second layers.

2. Atom scattering[69]

He atoms in an atomic beam having kinetic energies of ~60 meV are elastically scattered at a solid surface by interaction with the electron cloud that projects into vacuum outside the outermost layer of nuclei. Diffracted beam positions are determined by the same considerations as are applied to electron diffraction (Sec. 21.01.B.1). Intensities of the diffraction peaks are primarily determined by the periodic modulation of the repulsive part of the particle–solid potential function $V(\mathbf{r})$. This is expressed in terms of a *corrugation function* $\zeta(\mathbf{R})$ which describes the equipotential profile $V(\mathbf{r}) = E$ on which the classical turning points of the incident atom lie. In these expressions $\mathbf{r}$ is the three-dimensional position vector of the atom outside the surface, $\mathbf{R}$ the in-surface vector component of $\mathbf{r}$, and E the energy of the normally incident atom. Since the potenital well depth D (~10 meV) is small relative to E (~60 meV) it is usually neglected and the repulsive part of the potential approximated by a hard wall determined by $\zeta(\mathbf{R})$. Several methods exist for calculating diffraction intensities from assumed hard-wall models. Parametric forms are fitted to measured diffraction intensities. Certain aspects of surface structure can be inferred from the structure of the corrugation function. Recent theoretical work indicates that the He-surface interaction poetntial $V_{He}(\mathbf{r})$ is reasonably proportional to the electron charge density $\rho(\mathbf{r})$ outside the surface. Since $\rho(\mathbf{r})$ can be calculated by theories of the electronic structure of solids, a more direct connection between atom position parameters and diffraction corrugation functions is possible.

3. Atom-probe field ion microscopy[3,70,71] (FIM)

The field ion microscope has the capability of "seeing" individual atoms on surfaces, a feat accomplished by the projection to a distant phosphor screen of ions produced by field-evaporation and ionization of surface atoms. It is thus capable of determining in-plane positions of surface atoms but not their vertical positions. Combined with atom-probe *m/e* analysis of field- or laser-desorbed atoms, FIM has the unique capacity of chemical identification of individual atoms whose in-plane position on the surface has been determined. Atomic motions in diffusion or migration can also be directly observed. Several crystal facets are also observable on a single FIM image. Field desorption can provide atomically clean surfaces for investigation.

4. Structure from theory

Theoretical values of semiconductor surface parameters are sensitive to atomic structure. A method has been developed for covalent and ionic semiconductors that yields the change in the ground-state total energy of the crystal as a function of atom positions in the surface layers and minimizes this energy by varying surface parameters to predict a surface structure. Applying this method to the Si(111)2×1 surface Pandey[72] showed that a pi-bonding, chain model is energetically very favorable compared to a buckling model in which alternate rows of surface atoms are raised and lowered. In another application of energy minimization GaAs(111) was shown to have a 2×2 unit mesh with one-quarter monolayer of Ga vacancies.[73] Later, x-ray crystallography indicated that this structure occurs on InSb(111), a surface expected to have the same structure as GaAs(111).[74] The GaAs(100) structure predicted by total energy minimization[73] was shown to be correct by scanning tunneling microscopy.[75] Calculations of this type also suggest an asymmetrical dimer geometry with partially ionic bonds between surface atoms for the Si(100)2×1 surface. A structure for this surface based on *minimization of the surface elastic energy* is found to require atomic displacements from bulk position at least four more layers into the crystal.[76]

Self-consistent pseudopotential and empirical tight-binding calculations of the electronic structure of semiconductors are also sensitive to surface atom position, and when compared with photoemission electron energy distributions yield structural information.

Experiment and theory concerned with the structural and electronic properties of the surface of elemental semiconductors, particularly silicon,[77] and of tetrahedrally coordinated comound semiconductors[78] have been reviewed recently. Reference 23 is a review of experiment and theory of semiconductor surface structures. *Ab initio*, self-consistent, spin-density determinations of electronic structure and magnetism at surfaces and interfaces have been reviewed in Ref. 79.

H. Surface phase transitions

The structure of solid surfaces can undergo phase transitions. These can be transitions from one ordered phase to another such as Si(111)7×7 changing to Si(111)1×1 which has been studied by LEED[80,81] and by high-resolution electron microscopy.[82] As the temperature of a surface is raised toward the bulk melting point one observes surface melting and surface roughening.[83] A different mode of loss of lateral crystalline order occurring reversibly far below the bulk melting temperature has been observed.[84] Two-dimensional phase transitions in chemisorption systems such as H on Fe, W, and Ni have been studied.[85] Surface wetting and reversible disordering near melting have been analyzed by the computer method of molecular dynamics simulation.[86]

Monolayers of rare gas atoms adsorbed on the basal planes of graphite at low temperatures, studied by x-ray scattering, display phase transitions such as melting and lack of true crystallinity in two dimensions.[87] The structures and phase transformations of rare gases[88] and molecules, such a CO and N_2,[89] physisorbed on graphite have also been studied.

21.02. SURFACE CHEMICAL COMPOSITION AND REACTIONS

A. Compositional analysis[90]

Surface atoms have two identifying characteristics used in chemical compositional analysis. One is the core-state energy level structure, the other the atom's nuclear mass.

1. Core level methods[2,3]

In x-ray photoemission spectroscopy (XPS) there is sufficient photon energy to eject core electrons into vacuum outside the solid surface where the kinetic energy can be measured. The electronic transition of XPS is shown in FIgure 21.02.A.1.a, transition 1. The binding energy of the ejected core electron is

$$E_b = \hbar\omega - E_k - \phi_s$$

in the notation defined in the caption of Figure 21.02.A.1.a. In an XPS spectrum peaks from several such electronic transitions will be seen, enough to identify the atom in question. The shift in the binding energy of a core electron is a very sensitive probe of the chemical environment of the atom that undergoes the x-ray transition. Because of this the magnitude of the shift for a surface atom is also related to the validity of surface structural models. Reviews that discuss theory and interpretation of core-level binding shifts have been given in Refs. 91 and 92. Other papers discuss the various mechanisms that are responsible for surface core-level shifts and how they are related to surface electronic structure.[93] Tables of core-level binding energies and of shifts in various molecular environments are given in Appendices 1 and 3 of Ref. 94.

For XPS and other so-called emission *electron spectroscopies* whose basic transition processes can occur at distances into the solid that exceed the depth of

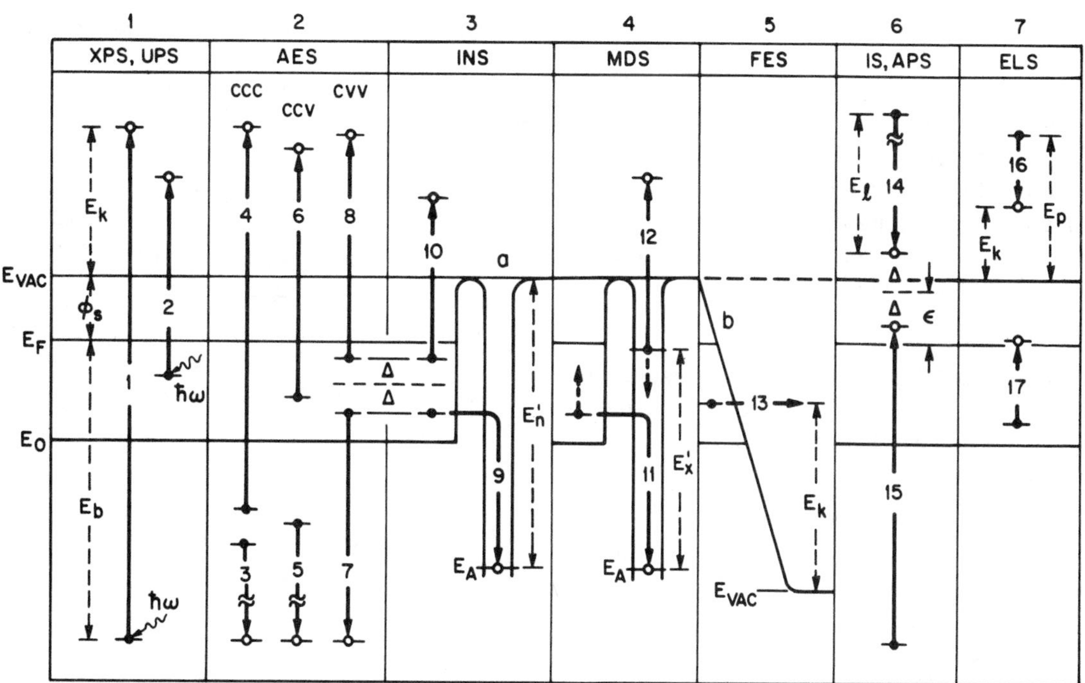

FIGURE 21.02.A.1.a. Electronic transitions basic to several electron spectroscopies. E_{VAC}, E_F, and E_0 indicate the vacuum level, the Fermi level, and the level of the bottom of the valence band, respectively. $\hbar\omega$, E_k, ϕ_s, and E_b are photon energy, electron kinetic energy just outside the surface, surface work function, and electron binding energy, respectively. In panel 2 the three letters involving *c* (core) and *v* (valence) designate the types of levels in which the three holes involved in the Auger process lie in order of decreasing binding energy.

the surface region by an order of magnitude or more, *surface sensitivity or specificity* is determined by the mean free path λ with respect to inelastic scattering.[95] This mean free path defines the *volume of sensitivity* or *information depth* of the spectroscopy. Rough values for λ in Å are 10, 5, 10, and 20 at electron energies in eV 8, 50, 600, and 2000, respectively.

Auger electron spectroscopy (AES) is based on the three two-electron transitions (3–4, 5–6, and 7–8 in column 2 of Figure 21.02.A.1.a) by which an excited core hole can decay. The most common means of producing the initial core hole is by impact of electrons having kinetic energy 3–5 keV. For elemental identification, kinetic energy spectra are differentiated to enhaced the relative sharp Auger features and then compared with standard plots taken from samples of known composition. Representative Auger spectra of the elements are given in Ref. 96. Reference 97 provides a review of the field.

Ionization spectroscopy (IS) detects the edges that occur in the kinetic energy spectrum of scattered electrons at energies equal to the energy of "ionization" of the core electron to the lowest unfilled levels at the vacuum level. In this case transitions 14 and 15 in Figure 21.02.A.1.a both terminate at the Fermi level E_F corresponding to $\epsilon = 0$.

Appearance potential spectroscopy (APS) when used for chemical analysis also determines the minimum energies of incident electron required to create core vacancies. In two APS variants the core hole creation is detected by measuring either the soft x rays or Auger electrons produced when the core hole is subsequently filled.

2. Nuclear mass methods

Mass determination by ion back scattering of He^+ or Ne^+ at an atom in the top layer of the solid is based on the scattering equation:

$$E_s/E_i = (1 + M_2/M_1)^{-2} \times \{\cos\theta + [(M_2/M_1)^2 - \sin^2\theta]^{1/2}\}^2$$

in which the energy of the scattered ion E_s outside the crystal relative to the ion's incident energy E_i is expressed in terms of the ion's mass M_1, the surface atom mass M_2, and the laboratory scattering angle θ. For scattering atoms of mass greater than that of the scattered ion, distinct peaks for each type of surface atom apear in the kinetic energy spectrum. In the high-energy regime (~1 MeV) ion scattering applied to single crystals in the channeling mode provides nondestructive, quantiative determination of atom surface density.

Other nuclear mass methods are based on the removal of surface atoms. In secondary ion mass spectroscopy (SIMS) the mass of sputtered ions is determined by conventional mass analysis techniques.[98] There are quantitative uncertainties as to the probability that the sputtered particle will be ionized. The *atom-probe field ion microscope* determines the mass of individually field-desorbed ions in a time-of-flight mass sepctrometer.[70,71] A modification of this technique has made possible observation of mass-selected ionic species desorbed from all the crystallographic planes available on the field emission tip.

B. Adsorption

Physisorption involves the van der Waals interaction that results from mutually induced fluctuating dipole moments in adsorbate and substrate. Bonding energies are less than about 0.5 eV. Chemisorption involves valence bonding for which bonding energies exceed 0.5 eV.

1. Chemisorption[99]

The sticking probability s is the rate of adsorption divided by the rate of arrival of gas molecules, usually given as a function of coverage in monolayers θ. The molecular arrival rate ν is

$$\nu = 3.513 \times 10^{22} P(MT)^{-1/2}\ \mathrm{cm}^2\ \mathrm{s}^{-1},$$

where P is pressure in Torr, M molecular mass in grams, and T temperature in kelvins. The form of $s(\theta)$ points to the kinetic mechanism of adsorption. Adsorption into immobile sites yields $s(\theta) \propto 1 - \theta$, with s falling approximately linearly as the surface covers. $s(\theta)$ independent of θ at lower coverages indicates the existence of a physisorbed precursor state in which the incident atom or molecule can find a vacant chemisorption site before desorption occurs.

The effects on adsorption on simple metals of the *coadsorption* of alkalis is reviewed in Ref. 100.

2. Physisorption

Underlying the phenomenon of physisorption is the particle–surface interaction potential. Consider, as an example of work in this field, the determination of this potential from atom beam scattering experiments.[101] In the phenomenon of *selective adsorption* an incident atom (He, say) having the requisite energy and momentum can be diffracted into a state in which the atom moves essentially as a free particle parallel to the surface. The atom gains total kinetic energy by the amount with which it is bound normal to the surface in the well-shaped particle-surface potential.

Resonant transitions into such bound states are evidenced by sharp "anomalies" in the intensities of various elastic beams as functions of incidence conditions of momentum and energy. Ground-state energies of the interaction potential determined by atom diffraction and by the measurement of thermodynamic binding energy of He atoms incident on graphite are in excellent agreement.

See Sec. 21.01.H for references to work on physisorption of atoms and molecules on graphite at low temperature.

C. Stimulated desorption

1. Thermal desorption[3,99]

Increasing substrate temperature, usually in a programmed manner, releases adsorbed molecules or their reaction products in specific temperature ranges. For a first-order desorption process the desorption activation energy E_m is proportional to the absolute temperature T_p at which the maximum of the desorption peak occurs. The residence time τ of an adsorbate on a surface can be written $\tau = A \exp(E_m/RT)$, where A is a constant, R the gas content, and T the temperature. If A is assumed to be 10^{-13} s, a condition met by many but not all adsorbate–substrate systems, the following expression holds:

$$E_m\quad (\mathrm{kJ\ mol}^{-1}) \cong 0.25 T_p\quad (\mathrm{K}).$$

In nonactivated adsorption on clean metals E_m is the differential heat of adsorption, an important adsorption parameter. Integration of the desorption spectrum yields relative coverages. Absolute coverages can be obtained from the pumping characteristics of the vacuum apparatus.

2. Electron- and photon-stimulated desorption[2,3,102–105] *(ESD, PSD)*

Neutral molecular and atomic fragments, ions, and metastable species can be desorbed from adsorbed layers on surfaces by bombardment of electrons with 10–1000 eV kinetic energy. The cross sections for the desorption of ions is smaller (10^{-1}) than those for neutral desorption, and each of these is usually much smaller (10^{-2}–10^{-7}) than the cross sections of comparable gas phase processes and sensitive to the mode of bonding. Transfer of momentum from incident electron to adsorbed species is insufficient to desorb it. The *Menzel–Gomer–Redhead (MGR) mechanism* is based on the electron-impact-induced transition of a valence electron from a bonding to an antibonding state. The repulsive potential of this excited electronic state then accelerates the desorbed atoms away from the surface, possibly in an ionized state. Qualitative understanding of many ESD observations has been achieved with this one-dimensional model.

Atomic and molecular ions desorbed by electron impact have been shown to leave the surface in discrete cones of emission whose axes are determined by the orientation of the surface molecular bonds that are "broken" in the desorption process. This method is known as *ESDIAD (electron-stimulated desorption ion angular distribution)*.

The *Knotek–Feibelman (KF) mechanism* of stimulated desorption explains the characteristic features of O^+ desorption from surfaces in which the oxygen is ionically bonded, a case difficult to understand on the basis one-electron Franck–Condon excitation. In the maximal valency ionic compound TiO_2 the cation is ionized to the noble-gas configuration Ti^{4+}, with the result that the highest occupied level, Ti(3*p*), lies ~34 eV below the conduction band minimum. If this level is ionized the dominant decay mode is an interatomic Auger process in which an O(2*p*) valence electron decays into the Ti(3*p*) hole and the ~31 eV thus released ejects one or two additional electrons from the O(2*p*) states. This amount of charge transfer turns O^{2-} into O^0 or O^+, which is desorbed by Coulomb repulsion. The KF mechanism is independent of how the shallow core hole in the cation is produced, and experiment has shown that both electrons and photons can initiate it. Identical angle-resolved adsorbate ejection patterns are observed by ESD and PSD. Deep core excitations by electrons and photons in adsorbate atoms have also been shown to initiate desorption.

D. Gas-surface interactions

Recent developments in atomic and molecular beam scattering techniques and in the use of ultra-high vacuum methods of surface preparation and characterization have produced rapid growth in our knowledge of the kinetics and dynamics of gas-phase interactions of atoms and molecules with surfaces. The basic phenomena are inelastic scattering, rotational and vibrational energy exchange, and chemical reactions. Recent reviews of the field have been given.[106–109] The quantum states of interacting particles can now be tailored and those of scattered particles determined. The current level of experiment and theory is leading to a unique description of the gas–surface interaction potential and the dynamics that occur on this potential. Reference 109 presents a review of reactive scattering.

21.03. SURFACE ELECTRONIC STRUCTURE

A. Work function ϕ[3,110–112]

The true work function of a uniform surface of an electronic conductor is the potential difference between the Fermi level (the electrochemical potential of the electrons inside the solid) and the potential at the near-surface vacuum level defined as the potential at the point at which the image force on an emitted electron has become negligible. Five methods of determining ϕ are described.

Thermionic emission of electrons (method T of Table 21.03.A.1) is governed by the Richardson–Dushman equation

$$j = A(1-r)T^2 \exp(-e\phi/kT),$$

in which j is the saturation current in a weak field, r the reflection coefficient of the surface for zero applied field (small for clean surfaces), and A a universal constant equal to $4\pi mk^2e/h^3$ (120 A cm^{-2} K^{-2}). The *Richardson plot* of $\ln(j/T^2)$ against $1/T$ yields the apparent work function ϕ^* as

$$\phi^* = (k/e)[d/d(1/T)]\ln(j/T^2) = \phi - T(d\phi/dT).$$

The method of *photoelectron yield near threshold* (method P of Table 21.03.A.1) is based for metals on the *Fowler equation*

$$\ln(j/T^2) = B + \ln f[(\hbar\omega - \phi)/kT],$$

in which $j(\omega)$ is the quantum yield for $\hbar\omega > \phi$, B a constant independent of ω and T, and $f(x)$ a tabulated function. The *Fowler plot* of $\ln(j/T^2)$ vs $\hbar\omega/kT$ is fitted to the function ln $f(x)$ by a shift of ϕ along the $\hbar\omega/kT$ axis and a shift of B along the $\ln(j/T^2)$ axis. In the limit $T \to 0$ the Fowler equation reduces to $j(\omega) \propto (\hbar\omega - \phi)^2$.

In the method of *field emission retarding potential* (method FERP in Table 21.03.A.1) field-emitted electrons that pass through an aperture in the anode reach the surface of the metal sample whose true ϕ is to be determined. The fastest electrons reaching the collecting sample are those that tunnel through the surface barrier at the Fermi level of the emitter. The retarding voltage V_r that reduces the collector current I_c to zero equals the work function to within the extent of a Boltzmann tail at finite temperature.

The *Fowler–Nordheim plot* (method F of Table 21.03.A.1) is of principal use in determining work functions of individual crystallographic planes on a field emission tip relative to one another or relative to the average work function of the entire emitting tip. The Fowler–Nordheim plot of $\ln(i/V^2)$ vs $1/V$ has slope $B\phi^{3/2}$, yielding relative ϕ values. B can be estimated for clean surfaces for which ϕ is known and then used in determining unknown ϕ of other surfaces.

The *contact potential difference* (method CPD of Table 21.03.A.1) between two solid surfaces is the difference $\phi_1 - \phi_2$ of the surface work functions. Thus ϕ of a metal or semiconductor surface is determined by the measured CPD to a metal (usually W) of known ϕ. In the *Kelvin method* the two surfaces are vibrated relative to one another, causing a time modulation $dC(t)/dt$ of the intersurface capacitance and a consequent current flow in the external circuit. In the null mode a voltage V_a applied to the condenser is varied till the circuit current

$$i(t) = [dC(t)/dt](\phi_1 - \phi_2 + V_a)$$

is zero, yielding $\phi_1 - \phi_2 = -V_a$.

Table 21.03.A.1 presents selected values for ϕ of the elements. A more recent compilation[111] contains some alternative values but no data later than 1976.

B. Electronic states

1. Photoemission spectroscopy[2,3,113] (PES)

Photoemission spectroscopy is done in two frequency ranges: ultraviolet photoemission spectroscopy (UPS) and x-ray photoemission spectroscopy (XPS) (Column 1 of Figure 21.02.A.1.a). In one mode of operation kinetic energy distributions of electrons ejected from filled initial states are measured at a series of constant $\hbar\omega$. For $\hbar\omega < 30$–50 eV these distributions are dominated by the joint density of initial and final states. Above this energy they reflect more closely the local density of states modified by matrix element effects. In the angle-integrated model (AIUPS) all electrons are detected along with energy-degraded and secondary electrons that distort the distribution near the vacuum level.

In the angle-resolved mode (ARUPS) only those electrons are detected whose exit momenta lie in a small range about a chosen exit angle. Since the parallel component of momentum,

$$k_\parallel = (2mE_k/\hbar^2)^{1/2} \sin\theta,$$

is conserved during the emission process and the initial state energy, specified by

TABLE 21.03.A.1. Selected values of electron work functions. No Miller index notation indicates polycrystalline sample. Methods: P, photoemission; CPD, contact potential difference; FERP, field emission retarding potential; F, Fowler–Nordheim plot; T, thermionic emission. From Ref. 112.

Element	ϕ (eV)	Method
Ag	4.26	P
(100)	4.64	P
(110)	4.52	P
(111)	4.74	P
Al	4.28	P
(100)	4.41	P
(110)	4.06	P
(111)	4.24	P
Au	5.1	P
(100)	5.47	P
(110)	5.37	
(111)	5.31	
Ba	2.52	CPD
Be	4.98	P
C	5.0	CPD
Ce	2.9	P
Co	5.0	P
Cr	4.5	P
Cs	2.14	P
Cu	4.65	P
(100)	4.59	P
(110)	4.48	P
(111)	4.98	P
(112)	4.53	P
Eu	2.5	P
Fe	4.5	P
(100)	4.67	P
α(111)	4.81	P
Ga	4.2	CPD
Ge	5.0	CPD
(111)	4.80	P
Gd	3.1	P

Element	ϕ (eV)	Method
Hf	3.9	P
In	4.12	P
Ir(110)	5.42	FERP
(111)	5.76	FERP
(111)	5.67	F
(210)	5.00	F
K	2.30	P
La	3.5	P
Mn	4.1	P
Mo	4.6	P
(100)	4.53	P
(110)	4.95	P
(111)	4.55	P
(112)	4.36	P
(114)	4.50	P
(332)	4.55	P
Na	2.75	P
Nb	4.3	P
(001)	4.02	T
(110)	4.87	T
(111)	4.36	T
(112)	4.63	T
(113)	4.29	T
(116)	3.95	T
(310)	4.18	T
Nd	3.2	P
Ni	5.15	P
(100)	5.22	P
(110)	5.04	P
(111)	5.35	P
Pb	4.25	P
Pd	5.12	P
(111)	5.6	P

Element	ϕ (eV)	Method
Pt	5.65	P
(111)	5.7	P
Re	4.72	T
(1011)	5.75	F
Rh	4.98	P
Ru	4.71	P
Sc	3.5	P
Si *n*	4.85	CPD
p(111)	4.60	P
Sm	2.7	P
Sr	2.59	T
Ta	4.25	T
(100)	4.15	T
(110)	4.80	T
(111)	4.00	T
Ti	4.53	P
U	3.63	P, CPD
(100)	3.73	P, CPD
(110)	3.90	P, CPD
(113)	3.67	P, CPD
V	4.3	P
W	4.55	CPD
(100)	4.63	FERP
(110)	5.25	FERP
(111)	4.47	FERP
(113)	4.18	CPD
(116)	4.30	T
Y	3.1	P
Zr	4.05	P

$$E_b = \hbar\omega - E_k - \phi_s,$$

is known, one obtains directly from experiment the two-dimensional band structure $E(k_{\parallel})$, the band dispersion, of the occupied initial state. θ is the polar angle measured from the surface normal. The intensity, tunability, and polarized nature of synchrotron radiation has made possible the extensive development of this technique in the study of metal and semiconductor surfaces.[114–117] The symmetry, dispersion, and orientation of orbitals in adsorbed molecules can be determined by angle-resolved, polarization-dependent photoelectron spectroscopy.[114,118]

2. Inverse photoemission spectroscopy[2,121] (IPES)

The basic phenomenon of inverse photoemission spectroscopy, originally known as bremsstrahlung isochromat spectroscopy (BIS), is the entrance of incident electrons of energy E into the electronic state structure at a solid surface that is followed by radiative or nonradiative decay. IPES detects the photon emitted in radiative decay. In the isochromat mode the energy of the photon is fixed and the spectrum is obtained by variation of the energy window of an electron spectrometer. IPES is thus the inverse of PES in which monoenergetic photons produce electrons over a range of energies. IPES complements PES in that it permits investigation of unoccupied electronic states of a solid and its surface. With it one can probe the energy range between the Fermi level, E_F, and the vacuum level, E_{vac}, that is inaccessible to PES. Other so-called "empty-state spectroscopies" are appearance potential spectroscopy (Sec. 21.03.B.5), near-edge x-ray absorption spectroscopy (Sec. 21.01.D.4), and scanning tunneling spectroscopy (Sec. 21.03.B.3). IPES has the unique advantage of the possibility of momentum resolution forming the basis of *k*-resolved PIES (KRIPES). X-ray BIS is reviewed in Ref. 119 and experiment and theory of IPES and KRIPES in Refs. 120–122.

3. Scanning tunneling spectroscopy[38–41,44,45] *(STS)*

The spectroscopy mode of scanning tunneling microscopy makes possible spatial resolution of the electronic structure of surfaces on an atomic scale because tunneling current depends on wave function overlap as well as tip-to-sample distance. The Si(111)7×7 surface affords an instructive example of this. When the voltage bias on the semiconductor is swept positively relative to the tip, steplike increases in tunneling conductance occur at energies of empty states in the semiconductor above its Fermi level. Sweeping bias negatively similarly reveals filled surface states below E_F. Figure 21.01.E.1.a is a gray-scale image at + 2 eV bias. Figure 21.03.B.3.a shows two images of electrons tunneling from filled surface states that lie in two different energy ranges below E_F. The higher-lying surface states imaged at *A* in this figure are localized on the adatoms while the deeper-lying states imaged at *B* are localized on the broken-bond orbitals in the corner holes and on the "rest atoms" that lie between the adatoms in the topmost layer of the underlying crystal. The intensity asymmetry between the two, triangular halves of the rhombic surface unit mesh in image *A* results from the difference in density of states between the two halves under one of which there is a stacking fault in the underlying crystal.[44,123] This stacking fault also gives rise to the mirror symmetry about the shorter diagonal of the unit mesh, observed in image *B*. All of these conclusions by STS are in agreement with the model of the surface derived by Takayanagi *et al.*[55] Filled and unfilled surface states can be observed by STS over the range − 4 to + 4 eV relative to E_F,[124] outside which the spectrum is dominated by barrier resonances.[125]

4. Tunneling spectroscopies involving incident ions, incident metastable atoms, field emission[2,4]

In columns 3–5 of Figure 21.02.A.1.a are depicted three *tunneling spectroscopies* that yield transition densities related to the local density of initial states. *Ion-neutralization spectroscopy (INS)* and *metastable deexcitation spectroscopy (MDS)*, also termed *Penning ionization*, are each based on two-electron processes whose transition probability depends upon the overlap of the evanescent tail of the surface electron with the vacant ground state of an incident, usually He, ion or metastable (transitions 9 and 11 in Figure 21.02.A.1.a). INS is a bona fide two-electron process since the initial states of both "up" and "down" electrons may vary in energy throughout the valence band. The initial-state transition density, related to the local denisty of states outside the surface, is obtained by taking the convolution square root of the meaured kinetic energy spectrum (the final-state transition density).[126] The process underlying MDS is quasi-one-electron since one electron equivalently moves between discrete atomic levels; so no deconvolution is needed.[127] For clean surfaces of work function > 4 eV the INS process occurs even for incident metastables because the metastable level, lying above the Fermi level E_F, if resonance ionized by thee solid.[128] Similarly for clean surfaces of work function < 2 eV the MDS process occurs for incident ions because they are resonance neutralized to the metastable level.[129] Each of these spectroscopies detects surface electronic structure down to about 12 eV below E_F.

Field emission spectroscopy[157] *(FES)* is based on direct electron tunneling through the surface barrier created by the high field at the field emitter tip. It is carried out in a field emission microscope equipped to measure electron kinetic energy (panel 5 of Figure 21.02.A.1.a). The measured electron energy distribution divided by the probability distribution for free-electron tunneling is a measure of the local density of states at the internal classical turning point of the barrier. Energy levels in adsorbed atoms are detected via their effect on tunneling probability. The range is limited to about 2 eV below E_F.

5. Other spectroscopies and surface probes[2,4]

The *ccv* and *cvv Auger processes* (panel 2 of Figure 21.02.A.1.a) resemble MDS and INS, respectively, and yield similar spectra, localized to the vicinity of the core-

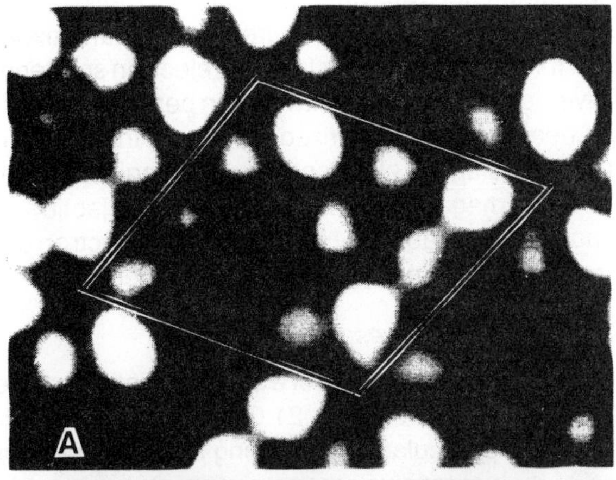

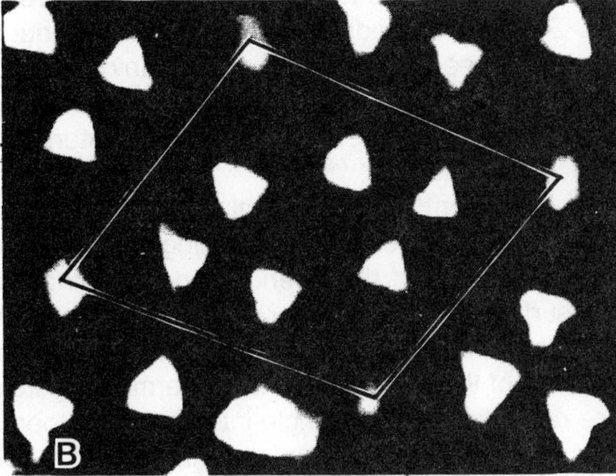

FIGURE 21.03.B.3.a. Gray-scale STM images of surface states of the Si(111)7×7 surface. A: states between E_F and 0.35 eV below E_F, located on the adatoms. B: states between 0.6 and 1.0 eV below E_F, located on the rest atoms. This figure is Fig. 19.2 of Ref. 41.

ionized atom, except in cases complicated by final-state multiplet structure and many-body effects. *Appearance potential spectroscopy (APS)* (panel 6 of Figure 21.02.A.1.a) yields an energy loss spectrum of scattered electrons (E_l) related to the convolution of the density of unfilled states above E_F. *Energy loss spectroscopy (ELS)* (panel 7 of Figure 21.02.A.1.a) can be used to detect transitions between filled and unfilled states via the measured energy loss $E_p - E_k$ of a scattered electron. Only the energy difference between the states is determined. AES, APS, and ELS, unlike the tunneling spectroscopies, have surface sensitivities dependent on electron escape depth.

Recent advances in lasers and systems of optical detection have resulted in new types of surface studies. Among these, second harmonic generation (SHG) is asimple and sensitive probe.[130] Chemisorption and coadsorption of atoms and molecules, reconstruction of and electronic transitions in surface monolayers, and surface melting have been studied by this method. Surface-enhanced Raman scattering (SERS) has been reviewed in Ref. 131.

6. Methods involving spin-polarized electrons

Electron spectroscopies of both magnetic and nonmagnetic surfaces involve spin-dependent interactions of two types: spin-orbit interaction with ion cores and exchange interaction with valence electrons. Refs. 132 and 133 review the extensive literature on theory and experiment in this field. Electron diffraction, electron scattering, and inverse photoemission have been performed with incident beams of spin-polarized electrons. Incident photons and unpolarized electrons eject spin-polarized electrons from magnetic surfaces. Electron ejection by spin-polarized metastable $He^*(2^3S)$[134] and electron capture by 150 keV deuterons[135] have also been studied.

7. Some examples of results

The electronic structure of a surface is intimately tied to its atomic structure. (Sec. 21.01.G.4) Semiconductor surfaces are particularly challenging in this respect because of their extensive reconstruction as compared to metals. Several methods of modeling semiconductor surfaces have been employed.[136,137] In many cases success depends on finding the correct surface model. These efforts are an interesting chapter in the interplay of experiment and theory.[77,78,23]

Self-consistent pseudopotential theories of clean semiconductor surfaces predict dangling-orbital surface states near the top of the valence band resulting from broken surface bonds. Also predicted are back-bond states below and within the valence band resulting from modification of the bonds within the outermost layers as surface relaxation occurs. Angle-integrated photoemission has verified the existence of dangling-orbital surface states on several silicon, germanium, and compound semiconductor surfaces. Angle-resolved photoemission provides enhanced intensity in surface state peaks in specific exit angle ranges, enabling the identification of surface states and the measurement of energy dispersion $E(k_{\parallel})$.

Theory and experiment have also combined to elucidate the chemisorption geometry on semiconductor surfaces.[138] Exposure of annealed Si(111)7×7 to atomic hydrogen produces the monohydride surface Si(111):H for which self-consistent pseudopotential and empirical tight-binding theories yield local densities of states in agreement with experimental angle-integrated spectra. Tight-binding theory further explains the results for saturated H absorption on quenched, disordered Si(111), leading to the conclusion that the top Si layer has been corroded away leaving a trihydride surface, Si(111):SiH_3, with three hydrogen atoms bonded to each second-layer Si atom. The observations with *s*- and *p*-polarized radiation on the Si(111)Cl and Ge(111)Cl surfaces when compared with calculated local densities of states demonstrate that Cl absorbs in the on-top, onefold site on Si(111) and in the hollow, threefold site on Ge(111).

The electronic structures of metal surfaces and of metal–adsorbate systems have been treated by the density functional method.[139] Chemisorption on metals has been extensively studied by electron spectroscopic techniques.[113] Orbital symmetry in adsorbed molecules. CO on Ni(100) for example, has been determined by observing the dependence of orbital peaks in angle-resolved photoemission on the polarization of the light. Orbital energy shifts on adsorption have been measured and calculated theoretically. Binding energies and induced dipole moments of a series of atomic adsorbates on jellium have been calculated using density functional theory, yielding results in good agreement with experimental work on metals.

21.04. ATOMIC MOTIONS AT SURFACES

A. Surface vibrational spectroscopies

The three principal vibrational spectroscopies for which entries into the literature are provided here are electron energy loss spectroscopy, inelastic atomic scattering spectroscopy, and surface infrared spectroscopy. Comparative discussions of these methods are to be found in Sec. 1.2 of Ref. 140, Ref. 141, Table 1 of Ref. 142, and Ref. 143.

1. Electron energy loss spectroscopy (EELS)

There are two modes of electronic interaction with surfaces that are bases for experimental methods. In one, electrons of energy less than 10 eV, on interacting with the dipole field of atoms or molecules vibrating normally to the surface, are scattered through snall angles peaked in the specular direction. The so-called impact mode is a short range, large angle, inelastic scattering of electrons of energy 100–200 eV. The dispersion of phonons excited in this interaction is measured. Energy resolution in each of these modes is 3–7 meV. Recent studies employing impact scattering are discussed in Refs. 141 and 144. A tabulation of EELS data has also been published.[145]

TABLE 21.04.A.1.a. Comparison of theoretical and experimental frequencies of various vibrational modes of H adsorbed on transition metal surfaces. Adapted from Ref. 149. Adsorbate coverage is one monolayer (ml) except for W(001) which is 2ml as indicated. Theoretical results are by D. R. Hamann, P. J. Feibelman, and R. Biswas in papers identified in the "Site" column by their reference numbers in Ref. 149. The experimental data, referenced in Ref. 149 by the letter superscripts in the "Experiment" column, were obtained by electron energy loss spectroscopy (a-e, h, i, k) and surface infrared spectroscopy (f, j).

Surface	Site	Mode	Theory (meV)	Experiment (meV)
Ru(0001)	3-fold[17]	as	100	138,[a] 102[b]
	fcc	ss	140	105,[a] 141[b]
W(001)	bridge[18]	wag	77	80[d,e]
	(2ml)	as	169	160,[c,d] 118[e]
		ss	141	130,[c,d,e] 133[f]
		opt	150	118[e]
Pt(111)	3-fold[19]	as	114	152,[h] 67[i]
	fcc	ss	166	68,[h] 112,[i] 155[j]
Rh(001)	4-fold[21]	as	67	138[k]
		ss	92	82[k]
		mix	142	
		mix	148	152[k]

Recent instrumental improvements make possible the study of surface rate processes by the recording of vibrational spectra in real time with resolution of a millisecond.[146–148] Theoretical and experimental frequencies of various vibrational modes of H adsorbed on transition metal substrates are compared in Table 21.04.A.1.a.[149]

2. Inelastic atomic scattering

The excitation of surface phonons and the measurements of surface phonon dispersion is also possible by the method of high resolution, inelastic, atomic scattering. This is the analog for an atomic beam of the impact mode of electron scattering. Nearly monochromatic beams of He^+ ions can be prepared with most probable energies in the range 8 to 60 meV. The energy half-widths of these beams are between 0.2 and 1 meV. This energy resolution is about an order of magnitude better than that achieved with EELS. A presentation of the history, theory, and experiment of inelastic atomic scattering has been published.[141]

3. Surface infrared spectroscopy (SIRS)

A comprehensive review of theory, experimental methods, and selected results for several variants of SIRS in given in Ref. 142 where also are referenced a number of other reviews of the field. A tabulation of the IR data has been published recently.[150] The technique of IR spectroscopy is best at measuring the differences between two states of a surface as, for example, adsorbate-covered and clean. The information obtainable relates to the chemical nature, and the geometrical and electronic structure of adsorbate and substrate atoms, as well as the dynamics and kinetics of gas–surface interactions, vibrations, and diffusion.

B. Surface diffusion[151–154]

1. Basic equations

Mobility of surface atoms involves jumping from the potential minimum across a barrier of height E_m, the activation energy for the jump. In a one-dimensional concentration gradient $\partial c/\partial x$, the diffusion flux J is given by Fick's first law,

$$J = -D(\partial c/\partial x),$$

in which D is the diffusion coefficient. Fick's second law,

$$\partial c/\partial t = \partial(D\,\partial c/\partial x)/\partial x,$$

expresses the result of diffusion in terms of the time rate of change of surface concentration, a convenience since J is difficult to measure. In the absence of a gradient ($\partial c/\partial x = 0$) an isolated adatom undergoes unrestricted random walk, jumping at a rate Γ over the activation barrier E_m and moving an average distance λ per jump. Diffusion theory interrelates D, Γ, λ, E_m, the mean-square displacement in one dimension $\langle \Delta x^2 \rangle$, and the time interval τ over which the experiment extends by the relation

$$\langle \Delta x^2 \rangle = 2\Gamma\tau\lambda^2,$$

the Einstein relation

$$\langle \Delta x^2 \rangle = 2D\tau,$$

and the Arrhenius relation

$$D = D_0 \exp(-E_m/kT).$$

2. Experimental methods

The techniques devised for measuring the diffusion parameters D_0 and E_m are of three kinds. In the *concentration gradient method* an initial deposit of the adsorbate is established on the surface at low temperatures with a sharp boundary between clean and covered regions. A probe, such as the field emission microscope, sensitive to concentration and having lateral resolution, is used to follow diffusion across the boundary at higher temperatures.

In the *homogeneous concentration method* the field emission microscope (FEM) is used with a probe hole to measure the emission current from a small area (50–100 Å diameter) centered on a given crystal plane. In the temperature range of mobility, current fluctuations indicate adatom concentration fluctuations within the probed area even though no net diffusion occurs. Such random fluctuations about equilibrium coverage, and the emission current, build up and decay with a relaxation time $\tau = \lambda^2/4D$, where λ is the radius of the region examined. τ is related to the time autocorrelation function of the concentration or current fluctuations. Comparison of the theoretical and experimental correlation functions yields τ and thus D.

The *method of direct adatom observation* employs the field ion microscope (FIM) in which individual atomic displacements are observed. For diffusion of an atom on a single-crystal plane, field ion patterns are obtained after each of a series of time intervals τ during which the surface is held at a given elevated temperature. These determine the mean-square displacement $\langle \Delta x^2 \rangle$ and thus the diffusion coefficient D. A few hundred heating periods to various temperatures are needed for an Arrhenius plot.

Formation and migration of atomic clusters can be observed in the FIM. Two W atoms in adjacent channels on W(211) form an interacting pair that exists in two stable configurations: two atoms abreast in the straight configuration or slanted at an angle to the rows in the straggered configuration. The dimer moves by hops of one atom at a time, alternating between straight and staggered configurations. Measured diffusion parameters, E_m and D, for metal monomers and dimers on metals are to be found in Refs. 151 and 152.

21.05. REFERENCES

Since the principal goal of this chapter is to provide an entry into the literature of surface physics, not a history of its development, publications cited are with few exceptions reviews and papers with review sections.

[1]A. Zangwil, *Physics at Surfaces* (Cambridge Univ. Press, 1988).

[2]G. Ertl and J. Küppers, *Low Energy Electrons and Surface Chemistry* (VCH Verlagsgesellschaft, 2nd ed. 1985).

[3]R. L. Park and M. G. Lagally, *Solid State Physics: Surfaces*, Vol. 22 of Methods of Experimental Physics, edited by R. Celotta and J. Levine (Academic Press, 1985).

[4]D. P. Woodruff and T. A. Delchar, *Modern Techniques of Surface Science* (Cambridge Solid State Science Series, 1988).

[5]E. A. Wood, J. Appl. Phys. **35**, 1306 (1964).

[6]P. J. Estrup and E. G. McRae, Surf. Sci. **25**, 1 (1971).

[7]G. A. Somorjai and M. A. Van Hove, *Adsorbed Monolayers on Solid Surfaces,* Vol. 38 of Structure and Bonding (Springer, Berlin, 1979).

[8]M. A. Van Hove, W. H. Weinberg, and C. M. Chan, *Low Energy Electron Diffraction: Experiment, Theory and Surface Structure*, Vol. 6 of Springer Series in Surface Sciences (Springer, 1986).

[9]M. A. Van Hove and S. Y. Tong, *Surface Crystallography by LEED,* Vol. 2 in Springer Series in Chemical Physics (Springer, Berlin, 1979).

[10]M. Henzler, in *Electron Spectroscopy for Surface Analysis,* edited by H. Ibach, Vol. 4 of Topics in Current Physics (Springer, Berlin, 1977), Chap. 4, p. 117.

[11]M. A. Van Hove, in *The Nature of the Surface Chemical Bond,* edited by T. N. Rhodin and G. Ertl (North Holland, Amsterdam, 1979), Chap. 4, p. 275.

[12]F. Jona, J. Phys. C **11**, 4271 (1978).

[13]M. G. Lagally and J. A. Martin, Rev. Sci. Instrum. **54**, 1273 (1983).

[14]E. G. McRae, R. A. Malic, and D. A. Kapilow, Rev. Sci. Instrum. **56**, 2077 (1985).

[15]J. M. MacLaren, J. B. Pendry, P. J. Rous, D. K. Saldin, G. A. Somorjai, M. A. Van Hove, D. D. Vvedensky, *Surface Crystallographic Information Service, A Handbook of Surface Structures* (Reidel, Dordrecht, 1987).

[16]Surface Crystallographic Information Service Database and Graphing Programs (Version 1.1, Jan. 1987) with Users Manual, edited by J. B. Pendry (Reidel, 1987).

[17]P. Estrup, in *Chemistry and Physics of Solid Surfaces,* edited by R. Vanselow, Vol. 5, page 205 (Springer, 1984).

[18]J. E. Inglesfield, in *Progress in Surface Science* (Pergamon Press, 1985),Vol. 20, p. 105.

[19]M. B. Webb and E. R. Moog, in *Springer Series in Surface Sciences*, Vol. 2 (Springer, 1985), p. 397.

[20]C. B. Duke, in *Chemistry and Physics of Solid Surfaces,* edited by R. Vanselow (Chemical Rubber, Cleveland, 1979), Vol. 2, p. 373.

[21]D. E. Eastman, J. Vac. Sci. Technol. **17**, 492 (1980).

[22]W. Mönch, Surf. Sci. **86**, 672 (1979).

[23]A. Kahn, Surf. Sci. Reports **3**, 193 (1983).

[24]I. K. Robinson, in *Handbook of Synchrotron Radiation,* edited by D. E. Moncton and G. S. Brown, Vol. 3 (North Holland, 1989).

[25]P. Eisenberger and W. C. Marra, Phys. Rev. Lett. **46**, 1081 (1981).

[26]R. Feidenhans'l, J. Bohr, M. Nielsen, M. Toney, R. L. Johnson, F. Gray, and I. K. Robinson, Festkörperprobleme **25**, 545 (1985).

[27]I. K. Robinson, W. K. Waskiewicz, P. H. Fuoss, and L. J. Norton, Phys. Rev. B **37**, 4325 (1988).

[28]B. W. Batterman, Phys. Rev. **133**, A759 (1964); J. A. Golovchenko, B. W. Batterman, and W. L. Brown, Phys. Rev. B **10**, 4239 (1974).

[29]J. R. Patel, P. E. Freeland, M. S. Hybertsen, D. C. Johnson, and J. A. Golovchenko, Phys. Rev. Lett. **59**, 2180 (1987).

[30]J. R. Patel, J. A. Golovchenkko, P. E. Freeland, and H.-J. Gossmann, Phys. Rev. B **36**, 7715 (1987).

[31]J. Stöhr, in *X-Ray Absorption–Principles, Applications, Techniques of EXAFS, SEXAFS, and XANES,* edited by D. C. Koningsberger and R. Prins, Chap. 10 (Wiley, 1988).

[32]P. H. Citrin, J. de Physique **47**, C8-437 (1986).

[33]F. Sette, C. T. Chen, J. E. Rowe, and P. H. Citrin, Phys. Rev. Lett. **59**, 311 (1987).

[34]A. Bianconi, in *X-ray Absorption–Principles, Applications, Techniques of EXAFS, SEXAFS, and XANES,* edited by D. C. Koningsberger and R. Prins, Chap. 11 (Wiley, 1988).

[35]F. Sette, in *EXAFS and Near-Edge Structure III,* edited by K. O. Hodgson, B. Hedman, and J. E. Penner-Hahn (Springer, 1984), p. 250.

[36]G. Binnig, H. Rohrer, Ch. Gerber, and E. Weibel, Phys. Rev. Lett. **49**, 57 (1982); **50**, 120 (1983).

[37]J. Tersoff and D. R. Hamann, Phys. Rev. Lett. **50**, 1998 (1983).

[38]N. D. Lang, Phys. Rev. Lett. **56**, 1164 (1986); Comments Condensed Matter Phys. **14**, 253 (1989); Phys. Rev. Lett. **58**, 45 (1987).

[39]R. M. Tromp, R. J. Hamers, and J. E. Demuth, Science **234**, 304 (1986).

[40]P. K. Hansma and J. Tersoff, J. Appl. Phys. **61**, R1 (1987).

[41]R. M. Tromp, in *Springer Series in Surface Science,* Vol. 10 (Springer, 1988), p. 547.

[42]J. Tersoff and D. R. Hamann, Phys. Rev. B **31**, 2 (1985).

[43]R. M. Tromp, R. J. Hamers, and J. E. Demuth, Phys. Rev. B **34**, 1388 (1986).

[44]R. S. Becker, J. A. Golovchenko, D. R. Hamann, and B. S. Swartzentruber, Phys. Rev. Lett. **55**, 2032 (1985).

[45]R. S. Becker, J. A. Golovchenko, E. G. McRae, and B. S. Swartzentruber, Phys. Rev. Lett. **55**, 2028 (1985).

[46]G. Binnig and H. Rohrer, Sci. Amer. **253**, 50 (1985); IBM J. Res. Develop. **30**, 355 (1986).

[47]J. A. Golovchenko, Science **232**, 48 (1986).

[48]Y. Kuk and P. J. Silverman, Rev. Sci. Instrum. **60**, 165 (1989).

[49]G. Binnig, C. F. Quante, and Ch. Gerber, Phys. Rev. Lett. **12**, 930 (1986).

[50]G. K. Binnig, Physica Scripta T **19**, 53 (1987).

[51]G. Binnig, Ch. Gerber, E. Stoll, T. R. Albrecht, and C. F. Quate, Surf. Sci. **189/190**, 1 (1987).

[52]J. M. Gibson, J. L. Batstone, R. T. Tung, and F. C. Unterwald, Phys. Rev. Lett. **60**, 1158 (1988).

[53]D. J. Smith, in *Chemistry and Physics of Solid Surfaces VI,* edited by R. Vanselow and R. Howe (Springer, Berlin, 1986), p. 413.

[54]See Table 15.3 of Ref. 53.

[55]K. Takayanagi, Y. Tanishiro, M. Takahashi, and S. Takahashi, J. Vac. Sci. Technol. A **3**, 1502 (1985); Surf. Sci. **164**, 367 (1985).

[56]K. Yagi, J. Appl. Cryst. **20**, 140 (1987).

[57]W. Telieps, Appl. Phys. A **44**, 55 (1987).

[58]L. C. Feldman, *Ion Scattering from Surfaces and Interfaces in Ion Beams for Materials Analysis,* edited by J. R. Bird and J. S. Williams (Academic Press, 1988).

[59]J. F. Van der Veen, Surf. Sci. Reports **5**, 199 (1985)

[60]W. N. Unertl, Appl. of Surf. Sci. **11/12**, 64 (1982).

[61]L. C. Feldman, J. W. Mayer, and S. T. Picraux, *Materials Analysis by Ion Channeling* (Academic Press, 1982).

[62]M. Aono, C. Oshima, S. Zaima, S. Otani, and Y. Ishizawa, Japan, J. Appl. Phys. **20**, L829 (1981).

[63]M. Aono, Y. Hou, C. Oshima, and Y. Ishizawa, Phys. Rev. Lett. **49**, 567 (1982).

[64]H. Niehus, Nucl. Instr. and Methods **218**, 230 (1983).

[65]T. M. Buck, G. H. Wheatley, and D. P. Jackson, Nucl. Instr. and Methods **218**, 257 (1983).

[66]B. J. J. Koeleman, S. T. de Zwart, A. L. Boers, B. Poelsema, and K. Verhey, Nucl. Instr. and Methods **218**, 225 (1983).

[67]W. C. Turkenburg, W. Soszka, F. W. Saris, H. H. Kersten, and B. G. Colenbrander, Nucl. Instr. and Methods **132**, 587 (1976).

[68]R. M. Tromp and J. F. Van der Veen, Surf. Sci. **133**, 159 (1983).

[69]T. Engel and K. Rieder, in *Advances in Surface Structure Investigations,* edited by G. Höhler and E. Niekisch, Vol. 91 of Springer Tracts in Modern Physics (Springer, Berlin, 1982) p. 55.

[70]T. T. Tsong, Surf. Sci. Reports **8**, 127 (1988).

[71]T. Sakurai, A. Sakai, and H. W. Pickering, Adv. in Electronics and Electron Physics, Supplement 20 (Academic Press, 1989).

[72]K. C. Pandey, Phys. Rev. Lett. **47**, 1913 (1981); **49**, 223 (1982); Proc. Indian Nat. Sci. Acad. A **51**, 17 (1985).
[73]D. J. Chadi, Phys. Rev. Lett. **52**, 911 (1984); J. Vac. Sci. Technol. A **5**, 834 (1987).
[74]J. Bohr, R. Feidenhans'l, M. Nielsen, M. Toney, R. L. Johnson, and I. K. Robinson, Phys. Rev. Lett. **54**, 1275 (1985).
[75]M. D. Pashley, K. W. Haberern, W. Friday, J. M. Woodall, and P. D. Kirchner, Phys. Rev. Lett. **60**, 2176 (1988).
[76]J. A. Appelbaum and D. R. Hamann, Surf. Sci. **74**, 21 (1978).
[77]M. Schluter, in The Chem. Phys. of Solid Surfaces and Heterogeneous Catalysis, edited by D. A. King and D. P. Woodruff (Elsevier, 1988) Vol. 5, Chap. 2, page 37.
[78]C. B. Duke, *ibid.*, Chap. 3, page 69.
[79]A. J. Freeman and C. L. Fu, Springer Prog. Phys. **14**, 16 (1986).
[80]P. A. Bennett and M. B. Webb, Surf. Sci. **194**, 74 (1981).
[81]E. G. McRae and R. A. Malic, Surf. Sci. **161**, 25 (1985).
[82]N. O. Osakabe, Y. Tanishiro, K. Yagi, and G. Honjo, Surf. Sci. **109**, 353 (1981).
[83]T. Engel, in *Chemistry and Physics of Solid Surfaces,* edited by R. Vanselow and R. Howe, Vol. 10, page 407, Springer Series in Surf. Sci. (Springer, 1988).
[84]E. G. McRae and R. A. Malic, Phys. Rev. Lett. **58**, 1437 (1987).
[85]L. Roelofs and P. Estrup, Surf. Sci. **125**, 51 (1983).
[86]E. G. McRae, J. M. Landwehr, J. E. McRae, G. H. Gilmer, and M. H. Grabow, Phys. Rev. B **38**, 13178 (1988).
[87]R. J. Birgenau and P. M. Horn, Science **232**, 329 (1986).
[88]M. B. Webb and L. W. Bruch, in *Interfacial Aspects of Phase Transitions,* edited by B. Mutaftschief, p. 365 (Reidel, Dordrecht 1982); M. B. Webb and E. R. Moog, Springer Ser. in Surf. Sci. **2**, 397 (1985).
[89]S. C. Fain, Jr., Berichte Bunsengesellschaft Phys. Chem. **90**, 211 (1986); H. You and S. C. Fain, Jr., Phys. Rev. B **34**, 2840 (1986); M. Toney and S. C. Fain, Jr., Phys. Rev. B **36**, 1248 (1987).
[90]R. L. Park, in *Surface Physics of Materials,* edited by J. M. Blakely (Academic, New York, 1975), Vol. 2, Chap. 8, p. 377.
[91]D. Spanjaard, C. Guillot, M. C. Desjonqueres, G. Treglia, and J. Lecante, Surf. Sci. Reports **5**, 1 (1985).
[92]W. F. Egelhoff, Jr. Surf. Sci. Reports **6**, 253 (1987).
[93]P. H. Citrin and G. K. Wertheim, Phys. Rev. B **27**, 3176 (1983); G. K. Wertheim, Appl. Phys. A **41**, 75 (1986).
[94]T. A. Carlson, Photoelectron and Auger Spectroscopy (Plenum, New York, 1975).
[95]C. R. Brundle, J. Vac. Sci. Technol. **11**, 212 (1974).
[96]L. E. Davis, N. C. MacDonald, P. W. Palmberg, G. E. Riach, and R. E. Weber, Handbook of Auger Electron Spectroscopy, Sec. Ed. (1976), Physical Electronics Div., Perkin-Elmer Corp., 6509 Flying Cloud Road, Eden Prairie, MN 55343.
[97]Auger Electron Spectroscopy, edited by C. L. Bryant and R. P. Messmer (Academic Press, 1988).
[98]A. Benninghoven, F. G. Rüdenauer, and H. W. Werner, Secondary Ion Mass Spectroscopy (Wiley, 1987).
[99]D. A. King, in *Chemistry and Physics of Solid Surfaces,* edited by R. Vanselow (Chemical Rubber, Cleveland, 1979), Vol. 2, p. 87.
[100]H. P. Bonzel, Surf. Sci. Reports **8**, 43 (1987).
[101]M. W. Cole, D. R. Frankl, and D. L. Goldstein, Rev. Mod. Phys. **53**, 199 (1981).
[102]T. E. Madey, in *Inelastic Particle-Surface Collisions,* edited by W. Heiland and G. Taglauer, page 8, vol. 10 of Springer Series in Chemical Physics (Springer 1981); Science **234**, 316 (1986).
[103]T. E. Madey, W. L. Johnson, and S. A. Joyce, Vacuum **38**, 579 (1988).
[104]M. L. Knotek and P. J. Feibelman, Surf. Sci. **90**, 78 (1979).
[105]Desorption Induced by Electronic Transisions, DIET III, edited by R. H. Stulen and M. L. Knotek, Springer Series in Surface Sciences, Vol. 13 (Springer, 1988).
[106]J. C. Tully, Ann. Rev. Phys. Chem. **31**, 319 (1980).
[107]M. J. Cardillo, Ann. Rev. Phys. Chem. **32**, 331 (1981); Langmuir **1**, 4 (1985).
[108]J. A. Barker and D. J. Auerbach, Surf. Sci. Reports **4**, 1 (1985).
[109]M. P. D'Evelyn and R. J. Madix, Surf. Sci. Reports **3**, 413 (1983).
[110]M. Cardona and L. Ley, in *Photoemission in Solids I,* edited by M. Cardona and L. Ley (Springer, Berlin, 1978), Sec. 1.2, Chap. 1.
[111]J. Hölzl and F. K. Schulte, in *Solid State Physics,* edited by G. Höler, Vol. 85 of Springer Tracts in Modern Physics (Springer, Berlin, 1979), p. 1.
[112]H. B. Michaelson, J. Appl. Phys. **48**, 4729 (1977).
[113]*Photoemission and the Electronic Properties of Surfaces,* edited by B. Feuerbacher, B. Fitton, and R. F. Willis (Wiley, New York, 1978).
[114]W. Eberhart and E. W. Plummer, Adv. Chem. Physics **49**, 533 (1982).
[115]F. J. Himpsel, Adv. Physics **32**, 1 (1983).
[116]*Angle-Resolved Photoemission,* edited by S. D. Kevan (Elsevier, 1989).
[117]G. V. Hansson and R. I. G. Uhrberg, Surf. Sci. Reports **9**, 197 (1988).
[118]R. D. Schnell, D. Rieger, A. Bogen, F. J. Himpsel, K. Wandelt, and W. Steinmann, Phys. Rev. B **32**, 8057 (1985).
[119]J. K. Lang and Y. Baer, Rev. Sci. Instrum. **50**, 221 (1979).
[120]V. Dose, Surf. Sci. Reports **5**, 337 (1985).
[121]N. V. Smith, Rep. Prog. Phys. **51**, 1227 (1988).
[122]G. Borstel and G. Thörner, Surf. Sci. Reports **8**, 1 (1988).
[123]J. E. Northrup, Phys. Rev. Lett. **57**, 154 (1986).
[124]J. A. Stroscio, R. M. Feenstra, and A. P. Fein, Phys. Rev. Lett. **57**, 2579 (1986).
[125]R. S. Becker, J. A. Golovchenko, and B. S. Swartzentruber, Phys. Rev. Lett. **55**, 987 (1985).
[126]H. D. Hagstrum, in *Chem. and Phys. of Solid Surfaces VII,* edited by R. Vanselow and R. Howe, Springer, Ser. in Surf. Sci. Vol. 10 (Springer, Berlin, 1988), p. 341; Phys. Rev. **150**, 495 (1966).
[127]W. Sesselmann, B. Woratschek, J. Küppers, G. Ertl, and H. Haberland, Phys. Rev. B **35**, 1547 and 8348 (1987).
[128]B. Woratschek, W. Sesselmann, J. Küppers, G. Ertl, and H. Haberland, Phys. Rev. Lett. **55**, 1231 (1985). Surf. Sci. **180**, 187 (1987).
[129]H. D. Hagstrum, P. Petrie, and E. E. Chaban, Phys. Rev. B **38**, 10264 (1988).
[130]Y. R. Shen, Ann. Rev. Mater. Sci. **16**, 69 (1986).
[131]M. Moskovits, Rev. Mod. Phys. **57**, 783 (1985).
[132]J. Kirschner, *Polarized Electrons at Surfaces*, Vol. 106, Springer Tracts in Modern Physics (Springer, Berlin, 1985); Appl. Phys. A **44**, 3 (1987).
[133]*Polarized Electrons in Surface Physics,* edited by R. Feder (World Scientific, Singapore, 1985).
[134]M. Onellion, M. W. Hart, F. B. Dunning, and G. K. Walters, Phys. Rev. Lett. **52**, 380 (1984).
[135]C. Rau, Appl. Surf. Phys. **13**, 310 (1982).
[136]J. A. Appelbaum and D. R. Hamann, Rev. Mod. Phys. **48**, 479 (1976).
[137]M. Schluter, in *Festkörperprobleme* (Advances in Solid State Physics), edited by J. Treusch (Viewag, Beaunschweig, 1978), Vol. 18, p. 155.
[138]*Theory of Chemisorption,* edited by J. R. Smith, Vol. 19 of Topics in Current Physics (Springer, Berlin, 1980).
[139]N. D. Lang, in *Theory of the Inhomogeneous Electron Gas,* edited by S. Lundqvist and N. H. March (Plenum, 1983), p. 309.
[140]H. Ibach and D. L. Mills, *Electron Energy Loss Spectroscopy and Surface Vibrations* (Academic Press, 1982).
[141]J. P. Toennies, in *Solvay Conference on Surface Science,* Springer Series in Surface Science (Springer, Berlin, 1988), Vol. 14, p. 248.
[142]Y. J. Chabal, Surf. Sci. Reports **8**, 211 (1988).
[143]P. Masri, *ibid.* **9**, 293 (1988).
[144]S. Lehwald, F. Wolf, H. Ibach, B. M. Hall, D. L. Mills, Surf. Sci. **192**, 131 (1987).
[145]P. A. Thiry, J. Electron Spectroscopy and Related Phenom. **39**, 273 (1986).
[146]T. H. Ellis, L. H. Dubois, S. D. Kevan, and M. J. Cardillo, Science **230**, 256 (1985).
[147]W. Ho, J. Vac. Sci. Technol. A **3**, 1432 (1985).
[148]S. D. Kevan and L. H. Dubois, Rev. Sci. Instrum. **55**, 1604 (1984).
[149]D. R. Hamann, J. Electron Spectroscopy and Related Phenom. **44**, 1 (1987).
[150]J. Darville, *ibid.* **39**, 311 (1986).
[151]G. L. Kellogg, T. T. Tsong, and P. Cowan, Surf. Sci. **70**, 485 (1978).
[152]G. Ehrlich and K. Stolt, Ann. Rev. Phys. Chem. **31**, 603 (1980).
[153]G. Ehrlich, Critical Rev. Solid State Mat. Sci. **10**, 391 (1982).
[154]A. G. Nauvomets and Yu. S. Vedula, Surf. Sci. Reports **4**, 365 (1984).
[155]P. I. Cohen, T. R. Pukitė, J. M. Van Hove, and C. S. Lent, J. Vac. Sci. Technol. **4**, 1251 (1986).
[156]M. G. Lagally, D. E. Savage, and M. C. Tringides, in *Reflection High Energy Electron Diffraction and Reflection Electron Imaging of Surfaces*, edited by P. K. Larsen and P. J. Dobson (Plenum, 1988), p. 139.
[157]E. W. Plummer, in *Interactions on Metal Surfaces*, edited by R. Gomer, Vol. 4 of Topics in Applied Physics (Springer, 1975), p. 144.

22.00. Thermophysics

Yeram S. Touloukian†
Purdue University

CONTENTS

† Deceased.

22.01. INTRODUCTION; DATA SOURCES

The selection of material to be included within the highly limited number of pages of this chapter has been by necessity somewhat arbitrary, considering the wide scope of subject matter relating to thermophysics of interest to a physicist. Hence, the primary considerations have been compactness of presentation, selectivity of materials (elements being given preference), and the omission of figures which could have been most informative from a pedagogic standpoint.

Of the eleven tables in which data are reported, Tables 22.02–22.06 have been adapted in abridged, selective, and restructured form from Chap. 4 of *American Institute of Physics Handbook,* 3rd ed. (McGraw-Hill, New York, 1972); Tables 22.07 and 22.08 have been adapted in abridged and selective form from various volumes of the *McGraw-Hill/CINDAS Data Series* and other publications; and Tables 22.09–22.12 have been similarly adapted from AIP's *Temperature, Its Measurement and Control in Science and Industry* (Rheinhold, New York, 1941). Therefore, when convenient and necessary, the reader should refer to the more extensive original sources for these data. Except for Tables 22.07 and 22.08, no serious effort was made by this chapter editor to check on the validity of the information extracted from the other two main sources, both of which are AIP publications.

The presented data are rather self-explanatory. A special mention may be made concerning Table 22.05, which reports the second and third virial coefficients for certain fluids considered to be of interest to physicists. This table, together with the relevant thermodynamic relations, is provided as a substitute for a large number of tables of derived thermodynamic properties for these substances whose tabulation would have required extensive space. The general availability of rather sophisticated hand calculators makes such an approach feasible.

Concerning Tables 22.09–22.12, one must remember that *Temperature, Its Measurement and Control in Science and Industry* is published at irregular intervals. For more extensive information on temperature scales and fixed points, one may refer to Pt. 1 of Vol. 4, 1972 edition; for resistance, electronic, and magnetic thermometry to Pt. 2; and for thermocouples of new noble metal alloys and temperature measurements in biology, medicine, geophysics, and space to Pt. 3.

To the extent that information of only a very limited scope could be presented within this chapter, it was felt worthwhile to list below the roster of numerical data centers which could serve as sources for in-depth information on thermophysical properties of materials.

Name and address of center	Point of contact	Telephone number
Alloy Phase Diagram Data Center National Institute of Standards and Technology Gaithersburg, MD 20899	E. N. Pugh	(301) 975-5960
CINDAS/Purdue University 2595 Yeager Road W. Lafayette, IN 47906	W. H. Shafer	(317) 494-9393
Chemical Thermodynamics Data Center National Institute of Standards and Technology Gaithersburg, MD 20899	M. W. Chase	(301) 975-2526
Fluid Mixtures Data Center National Institute of Standards and Technology Boulder, CO 80302	W. M. Haynes	(303) 497-3257
Crystal Data Center National Institute of Standards and Technology Gaithersburg, MD 20899	A. D. Mighell	(301) 975-6254
Molten Salts Data Center Rensselaer Polytechnic Institute Troy, NY 12181	G. J. Janz	(518) 276-6337
National Center for Thermodynamic Data of Minerals USGS, 12201 Sunrise Valley Drive, MS 959 Reston, VA 22092	B. S. Hemingway	(703) 860-6740
Phase Diagrams for Ceramists Data Center National Institute of Standards and Technology Gaithersburg, MD 20899	S. Freiman	(301) 975-5761
Thermodynamics Research Center Texas A&M University College Station, TX 77843	K. N. Marsh	(409) 845-4971

22.02. TABLE: DEFINED FIXED POINTS ON THE TEMPERATURE SCALE[a]

Equilibrium state	Assigned temperature value of International Practical Temperature[b] (K)
Triple point of equilibrium hydrogen	13.81
Equilibrium between liquid and vapor phases of equilibrium hydrogen at pressure of 33 330.6 N/m^2	17.042
Boiling point of equilibrium hydrogen	20.28
Boiling point of neon	27.402
Triple point of oxygen	54.361
Boiling point of oxygen	90.188
Triple point of water[c]	273.16
Boiling point of water[c,d]	373.15
Freezing point of zinc	692.73
Freezing point of silver	1235.08
Freezing point of gold	1337.58

[a]Adapted from *American Institute of Physics Handbook*, 3rd ed., edited by Dwight E. Gray (McGraw-Hill, New York, 1972).
[b]IPTS-68. Except for the triple points and one equilibrium hydrogen point (17.042 K) the assigned values of temperature are for equilibrium states at a pressure $p_0 = 1$ atm (101 325 N/m^2).
[c]The water used should have the isotopic composition of ocean water.
[d]The freezing point of tin has the assigned value of $t_{68} = 231.9681$°C and may be used as an alternative to the boiling point of water.

22.03. TABLE: VAPOR PRESSURE OF THE ELEMENTS[a]

Element	Temperature (°C) 10^{-10} atm	10^{-8} atm	10^{-6} atm	10^{-5} atm	1 atm
Actinium, Ac	(1025)[b]	(1211)	(1459)	(1617)	3200
Aluminum, Al	744	889	1084	1209	2520
Americium, Am	625	767	958[b]	1085	2614
Antimony, Sb	308	381	476	534	1587
Argon, Ar	−245	−240	−234	−230	−186
Arsenic, As	123	170	231	269	612
Astatine, At	−10	21	62	87	335
Barium, Ba	360	466	614	712[b]	2125
Beryllium, Be	747	887	1074	1193[b]	2472
Bismuth, Bi	366	459	586	669	1564
Boron, B	1450	1620	1900[b]	2076	3802
Bromine, Br	−139	−122	−100	−86	58
Cadmium, Cd	91	143	213	257	767
Calcium, Ca	314	399	516	591	1484
Carbon, C	1750	1980	2267	2439	3827
Cerium, Ce	1091	1288	1552	1721	3426
Cesium, Cs	0	47	109	159	682
Chlorine, Cl	−184	−172	−157	−147	−34
Chromium, Cr	902	1056	1257	1384	2672
Cobalt, Co	995	1161	1379[b]	1517	2928
Copper, Cu	774	918	1109	1237	2566
Dysprosium, Dy	667	803	987	1150	2562
Erbium, Er	757	907	1111	1242	2863
Europium, Eu	588	404	524	602	1597
Fluorine, F					−188
Francium, Fa	−15	30	90	131	(674)
Gadolinium, Gd	987	1165[b]	1406	1564	3266
Gallium, Ga	617	748	926	1038	2247
Germanium, Ge	851	1013	1232	1372	2834
Gold, Au	863	1024[b]	1237	1374	2808
Hafnium, Hf	1608	1862	2197[b]	2417	4603
Helium, He					4.22
Holmium, Ho	703	844	1035	1159	2695
Hydrogen, H	−269	−268	−267	−266	−253
Indium, In	533	654	819	924	2070
Iodine, I	−86	−63	−32	−14	183
Iridium, Ir	1673	1921	2242	2441[b]	4389
Iron, Fe	948	1109	1321[b]	1455[b]	2862
Krypton, Kr	−234	−228	−220	−214	−153
Lanthanum, La	1102	1299	1564	1733	3457
Lead, Pb	404	478	616	706	1750
Lithium, Li	265	344	456	531	1324
Lutetium, Lu	1056	1239	1481	1635[b]	3395
Magnesium, Mg	204	273	365	424	1090
Manganese, Mn	586	700[b]	858	956	2062
Mercury, Hg	−60	−29	14	42	357
Molybdenum, Mo	1670	1925	2259	2469[b]	4610
Neodymium, Nd	796[b]	954[b]	1173	1318	3068
Neon, Ne	−265	−264	−262	−261	−246
Nickel, Ni	994	1155	1373[b]	1511	2914
Niobium, Nb	1833	2101	2448[b]	2671	4744
Nitrogen, N	−248	−244	−239[b]	−236	−196
Osmium, Os	2002	2288	2662	2888[b]	4987
Oxygen, O	−243	−239	−234	−230[b]	−183
Palladium, Pd	917	1085	1308	1450[b]	2964
Phosphorus, P	67	106	153	182	431
Platinum, Pt	1385	1600	1882	2058	3824
Plutonium, Pu	895	1076	1323	1485	3230
Polonium, Po	132	183	258	308	947
Potassium, K	39	88	157	203	758
Praseodymium, Pr	902[b]	1082	1331	1495	3512
Radium, Ra	275	359	474	547	1527
Rhenium, Re	2052	2362	2771	3032[b]	5687
Rhodium, Rh	1358	1567	1841[b]	2017	3727
Rubidium, Rb	16	59	123	165	694
Ruthenium, Ru	1592	1826	2128	2316[b]	4119
Samarium, Sm	407	507	641	728	1791
Scandium, Sc	879	1035	1241[b]	1372	2831
Selenium, Se	80	130	194[b]	237	679
Silicon, Si	1071	1244	1476	1636	3267
Silver, Ag	617	740	904[b]	1010	2163
Sodium, Na	95[b]	152	231	282	883
Strontium, Sr	270	350	458	528[b]	1375
Sulfur, S	1	35	77[b]	105[b]	445
Tantalum, Ta	2079	2384	2779[b]	3025	5365
Technetium, Tc	1664	1910	2233	2458	4627
Tellurium, Te	196	253	327	373	988
Terbium, Tb	950	1122	1352[b]	1507	3223
Thallium, Tl	321	410	530	609	1487
Thorium, Th	1527	1782	2134	2361	4788
Thulium, Tm	487	599	748	845	1947
Tin, Sn	736	888	1093	1225	2623
Titanium, Ti	1124	1310	1554[b]	1709	3289
Tungsten, W	2235	2550	2955	3205[b]	5555
Uranium, U	1320	1548	1855	2053	4134
Vanadium, V	1228	1420	1675	1827[b]	3409
Xenon, Xe	−219	−211	−200	−192	−108
Ytterbium, Yb	228	301	400	463	1194
Yttrium, Y	1040	1220	1458[b]	1616	3338
Zinc, Zn	148	208	286	338	911
Zirconium, Zr	1574	1822[b]	2156	2367	4409

[a]Adapted from *American Institute of Physics Handbook*, 3rd ed., edited by Dwight E. Gray (McGraw-Hill, New York, 1972).
[b]Phase transition.

22.04. TABLE: CRITICAL TEMPERATURE, PRESSURE, AND DENSITY OF ELEMENTS AND COMPOUNDS[a]

Elements and organic compounds

Element or compound	T_c (K)	P_c (atm)	ρ_c (g/cm^3)
Ammonia	405.51	111.3	0.235
Antimony trichloride	794.1	...	0.842
Argon	150.72	48.00	0.5308
Arsenic trichloride	311.29	...	0.720
Bismuth tribromide	1220	...	1.487
Bismuth trichloride	1178	...	1.210
Boron tribromide	573	...	0.90
Boron trichloride	452.0	38.2	...
Boron trifluoride	260.9	49.2	0.59
Bromine	584	102	1.18
Carbon dioxide	304.20	72.85	0.468
Carbon diselenide	612	69	0.850
Carbon disulfide	552	78	0.441
Carbon monoxide	133.0	34.5	0.301 0
Carbonyl sulfide	378	61	...
Cesium	2056	130.8	0.451
Chlorine	417.2	76.1	0.573
Chlorotrifluorosilane	307.64	34.20	...
Cyanogen	400	59	...
Deuterium (equilibrium)	38.26	16.28	0.066 8
Deuterium (normal)	38.35	16.43	...
Dichlorodifluorosilane	368.93	35.54	...
Dichlorosilane	470	44.7	0.515
Fluorine	144	55	...
Gallium	5410	250	1.58
Germanium tetrachloride	550.1	38	...
Helium-3	3.38	1.22	0.041
Helium-4	5.21	2.26	0.069 3
Hydrazone	653	145	...
Hydrogen (equilibrium)	32.94	12.77	0.030 8
Hydrogen (normal)	33.24	12.80	0.031 02
Hydrogen bromide	362.96	84.00	...
Hydrogen chloride	324.7	81.5	0.45
Hydrogen cyanide	456.7	53.2	0.195
Hydrogen deuteride	35.91	14.65	0.048 1
Hydrogen fluoride	461	64.1	0.29
Hydrogen iodide	423.2	80.8	...
Hydrogen selenide	411	88	...
Hydrogen sulfide	373.6	88.9	0.348 8
Iodine	785	116	...
Krypton	209.39	54.27	0.908 5
Lead	5400	850	2.2
Mercuric chloride	972	...	1.555
Monochlorosilane	409	47.5	0.444
Neon	44.44	26.86	0.483 5
Niobium pentabromide	1009	...	1.05
Niobium pentachloride	807	46	0.68
Nitric oxide	180.3	64.6	0.52
Nitrogen	126.3	33.54	0.311 0
Nitrogen dioxide	431	100	0.56
Nitrogen trifluoride	233.90	44.72	...
Nitrous oxide	309.59	71.596	0.452 5
Nitryl fluoride	349.5	...	...
Oxygen	154.78	50.14	0.41
Oxygen fluoride	215.2	48.9	0.553
Ozone	285.3	54.6	...
Perchloryl fluoride	368.4	53.0	0.637
Phosgene	455	56	0.52
Phosphine	324.5	64.5	...
Phosphonium chloride	322.3	72.7	...
Phosphorus	993.8	120.8	...
Phosphorus trichloride	793	...	0.520
Radon	377.16	62.0	...
Rubidium	2111	...	0.334
Silane	270	42.2	0.309
Silicon tetrachloride	506.8	37.1	0.584
Silicon tetrafluoride	259.01	36.66	...
Silver	7500	...	1.85
Stannic chloride	591.9	36.95	0.741 9
Sulfur	1313	116	...
Sulfur dioxide	430.7	77.808	0.525
Sulfur hexafluoride	318.71	37.11	0.751 7
Sulfur tetrafluoride	364.1	...	...
Sulfur trioxide	491.4	83.8	0.633
Tantalum pentabromide	973	...	1.26
Tantalum pentachloride	767	...	0.89
Titanium tetrachloride	...	45.7	...
Trichlorofluorosilane	438.42	35.33	...
Trichlorosilane	495	41.2	0.533
Tritium	40.0	...	0.109
Uranium hexafluoride	503.4	45.5	...
Water	647.4	218.3	0.326
Water (heavy)	644.1	216	...
Xenon	289.75	58.0	1.105

Organic compounds

Compound	T_c (K)	P_c (atm)	ρ_c (g/cm^3)
Acetic acid	594.8	57.1	0.351
Acetone	508.7	46.6	0.273
Acetonitrile	547.9	47.7	0.237
Acetylene	309.5	61.6	0.231
Aniline	698.8	52.3	0.340
Benzene	562.7	48.6	0.300
Bromobenzene	670.9	44.6	0.458
n-Butane	425.17	37.47	0.228
Butanol	560.11	48.60	0.270
1-Butene	419.6	39.7	0.234
2-Butene (cis)	428.2	40.5	0.236
2-Butene (trans)	433.2	41.5	0.240
Carbon tetrachloride	556.4	44.97	0.558
Chlorobenzene	632.4	44.6	0.365
Chlorodifluoromethane	369.6	48.48	0.525
Chloroform	536.6	54	0.496
Chlorotrifluoroethylene	379	40	0.55
Chlorotrifluoromethane	302.02	38.2	0.578
1-Chloro-1,1-difluoroethane	410.3	40.7	0.435
2-Chloro-1,1-difluoroethylene	400.6	44.0	0.499
Cyclohexane	554.2	40.57	0.273
Cyclopentane	511.8	44.55	0.27
1,2-Dichloroethane	561	53	0.44
1,1-Dichloro-1,2,2,2,-tetrafluoroethane	418.7	32.6	0.582
Dichlorodifluoromethane	384.7	39.6	0.555
Dichlorofluoromethane	451.7	51.0	0.522
Diethyl ether	467.8	35.6	0.265
Diethyl ketone	561.0	36.9	0.256
1,1-Difluoroethane	386.7	44.4	0.365
1,1-Difluoroethylene	303.3	43.8	0.417
2,2-Dimethylbutane	489.4	30.67	0.240
2,3-Dimethylbutane	550.3	30.99	0.241
Dimethyl ether	400.1	52.6	0.246
Dioxane	585	50.7	0.36
Ethane	305.43	48.20	0.203
Ethyl acetate	523.3	37.8	0.308
Ethyl alcohol	516	63.0	0.276
Ethyl bromide	503.9	61.5	0.507
Ethyl cyclopentane	569.5	33.53	0.262
Ethyl formate	508.5	46.8	0.323
Ethyl mercaptan	499	54.2	0.300
Ethyl methyl ether	437.9	43.4	0.272
Ethyl methyl ketone	533.7	39.46	0.252
Ethyl propyl ether	500.6	32.1	0.260
Ethyl sulfide	498.7	54.2	0.300
Ethylene	283.06	50.50	0.227
Ethylene oxide	469.0	70.97	0.32
Fluorobenzene	560.08	44.91	0.269
n-Hexane	507.9	29.94	0.234
Iodobenzene	721	44.6	0.581
Isobutane	408.14	36.00	0.221
Isopentane	461.0	32.9	0.234
Methane	191.1	45.80	0.162
Methyl acetate	506.9	46.3	0.325
Methyl alcohol	513.2	78.47	0.272
Methyl butyrate	554.5	34.3	0.300
Methyl chloride	416.28	65.93	0.353
Methyl cyclopentane	532.77	37.36	0.264
Methyl fluoride	317.71	58.0	0.300
Methyl formate	487.2	59.2	0.349
Methyl isopropyl ketone	553.4	38.0	0.278
2-Methylpentane	497.9	29.95	0.235
3-Methylpentane	504.4	30.83	0.235
Methyl *n*-propyl ketone	564.0	38.4	0.286
Methyl sulfide	503.1	54.6	0.309
Methylene chloride	510.2	59.97	...
Neopentane	433.76	31.57	0.238
Nitromethane	588	62.3	0.352
n-Octane	569.4	24.64	0.235
n-Pentane	569.78	33.31	0.232
Perfluorobutane	386.4	22.93	0.600
Perfluoro-*n*-heptane	474.8	16.0	0.584
Propane	370.0	42.01	0.220
Propene	365.0	45.6	0.233
Propionic acid	612	53	0.32
Propionitrile	564.4	41.3	0.240
n-Propyl acetate	549.4	32.9	0.296
n-Propyl alcohol	537.3	50.2	0.273
Propyl formate	538.1	40.1	0.309
Toluene	594.0	41.6	0.29
Trichlorotrifluoromethane	471.2	43.2	0.554
Trichlorotrifluoroethane	487.3	33.7	0.576
1,1,1-Trifluoroethane	346.3	37.1	0.434
Trimethylamine	433.3	40.2	0.233

[a]Adapted from *American Institute of Physics Handbook*, 3rd ed., edited by Dwight E. Gray (McGraw-Hill, New York, 1972).

22.05. TABLE: SECOND AND THIRD VIRIAL COEFFICIENTS OF VARIOUS SUBSTANCES[a]

B is in cm^3/mol and *C* in $cm^6/mol^2 \times 10^2$.

Temp. (K)	Air (dry, no CO_2)		Argon		Carbon dioxide		*n*-Deuterium		Helium		*n*-Hydrogen		Krypton		Methane		Neon		Nitrogen		Oxygen		Xenon	
	B	*C*	*B*	*C*	*B*	*C*	*B*	*C*	*B*	*C*	*B*	*C*	*B*	*C*	*B*	*C*	*B*	*C*	*B*	*C*	*B*	*C*	*B*	*C*
10									−21.7															
15									−8.7		−230													
20							−187	86	−2.2		−155													
25							−146	61	1.5		−106	14.0												
30							−118	45	3.8		−80.7	16.0												
35							−96	33	5.4		−63.2	14.3												
40							−76	26	6.6		−50.3	12.1												
45							−62	21	7.5		−40.8	10.7												
50							−50	17	8.2		−33.4	9.6												
55							−40	15	8.7		−27.6	8.9												
60							−32	13	9.2	2.7	−22.7	8.4						4						
70							−20	10	9.9	2.5	−15.2	7.4						4						
80			−288	7			−13	8.2	10.6	2.4	−9.9	6.9					−11.8	4	−242					
90			−229	9			−9.7	6.9	11.0	2.3	−5.7	6.4					−7.8	4	−196		−245			
100	−167		−187	12			−4.2	6.5	11.4	2.2	−2.5	6.1					−4.8	4	−160		−197			
110	−140		−156	16			−1.3	5.7	11.6	2.1	0.0	5.9	−366		−334		−2.3	3	−133		−164			
120	−118		−133	20			1.0	4.6	11.8	2.0	2.0	5.7	−308		−281		−0.4	3	−113		−139			
130	−101		−115	23			3.0	5	11.9	1.9	3.7	5.5	−264		−241		1.2	3	−96.3		−120			
140	−87.3	28	−99.8	25			4.6	5	12.0	1.8	5.1	5.4	−229		−209		2.6	3	−82.8		−104			
150	−75.9	26	−87.5	23			6.0	5	12.1	1.7	6.4	5.3	−201		−183		3.7	3	−71.7		−91.1			
160	−66.2	24	−77.2	22			7.1	5	12.3	1.6	7.6	5.2	−178		−162		4.8	3	−62.3	26	−79.9	23		
180	−50.9	21	−60.9	20			8.9	5	12.3	1.5	9.5	5.0	−143		−129		6.2	3	−47.3	21	−62.9	20		
200	−39.3	19	−48.7	18			10.2	5	12.3	1.3	10.8	4.8	−118		−105		7.6	3	−35.9	19	−50.0	17		
220	−30.2	18	−39.2	16			11.3	5	12.2	1.2	11.7	4.6	−97.9	33	−86.6		8.5	3	−26.9	17	−40.1	15	−231	
240	−22.9	17	−31.5	15			12.2	5	12.1	1.1	12.6	4.5	−82.4	30	−72.0		9.4	3	−19.7	16	−32.1	13	−196	
260	−16.9	16	−25.3	13	−167		12.8	5	12.0	1.1	13.3	4.4	−69.9	28	−60.2		10.0	3	−13.8	15	−25.6	12	−169	
280	−11.9	15	−20.1	12	−142	56	13.2	5	11.9	1.0	13.8	4.1	−59.6	26	−50.5	28	10.6	3	−8.9	15	−20.2	11	−147	59
300	−7.7	15	−15.7	11	−121	52	13.5	5	11.8	1.0	14.4	3.9	−51.0	24	−42.3	26	11.0	2	−4.7	14	−15.7	10	−128	54
320	−4.1	14	−11.9	11	−105	49	14.0	5	11.7	1.0	14.7	3.6	−43.7	23	−35.4	24	11.5	2	−1.2	14	−11.8		−113	50
340	−1.0	14	−8.7	10	−90.8	45	14.4	5	11.6	0.9	15.0	3.4	−37.4	21	−29.4	22	11.8	2	1.9	14	−8.4		−99.7	46
360	1.7		−5.8	9	−79.0	42	14.7	5	11.5	0.8	15.3	3.2	−31.9	20	−24.2	21	12.1	2	4.6	13	−5.5		−88.4	41
380	4.2		−3.4	9	−68.9	38	15.0	4	11.4	0.8	15.6	3.0	−27.1	19	−19.7	19	12.3	2	7.0	13	−2.9		−78.5	36
400	6.3		−1.1	9	−60.2	36	15.2	4	11.3	0.7	15.9	2.9	−22.9	18	−15.7	18	12.6	2	9.1	13	−0.6		−69.8	34
450	10.7		3.4	8	−42.8				11.1				−14.2	17	−7.4	16	12.9		13.5	12	4.1		−52.3	29
500	14.1		6.9	7	−30.0				10.8				−7.5	15	−1.1	15	13.3		16.8	12	7.7		−38.8	24
600	19.0		11.9	7	−12.4				10.4				2.0	13	8.1	13	13.8		21.7		12.9		−19.6	
700	22.3		15.4		−1.3				10.1				8.5	12	14.3		14.0		25.0		16.5		−6.6	

[a]Adapted from *American Institute of Physics Handbook*, 3rd ed., edited by Dwight E. Gray (McGraw-Hill, New York, 1972).

Enthalpy:

$$\bar{H}-\bar{H}^{0}=RT\left(\frac{B-B_1}{\bar{V}}+\frac{2C-C_1}{2(\bar{V})^2}+\cdots\right).$$

Specific heat at constant volume:

$$\bar{c}_v-\bar{c}_v^0=-R\left(\frac{2B_1+B_2}{\bar{V}}+\frac{2C_1+C_2}{2(\bar{V})^2}+\cdots\right).$$

Specific heat at constant pressure:

$$\bar{c}_p-\bar{c}_p^0=-R\left(\frac{B_2}{\bar{V}}-\frac{(B-B_1)^2-(C-C_1)-\frac{1}{2}C_2}{\bar{V}^2}+\cdots\right).$$

Entropy

$$\bar{S}-\bar{S}^0=-R\left(\ln P+\frac{B_1}{\bar{V}}+\frac{B^2-C+C_1}{2(\bar{V})^2}+\cdots\right).$$

Joule-Thomson coefficient:

$$\mu=\frac{1}{\bar{c}_p^0}\left((B_1-B)+\frac{2B^2-2B_1B-2C+C_1}{\bar{V}}+\frac{R}{\bar{c}_p^0}\frac{B_2(B_1-B)}{\bar{V}}+\cdots\right).$$

Nomenclature:

$\bar{\ }\equiv$ molar quantity,

$^0\equiv$ perfect gas, or zero pressure, state,

$B_1=T(dB/dT)$, $B_2=T^2(d^2B/dT^2)$,

$C_1=T(dC/dT)$, $C_2=T^2(d^2C/dT^2)$.

22.06. TABLE: PHASE TRANSITION TEMPERATURE AND PRESSURE AND HEATS OF TRANSFORMATION[a]

Element	Process	State Initial	State Final	P (mm Hg)	T (K)	ΔH (kJ/mol)
Actinium, Ac	fus	c	liq		1323	
	vap	liq	g	760	3473	
Aluminum, Al	fus	c	liq	2.66(E − 9)	933.2	10.79
	vap	liq	g	760	2793	293.43
Americium, Am	fus	c	liq	1.4(E − 3)	1268	12.1
Antimony, Sb	fus	c	liq		904	19.87
	vap	liq	g, equil.	760	1800	
Argon, Ar	fus	c	liq	516.8	83.81	1.188
	vap	liq	g	760	87.29	6.506
Arsenic, As	sub	c	g, equil.	760	885	
Barium, Ba	tr	c, α	c, β	58(E − 6)	648	
	fus	c, β	liq	0.0107	1002	
	sub	c, β	g	1.1(E − 3)	900	169.5
Beryllium, Be	tr	c, α	c, β		1527	2.556
	fus	c, β	liq	0.037	1560	12.21
	vap	liq	g	760	2745	292.41
Bismuth, Bi	fus	c	liq		544.52	11.30
	vap	liq	g, equil.	760	1837	
Boron, B	fus	c	liq		2340	20.9
	vap	liq	g	760	4075	
Cadmium, Cd	fus	c	liq	0.109	594.18	6.19
	vap	liq	g	760	1040	99.54
Calcium, Ca	tr	c, α	c, β		720	0.920
	fus	c, β	liq	6.0(E − 5)	1112	8.54
	vap	liq	g	760	1757	153.64
Carbon, C	sub	c, graphite	g, std.	760	298.15	716.682
	sub	c, graphite	g, equil.	760	4100	
Cerium, Ce	tr	c, α	c, β		125	
	tr	c, β	c, γ		350	
	tr	c, γ	c, δ		999	2.992
	fus	c, δ	liq		1071	5.460
	vap	liq	g	760	3699	414
Cesium, Cs	fus	c	liq	1.4(E − 6)	301.8	2.18
	vap	liq	g, equil.	760	955	
Chlorine, Cl_2	fus	c	liq	10.1	172.12	6.406
	vap	liq	g	760	239.05	20.410
Cobalt, Co	tr	c, α	c, β		700	0.452
	fus	c, β	liq		1768	16.19
	vap	liq	g	760	3201	376.6
Chromium, Cr	fus	c	liq	3.25	2130	16.93
	vap	liq	g	760	2945	344.3
Copper, Cu	fus	c	liq	4.49(E − 4)	1356.5	13.14
	vap	liq	g	760	2839	300.29
Dysprosium, Dy	tr	c, α	c, β		1657	3.996
	fus	c, β	liq	0.591	1682	11.06
	vap	liq	g	760	2835	230.1
Erbium, Er	fus	c	liq	0.317	1795	19.92
	vap	liq	g	760	3136	261.37
Europium, Eu	fus	c	liq	0.72	1090	9.21
	vap	liq	g	760	1870	143.49
Fluorine, F_2	tr	c	c		45.55	0.728
	fus	c	liq	1.66	53.54	0.5104
	vap	liq	g	760	85.02	6.535
Gadolinium, Gd	tr	c, α	c, β		1533	3.912
	fus	c, β	liq		1585	10.04
	vap	liq	g	760	3539	359.4

[a]Adapted from *American Institute of Physics Handbook*, 3rd ed., edited by Dwight E. Gray (McGraw-Hill, New York, 1972).

22.06. TABLE—*Continued*

Element	Process	State Initial	State Final	P (mm Hg)	T (K)	ΔH (kJ/mol)
Gallium, Ga	fus	c	liq		302.9	5.585
	vap	liq	g	760	2520	257.16
Germanium, Ge	fus	c	liq		1210.4	36.94
	vap	liq	g	760	3107	330.9
Gold, Au	fus	c	liq	2.15(E − 5)	1336	
	vap	liq	g	760	3081	335.03
Hafnium, Hf	tr	c, α	c, β		2013	6.736
	fus	c, β	liq	1.1(E − 3)	2500	24.06
	vap	liq	g	760	4876	573.2
Helium, He	fus	c	liq	22.5(E + 3)	1.764	0.0084
	tr	liq, II	liq, I	37.8	2.172	
	vap	liq, I	g	760	4.214	0.084
Holmium, Ho	tr	c, α	c, β		1701	4.686
	fus	c, β	liq		1743	12.17
	vap	liq	g	760	2968	241.0
Hydrogen, H_2	fus	c	liq	54.0	13.957	0.117
	vap	liq	g	54.0	13.957	0.9163
	vap	liq	g	760	20.38	0.9163
Indium, In	fus	c	liq		429.76	
	vap	liq	g	760	2343	231.8
Iodine, I_2	sub	c	g	0.31	298.15	62.467
	fus	c	liq	92.0	386.75	15.52
	vap	liq	g	760	458.39	41.80
Iridium, Ir	fus	c	liq		2716	2.64
	vap	liq	g		4662	612.3
Iron, Fe	tr	c, α	c, β		1033	0.0
	tr	c, β	c, γ		1184	0.8996
	tr	c, γ	c, δ		1665	0.837
	fus	c, δ	liq	0.026	1809	13.81
	vap	liq	g	760	3135	349.56
Krypton, Kr	fus	c	liq	549	115.78	1.640
	vap	liq	g	760	119.93	9.046
Lanthanum, La	tr	c, α	c, β		550	0.364
	tr	c, β	c, γ		1134	3.121
	fus	c, γ	liq		1193	6.196
	vap	liq	g	760	3730	413.7
Lead, Pb	fus	c	liq		600.45	4.7990
	vap	liq	g	760	2023	177.8
Lithium, Li	tr	c, II	c, I		77	
	fus	c, I	liq		453.69	3.000
	vap	liq	g	760	1597	148.13
Lutetium, Lu	fus	c	liq	0.011	1936	18.65
	vap	liq	g	760	3668	355.89
Magnesium, Mg	fus	c	liq	3.10	922	8.954
	vap	liq	g	760	1363	127.40
Manganese, Mn	tr	c, α	c, β		980	2.226
	tr	c, β	c, γ		1360	2.121
	tr	c, γ	c, δ		1410	1.879
	fus	c, δ	liq	1.03	1517	12.05
	vap	liq	g	760	2335	225.9
Mercury, Hg	fus	c	liq		234.29	2.292
	vap	liq	g	760	629.73	59.296
Molybdenum, Mo	fus	c	liq	0.031	2890	27.82
	vap	liq	g	760	4880	592.45
Neodymium, Nd	tr	c, α	c, β		1128	3.01
	fus	c, β	liq		1289	7.15
	vap	liq	g	760	3341	272.8
Neon, Ne	fus	c	liq	324	24.544	0.33
	vap	liq	g	324	24.544	1.803
	vap	liq	g	760	27.15	1.795

22.06. TABLE—*Continued*

Element	Process	State Initial	State Final	P (mm Hg)	T (K)	ΔH (kJ/mol)
Neptunium, Np	tr	c, III	c, II		533	4.4
	tr	c, II	c, I		850	
	fus	c, I	liq		910	
Nickel, Ni	fus	c	liq	3.1(E − 3)	1726	17.472
	vap	liq	g	760	3187	370.3
Niobium, Nb	fus	c	liq		2740	26.36
	vap	liq	g	760	5017	682.0
Nitrogen, N_2	tr	c, II	c, I		35.61	0.230
	fus	c, I	liq	93.9	63.15	0.719
	vap	liq	g	93.9	63.15	6.050
	vap	liq	g	760	77.35	5.586
Osmium, Os	fus	c	liq		3323	
	sub	c	g	6.2(E − 6)	2550	784.1
Oxygen, O_2	tr	c, III	c, II		23.85	0.0920
	tr	c, II	c, I		43.77	0.745
	fus	c, I	liq	1.14	54.363	0.4435
	vap	liq	g	1.14	54.363	7.648
	vap	liq	g	760	90.180	6.820
Palladium, Pd	fus	c	liq	0.031	1825	17.56
	vap	liq	g	760	3237	357.3
Phosphorus, P_4	tr	c, IV	c, III		195.35	2.092
	fus	c, III	liq		317.30	2.628
	vap	liq	g		317.30	55.731
	vap	liq	g	760	530	
Platinum, Pt	fus	c	liq		2043	19.7
	vap	liq	g	760	4097	509.6
Plutonium, Pu	tr	c, VI	c, V		395	3.35
	tr	c, V	c, IV		480	0.586
	tr	c, IV	c, III		588	0.544
	tr	c, III	c, II		730	0.084
	tr	c, II	c, I		753	1.84
	fus	c, I	liq		913	2.85
	vap	liq	g	760	3503	343.7
Polonium, Po	tr	c, II	c, I		327	
	fus	c, I	liq		527	12.5
	vap	liq	g	760	1235	
Potassium, K	fus	c	liq		336.4	2.351
	vap	liq	g	760	1031	80.23
Praseodymium, Pr	tr	c, α	c, β		1068	3.18
	fus	c, β	liq		1204	6.904
	vap	liq	g	760	3785	296.6
Radium, Ra	fus	c	liq		973	
Radon, Rn	fus	c	liq	502	202	2.89
	vap	liq	g	760	211	16.7
Rhenium, Re	fus	c	liq	0.024	3453	33.1
	vap	liq	g	760	5960	715.5
Rhodium, Rh	fus	c	liq		2233	21.55
	vap	liq	g	760	4000	493.7
Rubidium, Rb	fus	c	liq		312	2.26
	vap	liq	g, equil.	760	967	
Ruthenium, Ru	fus	c	liq		2700	25.9
	vap	liq	g	760	4390	589.9
Samarium, Sm	tr	c, II	c, I		1190	3.10
	fus	c, I	liq	3.18	1345	8.619
	vap	liq	g	760	2064	166.5
Scandium, Sc	tr	c, II	c, I		1608	4.02
	fus	c, I	liq	0.084	1812	14.10
	vap	liq	g	760	3104	314.2
Selenium, Se	tr	c, II	c, I		398	0.753
	fus	c, I	liq		494	5.230
	vap	liq	g, equil.	760	958	

22.06. TABLE—*Continued*

Element	Process	State: Initial	State: Final	P (mm Hg)	T (K)	ΔH (kJ/mol)
Silicon, Si	fus	c	liq		1685	50.62
	vap	liq	g, equil.	760	3540	
Silver, Ag	fus	c	liq		1234	11.30
	vap	liq	g	760	2436	250.63
Sodium, Na	fus	c	liq		370.98	2.601
	vap	liq	g	760	1156	98.01
Strontium, Sr	tr	c, α	c, β		505	
	tr	c, β	c, γ		893	
	fus	c, γ	liq	1.8	1043	
	vap	liq	g	760	1648	138.9
Sulfur, S	tr	c, rhomb.	c, monocl.	3.8(E − 3)	368.46	0.402
	tr	c, rhomb.	c, monocl.		374.15	0.0
	fus	c, monocl.	liq		388.33	1.711
	vap	liq	g, equil.	760	717.75	9.20
Tantalum, Ta	fus	c	liq		3250	31.4
	vap	liq	g	760	5638	761.91
Technetium, Tc	sub	c	g	2.0(E − 6)	2150	686.2
	fus	c	liq	2.0(E − 4)	2443	
Tellurium, Te	fus	c	liq	0.176	722.95	17.49
	vap	liq	g, equil.	760	1261	
Terbium, Tb	tr	c, II	c, I		1560	5.021
	fus	c, I	liq	8.1(E − 4)	1630	10.79
	vap	liq	g	760	3496	331.0
Thallium, Tl	tr	c, α	c, β		507	0.38
	fus	c, β	liq		577	4.10
	vap	liq	g	760	1760	164.8
Thorium, Th	tr	c, α	c, β		1636	2.72
	fus	c, β	liq		2028	16.11
	vap	liq	g	760	5061	514.63
Thulium, Tm	fus	c	liq		1818	16.82
	vap	liq	g	760	2220	190.8
Tin, Sn	tr	c, white	c, grey		286.2	2.092
	fus	c, grey	liq		505.06	6.987
	vap	liq	g	760	2896	296.2
Titanium, Ti	tr	c, α	c, β		1167	4.15
	fus	c, β	liq	4.4(E − 3)	1943	15.5
	vap	liq	g	760	3562	420.91
Tungsten, W	fus	c	liq	0.039	3653	35.40
	vap	liq	g	760	5828	824.25
Uranium, U	tr	c, α	c, β		941	2.791
	tr	c, β	c, γ		1048	4.7572
	fus	c, γ	liq		1405	8.5186
	vap	liq	g	760	4407	464.01
Vanadium, V	fus	c	liq	2.0(E − 3)	2175	20.92
	vap	liq	g	760	3682	451.87
Xenon, Xe	fus	c	liq	611	161.36	2.293
	vap	liq	g	760	165.03	12.640
Ytterbium, Yb	tr	c, α	c, β		1033	1.749
	fus	c, β	liq	19.8	1097	7.657
	vap	liq	g	760	1467	128.9
Yttrium, Y	tr	c, α	c, β		1752	4.9915
	fus	c, β	liq	2.2(E − 3)	1799	11.397
	vap	liq	g	760	3611	363.2
Zinc, Zn	fus	c	liq	0.15	692.65	7.3848
	vap	liq	g	760	1184	115.56
Zirconium, Zr	tr	c, α	c, β	1.8(E − 18)	1136	3.93
	fus	c, β	liq	1.2(E − 5)	2125	16.9
	vap	liq	g	760	4682	581.6
	sub	c, α	g	1.8(E − 18)	1136	605.42

22.07. TABLE: THERMAL CONDUCTIVITY, SPECIFIC HEAT, AND COEFFICIENT OF THERMAL LINEAR EXPANSION OF ELEMENTS WHICH ARE SOLID AT NTP, AT 300 K

Element	λ[a] ($W\,m^{-1}\,K^{-1}$)	c_p[b] ($kJ\,kg^{-1}\,K^{-1}$)	α[c] ($10^{-6}\,K^{-1}$)
Aluminum	237	0.904	23.2
Antimony	24.3	0.209	11.0
Arsenic	50.0	0.335	15.5
Barium	18.4	0.205	20.7
Beryllium	200	1.867	11.5
Bismuth	7.87	0.122	13.4
Boron	27.0	1.033	4.7
Cadmium	96.8	0.231	30.9
Calcium	190	0.649	22.4
Carbon (diamond)	2300 (type IIa)	0.516	1.1
Cerium	11.4	0.192	5.3
Cesium	35.9	0.237	...
Chromium	93.7	0.456	4.5
Cobalt	100	0.418	13.1
Copper	401	0.385	16.6
Dysprosium	10.7	0.172	9.6
Erbium	14.3	0.168	9.4
Europium	13.9	0.178	40
Gadolinium	10.6	0.226	5
Gallium	40.6	0.358	...
Germanium	59.9	0.322	5.8
Gold	317	0.129	14.2
Hafnium	23.0	0.144	5.9
Holmium	16.2	0.172	9.8
Indium	81.6	0.234	32.5
Iridium	147	0.132	6.4
Iron	80.2	0.444	11.9
Lanthanum	13.5	0.189	5.3
Lead	35.3	0.128	29.0
Lithium	84.7	0.357	47
Lutetium	16.4	0.153	8.3
Magnesium	156	1.025	25.0
Manganese	7.82	0.477	21.9
Mercury	8.34 (liquid)	0.139 (liquid)	...
Molybdenum	138	0.255	4.8
Neodymium	16.5	0.191	6.9
Nickel	90.7	0.446	13.5
Niobium	53.7	0.270	7.3
Osmium	87.6	0.127	5.1
Palladium	71.8	0.246	11.9
Phosphorus	0.235 (white)	0.753	...
Platinum	71.6	0.132	8.8
Plutonium	6.74	0.142	47.0
Potassium	102	0.761	...
Praseodymium	12.5	0.194	5.4
Rhenium	47.9	0.138	6.2
Rhodium	150	0.248	8.2
Rubidium	52.8	0.351	...
Ruthenium	117	0.234	6.4
Samarium	13.3	0.190	...
Scandium	15.8	0.571	10
Selenium	2.04	0.320	45.8
Silicon	148	0.715	2.6
Silver	429	0.236	19.0
Sodium	141	1.230	71
Strontium	49	...	22.5
Sulfur	0.269	0.707	...
Tantalum	57.5	0.140	6.3
Tellurium	2.35	0.202	18.8
Terbium	11.1	0.178	9.5
Thallium	46.1	0.133	30.0
Thorium	54.0	0.118	11.1
Thulium	16.8	0.160	11
Tin	66.6	0.227	22.2
Titanium	21.9	0.523	8.7
Tungsten	174	0.132	4.5
Uranium	27.6	0.116	14.0
Vanadium	30.7	0.490	8.5
Ytterbium	34.9	0.154	25.2
Yttrium	17.2	0.298	11.3
Zinc	116	0.389	30.3
Zirconium	22.7	0.276	5.7

[a]From C. Y. Ho, R. W. Powell, and P. E. Liley, *Thermal Conductivity of the Elements: A Comprehensive Review*, J. Phys. Chem. Ref. Data **3**, Suppl. 1 (1974).

[b]From R. Hultgren *et al.*, *Selected Values of the Thermodynamic Properties of the Elements* (ASM, Metals Park, 1973).

[c]From *Thermophysical Properties of Matter—The TPRC Data Series,* edited by Y. S. Touloukian and C. Y. Ho (IFI/Plenum, New York), Vols. 4, 5, 12, and 13.

22.08. TABLE: THERMAL CONDUCTIVITY, SPECIFIC HEAT, AND VISCOSITY OF ELEMENTS WHICH ARE FLUID AT NTP, AT 300 K[a]

Element	λ ($W\,m^{-1}\,K^{-1}$)	c_p ($kJ\,kg^{-1}\,K^{-1}$)	η (10^{-4} Pa s)
Argon	0.0179	0.521	0.229
Bromine	0.0047	0.226	0.155
Chlorine	0.0089	0.479	0.136
Deuterium	0.1397	7.248	0.126
Fluorine	0.0256	0.827	0.227
Helium-3	0.1781	5.191	0.172
Helium-4	0.155	5.193	0.199
Hydrogen, normal	0.183	14.31	0.090
Hydrogen, ortho	0.181	14.10	0.081
Hydrogen, para	0.188	14.85	...
Iodine	...	0.145	...
Krypton	0.0095	0.249	0.256
Neon	0.0491	1.030	0.318
Nitrogen	0.0258	1.042	0.180
Oxygen	0.0263	0.920	0.207
Ozone	...	...	0.134
Radon	0.0036	...	...
Tritium	0.1083	4.837	0.155
Xenon	0.0056	0.160	0.234

[a]*McGraw-Hill/CINDAS Data Series on Material Properties* (McGraw-Hill, New York, 1980), Vol. III-2.

22.09. TABLE: THERMAL emf OF MATERIALS RELATIVE TO PLATINUM (mV)

Reference junction at 0°C. The values below 0°C, in most cases, have not been determined on the same samples as the values above 0°C. Based upon the original table in American Institute of Physics, *Temperature, Its Measurement and Control in Science and Industry* (Reinhold, New York, 1941). Values of the emf have been adjusted to correspond to temperatures expressed on the International Practical Temperature Scale of 1968.

Material	Temperature (°C)										
	−200	−100	0	100	200	400	600	800	1000	1200	1400
Elements											
Aluminum	0.45	0.06	0	0.42	1.06	2.84	5.15				
Antimony			0	4.89	10.14	20.53	28.87				
Bismuth	12.39	7.54	0	−7.34	−13.57						
Cadmium	−0.04	−0.31	0	0.90	2.35						
Calcium			0	−0.51	−1.13						
Carbon			0	0.70	1.54	3.72	6.79	10.98	16.46		
Cerium			0	1.14	2.46						
Cesium	0.22	−0.13	0								
Cobalt			0	−1.33	−3.08	−7.24	−11.28	−13.99	−14.21	−10.70	
Copper	−0.19	−0.37	0	0.76	1.83	4.68	8.34	12.81	18.16		
Germanium	−44.00	−26.62	0	33.9	72.4	82.3	43.9				
Gold	−0.21	−0.39	0	0.78	1.84	4.63	8.12	12.26	17.05		
Indium			0	0.69							
Iridium	−0.25	−0.35	0	0.65	1.49	3.55	6.10	9.10	12.57	16.45	20.47
Lead	0.24	−0.13	0	0.44	1.09						
Lithium	−1.12	−1.00	0	1.82							
Magnesium	0.37	−0.09	0	0.44	1.10						
Mercury			0	−0.60	−1.33						
Molybdenum			0	1.45	3.19	7.57	13.13	19.83	27.74	36.86	
Nickel	2.28	1.22	0	−1.48	−3.10	−5.45	−7.04	−9.33	−12.11		
Palladium	0.81	0.48	0	−0.57	−1.23	−2.82	−5.03	−7.96	−11.61	−15.86	−20.40
Potassium	1.61	0.78	0								
Rhodium	−0.20	−0.34	0	0.70	1.61	3.91	6.77	10.14	14.02	18.39	22.99
Rubidium	1.09	0.46	0								
Silicon	63.13	37.17	0	−41.56	−80.57						
Silver	−0.21	−0.39	0	0.74	1.77	4.57	8.41	13.33			
Sodium	1.00	−0.29	0								
Tantalum	0.21	−0.10	0	0.33	0.93	2.91	5.95	10.02	15.15	21.37	
Thallium			0	0.58	1.30						
Thorium			0	−0.13	−0.26	−0.50	−0.45	0.22	1.72	4.03	
Tin	0.26	−0.12	0	0.42	1.07						
Tungsten	0.43	−0.15	0	1.12	2.62	6.70	12.26	19.25	27.73	37.72	
Zinc	−0.07	−0.33	0	0.76	1.89	5.29					
Thermocouple materials											
Alumel	2.39	1.29	0	−1.29	−2.17	−3.64	−5.28	−7.07	−8.78	−10.33	−11.77
Chromel P	−3.36	−2.20	0	2.81	5.96	12.75	19.61	26.20	32.47	38.48	44.04
Constantan	5.35	2.98	0	−3.51	−7.45	−16.19	−25.46	−34.81	−43.85		
Copper	−0.19	−0.37	0	0.76	1.83	4.68	8.34	12.81	18.16		
Iron	−2.92	−1.84	0	1.89	3.54	5.88	7.80	10.84	14.28		
Alloys											
Brass, yellow			0	0.60	1.49	3.85	6.96				
Copper-beryllium			0	0.67	1.62	4.19					
75Cu-25Ni (nickel coin)			0	−2.76	−6.01	−13.78	−22.59				
95Cu-4Sn-1Zn (copper coin)			0	0.60	1.48	3.91	7.14				
Gold-chromium			0	−0.17	−0.32	−0.55	−0.66				
Manganin			0	0.61	1.55	4.25	7.84				
60Ni-24Fe-16Cr			0	0.85	2.01	5.00	8.68	13.03	18.06		
80Ni-20Cr			0	1.14	2.62	6.25	10.53	15.41	20.87		
Phosphor bronze			0	0.55	1.34	3.50	6.30				
50Sn-50Pb (solder)			0	0.46							
96.5Sn-3.5Ag (solder)			0	0.45							
90Ag-10Cu (silver coin)			0	0.80	1.90	4.81	8.64				
Steel, spring			0	1.32	2.63	4.84	6.86				
Steel, 18-8 stainless			0	0.44	1.04	2.60	4.67	7.35			

22.10. TABLE: emf FOR COMMON THERMOCOUPLE MATERIALS AS A FUNCTION OF TEMPERATURE (ABSOLUTE mV)

A positive sign means that in a simple thermoelectric circuit the resultant emf given is in such a direction as to produce a current from the element to the platinum at the reference junction (0°C). The values below 0°C, in most cases, have not been determined on the same samples as the values above 0°C. Based upon the original table in American Institute of Physics, *Temperature, Its Measurement and Control in Science and Industry* (Reinhold, New York, 1941). Values of the emf have been adjusted to correspond to temperatures expressed on the International Practical Temperature Scale of 1968.

Temp.	Copper vs Constantan					Iron vs Constantan					Chromel vs Constantan					Chromel vs Alumel			
(°C)	0	20	40	60	80	0	20	40	60	80	0	20	40	60	80	0	20	40	
−200	−5.603	−5.889	−6.105	−6.232		−7.890					−8.824	−9.274	−9.604	−9.797		−5.891	−6.158	−6.344	−6
−100	−3.378	−3.923	−4.419	−4.865	−5.261	−4.632	−5.426	−6.159	−6.821	−7.402	−5.237	−6.167	−6.907	−7.631	−8.273	−3.553	−4.138	−4.669	−5
−0	0.00	−0.757	−1.475	−2.152	−2.788	0.00	−0.995	−1.960	−2.892	−3.785	0.00	−1.151	−2.254	−3.306	−4.301	0.00	−0.777	−1.527	−2
+0	0.00	0.789	1.611	2.467	3.357	0.00	1.019	2.058	3.115	4.186	0.00	1.192	2.419	3.683	4.983	0.00	0.798	1.611	2
100	4.277	5.227	6.204	7.207	8.235	5.268	6.359	7.457	8.560	9.667	6.317	7.683	9.078	10.501	11.949	4.095	4.919	5.733	6
200	9.286	10.360	11.456	12.572	13.707	10.777	11.887	12.998	14.108	15.217	13.419	14.909	16.417	17.942	19.481	8.137	8.938	9.745	10
300	14.860	16.030	17.217	18.420	19.638	16.325	17.432	18.537	19.640	20.743	21.033	22.597	24.171	25.754	27.345	12.207	13.039	13.874	14
400	20.869					21.846	22.949	24.054	25.161	26.272	28.943	30.546	32.155	33.767	35.382	16.395	17.241	18.088	18
500						27.388	28.511	29.642	30.782	31.933	36.999	38.617	40.236	41.853	43.470	20.640	21.493	22.346	23
600						33.096	34.273	35.464	36.671	37.893	45.085	46.697	48.306	49.911	51.513	24.902	25.751	26.599	27
700						39.130	40.382	41.647	42.922		53.110	45.703	56.291	57.873	59.451	29.128	29.965	30.799	31
800											61.022	62.588	64.147	65.700	67.245	32.277	34.095	34.909	35
900											68.783	70.313	71.835	73.350	74.857	37.325	38.122	38.915	39
1000											76.358					41.269	42.045	42.817	43
1300																53.398	53.093	53.782	54

22.11. TABLE: emf FOR PLATINUM VERSUS PLATINUM-RHODIUM THERMOCOUPLES (ABSOLUTE mV)

Reference junction at 0°C. Based upon the original table in American Institute of Physics, *Temperature, Its Measurement and Control in Science and Industry* (Reinhold, New York, 1941). Values of the emf have been adjusted to correspond to temperatures expressed on the International Practical Scale of 1968.

Temp.	Platinum–10% rhodium vs platinum					Platinum–13% rhodium vs platinum				
(°C)	0	20	40	60	80	0	20	40	60	80
0	0.000	0.113	0.235	0.365	0.502	0.000	0.111	0.232	0.363	0.501
100	0.645	0.795	0.950	1.109	1.273	0.647	0.800	0.959	1.124	1.294
200	1.440	1.611	1.785	1.962	2.141	1.468	1.647	1.830	2.017	2.207
300	2.323	2.506	2.692	2.880	3.069	2.400	2.596	2.795	2.997	3.201
400	3.260	3.452	3.645	3.840	4.036	3.407	3.616	3.826	4.039	4.254
500	4.234	4.432	4.632	4.832	5.034	4.471	4.689	4.910	5.132	5.356
600	5.237	5.442	5.648	5.855	6.064	5.582	5.810	6.040	6.272	6.505
700	6.274	6.486	6.699	6.913	7.128	6.741	6.979	7.218	7.460	7.703
800	7.345	7.563	7.782	8.003	8.225	7.949	8.196	8.445	8.696	8.949
900	8.448	8.673	8.899	9.126	9.355	9.203	9.460	9.718	9.978	10.240
1000	9.585	9.816	10.048	10.282	10.517	10.503	10.768	11.035	11.304	11.574
1100	10.754	10.991	11.229	11.467	11.707	11.846	12.119	12.394	12.669	12.946
1300	13.155	13.397	13.640	13.883	14.125	14.624	14.906	15.188	15.470	15.752
1500	15.576	15.817	16.057	16.296	16.534	17.445	17.726	18.006	18.286	18.564
1700	17.942	18.170	18.394	18.612		20.215	20.483	20.748	21.006	

22.12. TABLE: ELECTRICAL RESISTIVITY OF SOME ELEMENTS ($\mu\Omega$ cm)

Based upon the original table in American Institute of Physics, *Temperature, Its Measurement and Control in Science and Industry* (Reinhold, New York, 1941).

Temp.	Elements				
(°C)	Copper	Iron	Nickel	Platinum	Silver
100	1.431	1.650	1.663	1.392	1.408
200	1.862	2.464	2.501	1.773	1.827
300	2.299	3.485	3.611	2.142	2.256
400	2.747	4.716	4.847	2.499	2.698
500	3.210	6.162	5.398	2.844	3.150
600	3.695	7.839	5.882	3.178	3.616
700	4.207	9.785	6.326	3.499	4.093
800	4.750	12.003	6.749	3.809	4.584
900	5.332	12.788	7.154	4.108	5.089
1000	5.959	13.070	7.541	4.395	
1100				4.672	
1200				4.937	

INDEX

[Entries in italics refer to chapters]

Q

R

S

T

U

V

W

X

Y